DESIGN AND OPERATION OF CIVIL AND ENVIRONMENTAL ENGINEERING SYSTEMS

DESIGN AND OPERATION OF CIVIL AND ENVIRONMENTAL ENGINEERING SYSTEMS

Edited by

Charles ReVelle
Arthur E. McGarity

A WILEY-INTERSCIENCE PUBLICATION

JOHN WILEY & SONS, INC.

New York • Chichester • Weinheim • Brisbane • Singapore • Toronto

This text is printed on acid-free paper.

Library of Congress Cataloging-in-Publication Data

Design and operation of civil and environmental engineering systems
/ edited by Charles ReVelle, Arthur McGarity.
p. cm.
"A Wiley Interscience publication."
Includes index.
ISBN 0-471-12816-3 (cloth : alk. paper)
1. Civil engineering—Linear programming. 2. Environmental
engineering—Linear programming. 3. Engineering economy—Linear
programming. 4. Operations research. I. Revelle, Charles.
II. McGarity, Arthur E.
TA153.C55 1997
624—dc21 97-52

To
Penny and Jane

CONTENTS

PREFACE

In the period from 1939 to 1957, a span of only 18 years, a group of men and women from mathematics, engineering, and economics created a new and powerful set of mathematical tools that we call, in aggregate, constrained optimization or mathematical programming. From the fertile minds of Kantorovich, Koopmans, Dantzig, Charnes, Bellman, Land, and Doig came the methods that include linear programming, dynamic programming, and integer programming. Originally focused primarily on military logistics problems, the tools of optimization quickly found application in industry, commerce, and the public sector.

These tools were part of a toolkit of methodologies and approaches that came from the allied war effort to defeat Germany and Japan. The toolkit included, in addition to constrained optimization, the methodologies known as simulation, queueing, game theory, and other techniques. These methodologies had been used in the study of military operations and the process of application of these techniques became known as *operations research,* a name that has stuck despite a lack of much precision. Application of these additional tools, as with constrained optimization/mathematical programming, focused as well on industrial, commercial, and public problems.

Industrial engineers gradually abandoned their bread-and-butter time-and-motion studies and embraced the methodologies of operations research, thereby redefining and renewing their discipline. Mechanical engineers found these tools especially helpful in machine design and the design of mechanical components. Chemical engineers utilized the tools for plant and process design, and electrical engineers applied them to power systems design and operation. And civil and environmental engineers discovered in these tools, especially the tool of constrained optimization, an instrument that could actually transform much of their discipline, an instrument that could provide an overarching perspective for the multiple tasks and operations of civil and environmental engineering.

So transforming were these methods for civil and environmental engineers that most modern curricula in these departments today have an undergraduate required course in "systems." *Systems analysis* is the name that Charnes applied to the application of these tools in their various settings. *Systems engineering* is a parallel name to *operations research* for the application of optimization and descriptive models. Thus a substantial number of civil and environmental engineers are exposed to these tools in their undergraduate education. The courses that offer these tools are at the basic optimizing skill level. The applications that undergraduates encounter in these courses are generally small and limited to repetitions of the simpler published research, but the optimizing skills learned in these courses help engineers understand the concepts that underlie the more sophisticated applications that they may encounter later in their careers.

These optimizing skills are also the foundation for further study and research at the graduate level. At the graduate level, courses become deeper. The basics of optimization are assumed and techniques may be extended, often in the context of application. Sometimes, specialty courses are created, but often the courses offer the breadth of optimization applications, reflecting the breadth of interests of civil and environmental engineering. Until now, no text existed for such courses. Such a book is what we offer here.

Designed not only for the graduate student building advanced skills in systems, the book is also suited for the engineer in practice who wants exposure to a range of advanced applications. The range of application is magnificently broad. From air quality models to transportation and water resources, engineers have found hundreds of applications for the methodologies in the systems toolkit. The chapters that follow, authored by the most skilled people in the world in their respective areas, reflect that stunning range of concern and can lead students and practicing engineers to the forefront in these areas of research.

The chapters, although they survey the literature, are not tailored for the specialist seeking another literature review. Instead, the focus is on teaching the materials to those being exposed to these applications for the first time. Of necessity, much of the literature in each area must be mentioned, but the idea behind the text is education rather than a display of erudition. Classes can focus on subsets of the chapters, say in transportation or in water, or can sample the breadth of material offered. Practicing engineers can use the book in a similar fashion. The text also makes an excellent reference for ready use.

In Chapter 1, ReVelle treats the design and operation of water reservoirs and systems of reservoirs for single and multiple purposes using models derived from inventory management applications. The chapter starts with a single-reservoir problem having constant release, and builds in complexity, one step at a time, to include flood control, varying monthly releases, operating rules and simulation, joint operation of multiple reservoirs, synergistic effects, hydropower, and integrated services.

Chapter 2 begins with an overview of water pollution fundamentals and the derivation of descriptive models for the transport of pollutants in freshwater streams. McGarity then derives a recursive, multiple-discharge model for calculating oxygen depletion that employs the concept of partial deficit for separating the effects of various upstream discharges. He goes on to develop several optimization models that incorporate the descriptive model in a linear programming framework to generate solutions that achieve water quality standards and simultaneously, minimize the costs of pollution control.

In Chapter 3, Yeh and Sun describe the general framework for groundwater management and illustrate its application with a specific case study: the San Jacinto basin in southern California. The analysis includes objective recognition, data collection, system equations and numerical methods, parameter estimation, model verification, and model application. The authors then survey the following applications of groundwater systems analysis: stochastic parameter estimation, groundwater remediation design, health risk assessment, and soil vapor extraction system design.

In Chapter 4, Ellis presents three optimization-based applications for managing air quality which represent a broad range of mathematical programming techniques, from straightforward linear programs to highly nonlinear, nonconvex programming problems requiring specialized heuristic solvers. The applications include acid rain control, acid rain emissions trading, and air pollution monitoring network design.

In Chapter 5, Liebman deals with models that are useful for designing and operating systems to collect, transport, and dispose of solid waste. He presents network flow and facility siting models for transshipment operations, capacitated facilities, and fixed charge problems, including transfer and terminal facilities. He also covers a wide range of different vehicle routing problems and algorithms for their solution.

In Chapter 6, Turnquist and Nozick describe various classes of hazardous waste and present an overview of environmental risk assessment methods, including an example of exposure calculations. Modeling formulations and solutions are presented for problems involving location of storage, treatment, and disposal facilities as well as routing methods for transportation of hazardous wastes. The use of geographical information systems in hazardous waste management is also covered, including an example relating to the optimal location of hazardous waste facilities in the Boston area.

Boyce and Daskin deal with the problem of modeling flows of traffic in an urban area in Chapter 7. They present models forecasting future flows and evaluating the ability of road and transit systems to accommodate the flows, as well as models for route choice, transportation mode choice, and origin–destination choices. Solution algorithms are discussed and evaluated. An example applied to the Chicago area involves the application of a travel choice model and its solution. Model implementation issues are discussed and conventional transportation planning practices are evaluated.

In Chapter 8, Crainic and Laporte cover models for three levels of freight transportation planning: strategic, tactical, and operational. Strategic models include facility location, network design, and regional multimodal planning, and tactical models include simulation and optimization models for service network design and vehicle routing problems. Operational models cover scheduling of services, empty vehicle distribution or repositioning, crew scheduling, allocation of resources, and routing with uncertainties.

In Chapter 9, Ceder and Wilson present a wide range of methods for public transportation system planning at different levels. Network and route design issues are discussed, and a multiobjective mathematical program with a numerical example is included. Trip frequency determination and timetable-setting models for tactical planning problems are presented, including treatment of high-ridership corridors. Methods are also included for vehicle scheduling, assigning vehicles to trips to minimize cost and satisfy performance constraints, and for crew scheduling and generation of drivers' daily duties.

In Chapter 10, Barnhart and Talluri survey airline operation problems of two types: schedule planning and revenue management. Schedule generation, fleet assignment, periodic maintenance routing, and crew scheduling methods, including model formulations and solution techniques, are covered under schedule planning. Revenue management models address problems in flight booking, overbooking, and reservations, and include models for small, single-leg systems and for large hub-and-spoke networks.

In Chapter 11, Hobbs provides an in-depth review of the external environmental costs that are associated with the electric utility industry, including regulatory trends over the past 40 years. A linear programming model for integrated resource planning treating multiple objectives addresses the problem of minimizing greenhouse emissions. Alternative approaches to environmental planning are evaluated including command-and-control regulations, emissions caps, and market-based systems such as emissions allowances and trading.

In Chapter 12, Cohon and Rothley deal with a valuable methodology for handling the complexity of competing goals that usually arise in civil and environmental engineering systems. Their thorough introduction to the concepts of multiobjective decision making includes the basic philosophy and the contrasts with other methods, such as benefit-cost analysis. A variety of techniques are introduced and examples are provided for two classes of multiobjective problems: multiple-criteria decision making, for which the potential alternatives are explicitly defined, and multiobjective programming, for which the alternatives are not predefined. Sample applications are included for water resource planning and nuclear waste management.

In Chapter 13, Heaney analyzes conflicts between efficiency and equity in the context of water resource projects. He presents analytical techniques for solving the problem of allocating costs among the various users of water projects, including n-person game theory and activity-based cost accounting. Examples of the application of these techniques are provided for water, wastewater, and stormwater facilities.

Whitlatch introduces concepts of regionalization and economies of scale to motivate the study of regional facilities in Chapter 14, with particular attention given to wastewater treatment plants. Network and linear programming models are presented for facility siting based on meeting effluent standards and water quality standards. Cost-sharing models based on n-person cooperative game theory are also included.

In Chapter 15, Brill makes the case for innovative approaches to water quality management and provides several tools for implementing them. The economic efficiency and equity aspects of discharge permits are covered in depth, and a constrained optimization formulation is developed for solving such problems. The concept of the transferable discharge permit is introduced and its development is traced over the past 40 years. An example is developed illustrating the application of a transferable discharge permit program to the problem of dissolved oxygen management in a stream.

In the final chapter, Chapter 16, Jacobs presents methods for determining the most cost-effective structure that satisfies the requirements of functionality and safety for its application. Basic concepts of structural optimization are developed using examples beginning with a statically determinate three-bar truss and moving into more complex problems such as optimal design of indeterminate structures. Methodologies include linear and nonlinear programming, integer programming, and stochastic models.

We would like to acknowledge the following reviewers for contributing to the development of the book: David P. Ahlfeld, University of Connecticut, Hal Cardwell, U.S. Agency for International Development, J. Wayland Eheart, University of Illinois at Urbana, Peter Furth, Northeastern University, Yacov Y. Haimes, University of Virginia, Karla Hoffman, George Mason University, Mark Houck, George Mason University, Bruce N. Janson, University of Colorado at Denver, Elizabeth K. Judge, North Carolina State University, James W. Male, University of Massachusetts, Peter M. Meier, AsiaPower Ltd., Pitu Mirchandani, University of Arizona, Warren Powell, Princeton University, Davis Sample, University of Colorado, and Rick I. Zadoks, the University of Texas at El Paso.

Charles ReVelle
Arthur E. McGarity

CONTRIBUTORS

Cynthia Barnhart, Professor, Department of Civil Engineering, Massachusetts Institute of Technology, Cambridge, Massachusetts

David E. Boyce, Professor, Department of Civil and Materials Engineering, University of Illinois at Chicago, Chicago, Illinois

E. Downey Brill, Professor & Head of Civil Engineering, Department of Civil Engineering, North Carolina State University, Raleigh, North Carolina

Avishai Ceder, Professor, Q. Tivon, Israel

Jared Cohon, President, Carnegie Mellon University, Pittsburgh, Pennsylvania

Teodor Gabriel Crainic, Professor, Center for Transportation Research, University of Montreal, Montreal, Quebec, Canada

Mark S. Daskin, Professor, Department of Industrial Engineering and Management Sciences, Northwestern University, Evanston, Illinois

J. Hugh Ellis, Professor, Department of Geography and Environmental Engineering, The Johns Hopkins University, Baltimore, Maryland

James Heaney, Professor, Department of Civil, Environmental and Architectural Engineering, University of Colorado, Boulder, Colorado

Benjamin F. Hobbs, Professor, Department of Geography and Environmental Engineering, The Johns Hopkins University, Baltimore, Maryland

Timothy L. Jacobs, Ph.D., P.E., Consultant, SABRE Decision Technologies, Southlake, Texas

Gilbert Laporte, Professor, Center for Transportation Research, University of Montreal, Montreal, Quebec, Canada

Jon C. Liebman, Professor Emeritus of Environmental Engineering, Department of Civil Engineering, University of Illinois at Urbana-Champaign, Urbana, Illinois

Arthur E. McGarity, Professor, Department of Engineering, Swarthmore College, Swarthmore, Pennsylvania

Linda K. Nozick, School of Civil and Environmental Engineering, Cornell University, Ithaca, New York

Charles ReVelle, Professor, Department of Geography and Environmental Engineering, The Johns Hopkins University, Baltimore, Maryland

Kristina Rothley, Professor, School of Forestry and Environmental Studies, Yale University, New Haven, Connecticut

Yung-Hsin Sun, Postdoctoral Scholar, Sacramento, California

Kalyan T. Talluri, Assistant Professor, Department of Economics and Business, Universistat Pompeu Fabra, Barcelona, Spain

Mark A. Turnquist, Professor, School of Civil and Environmental Engineering, Cornell University, Ithaca, New York

Earl Whitlatch, Professor, Department of Civil and Environmental Engineering and Geodetic Science, Ohio State University, Columbus, Ohio

Nigel H. M. Wilson, Professor, Department of Civil Engineering, Massachusetts Institute of Technology, Cambridge, Massachusetts

William W-G. Yeh, Professor of Engineering and Applied Science, Department of Civil and Environmental Engineering, University of California, Los Angeles, Los Angeles, California

CHAPTER 1

Water Resources: Surface Water Systems

CHARLES REVELLE

1.A. INTRODUCTION

The focus of this chapter is the design and operation of the water reservoir and systems of reservoirs. Water resources is a fascinating area of study, rich in problems and challenges. Further, the mathematical form used to model the reservoir shows up again and again in numerous industrial and commercial settings where the management of inventory is at issue. Thus study of the reservoir model reveals not only the subject itself but illustrates a basic mathematical structure that can be utilized in many other applied settings.

The fundamental components that make up surface water systems include not only reservoirs but also their withdrawal structures and spillways together with pipelines, irrigation channels, and hydropower units, components that reflect the multiple functions of the reservoirs. These functions include water supply for municipal and industrial use, irrigation, flood control, recreation, hydropower, and flow maintenance for navigation or aquatic life. All these functions are modeled in this chapter. Regional wastewater treatment is discussed extensively in Chapter 15. Groundwater systems are discussed in detail in Chapter 3.

The water reservoir takes different structural forms depending on its design functions. If the reservoir is used only for flood control, it can be as simple in design as a bathtub, with a single limited-size unadjustable outlet structure as well as a spillway. Alternatively, the reservoir can deliver water through multiple hydropower units or through complex units such as a submerged tower with inlet ports at multiple levels of the reservoir.

The reservoir, for our purposes, will be an impoundment created by either an earthen dam or a concrete dam. It will have a spillway to get rid of excessively high and unexpected flows—if it should become absolutely necessary to do so. And it will have one or more outlet structures, at least one of which discharges water to the stream downstream of the reservoir, possibly through gates, possibly through a turbine-generator.

The water discharged to the stream may be planned release to the stream to maintain desired aquatic life, or the discharge may be used later for water supply withdrawal. Alternatively, water released to the stream may be "wasted" from the reservoir to keep the reservoir contents below the "full" level, thereby preventing use of the spillway. Since most spillways are never tested (for fear of their destruction), the latter practice is only common sense. If water used for water supply or irrigation is not drawn from the stream but from the reservoir, a separate structure will provide the means of withdrawal. The structure might be a submerged port connected to a pipeline that directs water to municipal and industrial use.

This basic reservoir structure provides the flexibility needed to begin to model the arrival of water at the reservoir and its dispatch to various uses—or to model the design process that determines a required reservoir capacity. The forms that we focus on in this chapter fall mainly in the category of deterministic optimization models, the simplest of which is the linear programming model. Indeed, we are fortunate that many of the hydrologic optimization models that have been built fall naturally in this category or in categories strongly related to linear programming. We are fortunate because these optimization models are the simplest to explain and the simplest to solve. We unapologetically emphasize linear programming models in this chapter because of the ease of problem statement and ease of solution. No specially written computer code is needed; multiple, easy-to-use solvers are available in the marketplace. Further, solutions obtained are optimal to the problems at hand.

To model the reservoir we need a record of streamflow inputs to the reservoir. The record of inflows that is selected for utilization reflects the intended functions that are to be considered in the optimization model. For instance, if the model will size a reservoir to be used only for water supply, the worst records of monthly inflows, usually no more than three or four years' worth, must be included within the record used for analysis. This worst record of inflows is referred to as the *critical period.*

Water demand and water supplies (inflows) are not matched in time—peak demands often correspond to seasons when inflows (and rainfall) are least. This is the reason that dams are built—to store water from seasons of abundance for seasons of shortage. And water demands and water supplies are not matched spatially. This is the reason that projects with pipelines and diversions are built—to provide water from areas that are water rich (if only temporarily) to areas where water is in short supply (if only temporarily).

1.B. WATER SUPPLY WITH CONSTANT RELEASE

With this background we can introduce the notation needed to structure the basic reservoir model. This initial model presumes a steady release to water supply month after month. A second model programs a consistent year-to-year release for each January, a consistent release for each February, and so on. We let

t, n = index of months and total length in months of the critical period

s_t = storage at the end of month t (billions of gallons), unknown

s_0 = starting end-of-period storage, unknown

q = steady month-to-month release for water supply (billions of gallons), specified in this first model, later a variable

w_t = water wasted to the stream from the reservoir in month t for lack of storage space, referred to as spill (billions of gallons), unknown

c = capacity of the reservoir to store water (billions of gallons), unknown in this first model, later known

i_t = historically recorded streamflow into the reservoir, month t, of the critical period (billions of gallons), known

The initial model seeks the smallest reservoir capacity needed to sustain a steady release of q throughout the duration of the critical period, at the same time requiring the ending reservoir condition to be no worse than the condition at the start of the critical period. A similar "inventory-type" linear programming formulation is presented in the context of water quality management in this book in Chapter 2 (Section 2.D.1).

The linear programming (LP) problem statement, originally due to Dorfman (1965), in nonstandard form is

$$\text{minimize } z = c$$

subject to

$$s_t = s_{t-1} + i_t - q - w_t \qquad t = 1,2,\ldots,n \tag{1.1}$$

$$s_t \leq c \qquad t = 1,2,\ldots,n \tag{1.2}$$

$$s_n \geq s_0 \tag{1.3}$$

$$s_t, w_t \geq 0 \qquad t = 1,2,\ldots,n$$

$$c, q \geq 0$$

The problem may be solved by any of the standard linear programming codes that are currently available. Problem size is such that solutions on a personal computer are not difficult. A simple specialized algorithm known as *sequent peak* that dates from the 1950s is also available to solve the problem (see, e.g., Potter, 1977; Viessman and Welty, 1985).

Constraints (1.1) are mass balance constraints, really identities in a sense. They say that the ending storage for a particular month is equal to the storage at the end of the preceding month plus any inflow during the month less any release either to water supply or to the stream itself. Constraints (1.2) prohibit any storage in excess of the storage capacity which is to be selected for the reservoir. Constraint (1.3) says that the storage at the end of the critical period must be at least as great as the unknown starting storage. This last constraint prevents "borrowing water" to artificially inflate the amount of water that can be delivered steadily throughout the course of the critical record. The objective, slightly unusual for linear programming problems, consists of just a single variable, the unknown capacity to be chosen by the model. The general technique of minimizing the upper bound of some function or variable is known as *minimizing the maximum.* The reader may ask: "Shouldn't the goal be to minimize costs?" Of course, minimum cost is the correct objective, but there is only one variable in the cost function, the reservoir capacity. Since the cost increases monotonically with capacity, it is sufficient to minimize capacity to minimize costs.

The problem can also be structured without the storage variable; partial sums replace the storage variable in the following formulation:

$$\text{minimize } z = c$$

subject to

$$s_0 + \sum_{t=1}^{k} i_t - \sum_{t=1}^{k} w_t - kq \leq c \qquad k = 1,2,\ldots,n \tag{1.4}$$

$$s_0 + \sum_{t=1}^{k} i_t - \sum_{t=1}^{k} w_t - kq \geq 0 \qquad k = 1,2,\ldots,n \tag{1.5}$$

$$\sum_{t=1}^{n} w_t + nq \leq \sum_{t=1}^{n} i_t \tag{1.6}$$

$$w_t \geq 0 \qquad t = 1,2,\ldots,n$$

$$c,q \geq 0$$

This alternative formulation has n fewer variables and the same number of constraints as the earlier formulation. Instead of a storage variable, the storage at the end of k is represented by

$$s_0 + \sum_{t=1}^{k} i_t - \sum_{t=1}^{k} w_t - kq$$

The storage at the end of k is made up of the initial storage plus all of the inflows up to and including the inflow in period k less all releases to the stream for want of further capacity less all planned releases to water supply in all periods up to and including period k. The mass balance constraints (1.1) are not needed here because they are replaced by the partial sums. However, this formulation needs constraints (1.5), which require nonnegative end-of-period storages. These constraints were not needed in the first formulation because the nonnegativity requirements of the basic LP formulation enforced nonnegativity without formal constraints.

These two formulations both seek the smallest capacity needed to deliver a stable flow through the critical record without "borrowing water." The formulations could be "turned around" with the capacity given and the largest sustainable flow sought as the objective. In this case, c is known and q is unknown. Thus, instead of minimizing the maximum needed capacity, the formulation would maximize the sustainable release to water supply (i.e., maximize q).

This largest sustainable release is often referred to as the *safe yield* of the reservoir. The term can be deceptive to the general public, however. It is "safe" to the water supply engineer because it was capable of delivery through the worst drought on record, but the engineer knows that worse droughts are entirely possible. The layperson assumes the term's conventional meaning, namely, without hazard. The delivery of this yield is not expected always to be safe. Hence the term should probably not be used. Probably *historical yield* could be used, or *critical period yield* might be substituted, so that people realize that the yield can be sustained only if the worst sequence of past inflows repeated precisely.

The use of the critical period for sizing the reservoir or for gauging the historical yield represents the engineer's practical approach to water supply. It is much like building a seaside cottage no nearer the sea than the highest recorded tides. In Florence, Italy, much of the famous statuary is presumably set on pedestals just above the height of the highest recorded floodwaters of the Arno River. These practices reflect the judgment that future conditions are very unlikely to be worse than past conditions—or, if they are, sufficient capacity to adapt is present.

In fact, water supply engineers have developed means to assess the reliability of delivering a stable water supply quantity from a reservoir of stated capacity. The methodology, known as *synthetic hydrology,* is statistical in nature and consists of generating a number of artificial records of streamflows of lengths comparable to the existing record. The records are generated in such a way that all of the relevant statistical parameters of the original streamflow sequence are maintained in the new records. The critical period yield may be tested on these records using the capacity for which the yield was derived. If, for instance, the critical period yield can be sustained in nine out of 10 of the records, the yield can be thought of as roughly 90% reliable or as having a 90% chance of being sustained in some future sequence of flows drawn from the same climactic regime. The reader interested in the methodology of synthetic hydrology is referred to Fiering and Jackson (1971). Hirsch (1979) provides an interesting investigation of the relations that can be built between synthetic hydrology and reliability.

Although synthetic sequences are not hard to generate, their use in assessing the reliability of a critical period yield has been questioned. Researchers have gradually discovered that the low flow sequences that make up the critical record typically may not be sufficiently well represented in the artificial records from synthetic hydrology. This leaves us unfortunately with only the historical critical period as the base data of water supply reservoir calculations, until further research can clarify the issue of reliability. The author of this chapter has a model of reservoir operation in preparation that should offer useful insights to the issue of reliability.

The concept of the critical period has allowed us to structure the basic capacity-minimizing or yield-maximizing linear programs for a water supply reservoir. Furthermore, the size of the problem we have needed to solve has been very small. The formulation using storage variables required only $2n + 2$ variables and $2n + 1$ constraints. Even if n were as large as 60 months, the problem would have only 122 variables and 121 constraints. However, the length and even the position of the critical period in the long record may shift as the yield is altered. When the yield is relatively small, the critical period is short. As the yield is pushed up, the length of the critical period increases and the position within the long record may move as well. In addition, the capacity required to deliver the specified yield also increases. The rate of increase of yield with capacity is constant as long as the length and position of the critical period do not change. However, if either the length of the critical period is increased or its position shifts, the rate of increase in yield with capacity will decline as capacity increases. The plot of yield versus capacity is known to water supply engineers as the *storage–yield curve.*

The fact that the critical period lengthens and shifts with an increasing water supply requirement and the fact that problem size is quite small for these programs—even for many years of record—suggests that the original linear program should be run over a very long portion of the record. This long subrecord should include within it the critical months for the initial particular water supply requirement. In this way the curve of yield versus capacity requires no recalibration of months of the critical period. That is, whatever the desired yield, the critical period will be within the total length of record being utilized for the calculation. Therefore, the number of months n should be much larger than the length of the critical period, and the positioning of these months should encompass the initial critical period, at the least.

1.C. WATER SUPPLY AND FLOOD CONTROL

Flood control, in contrast to water supply, requires free volume in the reservoir to catch and store floods. The captured floodwaters are gradually released at rates that do not cause damage to downstream users or at rates which minimize the damage that might be caused. We will consider a reservoir that provides both the water supply and flood control functions. Thus a portion of the reservoir is dedicated to water supply storage and another portion to emptiness. Methods of assessing the volume needed for flood storage in each month or throughout the year are designed to minimize expected damages or to achieve a sufficiently small probability of spillway use or of overtopping the reservoir.

One method to calculate flood storage pools involves translation of flood pool volumes into the maximum discharge rates that would result. These discharge rates yield stage levels (height) of the stream, which in turn determine damages. The portion of the reservoir devoted to the flood pool and the portion devoted to water supply are chosen so that the sum of water supply benefits and flood damages prevented are as large as possible. The goal in this section is to show how to determine the largest sustainable yield that can be achieved, given (1) a historical record of inflows, (2) a total storage capacity c for the reservoir, and (3) 12 flood storage volumes or flood pools, one for each month of the year. The difference between this model for water supply and the previous model is that in the previous model the maximum capacity available for water storage was the same month to month. In this model, storage available for water supply at the end of each month is reduced from the reservoir capacity by the flood storage volume. These numbers for flood pool volumes, which may be specific by month or constant throughout the year, are indicated as v_k, the flood pool volume to be maintained in the reservoir during month k of the year (billions of gallons), known. These numbers can now be incorporated into the basic reservoir model in which the reservoir capacity, c, is already known. Using the notation we introduced at the outset, we can seek the largest steady monthly supply that can be provided by the reservoir through some critical period or long record of stream flow.

The model of maximizing sustainable yield may be stated as

$$\text{maximize } z = q$$

subject to

$$s_t = s_{t-1} - q - w_t + i_t \qquad t = 1,2,\ldots,n \qquad (1.7)$$

$$s_t \leq c - v_k \qquad t = 1,2,\ldots,n \qquad (1.8)$$

$$k = t - 12[(t-1)/12]$$

$$s_n \geq s_0 \qquad (1.9)$$

$$s_t, w_t \geq 0 \qquad t = 1,2,\ldots,n$$

$$q \geq 0$$

$$[u] = \text{integer part of } u$$

Once again, constraints (1.7) represent the mass balance statement. Constraints (1.8) limit month k end-of-period storages to $c - v_k$, and constraint (1.9) prevents borrowing water from initial storage. Constraints (1.8) are written so that in each recurrence of month k in successive years the appropriate upper limit on water storage is always enforced. For instance, suppose that $t = 26$, and we are referring to the second month of the third year. Then $[(t - 1)/12] = 2$. The value of $k = 26 - 12 \times 2 = 2$. Hence

$$s_{26} \leq c - v_2$$

Since v_k is the volume needed during month k, the model should probably be modified so that the storage through the month is less than $c - v_k$. This might be an interesting exercise for the reader. It might be approached by constraining the average storage in the month less than that value. The reader should also be able to extend this model to a multiple reservoir system that serves both flood control and water supply functions (see, e.g., Kelman and Damazio, 1989). Later in this chapter you will see how the basic water supply model is extended to the multiple reservoir water supply system.

This model is for monthly water supply operation given a reservoir of a fixed size with specified monthly flood control volumes. The model does not guide a reservoir manager in operation during the hours or days of a flood. Such models are beyond the scope of this chapter. The reader interested in the question of operation during floods can consult articles listed in the "References on Operation for Flood Control."

1.D. WATER SUPPLY WITH MONTHLY VARYING RELEASE

When the release varies by month, the objective of the linear program can no longer be the simple maximum steady month-to-month release. Here then we are in the situation of assessing how much water can be delivered from a reservoir of known capacity when the quantity to be delivered is specific for each month of the year. There is a January release to water supply, a February release, a March release, and so on. We do not know what these numbers are, only that they are the same numbers year to year. Furthermore, since they are the same year to year, we can deduce that the January release is always the same proportion of the total of the 12 monthly releases; the February release is a constant fraction of the sum of the 12 releases; and so on. The number that is singular, that is, the same year to year, although still unknown, and which we can optimize, is the sum of the 12 monthly releases, namely the annual yield.

We introduce the following new notation to supplement earlier notation:

q_A = annual release to water supply, unknown

β_k = fraction of the annual yield that is released in month k ($k = 1,2,...,12$), known

The β_k positive fractions, which sum to 1, would be expected to be largest in the drier summer and fall months and least in the wetter winter and spring months. The record of inflows over which the model is solved should be the multiple years that encompass all possible critical periods. The maximum annual yield model can now be structured as

$$\text{maximize } z = q_A$$

subject to

$$s_t = s_{t-1} + i_t - w_t - \beta_k q_A \qquad t = 1,2,...,n \qquad (1.10)$$

$$k = t - 12[(t-1)/12]$$

$$s_t \le c \qquad t = 1,2,...,n \qquad (1.11)$$

$$s_n \ge s_0 \qquad (1.12)$$

$$s_t, w_t \ge 0 \qquad t = 1,2,...,n$$

$$q_A \ge 0$$

$$[u] = \text{integer part of } u$$

Constraints (1.10) are the mass balance equations modified to reflect different releases in each of the 12 months. To illustrate, suppose that $t = 26$. Then $k = 2$ and

$$s_{26} = s_{25} + i_{26} - w_{26} - \beta_2 q_A$$

That is, the release $\beta_2 q_A$ is taken from the reservoir contents toward the February requirement. Constraints (1.11) limit storage to the reservoir's capacity, and constraints (1.12) prevent borrowing water.

As in the basic model, a storage–yield curve may be constructed. The annual water delivery q_A will be found to increase in successively lower sloped linear segments as the reservoir capacity c is allowed to increase. That is, there are declining returns to scale. Each linear segment corresponds to either an expanded-in-length or shifted-in-position critical period.

1.E. WATER SUPPLY OPERATING RULES AND SIMULATION

The model that maximizes the steady month-to-month yield, given a capacity c, follows an *operating rule,* albeit a simple operating rule. An operating rule is an equation or a chart or a look-up table that specifies the amount to be released for various purposes as a function of system states and parameters. Such rules become most necessary when the reservoir is in a state of "stress" (i.e., when demands are becoming difficult to meet without threatening future supplies). Most often those system characteristics that determine release are state variables such as reservoir storage at the end of the preceding period, or stream flows, such as the flow in the immediately preceding month and the flow projected for the next month. Release may be composed of linear or nonlinear functions of these separate descriptors or may be a function of the sum of the descriptors or could even take other forms. We provide further discussion of possible operating rules later, but for now we return to consideration of the operating rule assumed by the linear programming reservoir model.

The LP model assumes that release is equal to the water supply level q when the sum of the previous reservoir storage and this month's inflow are within the following range. On the lower end of the range, the sum is q. If storage plus inflow sum to q or greater, q is released. However, if previous storage plus this month's inflow less q exceeds the capacity of the reservoir, spill w_t must take place. The amount of the spill is the difference between storage plus inflow less q and less the capacity c. If the storage plus inflow are less than q, clearly q cannot be released, as the formalism of release would lead to a negative storage. Since q is the largest possible steady release that can be achieved without a negative reservoir storage with a reservoir capacity of c over a critical record, never, over that record, will there be a need to release less than q. If, however, a quantity larger than q were attempted steadily over the critical record, at some moment storage would go negative. Hence, if a quantity larger than q were attempted, nevertheless, the reservoir operator would have to be prepared at times during the critical record to decrease the release to water supply. The rule implied by the linear program, however, never needs to make such a choice because the q chosen is always achievable over the record being investigated.

Let us assume that a release to water supply is chosen that is greater than q; say d, a number that is not always physically achievable through the record. Now a rule does need to be created for the situation in which the previous storage plus current inflow is less then the desired water supply release. The simplified rule we will use first, if the desired supply is not available, is to release the entire contents of the reservoir, storage plus inflow, toward water supply.

To recapitulate, in this new situation in which a water supply release of d is desired, the following responses occur. If storage plus this month's inflow are less than d, release storage plus inflow. If storage plus inflow exceed d

and the combined storage and inflow less d is less than capacity, release d precisely. If storage plus inflow less d exceed capacity, release the difference or amount of exceedance as spill to the stream, in addition to d. The relations discussed above are summarized in the following equations and in Figure 1.1.

New notation needed is:

$$x_t = \text{release to water supply in month } t$$

$$d = \text{desired water supply release}$$

The release to water supply in month t is

$$x_t = \begin{cases} s_{t-1} + i_t & s_{t-1} + i_t < d \\ d & s_{t-1} + i_t \geq d \end{cases} \tag{1.13}$$

The spill release to stream w_t, is

$$w_t = \begin{cases} 0 & s_{t-1} + i_t - d \leq c \\ s_{t-1} + i_t - d - c & s_{t-1} + i_t - d > c \end{cases} \tag{1.14}$$

Finally, the reservoir contents are given by

$$s_t = s_{t-1} + i_t - x_t - w_t \tag{1.15}$$

where x_t and w_t are determined by the previous equations. Alternatively, reservoir contents could also be written as

$$s_t = \begin{cases} 0 & s_{t-1} + i_t \leq d \\ s_{t-1} + i_t - d & d < s_{t-1} + i_t < c + d \\ c & s_{t-1} + i_t - d \geq c \end{cases}$$

The reservoir release rule and its consequences (1.13)–(1.15), constitute a

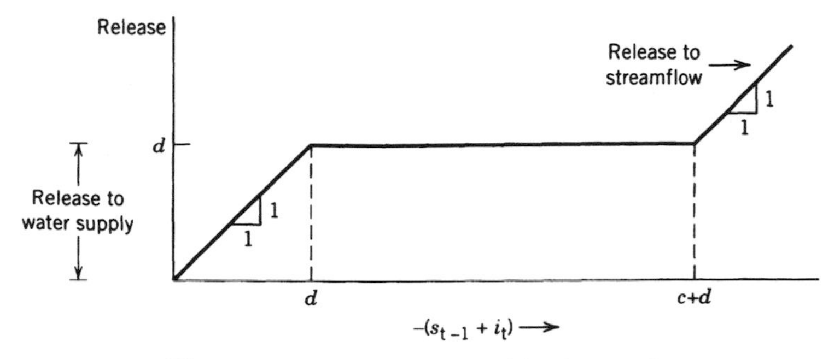

Figure 1.1 Water supply spill release rule.

pattern to guide monthly operation of a reservoir over many years of record. The choice of a water supply release quantity obviously influences the reservoir contents through time. A release set equal to q, the maximum sustainable steady release, will lead to a reservoir which, although it may empty at times, never needs to cut back its release for lack of storage and inflow if operation is performed over the historical record. A release smaller than q will lead to a reservoir that never empties over the record, has a higher average storage, and spills water to downstream flow more often. A release larger than q leads to a reservoir that has to cut back its release to some extent on some occasions, hits bottom more frequently than when the steady release is q, and has a lower average storage, spilling less frequently than when the steady release is q. The larger that the steady release is made, as long as it is above q, the more frequent and the greater in extent are release cutbacks. In addition, average storage decreases with increasing steady release, as does the frequency of release to downstream flow.

The process of adding inflows and subtracting releases month by month is a form of *simulation*. Simulation means repeated operation in an artificial as opposed to real environment. In contrast to optimization, which locates some optimal policy or design, simulation tests alternative policies and designs and compares them in as many perspectives as possible. Each repetition of a simulation alters some facet of the environment, such as the steady release or the capacity of the reservoir or perhaps the rate of release in the situation in which the full water supply quantity cannot be delivered.

These repeated tests can lead to a number of useful plots. For instance, for a given capacity, the frequency of needing to cut back from the otherwise steady delivery may be plotted against the target delivery rate. The same variables can be plotted at different capacity levels. Alternatively, for a given steady release, the frequency of cutbacks might be plotted against capacity. Reservoir simulation can be a useful technique when the system is simple (e.g., one reservoir) and the functions few (e.g., water supply and streamflow maintenance). When the system is more complex, analysis by simulation requires "art" to tease relevant information from the system in a useful form.

Simulation is a very popular method of water resources analysis, and a number of agencies and organizations have prepared their own simulation codes. These include the U.S. Army Corps of Engineers, the Colorado Water Resources Institute at Colorado State University, the Texas Water Development Board, and the Center for Advanced Decision Support for Water and Environment Systems (University of Colorado). The reader interested in water resources simulation can consult the literature in the section "References on Water Resources Simulation."

1.F. WATER SUPPLY WITH MULTIPLE RESERVOIRS

In this section we extend the single water supply reservoir model to a system of reservoirs in parallel. By *in parallel* we mean that each reservoir is on a parallel river above any point of juncture with the other rivers in the system. Reservoirs on the same stream (i.e., in series) do not require special modeling unless times of flow between the reservoirs are more than a few days or a week. Such reservoirs in series on a single stream may generally be treated as a single reservoir for water supply operational purposes.

The basic issue we will examine for operation of reservoirs on parallel rivers is how much yield, either a steady through-the-year monthly yield or an annual yield with variable yields as a function of month, can be obtained from the system of reservoirs. One overly simple and overly conservative answer to this question is that the system yield is the sum of historic or critical period yields of the individual reservoirs. This specific answer presumes that each reservoir will continue to act individually as though it were not a part of a larger system. In such a system, the only advantage to adding a new reservoir is its individual historic yield.

In fact, considerable yield advantage may be obtained from operating a system of reservoirs *as a system.* The differences in local watershed climates often produce seasonal differences in streamflows. These seasonal differences can be translated into different monthly contributions from the individual reservoirs, contributions that may vary by month across the reservoirs but sum to the desired total for each month.

System operation is not yet a finished science. There is no single optimal way to operate a system of reservoirs for water supply, especially in the face of an unknown future. There are not only multiple approaches to scientific operation. There is also the school of "experience," widely used still as we begin to transit into an era of scientific reservoir management. The school of "experience" draws on the accumulated years of operating wisdom of reservoir system managers. It is not to be dismissed quickly. Nonetheless, the mathematical approaches to be discussed next are designed to supplement expert knowledge as well as to provide rules against which expert judgment can be tested.

We will structure the problem of reservoirs in parallel as one of either maximizing the yield that can be sustained month to month or maximizing the annual yield. The reservoirs will be assumed to be in place with storage capacities at prespecified levels. The numbers we obtain represent a possible upper bound on the system yield, the best yield that consistent management practice can provide, given a repetition of the worst system flow conditions of the past.

What is the justification of seeking the maximum system yield? The justification of this objective is that demand almost always grows with time. Maximum yield tells us how long the physical system will be able to provide water before new sources of water supply or water conservation efforts will be needed. Of course, the cost of water conservation should be balanced against the full cost (economic and environmental) of additional water supplies.

There is an obverse or inverted form of this problem statement, just as in the case of the single reservoir. For the single reservoir, we can minimize capacity—as a surrogate for cost—given a specified yield, or we can maximize yield subject to a particular level of capacity. Both problems are straightforward. For the reservoir system, the inverted problem form to the one stated above is to assume that the reservoirs are not in place and to determine the individual capacities that minimize the cost of the reservoir system given a required water supply yield. Unfortunately, this inverted problem is not so easy to solve as the problem that maximizes yield given capacities. The inverted or cost-minimizing system problem will be structured after we have treated the yield-maximizing problem.

1.F.1. Maximizing Yield with Reservoirs in Place

To begin, we will structure the problem of maximizing yield given reservoirs in place. We will assume that a single reservoir is on each of three nonintersecting streams and that a contribution toward water supply can be drawn from each reservoir (see Palmer et al., 1982). The following notation is needed:

c_1, c_2, c_3 = capacities of each of three reservoirs

i_{1t}, i_{2t}, i_{3t} = streamflow sequences for months t on each of the three streams; the record length is given as n months; the record encompasses the worst flow sequences (i.e., those that define the system yield)

q_{1k}, q_{2k}, q_{3k} = release toward water supply from each of the three reservoirs for month k of the year, unknown

w_{1t}, w_{2t}, w_{3t} = unknown release to streamflow from each of the three reservoirs in month t

s_{1t}, s_{2t}, s_{3t} = end-of-month storages in each of the three reservoirs

q_M = largest cumulative draft that can be achieved month after month from the three reservoirs toward the system water supply, unknown

The linear program can be stated as

$$\text{maximize } z = q_M$$

subject to

$$s_{1t} = s_{1t-1} + i_{1t} - w_{1t} - q_{1k} \quad t = 1,2,\ldots,n \quad (1.16)$$

$$k = t - 12[(t - 1)/12]$$

$$s_{1t} \leq c_1 \quad t = 1,2,\ldots,n \quad (1.17)$$

$$s_{2t} = s_{2t-1} + i_{2t} - w_{2t} - q_{2k} \qquad t = 1,2,...,n \qquad (1.18)$$

$$k = t - 12[(t - 1)/12]$$

$$s_{2t} \leq c_2 \qquad t = 1,2,...,n \qquad (1.19)$$

$$s_{3t} = s_{3t-1} + i_{3t} - w_{3t} - q_{3k} \qquad t = 1,2,...,n \qquad (1.20)$$

$$k = t - 12[(t - 1)/12]$$

$$s_{3t} \leq c_3 \qquad t = 1,2,...,n \qquad (1.21)$$

$$s_{in} > s_{i0} \qquad i = 1,2,3 \qquad (1.22)$$

$$q_{1k} + q_{2k} + q_{3k} = q_M \qquad k = 1,2,...,12 \qquad (1.23)$$

Of course, all variables are constrained greater than or equal to zero.

Equations (1.16), (1.18), and (1.20) are the mass balance equations for each of the three reservoirs. Constraints (1.17), (1.19), and (1.21) limit storages to the known reservoir capacities. Constraint (1.22) compares to constraint (1.12), which is reservoir specific, requiring that the ending storage in each reservoir exceeds the beginning and unknown storage in each reservoir. Constraint (1.23) requires that the sum of releases in month k (e.g., February) equal the steady monthly yield q_M.

The crucial answers to this problem, as in the single-reservoir model, are not the reservoir storages through time or the spill quantities. The crucial answers from this program are the 36 q's, three for each month of the year. These are the allocations toward the steady monthly system yield to be drawn from each of the three reservoirs for each of the 12 months of the year. For any month, the three reservoir drafts sum to system yield, but for any particular reservoir the numbers may change by month through the year. In some months, one of the reservoirs may contribute a large share of the system yield, while in others, the same reservoir may contribute a small share. This contrasts to the situation in which the three reservoirs are operated independently of one another and the system yield (a smaller number) is obtained by merging the individual yields of the reservoirs.

The inconsistency of release may be disconcerting and may be eliminated by further programming. We might proceed by maximizing system yield with constraints for each reservoir i that limit the month-to-month change in that reservoirs contribution to some level. That is, maximize q subject to all of the original constraints and

$$q_{ik} - q_{i,k+1} \leq \Delta \qquad k = 1,2,...,11$$

$$q_{i,k+1} - q_{ik} \leq \Delta \qquad k = 1,2,...,11$$

These two constraints limit the increase or decrease to Δ. The reduction in system yield is likely to be small and as Δ is allowed to increase, smaller still.

The basic multiple-reservoir model can also be structured for the situation in which the system requirement cycles with month or season. In this case we seek to maximize the annual yield rather than monthly yield. Specifically, we

$$\text{maximize } z = q_A$$

subject to constraints (1.16)–(1.22) inclusive. Equations (1.23) are now modified to

$$q_{1k} + q_{2k} + q_{3k} = \beta_k q_A \qquad k = 1,2,\ldots,12 \tag{1.24}$$

where β_k is the portion of annual yield that is attributed to month k. Once again, the crucial result is not only the annual yield but the monthly contribution from each reservoir toward the monthly requirement $\beta_k q_A$.

The rules we have formulated thus far for the operation of the parallel reservoirs may be termed *zero-order rules.* They depend on no information which is current at the time that the individual reservoir releases are made. Another term applied to describe these rules is *once-and-for-all,* in the sense that individual releases are specified in advance and are not adjusted for any current conditions.

Other classes or orders of rules may be formulated that do utilize current information. For instance, a first-order rule might specify that the release from a particular reservoir is linearly proportional to storage or inflow or the sum of these for that reservoir. Such rules as these might be termed *here-and-now rules* in the sense that they adjust to or are attuned to current conditions. Another descriptive term for such operating guidelines is *real-time operating rules.* We suggest several approaches to yield maximization in real time. These approaches include simulation and optimization in real time.

1.F.2. Simulation of Water Supply from Multiple Reservoirs

Heuristic Operating Rules and System Yield Maximization Using Simulation. One way to deal with the issue of yield maximization using rules that adjust reservoir contributions to current conditions is to specify the rules in advance. The specification clearly is a heuristic, since the best self-adjusting rules cannot be known a priori. Suppose that the rules are specified. Suppose further that a value for maximum yield has been estimated. The estimate of maximum yield and the rules can be tested together using a long record that includes the worst system droughts experienced over the years of flow measurement. The record should begin in years of ordinary rather than drought flows so that the reservoirs begin their years of crucial operation with realistic opening storage conditions. When the long record opens, the reservoirs will be assumed full, a reasonable assumption during years of ordinary flow.

The testing, of course, is by a simulation, much like that described for the single water supply reservoir. We will offer one rule for allocating releases amongst reservoirs. A number of others are possible. We need to define three additional parameters to structure the rule:

$\hat{i}_{it}$ = projected flow for month t into reservoir i, a known number

α_{it} = unknown proportion of month t's demand allocated to reservoir i

T = estimate of system yield to be tested

The projected flow number may be based on a correlation of this month's streamflow with last month's, or it may be based on snowpack in a mountainous area, or it may be estimated by some other means. This is a projected as opposed to actual flow for month t. Both actual and projected flows will be carried. The projection is carried and used to calculate system operation (the month's calculated releases) at the beginning of a month. The actual flows are used to update the reservoir contents after the month's calculated releases have been made.

The rule we propose here is

$$\alpha_{it} = \frac{(s_{it-1} + \hat{i}_{it})/c_i}{\sum_{i=1,2,3} (s_{it-1} + \hat{i}_{it})/c_i} \tag{1.25}$$

and

$$x_{it} = \alpha_{it} T \tag{1.26}$$

The proportion, α_{it}, of system yield to be taken from reservoir i in month t is given by that portion of system fullness that resides in reservoir i at the beginning of the month. Individual reservoir fullness is given as the ratio of reservoir storage plus projected inflow divided by the capacity of the reservoir. The testing of an estimated system yield T proceeds by operating each reservoir in the system through time using the allocation rule (1.25) and (1.26) along with the following mass balance equations:

$$s_{it} = s_{it-1} + i_{it} - x_{it} \qquad 0 \le s_{it-1} + i_{it} - x_{it} \le c_i \tag{1.27}$$

$$s_{it} = c_i \qquad c_i < s_{it-1} + i_{it} - x_{it} \tag{1.28}$$

$$s_{it} = \text{inadmissible/stop} \qquad s_{it-1} + i_{it} - x_{it} < 0 \tag{1.29}$$

These equations are written for each reservoir i in the system. If at any time, using the capacities, inflows, allocation rules, and estimated system yield, one of the reservoir storages is an inadmissible value (i.e., negative), the system yield T is not attainable with the specified allocation rule.

The means to determine the maximum system yield is to start with a conservative estimate of T, one that is likely to be achievable with the operating rule at hand. The value of T is increased incrementally until system operation, according to (1.25)–(1.29), leads to a negative storage in some reservoir at some time. When this occurs, the largest T that does not lead to a negative storage anywhere in the system over the entire record is the maximum system yield. The reader can imagine and create other plausible and useful operating rules, some of which might, under some records, provide a larger system yield than the rule described here. The apparent quality of one rule vis-à-vis another could be record and capacity dependent, however.

Real-Time Operation Using Optimization and System Yield Maximization Using Simulation. In the previous model we utilized a heuristic rule to decide on individual reservoir contributions toward a system target. Then we increased the target incrementally until one of the reservoir storages went negative, when the reservoirs were operated according to the heuristic rule. In this model we once again decide on individual reservoir contributions toward the target and increase the target until some reservoir reaches a negative storage, when operation follows a set of optimized monthly decisions. The key difference here between this and the previous model is that the method for deciding each month on reservoir contributions will be by an optimization model rather by a heuristic rule.

Real-Time Operation. The optimization model, which we will assume to use monthly time steps, requires the following notation. We let

s_i = storage in reservoir i at the beginning of the month in which operation is to be taken (also, the storage at the end of the preceding month)

x_i = contribution of reservoir i toward the system target

ei_i = expected inflow in the month to reservoir i

w_i^+ = expected spill (water wasted) during the month from reservoir i

w_i^- = expected empty space in reservoir i at the end of the month of operation

c_i = capacity of reservoir i

The expected inflow might be established by use of a regression model which, using historic flows for each reservoir, regresses the flow in month i

(say, September) with the flow in month $i - 1$ (say, August). Then the expected value of the flow in month i is calculated by the regression equation using the flow that actually occurred in month $i - 1$.

The objective of the optimization model is to deliver the target yield in such a way that when the month is over the total volume of inflow that is retained in all the reservoirs is as large as possible. Put another way, maximizing the volume retained is equivalent to minimizing the volume spilled or wasted and places the reservoir system in a strong position to meet future demands. Note that we say "strong position" as opposed to "the strongest possible position" even though we are optimizing for the month. This is because the objective of maximum water retention is a short-term goal. If we are to optimize with a view to the long term, we need a modified objective or a new constraint or set of constraints. We will introduce such a modification once we have structured the basic model. This model, the one that optimizes monthly operation with an eye to the future, will then be used in conjunction with multiple system simulations [see equations (1.25)–(1.29)] to establish a maximum system yield.

Note: The reader should take note that the model which allocates reservoir contributions in order to maximize expected water retained/minimize water wasted—with an eye to future needs—is a fundamental model in its own right. Once a system target has been established and the reservoirs need to be operated through real, present time, the allocation model can be used as the means to make decisions about individual contributions in each month, as that month occurs. That is, the allocation model falls also in the class of *real-time decision models.*

The allocation model that minimizes spill may be written as

$$\text{maximize } z = \sum_{i=1,2,3} w_i^+$$

subject to

$$s_i + ei_i - x_i - c_i = w_i^+ - w_i^- \qquad i = 1,2,3 \tag{1.30}$$

$$\sum_{i=1,2,3} x_i = T \tag{1.31}$$

$$x_i, w_i^+, w_i^- \geq 0 \qquad i = 1,2,3$$

In the standard form the model is

$$\text{minimize } z = \sum_{i=1,2,3} w_i^+$$

subject to

$$w_i^+ - w_i^- + x_i = s_i + ei_i - c_i \qquad i = 1,2,3 \tag{1.32}$$

$$\sum_{i=1,2,3} x_i = T \tag{1.33}$$

$$x_i, w_i^+, w_i^- \geq 0 \qquad i = 1,2,3$$

By example, we can see that there is often an obvious solution to the problem of minimizing spill. For instance, suppose that the three equations (1.32) come out to be

$$w_1^+ - w_1^- + x_1 = 5$$

$$w_2^+ - w_2^- + x_2 = -2$$

$$w_3^+ - w_3^- + x_3 = 7$$

and that

$$x_1 + x_2 + x_3 = 10$$

There are multiple obvious solutions to the problem of minimizing $(w_1^+ + w_2^+ + w_3^+)$. They include the following (x_i, x_3) pairs: (5,5), (4,6)(3,7), with $x_2 = 0$. The methodology is described in Hirsch et al. (1977). All three of the solutions shown can provide zero spill (i.e., $w_1^+ = w_2^+ = w_3^+ = 0$).

The possibility of multiple alternative optima always raises the possibility of imposing other objectives or other constraints. In this case the constraints we will add are designed to select from among the multiple optima those that distribute emptiness in an advantageous way. Early on, almost predating the era of water resources systems, Clark suggested a rule for allocating emptiness across parallel reservoirs that serve a common supply (see Maas et al., 1962). That rule, known as the *space rule,* suggests that the fraction of total system emptiness that resides in an individual reservoir should reflect the fraction of the total system inflow which will occur to that reservoir in some future period of time as measured from the present. Of course, future inflows are predicted inflows, again likely from regression equations.

To achieve Clark's space rule within the context of minimizing the sum of spills, we need to define the fraction of future inflow to reservoir i. This requires a statement of the horizon of concern. If the reservoir refills each year, the horizon may be taken as the number of months prior to all reservoirs in the system refilling. We define

$$\alpha_i = \frac{f_i}{\Sigma_{i=1,2,3} f_i} = \text{fraction of future inflow to reservoir } i$$

where f_i is the anticipated inflow into reservoir i over the planning horizon. Then the constraints that distribute emptiness are

$$\frac{w_i^-}{\sum_{i=1,2,3} w_i^-} = \alpha_i \qquad i = 1,2,3 \tag{1.34}$$

The mathematical program that minimizes the expected system spill for the month and distributes emptiness at the end of the month is, in standard form,

$$\text{minimize } z = \sum_{i=1,2,3} w_i^+$$

subject to

$$w_i^+ - w_i^- + x_i = s_i + ei_i - c_i \qquad i = 1,2,3 \tag{1.35}$$

$$\sum_{i=1,2,3} x_i = T \tag{1.36}$$

$$w_i^- - \alpha_i \sum_{i=1,2,3} w_i^- = 0 \qquad i = 1,2,3 \tag{1.37}$$

$$x_i, w_i^+, w_i^- \geq 0 \qquad i = 1,2,3$$

System Yield Maximization Given Real-Time Optimization. Now that a new mechanism has been carved out for real-time operation, we can return to the maximization of system yield. Once again we will rely on equations (1.25)–(1.29) for operation of the reservoir through the historical record. Decisions on individual reservoir contributions for each month are made by the model that minimizes system spill subject to (1.35)–(1.37). Then the actual end-of-period storages are calculated using the historic flows that did occur using (1.25)–(1.29). This operation is performed for a specific value of system yield T. Presumably, the first estimate of T has been conservative, so that no reservoir ever goes negative in storage when real-time operation is practiced and actual historic flows occur. Then T is incremented just until some reservoir storage goes negative over the record of inflows.

1.F.3. Cost Minimization Given a System Yield Requirement

We indicated earlier that we were treating first the simpler problem (i.e., yield maximization given reservoirs in place) and that we would return to the obverse or inverted problem. Although there is much more that can be said about system yield maximization, the interested reader can refer to the literature for enlightenment. For now, we proceed to the problem of minimizing cost to obtain a required yield. Because this is already a difficult problem, our treatment will not be as extensive as for the yield-maximization problem. In particular, we will need to make some early assumptions regarding reservoir operations. We will consider two possibilities. The first is that the reservoirs are operated entirely independent of one another. The second is that they will operate in concert with another. The latter assumption is the same as utilized in model (1.16)–(1.23) but leads potentially to a more difficult model to solve. The first leads to a dynamic program of a standard sort.

Reservoirs Operated Independent of One Another. The problem we are stating here is one in which reservoirs are not in place but are to be built, if economically efficient to do so, to provide a specified level of water supply. As in the earlier system models, we assume three parallel streams and an associated possible reservoir on each of these stream (see Wathne et al., 1975). We define

q_T = needed monthly supply of the system, known

q_1, q_2, q_3 = unknown but consistent monthly supplies to be provided by reservoirs 1, 2, and 3

i_{1t}, i_{2t}, i_{3t} = long record of streamflows by month t for each of the three reservoirs

c_1, c_2, c_3 = unknown capacities of each of the three reservoirs

s_{1t}, s_{2t}, s_{3t} = unknown storage at the end of month t in each of the three reservoirs

w_{1t}, w_{2t}, w_{3t} = spills in month t for each of the three reservoirs

$g_1(c_1), g_2(c_2), g_3(c_3)$ = cost of each reservoir as a function of its capacity

For each reservoir i, we solve for the smallest capacity subject to equations (1.1)–(1.3) where q_i is initially given; that is,

$$\text{minimize } z = c_i$$

subject to

$$s_{it} = s_{it-1} + i_{it} - q_i - w_{it} \qquad t = 1,2,\ldots,n \qquad (1.1)$$

$$s_{it} \leq c_i \qquad t = 1,2,\ldots,n \qquad (1.2)$$

$$s_{in} \geq s_{i0} \qquad (1.3)$$

This is, of course, the fundamental model with which we began the chapter. For each reservoir i we create a storage–yield curve, solving for the needed capacity at increasing levels of reservoir yield q_i. The curve of capacity versus reservoir yield is not smooth, as we indicated earlier, but is composed of linear segments. Using the cost function $g_i(c_i)$ given above, the capacity versus yield curve is translated to a cost–yield curve. The new function for reservoir i represented by the curve is $f_i(q_i)$. That is, $f_i(q_i)$ is the cost of the reservoir i needed to deliver a steady monthly supply of q_i.

Now if q_T is demanded from the system each month and each reservoir is to contribute the same quantity each month of the year toward the joint supply, the problem is

$$\text{minimize } z = f_1(q_1) + f_2(q_2) + f_3(q_3)$$

subject to

$$q_1 + q_2 + q_3 = q_T$$

This, of course, is in the standard form of the dynamic program (see, e.g., Dreyfus and Law, 1977). While the problem stated here is of interest, the assumption of independent operation does not allow for the synergy of joint operation.

Reservoirs Operated in Concert. We choose next to study the problem with the once-and-for-all or zero order rules that were used in model (1.16)–(1.23). These rules allow the contribution of each reservoir toward the common supply to vary by month even though they are the same for a particular month from year to year. In some months, one or two reservoirs may make the dominant contributions, while in others the remaining reservoir(s) may provide the larger amounts toward the system requirement. In this model as in the preceding model, which used dynamic programming, we avoid entirely the simulation step we have been utilizing in the earlier models. Our notation then will be precisely the notation used in the model (1.16)–(1.23) except that the capacities c_1, c_2, c_3 will be unknown. We will use, in addition, the cost functions $g_i(c_i)$, which indicate reservoir cost as a function of capacity.

The cost of a reservoir increases as the height of the dam is increased. Initially, there are large economies of scale and then marginal costs may decline a little. Eventually, at larger capacities, the cost curve could become convex. For purposes of this exercise, we approximate the cost–capacity curve as a fixed-charge cost function (see Figure 1.2). In this book, fixed-charge cost functions are also used in problem formulations in Chapter 5 (Section 5.C; solid waste), Chapter 8 (Section 8.B.1; location of freight terminal facilities), and Chapter 14 (Section 14.E.3; wastewater treatment plant siting). That is, $g_i(c_i)$ is replaced by $b_iy_i + a_ic_i$. The fixed-charge cost function (which will be minimized) states that

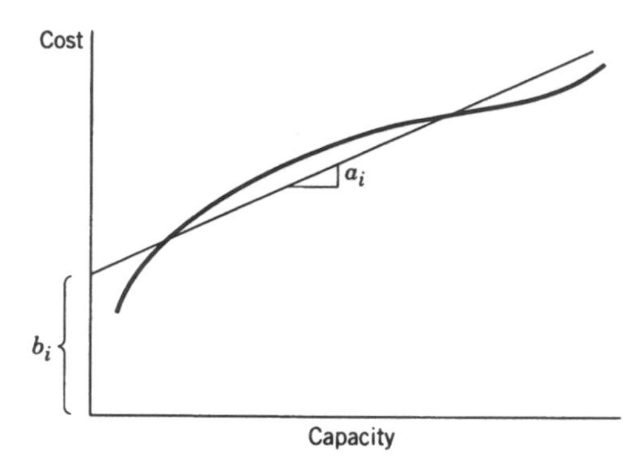

Figure 1.2 Fixed-charge approximation of the reservoir cost function.

$$\text{cost} = \begin{cases} b_i y_i + a_i c_i & c_i > 0 \\ 0 & c_i = 0 \end{cases}$$

where $y_i = 0, 1$; it is 1 if reservoir i is built to any level and zero otherwise, unknown. The 0, 1 values of the y_i are achieved in most computer codes by the widely known procedure of branch and bound. The b_i and a_i are known constants that define each of the fixed-charge cost functions. Thus, if reservoir i is built to any capacity level other than zero, the cost is $b_i + a_i c_i$. However, if the reservoir is not built, no cost is incurred. In making decisions on what capacities to build to, it is entirely possible that a particular reservoir will not be built at all, and this circumstance must be accounted for correctly by the cost function.

The fixed-charge cost function needs to be enforced by some programming mechanism or constraint. The most common way to enforce this unusual cost function is by introducing an upper bound on the capacity of each reservoir. This upper bound is the largest conceivable capacity that the reservoir could take. We define

m_i as the maximum possible capacity for the ith reservoir.

The constraint that enforces the fixed charge for the ith reservoir is

$$c_i \le m_i y_i \qquad i = 1,2,\ldots,3 \tag{1.38}$$

This constraint may also be written in a slightly more understandable form as

$$y_i \ge \frac{c_i}{m_i} \qquad i = 1,2,3$$

The constraint in combination with a minimizing objective forces y_i to be positive if c_i is positive and zero if c_i is zero. If y_i is forced to be positive by a positive c_i in constraint (1.38), y_i will have to be 1 since it can only assume the two possible values, zero or 1. If c_i is zero, y_i need only be greater than or equal to zero, and y_i will be pushed down to zero because the objective minimizes the $b_i y_i$ term.

To proceed to minimizing the sum of reservoir costs subject to a monthly system requirement, given once-and-for-all monthly drafts from each of the reservoirs, which drafts may differ by month, we structure the following mixed-integer program. By *mixed-integer program* we mean that some of the variables are integer (in this case, zero–one) and some are continuous. Thus the problem is

$$\text{minimize } z = \sum_{i=1,2,3} (b_i y_i + a_i c_i)$$

subject to

$$s_{it} = s_{it-1} + i_{it} - w_{it} - q_{ik} \qquad i = 1,2,3 \qquad (1.39)$$

$$t = 1,2,\ldots,n$$

$$k = t - 12[(t - 1)/12]$$

$$s_{it} \leq c_i \qquad i = 1,2,3 \qquad (1.40)$$

$$t = 1,2,\ldots,n$$

$$s_{in} \geq s_{i0} \qquad i = 1,2,3 \qquad (1.41)$$

$$\sum_{i=1,2,3} q_{ik} = q_T \qquad k = 1,2,\ldots,12 \qquad (1.42)$$

$$c_i \leq m_i y_i \qquad i = 1,2,\ldots,3 \qquad (1.43)$$

As in model (1.16)–(1.22) and (1.24), if the system requirement were variable by month, q_T would be replaced by q_A, the known annual requirement, multiplied by β_k, the portion of annual yield needed in month k. That is

$$\sum_{i=1,2,3} q_{ik} = \beta_k q_A \qquad k = 1,2,\ldots,12 \qquad (1.44)$$

and the minimization of costs is carried out subject to all of the foregoing constraints except (1.42), which is replaced by (1.44).

Although it is relatively easy to invoke the mixed-integer programming option of most optimization codes, a problem as small as this can be optimized by a combination of linear programming and enumeration. With three reservoirs there are only seven system possibilities of open/not open reservoirs. These are triplets of the y_i, namely (y_1, y_2, y_3). The possible combinations are (1, 0, 0), (0, 1, 0), (0, 0, 1), (1, 1, 0), (1, 0, 1), (0, 1, 1), and (1, 1, 1).

That is, one at a time, one of these combinations is examined and the linear portion, as opposed to the integer portion, of the objective is minimized. For instance, when (1, 0, 1) is examined, the objective is $a_1c_1 + a_3c_3$ and the constants b_1 and b_3 are added to the value of the optimum linear objective. The least costly of these seven combinations is the answer to the minimization of the fixed-charge objective function. Of course, systems with many reservoirs become more difficult to examine in this enumerative fashion, and the mixed-integer programming option becomes the methodology of choice.

With this cost-minimizing model, we conclude our consideration of water-supply-only models. The justification for spending so much text on these models is that water supply is probably the reservoir use that is first in importance worldwide. Water for municipal and industrial use as well as for agricultural irrigation are among the chief reasons that reservoirs are built. Indeed, there is a great deal we have still not said about water supply.

We have not said anything about how much or how frequently water supply releases would be cut back in the event of a shortfall. The study of the reliability of water supplies is ongoing. The consideration of the stochastic or random nature of stream flow began with the work of Fiering and others (Fiering, 1967; Fiering and Jackson, 1971) referred to earlier. Fiering's ideas and time-series notions, in general, were applied by Hirsch (1979) in an examination of reservoir reliability in real time. In addition, the behaviors or responses of reservoirs in the face of random flows—their reliability, resilience, and vulnerability—have been examined by a number of investigators, including Hashimoto et al. (1982), Moy et al. (1986), and most recently, Vogel and Bolognese (1995). Recently, Shih and ReVelle (1994, 1995) began an investigation to identify the signals to be used for beginning various phases of water rationing, and we have touched only lightly on the issue of reservoir operating rules (Yeh, 1985). Nor have we been able to say anything about water pricing (see, e.g., Howe and Linaweaver, 1967). Finally, we have not offered any modifications to incorporate evaporation or seepage loss. Frankly, in this brief introduction to water resources systems, we are unable to elaborate on some truly interesting and valuable research and modeling in water supply planning. This is because there is much still to say about the other functions of water supply reservoirs in addition to water supply and flood control.

The other functions we want to discuss include recreation, hydropower, and streamflow maintenance for navigation or pollution control. Often, reservoirs may serve all of these functions and water supply and flood control purposes as well. The challenge is to operate the reservoirs in a manner that serves all these adequately. The multiple modifications to the basic model that are necessary to perform these functions are not straightforward and so are introduced in stages. We begin by adding the hydropower function to the basic water supply/flood control model.

1.G. HYDROPOWER FUNCTION

The energy in falling water was one of the first forms of energy to be captured and converted to work. The old grist mill has now given way to the modern hydroelectric power plant and the energy that had been used locally may now be transported many hundreds of miles as electricity. Just as at the grist mill, though, water turns a "wheel," now called a *turbine*. The energy that goes into turning the turbine depends on the product of two variables: the volumetric water flow through the turbine and the height of water from the water surface to the turbine. This height is referred to as *head*.

In the case of the water supply reservoir, the question asked was: How much water can be reliably delivered month after month from a reservoir of stated capacity through a long record that includes droughts? A perfectly parallel question can be asked for a reservoir devoted to hydropower: What is the largest amount of power that can be delivered steadily month after month from a reservoir of stated capacity through a long record that includes droughts? This is the question that we address first in this section.

In addition to the notation introduced for the basic reservoir model, we need to add

h_t = head (height in feet) of water above the turbines at the end of month t

$\bar{h}_t$ = average head in month t

$\bar{s}_t$ = average storage volume during month t

x_t = amount of water released through the turbines in month t

The height of water above the turbines is obviously a function of the volume of water stored in the reservoir. The function is represented by the solid curve in Figure 1.3 and is approximated by the dashed line, which may be written as the linear function

$$h = p + ms$$

where h and s are the instantaneous height and storage. Thus the average height in month t may be written as a function of the average storage in month t:

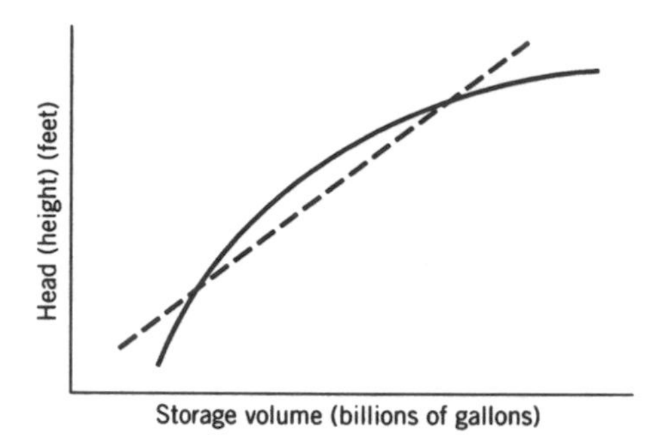

Figure 1.3 Head as a function of storage volume.

$$\bar{h}_t = p + m\bar{s}_t \tag{1.45}$$

The average storage in month t may be well approximated by

$$\bar{s}_t = \frac{s_{t-1} + s_t}{2} = \frac{1}{2} s_{t-1} + \frac{1}{2} s_t \tag{1.46}$$

The power generated in month t is proportional to the product of average head during t and release throughout t:

$$\pi_t = \alpha \bar{h}_t x_t$$

where π_t is the power generated in month t. Substituting equation (1.45) for $\bar{h}_t$ gives

$$\pi_t = \alpha(p + m\bar{s}_t)x_t \tag{1.47}$$

In equation (1.47), $\bar{s}_t$ can be replaced by its equivalent in equation (1.46), so that power production in t can be written as

$$\pi_t = \alpha \left(p + \frac{m}{2} s_{t-1} + \frac{m}{2} s_t \right) x_t \tag{1.48}$$

It is this equation for power production that is used in conjunction with the basic reservoir model (1.1)–(1.3).

The question that is being asked, the maximum amount of firm, steadily deliverable power that can be provided, cannot be approached directly with linear programming, to our knowledge. Nonetheless, by solving a linear program iteratively, that value, the firm power, can be determined relatively precisely. We begin by prespecifying an amount of firm power delivery at a level we are confident can be achieved: d, the minimal monthly power production. We then write the basic reservoir constraints and append to them a set constraints that require that the prespecified power be produced in each month. These appended constraints are nonlinear. The model constraints take the form

$$s_t = s_{t-1} + i_t - x_t - w_t \qquad t = 1,2,...,n \tag{1.49}$$

$$s_t \leq c \qquad t = 1,2,...,n \tag{1.50}$$

$$s_n \geq s_0 \tag{1.51}$$

$$\alpha\left(p + \frac{m}{2}s_{t-1} + \frac{m}{2}s_t\right)x_t \geq d \qquad t = 1,2,...,n \tag{1.52}$$

The challenge is to find a feasible solution to (1.49)–(1.52), which produces power at a rate d. The problem is difficult because constraints (1.52) involve product terms. This model differs from the basic model (1.1)–(1.3) not only because of the presence of the nonlinear power production constraints (1.52) but also in the release quantity x_t, which is allowed to differ for each month of the entire record. It should be noted that this release x_t is not for water supply, but for power production and that the quantity of water that flows through the turbines becomes downstream flow. To find a feasible solution to (1.49)–(1.52), we first transform (1.52) into linear constraints. Constraints (1.52) can also be written

$$\alpha p + \frac{\alpha m}{2}s_{t-1} + \frac{\alpha m}{2}s_t \geq \frac{d}{x_t} \qquad t = 1,2,...,n \tag{1.53}$$

The term d/x_t is a nonlinear function whose shape is illustrated in Figure 1.4. It is an hyperbola that approaches zero as x_t becomes large and that approaches infinity as x_t goes to zero.

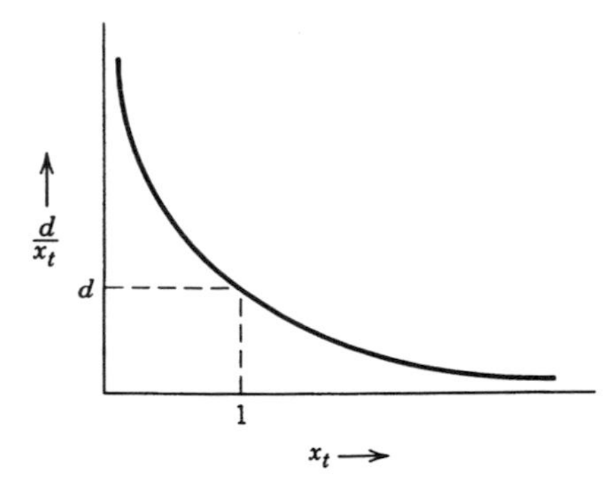

Figure 1.4 The equation d/x_t as a function of x_t.

The reader notes that the function is convex because a line segment joining any two points on the curve falls entirely above the curve except for its endpoints. Constraint (1.53) can be reorganized to read

$$\frac{d}{x_t} - \frac{\alpha m}{2} s_{t-1} - \frac{\alpha m}{2} s_t \leq \alpha p \tag{1.54}$$

This constraint has two linear terms and a convex term and is less than or equal to a right-hand-side value. The linear terms are standard issue for the application of linear programming. The convex term, because the sense of the constraint is "less than," can be piecewise approximated by line segments and the segments will enter in the proper order. In Figure 1.5 we show the piecewise approximation of d/x_t in three segments. The approximation will be the same for all months.

In the figure, the following new notation is introduced:

r_0 = value of d/x_t at first point of approximation

r_i = slope of the ith segment

x_{it} = amount of the ith segment filled or occupied

l_i = length of segment i

With this new notation constraints (1.54) are replaced by constraints (1.55):

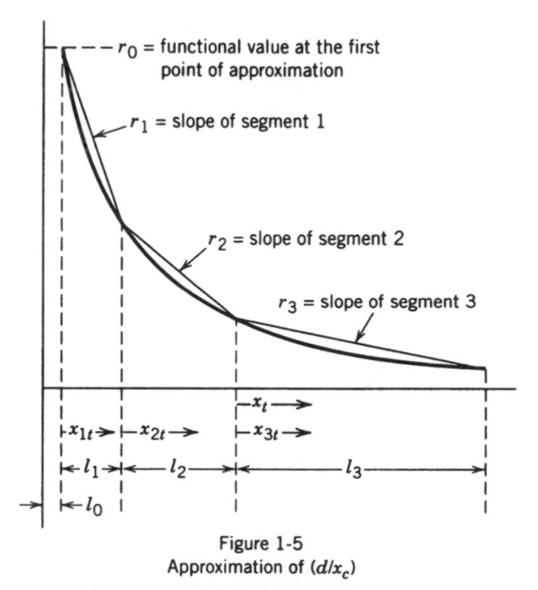

Figure 1.5 Approximation of d/x_t.

$$r_0 - \sum_{i=1}^{3} r_i x_{it} - \frac{\alpha m}{2} s_{t-1} - \frac{\alpha m}{2} s_t \leq \alpha p \qquad t = 1,2,\ldots,n \qquad (1.55)$$

$$x_t - \sum_{i=1}^{3} x_{it} = l_0 \qquad t = 1,2,\ldots,n \qquad (1.56)$$

$$x_{it} \leq l_i \qquad t = 1,2,\ldots,n \qquad (1.57)$$
$$i = 1,2,3$$

in which constraints (1.55) force hydropower production in each month to be greater than or equal to d; constraints (1.56) define x_t in terms of the x_{it}; and constraints (1.57) limit the x_{it} to the appropriate length.

The new constraint set consists of (1.49)–(1.51) and (1.55)–(1.57). The value of d was chosen at a conservative level, a level we were confident could be delivered. Note that no objective function has been articulated other than the maximum firm power, but the firm power is temporarily a specification rather than the objective. To be certain that the initial value of d can be delivered, we need to solve (1.49)–(1.51) and (1.55)–(1.57) under some objective. Any reasonable objective will do. As an example, we might maximize the average storage, that is,

$$\text{maximize } z = \sum_{t=1}^{n} s_t$$

If the solution to this problem is feasible, that is, a power level of d can be delivered, we will obtain an answer that consists of n values of the x_t, $(n + 1)$ of s_t, and n of w_t.

If it is feasible to deliver a firm power of d, the next step is to increase d by a small increment. For this new value of d, the function d/x_t will need to be reapproximated. Probably the segment lengths can remain the same, but the slopes of the approximating line segments will need to be recalculated for each new value of d. The firm power level is increased in steps until the linear program is no longer feasible. The highest firm power level that can be produced through the historical record from a reservoir of capacity c is the last value of d attempted before the next increment produces an infeasible solution. For each value of capacity, a maximum firm power level can be calculated in this manner.

The problem may also be approached efficiently in the following way. Set a firm power level d. Using piecewise approximation, write the constraints requiring that power. Minimize the capacity necessary to achieve this. Of course, if d is chosen too high, no capacity will be capable of providing it and the problem will be infeasible. Let's assume, though, that we have specified a value of d that can be achieved. If that is the case, we can calculate the smallest needed capacity to accompany the value of d. We then specify multiple values of d from small to large. For each value of d, each of which requires approximation of d/x_t, a capacity is determined. A figure can be created in which d is plotted against c, firm power against capacity. The plot is analogous to the storage–yield curve we described in the first portion of the chapter. It might be called a storage–power curve.

The problem of maximizing the sum of hydropower production rates from a single reservoir is not so straightforward as maximizing the firm yield. Many investigators have studied the problem (see "References on Hydropower Optimization"). Nor is the problem of optimizing firm electric power yield from multiple reservoirs, in which the power constraint consists of the sum of product terms, an easy problem.

1.H. INTEGRATING RESERVOIR SERVICES

To this point we have shown how to model and optimize the water supply function. We have also shown how to incorporate flood control into water supply reservoir models. We did this by modifying the capacity limits of the water supply reservoir by month by subtracting the storage volume that was needed for capturing floods in that month. Then we modeled hydropower production, structuring nonlinear constraints requiring a specified firm power. We were able, by piecewise approximation, to convert the nonlinear constraints to linear constraints.

The first function, water supply, is probably the most basic service the reservoir provides. Flood control is probably the second most basic service. Hydropower is both very important and the most difficult of the services to model.

The principal reservoir services remaining are generally easier to model. One such service remaining is flow maintenance for (1) navigation or (2) for dilution of pollution or (3) for environmental purposes such as enhancement of conditions for fish spawning. Another service is recreation. Whereas flow maintenance implies sustaining the flow in the stream, possibly even through the dry season, recreation requires a relatively steady, low-fluctuation storage volume. Neither of these services is particularly hard to model.

The last reservoir service we need to discuss is irrigation. Although superficially, irrigation might seem to be very similar to water supply in its exercise, in fact, irrigation needs are often more flexible. Water supply implies use by cities for people and industry. Irrigation means application of water in agriculture. Water supply needs are more firm; agricultural needs have targets with some "give" to them. In addition, a shortfall of irrigation water in one early season may decrease needs later because there are fewer crops to water.

Thus, to conclude our tour of reservoir modeling, we assign all these functions to a single reservoir: water supply, flood control, hydropower production, flow maintenance, recreation, and irrigation, and we show how to model these services simultaneously. In Chapter 13, methods are discussed for allocating the cost of reservoirs among the various services that they provide.

We begin by assuming we have a reservoir of capacity c and a long flow record to accompany it. From this long flow record, the worst drought is abstracted for our analysis. That is, we will look at how the reservoir is operated for these multiple functions in the worst water availability environment that has been experienced.

Initially, we will need to add one more variable and a number of parameters to our lengthy set of notation. As we proceed, a few more variables will be needed. The new variables and parameters we need now are

z_k = amount of water to be delivered for irrigation, in month k (k = 1,2,...,12)

t_k = target or ideal irrigation delivery in month k (k = 1,2,...,12)

q_k = water supply need that must be provided in month k (k = 1,2,...,12)

d_k = power requirement for month k (k = 1,2,...,12)

u = maximum monthly flow through the turbines

f_k = streamflow maintenance requirement, month k (k = 1,2,...,12)

The value of the irrigation target, t_k, may be larger during this period of low streamflows than during a period of normal flows. This condition may obtain because the time of low streamflow may reflect rainfall drought in the irrigation region as well as in the watershed, resulting in greater irrigation needs by crops.

The model needs some overarching purposes for us to provide an appropriate framework for modeling. Thus we assume that the water supply amounts, q_k, must be provided. Additionally, the hydropower needs in month k must be met, and finally, the flood storage volumes, the v_k introduced earlier, are essential to maintain.

Our first modeling step is to restructure all the basic reservoir equations:

$$s_t = s_{t-1} + i_t - w_t - q_k - x_t - z_k \quad t = 1,2,\ldots,n \quad (1.58)$$
$$k = t - 12[(t-1)/12]$$

$$s_t \leq c - v_k \quad t = 1,2,\ldots,n \quad (1.59)$$

$$s_n \geq s_0 \quad (1.60)$$

$$x_t \leq u \quad t = 1,2,\ldots,n \quad (1.61)$$

$$\frac{d_k}{x_t} - \frac{\alpha m}{2} s_{t-1} - \frac{\alpha m}{2} s_t \leq \alpha p \quad t = 1,2,...,n \tag{1.62}$$

$$k = t - 12[(t - 1)/12]$$

$$x_t + w_t \geq f_k \quad t = 1,2,...,n \tag{1.63}$$

$$k = t - 12[(t - 1)/12]$$

The mass balance equation (1.58) now contains all of its prior terms plus more removal terms. As before streamflow, i_t, enters during month t. As it was earlier, spill, w_t, is simply the result of an inability to store water because of capacity limitations or, in this case, to use excess water by passing it through the turbines because of a flow limitation through the turbines. The water supply requirement, q_k, is drawn off in each of the 12 months of the year. Release through the turbines to the stream, x_t, is used to generate power. Finally, a quantity of water, z_k, is drawn off for irrigation. This is a busy reservoir.

In addition, constraints (1.59) limit water storage to the portion of the reservoir that can be devoted to water storage in month k, $c - v_k$. Constraints (1.60) prevent water from being borrowed. Constraints (1.61) limit flow through the turbines. Constraints (1.62) require that a power level d_k be provided in month k. The constraints are nonlinear, but each can be linearized by piecewise approximation. This step will be assumed. Constraints (1.63) are the flow maintenance requirements. These constraints may force spill to be positive in some months even when storage does not reach the storage limit $c - v_k$.

At this point, four services have all been accounted for: water supply, flood control, hydropower production, and flow maintenance. The irrigation function is partially accounted for by the z_k term in (1.58), but the recreation function remains completely unaccounted for. The four services that have been fully modeled may be thought of as services with requirements that must be met.

Irrigation and recreation may be thought of as services that are needed but for which strict requirements are not applicable. Meeting these additional needs as well as possible, given that the other needs are met precisely, may be sufficient. Meeting needs "as well as possible" suggests that the measures of how well these needs are met can be placed in the objective function or can be traded off against one another, or at least are not "hard constraints."

We will treat irrigation and recreation services as objective functions that can be traded off against one another. For irrigation, our objective will be to minimize the maximum fractional deviation from the target for irrigation flow over all months. We will assume more water than the target will not be provided. That is,

$$z_k \leq t_k \quad k = 1,2,...,12 \tag{1.64}$$

We choose to minimize the maximum because irrigation losses are probably convex with increasing shortfalls from the target. Objective functions that involve minimizing the maximum of a function are not unusual in the field of environmental systems. In this book other formulations also use this type of objective function in Chapter 2 (Section 2.D.3; water quality management) and in Chapter 6 (Section 6.D.2; hazardous waste management).

The fractional deviation from the irrigation target in month t is

$$\frac{t_k - z_k}{t_k} \qquad k = 1,2,...,12$$

It is this term for which we minimize the maximum. Thus we write

$$\frac{t_k - z_k}{t_k} < U \qquad k = 1,2,...,12 \tag{1.65}$$

where U is the unknown upper bound of fractional deviations (shortfalls) to be minimized.

The recreation objective will be served if fluctuations in storage and thus in the height of the shoreline can be kept in the narrowest possible range. This translates into keeping the storage in the tightest possible range. We can achieve this by adding constraints to the problem which define the unknown upper bound and the unknown lower bound for storage. Thus we write

$$s_t \leq s_U \qquad t = 1,2,...,n \tag{1.66}$$

$$s_t \geq s_L \qquad t = 1,2,...,n \tag{1.67}$$

where s_U is the unknown upper bound on storage and s_L is the unknown lower bound on storage. The difference between s_U and s_L is to be minimized.

The integrated reservoir problem can be summarized as

$$\text{minimize } z_1 = U$$

$$z_2 = s_U - s_L$$

subject to (1.58)–(1.67) and nonnegativity constraints. The trade-off between the objectives can be developed by the constraint or weighting methods of multiobjective analysis as described in detail in Chapter 12.

EXERCISES

1.1. In Section 1.C we discussed water supply and flood control. We suggested that to maintain a flood volume of v_k in the kth month for the year, the constraint needed would be

$$s_t \leq v_k \qquad k = t - 12[(t - 1)/12]$$

We noted, however, that this constraint really only assures that the volume v_k is free at the end of the month. There is something else wrong with this constraint as well. It assumes that ends of periods are critical, but a later moment, at the beginning of the next period, the value of v_{k+1} may have changed in value from v_k.

(a) Show how you could modify these constraints so that the storage volume left free of water is more nearly v_k throughout the period. A hint is that the average storage in the period might be constrained to assure this free volume. Show such a constraint in linear form.

(b) Suppose that we have v_k needed free in month k and v_{k+1} needed free in the next month. What might be a good value to use for end-of-month storage? Write the constraint.

1.2. In Section 1.F.3 we discussed cost minimization of a water supply reservoir system given a system yield requirement. We focused principally on the situation in which the reservoirs were operated jointly (as opposed to independently) with zero-order rules (independent of current reservoir conditions). The idea was that the system was being built from scratch; no prior reservoirs were in place.

Suppose that the state of the system were slightly different. Suppose that the system could consist of as many as four reservoirs on parallel, nonintersecting streams. Two of the reservoirs are not yet built and two are already built to their maximum size. The two that are built are reservoirs 1 and 4. Reservoirs 2 and 3 may be built.

Community A is a fast-growing metropolis demanding large quantities of water; it is served by reservoir 1 (already built) and can also be served by reservoirs 2 and 3. Reservoirs 1, 2, and 3 are upland sources, so water sent to community A will not require pumping. Reservoir 4 presently supplies only community B, but the yield of reservoir 4 is well in excess of community B's needs. Community B is willing to sell water from reservoir 4 to community A at a flat price of P dollars per billion gallons (assume that any pumping costs are inside P). The maximum quantity they are willing to sell per month is q_M. Community B still requires q_B billion gallons/month, every month.

For community B to sell to A requires a pipeline (aqueduct) to convey water from reservoir 4 to the water treatment plant for community A, where water from reservoir 1 is being merged (potentially) with water from reservoirs 2 and 3.

If water from reservoir 4 is to be used for community A, the pipeline would be built large enough to convey the maximum contribution that reservoir 4 could make, namely q_M. If water from reservoir 4 is used in any quantity, the pipeline would be built to carry q_M. If water from reservoir 4 is not used, the pipeline would not be built.

The cost functions for reservoirs 2 and 3 are represented as fixed-charge cost functions where

$$g_2(c_2) = a_2y_2 + b_2c_2$$

$$g_3(c_3) = a_3y_3 + b_3c_3$$

The units of cost are the annual costs needed to repay the loans for reservoirs of the stated size, where an interest rate and payback period are known. These functions already include the cost of a pipeline from each reservoir to the water treatment plant.

As a technical consultant employed by the water department of community A, you are asked to recommend the capacities to which reservoirs 2 and 3 are to be built, whether a pipeline is to be built from reservoir 4 to A, and how the reservoirs are to be operated. That is, you should derive the monthly contribution of each reservoir toward the supply needed by community A, which we will call q_A, a number that we assume is the same for each of the 12 months of the year. The goal of your optimization problem will be to minimize the annual costs, which are the sum of building and operating the system. Use the variable names for storage, inflow, release, spill and capacity that are standard within the chapter. Assume that long flow records exist for each of the four parallel streams. Releases from each reservoir are specific by month of the year.

REFERENCES AND BIBLIOGRAPHY

References on the Single Water Supply Reservoir

Dorfman, R., 1965. Formal Models in the Design of Water Resource Systems, *Water Resources Research,* 1(3), 329–336.

Fiering, M., 1967. *Synthetic Hydrology,* Harvard University Press, Cambridge, MA.

Fiering, M., and B. Jackson, 1971. *Synthetic Streamflows,* Water Resources Monograph 1, American Geophysical Union, Washington, DC.

Hashimoto, T., J. Stedinger, and D. Loucks, 1982. Reliability, Resilience, and Vulnerability Criteria for Water Resource System Performance Evaluation, *Water Resources Research,* 18(1), 14–20.

Hirsch, R., 1979. Synthetic Hydrology and Water Supply Reliability, *Water Resources Research,* 15(6), 1603–1615.

Howe, C., and F. Linaweaver, 1967. The Impact of Price on Residential Water Demand and Its Relation to System Design and Price Structure, *Water Resources Research,* 3(1), 12–32.

Moy, W.-S., J. Cohon, and C. ReVelle, A Programming Model for Analysis of the Reliability, Resilience, and Vulnerability of a Water Supply Reservoir, *Water Resources Research,* 22(4), 489–498.

Potter, K., 1977. Sequent Peak Procedure: Minimum Reservoir Capacity Subject to Constraint on Final Storage, *Water Resources Bulletin,* 13(3), 521–528.

Shih, J.-S., and C. ReVelle, 1994. Water Supply Operations During Drought: Continuous Hedging Rule, *ASCE Journal of Water Resources Planning and Management,* 120(5), 613–629.

Shih, J.-S., and C. ReVelle, 1995. Water Supply Operations During Drought: A Discrete Hedging Rule, *European Journal of Operational Research,* 82, 163–175.

Viessman, W., and C. Welty, 1985. *Water Management: Technology and Institutions,* Harper & Row, New York.

Vogel, R., and R. Bolognese, 1965. Storage–Reliability–Resilience–Yield Relations for Over-Year Water Supply Systems, *Water Resources Research,* 31(3), 645–654.

Yeh, W., 1985. Reservoir Management and Operations Models: A State-of-the-Art Review, *Water Resources Research,* 21(12).

References on Operation for Flood Control

Karbowski, A., 1993. Optimal Flood Control in Multireservoir Cascade Systems with Deterministic Inflow Forecasts, *Water Resources Management,* 7, 207–223.

Kelman, J., and J. M. Damazio, 1989. The Determination of Flood Control Volumes, *Water Resources Research,* 25(3), 337–344.

Marien, J. L., J. M. Damazio, and F. S. Costa, 1994. Building Flood Control Rule Curves for Multipurpose Multireservoir Systems Using Controllability Conditions, *Water Resources Research,* 30(4), 1135–1144.

Pytlak, R., and K. Malinowski, 1989. Optimal Scheduling of Reservoir Releases During Flood: Deterministic Optimization Problem: 1. Procedure, *Journal of Optimization Theory and Applications,* 61(3), 409–449.

Unver, O., and L. Mays, 1990. Model for Real-Time Optimal Flood Control Operation of a Reservoir System, *Water Resources Management,* 4, 21–46.

Wasimi, S., and P. K. Kitanidis, 1983. Real-Time Forecasting and Daily Operation of a Multireservoir System During Floods by Linear Quadratic Gaussian Control, *Water Resources Research,* 19(6), 1511–1522.

References on Water Resources Simulation

Browder, L. E., 1978. *RESOP-II Reservoir Operating and Quality Routing Program, Program Documentation and User's Manual,* UM-20, Texas Department of Water Resources (now renamed Texas Water Development Board), Austin, TX.

Feldman, A. D., 1981. HEC Models for Water Resources System Simulation: Theory and Experience, in V. T. Chow (ed.), *Advances in Hydroscience,* Vol. 12, Academic Press, San Diego, CA.

Kuczera, G., and G. Diment, 1988. General Water Supply System Simulation Model: Wasp, *ASCE Journal of Water Resources Planning and Management,* 114(4).

Labadie, J. W., A. M. Pineda, and D. A. Bode, 1984. *Network Analysis of Raw Supplies Under Complex Water Rights and Exchanges: Documentation for Program MODSIM3,* Colorado Water Resources Institute, Fort Collins, CO, March.

Sigvaldason, O. T., 1976. A Simulation Model for Operating a Multipurpose Multi-reservoir System, *Water Resources Research,* 12(2), 263–278.

Strzepek, K. M., L. A. Garcia, and T. M. Over, 1989. *MITSIM 2.1 River Basin Simulation Model, User Manual,* draft, Center for Advanced Decision Support for Water and Environmental Systems, University of Colorado, Boulder, CO, May.

U.S. Army Corps of Engineers, Hydrologic Engineering Center, 1982. *HEC-5 Simulation of Flood Control and Conservation Systems, User's Manual,* April.

U.S. Army Corps of Engineers, Hydrologic Engineering Center, 1989. *HEC-5 Simulation of Flood Control and Conservation Systems, Exhibit 8, Input Description,* January.

References on Multiple Reservoir Models

Dreyfus, S., and A. Law, 1977. *The Art and Theory of Dynamic Programming,* Academic Press, San Diego, CA.

Hirsch, R., J. Cohon, and C. ReVelle, 1977. Gains from Joint Operation of Multiple Reservoir Systems, *Water Resources Research* 13(2), 239–245.

Maas, A., M. Hufschmidt, R. Dorfman, H. Thomas, S. Marglin, and G. Fair, 1962. *The Design of Water-Resource Systems,* Harvard University Press, Cambridge, MA.

Major, D., and R. Lenton, 1979. *Applied Water Resource Systems Planning,* Prentice Hall, Upper Saddle River, NJ.

Palmer, R., J. Smith, J. Cohon, and C. ReVelle, 1982. Reservoir Management in the Potomac River Basin, *ASCE Journal of Water Resources Planning and Management,* 108(1), 47–66.

Wathne, M., J. Liebman, and C. ReVelle, 1975. Capacity Determination for Water Supply Reservoirs in Series and Parallel, *Water Resources Bulletin,* June.

References on Hydropower Optimization

Grygier, J. C., and J. R. Stedinger, 1985. Algorithms for Optimizing Hydropower System Operation, *Water Resources Research,* 21(1), 1–11.

Pereira, M., and L. Pinto, 1985. Stochastic Optimization of Multi-reservoir Hydro-Electric System, *Water Resources Research,* 21(6), 779–792.

Reznicek, K., and S. Siminovic, 1990. An Improved Algorithm for Hydropower Optimization, *Water Resources Research,* 26(2), 189–198.

Rosenthal, R. E., 1981. A Nonlinear Network Flow Algorithm for Maximization of Benefits in a Hydroelectric Power System, *Operational Research,* 29(4), 763–786.

Trezos, T., and W. Yeh, 1987. Use of Stochastic Dynamic Programming for Reservoir Management, *Water Resources Research,* 23(6), 983–996.

Turgeon, A., 1981. Optimum Short-Term Hydro Scheduling from the Principle of Progressive Optimality, *Water Resources Research,* 17(3), 481–486.

Yeh, W., L. Becker, and W. Chu, 1979. Real-Time Hourly Reservoir Operation, *ASCE Journal of Water Resources Planning and Management,* 105(2), 187–203.

CHAPTER 2

Water Quality Management

ARTHUR E. McGARITY

2.A. INTRODUCTION

Water quality management models are mathematical formulations that can be solved to help decision makers generate cost-effective pollution control strategies that focus on specific measures of the quality of water in receiving waters. Such models are particularly useful for management efforts that include entire watersheds because they are capable of accounting for the many interactions that occur among the various sources of pollution in a watershed, the effects of pollution on water quality, and the costs of options for reducing pollution discharges.

In this chapter, an overview of water pollution fundamentals is presented followed by the derivation of descriptive models for transport of pollutants and calculation of water quality measures. Then a set of optimization models are presented which can generate solutions that achieve water quality standards and simultaneously, minimize the costs of pollution control. Thus these formulations bring together two goals that are often in conflict: environmental quality and economic efficiency. This chapter focuses most attention on formulations that are applied to freshwater streams and estuaries, although discharges into lakes and the ocean are also important.

Research on water quality management modeling had its origins in the 1920s with the pioneering work of Streeter and Phelps (1925), who developed an analytical solution to the differential equations describing the depletion of dissolved oxygen downstream from a discharge of wastewaters containing organic wastes that are metabolized by microbes in the stream. The usefulness of descriptive models of this sort for management purposes was limited, however, until the advent of the widespread availability of digital computers in the 1950s and parallel developments in the field of operations research led to new possibilities for combining descriptive models with economic considerations. Constrained optimization formulations appeared in the literature in the 1960s, reporting on research by ReVelle et al. (1967, 1968) and others. The models developed in this chapter build on the work of these and other researchers who have made significant contributions over the past 30 years.

This chapter demonstrates that research on theoretical optimization model formulations for water quality management has reached a fairly mature stage of development. However, application of these models, including full implementation, to actual watershed management problems is rare, in part, because of the regulatory approach adopted in the United States toward water quality management and pollution control.

The water pollution regulatory approach in the United States is based primarily on regulation of point-source discharges and determined by the availability of pollution control technology. This approach has not been conducive to the "systems approach" that is inherent in optimization management models. Fortunately, recent trends in regulation point toward much greater emphasis on the entire watershed system, explicit consideration of economic factors, greater participation by local governments, and consideration of alternatives that reduce the amount of pollution generated. Water quality management models based on optimization, which are the primary focus of this chapter, have great potential for providing the types of analyses that are necessary for effective management in this emerging context.

In parts of the world where initial efforts at environmental management are occurring only now, methods of optimization management modeling can be applied to the development of a systems approach to water quality management from the very beginning. In countries of Eastern Europe, for example, where more than 50% of surface water is unsuitable as a source for treated drinking water, watershed management is an urgent matter. The emphasis that water quality management models places on balancing environmental quality concerns with economic considerations is particularly important in such countries where capital to support pollution control projects is in short supply.

2.B. WATER POLLUTION FUNDAMENTALS

This section provides background necessary to understand the various sources of water pollution and the effects that they have on natural receiving waters.

2.B.1. Sources, Characteristics, and Transport

There are many sources of the contaminants that pollute streams, such as industrial discharges, domestic sewage treatment plants, and uncontrolled stormwater runoff from roads, parking lots, and construction sites. Sewer lines in populated areas usually follow streambeds, sometimes crossing them several times and exposing the streams to risks from overflows and broken pipes that leak of raw sewage. Runoff from rural farms and suburban lawn and golf courses includes excess nutrients from fertilizers and pesticides applied to control weeds and insects in seasonally varying amounts. Mining activities expose naturally occurring substances such as arsenic, lead, sulfur, and various salts to leaching by surface water runoff, and sometimes require the removal and discharge of saline groundwater to make coal and ores accessible. Contaminated leachates are also generated by domestic solid waste landfills and abandoned industrial sites.

There are many parameters that are useful for characterizing wastewaters and for measuring the severity of their effects on streams. Physical characteristics of water and wastes include the turbidity (the degree to which light is absorbed as it passes through the water), the suspended solids content (including total solids and the organic and mineral components, some that settle into the streambed and others that remain in colloidal or dissolved form), and other properties, such as odor and taste. Chemical characteristics include concentrations of a host of different elements and compounds, including quality enhancers such as oxygen gas [the dissolved oxygen (DO)] and buffering salts of calcium and magnesium; contaminants such as nitrogen in its ionic forms of ammonium, nitrite, and nitrate; phosphorus in various forms; excess carbonaceous material from organic waste, including excretions from humans, livestock, and waterfowl; excess sulfite ion; oils and grease; chlorinated organic compounds from pesticides and industrial solvents; and metals such as lead, mercury, and cadmium. Biological characteristics include microbiological contamination by pathogens, indicated by the presence of fecal coliform bacteria; excessive populations of all kinds of microbes and algae that deplete dissolved oxygen when stimulated by discharges of organic wastes and nutrients; and macrobiological indicators of water quality, such as an overabundance of pollution-tolerant species that point to long-term impact of contaminated waters.

Pollutants are transported and transformed by many different mechanisms. Physical processes include *advection,* which is the entrainment of the pollution in the flowing water, and *dispersion,* which is the movement of pollution driven by a concentration gradient out from zones of high concentration into areas of lower concentration as a result of diffusion and flow turbulence. Most pollutants are subject to some combination of removal and/or decay processes. Solid pollutants will tend to settle out onto the bottom of the stream and may be immobilized for a period of time in the streambed sediments but later move downstream when the bottom is scoured by high-velocity flow following a heavy rainfall. Dissolved metals may precipitate or reenter solution as the stream's pH and temperature change. The biota of a stream are responsible for some of the more complex transport and transformation mechanisms. Excess nutrients, metals, and synthetic organic molecules from pollution discharges work their way, primarily downstream, through the stream ecosystem by incorporation in the cells of microbes and the tissues of plants, macroinvertibrates, and fish. In estuaries and bays, tidal forces transport pollutants both upstream and downstream as the water sloshes back and forth on its way toward the sea.

In this chapter much attention is focused on the transport and transformation of the organic components of wastewaters through microbial action.

When such wastes are discharged, the microbes in the stream, or in the waste if the source is a sewage discharge (treated or untreated), will increase in population as they consume the waste, converting the suspended and dissolved forms of carbon and nutrients into cell material and energy. The respiration of these microbes places a demand on the meager dissolved oxygen content of the water (less than 14 parts per million under the best of circumstances, i.e., when saturated), which can temporarily deplete the oxygen and deprive other aquatic species, such as fish, of the oxygen they need to survive. As the waste and the associated exploding population of microbes moves downstream, the oxygen deficit grows until a critical point is reached where the decaying quantity of waste can no longer support the high microbe population and the rate of oxygen consumption equals the rate of natural replenishment of oxygen from the atmosphere. At this point the oxygen deficit begins to decline, and unless the stream encounters additional discharges of organic waste, the dissolved oxygen concentration will gradually approach a level near the saturation point.

2.B.2. Measurement and Monitoring of Pollutants

Methods and Units of Measurement. The severity of a pollution discharge is measured both by the total mass of pollutant discharged and by the resulting concentration of the pollutant in the receiving waters. Pollutant mass *flow rates* are also important and are calculated by multiplying the volumetric flow rate by the concentration. Total mass and mass flow rates are typically measured in units of pounds or tons (kilograms or metric tonnes outside the United States) and in pounds per day or tons per year (kilograms per day or tonnes per year). Volumetric flow rates of pollution discharges and streamflows are expressed as either million gallons per day or cubic feet per second (cubic meters per second outside the United States). Concentrations are typically measured in units of milligrams per liter, which is conveniently equivalent to parts per million by weight since 1 liter of water weighs 1000 grams (1 million milligrams).

A wide variety of methods are available to measure the concentration of pollutants in wastewaters and streams. Conventional methods of analytical chemistry, including titrations and colorimetric techniques, are used to measure concentrations of compounds, ions, and metals in the range of low parts per million. Many water quality parameters can be measured by electronic probes, including dissolved oxygen, conductivity (which correlates with salinity), nitrate and ammonium ion, and turbidity. Microbiological contamination is measured by either membrane filtration or multitube fermentation tests that indicate the number of fecal coliform bacteria present in a volume of sample, typically 100 mL.

Water Quality Monitoring. Occasionally sampling from streams followed by laboratory analysis provides useful "snapshots" of quality parameters. However, many important events that degrade water quality are missed by grab samples taken once per year or even once per month. Storm-related discharges from parking lots, roads, lawns, farms, and sanitary sewer overflows will flush a variety of contaminants into streams.

One way to estimate the integrated effects of all the various discharge and runoff events is by examining the kinds of biota that live in a stream. Macrobiological sampling and classification of invertebrates by species in stream riffles can determine whether these habitats are occupied primarily by pollution-intolerant species (indicating good water quality) or by pollution tolerant species (indicating poor water quality). Biological monitoring has been used frequently in recent years to evaluate the overall health of streams, particularly in areas where the streams are stressed by continuous discharges, sewer overflows, and storm runoff.

A shortcoming of biological monitoring, though, is the lack of data on the specific physical and chemical causes of poor water quality conditions. One solution is to deploy electronic monitoring devices or *sondes* in a stream to monitor certain parameters once per hour or even more frequently. Data are stored in a computer embedded in the battery-powered sonde or encoded in telemetry and transmitted to a central site, sometimes by satellite.

Another approach is to sample from streambed sediments and analyze the samples for toxic metals and organic compounds. Atomic absorption spectrophotometry is commonly used for measurement of metals such as cadmium, lead, arsenic, and mercury. Organic compounds such as pesticides and polychlorinated biphenyls (PCBs) are routinely measured using gas and liquid chromatography. Standard methods have been developed for such measurements and are promulgated by professional societies and government agencies such as the U.S. Environmental Protection Agency (EPA).

Measurement of Biochemical Oxygen Demand (BOD). Special attention is given here to measurement of BOD because the method simulates the actual oxygen depletion and BOD decay process in a stream exposed to oxygen-consuming organic wastes. The descriptive models of oxygen sag in a stream developed later in this chapter contain mathematical representations of these processes that are similar to the formulas used to calculate BOD.

BOD is a convenient way of measuring the effective concentration of a host of different organic compounds in water and wastewater. These compounds take many different forms, including suspended solid matter, dissolved solids and liquids, and living microorganisms. The decomposition of these compounds by microbes consumes oxygen gas molecules dissolved in the water, and the amount of oxygen consumed is a measure of the potency of the waste (i.e., its potential to do damage to the stream environment). The decomposition process is a slow one, taking up to 2 weeks, so measurement of BOD takes a long time, and the standard test requires 5 days (5-day BOD).

The 5-day BOD test can be performed on a stream sample by (1) aerating the sample to boost its oxygen content until it reaches the saturation concentration, (2) pouring the sample into a bottle which is then tightly sealed to avoid contamination from atmospheric oxygen, and (3) placing the bottle in a dark chamber held at a constant temperature at 20°C for 5 days. The oxygen concentration of the sample is measured at both the beginning and end of the period, and the difference in these concentrations [the total decrease in oxygen gas (O_2) concentration] is equal to the 5-day BOD in mg/L. If the ending concentration is near zero, it is likely that some of the organic material remains and the test must be repeated with a diluted sample. Typically, multiple dilutions are run for each sample, particularly for wastewaters that can have BOD concentrations of 200 mg/L or higher. The difference between the initial and final O_2 concentrations is multiplied by a dilution factor to obtain the BOD of the undiluted sample.

The kinetics of the BOD decay process is usually modeled as a first-order decay process:

$$L_t = Le^{-k_1 t} \tag{2.1}$$

where

t = elapsed time since beginning of test (days)

k_1 = biooxidation rate constant (day^{-1})

L = ultimate BOD of the sample (mg/L), the BOD remaining at the beginning of the test and the total BOD exerted if the test were allowed to run until all organic material is decomposed

L_t = amount of BOD remaining at time t

The 5-day BOD is the amount of oxygen consumed over a 5-day period and is equal to the difference between the ultimate BOD, L, and the BOD remaining after 5 days, L_5. In general,

$$y_t = L - L_t \tag{2.2}$$

where y_t is the amount of oxygen consumed (BOD exerted, mg/L) over t days. L and y are often confused by students learning about BOD for the first time. It may be helpful to think of L as representing the effective concentration of biodegradable organic compounds and to think of y as the amount of dissolved oxygen consumed per liter of water by the biooxidation process.

The biooxidation rate constant k_1 varies with the composition of the organic waste and with the water temperature. Water quality management models require an accurate estimate of k_1 for the actual chemical composition and temperature of the stream. Greatest accuracy is obtained by laboratory tests with polluted stream water conducted at the actual temperature of the stream. However, reasonably accurate estimates can be obtained by laboratory tests at 20°C and adjusted for the actual stream temperature using the *Hoff–Arrhenius formula:*

$$k_1(T) = k_{(1,20^\circ)}\theta^{T-20} \tag{2.3}$$

where T is the stream water temperature (°C), θ an empirically determined constant which depends on the type of organic waste, $k_{1,20^\circ}$ a biooxidation constant determined from laboratory tests at 20°C, and $k_1(T)$ a biooxidation constant at temperature T.

2.C. DESCRIPTIVE MODELS FOR FRESHWATER STREAMS

In this section we present models that are useful for describing important physical, chemical, and biological processes in freshwater streams that receive discharges of polluted water. In following sections, these descriptive models are incorporated into linear programming management models that impose objectives and constraints on the larger system that is comprised of the natural stream processes and the technological pollution control processes.

2.C.1. Dilution Processes

The concentration of any material dissolved or suspended in water changes in direct proportion to the amount of water in the stream through dilution processes. Streamflow rates change as wastewater discharges and tributaries join the flow, as runoff from the watershed varies with rainfall amounts, and as groundwater levels vary with the seasons when the streambed intersects the water table. The concentration of any pollutant can be reduced to an arbitrarily low level by mixing enough pure water with the polluted water. The dispersal of pollutants, as opposed to their removal, is often called the *dilution solution to pollution.*

Dilution processes are modeled by a mass balance on each component or characteristic of interest in water quality assessment. Applying the mass balance can be complicated by flow phenomena downstream from a discharge that inhibit complete mixing. A significant concentration gradient can appear vertically and transverse to the streamflow for many miles downstream of a discharge when the stream is deep and moving slowly with little turbulence. Often, special equipment is installed at the point of wastewater discharges to force mixing and rapid dilution of the waste.

Most management models make the simplifying assumption of "perfect mixing" at points of discharge of wastewater and tributaries. If X represents the concentration of a chemical component (such as O_2 or BOD) or a physical property (such as temperature or turbidity), Q represents volumetric flow rates in the stream, and q represents the volumetric flow rate of the pollution or tributary discharge, the perfect mixing assumption leads to the following mass balance relationship:

$$X_uQ_u + X_jq_j = X_0Q_0 \tag{2.4}$$

where the following subscripts indicate the specific locations in the proximity of the discharge: u, indicating in the stream, just upstream of the discharge point; j, indicating in the pollution discharge (or tributary) just before mixing with stream; and 0, indicating in the stream, just downstream of the discharge mixing point. Equation (2.3) is typically applied in situations where X_0 is calculated from the other quantities using the formula

$$X_0 = \frac{Q_u}{Q_u + q_j} X_u + \frac{Q_j}{Q_u + q_j} X_j \tag{2.5}$$

which includes the water flow balance: $Q_0 = Q_u + q_j$.

2.C.2. Pollution Decay and Oxygen Depletion Processes

The concentration of a pollutant also changes as a result of decay processes that remove it from solution such as (1) conversion by chemical reactions or biological processes to a different compound; (2) precipitation and coagulation, creating settleable particles; (3) bottom deposition; and (4) bubble formation followed by escape to the atmosphere. Of course, decay of one type of pollutant can result in formation of another type of pollutant if the decay products are considered harmful to water quality.

Descriptive models that determine the concentration of pollutants in a body of water at different times and at different distances away from the points of discharge must account for the rates of decay and formation. Frequently, a simple first-order representation of the decay process is adequate:

$$r_d = -k_d X \tag{2.6}$$

where r_d is the rate of decay in pollutant concentration (mg/L per day), k_d a decay coefficient (day^{-1}), and X the pollutant concentration (mg/L).

For example, when the concern is organic wastes that exert a BOD on the stream, the rate of decay in BOD is expressed by

$$r_1 = -k_1 L \tag{2.7}$$

where r_1 the rate of BOD decay (mg/L per day), k_1 a BOD decay (biooxidation) coefficient (day^{-1}) as in equation (2.1), and L the amount of BOD remaining (mg/L). The subscript 1 is used because BOD decay is one of two processes (aeration is the other) involved in the model developed below to calculate the dissolved oxygen concentration in a polluted stream.

When there are no other processes affecting BOD concentration, such as bottom scour, which can add BOD to the stream, the rate of decay is also the rate of change in the BOD concentration:

$$\frac{dL}{dt} = -k_1 L \qquad \text{which has the solution} \qquad L = L_0 e^{-k_1 t} \tag{2.8}$$

where L_0 is the initial BOD concentration in the stream just after complete mixing of the stream with waste (mg/L) and t is the time of flow downstream from the discharge point (days).

The depletion of stream-dissolved oxygen is directly linked to BOD decay because the amount of oxygen removed from the stream is identical to the amount of BOD exerted in the stream:

$$y = L_0 - L \tag{2.9}$$

where y is the oxygen depletion (mg/L), and the *rate* of oxygen depletion is

$$\frac{dy}{dt} = -\frac{dL}{dt} = -r_1 = k_1 L \tag{2.10}$$

The formulas for first-order decay in a stream are very similar to those describing the kinetics of BOD measurement in the laboratory. In fact, it is this similarity between the laboratory measurement process and the actual stream processes that is responsible for the widespread use of BOD as a measure of the effects of wastewater discharges on water quality. Of course, the biooxidation constant k_1 in the stream will not be the same in the stream as it is in the BOD bottle in the laboratory. However, adjustments for the stream water temperature can be made using equation (2.3), and additional stream conditions as described in Problem 2.1.

2.C.3. Aeration Processes

Streams and other bodies of water having an upper surface in contact with the atmosphere obey *Henry's law,* which states that the concentration of an atmospheric gas dissolved in water in equilibrium is directly proportional to the partial pressure of that gas in the atmosphere. Equilibrium conditions exist when the rate of movement of the gas into solution equals its rate of movement out of solution (back to the atmosphere) and corresponds to the condition normally referred to as *saturation.* Henry's law is usually stated as

$$X_g = HP_g \tag{2.11}$$

where X_g is the mole fraction of the gas in water (dimensionless), P_g the partial pressure of the gas in the atmosphere (atm), and H is Henry's constant (atm^{-1}). H varies strongly with the temperature of the water and is also affected somewhat by the concentration of dissolved ionic solids (salinity) in the water.

In water quality modeling, we are particularly interested in the concentration of O_2 molecules in the water. A more convenient form of Henry's law specifically for oxygen expresses the saturation concentration of water as a function of its temperature. Tabular representations of this function are widely published.

A polynomial fit to tabulated data can also be useful for computer programs that simulate aeration processes in streams:

$$C_s = 14.652 - 0.41022T + 0.00799T^2 - 0.000777774T^3 \quad (2.12)$$

where C_s is the dissolved oxygen saturation concentration (mg/L) and T is the water temperature (°C). Equation (2.12) applies to water that is low in salinity, as in a freshwater stream. Similar representations can be developed for saline waters.

When water is polluted with oxygen-depleting compounds, the dissolved oxygen concentration is usually depressed below the saturation value and an oxygen *deficit* exists:

$$D = C_s - C \quad (2.13)$$

where D is the dissolved oxygen deficit (mg/L) and C is the dissolved oxygen concentration (mg/L). When the deficit is greater than zero, a *reaeration process* occurs that is driven by diffusion of oxygen molecules across the air–water interface. The rate of reaeration is directly proportional to the deficit:

$$r_2 = k_2 D \quad (2.14)$$

where r_2 is the rate of oxygen diffusion into water per unit volume (mg/L per day) and k_2 is the reaeration rate coefficient (day^{-1}).

The proper value of k_2 to use in a specific stream modeling situation depends on several factors, including the amount of turbulence (which affects the surface area exposed to the atmosphere per unit volume of water) and the water temperature. Much research has been done on the problem of calculating k_2 directly from easily measured stream parameters. The formula of O'Connor and Dobbins (1958) is based on theoretical considerations and compares well with empirical formulas:

$$k_2 = \frac{(D_L u)^{1/2}}{h^{3/2}} \quad (2.15)$$

where D_L is the oxygen diffusivity at 20°C (0.000081 ft^2/hr), u the average stream velocity (ft/hr), and h the average stream depth (at a point, h = cross-sectional area ÷ width, and over an entire reach, h = total volume ÷ surface area).

2.C.4. Example: Dissolved Oxygen Monitoring Data

An example of data covering 6 years at sites on Crum Creek, a small stream in urbanized Delaware County, Pennsylvania, is included here to illustrate dissolved oxygen depletion problems on freshwater streams. Dissolved oxygen data, shown in Figure 2.1, were obtained by a group of volunteer stream monitors called the Crum–Ridley–Chester monitors. The author of this chapter serves as a technical advisor to this group. Most of the data were taken at 1-month intervals.

The measured points are plotted directly from the volunteer monitors' data. The saturated points indicate what the dissolved oxygen concentrations would have been if the stream water had been saturated with oxygen. These points are calculated from water temperature measurements that were taken simultaneously with the dissolved oxygen measurements.

These results suggest that there are times, particularly in the summer months, but also during the winter, when significant dissolved oxygen deficits occur at certain points on the stream. However, little more can be determined from the data alone. Additional information is needed, particularly stream flow rates, and these data, along with information on the biochemical oxygen demand of known discharges, biooxidation decay coefficients, and stream reaeration rates, can be incorporated into a descriptive model that calculates

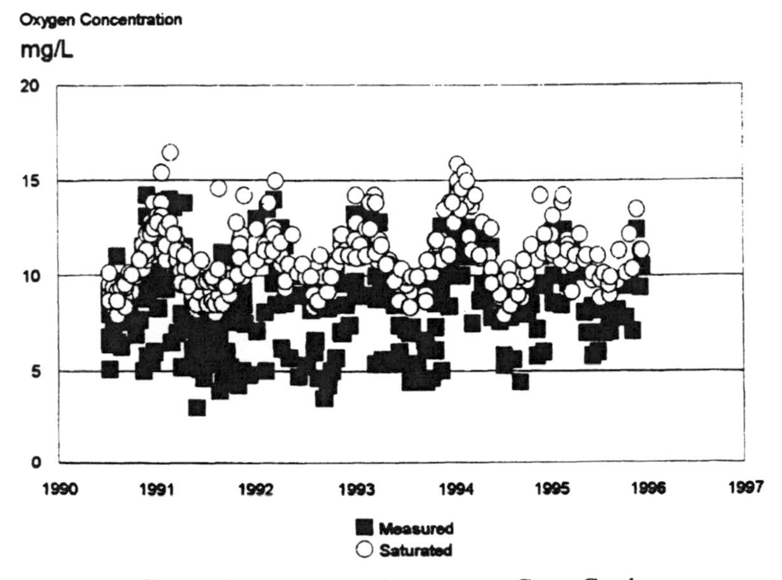

Figure 2.1 Dissolved oxygen on Crum Creek.

the oxygen deficit at the points monitored. One such model, a *dissolved oxygen sag model,* is presented in the next section.

2.C.5. Streeter–Phelps Analytical Dissolved Oxygen Sag Model

The BOD decay, oxygen depletion, and aeration processes described above are now combined into a single model for calculating the dissolved oxygen (DO) concentration in a freshwater stream. The model is considered *one-dimensional* because variations in DO are allowed in only one dimension, the flow distance downstream from the points of pollution discharge. In effect, we follow an element of volume having the same depth and width as the stream as it moves along down the stream. We assume that the volume element is perfectly mixed and that its DO concentration (and all other properties as well) can be represented by a single value.

The processes affecting DO concentration acting on the volume element must also obey the law of conservation of mass. The *oxygen balance* representing this physical law is expressed as

$$\frac{dC}{dt} = -k_1 L + k_2(C_s - C) \tag{2.16}$$

When this equation is expressed as an oxygen deficit using equation (2.13), and L is substituted using equation (2.8), we have the linear, first-order differential equation

$$\frac{dD}{dt} + k_2 D = k_1 L_0 e^{-k_1 t} \tag{2.17}$$

with analytical solution

$$D = \frac{k_1}{k_1 - k_2}(e^{k_1 t} - e^{k_2 t})L_0 + (e^{-k_2 t})D_0 \tag{2.18}$$

where D_0 is the DO deficit just after the discharge point. This is the well-known *Streeter–Phelps equation* (Streeter and Phelps, 1925). Figure 2.2 illustrates a typical solution to equation (2.18) displayed as DO concentration for the set of parameters k_1, k_2, C_s, D_0, and L_0, as shown.

The downstream flow time is related to the downstream distance by the velocity of flow. The average velocity over a cross section of the stream transverse to the flow is the volumetric flow rate divided by the cross-sectional area. Over a section of a stream, called a *reach,* the volumetric flow rate remains fairly constant, but the cross-sectional area may vary greatly, usually as a result of changes in water depth as the stream alternates between deep pools and shallow riffles. If an average flow velocity over each reach can be

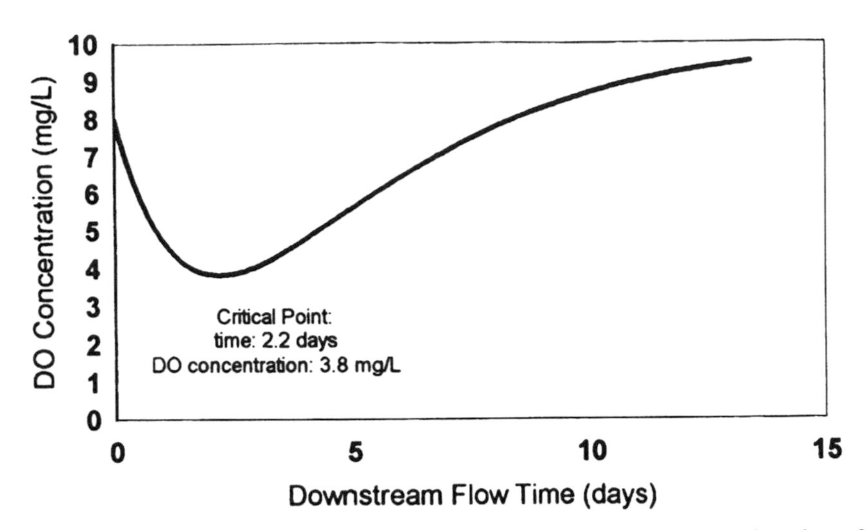

Figure 2.2 Analytical solution to the Streeter–Phelps oxygen sag equation. $k_1 = 0.3$ day^{-1}, $k_2 = 0.5$ day^{-1}, $C_S = 10$ mg/L, $D_0 = 2$ mg/L, $L_0 = 20$ mg/L.

determined, the relationship between the distance downstream from a discharge and the flow time is approximated

$$x = \bar{v}t \tag{2.19}$$

where $\bar{v}$ is the average velocity over the length of the reach and x is the distance downstream. In the model derivations that follow, flow time rather than downstream distance will be used since the pollutant decay and reaeration processes are fundamentally dependent on time rather than distance. Whenever a distance reckoning is required, equation (2.19) can be used for the conversion.

Figure 2.2 reveals that there is a "critical point" where the concentration falls to a minimum. The critical time, t_c, the downstream flow time associated with the critical point, is determined analytically by setting $dD/dt = 0$ and solving for $t = t_c$. The resulting analytical formula is

$$t_c = \frac{1}{k_2 - k_1} \ln \left[\frac{k_2}{k_1} \left(1 - \frac{k_2 - k_1}{k_1} \frac{D_0}{L_0} \right) \right] \tag{2.20}$$

The critical deficit, D_c, can be calculated from t_c:

$$D_c = \frac{k_1 L_0}{k_2} e^{-k_1 t_c} \tag{2.21}$$

or directly from the stream parameters by substituting for t_c:

$$D_c = \frac{k_1 L_0}{k_2} \left[\frac{k_2}{k_1} \left(1 - \frac{k_2 - k_1}{k_1} \frac{D_0}{L_0} \right) \right]^{-[k_1/(k_2-k_1)]} \tag{2.22}$$

and the critical DO concentration, C_c, is calculated directly from the saturation concentration, C_s, and the critical deficit:

$$C_c = C_s - D_c \tag{2.23}$$

Figure 2.2 displays the critical time and DO concentration calculated from these formulas for the example DO sag problem.

The Streeter–Phelps equation represents in the simplest and most basic form the dynamics of the competition between the oxygen-depleting and oxygen-renewing processes in a stream, and it is used in the models that follow in this chapter. Extensions of this equation have been developed to account for factors such as settling of suspended organic particulate matter onto the bottom in low-velocity reaches of streams, and the reverse, scouring of bottom sediments in high-velocity reaches. Settling (sedimentation) decreases the BOD in the water column, whereas scouring increases the BOD. Although settled organic solids in the stream sediments are not included in the water's BOD concentration, they do contribute to oxygen depletion in a different way called *benthal oxygen demand.* The presence of photosynthetic plant life in the stream, especially algae, contributes oxygen to the water, particularly in well-illuminated reaches. Nonpoint sources of runoff can also contribute BOD and nutrients to a stream. Straightforward extensions of the Streeter–Phelps equation incorporating all of the above-mentioned factors have been developed (Camp, 1963; Dobbins, 1964). Others have extended the model to include the "nitrogenous" oxygen demand exerted by *Nitrosomonas* and *Nitrobacter* bacteria in the conversion of ammonia into nitrite and nitrate. The effects of algae, including photosynthetic production of oxygen and oxygen depletion through respiration, can also be incorporated into this analytical solution. These extensions are not necessary, however, to derive the descriptive and management models in this chapter, and their inclusion in these models is left as an exercise for the reader.

2.C.6. Multireach–Multidischarge Models

In this section a general model is developed for calculating pollutant concentrations and water quality parameters in streams having several reaches distinguished by different flow, biooxidation, and reaeration parameters and subjected to discharges from several different point sources. Typically, a section of the stream is designated as a separate reach whenever a significant inflow, such as a wastewater discharge or a tributary, occurs. However, significant changes in the average depth of the stream or other streambed characteristics, which affect the velocity and velocity-dependent parameters such as the reaeration coefficient, could also lead to the designation of a separate reach.

Consider the sequence of three reaches shown in Figure 2.3. The reaches are indexed by i. At the top of each reach is a discharge, indicated by a downward-pointing arrow and indexed by j. These indices correspond exactly (i.e., discharge 1 joins the stream at the top of reach 1, and so on). Each discharge can affect the water quality in all downstream reaches because of transport of pollutants by the stream from one reach to another. Different indices are used for the reaches and discharges so that pollution transport coefficients can be defined to indicate the effects of a discharge on points in any downstream reach. Each discharge has an associated withdrawal, indicated by an upward-pointing arrow, to allow for the case of a town or industry that uses a significant portion of the stream's flow for treated water supply. The withdrawal will sometimes be zero, for example, when the discharge is a tributary.

Initial, Critical, and Endpoint Calculations for Each Reach. There are three particularly significant flow times on each reach. The first, indicated by $t_{i,0}$, is just downstream from the point where the discharge $j = i$ mixes with the stream. The next point on each stream is associated with the critical time, indicated by $t_{i,c}$, when the oxygen deficit is greatest. The last point is the endpoint, just before the next discharge mixes with the stream, indicated by $t_{i,e}$. Pollutant concentrations and water quality parameters must be calculated at points $t_{i,0}$ and $t_{i,e}$ to account for the dilution processes at each point of discharge. Similar calculations are necessary at $t_{i,c}$ to check whether water quality standards are being violated. There could also be other points in a reach, such as at the location of a water intake, where the calculation of water quality parameters is required, and in the formulas that follow, the flow time $t_{i,s}$ associated with that point can be substituted for the critical time $t_{i,c}$.

Organic waste pollution, which exerts BOD, is used here as an example of a pollutant that decays with time downstream of the discharge point, and DO concentration (deficit) is the water quality parameter to which a standard is applied. If we let $L_{i,s}$ and $D_{i,s}$ represent the BOD concentration and DO deficit, respectively, at points $s = 0$ and e, equation (2.5) applied to the

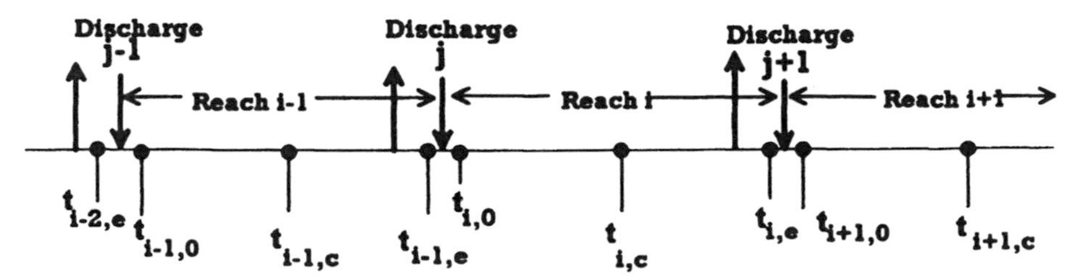

Figure 2.3 Schematic diagram of multireach–multidischarge stream model.

dilution process at discharge $j = i$ is used to calculate L and D at the top of reach i:

$$L_{i,0} = \frac{Q_{i-1} - q_{j,\text{out}}}{Q_i} L_{i-1,e} + \frac{q_{j,\text{in}}}{Q_i} l_j \tag{2.24}$$

and

$$D_{i,0} = \frac{Q_{i-1} - q_{j,\text{out}}}{Q_i} D_{i-1,e} + \frac{q_{j,\text{in}}}{Q_i} d_j \tag{2.25}$$

where Q_i is the flow rate in reach i (m^3/s), $q_{j,\text{out}}$ the flow rate of withdrawal j (m^3/s), $q_{j,\text{in}}$ the flow rate of discharge j (m^3/s), l_j the BOD concentration in discharge j (mg/L), and d_j the DO deficit in discharge j (mg · L).

The BOD decay processes, expressed in equation (2.8), determine the BOD concentrations at points c and e downstream on reach i as follows:

$$L_{i,s} = \rho^i_{i,s} L_{i,0} \tag{2.26}$$

where $\rho^i_{i,s} = e^{-k_{1i}t_{i,s}}$ is the BOD decay factor for discharge $j = i$ at point $s = c$ or e on reach i.

Dissolved oxygen depletion and recovery processes, expressed in equation (2.18), determine the DO concentrations at points c and e downstream on reach i as follows:

$$D_{i,s} = \alpha^i_{i,s} L_{i,0} + \beta^i_{i,s} D_{i,0} \tag{2.27}$$

where $\alpha^i_{i,s} = [k_{1i}/(k_{1i} + k_{2i})](e^{-k_{1i}t_{i,s}} - e^{-k_{2i}t_{i,s}})$ is the BOD exertion factor for discharge $j = i$ at point $s = c$ or e on reach i, and $\beta^i_{i,s} = e^{-k_{2i}t_{i,s}}$ is the initial deficit decay factor for discharge $j = i$ at point $s = c$ or e on reach i.

Recursive Formulas for a Two-Reach Model. The derivations above relate water quality on a reach to the discharge at the top of the reach. The effects of other upstream discharges are considered only indirectly through the concentrations $L_{i-1,e}$ and $D_{i-1,e}$ at the end of the previous reach. Here, coefficients are derived that directly express the effects of each upstream discharge on the water quality at the critical point on a reach. These coefficients can be obtained for each reach recursively from coefficients that apply to the previous reach. The derivation is shown here for two reaches, $i = 1$ and 2, and is then generalized for an arbitrary number of reaches.

The derivation of the recursive formulas for dissolved oxygen deficit proceeds by combining the dilution processes [equation (2.25)] at the top of each reach with the oxygen sag processes [equation (2.27)] over each reach in a single formula. Dilution at discharge 1 is expressed by applying equations (2.24) and (2.25) for the case of $i = 1$ in equations (2.28) and (2.29).

$$L_{1,0} = \frac{Q_0 - q_{1,\text{out}}}{Q_1} L_{0,e} + \frac{q_{1,\text{in}}}{Q_1} l_1 \tag{2.28}$$

$$D_{1,0} = \frac{Q_0 - q_{1,\text{out}}}{Q_1} D_{0,e} + \frac{q_{1,\text{in}}}{Q_1} d_1 \tag{2.29}$$

The deficit at the critical point on reach 1 is

$$D_{1,c} = \alpha_{1,c}^1 L_{1,0} + \beta_{1,c}^1 D_{1,0} \tag{2.30}$$

Combining equations (2.28)–(2.30), and separating the terms related to the upstream parameters from those of the discharge parameters, we obtain the following expression for the critical deficit that combines dilution and oxygen sag:

$$D_{1,c} = \frac{Q_0 - q_{1,\text{out}}}{Q_1} (\alpha_{1,c}^1 L_{0,e} + \beta_{1,c}^1 D_{0,e}) + \frac{q_{1,\text{in}}}{Q_1} (\alpha_{1,c}^1 l_1 + \beta_{1,c}^1 d_1) \tag{2.31}$$

At the end of the first reach, the deficit is expressed similarly:

$$D_{1,e} = \frac{Q_0 - q_{1,\text{out}}}{Q_1} (\alpha_{1,e}^1 L_{0,e} + \beta_{1,c}^1 D_{0,e}) + \frac{q_{1,\text{in}}}{Q_1} (\alpha_{1,e}^1 l_1 + \beta_{1,e}^1 d_1) \tag{2.32}$$

and the BOD concentration at the endpoint is

$$L_{1,e} = \frac{Q_0 - q_{1,\text{out}}}{Q_1} \rho_{1,e}^1 L_{0,e} + \frac{q_{1,\text{in}}}{Q_1} \rho_{1,e}^1 l_1 \tag{2.33}$$

The expression for the critical deficit on the second reach, analogous to equation (2.31) for the first reach, is

$$D_{2,c} = \frac{Q_1 - q_{2,\text{out}}}{Q_2} (\alpha_{2,c}^2 L_{1,e} + \beta_{2,c}^2 D_{1,e}) + \frac{q_{2,\text{in}}}{Q_2} (\alpha_{2,c}^2 l_2 + \beta_{2,c}^2 d_2) \tag{2.34}$$

Now, $L_{1,e}$ and $D_{1,e}$ are replaced using equations (2.33) and (2.32), respectively, and terms are regrouped to separate those relating to stream parameters above the first discharge from those of the first discharge and those of the second discharge:

$$D_{2,c} = \frac{Q_1 - q_{2,\text{out}}}{Q_2} \frac{Q_0 - q_{1,\text{out}}}{Q_1} [(\alpha^2_{2,c}\rho^1_{1,e} + \beta^2_{2,c}\alpha^1_{1,e})L_{0,e} + \beta^2_{2,c}\beta^1_{1,e}D_{0,e}]$$

$$+ \frac{Q_1 - q_{2,\text{out}}}{Q_2} \frac{q_{1,\text{in}}}{Q_1} [(\alpha^2_{2,c}\rho^1_{1,e} + \beta^2_{2,c}\alpha^1_{1,e})l_1 + \beta^2_{2,c}\beta^1_{1,e}d_1]$$

$$+ \frac{q_{2,\text{in}}}{Q_2} (\alpha^2_{2,c}l_2 + \beta^2_{2,c}d_2) \tag{2.35}$$

This equation shows clearly that the critical deficit on the second reach can be related directly to a linear combination of the initial stream parameters and the parameters of the first and second discharges. The linearity of this relationship will be exploited in the following sections, where linear programming formulations for water quality management are developed. First, though, a more compact version of equation (2.35) is developed which incorporates a recursive relationship for the coefficients.

We define the following coefficients for the effects of initial stream and first discharge parameters on deficit at the critical point on reach 2:

$$\alpha^1_{2,c} = \alpha^2_{2,c}\rho^1_{1,e} + \beta^2_{2,c}\alpha^1_{1,e} \tag{2.36}$$

= BOD exertion factor for discharge 1 on reach 2

$$\beta^1_{2,c} = \beta^2_{2,c}\beta^1_{1,e} \tag{2.37}$$

= initial deficit decay factor for discharge 1 in reach 2

To simplify equation (2.35) still further, the concept of a *partial deficit* is introduced. Conceptualize the flow in the stream as a bundle of strands, starting with a single strand before the first discharge point. Each discharge adds an additional strand of BOD and DO deficit to the bundle. Let the *undiluted partial deficit,* $D^j_{i,c}$, be the DO deficit in the strand emanating from discharge j, unaffected by dilution from the other strands. The overall deficit at the critical point of reach i is a weighted sum of the undiluted partial deficits from all upstream discharges and the initial stream deficit. The weighting factors account for the dilution effects of the withdrawals and discharges, and each term in the summation can be thought of as a *diluted partial deficit.*

The undiluted partial deficits comprising the deficit at the critical point on reach 2 are

$$D^0_{2,c} = \alpha^1_{2,c}L_{0,e} + \beta^1_{2,c}D_{0,e} \tag{2.38a}$$

$$D^1_{2,c} = \alpha^1_{2,c}l_1 + \beta^1_{2,c}d_1 \tag{2.38b}$$

$$D^2_{2,c} = \alpha^2_{2,c}l_2 + \beta^2_{2,c}d_2 \tag{2.38c}$$

Now, the critical deficit in reach 2 can be expressed as

$$D_{2,c} = \frac{Q_1 - q_{2,\text{out}}}{Q_2} \frac{Q_0 - q_{1,\text{out}}}{Q_1} D_{2,c}^0 + \frac{Q_1 - q_{2,\text{out}}}{Q_2} \frac{q_{1,\text{in}}}{Q_1} D_{2,c}^1 + \frac{q_{2,\text{in}}}{Q_2} D_{2,c}^2 \tag{2.39}$$

The dilution weighting factors multiplying the undiluted partial deficits are now transformed to a form that is more readily generalized:

$$\frac{Q_1 - q_{2,\text{out}}}{Q_2} \frac{Q_0 - q_{1,\text{out}}}{Q_1} = \left(1 - \frac{q_{2,\text{out}}}{Q_1}\right)\left(1 - \frac{q_{1,\text{out}}}{Q_0}\right) \frac{Q_c}{Q_2} \tag{2.40a}$$

$$\frac{Q_1 - q_{2,\text{out}}}{Q_2} \frac{q_{1,\text{in}}}{Q_1} = \left(1 - \frac{q_{2,\text{out}}}{Q_1}\right) \frac{q_{1,\text{in}}}{Q_2} \tag{2.40b}$$

Now, define *flow reduction factors* for withdrawals:

$$\gamma_2^0 = \left(1 - \frac{q_{2,\text{out}}}{Q_1}\right)\left(1 - \frac{q_{1,\text{out}}}{Q_0}\right) \tag{2.41a}$$

$$\gamma_2^1 = 1 - \frac{q_{2,\text{out}}}{Q_1} \tag{2.41b}$$

$$\gamma_2^2 = 1 \tag{2.41c}$$

and *dilution factors:*

$$\delta_2^0 = \gamma_2^0 \frac{Q_0}{Q_2} \tag{2.42a}$$

$$\delta_2^1 = \gamma_2^1 \frac{q_{1,\text{in}}}{Q_2} \tag{2.42b}$$

$$\delta_2^2 = \gamma_2^2 \frac{q_{2,\text{in}}}{Q_2} \tag{2.42c}$$

Then the critical deficit on reach 2 can be calculated from a sum of the undiluted partial deficits weighted by the dilution factors:

$$D_{2,c} = \delta_2^0 D_{2,c}^0 + \delta_2^1 D_{2,c}^1 + \delta_2^2 D_{2,c}^2 \tag{2.43}$$

This compact form can also be applied to reach 1 by defining undiluted partial deficits,

$$D_{1,c}^{0} = \alpha_{1,c}^{1} L_{0,e} + \beta_{1,c}^{1} D_{0,e} \tag{2.44a}$$

$$D_{1,c}^{1} = \alpha_{1,c}^{1} l_{1} + \beta_{1,c}^{1} d_{1} \tag{2.44b}$$

flow reduction factors,

$$\gamma_2^0 = 1 - \frac{q_{1,\text{out}}}{Q_0} \quad \text{and} \quad \gamma_1^1 = 1 \tag{2.44c}$$

and dilution factors,

$$\delta_1^0 = \gamma_1^0 \frac{Q_0}{Q_1} \quad \text{and} \quad \delta_1^1 = \gamma_1^1 \frac{q_{1,\text{in}}}{Q_1} \tag{2.44d}$$

such that

$$D_{1,c} = \delta_1^0 D_{1,c}^0 + \delta_1^1 D_{1,c}^1 \tag{2.44e}$$

Generalized Recursive Model for an Arbitrary Number of Reaches. Now the number of reaches is allowed to be any value, m. The flow rate in any reach i is

$$Q_i = Q_0 + \sum_{k=1}^{i} q_{i,\text{net}} \tag{2.45}$$

where the net inflow, $q_{i,\text{net}}$, is defined as

$$q_{i,\text{net}} = q_{i,\text{in}} - q_{i,\text{out}} \tag{2.46}$$

The general form for the withdrawal factors used in calculating the effects of discharge j on downstream reach i is

$$\gamma_i^j = \prod_{k=j+1}^{i} \left(1 - \frac{q_{k,\text{out}}}{Q_{k-1}}\right) \qquad j = 0,1,\ldots,i-1, \quad i = 1,2,\ldots,m \tag{2.47a}$$

$$\gamma_i^i = 1 \qquad i = 1,2,\ldots,m \tag{2.47b}$$

Note that if all withdrawals are zero, all withdrawal factors are 1. This special case is considered in the stochastic models referenced in Section 2.D.

The general form for the dilution factors for calculating the effects of discharge j on downstream reach i is

$$\delta_i^j = \gamma_i^j \frac{q_{j,\text{in}}}{Q_i} \qquad j = 0,1,\ldots,i \tag{2.48}$$

This generalized form requires that we consider $q_{0,\text{in}} = Q_0$, the initial flow in the stream.

The dissolved oxygen concentration at the critical point in a reach depends, in part, on the BOD discharged into the reach plus the remaining BOD from other upstream discharges, which equals the BOD remaining at the end of the preceding reach. The general form for the BOD concentration at the end of each reach is

$$L_{i,e} = \sum_{j=0}^{i} \delta_i^j L_{i,e}^j \tag{2.49}$$

where

$$L_{i,e}^j = \rho_{i,e}^j l_j \qquad (\text{defining } l_0 = L_{0,e}) \tag{2.50}$$

which uses the recursive definition

$$\rho_{i,e}^j = e^{-k_{1i}t_{i,e}} \rho_{(i-1,e)}^j = \rho_{i,e}^i \rho_{(i-1,e)}^j \tag{2.51}$$

Now the general form for the BOD exertion factor of upstream discharge j at the critical point on reach i can be expressed:

$$\alpha_{i,c}^j = \alpha_{i,c}^i \rho_{(i-1,e)}^j + \beta_{i,c}^i \alpha_{(i-1,e)}^j. \tag{2.52}$$

Similarly, the initial deficit decay factor for upstream discharge j at the critical point on reach i is expressed in general as

$$\beta_{i,c}^j = \beta_{i,c}^i \beta_{(i-1,e)}^j \tag{2.53}$$

The undiluted partial deficit at the critical point on reach i resulting from the discharge at j is given by

$$D_{i,c}^j = \alpha_{i,c}^j l_j + \beta_{i,c}^j d_j \tag{2.54}$$

which is generalized by letting $l_0 = L_{0,e}$ and $d_0 = D_{0,e}$.

Now the deficit at the critical point on reach i is expressed in general as the sum of the diluted partial deficits:

$$D_{i,c} = \sum_{j=0}^{i} \delta_i^j D_{i,c}^j \tag{2.55}$$

This derivation for the deficit at the critical point is also valid for calculating the deficit at any other point of interest on the reach.

The equations derived in this section comprise analytical solutions to the coupled algebraic and differential equations that arise when modeling a multireach–multidischarge stream system with a first-order differential equation such as the Streeter–Phelps model or one its various extensions. The generalized models presented here can be extended to handle estuaries as well. Estuary models require more complex descriptive models which are beyond the scope of this chapter.

First-order models of this sort are useful for developing an understanding of the fundamental processes affecting stream quality without having to specify a great many parameters. Such models are also useful for representing the physical processes in a stream in optimization models that focus primarily on watershed management issues involving trade-offs between treatment costs and water quality or equity issues among multiple dischargers. In Section 2.D we derive management models that incorporate the analytical descriptive model derived above.

There are other descriptive models, though, that simulate the physical processes in a stream using much greater detail than the first-order DO/BOD model that is used here, and some discussion of the popular "model packages" is in order before proceeding to management modeling.

2.C.7. Simulation Model Packages

Several computer simulation programs have been developed in recent years that enable the user to examine closely the effects of pollution discharges and contaminated runoff flows on many different aspects of water quality. Specialized software packages are available, often at no cost from government sources, for calculating: concentrations of toxic chemicals in the water and accumulated in the bottom sediments, elevated temperatures resulting from thermal discharges such as electric generating stations, effects of nutrient (nitrogen and phosphorus) discharges on lakes and estuaries, and effects of stormwater discharges and combined sewer overflows. An increasing number of these packages can be obtained by downloading them directly onto a microcomputer over an Internet link, and several have incorporated user-friendly graphical user interfaces. In this section we summarize the capabilities of four useful packages that are available at no charge from the EPA.

QUAL2E: Enhanced Stream Water Quality Model. This model simulates the major reactions of nutrient cycles, algal production, benthic and carbonaceous oxygen demand, atmospheric reaeration, and the effects of these processes on the dissolved oxygen balance. It can predict the concentrations of up to 15 water quality constituents. It is intended as a water quality planning tool for developing total maximum daily loads (TMDLs) and can be used in conjunction with field sampling for identifying the magnitude and quality characteristics of nonpoint sources. It is also possible to study dissolved oxygen cycles that occur on a diurnal basis as a result of algal growth. The model has a user interface for Microsoft Windows, which provides input screens for convenient entry of model inputs and for graphical views of input data and model results. The model was developed by EPA's Environmental Research Laboratory in Athens, Georgia (U.S. EPA, 1995).

WASP5: Water Quality Analysis Simulation Program. This package provides a generalized framework for modeling the fate and transport of contaminants in surface waters and has been used to study biochemical oxygen demand and dissolved oxygen dynamics, nutrients and eutrophication, bacterial contamination, and toxic contamination by organic chemicals and heavy metals. The program can be extended through links to user-written subcomponent models. Two such models are provided by EPA with TOXI5, a toxics model that simulates the transport and transformation of up to three chemicals and up to three types of particulate material, and EUTRO5, a dissolved oxygen/eutrophication model that simulates the transport and transformation of variables in the water column and sediment bed, including dissolved oxygen, carbonaceous biochemical oxygen demand, phytoplankton carbon and chlorophyll *a*, ammonia, nitrate, organic nitrogen, organic phosphorus, and orthophosphate. The model was developed by EPA's Center for Exposure Assessment Modeling (CEAM) in Athens, Georgia (U.S. EPA, 1993a).

HSPF: Hydrological Simulation Program–FORTRAN. This model integrates stream water quality modeling with watershed hydrology modeling that enables simulation of land and soil contaminant runoff processes and their interactions with in-stream hydraulics and the chemical composition of bottom sediments. The model calculates a time history of the runoff flow rate, sediment load, and nutrient and organic chemical concentrations, as well as a time history of water quantity and quality in the stream that receives the runoff. The data requirements for this model are extensive. At a minimum, continuous rainfall records are required to drive the runoff model, and additional records of evapotranspiration, temperature, and solar intensity are desirable. HSPF was also developed by EPA's Center for Exposure Assessment Modeling (CEAM) in Athens, Georgia (U.S. EPA, 1996).

SWMM: Stormwater Management Model. The purpose of this model is simulation of the movement of water and pollution carried by precipitation runoff from the ground surface through pipe and channel networks, storage treatment units, and finally, to receiving waters. The simulations can be structured as either single events or continuous flows over an arbitrary period of time. The model is useful for a general assessment of the urban runoff problems and options for their solution. Single-event simulations can be used for detailed design of catchments having storm sewers and natural drainage. Like QUAL2E, this model has a Microsoft Windows user interface and was developed by EPA's Environmental Research Laboratory in Athens, Georgia (U.S. EPA, 1993b).

2.D. MANAGEMENT MODELS FOR FRESHWATER STREAMS

In this section we describe a group of models that have been developed primarily for managing water quality on freshwater streams. These models use techniques from the field of operations research, linear and nonlinear programming, to identify optimal solutions to management problems. The highly detailed simulation models discussed in the preceding section have also been used successfully in water quality management. The complexity of these models, which makes them so useful for accurate analysis of a specific management alternative, is also their main weakness. Their complexity can make it very difficult to incorporate such models into an optimization study that requires a search for solutions that best satisfy management objectives. Such objectives include cost minimization, environmental quality maximization, and equity in the distribution of treatment costs among multiple dischargers.

The optimization models derived in this section incorporate the simpler first-order descriptive models derived in Section 2.C. Although these management models must use simplifying assumptions that are not required in complex simulation models, they are capable of searching among an essentially infinite number of alternatives to identify those management options that do the best job of satisfying the goals of the watershed planners and managers. Also, the simplicity of the embedded descriptive model, and the relatively few stream parameters required to implement it, makes it possible to use the optimization model to gain a basic understanding of the salient issues in watershed management without expending a greal deal of time and money to obtain data.

An ideal strategy for water quality management modeling is combined use of optimization and simulation models. In situations where sufficient computing and data resources are available for implementing a detailed simulation model, an optimization model can be run first to "screen out" a great many inferior alternatives, leaving only the more promising solutions for close scrutiny by the simulation model.

2.D.1. Dissolved Oxygen Management with Cost Minimization

In this section we derive a model for dissolved oxygen management in a freshwater stream that illustrates the basic technique of optimization management modeling. Here the model is shown in its full form, which displays (1) the individual components of an embedded descriptive model based on Section 2.C, (2) a nonlinear objective function for cost minimization, (3) constraints representing the requirements of water quality regulators, and (4) constraints representing the capabilities of available technology. Example solutions of the model are also presented.

The theoretical development of optimization models for watershed management originated in the 1960s. Initial efforts used dynamic programming, such as the model solved by Liebman and Lynn (1966), which determined the optimal removal of BOD at wastewater treatment plants, and the model by Converse (1972), which solved for the required number of treatment plants and their locations for a scheme of regional wastewater treatment plants.

ReVelle et al. (1967, 1968) used linear programming and separable convex nonlinear programming to find cost-minimizing wastewater treatment plant removal efficiencies for discharges into a single stream. Much work in the field has built upon this constrained optimization methodology, as does the current chapter.

Treatment Efficiency and Cost. Figure 2.4 shows schematically the overall effect of a wastewater treatment plant on a discharge of water polluted with BOD. The notation used in Figure 2.4 is defined as follows:

l_j^u = BOD concentration in untreated wastewater discharge j (mg/L)

l_j = BOD concentration in treated wastewater discharge j (mg/L)

$$x_j = \text{efficiency of BOD removal at discharge } j = \frac{l_j^u - l_j}{l_j^u} \text{ (dimensionless)} \tag{2.56}$$

On a stream having m discharges of treated wastewater, the treatment efficiencies, x_j, $j = 1,\ldots,m$, are the main decision variables in the management problem. The BOD concentrations in the treated discharges are also decision variables, linked to the x_j through the following equation, which is written in linear programming standard form, with terms containing decision variables on the left and constants on the right:

$$x_j + \frac{1}{l_j^u} l_j = 1 \qquad j = 1,\ldots,m \tag{2.57}$$

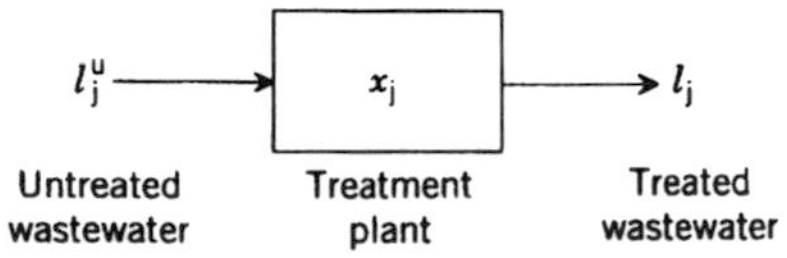

l_j^u → x_j → l_j

Untreated wastewater — Treatment plant — Treated wastewater

Figure 2.4 Schematic of wastewater treatment plant.

The cost of wastewater treatment varies in a nonlinear way with the treatment efficiency, as shown in Figure 2.5. Three regions are shown on the cost curve, corresponding to different classes of treatment technology: primary, secondary, and tertiary. Primary treatment consists mainly of suspended solids removal, typically by screens and sedimentation tanks which removes the first 35% of the BOD. The costs for this portion of the curve include land acquisition and site-preparation costs, which contribute to the high initial slope. Secondary treatment involves removal of dissolved organic contaminants by biological processes, and costs are fairly linear with treatment efficiency at a reduced slope. Combined primary and secondary treatment can easily remove up to 80% of the BOD. Tertiary treatment involves a variety of advanced processes, such as nutrient removal and activated carbon filtration, and are characterized by rapidly increasing marginal costs as the treatment efficiency approaches 100%.

The treatment costs at each plant discharging into a stream can be represented by a cost function $c_j(x_j)$ having characteristics similar to those shown in Figure 2.5. Furthermore, the total cost of wastewater treatment for the entire stream is the sum of these cost functions for all dischargers:

$$\text{total cost} = \sum_{j=1}^{m} c_j(x_j) \tag{2.58}$$

The management model presented in this section has the objective of finding a set of optimal treatment efficiencies that minimize the total cost. The nonlinear function for the total cost in equation (2.58) has two important char-

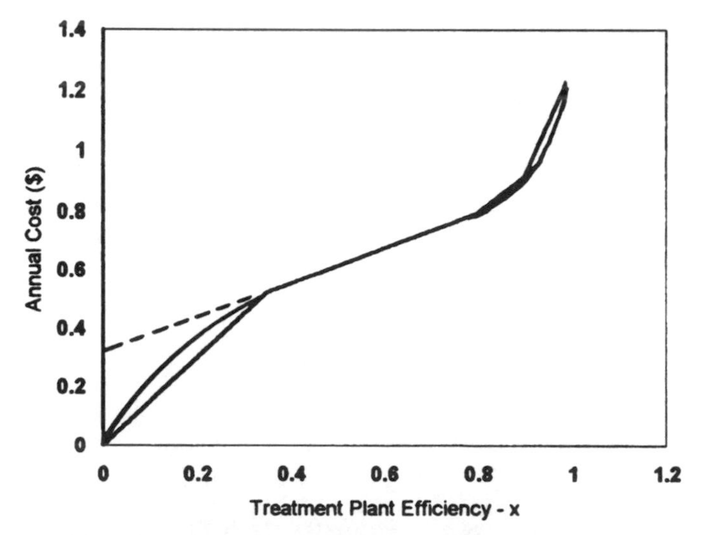

Figure 2.5 Treatment plant annual costs versus treatment efficiency.

acteristics: (1) it is separable (i.e., it consists of a sum of m separate nonlinear functions), and (2) each separable function has both concave and convex sections. The second characteristic complicates the search for the optimal solution. This problem is avoided, though, by assuming that treatment plants will use at least primary treatment. This assumption is quite reasonable considering the host of problems that are caused by discharging completely untreated wastewater into streams. Thus, only the convex portions of the curves above 35% efficiency are relevant in the solution, and the cost functions can be modified to make them convex by extending the secondary treatment portion of the curve back to the origin using a straight-line segment (shown as a dashed line in Figure 2.5). With this modification, the function for the total cost is both separable and convex, and the technique of *separable convex programming* can be used to solve the management problem using efficient linear programming algorithms.

The linear programming formulation requires that the convex portions of the cost curves be approximated by piecewise linear segments, as shown in Figure 2.5. The linear approximation to the total cost function is given by

$$\text{total cost (approximate)} = z = c_{j0} + c_{j1}x_{j1} + c_{j2}x_{j2} + c_{j3}x_{j3} \quad (2.59)$$

where

c_{j0} = intercept of the convex approximation with the cost axis (\$/year)

c_{j1} = slope of linear approximation to the secondary treatment portion (\$/year/efficiency fraction)

c_{j2} = slope of linear approximation to the first tertiary treatment segment (\$/year/efficiency fraction)

c_{j3} = slope of linear approximation to the second tertiary treatment segment (\$/year/efficiency fraction)

x_{j1} = efficiency segment from zero to η_{j1}, the upper limit of the secondary segment

x_{j2} = efficiency segment from η_{j1} to η_{j2}, the upper limit of the first tertiary segment

x_{j3} = efficiency segment from η_{j2} to η_{j3}, the upper limit of the third tertiary segment

Note that $x_j = x_{j1} + x_{j2} + x_{j3} = \Sigma_{k=1}^{3} x_{jk}$, where k is used to index the cost curve segments.

In general, if p_j linear segments are required to represent the cost curve for discharge j, the following equations are used for the total cost objective function and the efficiency definitions:

$$z = \sum_{j=1}^{m} \left(c_{j0} + \sum_{k=1}^{p_j} c_{jk} x_{jk} \right) \tag{2.60}$$

$$\sum_{k=1}^{p_j} x_{jk} + \frac{1}{l_j^u} l_j = 1 \qquad j = 1,\ldots,m \tag{2.61}$$

$$\eta_{j,\min} \le x_{j1} \le \eta_{j1} \qquad j = 1,\ldots,m \tag{2.62a}$$

$$0 \le x_{jk} \le \eta_{jk} - \eta_{j,k-1} \qquad k = 2,\ldots,p_j, \quad j = 1,\ldots,m \tag{2.62b}$$

where $\eta_{j,\min}$ is the lowest removal efficiency allowed at discharge j. This level will typically be established by a government regulatory agency, and it must be at least as high as the point where the cost curve becomes convex if separable convex linear programming is to be used to solve the problem.

Mixing and Dilution at Discharge Points. The management model uses the same multireach–multidischarge schematic (Figure 2.3) used to develop the general descriptive model in Section 2.C.5. The volumetric flow rates of the withdrawals, $q_{j,\text{out}}$, and the discharges, $q_{j,\text{in}}$, are independent of the pollution removal efficiencies, so the streamflow in each reach can be calculated separately from the management model using specified values for the net discharges, $q_{j,\text{net}} = q_{j,\text{in}} - q_{j,\text{out}}$ and the initial flow upstream of the first discharge, Q_0:

$$Q_i = Q_0 + \sum_{j=1}^{i} q_{j,\text{net}} \tag{2.63}$$

The management model presented here uses equality constraints in the linear program to deal explicitly with the continuity equations expressing the mixing and dilution processes at each discharge point. Thus the BOD concentration and the DO deficit at the top of each reach are calculated as auxiliary decision variables and are available in the model for use by other constraints that calculate BOD decay and oxygen sag. (The generalized management model presented below in Section 2.D.2 incorporates the recursive model developed in Section 2.C.5, which folds the equations for mixing, dilution, and BOD decay into a single set of constraints for the oxygen deficit.)

Applying equation (2.4) to mixing and dilution for both BOD and oxygen deficit yields two sets of equality constraints:

$$(Q_{i-1} - q_{j,\text{out}})L_{i-1,e} + q_{j,\text{in}} l_j - Q_i L_{i,0} = 0 \qquad i = 1,\ldots,m, \quad j = i \tag{2.64}$$

$$(Q_{i-1} - q_{j,\text{out}})D_{i-1,e} + q_{j,\text{in}} d_j - Q_i D_{i,0} = 0 \qquad i = 1,\ldots,m, \quad j = i \tag{2.65}$$

Note that the wastewater discharges, indexed by j, and the reaches of the stream, indexed by i, correspond exactly here because, in this model, the effects of each discharge are calculated directly only for the reach into which the discharge is made. Effects on the other downstream reaches are calculated indirectly through these mixing and dilution constraints written for each discharge point on the stream.

BOD Decay, Oxygen Deficit at the End of Each Reach. Equations (2.64) and (2.65) show that auxiliary decision variables for the BOD concentration and oxygen deficit at the end of each reach are also needed. These variables are defined in the model by applying equations (2.26) and (2.27) to the endpoint of each reach (i.e., $s = e$), as shown in a set of equality constraints written for all reaches except the last:

$$L_{i,e} - \rho^{i}_{i,e} L_{i,0} = 0 \qquad i = 1,2,\ldots,m-1 \tag{2.66}$$

$$D_{i,e} - \alpha^{i}_{i,e} L_{i,0} - \beta^{i}_{i,e} D_{i,0} = 0 \qquad i = 1,2,\ldots,m-1 \tag{2.67}$$

Note that the coefficients ρ, α, and β are for discharge $j = i$ into reach i.

Water Quality Constraints. The indicator of water quality in this model is the dissolved oxygen concentration. A lower limit on the DO concentration is imposed by setting an upper bound on the DO deficit at the critical point on each reach. Applying equation (2.27) to the critical point on each reach yields the following set of equality constraints:

$$\alpha^{i}_{i,c} L_{i,0} + \beta^{i}_{i,c} D_{i,0} \leq D_{i,\max} \qquad i = 1,\ldots,m \tag{2.68}$$

where $D_{i,\max}$ is the maximum allowed deficit that will satisfy water quality standards.

Equation (2.68) is difficult to implement, however, because, as equation (2.20) shows, the critical time $t_{i,c}$, and therefore the parameters $\alpha^{i}_{i,c}$ and $\beta^{i}_{i,c}$, depend on the values of the variables $L_{i,0}$ and $D_{i,0}$. In other words, it is not possible to know in advance of solving the management model where the critical points are because they depend on the results of the same model.

One solution to this problem has been proposed by Arbabi et al. (1974). A linear approximation to equation (2.22) for the critical deficit, $D_{i,c}$, can be obtained having the following form:

$$\hat{D}_{i,c} = a_{0,i} + a_{1,i} L_{i,0} + a_{2,i} D_{i,0} \qquad i = 1,\ldots,m \tag{2.69}$$

where $\hat{D}_{i,c}$ is a linear approximation of $D_{i,c}$ and $a_{0,i}$, a_{1i}, and $a_{1,i}$ are constants determined by a best-fit procedure applied to each reach. This equation can replace the left-hand side of the water quality constraints in equation (2.68).

The procedure required to determine values for the constants in equation (2.69) is fairly complicated, and since the result is an approximation, errors are involved. However, for large models involving many discharges that will be run several times, it may be worthwhile to calculate the constants and use the approximate form. However, for small models, such as the one developed for the example below, another method, which is simpler to implement, can be used to deal with the problem of not knowing in advance where the critical deficit occurs. The method involves dividing each reach into several segments by specifying fixed flow times at which the DO deficit is evaluated. Each reach i is divided into a total of n_i segments, and flow times at the end of each segment are specified as $t_{i,s}$, $i = 1,...,m$; $s = 1,...,n_i$. These points at the end of each segment are sometimes called *mesh points* in the literature. Then values of the parameters $\alpha^i_{i,s}$ and $\beta^i_{i,s}$ are calculated for each of these flow times, and equation (2.68) is written n_i times for each reach in the modified form

$$\alpha^i_{i,s} L_{i,0} + \beta^i_{i,s} D_{i,0} \leq D_{i,\max} \qquad i = 1,2,...,m, \quad s = 1,2,...,n_i \qquad (2.70)$$

If a sufficiently large number of segments are specified for each reach, one of these water quality constraints for each reach is certain to be near the critical point for any values of $L_{i,0}$ and $D_{i,0}$.

Complete Linear Programming Management Model. The complete formulation of the management model is shown in Table 2.1 in standard linear programming form. The constant fixed-cost terms $c_{j,0}$ have been dropped because adding a constant term to the objective function has no effect on the optimal values of the decision variables. These costs can be added to the optimal value of the objective function after the problem is solved if total costs are desired.

Example A. An example is presented for a stream having three reaches, with a significant BOD discharge at the top of each reach. The problem is to determine the optimal efficiencies for wastewater treatment plants at each discharge point so as to minimize cost while keeping the dissolved oxygen deficit below 4.5 mg/L. Regulations require that each plant provide primary treatment with at least 35% BOD removal. Secondary treatment provides up to 80% removal. The first stage of tertiary treatment provides up to 90% removal, and the second stage up to 99%. Just upstream of the first discharge, the flow rate in the stream is 400 million gallons per day (mgd), the BOD concentration is 8.0 mg/L, and the DO deficit is 2.0 mg/L. Just upstream of each discharge is a withdrawal having the same flow rate as that of the discharge. The DO deficit in each discharge is 7 mg/L. The parameters of each discharge are shown in Table 2.2 and the parameters of each reach are shown in Table 2.3.

TABLE 2.1 Complete Linear Programming Management Model

Minimize	$z = \sum_{j=1}^{n} \sum_{k=1}^{p_j} c_{jk} x_{jk}$		(2.71)
subject to			
	$\sum_{k=1}^{p_j} x_{jk} + \frac{1}{l_j^u} l_j = 1$	$j = 1,2,\ldots,m$	(2.72)
	$\eta_{j,\min} \le x_{j1} \le \eta_{j1},$	$j = 1,2,\ldots,m$	(2.73)
	$0 \le x_{jk} \le (\eta_{jk} - \eta_{j,k-1})$	$k = 2,3,\ldots,p_j$ $j = 1,2,\ldots,m$	(2.74)
	$(Q_{i-1} - q_{j,\text{out}})L_{i-1,e} + q_{j,\text{in}} l_j - Q_i L_{i,0} = 0$	$i = 1,2,\ldots,m$ $j = i$	(2.75)
	$(Q_{i-1} - q_{j,\text{out}})D_{i-1,e} + q_{j,\text{in}} d_j - Q_i D_{i,0} = 0$	$i = 1,2,\ldots,m$ $j = i$	(2.76)
	$L_{i,e} - \rho_{i,e}^{i} L_{i,0} = 0$	$i = 1,2,\ldots,m - 1$	(2.77)
	$D_{i,e} - \alpha_{i,e}^{i} L_{i,0} - \beta_{i,e}^{i} D_{i,0} = 0$	$i = 1,2,\ldots,m - 1$	(2.78)
	$\alpha_{i,s}^{i} L_{i,0} - \beta_{i,s}^{i} D_{i,0} \le D_{i,\max}$	$i = 1,2,\ldots,m$ $s = 1,2,\ldots,n_i$	(2.79)
	$D_{i,0} \le D_{i,\max}$	$i = 1,2,\ldots,m$	(2.80)
	$l_j \ge 0$	$j = 1,2,\ldots,m$	(2.81)
	$L_{i,e} \ge 0$	$i = 1,2,\ldots,m$	(2.82)

The linear programming management model of Table 2.1 was solved using the general mathematical programming language AMPL (Fourer et al., 1993). The program and the associated data file are shown in the Appendix. Figure 2.6 displays for each segment on each reach the BOD, dissolved oxygen, and deficit concentrations, which are calculated from the optimal values of the decision variables after the linear program is solved. Note that the maximum deficit is reached once in the third reach. Table 2.4 shows the optimal treatment efficiencies at each discharge.

TABLE 2.2 Discharge Parameters for Example A

			Cost Slope (millions/year)		
Discharge Number	Discharge Flow Rate (mgd)	Untreated BOD Concentration (mg/L)	Secondary Segment	First Tertiary Segment	Second Tertiary Segment
1	40	275	$0.5	$1	$3
2	16	160	0.7	1.5	4
3	35	390	0.6	1.2	3.5

TABLE 2.3 Stream Reach Parameters for Example A

Reach Number	Number of Segments	Flow Time (days)	k_1 (day^{-1})	k_2 (day^{-1})
1	10	1	0.31	0.39
2	15	2.2	0.27	0.44
3	10	1.4	0.28	0.63

A discharge, j, having no treatment plant can be handled in this model by setting all the efficiency bounds to zero (i.e., $\eta_{j,\min} = \eta_{j,1} = \eta_{j,2} = \cdots = \eta_{j,p_i} = 0$). A tributary is a common example of a discharge that will have initial BOD and DO deficit concentrations but no possibility of BOD removal. Withdrawals from the stream, without corresponding discharges, can be handled similarly.

2.D.2. Generalized Models for Cost Minimization

In this section we incorporate the general descriptive model developed in Section 2.C.5 into the management model of Section 2.D.1 to create a compact model that is more easily generalized to other types of waterways, such as an estuary, and to other types of pollutants, such as excess nutrients or heavy metals.

Compact Model for Dissolved Oxygen Management in a Freshwater Stream. The management model developed here, based on the generalized recursive multireach–multidischarge model of Section 2.C.5, is equivalent to

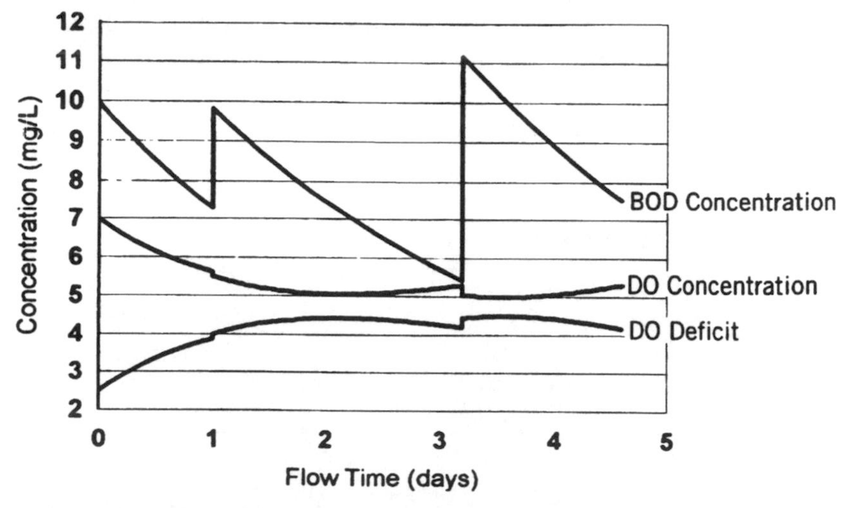

Figure 2.6 Water quality parameters versus time for the optimal solution.

TABLE 2.4 Optimal Treatment Efficiencies

Discharge Number	BOD Removal Efficiency
1	0.9
2	0.56
3	0.82

the full-form model of Table 2.1 but has fewer variables and constraints and is therefore called *compact*. In effect, the equality constraints are used to solve for certain variables, and then these variables are eliminated from other constraints by substitution of the expressions obtained from the equality constraints. The generalized recursive descriptive model was developed in Section 2.C.5 using essentially the same procedure.

Equation (2.55) and its expanded form are shown below applied to the point at the end of segment s in reach i:

$$D_{i,s} = \sum_{j=0}^{i} \delta_i^j D_{i,s}^j = \sum_{j=0}^{i} \delta_i^j(\alpha_{i,s}^j l_j + \beta_{i,s}^j d_j) \tag{2.83}$$

As before, $l_0 = L_{0,e}$ and $d_0 = D_0$. The definition of removal efficiency from equation (2.56) is rewritten as

$$l_j = l_j^u(1 - x_j) \tag{2.84}$$

which is used to substitute for l_j in equation (2.83) to yield

$$D_{i,s} = \sum_{j=0}^{i} \delta_i^j(\alpha_{i,s}^j l_j^u + \beta_{i,s}^j d_j) - \sum_{j=0}^{i} (\delta_i^j \alpha_{i,s}^j l_j^u)x_j \tag{2.85}$$

The first term in this formula is the deficit that would result at point s in reach i if all upstream discharges are untreated, and this quantity is defined separately as

$$D_{i,s}^{\text{ut}} = \sum_{j=0}^{i} \delta_i^j(\alpha_{i,s}^j l_j^u + \beta_{i,s}^j d_j) \tag{2.86}$$

where $D_{i,s}^{\text{ut}}$ is the deficit occurring at point s on reach i if all upstream waste discharges are untreated. The water quality constraint (2.68) limiting the DO deficit is now written as

$$D_{i,s}^{\text{ut}} - \sum_{j=0}^{i} (\delta_i^j \alpha_{i,s}^j l_j^u) x_j \leq D_{i,\max} \tag{2.87}$$

which can be rearranged to the form

$$\sum_{j=0}^{i} (\delta_i^j \alpha_{i,s}^j l_j^u) x_j \geq D_{i,s}^{\text{ut}} - D_{i,\max} = D_{i,s}^{\text{red}} \tag{2.88}$$

where $D_{i,s}^{\text{red}}$ is the reduction in DO deficit required to satisfy the water quality constraint at point s in reach i.

The notation can be simplified still further by defining

$$a_{i,s}^j = \delta_i^j \alpha_{i,s}^j l_j^u \tag{2.89}$$

where $a_{i,s}^j$ is the transfer coefficient for water quality improvement at point s in reach i per unit increase in x_j. Now the management model can be written in compact form with a general nonlinear cost function as

$$\text{minimize} \sum_{j=1}^{m} c_j(x_j) \tag{2.90a}$$

subject to

$$\sum_{j=0}^{i} a_{i,s}^j x_j \geq D_{i,s}^{\text{red}} \qquad i = 1,2,\ldots,m, \quad s = 1,2,\ldots,n_i \tag{2.90b}$$

$$\eta_{j,\min} \leq x_j \leq \eta_{j,\max} \qquad j = 1,2,\ldots,n \tag{2.90c}$$

The approximate linear programming model is

$$\text{minimize } z = \sum_{j=1}^{m} \sum_{k=1}^{p_j} c_{jk} x_{jk} \tag{2.91a}$$

subject to

$$\sum_{j=0}^{i} a_{i,s}^j \sum_{k=1}^{p_j} x_{jk} \geq D_{i,s}^{\text{red}}, \qquad i = 1,2,\ldots,m, \quad s = 1,2,\ldots,n_i \tag{2.91b}$$

$$\eta_{j,\min} \leq x_{j1} \leq \eta_{j,1}, \qquad j = 1,2,\ldots,m \tag{2.91c}$$

$$0 \leq x_{jk} \leq \eta_{jk} - \eta_{j,k-1}, \qquad j = 1,2,\ldots,m \quad k = 2,3,\ldots,p_j \tag{2.91d}$$

General Water Quality Management Model. In its most general form, the management model for minimization of pollution treatment costs can be expressed as

$$\text{minimize} \sum_{j=1}^{m} c_j(x_j) \tag{2.92a}$$

subject to

$$\sum_{j=1}^{m} a_{i',j} x_j \geq b_{i'} \qquad i' = 1,\ldots,n \tag{2.92b}$$

$$l_j \leq x_j \leq u_j \qquad j = 1,\ldots,m \tag{2.92c}$$

where

x_j = pollution removal efficiency at discharge j

m = number of dischargers on stream

$c_j(x_j)$ = cost function for pollution control at discharge j

$a_{i'j}$ = transfer coefficient indicating the water quality improvement in the stream at point i' as a result of pollution removal at discharge j

$b_{i'}$ = improvement in water quality in the stream at checkpoint i' required to satisfy water quality standards

n = number of points in stream where water quality is monitored in the model

l_j = lower limit on pollution removal efficiency at discharge j, usually imposed by regulators

u_j = upper limit on pollution removal efficiency at discharge j, usually imposed by limitations of technology

A new index i' is introduced that combines indices i and s from the previous model. The end of each segment s on each reach i is assigned a unique "checkpoint" identification number for a total of n checkpoints indexed by i'. The coefficients $b_{i'}$, $i' = 1,2,\ldots,n$, are equivalent to $D_{i,s}^{\text{red}}$ in the dissolved oxygen deficit model above. In general, the $b_{i'}$ will be the minimum improvement in an arbitrary water quality parameter of interest to the modeler. Of course, improvements in this parameter must be effected by pollution removal at the treatment plants, and the coefficients $a_{i',j}$ must be determined for this parameter for each checkpoint i' and discharge j. Typically, much descriptive modeling is required to obtain the transfer coefficients $a_{i'}$ by modeling the physical, chemical, and biological processes. In Chapter 15 these transfer coefficients are called *impact coefficients* and are used in a model for evaluating the concept of *transferrable BOD discharge permits*. Substantial effort is also required in the area of pollution control economics to obtain the cost functions $c_j(x_j)$ for each pollution discharge. In the most general case, these functions can represent pollution prevention costs as well as pollution removal costs.

Example B. Example A of Section 2.D.1, with parameters as shown in Tables 2.2 and 2.3, is now solved using the generalized model of the current section. Values for the parameters $a^j_{i,s} = a_{i'j}$ are shown in Table 2.5. Also shown are the dissolved oxygen deficits from the optimal solution and the indices for both the reach-segment indexing and the checkpoint indexing methods. A plot of the transfer coefficients and the DO deficit versus time is shown in Figure 2.7. The results for optimal treatment plant efficiencies, BOD levels, and DO concentrations and deficits are identical to those obtained with the formulation shown in Table 2.1, as shown in Table 2.5 and Figure 2.7. As in Example A, the problem was solved using the AMPL modeling language, and the program files are shown in the Appendix.

2.D.3. Equity Objectives and Multiobjective Trade-offs

Solutions to the management model formulations above can contain inequities among the different dischargers on a single stream, with one or a few dischargers bearing most of the cost of water quality improvements. This situation may be acceptable if one government entity, such as the federal government, is bearing most of the cost for all the plants. In fact, this was the situation during much of the 1970s in the United States for the treatment of domestic wastes. However, more recently, greater portions of the cost burden are placed on local governments and private industries. Thus there is a need to address the issue of inequities among the various dischargers.

A model that incorporates equity objectives was developed by Brill et al. (1976) to address this shortcoming of models that minimize cost exclusively. Two such models are presented here. One model has the objective of minimizing the sum of the absolute values of the deviations from the average treatment efficiency. The other minimizes the range of efficiencies. These objectives are then combined with the cost-minimization objective in a multiobjective analysis that treats total costs and equity simultaneously.

Minimizing Deviations from the Average Treatment Efficiency. This model minimizes the sum of the absolute deviations of individual treatment plant efficiencies from the average treatment efficiency. Constraints are imposed to achieve water quality goals and to keep the total cost below a budget limit, designated by c_{total}, which is an upper limit on the sum of annualized

TABLE 2.5 Transfer Coefficients and Dissolved Oxygen Deficit for Example B

Reach, Segment $[i,s]$	Checkpoint i'	Time $t_{i'}$ (days)	Deficit, $D_{i'}$ (mg/L)	Transfer Coefficients $a_{i',0}$ (mg/L)/ efficiency	$a_{i',1}$ (mg/L)/ efficiency	$a_{i',2}$ (mg/L)/ efficiency	$a_{i',3}$ (mg/L)/ efficiency
[1,0]	1	0	2.5	0	0	0	0
[1,1]	2	0.1	2.7	0.21	0.82	0	0
[1,2]	3	0.2	2.89	0.41	1.59	0	0
[1,3]	4	0.3	3.06	0.6	2.3	0	0
[1,4]	5	0.4	3.21	0.77	2.96	0	0
[1,5]	6	0.5	3.35	0.93	3.58	0	0
[1,6]	7	0.6	3.48	1.09	4.15	0	0
[1,7]	8	0.7	3.59	1.22	4.67	0	0
[1,8]	9	0.8	3.69	1.35	5.16	0	0
[1,9]	10	0.9	3.79	1.47	5.6	0	0
[1,10]	11	1	3.87	1.57	6.01	0	0
[2,0]	12	1	3.99	1.51	5.77	0	0
[2,1]	13	1.15	4.11	1.61	6.14	0.24	0
[2,2]	14	1.29	4.21	1.69	6.45	0.45	0
[2,3]	15	1.44	4.29	1.76	6.72	0.65	0
[2,4]	16	1.59	4.35	1.82	6.95	0.82	0
[2,5]	17	1.73	4.39	1.87	7.13	0.97	0
[2,6]	18	1.88	4.42	1.91	7.29	1.11	0
[2,7]	19	2.03	4.43	1.94	7.41	1.23	0
[2,8]	20	2.17	4.44	1.96	7.49	1.34	0
[2,9]	21	2.32	4.43	1.98	7.56	1.43	0
[2,10]	22	2.47	4.41	1.99	7.59	1.51	0
[2,11]	23	2.61	4.38	1.99	7.61	1.58	0
[2,12]	24	2.76	4.35	1.99	7.6	1.63	0
[2,13]	25	2.91	4.31	1.99	7.58	1.68	0
[2,14]	26	3.05	4.26	1.98	7.54	1.72	0
[2,15]	27	3.2	4.2	1.96	7.49	1.75	0
[3,0]	28	3.2	4.45	1.79	6.83	1.6	0
[3,1]	29	3.34	4.48	1.73	6.62	1.58	1.25
[3,2]	30	3.48	4.5	1.68	6.4	1.56	2.36
[3,3]	31	3.62	4.5	1.62	6.19	1.54	3.32
[3,4]	32	3.76	4.49	1.57	5.99	1.51	4.15
[3,5]	33	3.9	4.46	1.52	5.79	1.49	4.88
[3,6]	34	4.04	4.42	1.47	5.6	1.46	5.5
[3,7]	35	4.18	4.37	1.42	5.41	1.43	6.03
[3,8]	36	4.32	4.31	1.37	5.22	1.4	6.47
[3,9]	37	4.46	4.25	1.32	5.04	1.37	6.84
[3,10]	38	4.6	4.18	1.27	4.87	1.34	7.15

capital costs and annual operating and maintenance costs for all the treatment plants.

The equity objective minimizing deviations from the average, called equity1, is expressed in its basic form as

$$\text{minimize equity1} = \sum_{j=1}^{m} |x_j - \overline{\eta}| \tag{2.93}$$

where

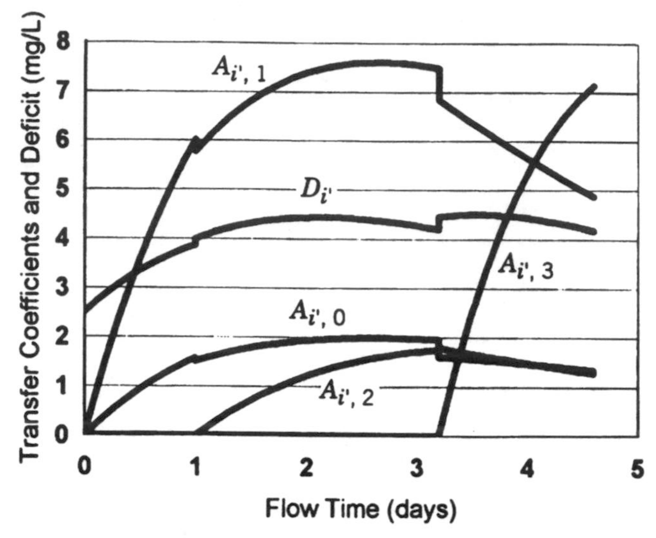

Figure 2.7 Transfer coefficient and O_2 deficit versus time.

$$\bar{\eta} = \frac{1}{m} \sum_{j=1}^{m} x_j = \text{average efficiency} \tag{2.94}$$

Equation (2.93) cannot be used as the objective function in a linear program, however, because the absolute value function is nonlinear. Fortunately, a technique is available for converting it into a linear objective function. Two auxiliary variables, u_j and v_j, are defined to represent the deviations from the average such that u_i handles the positive deviations and v_j handles the negative deviations. This definition is accomplished by the following two equations, which can be added as constraints in the model:

$$u_j - v_j = x_j - \bar{\eta} \qquad j = 1,2,\ldots,m \tag{2.95}$$

$$u_j v_j = 0 \qquad j = 1,2,\ldots,m \tag{2.96}$$

Equation (2.96) assures that when $u_j > 0$, because $x_j > \bar{\eta}$ (positive deviation), then $v_j = 0$ and $u_j = x_j - \bar{\eta}$. Similarly, when $v_j > 0$, because $x_j < \bar{\eta}$ (negative deviation), then $u_j = 0$ and $v_j = \bar{\eta} - x_j$. Now, the objective function can be written as the sum of these two auxiliary variables:

$$\sum_{j=1}^{m} |x_j - \bar{\eta}| = \sum_{j=1}^{m} u_j + v_j \tag{2.97}$$

In fact, this transformation of the problem does not eliminate the nonlinearities; it merely moves them from the objective function (nonlinear absolute value functions) to the constraint set [nonlinear constraints, equation (2.96), containing product terms]. Fortunately, in this case, the linear form of the objective function [equation (2.97)] assures that these nonlinear constraints are satisfied automatically, and they can be dropped from the formulation and replaced by lower bounds of zero (nonnegativity) on the deviation variables u_j and v_j, yielding a linear constraint set. The reason this works is that the objective function is being *minimized,* and it is not possible for an optimal solution to have both of the deviation variables positive simultaneously. Any solution having both u_j and v_j positive for some value of j is clearly not optimal because one of these variables can be decreased to zero without violating the constraints, yielding a lower value of the objective function. Thus since there can be no feasible solution having an objective function lower than the optimal solution, by definition of optimality, an optimal solution must satisfy equation (2.96) automatically. The complete formulation, with piecewise linear cost functions, as before, is shown in Table 2.6.

Minimizing the Range of Treatment Plant Efficiencies. The equity objective expressed above forces treatment efficiencies toward the overall average but does not rule out the possibility of a solution that assigns two individual treatment plants very different efficiencies, leaving an unacceptable inequity. The second equity model, presented below, addresses this concern by minimizing range of efficiencies, the difference between the greatest treatment plant efficiency and the smallest treatment plant efficiency. As before, additional constraints are imposed to achieve water quality goals and to keep the total cost below a budget limit, c_{total}.

TABLE 2.6 Equity Model 1: Minimize Sum of Absolute Deviations from Average

Minimize	$\text{equity1} = \sum_{j=1}^{m} (u_j + v_j)$		(2.98)
subject to			
	$u_j - v_j - x_j + \bar{\eta} = 0$	$j = 1,2,\ldots,m$	(2.99)
	$\bar{\eta} - \frac{1}{m}\sum_{j=1}^{m}\sum_{k=1}^{p_j} x_{jk} = 0$		(2.100)
	$\sum_{j=1}^{m} a_{i'j} \sum_{k=1}^{p_j} x_{jk} \geq b_{i'}$	$i' = 1,2,\ldots,n$	(2.101)
	$\sum_{j=1}^{m}\sum_{k=1}^{p_j} c_{jk}x_{jk} \leq c_{\text{total}}$		(2.102)
	$\eta_{j,\min} \leq x_{j1} \leq \eta_{j1}$	$j = 1,2,\ldots,m$	(2.103)
	$0 \leq x_{jk} \leq (\eta_{jk} - \eta_{j,k-1})$	$j = 1,2,\ldots,m;\ k = 2,3,\ldots,p_j$	(2.104)
	$\bar{\eta} \geq 0;\ u_j \geq 0;\ v_j \geq 0$	$j = 1,2,\ldots,m$	(2.105)

The equity objective minimizing range of efficiencies, called equity2, is expressed in its basic form as

$$\text{minimize equity2} = \max_{j=1,2,\ldots,m} (x_j) - \min_{j=1,2,\ldots,m} (x_j) \tag{2.106}$$

This objective function is also nonlinear because both the max(·) and the min(·) functions are nonlinear. However, as before, there is a way to convert this to a linear objective function by defining auxiliary variables and constraints. The first step is simply to define the auxiliary variables in their basic form:

$$x_{\max} = \max_{j=1,2,\ldots,m} (x_j) \quad \text{and} \quad x_{\min} = \min_{j=1,2,\ldots,m} (x_j) \tag{2.107}$$

and equity2 becomes simply $(x_{\max} - x_{\min})$. The definitions in equation (2.107) are implemented by the following sets of linear constraints which are added to the model:

$$\sum_{k=1}^{p_j} x_{jk} \leq x_{\max} \qquad j = 1,2,\ldots,m \tag{2.108}$$

$$x_{\min} \leq \sum_{k=1}^{p_j} x_{jk} \qquad j = 1,2,\ldots,m \tag{2.109}$$

The solution of this problem is accomplished by setting treatment efficiencies that force $x_{\max}$ and $x_{\min}$ as close together as possible by pushing down on $x_{\max}$ [equation (2.108)] and by pushing up on $x_{\min}$ [equation (2.109)]. The complete linear programming formulation is shown in Table 2.7.

Multiobjective Models Combining Cost Minimization and Equity. The objectives of cost minimization and equity among dischargers are usually in conflict. A solution that minimizes total treatment costs is likely to contain some striking inequities. Moreover, each of the two equity formulations above produces steadily decreasing measures of inequity as the total cost constraint [equation (2.102) in model equity1 and equation (2.114) in model equity2] is relaxed by increasing c_{total}. In fact, in both formulations, inequities can be eliminated altogether (identical treatment efficiencies at all plants) by increasing c_{total} to a sufficiently high value.

The trade-off between cost minimization and equity objectives can be examined directly through the use of multiobjective analysis, which involves the generation and evaluation of a trade-off curve displaying solutions that are "noninferior" with respect to cost minimization and one of the equity objectives. Chapter 12 of this book provides an overview of multiobjective

TABLE 2.7 Equity Model 2: Minimize Range of Treatment Plant Efficiencies

Minimize $\text{equity2} = (x_{\max} - x_{\min})$ (2.110)

subject to

$$\sum_{k=1}^{p_j} x_{jk} - x_{\max} \le 0 \qquad j = 1,2,\ldots,m \qquad (2.111)$$

$$\sum_{k=1}^{p_j} x_{jk} - x_{\min} \ge 0 \qquad j = 1,2,\ldots,m \qquad (2.112)$$

$$\sum_{j=1}^{m} a_{i'j} \sum_{k=1}^{p_j} x_{jk} \ge b_{i'} \qquad i' = 1,2,\ldots,n \qquad (2.113)$$

$$\sum_{j=1}^{m} \sum_{k=1}^{p_j} c_{jk} x_{jk} \le c_{\text{total}} \qquad (2.114)$$

$$\eta_{j,\min} \le x_{j1} \le \eta_{j1} \qquad j = 1,2,\ldots,m \qquad (2.115)$$

$$0 \le x_{jk} \le \eta_{jk} - \eta_{j,k-1} \qquad j = 1,2,\ldots,m;\ k = 2,3,\ldots,p_j \qquad (2.116)$$

$$\overline{\eta} \ge 0;\ u_j \ge 0;\ v_j \ge 0 \qquad j = 1,2,\ldots,m \qquad (2.117)$$

analysis and programming techniques with an extensive list of references. The constraint method of multiobjective programming can easily be applied to either of the equity formulations above by solving the model repeatedly for several different values of c_{total}, beginning with the minimum total cost obtained from the cost-minimizing model (from Table 2.1) and increasing the total cost until all treatment efficiencies are the same. The trade-off curve is then created by plotting the equity objective versus cost.

Example C. The three-reach, three-discharge model used in Examples A and B now serves as the basis for an example of multiobjective applied to water quality management. Table 2.8 displays the values of the objective functions for the three models (minimum cost, equity1, and equity2) solved separately, with c_{total} set equal to the optimal solution to the minimum-cost model. Clearly, the minimum-cost solution has significant inequities according to both measures of inequity. Now, the multiobjective trade-off curves are generated by reducing c_{total} in increments of \$0.05 million (\$50,000) per year in both equity1 and equity2 until the objective functions reach zero, indicating identical treatment efficiencies at all three plants. Both trade-off curves are shown in Figure 2.8. Each of the curves in the figure indicates the values of

TABLE 2.8 Minimum Cost, Equity1 and Equity2 Models Solved Separately

Minimum Cost, z (millions/year)	Equity1 (%)	Equity2 (%)
\$1.39	39.1	33.5

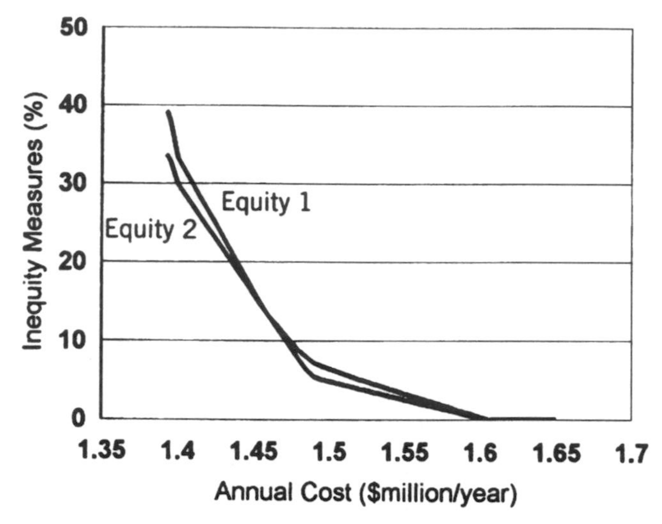

Figure 2.8 Multiobjective trade-off curves.

a noninferior set of solutions. Solutions above the curves are inferior with respect to minimization of cost and minimization of the particular equity objective. Solutions below the curves are infeasible.

Note the sharp bends in both curves at an annual cost of about $1.48 million per year. These points may be particularly attractive candidates for a compromise solution that achieves a great improvement in equity, according to both measures of equity, without sacrificing too much in increased costs. The equity1 objective also has a slight bend at about $1.45 million per year, which indicates a possibly attractive alternative solution.

2.D.4. Extensions

All of the formulations discussed so far are deterministic. They require the modeler to select a set of data on stream parameters such as flow rates, reaeration rates, pollution decay rates, and pollution transfer coefficients that are representative of certain conditions, usually worst-case water quality conditions. These data are selected from a limited record of observations on the stream and may not actually represent worst-case conditions. Moreover, it may not be prudent to base a management plan on worst-case conditions if the occurrence of such conditions is extremely rare. Stochastic models can explicitly account for uncertainty and provide a way for watershed managers to express reliability levels as the specific probability that a constraint may be violated. Fugiwara et al. (1986, 1987) and Ellis (1987) have published formulations for stochastic (chance constrained) versions of the generalized cost-minimization model of Section 2.D.2. However, implementation of stochastic models can be quite difficult because of the multiple covariances that exist among the parameters of the model, and a great deal of data analysis is required to establish the covariance matrices. Ongoing research in this area may lead to simplified chance constrained models that are applied more easily.

In Chapter 14 of this book, Whitlatch presents models for siting regional wastewater treatment facilities based on goals for the water quality in the receiving stream. In Chapter 15, Brill presents models that can be used to determine optimal effluent charges to assess dischargers into a stream or estuary. Chapter 15 also extends considerations of equity by exploring the idea of transferable permits for discharges and presents model formulations that can help to establish a market for the trading of discharges.

2.E. CONCLUSIONS

In this chapter we have presented derivations and example applications of several models for water quality management. All the models are based on research literature that has appeared over the past 30 years. Descriptive models of pollution transport and water quality effects are developed and generalized. Simple first-order models are contrasted with detailed simulation models, and the strengths and weaknesses of each are discussed. Optimization management models using linear programming formulations are derived and applied. A formulation for minimizing cost subject to environmental quality constraints is combined with two different formulations for determining equitable solutions in a multiobjective analysis.

The challenge for the future is to identify real planning problems for watersheds that are in need of better management and to implement these formulations. Regulatory agencies are beginning to realize the importance of a whole watershed approach to water quality management, and the models presented here have great potential for development into innovative and useful tools that managers can put to good use on a more routine basis. Hopefully, the presentation here will help further the goal of better watershed management through more widespread application of these methods.

EXERCISES

2.1. *Simulation of dissolved oxygen in a freshwater stream.* Consider the problem involving discharge of municipal wastewater into a freshwater stream. A town discharges its waste into a stream after treatment in a sewage treatment plant. The stream is rapidly moving and shallow for the first 40 miles downstream from the discharge point (reach 1). Then it widens into a slowly moving and deep stream for the next 40 miles (reach 2). In reality, this situation could occur with a dam at the end of reach 2. We model reach 2 as a stream, even though it may have some characteristics of a lake.

The problem is to determine the percent removal of BOD that is necessary to meet a stream standard for dissolved oxygen of 4.5 mg/L everywhere on the stream in reaches 1 and 2. You will do this for two different streamflows (volumetric flow rate in m^3/s). This will involve plotting the DO sag curve for a range of different treatment levels for both flow rates. You will need to calculate the parameters for the wastewater and the stream for each reach. The temperature-adjusted laboratory rate constant (k_T) should be corrected to account for the difference in laboratory and stream mixing conditions and microorganisms using the formula

$$k_1 = k_T + \frac{v}{h} \eta$$

where v is the stream velocity (m/s), h the average depth of stream (m), and η the bed activity coefficient (s/day).

Parameters upstream of discharge. First, assume that the flow upstream of the discharge is 2 m^3/s (normal). Then assume that this flow is 0.75 m^2/s (dry weather). In both cases, use upstream DO concentration = 9 mg/L and upstream BOD concentration = 0 mg/L.

Untreated waste parameters:

Biooxidation rate constant $k_{1,\text{lab}} = 0.4077\ \text{day}^{-1}$ (base e, in laboratory at 20°C)

Hoff–Arrhenius $\theta = 1.047$ (for temperature adjustment of k_1)

BOD_5 = 200 mg/L (5-day BOD measured in laboratory at 20°C; *note:* must convert to *ultimate* BOD)

Dissolved oxygen concentration = 0 mg/L

Flow rate = 0.2 m^3/s (same for treated waste)

Parameters downstream of discharge:

Parameter	Reach 1	Reach 2
Average normal flow depth (m)	3	6
Average dry weather depth (m)	2	5
Average width (m)	20	65
Bed activity coefficient (s/day)	0.6	0.2
Average water temperature (°C)	10	18

2.2. *Management model for dissolved oxygen.* Consider the problem involving discharge of municipal wastewater into a freshwater stream from two towns and a paper mill. The geography of the problem is shown in Figure 2.9. The problem is identical to Example A in Section 2.D.1 except that

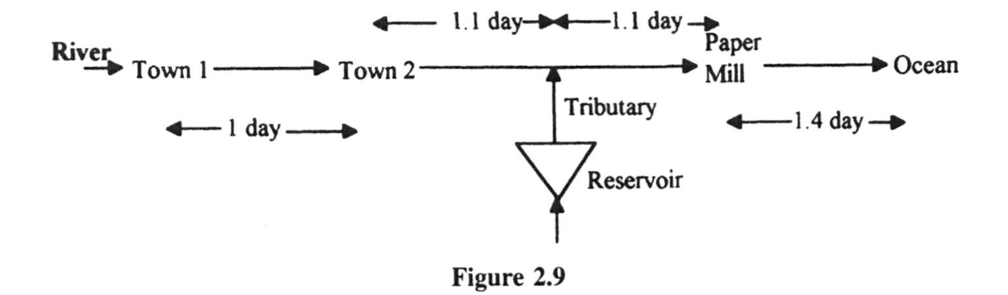

Figure 2.9

now, a tributary flows in midway between the second and third BOD discharges. The first two discharges are from municipal wastewater treatment plants, and the third discharge is from a paper mill.

The reservoir does not yet exist. It has been proposed as a measure to increase the level of the dry-weather flow in the stream and thereby improve the quality of the main stream below the point where the tributary joins it. We do not yet know whether it will be more expensive to build the reservoir or to remove more BOD at the towns and paper mill. Also, the size of the reservoir is not yet specified. A small reservoir would provide only a small increase in the dry-weather flow, but a large reservoir would increase it much more.

The problem is to create a plot of optimal BOD removal costs (i.e., cost-minimizing treatment plant costs at the towns and paper mill) versus flow rate from the tributary. Treatment costs should decrease as the reservoir flow rate increases because the amount of BOD removal necessary to meet the water quality standard goes down as the flow rate goes up. In other words, you will be determining the savings that can result from the "dilution solution to pollution." We will not examine the costs of the reservoir in this exercise, though, so we will not be able to determine whether dilution is more or less expensive than waste treatment. The environmental impacts of reservoirs are also significant and would have to be included in a complete assessment.

Assume that the dry-weather flow from the tributary without the reservoir is 25 mgd (million gallons per day). (*Note:* All flows in the model are in units of mgd, but it does not matter what units are used as long as they are the same for all flows.) Increase the flow to a maximum of 250 mgd, which is the greatest possible dry-weather flow corresponding to the largest possible reservoir.

Obtain the minimum-cost solution for treatment efficiencies at the towns and paper mill for each tributary flow rate. Use the same cost values as are used in Example A. The units of the objective function are $ million/year annualized combined capital and operating costs. Also assume that the BOD concentration in the tributary at the point it joins the main stream is 10 mg/L when the flow is 25 mgd. (It will decrease as the tributary flow increases because of additional dilution.)

Note: The times of flow given in Figure 2.9 are for a tributary flow of 25 mgd. Increasing the flow below the tributary will change the stream velocity downstream, and the flow times will tend to decrease. However, it is difficult to calculate the new velocities because the depth and width of the stream will also change. The velocity of flow in an open channel, v, can be calculated as a function of the hydraulic radius R and the stream slope S using Manning's formula, $v = (1/n)R^{2/3}S^{1/2}$, where n, called Manning's n, depends on the nature of the streambed. The hydraulic radius increases with the volumetric flow rate in a way that depends on the geometry of the streambed. We will not attempt to solve Manning's equation for this exercise. Rather, we will make the simplifying assumption that the flow times given in the figure are accurate even at the higher flow rates. If we were to apply this water quality management model to a real stream, we should calculate more realistic flow times for each different flow rate below the tributary.

REFERENCES

Arbabi, M., J. Elzinga, and C. ReVelle, 1974. The Oxygen Sag Equation: New Properties and a Linear Equation for the Critical Deficit, *Water Resources Research,* 10(5).

Brill, E. D., Jr., et al., 1976. Equity Measures for Exploring Water Quality Management Alternatives, *Water Resources Research,* 12(5).

Camp, T. R., 1963. *Water and Its Impurities,* Rheinhold Press, New York.

Converse, A. O., 1972. Optimum Number and Location of Treatment Plants, *Journal of the Water Pollution Control Federation,* 44(8).

Dobbins, W. E., 1964. BOD and Oxygen Relationships in Streams, *Journal of the Sanitary Engineering Division, ASCE,* 90(3), 53.

Ellis, J. H., 1987. Stochastic Water Quality Optimization Using Imbedded Chance Constraints, *Water Resources Research,* 23(12).

Fourer, R. M., D. M. Gay, and B. W. Kernighan, 1993. *AMPL: A Modeling Language for Mathematical Programming,* Academic Press, San Diego, CA.

Fujiwara, O., S. K. Gnanendran, and S. Ohgaki, 1986. River Quality Management Under Stochastic Streamflow, *ASCE Journal of Environmental Engineering,* 112(4).

Fujiwara, O., S. K. Gnanendran, and S. Ohgaki (1987). Chance Constrained Model for River Water Quality Management, *ASCE Journal of Environmental Engineering,* 113(5).

Liebman, J. C. and W. R. Lynn, 1966. The Optimal Allocation of Stream Dissolved Oxygen, *Water Resources Research,* 2(3).

O'Connor, D. J., and W. E. Dobbins, 1958. Mechanisms of Reaeration in Natural Streams, *Transactions of the American Society of Civil Engineers,* 123, 655.

ReVelle, C. S., D. P. Loucks, and W. R. Lynn, 1967. A Management Model for Water Quality Control, *Journal of the Water Pollution Control Federation,* 39(4).

ReVelle, C. S., D. P. Loucks, and W. R. Lynn, 1968. Linear Programming Applied to Water Quality Management, *Water Resources Research,* 4(1).

Streeter, H. W. and E. B. Phelps, 1925. A Study of the Pollution and Natural Purification of the Ohio River, *U.S. Public Health Bulletin 146,* February.

U.S. EPA, 1993a. Internet file URL: http://www.epa.gov/earth100/records/wasp.html, U.S. Environmental Protection Agency, Center for Exposure Assessment Modeling (CEAM), Athens, GA.

U.S. EPA, 1993b. Internet file URL: http://www.epa.gov/docs/SWMM_WINDOWS/metadata.txt.html, U.S. Environmental Protection Agency, Watershed Modeling Section, Washington, DC.

U.S. EPA, 1995. Internet file URL: http://www.epa.gov/docs/QUAL2E-WINDOWS/metadata.txt.html, U.S. Environmental Protection Agency, Watershed Modeling Section, Washington, DC.

U.S. EPA, 1996. Internet file URL: http://www.epa.gov/epa_ceam/wwwhtml/hspf.htm, U.S. Environmental Protection Agency, Center for Exposure Assessment Modeling (CEAM), Athens, GA.

APPENDIX

AMPL Computer Code of Water Quality Management Models

I. Full-Form Model

Model File: QUAL.MOD

```
# Ampl water quality management model
param m > 0 ; # number of reaches (discharges)
set REACH := {1..m}; # indexed by i
param p{REACH} > 0 ; # number of breakpoints for each
    reach
set BREAKPOINT{i in REACH} := {1..p[i]}; # indexed by k

param n{REACH} > 0 ; # number of segments in each reach
set SEGMENT{i in REACH} := {1..n[i]}; # indexed by s

param Q0; # Flowrate in stream above first discharge
    (MGD)
param Dinit; # Oxygen deficit just upstream of first
    discharge
param Linit; # BOD concentration just upstream of first
    discharge
param eta_min{REACH}; # minimum treatment efficiency
    breakpoint
param eta{i in REACH,breakpoint[i]}; # tertiary
    treatment breakpoints
```

```
param qout{REACH}; # flowrate of each withdrawql (MGD)
param qin{REACH}; # flowrate of each discharge (MGD)
param k1{REACH}; # bioxidation constant for each reach
    (1/day)
param k2{REACH}; # reaeration constant for each reach
    (1/day)
param Lu{REACH}; # BOD concentration entering each plant
    (mg/L)
param d{REACH}; # oxygen deficit in discharge from each
    plant (mg/L)
param Dmax{REACH}; # greatest oxygen deficit allowed on
    each reach (mg/L)
param Csat{REACH}; # saturation DO concentration for
    each reach (mg/L)
param c{i in REACH,BREAKPOINT[i]};# slopes of cost vs.
    efficiency
param trun{REACH}; # flow time for each reach (day)
param Q{i in REACH union {0}}; # streamflow on each reach
    (MGD)

# Calculated Parameters:
param t{i in REACH,s in SEGMENT[i]} := s*trun[i]/n[i];
param A{i in REACH,s in SEGMENT[i]}
     := (k1[i]/(k2[i]-k1[i]))*(exp(-k1[i]*t[i,s]) -
         exp(-k2[i]*t[i,s]));
param B{i in REACH,s in SEGMENT[i]} :=
    exp(-k2[i]*t[i,s]);
param R{i in REACH} := exp(-k1[i]*t[i,n[i]]);

data QUAL.DAT;
model;

let Q[0] := Q0;
let {i in REACH} Q[i] := Q[i-1] + qin[i] - qout[i];

var x{i in REACH,BREAKPOINT[i]} >= 0; # treatment
    efficiency
#var D{ {i in REACH,SEGMENT[i]} union {(0,n[0])} } >=
    0; # oxygen deficit
var L0{REACH} >= 0; # BOD at beginning of each reach
var Le{REACH union {0} } >= 0; # BOD at end of each
    reach
var D0{REACH} >= 0; # DO at beginning of each reach
```

```
var De{REACH union {0} } >= 0; # DO at end of each
    reach
var l{REACH} >= 0; # BOD of each discharge

minimize total_cost:
    sum {i in REACH,k in BREAKPOINT[i]} c[i,k]*x[i,k];
subject to efficiency{i in REACH}:
    sum{k in BREAKPOINT[i]}x[i,k] + (1/Lu[i])*l[i] =
        1.0;
subject to primary_secondary{i in REACH}:
    eta_min[i] <= x[i,1] <= eta[i,1];
subject to tertiary{i in REACH, k in (BREAKPOINT[i] diff
    {1})}:
    x[i,k] <= (eta[i,k] - eta[i,k-1]);
subject to bodmix{i in REACH}:
    (Q[i-1]-qout[i])*Le[i-1] + qin[i]*l[i] -
        Q[i]*L0[i] = 0;
subject to oxymix{i in REACH}:
    -(Q[i-1]-qout[i])*De[i-1] + Q[i]*D0[i] =
        qin[i]*d[i];
subject to bodend{i in REACH}:
    Le[i] - R[i]*L0[i] = 0;
subject to oxyend{i in REACH}:
    De[i] - A[i,n[i]]*L0[i] - B[i,n[i]]*D0[i] = 0;
subject to oxysag{i in REACH,k in SEGMENT[i]}:
     A[i,k]*L0[i] + B[i,k]*D0[i] <= Dmax[i];
subject to oxystart{i in REACH}:
     D0[i] <= Dmax[i];
subject to Initial_Deficit:
    De[0] = Dinit;
subject to Initial_BOD:
    Le[0] = Linit;
```

Date File: QUAL.DAT

```
param m := 3;
param p := 1 3
           2 3
           3 3;
param n := 1 10
           2 15
           3 10;
param Q0 := 400;
param Dinit := 2.0;
```

```
param Linit := 8.0;
param eta_min := 1  0.35
                 2  0.35
                 3  0.35 ;
param eta :=
      :    1     2     3    :=
      1  0.80  0.90  0.99
      2  0.80  0.90  0.99
      3  0.80  0.90  0.99  ;

param qout := 1  40
              2  16
              3  35 ;

param qin := 1  40
             2  16
             3  35  ;

param k1 := 1  0.31
            2  0.27
            3  0.28  ;

param k2 := 1  0.39
            2  0.44
            3  0.63  ;

param Lu := 1  275
            2  160
            3  390  ;

param d := 1  7
           2  7
           3  7  ;

param Dmax := 1  4.5
              2  4.5
              3  4.5  ;

param Csat := 1  9.5
              2  9.5
              3  9.5  ;

param c :=
       :    1    2    3   :=
       1   0.5  1.0  3.0
       2   0.7  1.5  4.0
       3   0.6  1.2  3.5   ;
```

```
param trun := 1  1.0
              2  2.2
              3  1.4  ;
```

II. Compact Model

Model File: GENQUAL.MOD

```
var x{j in DISCHARGE,k in BREAKPOINT[j]} >= l[j,k] <=
    u[j,k];
# treatment efficiency

minimize total_cost:
        sum {j in DISCHARGE, k in BREAKPOINT[j]}
            c[j,k]*x[j,k];

subject to deficit{i in REACH, s in SEGMENT[i]}:
    sum{j in UPSTREAM[i]} A[i,s,j] * (sum{k in
        BREAKPOINT[j]} x[j,k])
                                        >= Dred[i,s];
```

Parameter File: GENQUAL.PAR

```
param m > 0 ; # number of discharges
set REACH := {1 .. m}; # indexed by i
set DISCHARGE := {0 .. m}; # indexed by j
set UPSTREAM{i in REACH} := {0..i}; # indexed by j -
    discharges influencing each reach i
param p{DISCHARGE} > 0 ; # number of breakpoints
set BREAKPOINT{k in DISCHARGE} := {1..p[k]}; # indexed
    by k
param n{({0} union REACH)} > 0 ; # number of segments
    in each reach
let n[0] := 1;
set SEGMENT{i in REACH} := {0..n[i]}; # indexed by s
param Q0; # Flowrate in stream above first discharge
    (MGD)
param eta_min{DISCHARGE}; # minimum treatment efficiency
    breakpoint
param eta{j in DISCHARGE,BREAKPOINT[j]}, # tertiary
    treatment breakpoints
param l{j in DISCHARGE,BREAKPOINT[j]};
param u{j in DISCHARGE,BREAKPOINT[j]};
```

```
param Q{DISCHARGE}; # flow in stream downstream from
    discharge k
param qout{DISCHARGE}; # flowrate of each intake (MGD)
param qin{DISCHARGE}; # flowrate of each discharge (MGD)
param k1{REACH}; # bioxidation constant for each reach
    (1/day)
param k2{REACH}; # reaeration constant for each reach
    (1/day)
param Lu{DISCHARGE}; # BOD concentration entering each
    plant (mg/L)
param d{DISCHARGE}; # oxygen deficit in discharge from
    each plant (mg/L)
param Dmax{REACH}; # greatest oxygen deficit allowed on
    each reach (mg/L)
param Csat{REACH}; # saturation oxygen conc in each
    reach (mg/L)
param c{j in DISCHARGE,BREAKPOINT[j]};
# slopes of cost vs. efficiency by discharge
param trun{REACH}; # flow time for each reach (day)
# Calculated Parameters:
param t{i in REACH,s in SEGMENT[i]} := s*trun[i]/n[i];
param f{i in REACH,s in SEGMENT[i]}
     := (k1[i]/(k2[i]-k1[i]))*(exp(-k1[i]*t[i,s]) -
         exp(-k2[i]*t[i,s]));
param g{i in REACH,s in SEGMENT[i]} :=
    exp(-k2[i]*t[i,s]);
param h{i in REACH,s in SEGMENT[i]} :=
    exp(-k1[i]*t[i,s]);
param alpha{i in REACH, SEGMENT[i], UPSTREAM[i]};
param beta{i in REACH, SEGMENT[i], UPSTREAM[i]};
param rho{i in REACH, SEGMENT[i], UPSTREAM[i]};
param A{i in REACH, SEGMENT[i], UPSTREAM[i]};
param gamma{i in REACH, UPSTREAM[i]};
param delta{i in REACH, UPSTREAM[i]};
param y{DISCHARGE diff {0}}; # (1-qout[j]/Q[j-1])
param Dunt{i in REACH, s in SEGMENT[i]}; # Deficit from
    untreated discharges
param Dred{i in REACH, s in SEGMENT[i]}; # Deficit
    reduction required

data genqual.dat;
model;
param count;
param product;
option display_precision 3;
```

```
option print_precision 3;
let Q[0] := Q0;
let qin[0] := Q0;
let {j in (DISCHARGE diff {0})} Q[j] := Q[j-1] -
    qout[j] + qin[j];
let {j in DISCHARGE diff {0})} y[j] := 1 - qout[j]/
    Q[j-1];
let {j in DISCHARGE} 1[j,1] := eta_min[j];
let {j in DISCHARGE, k in (BREAKPOINT[j] diff {1})}
    1[j,k] := 0;
let {j in DISCHARGE} u[j,1] := eta[j,1];
let {j in DISCHARGE, k in (BREAKPOINT[j] diff {1})}
                    u[j,k] := (eta[j,k] - eta[j,k-1]);
for {i in REACH} {
  for {j in UPSTREAM[i]} {
    if (j == i) then {
      let gamma[i,j] := 1;
    } else {
      let gamma[i,j] := prod{k in {(j+1)..i}} y[k];
    }
    let delta[i,j] := gamma[i,j]*(qin[j]/Q(i]);
  }
}
for {i in REACH} {
  for {j in (UPSTREAM[i] diff {0})} {
    if(j = = i) then {
      let {s in SEGMENT[i]} rho[i,s,j] := h[i,s];
      let {s in SEGMENT[i]} alpha[i,s,j] := f[i,s];
      let {s in SEGMENT[i]} beta[i,s,j] := g[i,s];
    } else {
      let {s in SEGMENT[i]} rho[i,s,j] :=
          h[i,s]*rho[i-1,n[i-1],j];
      let {s in SEGMENT[i]}
      alpha[i,s,j] := f[i,s]*rho[i-1,n[i-1],j] +
          g[i,s]*alpha[i-1,n[i-1],j];
      let {s in SEGMENT[i]} beta[i,s,j] :=
          g[i,s]*beta[i-1,n[i-1],j];
    }
  }
}
let {i in REACH, s in SEGMENT[i]} rho[i,s,0] :=
    rho[i,s,1];
let {i in REACH, s in SEGMENT[i]} alpha[i,s,0] :=
    alpha[i,s,1];
```

```
let {i in REACH, s in SEGMENT[i]} beta[i,s,0] :=
    beta[i,s,1];
let {i in REACH, s in SEGMENT[i], j in UPSTREAM[i]}
  A[i,s,j] := delta[i,j]*alpha[i,s,j]*Lu[j];
let {i in REACH, s in SEGMENT[i]}
    Dunt[i,s] := sum{j in UPSTREAM[i]}
      delta[i,j]*(alpha[i,s,j]*Lu[j]+beta[i,s,j]*d[j]);
let {i in REACH, s in SEGMENT[i]}
   Dred[i,s] := Dunt[i,s] - Dmax[i];
```

Date File: GENQUAL.DAT

```
data;
param m := 3;
param p := 0  3
           1  3
           2  3
           3  3 ;
param n := 1  10
           2  15
           3  10;
param eta_min := 0  0.0
                 1  0.35
                 2  0.35
                 3  0.35 ;
param eta :=
      :   1      2      3    :=
        0   0.00   0.00   0.00
      1   0.80   0.90   0.99
      2   0.80   0.90   0.99
      3   0.80   0.90   0.99  ;

param Q0 := 400;

param qin := 1  40
             2  16
             3  35  ;

param qout := 1  40
              2  16
              3  35  ;

param k1 := 1  0.31
            2  0.27
            3  0.28  ;
```

```
param k2 := 1  0.39
            2  0.44
            3  0.63  ;

param Lu := 0    8
            1  275
            2  160
            3  390  ;

param d := 0  2
           1  7
           2  7
           3  7  ;

param Dmax := 1  4.5
              2  4.5
              3  4.5  ;

param Csat := 1  9.5
              2  9.5
              3  9.5  ;

param c  :=
       :    1    2    3   :=
       0   0.0  0.0  0.0
       1   0.5  1.0  3.0
       2   0.7  1.5  4.0
       3   0.6  1.2  3.5  ;

param trun := 1  1.0
              2  2.2
              3  1.4  ;
```

CHAPTER 3

Groundwater Systems

WILLIAM W-G. YEH and YUNG-HSIN SUN

3.A. INTRODUCTION

In the hydrologic cycle, groundwater occurs whenever surface water occupies and saturates the pores or interstices of the rocks and soils beneath the earth's surface. The geologic formations that store and transmit the subsurface water are known as *aquifers*. Aquifers, *aquitards* (semipermeable formations), or *aquicludes* (nonpermeable formations) may underlie a geographic area, watershed, or drainage basin, and all may hold water. But drawing water from aquitards and aquicludes is impractical and economically prohibitive, whereas groundwater in aquifers can be removed economically and is often a dependable source of water supply (Todd, 1980). Most aquifers can be considered as underground storage reservoirs that receive recharge from both natural and artificial sources. Depending on local geological formation and boundary conditions, groundwater may flow out of an aquifer, contributing to surface runoff. In most cases, each aquifer formation has spatially varying properties, such as transmissivity, storage capacity, and water quality, which affect the basin's response to artificial pumping and recharge. These formations are referred to collectively as a *groundwater reservoir* or *groundwater system* (Willis and Yeh, 1987). Throughout human civilization, groundwater has been relied upon as a low-cost water source for domestic, municipal, industrial, and agricultural consumption.

Groundwater aquifers can be classified as *confined* or *unconfined*, depending on the existence of a water table (see Figure 3.1). A leaky confined aquifer represents a geological formation that leaks and allows water to flow through the confining layer. However, a groundwater system can be defined more precisely by the following components (Willis and Yeh, 1987):

1. The controlled and partially controlled system inputs, such as subsurface flows, natural recharge, precipitation, and replenishment from irrigation return flows, streams, and artificial recharge practices

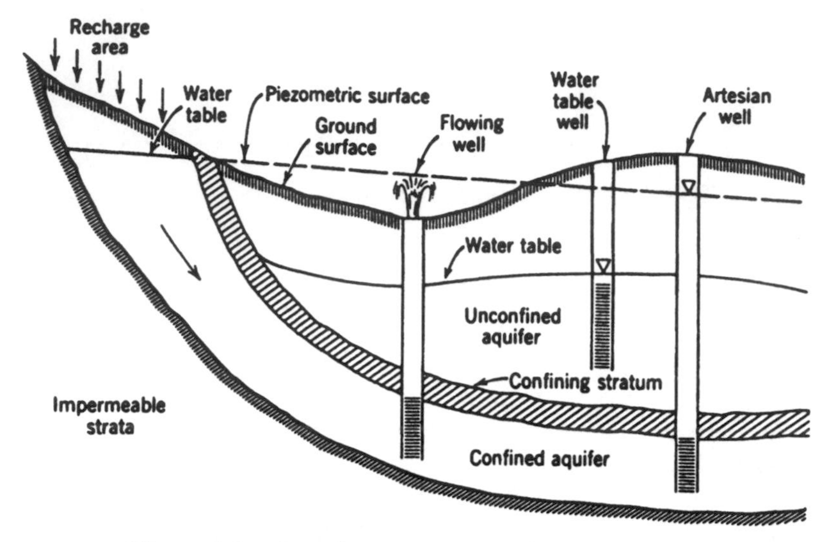

Figure 3.1 Groundwater system. (After Todd, 1980.)

2. The system outputs, which include subsurface outflows, discharges to surface waters, natural springs, and evapotranspiration losses
3. The system parameters, which define the flow, quality, and thermal properties of the aquifer system (e.g., storativities, transmissivities, and dispersion parameters)
4. The control or decision variables, which represent the artificial extraction, injection, and recharge of the groundwater system
5. The state variables, which characterize the condition of the system (e.g., the hydraulic head, constituent concentrations, and temperature in the groundwater system)

Since the discovery of numerous hazardous-waste sites in the late 1970s, groundwater contamination and groundwater quality have become major issues in the management of groundwater systems (Bedient et al., 1994). Public health is under the threat of possible long-term effects caused by various contamination sources, which include industrial spills, infiltration through landfills and surface impoundment, leakage of storage/septic tanks and pipelines, artificial recharge of poor-quality water, saltwater intrusion, pesticide use, and improper waste disposal (Pettyjohn, 1987). Figure 3.2 shows how waste disposal practices contaminate the groundwater system. The complex underground geological formations contribute significantly to diversifying the patterns of contaminant migration and to reaching pathways of human exposure. Among the more than 1500 contaminated sites nationwide on the National Priority List from the U.S. Environmental Protection Agency (EPA), the main contaminants of concern include petroleum hydrocarbons, chlori-

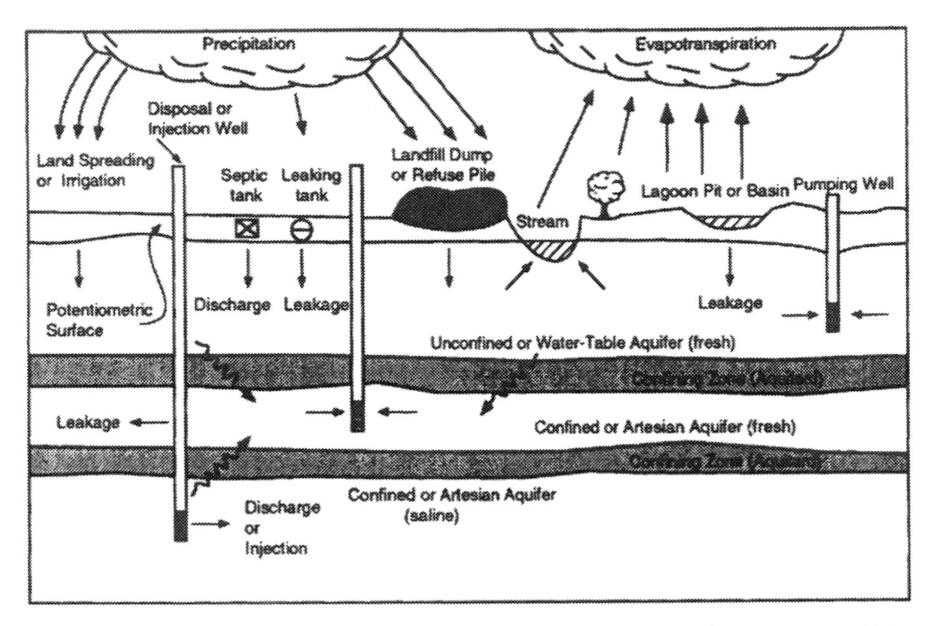

Figure 3.2 How waste-disposal practices contaminate the groundwater system. (After Bedient et al., 1994.)

nated organics (e.g., trichloroethylene), heavy metals (e.g., lead, zinc, and chromium), and inorganic salts. Various actions are now being taken to preserve and restore groundwater (Bedient et al., 1994). Groundwater system analysis has never been more important than today, for system analysis systematically investigates groundwater resources and contamination and will provide long-range plans for groundwater resources management that is both technically and environmentally sound.

3.A.1. System Analysis in Groundwater Management

System analysis identifies the basic components of an integrated system and, through mathematics, describes the interrelationships that link the components into the active system. In the analysis of groundwater systems, Willis and Yeh (1987) consider three classes of management problems: the instrument or detection problem, the inverse or parameter estimation problem, and the prediction problem. These three types of problems focus on different components of a groundwater system. The *instrument or detection problem* focuses on retrieval of system inputs. Examining the response properties of the aquifer system to identify recharge or leakage in semiconfined or unconfined aquifers is an example of the instrument problem.

Assuming that historical concurrent input and output observations are available, the *inverse or parameter estimation problem* seeks to identify unknown model parameters embedded in the partial differential equations that govern the groundwater flow and solute transport in the groundwater system and, possibly, some of the unknown initial or boundary conditions. The inverse problem is extremely important in groundwater management because the identified parameters will be used later for prediction in groundwater models. The availability and reliability of field observations are of great importance for the inverse problem and, consequently, for the successful execution of a management model.

The *prediction problem* focuses on the dynamic behavior of the groundwater system in response to human activities, usually with the assistance of an analytical or numerical simulation model. The parameters estimated in the inverse problem are used to characterize the aquifer system and therefore are assumed to be known in the prediction problem. The prediction problem often occurs as part of the simulation or optimization framework of a groundwater system for examining groundwater flows and/or contaminant fates under a given set of operational rules (the control variables).

Although in groundwater management the instrument, inverse, and prediction problems seem to occur only in sequence, in fact they usually have a complex interdependency, that renders groundwater system analysis a highly dynamic process. Their relationships are discussed further below.

3.A.2. Framework of System Analysis in Groundwater Management

The management of a groundwater system is rarely an isolated problem. A system analyst often finds it necessary to integrate groundwater system analysis into a broad, regional water resources planning and management framework. To assist in groundwater system analysis, a simulation model will translate the physical groundwater system and operational strategies into a mathematical representation. Translating adequately from the real system to the mathematical representation is essential; otherwise, groundwater management alternatives will be evaluated incorrectly.

The general framework of system analysis in groundwater management consists of six major procedures; each poses one or more questions to the system analyst:

1. *Objective recognition:* What is the system analysis for? What type of groundwater model is required?
2. *Data collection:* What do we know about the groundwater system and its management strategies?
3. *Groundwater model composition:* What components does the model need to represent the groundwater system properly? How can we solve this resulting model?
4. *Parameter estimation:* How can model parameters be quantitatively determined? Are the estimated parameters reliable?

5. *Model verification:* Is the calibrated model reliable?
6. *Model application:* How can the groundwater model help us? Is it adequate for the purpose stipulated?

In this framework, if the analyst is not satisfied with the results of one procedure, it may be necessary to trace backward and to find ways to improve the results. For example, unreliable parameter estimation may require experimental design to guide additional data collection. If the calibrated model is proven to be unreliable for the purpose stipulated, ways must be sought to improve conceptual model formulation, parameter estimation, and data collection. The three types of management problems mentioned previously are blended in various parts of this framework, resulting in a highly complex interrelationship problem of groundwater system planning and management. It should be noted that a system analyst's key role is to integrate the various aspects of the framework, since it is possible and even desirable to rely on specialized experts to carry out the required analysis of some of the procedures.

In this chapter the framework of groundwater system analysis is introduced through a case study of conjunctive use of surface water and groundwater, followed by four additional applications with their own distinctive significance for the modern practice of groundwater system analysis. Through these applications we wish to demonstrate the versatility of groundwater system analysis under a single consistent framework.

3.B. SYSTEM ANALYSIS IN GROUNDWATER MANAGEMENT: A CASE STUDY

The framework of groundwater system analysis will be illustrated by using the model development for conjunctive use study of surface water and groundwater at the San Jacinto Basin in southern California (Yeh et al., 1991; Wang et al., 1995). Water demands in southern California reached 4 million acre-feet in 1990. This large amount of water consumption relies heavily on imported water from both northern California and the Colorado River. However, increasing public awareness of environmental issues and changes in regulatory decisions have reduced the reliability of imported water. Thus the Metropolitan Water District of Southern California (MWD), the regional water wholesaler, in collaboration with local member agencies and representatives from the business, agricultural, and environmental communities, is actively developing an integrated resources plan to make best use of existing and potential water supplies and through this plan to assure reliable water service that addresses environmental and institutional concerns at the lowest possible long-term cost.

One of the most important components of the integrated resources plan is the expanded development of conjunctive use of surface water and groundwater. Under the conjunctive use program, surplus water imported during a period of abundance—water that would otherwise be lost to the ocean—would recharge local groundwater basins by direct or indirect means. This water would remain in storage until needed during seasons of shortage or years of drought. San Jacinto Basin (see Figure 3.3a) was included in the conjunctive use program due to its location and excellent recharge capacity. The California Department of Water Resources estimated that more than 300,000 acre-ft of water could be stored in the San Jocinto Basin before reaching its capacity. A modest recharge plan was proposed to recharge the

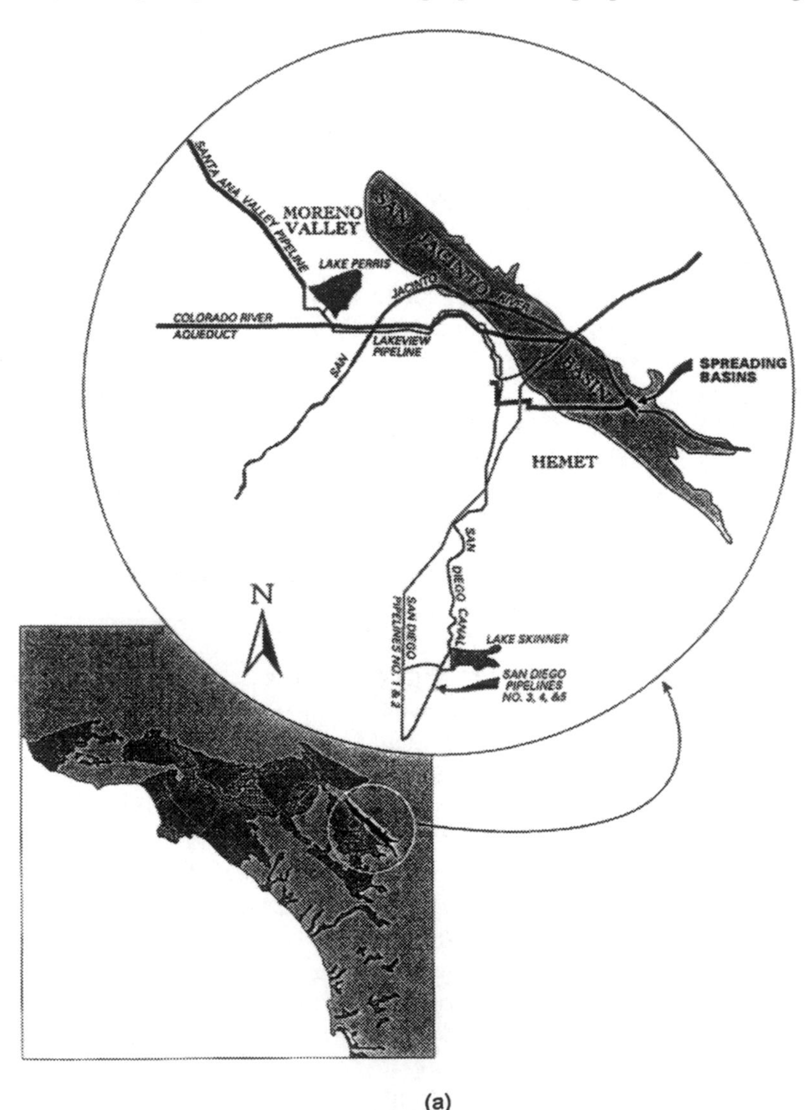

(a)

Figure 3.3 (a) San Jacinto Basin, Riverside, California. (After Wang et al., 1995.) (b) Re-creation of water heads and aquitard distribution. (After Yeh et al., 1991.) (c) Plan view of geological cross-sections. (Yeh et al., 1991.)

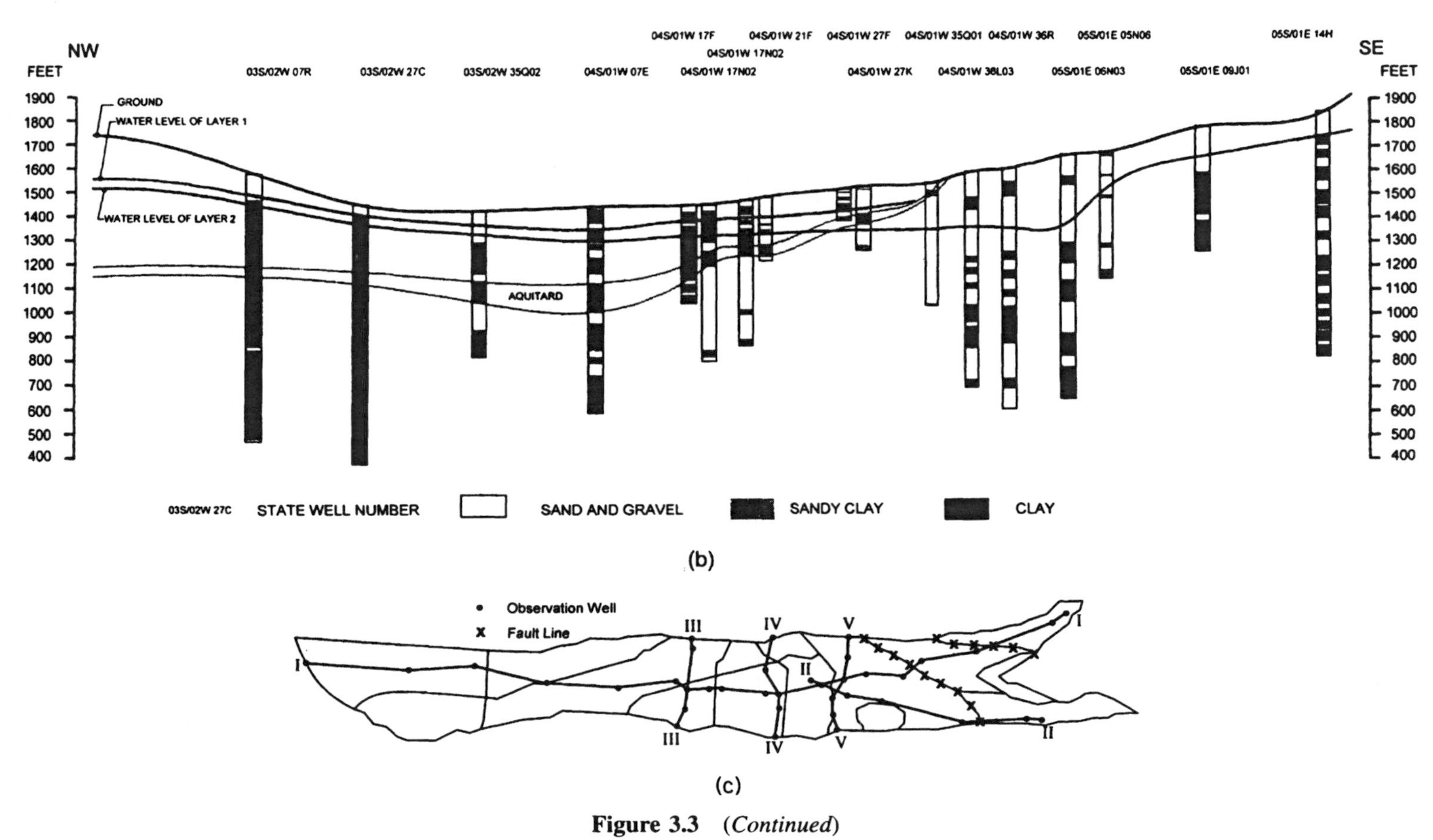

Figure 3.3 *(Continued)*

basin at the rate of 80 acre-ft/day from October 1 to January 31 each year. To assist in the validation of the hydrologic feasibility and efficiency of the proposed plan, Yeh et al. (1991), in collaboration with MWD, developed a groundwater flow and transport model for the San Jacinto Basin.

3.B.1. Objective Recognition

In groundwater system analysis, it is important to recognize the objective, because the objective often determines those requirements necessary for subsequent procedures. The system analysis for regional planning and management does not require a groundwater model as sophisticated as that needed for local aquifer study.

The objective of the conjunctive use study by Yeh et al. (1991) is straightforward: to ascertain if the proposed recharge plan is feasible and efficient. A groundwater model of high resolution is necessary for this local aquifer study. The control of water in and out of the aquifer is especially important to a recharge study. In particular, the change of total storage, river recharge, and possible aquifer outflow should be monitored constantly. These understandings were reflected in the following study.

3.B.2. Data Collection

Before a groundwater model can be used for prediction, it should be calibrated based on available observations, historical operation, and boundary and initial conditions. Without doubt, the more data we have, the more reliable the calibrated model will be. But data deficiency is always a serious impediment to implementing a groundwater system analysis. Data should at the very least support water balance in the study area.

The data collected, compiled, and reviewed for the determination and formulation of a conceptual groundwater model (flow and solute transport) for the San Jacinto Basin include (Yeh et al., 1991) the following:

1. Topographical and geographical maps, which assisted in defining the region of groundwater basin and boundary types.
2. Well logs, which contained the vertical distribution of geological formations, including depth, color, character, size of material, and structure of the strata. With these data, Yeh et al. (1991) recreated the aquifer strata, which assisted in composing the groundwater model and estimating parameters. Figure 3.3b shows the re-creation of water heads and aquitard distribution along section I–I (see Fig. 3.3c). A layer of aquitard, which consists mostly of clay, divides the San Jacinto Basin into two layers: one unconfined, and the other, partially confined and partially unconfined.
3. Groundwater level and quality data, which were the essential data in model calibration (parameter estimation). As often happens in a field

study, available data were scattered in time and space. Determining a usable sequence of data requires consulting the history of local activity and data integrity. Recent observations were considered more reliable for calibrating the current system, and the starting data of calibration was selected so that the concurrent water level and quality could be restored for the entire basin with confidence.

4. Historical precipitation and streamflow data, which were part of the boundary conditions.
5. Groundwater extraction records, which were incomplete. Missing records were estimated by personnel of the related water agencies based on their experience.
6. Land use data, which included detailed acreages of human activities and corresponding extraction factor, return factor, and net unit waste increment. Human activities included residential, commercial, industrial, and agricultural uses. Extraction factor and return factor were necessary to estimate the yearly groundwater consumption and return flow, and net unit waste increment was used to estimate the increment of the total dissolved solids (TDS) in groundwater caused by human activities.

The problem of data deficiency and the need to design an efficient and effective strategy for additional data collection are discussed later in the chapter. It should be noted that data requirements are problem specific and highly dependent on the associated management objective. The data categories above are for a conjunctive use study, with particular concern for TDS. If the data collected are to be used in a groundwater management model, additional information on financial, institutional, and other socioeconomic conditions may be required.

3.B.3. Groundwater Model Composition

A groundwater model is a physically based mathematical model. Model composition involves (1) choosing the building blocks that represent the groundwater system as closely as possible, and (2) selecting the proper method to solve the resulting mathematical model.

Determination of System Equations. Yeh et al. (1991) and Wang et al. (1995) analyzed the geological data of the basin. It was concluded that a two-layered leaky aquifer system was necessary and sufficient for characterizing the groundwater system. The confined aquifer portion of layer 2 is separated from layer 1 (unconfined aquifer) by an aquitard with a hydraulic conductivity value much lower than those of the upper and lower aquifers. The flow in the aquitard is assumed to be vertical.

In the following mathematical representation, subscript 1 is used for variables associated with layer 1 and the unconfined portion of layer 2, and subscript 2 for variables associated with the confined portion of layer 2. The hydraulics of groundwater flow can be described by Darcy's laws, mass conservation, and Dupuit assumptions. Flows in the unsaturated and the saturated zones are represented by the equations (Bear, 1979)

$$S_y \frac{\partial h}{\partial t} = \frac{\partial}{\partial x}\left(K_1 m_1 \frac{\partial h}{\partial x}\right) + \frac{\partial}{\partial y}\left(K_1 m_1 \frac{\partial h}{\partial y}\right) + P + R + I - Q_1 - \frac{K_z}{b'}(h - \phi) \tag{3.1}$$

$$S \frac{\partial \phi}{\partial t} = \frac{\partial}{\partial x}\left(T \frac{\partial \phi}{\partial x}\right) + \frac{\partial}{\partial y}\left(T \frac{\partial \phi}{\partial y}\right) - Q_2 + \frac{K_z}{b'}(h - \phi) \tag{3.2}$$

in which

K = hydraulic conductivity [L/T]

m = saturated thickness [L]; $m_1 = h - b'$

$T = K_2 m_2$, the transmissivity [L^2/T]

h, ϕ = water heads in the unconfined and the confined aquifers, respectively [L]

S_y, S = storativity of the unconfined aquifer, and the storage coefficient of the confined aquifer, respectively, dimensionless

Q = extraction rate per unit area [L/T]

K_z = vertical hydraulic conductivity of the aquitard [L/T]

b' = thickness of the aquitard [L]

P, R, I = rates of replenishment, river leakage, and return water per unit area [L/T]

Solute transport in groundwater consists of two parts: advection and dispersion. *Advection* refers to the transport of contaminants at the same speed as the average linear velocity of groundwater. *Dispersion* is caused by the concentration gradient of each constituent and can be described by Fick's law (Bear, 1972). Readers should refer to Freeze and Cherry (1979) or Willis and Yeh (1987) for a detailed derivation of the solute transport equation. Solute transport in the unconfined and the confined aquifers are governed by

$$\frac{\partial(m_1 C_1)}{\partial t} = \nabla \cdot (m_1 D_1 \nabla C_1 - V_1 m_1 C_1) + PC_P + RC_R + IC_I - Q_1 C_1 - \frac{K_Z}{b'}(h - \phi)\overline{C} \tag{3.3}$$

$$\frac{\partial(m_2C_2)}{\partial t} = \nabla \cdot (m_2D_2\nabla C_2 - V_2m_2C_2) - Q_2C_2 + \frac{K_Z}{b'}(h - \phi)\overline{C} \quad (3.4)$$

in which

C = solute concentrations

D = hydrodynamic dispersion coefficient tensor [L^2/T]

V = average flow velocities [L/T]

C_P, C_R, C_I = solute concentrations of the replenishment, river leakage, and return water, respectively

$\overline{C}$ = solute concentration in the aquitard

The groundwater flow velocity (pore velocity) can be obtained by using Darcy's law. The components of V_1 and V_2 in the x and y directions can be represented by

$$V_{1x} = \frac{K_1}{n_1}\frac{\partial h}{\partial x} \qquad V_{1y} = -\frac{K_1}{n_1}\frac{\partial h}{\partial y} \quad (3.5a)$$

$$V_{2x} = -\frac{K_2}{n_2}\frac{\partial \phi}{\partial x} \qquad V_{2y} = -\frac{K_2}{n_2}\frac{\partial \phi}{\partial y} \quad (3.5b)$$

in which n is the porosity. The dispersion coefficient tensor D, ignoring molecular diffusion, can be calculated as follows (Bear, 1979):

$$D_{ij} = \alpha_T V\delta_{ij} + (\alpha_L - \alpha_T)\frac{V_iV_j}{V} \qquad i,j = x,y \quad (3.6)$$

where α_L and α_T are the longitudinal and transverse dispersivities, respectively, and δ_{ij} is the Kronecker delta ($\delta_{ij} = 1$ if $i = j$; $\delta_{ij} = 0$ otherwise).

The flow equations of the confined and the unconfined aquifers are coupled by the leakage term, and the solute transport equations and the flow equations by velocity terms. Significant difficulties arise in solving these coupled equations. Due to the thin layer of the aquitard, it is further assumed that $\overline{C} = C_1$ if $h > \phi$ and that $\overline{C} = C_2$ if $h < \phi$. The approximation helps to decouple equations (3.3) and (3.4). However, it should be noted that this type of approximation always requires verification.

In the studies by Yeh et al. (1991) and Wang et al. (1995), TDS is the target concentration in the solute transport equations, (3.3) and (3.4). The solute is assumed to travel with the flow of groundwater without retardation. However, if contaminants tend to sorb onto soil particles, the resulting solutes would move more slowly through the aquifer than the groundwater that transports them (Fetter, 1993). Benzene, TCE (trichloroethene), and PCE (tetrachloroethene) are examples. The sorption process has a significant impact on solute transport and thus on the efficiency of groundwater remediation. This is discussed further later in the chapter.

Selection of Numerical Methods. Analytical solutions of groundwater flow and solute transport have been reported for idealized groundwater systems; however, a more common approach for solving the distributed-parameter, time-dependent partial differential equations that govern groundwater flow and solute transport in the groundwater system is through numerical techniques such as finite difference or finite element methods (Willis and Yeh, 1987). These techniques transform the partial differential equations into a system of algebraic equations. The solution of these equations determines the state variables at a predetermined set of discrete nodal points within the aquifer system.

Finite difference approximation is based on a Taylor series representation of the time and spatial derivatives. It is conceptually more straightforward than finite element approximation, which is based on the method of weighted residuals. For many groundwater problems, the finite element method is superior to the classical finite difference method. Medium heterogeneity and irregular boundary conditions are handled naturally by the finite element method. This contrasts with finite difference approximation, which requires complicated interpolation schemes to approximate complex boundary conditions. Moreover, in the finite element method, the size of the elements can easily be varied to reflect rapidly changing state variables or parameter values. The piecewise continuous representation of the dependent variables and, possibly, the parameters of the groundwater system can also increase the accuracy of numerical approximation (Willis and Yeh, 1987).

Since these two types of numerical methods have been applied to many fields, references are abundant in the literature. Willis and Yeh (1987) and N.-Z. Sun (1994b) have detailed derivations of how these two methods are to be applied to groundwater modeling. Yeh et al. (1991) solved the mathematical model above by the multiple-cell balance method (MCB; Sun and Yeh, 1983). MCB is a variation of the finite element method. An advantage of MCB is its superior mass balance in both local and global scales (N.-Z. Sun, 1994b). It is also worth noting that many established groundwater modeling softwares are available in the public domain. Thus system analysts do not have to code their own versions of groundwater model. Bedient et al. (1994) provided a summary list of existing numerical models of groundwater flow and solute transport.

3.B.4. Parameter Estimation

The accuracy of model prediction depends on the reliability of the estimated parameters as well as on the accuracy of prescribed initial and boundary conditions. Parameters used in deriving the governing equations are not directly measurable from the physical point of view. In practice, model parameters are required to be estimated from historical input–output observations using an inverse procedure of parameter estimation.

Solution algorithms for parameter identification in groundwater system analysis are usually classified as being part of either the equation error criterion method or the output error criterion method (Yeh, 1986). The *equation error criterion method* requires using either finite difference or finite element approximations to obtain a system of algebraic equations in terms of the unknown parameters; it also requires that observations be available (observed or interpolated) at each node of the discretized grids used in the associated numerical scheme. The unknown parameters are determined by minimizing the least squares error (or residual sum of squares) of the equations. Major disadvantages of this approach are as follows: (1) Observation data are not available at all grid points, thus estimation by interpolation or extrapolation is often needed; (2) the solution is highly dependent on the level of discretization used in numerical solution of the governing equation, especially in the presence of measurement errors; and (3) data interpolation or extrapolation introduces collinearity among parameters, which violates the independence assumption associated with the least squares estimation; thus the coefficient matrix of the resulting algebraic system equation may have rank deficiency, causing numerical instability in the least squares estimation without proper parameterization (Yeh, 1986).

Yeh et al. (1991) formulated the inverse problem as an optimization problem using the *output error criterion method.* The objective of the optimization model is to minimize a given output error criterion. Yeh et al. (1991) used a standard least squares error criterion in identification of the flow model parameters [see equations (3.1) and (3.2)]. That is,

$$\text{minimize } \mathrm{Err}(\hat{\mathbf{P}}) = [\mathbf{h}_{\mathrm{D}}(\hat{\mathbf{P}}) - \mathbf{h}_{\mathrm{D}}^{*}]^{\mathrm{T}}[\mathbf{h}_{\mathrm{D}}(\hat{\mathbf{P}}) - \mathbf{h}_{\mathrm{D}}^{*}] \tag{3.7}$$

where $\mathbf{h}_{\mathrm{D}}^{*}$ is the vector of observed heads, $\mathbf{h}_{\mathrm{D}}(\hat{\mathbf{P}})$ the corresponding model prediction based on the current vector of parameter estimates $\hat{\mathbf{P}}$, and the superscript T represents the transpose operator. The constraints applied to the optimization objective [equation (3.1)] include the governing equations of groundwater flow [equations (3.1) and (3.2)] and the upper and lower bounds of parameters. Parameter identification in a leaky aquifer is complicated due to the water transfer between layers and the presence of fault lines (Yeh and Sun, 1990). The parameters to be identified by Yeh et al. (1991) include hydraulic conductivities, leakage coefficients, specific yields, the precipitation-infiltration coefficient, the irrigation-infiltration coefficient, recharge boundary conditions, longitudinal fault coefficients, transverse fault coefficients, river recharge parameters, and the infiltration delay time. The total number of unknown parameters is 91. The time period used in the parameter identification is from April 1963 to December 1983, and a total of 1117 head observations collected from 25 observation wells are used (Yeh et al., 1991). The optimization problem was then solved by the Gauss–Newton algorithm.

The Gauss–Newton algorithm has been widely used in solving parameter identification problems (Yeh, 1986). The algorithm is attractive because it does not require the calculation of the Hessian matrix, and the rate of convergence is faster than with classical gradient search procedures (Luenberger, 1984). The algorithm is developed for unconstrained minimization problems, but constraints, such as upper and lower bounds, can be incorporated in the algorithm with minor modifications. The algorithm starts with a set of initial estimates of parameters and converges to a local optimum. If the objective function is convex, local optimum is the global optimum. However, in general, global convergence cannot be guaranteed.

The Gauss–Newton algorithm solves the optimization by iterations. The solution is improved from iteration k to iteration $k + 1$ as follows:

$$\hat{\mathbf{P}}^{k+1} = \hat{\mathbf{P}}^{k} - \rho^{k}\mathbf{d}^{k} \tag{3.8}$$

where

$\mathbf{d}^k$ = Gauss–Newton direction vector; $\mathbf{A}^k\mathbf{d}^k = \mathbf{g}^k$

$\mathbf{A}^k = [\mathbf{J_D}(\hat{\mathbf{P}}^k)]^{\mathrm{T}}[\mathbf{J_D}(\hat{\mathbf{P}}^k)]$, which is the first-order approximation of the Hessian matrix

$\mathbf{g}^k = [\mathbf{J_D}(\hat{\mathbf{P}}^k)]^{\mathrm{T}}[\mathbf{h_D}(\hat{\mathbf{P}}^k) - \mathbf{h}^*_{\mathbf{D}}]$

$\mathbf{J_D}(\hat{\mathbf{P}}^k)$ = Jacobian matrix of the head with respect to $\hat{\mathbf{P}}$

ρ^k = step size (scalar)

The step size can be determined by a quadratic interpolation scheme such that $\mathrm{Err}(\hat{\mathbf{P}}^{k+1}) < \mathrm{Err}(\hat{\mathbf{P}}^{k})$, or simply by a trial-and-error procedure. The elements of the Jacobian matrix are represented by sensitivity coefficients as follows:

$$\mathbf{J}_D = \begin{bmatrix} \dfrac{\partial h_1}{\partial P_1} & \dfrac{\partial h_1}{\partial P_2} & \cdots & \dfrac{\partial h_1}{\partial P_L} \\ \dfrac{\partial h_2}{\partial P_1} & \dfrac{\partial h_2}{\partial P_2} & \cdots & \dfrac{\partial h_2}{\partial P_L} \\ \vdots & \vdots & \ddots & \vdots \\ \dfrac{\partial h_M}{\partial P_1} & \dfrac{\partial h_M}{\partial P_2} & \cdots & \dfrac{\partial h_M}{\partial P_L} \end{bmatrix} \tag{3.9}$$

in which M is the total number of observations and L is the total number of parameters. In the current problem, M and L are equal to 1117 and 91, respectively. However, since the analytical form of sensitivity coefficient is not available, the elements are calculated by the influence coefficient method (Becker and Yeh, 1972), which used the concept of parameter perturbation:

$$\frac{\partial h_i}{\partial P_l} \approx \frac{h_i(\hat{\mathbf{P}} + \Delta P_l \mathbf{e}_l) - h_i(\hat{\mathbf{P}})}{\Delta P_l} \qquad i = 1,\ldots,1117 \tag{3.10}$$

where ΔP_l is a small increment of P_l and $\mathbf{e}_l$ is the lth unit vector. Besides the influence coefficient method, both the sensitivity equation method (Yeh and Yoon, 1981) and the variational method (Sun and Yeh, 1985) can be used to generate the sensitivity coefficients of the Jacobian matrix. N.-Z. Sun (1994a) provides an extensive discussion about techniques for parameter estimation.

One may ask: How reliable are these parameter estimates? Conceptual models and observations introduce various types of errors. For example, natural aquifer strata are rarely homogeneous. Hydraulic and transport parameters, such as transmissivity and dispersivity, are continuous functions of the spatial domain of the groundwater system. These continuous functions may be identified in an idealized situation but not with limited and scattered point measurements (observations made at a limited number of locations in the spatial domain). Therefore, these continuous functions must be approximated by finite-dimensional functions to have a solvable inverse problem. The reduction of parameter dimension is generally accomplished by parameterization using either the zonation approach or the interpolation method (Yeh, 1986).

With the zonation approach, the flow region is divided into a number of subregions or zones and a constant parameter value characterizes each zone. The unknown parameter function is then represented by the collection of constants for corresponding zones (Yeh and Yoon, 1976). Figure 3.4 shows the distribution of zones in the San Jacinto Basin used by Yeh et al. (1991) and Wang et al. (1995). In the interpolation method, the distribution of unknown parameters is assumed to follow one particular function [e.g., geostatistical models (Kitanidis and Vomvoris, 1983; Tong and Yeh, 1995; Tong, 1996)] or a combination of its kind [e.g., the finite element basis functions (Yoon and Yeh, 1976; Yeh and Yoon, 1981)]. In any case, approximation introduces errors.

There are two sources of error that affect the reliability of parameter estimation: measurement error and system error (TASC, 1974). *Measurement error* is introduced by imperfect measurements performed in the field. Although measurement techniques have been studied and improved continually over the years, it is still impossible to avoid completely this type of error. Aware of data deficiency, the system analyst may well acknowledge mea-

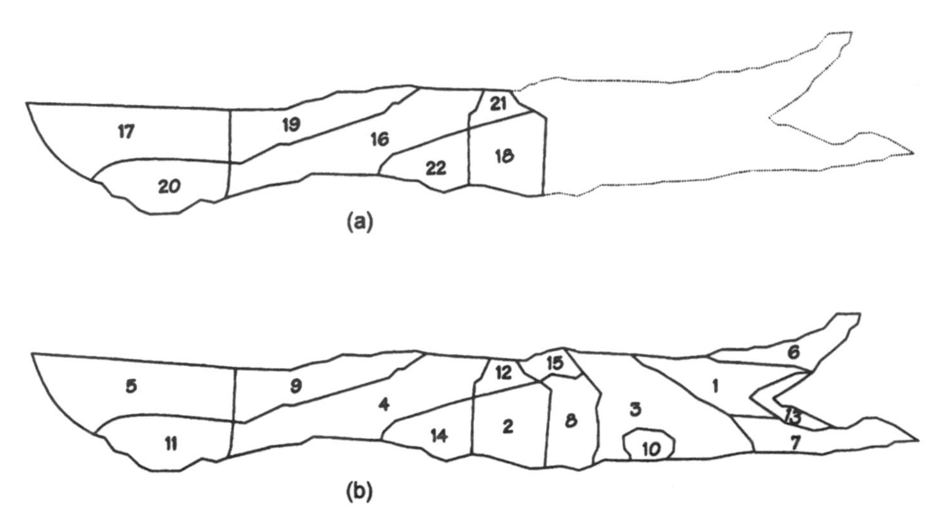

Figure 3.4 Soil zone distribution in the San Jacinto Basin, Riverside, California. (After Wang et al., 1995.)

surement errors, but such errors are not necessary expressed directly in the inverse problem. In other words, measurement errors are generally accounted for together with the system error.

Two types of *system error* are often discussed in the inverse problem: the system modeling error and the error associated with parameter uncertainty (Yeh, 1986). It is rather difficult to distinguish one from the other. The *system modeling error* stems from an imperfect conceptual model for groundwater flow and contaminant transport, or parameterization. In general, system modeling error can be reduced by improving the model structure (improving the model composition procedure) or increasing the number of model parameters (increasing the model resolution).

Parameter uncertainty is a consequence of system modeling error and of deficient observations. Parameter uncertainty can be reduced by increasing the number of observations but it cannot be totally eliminated. Thus a parameter can be identified only as a range of values with different occurrence probabilities. In other words, the parameter estimate is a random variable. The inverse problem results in the expected value of the parameter estimated being based on the available data. If the estimation is unbiased, with a sufficiently large sample size, the expected value of parameter estimate should be equal to the true value of the parameter (Haan, 1977).

Parameter variance is a measure of the dispersion of the parameter estimate, or how extensive $\hat{P}_i$ deviates from its true value, P_i. Intuitively, if a random variable can vary in a wider range, its mean (the expected value) would be less representable. Therefore, the variance of parameter estimate is often used to define parameter uncertainty quantitatively. The variance of parameter estimate can be written as follows:

$$\mathrm{Var}(\hat{P}_i) = \mathrm{E}[(\hat{P}_i - P_i)^2] \tag{3.11}$$

where E represents the expectation operator, and the square root of $\mathrm{Var}(\hat{P}_i)$ is the standard deviation. Statistically speaking, if the estimation is consistent, $\mathrm{Var}(\hat{P}_i)$ can be as small as desired with a sufficiently large sample size. In other words, we gain more confidence in the parameter estimate as more data are included in the estimation. If a large sample size is not available, a consistent estimator resulting in less error variance should be used (Haan, 1977). In multidimensional parameter estimation, the concept is extended to the covariance matrix of parameter estimates, which can be written as follows:

$$\mathrm{Cov}(\hat{\mathbf{P}}) = \mathbf{E}[(\hat{\mathbf{P}} - \mathbf{P})(\hat{\mathbf{P}} - \mathbf{P})^{\mathrm{T}}] \tag{3.12}$$

Individual parameter variances appear as the diagonal elements in the covariance matrix. The additional information of parameter covariances appear as the off-diagonal elements. The covariance between two parameters often relates to the significance of their interdependency. In the inverse problem, since there is no knowledge of true parameter values, an approximation of the covariance matrix can be represented by the equation (Yeh and Yoon, 1981)

$$\mathrm{Cov}(\hat{\mathbf{P}}) = \frac{\mathrm{Err}(\hat{\mathbf{P}})}{M - L}\,[\mathbf{J}_{\mathbf{D}}^{\mathrm{T}}\mathbf{J}_{\mathbf{D}}]^{-1} \tag{3.13}$$

The covariance matrix has been used to evaluate the goodness of model fitting and the interdependency among parameters and thereby to determine the appropriate level of parameterization (Yeh and Yoon, 1981; Yeh and Sun, 1984; McCarthy and Yeh, 1990). It is also possible to look at parameter uncertainty from a statistical point of view.

Parameter variance determines the significance level of the parameter estimate in linear models. It provides a reference for nonlinear parameter estimation. The significance is tested against the hypothesis that the true value of the parameter is zero. Since the true value of any hydraulic parameter cannot be zero, an insignificant estimate that is not distinguishable from zero implies that the parameter estimate is insufficient. Possible causes of the insufficient estimation are parameter redundancy or interdependency, inappropriate model specification, insufficient data, and other human errors. The ratio

$$t = \frac{\hat{P}_i}{\sqrt{\mathrm{Var}(\hat{P}_i)}}$$

has a distribution approaching the standard normal distribution as sample size becomes large and can be used to define parameter significance quantitatively (Haan, 1977). For example, for zone 3 shown in Figure 3.4, the estimated precipitation-infiltration parameter is 0.650 with a variance of 0.0082, and the estimated recharge boundary condition is 8.820 with a variance of 10.6076 (Yeh et al., 1991). The t ratios of these two parameters are 2.27 and 2.71, respectively. Thus these two parameter estimates are said to be significant up to 97.7% and 99.3% confidence levels, respectively. The hypothesis that these parameters are truly zero can be rejected in high confidence. However, the hydraulic conductivity of zone 13 has the estimate of 1.650 with a variance of 1.4235. The estimate is significant only to a 83.2% confidence level, based on its t ratio of 1.38. A quick review of the t statistics of parameter estimates can provide good knowledge regarding parameter uncertainty and will indicate how well the parameter estimates are supported by available data. A parameter estimate that has a low significant level can be a result of overparameterization, or data deficiency, or both.

Yeh et al. (1991) and Wang et al. (1995) adapted the normalized sensitivity coefficients in defining parameter uncertainty. The same definition is used later in model verification to define the reliability of model prediction. The normalized sensitivities,

$$\frac{\partial \text{Err}}{\partial \hat{P}_i} \frac{\hat{P}_i}{\text{Err}} \qquad i = 1,2,\ldots,91 \tag{3.14}$$

are computed by the method of perturbation. The absolute value of the normalized sensitivity shows the relative sensitivity of the parameter estimate to the least squares error. The more sensitive the parameter estimate, the more reliable it is in the model calibration. The sensitivity analysis and the statistical inference above result in parallel conclusions in parameter reliability.

Parameter collinearity serves an index for overparameterization, and it is more conveniently observed from the correlation matrix

$$\mathbf{R} = \begin{bmatrix} \frac{c_{11}}{\sqrt{c_{11}c_{11}}} & \frac{c_{12}}{\sqrt{c_{11}c_{22}}} & \cdots & \frac{c_{1L}}{\sqrt{c_{11}c_{LL}}} \\ \frac{c_{21}}{\sqrt{c_{22}c_{11}}} & \frac{c_{22}}{\sqrt{c_{22}c_{22}}} & \cdots & \frac{c_{2L}}{\sqrt{c_{22}c_{LL}}} \\ \vdots & \vdots & \ddots & \vdots \\ \frac{c_{L1}}{\sqrt{c_{LL}c_{11}}} & \frac{c_{L2}}{\sqrt{c_{LL}c_{22}}} & \cdots & \frac{c_{LL}}{\sqrt{c_{LL}c_{LL}}} \end{bmatrix} \tag{3.15}$$

where c_{ij} is the covariance of parameters i and j in the covariance matrix. It can be seen that all diagonal elements in the correlation matrix are equal to 1. Correlation coefficient ($\mathbf{R}_{ij}$) is a normalized statistic with a range from -1 to 1, and is a measure of the collinearity of two parameters. A perfect collinearity between two parameters is shown by a correlation coefficient of 1 or -1, depending on whether a sign conversion is needed. Parameter collinearity implies parameter redundancy and causes numerical instability or induces errors in the execution of parameter identification. A correlation coefficient of zero implies that no collinearity can be found in these two parameters; however, it does not exclude the possibility for having a nonlinear relationship between these two parameters. It is extremely important to recognize the limitations of statistical diagnosis.

One of the major difficulties associated with the inverse problem is determining the proper level of parameterization in the groundwater system model. The foregoing methods for reliability analysis cannot clearly identify the optimum parameter dimension, due to the deficiency of observation data. Yeh and Yoon (1981) demonstrated that an increase in parameter dimension (the number of unknown parameters associated with parameterization) can reduce resulting system modeling error but may increase the parameter uncertainty at some point (see Figure 3.5). These observations can be confirmed by examining equation (3.13).

The ill-posedness of the inverse problem, which is characterized by the nonuniqueness and instability of the identified parameters, has recently inspired much research on methodologies that evaluates and remedies the ill-

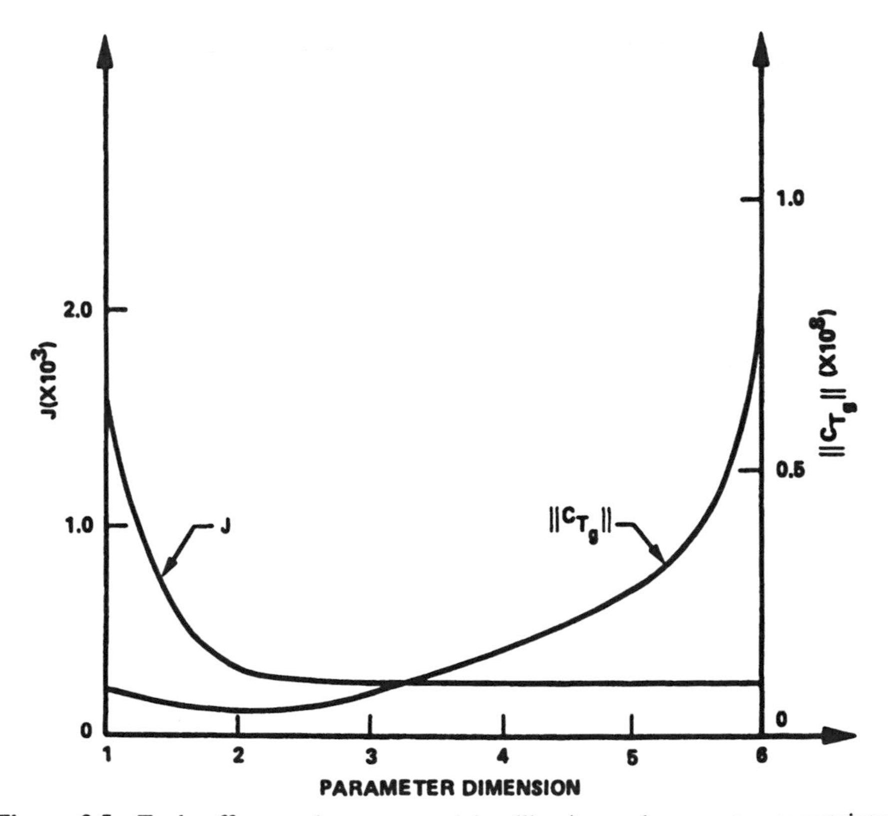

Figure 3.5 Trade-off curve between model calibration and parameter uncertainty. (After Yeh and Yoon, 1981.)

posedness. Statistical methods, such as regression analysis (Cooley, 1982) and geostatistical methods (Kitanidis and Vomvoris, 1983), have been used to remedy the rather subjective parameterization in the zonation approach. Computation efficiency can be improved by applying the sensitivity equations or the adjoint state theory to computation of the Jacobian matrix (Sun and Yeh, 1985; Yeh, 1986).

In view of the various uncertainties involved, a groundwater model can only approximate the physical mechanism of the aquifer system under study. Parameter identifiability, which is related to the question of whether the parameters can be determined uniquely by the available data, has also been studied by Yeh and Sun (1984). The δ-identifiability criterion presented by Yeh and Sun (1984) represents the first attempt in the development of an extended identifiability criterion for experimental design for parameter estimation. Rather than seeking the uniqueness of the identified parameters, the criterion emphasizes the reliability of model prediction. A δ-identifiable pumping test is an experiment producing data sufficient to guarantee that the parameter estimates of the groundwater model yield predictions accurate enough for the overall management objective. McCarthy and Yeh (1990) applied this concept to the optimum design of a pumping test for aquifer parameter identification. The integrated approach, which includes stochastic parameter estimation, reliability analysis, and experimental design, has been reported by Sun and Yeh (1990a,b), Tong and Yeh (1995) and Tong (1996).

3.B.5. Model Verification

Due to the uncertainties involved in modeling and parameter estimation, the calibrated groundwater model should be verified before being applied. A typical verification exercise should include the out-sample prediction for water level and solute concentration; in other words, the observations that were not used in parameter estimation are compared with the results of model simulation to evaluate the accuracy of model prediction. However, as mentioned previously, groundwater system analysis generally suffers greatly from data deficiency, and too often model verification by an out-sample prediction is simply impossible. Under this circumstance, the unverified model should be used only after sensitivity analyses for both model calibration and prediction are performed and evaluated (Bedient et al., 1994). Nonetheless, the groundwater model should at the very least support the mass balance calculation for groundwater flow and solute transport. The calibrated model for the San Jacinto Basin supported the basic mass balances of flow and solute transport (Yeh et al., 1991), and the sensitivity analysis for model prediction is performed similarly to that of parameter estimation mentioned previously.

Since the feasibility analysis of the recharge plan is the objective of the groundwater system analysis, Yeh et al. (1991) performed the sensitivity analysis against the predictions of river recharge, change in storage, and outflow from the given head boundary within 10 years. A base case using a 10-year hydrology corresponding to the historical dry period, initial and boundary conditions, land use data, and pumping policy is simulated by the calibrated model to compute reference values for river recharge, change in storage, and outflow from the given head boundary. The results of the sensitivity analysis indicate that these quantities are insensitive to all unreliable (insignificant) parameters. Thus the calibrated model is considered reliable enough for the current feasibility analysis of the proposed recharge plan.

3.B.6. Model Application

The calibrated and verified groundwater model is now ready for predicting how the system will respond to the proposed recharge plan. In the recharge plan, 100 acres of land at the downstream of the confluence of the San Jacinto River and Bautista Creek are used for water spreading (Yeh et al., 1991). The planned recharge rate is 80 acre-ft/day, and the operation period is from October 1 through January 31 for 10 years. Two historical hydrologies corresponding to a dry period and a wet period were used in the analysis. The average annual increase in storage was compared with the annual replenishment to evaluate the recharge efficiency. The simulation results indicate that the planned recharge can infiltrate into the aquifer almost completely during the dry period, but much will be lost during the wet period due to the high groundwater level and the plentiful precipitation.

A groundwater model can be linked with a management model for policy evaluation. A groundwater management model is often represented as an optimization problem:

$$\underset{\mathbf{x}\in\mathbf{X}}{\text{minimize}}\ f = F(\mathbf{x}) \tag{3.16a}$$

subject to

$$\boldsymbol{\Psi}(\mathbf{X}) \geq 0 \tag{3.16b}$$

where $\mathbf{x}$ is the vector of decision variables and $\boldsymbol{\Psi}$ is the function vector of constraints. Usually, the objective of the management model is to determine the optimum operational policy under a given set of constraints, focusing on various aspects of groundwater management, such as experimental design, water supply management, and contamination remediation. The groundwater model is incorporated as part of the physical constraint set. Due to the complexity of the computation, the groundwater model is often linked to the management model (instead of being embedded) as a simulation module to provide information of how the system responds to the change of operational policies. The solution methods for the management model that have been reported in the literature include linear programming, nonlinear programming, integer and mixed-integer programming, and dynamic programming. Gorelick

(1983) and Yeh (1992) presented extensive reviews of groundwater management models. Additional examples of these two types of prediction in modern practice of groundwater system analysis are introduced in the next section.

3.C. APPLICATIONS OF GROUNDWATER SYSTEM ANALYSIS

Four applications of groundwater system analysis are introduced in this section to represent some of the ongoing studies in groundwater system management. These applications involve research of chemical properties, public health, risk assessment, mathematics, and management philosophy. The quality of groundwater is today much emphasized, for public awareness of contamination has greatly increased in recent years. The significance of each application in groundwater system analysis is summarized as follows:

1. *Stochastic parameter estimation.* A geostatistical approach is used to improve the estimation of hydraulic conductivity, and eventually to unify the efforts of parameter estimation, reliability analysis, and experimental design in groundwater modeling.
2. *Groundwater remediation design.* The groundwater remediation design is accomplished by linking a groundwater model to a management model, resulting in a systematic, thorough investigation on the interaction between groundwater operation and system response.
3. *Simulation framework of groundwater-related health risk problems.* The groundwater model is present as part of an integrated, interdisciplinary simulation framework so that the relationship between public health risk and groundwater management can be better observed.
4. *Soil vapor extraction system design.* A comparison of decision rules is inspired as groundwater remediation is reinforced by cleaning up the contamination in the unsaturated zone.

3.C.1. Stochastic Parameter Estimation

As mentioned previously, the inverse problem of a groundwater system is often ill-posed in reality, due to data deficiency. A useful guideline in groundwater modeling is the principle of parameter conservation, which means that a good groundwater model should have the least number of parameters and yet be able to capture the response of the groundwater system.

Kitanidis and Vomvoris (1983) proposed a geostatistical approach for the estimation of hydraulic parameters. Hydraulic conductivity was assumed to be a random variable, following an empirical, statistical model. Observations were assumed to be normally distributed, and the associated stochastic parameters were estimated by the maximum likelihood method. Parameter estimation was further enhanced by Gaussian conditional mean, or *cokriging.*

Head observations were used as the secondary variable to assist in the reduction of parameter uncertainty by cokriging. Sun and Yeh (1992) improved the geostatistical approach by using the adjoint state method in computating the sensitivity coefficient. They also combined parameter estimation, reliability analysis, and experimental design in a unified framework. Tong (1996) extended the results of Sun and Yeh (1992) and further incorporated observations of water quality to assist in the estimation of hydraulic conductivity. The following discussion is based largely on Tong (1996).

Geostatistical Approach. Freeze (1975) reported that hydraulic conductivity (K) fits a log-normal distribution; that is, $Y = \ln(K)$ is normally distributed. It is further assumed that the random field of Y is characterized by a constant mean and an isotropic, exponential covariance (Hoeksema and Kitanidis, 1985; Sun and Yeh, 1992).

$$\mathrm{E}(Y) = \mu \tag{3.17a}$$

$$\mathrm{Cov}(Y_j,Y_j) = \sigma^2 \exp\left(\frac{-d_{ij}}{l}\right) \tag{3.17b}$$

where Y_i is the value of Y at location x_i with a variance of σ^2, l the log hydraulic conductivity correlation scale, d_{ij} the distance between points x_i and x_j, and $\mathrm{Cov}(Y_i,Y_j)$ the covariance of Y_i and Y_j. Since the covariance of Y is a function of distance only, the Y field can be said to be weakly stationary or second-order stationary. By taking the geostatistical approach, the parameters needed to be identified are now reduced to mean, variance, and correlation scale of the log-hydraulic conductivity field.

The occurrence probability of observations $(\mathbf{z}_1,\mathbf{z}_2,\ldots,\mathbf{z}_{\mathrm{TM}})$ under a given set of model parameters can be written as $\mathrm{prob}(\mathbf{z}_1,\mathbf{z}_2,\ldots,\mathbf{z}_{\mathrm{TM}}|\mathbf{P})$, where $\mathbf{P}^{\mathrm{T}} = (\mu,\sigma^2,l)$; $\mathbf{z}^{\mathrm{T}} = [\mathbf{y}^{\mathrm{T}},\mathbf{h}_1^{\mathrm{T}},\ldots,\mathbf{h}_{lh}^{\mathrm{T}},\mathbf{c}_1^{\mathrm{T}},\ldots,\mathbf{c}_{lc}^{\mathrm{T}}]$; $\mathbf{h}_i$ is the vector of head observations at location i from time 1 to time kh; $\mathbf{c}_j$ is the vector of concentration observations at location j from time 1 to time kc; $\mathbf{y}$ is the vector of the direct measurement of Y at ly locations; and $\mathrm{TM} = [kh \times lh + kc \times lc + ly]$. These parameters can be estimated by maximizing the occurrence probability of this set of observations; thus the parameter estimates are called maximum likelihood estimates (MLEs).

The observations are assumed to fit a joint normal distribution (Kitanidis and Vomvoris, 1983) as follows:

$$\mathrm{prob}(\mathbf{z}|\mathbf{P}) = \frac{1}{\sqrt{(2\pi)^{\mathrm{TM}}|\Omega|}} \exp\left[-\tfrac{1}{2}(\mathbf{z} - \bar{\mathbf{z}})^{\mathrm{T}}\Omega^{-1}(\mathbf{z} - \bar{\mathbf{z}})\right] \tag{3.18}$$

where $\bar{\mathbf{z}} = \mathrm{E}(\mathbf{z}|\mathbf{P})$ and $\Omega = \mathrm{E}[(\mathbf{z} - \bar{\mathbf{z}})(\mathbf{z} - \bar{\mathbf{z}})^{\mathrm{T}}|\mathbf{P}]$, which is the covariance matrix of observations. It is possible to use observations at different times as independent observations; thus the computational requirement can be reduced. However, the simultaneous usage of all observations incorporates the temporal correlations of observations. Without loss of generality, MLE can be obtained by minimizing the negative logarithm of likelihood function (occurrence probability):

$$L(\mathbf{P}) = \frac{TM}{2} \ln (2\pi) + \tfrac{1}{2} \ln |\Omega| + \tfrac{1}{2}(\mathbf{z} - \bar{\mathbf{z}})^{T}\Omega^{-1}(\mathbf{z} - \bar{\mathbf{z}}) \tag{3.19}$$

MLE requires determination of the covariance matrix Ω. The first-order approximation of Ω can be obtained by using the covariance matrix of Y field. For example, the covariance of head observations $h_{i,1}$ and $h_{j,2}$ can be written as

$$\begin{aligned} \mathrm{Cov}(h_{i,1}, h_{j,2}) &= \mathrm{E}[(h_{i,1} - \bar{h}_{i,1})(h_{j,2} - \bar{h}_{j,2})] \\ &= \mathrm{E}\left[\left(\frac{\partial h_{i,1}}{\partial \mathbf{Y}}\right)^{T} (\mathbf{Y} - \bar{\mathbf{Y}})(\mathbf{Y} - \bar{\mathbf{Y}})^{T} \frac{\partial h_{j,2}}{\partial \mathbf{Y}}\right] \\ &= \mathbf{J}_{\mathbf{Y}}^{T}(h_{i,1})\Omega_{\mathbf{Y}}\mathbf{J}_{\mathbf{Y}}(h_{j,2}) \end{aligned} \tag{3.20}$$

where $\bar{h}$ is the true value of the head observation, $\mathbf{Y}$ the vector of log hydraulic conductivity at all nodes, and $\bar{\mathbf{Y}}$ the vector of the true value of $\mathbf{Y}$. $\mathbf{J}_{\mathbf{Y}}(h)$ is the Jacobian matrix of head observations respect to $\mathbf{Y}$, and $\Omega_{\mathbf{Y}}$ is the covariance matrix of $\mathbf{Y}$, which is determined by equation (3.17b). Thus

$$\Omega = \mathbf{J}_{\mathbf{Y}}^{T}(\mathbf{z})\Omega_{\mathbf{Y}}\mathbf{J}_{\mathbf{Y}}(\mathbf{z}) \tag{3.21}$$

The Gauss–Newton method mentioned previously in parameter estimation was used in the inverse problem. Hessian matrix is approximated, and the elements can be calculated by using Gaussian moment factoring (Kitanidis and Vomvoris, 1983):

$$\mathrm{E}\left(\frac{\partial L}{\partial P_i}\frac{\partial L}{\partial P_j}\right) = \frac{1}{2}\,\mathrm{Tr}\left(\Omega^{-1}\frac{\partial \Omega}{\partial P_i}\Omega^{-1}\frac{\partial \Omega}{\partial P_j}\right) + \left(\frac{\partial \bar{z}}{\partial P_i}\right)^{T}\Omega^{-1}\frac{\partial \bar{z}}{\partial P_j} \tag{3.22}$$

By applying the following properties,

$$\frac{\partial}{\partial P_j}\ln|\Omega| = \mathrm{Tr}\left(\Omega^{-1}\frac{\partial \Omega}{\partial P_j}\right) \tag{3.23a}$$

$$\frac{\partial}{\partial P_j}\Omega^{-1} = -\Omega^{-1}\frac{\partial \Omega}{\partial P_j}\Omega^{-1} \tag{3.23b}$$

the gradient evaluation can be obtained as follows:

$$\frac{\partial L}{\partial P_j} = \frac{1}{2} \operatorname{Tr}\left(\Omega^{-1} \frac{\partial \Omega}{\partial \theta_j}\right) - \frac{1}{2} (\mathbf{z} - \bar{\mathbf{z}})^{\mathrm{T}} \Omega^{-1} \frac{\partial \Omega}{\partial P_j} \Omega^{-1} (\mathbf{z} - \bar{\mathbf{z}})$$
$$- (\mathbf{z} - \bar{\mathbf{z}})^{\mathrm{T}} \Omega^{-1} \frac{\partial \bar{\mathbf{z}}}{\partial P_j} \qquad (3.24)$$

The symbol "Tr" denotes the trace of the corresponding matrix, which is the summation of the diagonal elements of the matrix. After the parameters for defining the random field of log-hydraulic conductivity have been estimated by MLE, the Y field can be re-created. However, the error associated with the estimation of hydraulic conductivity can be improved by using cokriging.

Cokriging improves the estimates of two or more random fields by using their measurements together if these random variables are correlated (N.-Z. Sun, 1994a). It is an extension of kriging. Cokriging has been applied to the inverse problem in groundwater systems by introducing head observations as the secondary variables when estimating hydraulic conductivity field (Kitanidis and Vomvoris, 1983; Hoeksema and Kitanidis, 1984; Sun and Yeh, 1992). Tong (1996) introduced an additional secondary variable of concentration observation to improve further the estimation of hydraulic conductivity.

For an arbitrary point 0 in the field, the cokriging estimate of Y_0 can be written as

$$\hat{Y}_0 = \sum_{m=1}^{ly} \alpha_m^0 y_m + \sum_{i=1}^{lh} \sum_{k=1}^{kh} \beta_{i,k}^0 \varepsilon_{i,k}^h + \sum_{j=1}^{lc} \sum_{k=1}^{kc} \gamma_{j,k}^0 \varepsilon_{j,k}^c \qquad (3.25)$$

where α^0, β^0, and γ^0 are cokriging coefficients for the estimate at point 0; $\varepsilon^h = h - \bar{h}$ and $\varepsilon^c = c - \bar{c}$ are two new random fields based on the results of MLE. ε^h and ε^c have zero mean and measurements $\varepsilon_{i,k}^h = h_{i,k} - \bar{h}_{i,k}$ ($i = 1,2,\ldots,lh$; $k = 1,2,\ldots,kh$) and $\varepsilon_{j,k}^c = c_{i,k} - \bar{c}_{i,k}$ ($j = 1,2,\ldots,lc$; $k = 1,2,\ldots,kc$), respectively. Thus the means of hydraulic heads and solute concentrations obtained from MLE are considered to be the true mean values. It should be noted that equation (3.25), which relates the fields of observation errors to the point estimate of hydraulic conductivity, can be written in other forms, such as replacing the Y's by $(Y - \mu)$'s to relate the deviations from the expected value to the observation errors. If we have the perfect knowledge about the true μ, the different forms should have the same effect. Since this is not possible in practical cases, equation (3.25) is used to avoid additional errors introduced by using the estimated μ.

The two most important properties of a parameter estimator are unbiasedness and consistency for a sufficiently large sample size (Haan, 1977). Unbiasedness means that the expected value of the estimated parameter is equal to the true value. Consistency implies that when sample size becomes sufficiently large, the probability of the absolute error of the parameter estimate being greater than a given small number can be as small as desired. When a small sample is used, the estimator with the least variance of estimation error should be used. These two properties determine the cokriging coefficients. The expected value and the variance of estimation error can be written as follows:

$$\mathrm{E}[\hat{Y}_0 - Y_0] = \sum_{m=1}^{ly} \alpha_m^0 \mathrm{E}[y_m] + \sum_{i=1}^{lh} \sum_{k=1}^{kh} \beta_{i,k}^0 \mathrm{E}[\varepsilon_{i,k}^h]$$

$$+ \sum_{j=1}^{lc} \sum_{k=1}^{kc} \gamma_{j,k}^0 \mathrm{E}[\varepsilon_{j,k}^c] - \mathrm{E}[Y_0]$$

$$= \left(\sum_{m=1}^{ly} \alpha_m^0 - 1 \right) \mu \tag{3.26a}$$

$$\mathrm{Var}(\hat{Y}_0 - Y_0) = \mathrm{E}\left[\left(\sum_{m=1}^{ly} \alpha_m^0 y_m + \sum_{i=1}^{lh} \sum_{k=1}^{kh} \beta_{i,k}^0 \varepsilon_{i,k}^h \right.\right.$$

$$\left.\left. + \sum_{j=1}^{lc} \sum_{k=1}^{kc} \gamma_{j,k}^0 \varepsilon_{j,k}^c - Y_0 \right)^2 \right] \tag{3.26b}$$

An unbiased cokriging estimator requires that

$$\sum_{m=1}^{ly} \alpha_m^0 = 1 \tag{3.27}$$

Thus the variance of the cokriging estimator of Y_0 can be written as follows:

$$\mathrm{Var}(\hat{Y}_0 - Y_0) = \mathrm{E}\left[\left(\sum_{m=1}^{ly} \alpha_m^0 \varepsilon_m^Y + \sum_{i=1}^{lh} \sum_{k=1}^{kh} \beta_{i,k}^0 \varepsilon_{i,k}^h \right.\right.$$

$$\left.\left. + \sum_{j=1}^{lc} \sum_{k=1}^{kc} \gamma_{j,k}^0 \varepsilon_{j,k}^c - \varepsilon_0^Y \right)^2 \right]$$

$$= \mathrm{E}[(\mathbf{A}_0^{\mathrm{T}} \mathbf{E}_Y + \mathbf{B}_0^{\mathrm{T}} \mathbf{E}_h + \mathbf{G}_0^{\mathrm{T}} \mathbf{E}_c - \varepsilon_0^Y)^2]$$

$$= \mathrm{E}[\mathbf{W}_0^{\mathrm{T}} \mathbf{E} \mathbf{E}^{\mathrm{T}} \mathbf{W}_0 - 2\mathbf{W}_0^{\mathrm{T}} \mathbf{E} \cdot \varepsilon_0^Y + (\varepsilon_o^Y)^2]$$

$$= \mathbf{W}_0^{\mathrm{T}} \Omega \mathbf{W}_0 - 2\mathbf{W}_0^{\mathrm{T}} \,\mathrm{Cov}(\mathbf{z}, Y_0) + \sigma^2 \tag{3.28}$$

where $\varepsilon^Y = Y - \mu$ is the random filed of the deviation with a zero mean and measurements $\varepsilon_m^Y = y_m - \mu$ $(m = 1,2,\ldots,ly)$; and

$$\mathbf{A}_0^{\mathrm{T}} = [\alpha_1^0,\ldots,\alpha_{ly}^0]_{(1\times ly)} \qquad \mathbf{B}_0^{\mathrm{T}} = [\beta_1^0,\ldots,\beta_{lh,kh}^0]_{(1\times(lh*kh))}$$

$$\mathbf{G}_0^{\mathrm{T}} = [\gamma_1^0,\ldots,\gamma_{lc,kc}^0]_{(1\times(lc*hc))}$$

$$\mathbf{E}_Y^{\mathrm{T}} = [\varepsilon_1^0,\ldots,\varepsilon_{ly}^0]_{(1\times ly)} \qquad \mathbf{E}_h^{\mathrm{T}} = [\varepsilon_1^h,\ldots,\varepsilon_{lh,kh}^h]_{(1\times(lh*kh))}$$

$$\mathbf{E}_c^{\mathrm{T}} = [\varepsilon_1^c,\ldots,\varepsilon_{lc,kc}^c]_{(1\times(lc*hc))}$$

$$\mathbf{W}_0^{\mathrm{T}} = [\mathbf{A}_0^{\mathrm{T}} \quad \mathbf{B}_o^{\mathrm{T}} \quad \mathbf{G}_0^{\mathrm{T}}] \quad \mathbf{E}^{\mathrm{T}} = [\mathbf{E}_Y^{\mathrm{T}} \quad \mathbf{E}_h^{\mathrm{T}} \quad \mathbf{E}_c^{\mathrm{T}}] = (\mathbf{z} - \bar{\mathbf{z}})^{\mathrm{T}}$$

To ensure that the unbiasedness condition is satisfied in the minimization of estimation variance, Eq. (3.27) is incorporated in the objective function by a Lagrangian multiplier 2ν. The resulting optimization is unconstrained with the objective function as follows:

$$F = \mathrm{Var}(\hat{Y}_0 - Y_0) + 2\nu(\mathbf{W}_0^{\mathrm{T}}\Lambda - 1) \tag{3.29}$$

where $\Lambda^{\mathrm{T}} = [\Phi^{\mathrm{T}} \quad \mathbf{O}^{\mathrm{T}}]$, Φ is a vector with ly entries corresponding to α^0 all equal to 1, and $\mathbf{O}$ is a vector with $(lh \times kh + lc \times kc)$ entries all equal to zero. The necessary conditions for the objective function to have a minimum are

$$\frac{\partial F}{\partial \mathbf{W}_0} = \frac{\partial\ \mathrm{Var}(\hat{Y}_0 - Y_0)}{\partial \mathbf{W}_0} + \nu\Lambda \tag{3.30a}$$

$$= 2\Omega\mathbf{W}_0 - 2\ \mathrm{Cov}(\mathbf{z},Y_0) + 2\nu\Lambda = 0$$

$$\frac{\partial F}{\partial \nu} = 2\mathbf{W}_0^{\mathrm{T}}\Lambda - 2 = 0 \tag{3.30b}$$

That is,

$$\begin{bmatrix} \Omega & \Lambda \\ \Lambda^{\mathrm{T}} & 0 \end{bmatrix} \begin{bmatrix} \mathbf{W}_0 \\ \nu \end{bmatrix} = \begin{bmatrix} \mathrm{Cov}(\mathbf{z},Y_0) \\ 1 \end{bmatrix} \tag{3.31}$$

Thus solution methods such as the Gauss–Newton algorithm can be used to solve for the cokriging coefficients and the Lagrangian multiplier ν. In addition, using the relationships shown in equation (3.30), the error of estimation error can be simplified as follows:

$$\mathrm{Var}(\hat{Y}_0 - Y_0) = \sigma^2 - \mathbf{W}_0^{\mathrm{T}}\ \mathrm{Cov}(\mathbf{z},Y_0) - \nu \tag{3.32}$$

Similarly to equation (3.20), the first-order approximation of $\mathrm{Cov}(\mathbf{z},Y_0)$ can be obtained by using the Jacobian matrix $\mathbf{J}_{\mathrm{Y}}(\mathbf{z})$.

$$\mathrm{Cov}(\mathbf{z},Y_0) = \mathbf{J}_Y^T(\mathbf{z})\,\mathrm{Cov}(\mathbf{Y},Y_0) \tag{3.33}$$

where $\mathrm{cov}(\mathbf{Y},Y_0)$ can be determined by equation (3.17).

Tong (1996) took a hypothetical problem to illustrate the improved geostatistical approach. Figure 3.6a shows the hypothetical, confined aquifer, for which observations of head (**h**), concentration (**c**), and hydraulic conductivity (**y**) are available at points specified. The true log K field shown in Figure 3.6b is generated by equation (3.17) by the turning bands method (Mantoglou and Wilson, 1982; Sun and Yeh, 1992) with a given set of statistical parameter values, including mean, variance, and correlation scale. The Y field was estimated by the proposed geostatistical method with three levels of data use: observations of **h** and **y**; observations of **c** and **y**; and observations of **h**, **c**, and **y**. Figure 3.6c shows the Y field estimated by using observations of **h**, **c**,

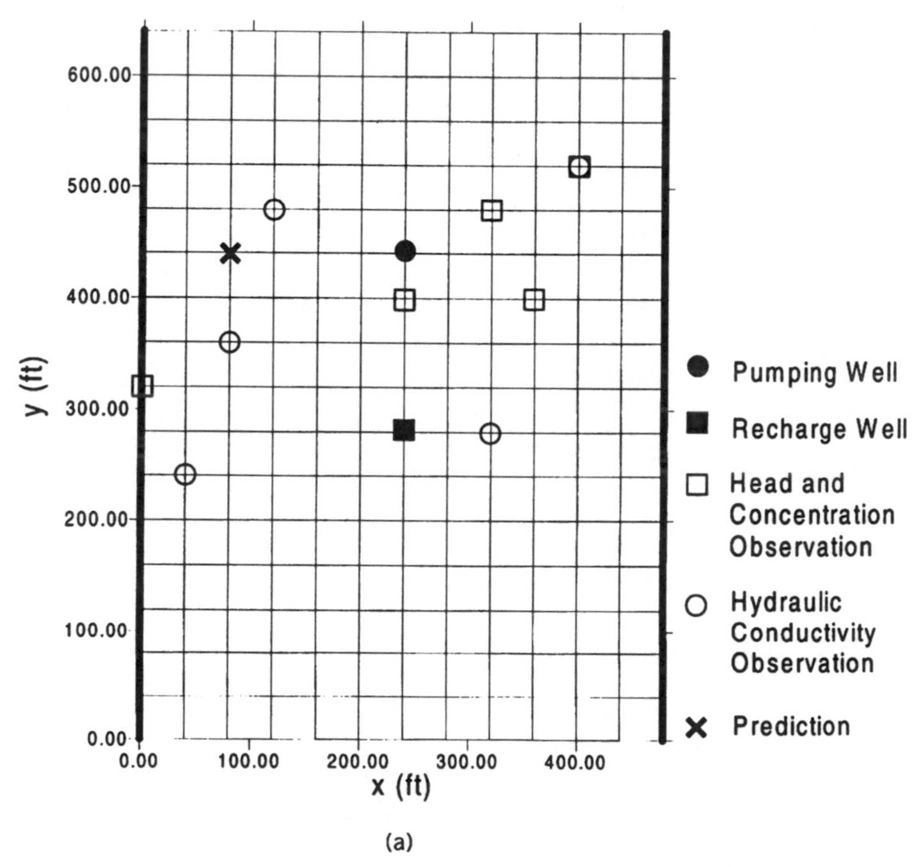

Figure 3.6 (a) Plan view of the hypothetical, confined aquifer; (b) true log hydraulic conductivity field; (c) estimated log hydraulic conductivity field; (d) sensitivity of concentration with respect to hydraulic conductivity. (After Tong, 1996.)

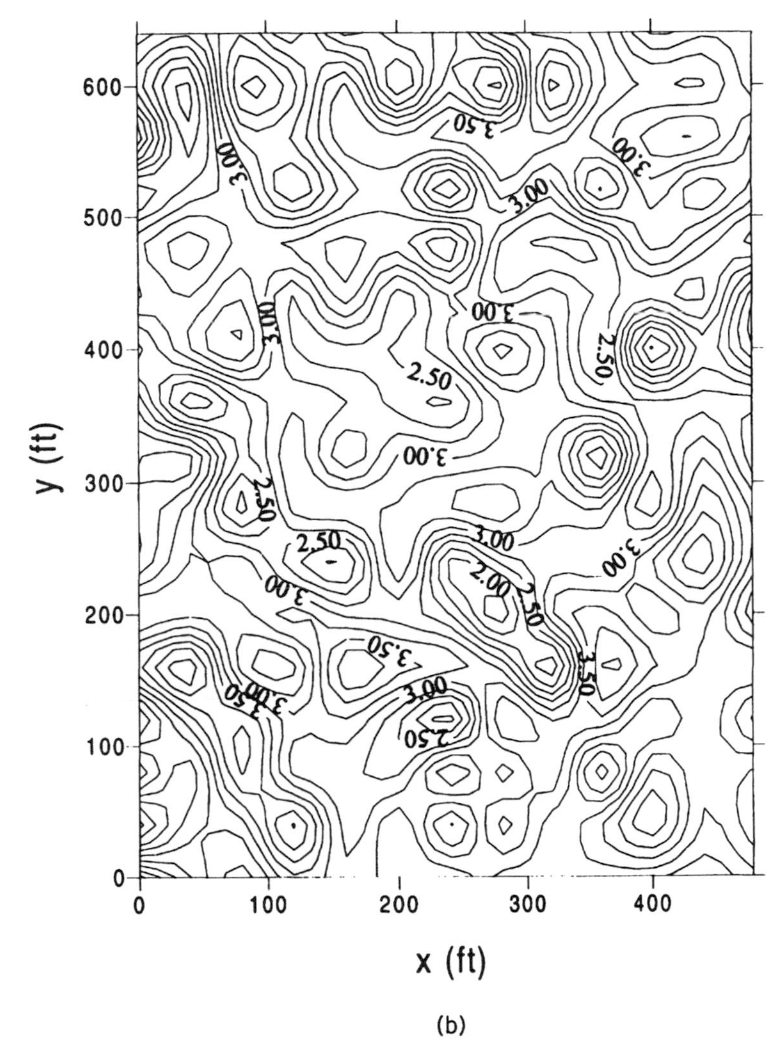

(b)

Figure 3.6 (*Continued*)

and y. The Y fields estimated with three levels of data use have a mean-square error of 0.422, 0.499, and 0.389, respectively.

The improved estimation of hydraulic conductivity also enhances the predictions of nodal concentration and areal concentration distribution. For the nodal concentration at prediction point X (see Figure 3.6a), the relative error of nodal concentration predictions by three estimated Y fields are 28.4%, 86.6%, and 11.9%. The corresponding mean-square errors of predictions for areal concentration distribution are 2.332, 7.104, and 0.976. Significant improvement on both parameter estimation and model prediction can be obtained by incorporating the concentration data in the inverse solution.

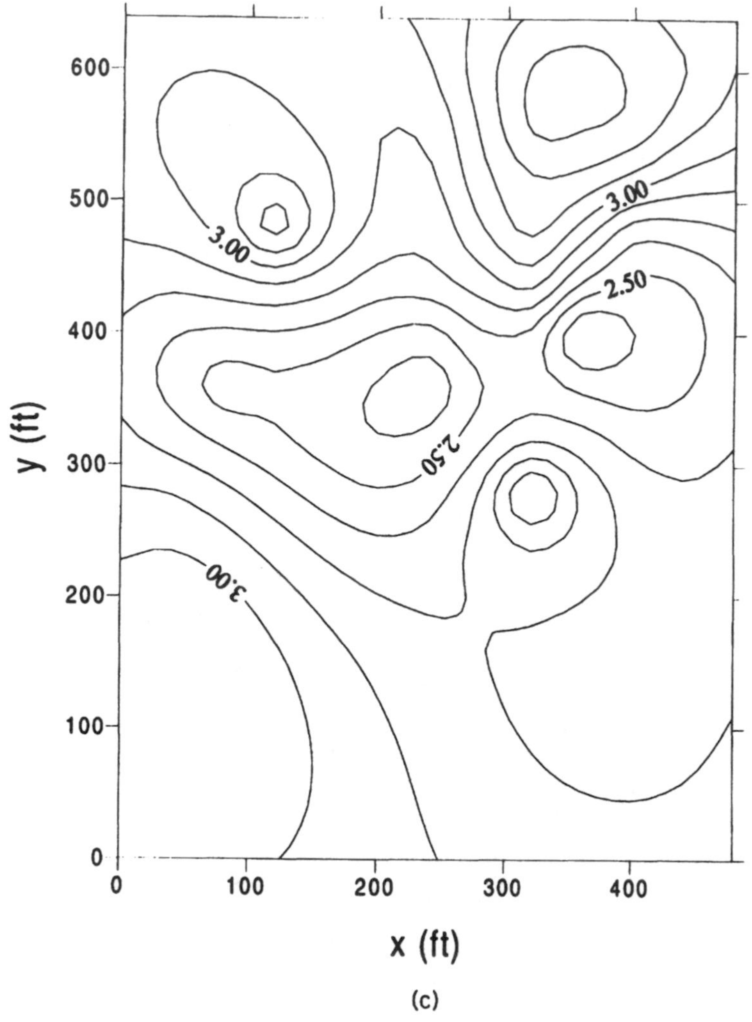

(c)

Figure 3.6 *(Continued)*

Sensitivity Coefficients. The implementation of MLE and cokriging estimation requires the estimate of the Jacobian matrix (sensitivity coefficient matrix) of observations with respect to Y at all nodes. The Jacobian matrix can be evaluated by the influence coefficient method (Becker and Yeh, 1972); however, it would require intensive computation. To reduce the computation requirements, Tong (1996) used the adjoint state method in the estimation of the Jacobian matrix.

The adjoint state method, which is based on the variational principle, has been shown to facilitate the solution of inverse problems and groundwater remediation design (Neuman, 1980; Sun and Yeh, 1985; Ahlfeld et al., 1988a; Yeh and Sun, 1990; Sun and Yeh, 1990a,b; N.-Z. Sun, 1994a). Sun and Yeh

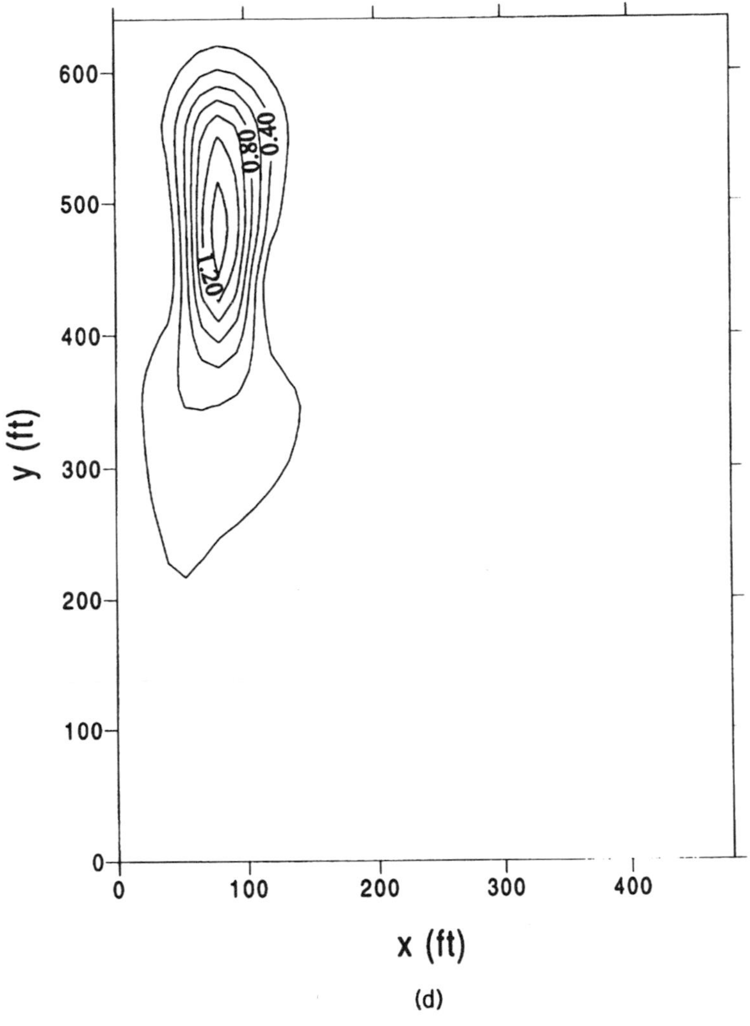

(d)

Figure 3.6 (*Continued*)

(1992) applied the adjoint state method to identifying stochastic parameters for a transient groundwater flow problem. Tong (1996) further incorporated the adjoint state equation for transient solute transport when identifying hydraulic parameters.

One important merit of using the adjoint state equations is that these equations have the identical forms of the original groundwater flow and solute transport equations; thus the same numerical method developed for solving problems of groundwater flow and solute transport can be used to solve the adjoint state equations. The number of simulations required for generating the Jacobian matrix is equal to the number of observation wells plus one, and the accuracy of sensitivity coefficients is also improved.

Beside the model-fitting error, sensitivity coefficients can be specified for any model performance criterion, such as sensitivity of the total volume of groundwater storage mentioned previously in the study of Yeh et al. (1991). The particular sensitivity of interest can be obtained by using a proper performance function and by solving the resulting adjoint state equations (Sun and Yeh, 1992). Figure 3.6d shows the sensitivity associated with the prediction of nodal concentration at point X. Thus the experimental design for future data collection designed for the improvement of nodal concentration prediction at X can be obtained.

3.C.2. Groundwater Remediation Design

A contaminant as defined by the Safe Drinking Water Act is "any physical, chemical, biological, or radiological substance or matter in water." However, groundwater contamination usually refers to situations where the concentration of contaminants introduced by human activities attains an objectionable level (Freeze and Cherry, 1979; Barcelona et al., 1988). Once polluted, a groundwater system remains contaminated for decades or even centuries because the mixing and dilution of contaminants are largely delayed by the low velocity of groundwater movement.

The migration of contaminants is subject to the chemical properties of contaminants, the geophysical conditions of aquifer, and remediation efforts. Groundwater remediation usually includes such basic steps as site investigation, free product recovery, remediation system design, and operation and maintenance. It is advised that all subsurface contaminant problems be viewed as two distinct zones of contamination: (1) contaminant source materials and contaminated soils in the unsaturated soil (or rock) zone, and (2) nonaqueous phase liquids (NAPLs) and/or groundwater containing dissolved contaminants within the saturated zone (Connor, 1994). While the dissolved contaminants move with groundwater flow, NAPLs tend to migrate vertically under the control of gravitational and capillary forces (Bedient et al., 1994).

The existence of NAPLs results in a multiphase flow in both the unsaturated and the saturated zones. Large amount of NAPLs can be recovered through extraction; however, NAPLs trapped in soil matrix are difficult to remove and likely to become a continuous source of contamination, even after the source material has been removed. NAPLs capacitate the transfer of contaminant mass to the ambient environment through volatilization, in situ biodegradation, and dissolution. Thus, to define properly the extent and concentration distribution of contamination, site investigation before groundwater remediation often proceeds in a two-stage process: (1) delineating the source in the unsaturated zone, comprised of chemical waste, product mass, and the associated contaminated soils, and (2) investigating the presence and extent of contaminant migration within the underlying groundwater system. Connor (1994) provided step-by-step procedures for these two tasks.

Accepted remediation alternatives of subsurface contamination problems include excavation and disposal, physical containment, hydraulic control and "pump and treat," in situ biological or chemical treatment, and soil vapor extraction (Bedient et al., 1994). Although it guarantees complete removal, excavation is economically feasible only for contamination that is small in extent and shallow in depth. Physical containment using barriers or slurry walls delays the spreading of contamination but leaves the problem unresolved. Aquifer restoration is usually implemented through the other three methods, used either independently or conjunctively (U.S. EPA, 1989; Pedersen and Curtis, 1991).

The pump-and-treat technique is the most common remedial technology for treating contaminated aquifers in the saturated zone. With a pump-and-treat system, contaminated groundwater is extracted from the aquifer through recovery wells, treated with activated carbon or other techniques, and then discharged. The recovered contaminants can then be disposed of properly. U.S. EPA (1989) reviewed 19 pump-and-treat systems installed to remove groundwater contamination and suggested that the design of a pump-and-treat system should be based on thorough analysis of the groundwater flow and contaminant transport. The need for efficient system design has inspired the development of many design strategies (Gorelick et al., 1984; Ahlfeld et al., 1986, 1988a,b; Ahlfeld and Sawyer, 1990; Dougherty and Marryott, 1991; McKinney and Lin, 1994; Haggerty and Gorelick, 1994; Ahlfeld and Heidari, 1994; Ahfeld et al., 1995; Barlow et al., 1996; Xiang et al., 1996). The design of a pump-and-treat system will be illustrated by an extensive study at a superfund site by Ahfeld et al. (1988a,b).

Groundwater Flow and Contaminant Transport Model. A contaminated site or a continuous plume is usually considered a problem of local scale. Before any systematic design process can be taken, an accurate groundwater flow and contaminant transport model capable of simulating the aquifer response to alternative system designs is required. Ahlfeld et al. (1988a) assumed a steady-state confined-flow transient contaminant transport model that consists of the following equations to represent the two-dimensional contaminant transport under the operation of a pump-and-treat system.

Steady-State Flow Equation

$$\nabla \cdot (b\mathbf{K} \cdot \nabla \phi) + f + \sum_{k=1}^{np} q_k \delta(x_k, y_k) = 0 \tag{3.34}$$

where b is the aquifer thickness, $\mathbf{K}$ the hydraulic conductivity tensor, ϕ the vertical average hydraulic head, f a leakage term, n_p the total number of pumps, q_k the extraction/injection rate at location (x_k,y_k), and $\delta(x_k,y_k)$ the Direc delta function evaluated at location (x_k,y_k).

Transient Transport Equation. In general, pump and treat is effective in hydraulic containment of plumes. Aquifer restoration thus becomes a long-term effort, possible only when site conditions are favorable and the extraction system properly designed and operated. The expectation of long-term efforts for aquifer restoration results from the retardation effect, which is a result of the sorption process and the existence of NAPLs (U.S. EPA, 1989; Bedient et al., 1994).

Sorption processes refer to the processes whereby contaminants originally in solution become distributed between the solution and the solid phase. In a natural groundwater system, sorption processes include adsorption, chemisorption, absorption, and ion exchange, forming a complex mechanism of mass transfer. However, these processes are often referred to only as partitioning (Fetter, 1993). Because of partitioning, contaminants are transported at a rate slower than the rate of groundwater flow. The partitioning of contaminant transport is often accommodated by using the retardation factor. The retardation factor can be seen as the ratio of the average pore velocity of the groundwater flow to the average velocity of the solute front where the concentration is one-half of the original. The implicit assumption for using the retardation factor is that the sorption process can be represented by a linear relationship between the amount of a solute sorbed onto solid C^* and the concentration of the solute C (i.e., a linear sorption isotherm as follows) (Freeze and Cherry, 1979; Fetter, 1993):

$$C^* = K_d C \tag{3.35}$$

where K_d is the distribution coefficient [L^3/M]. Thus the retardation factor (R) is defined as follows:

$$R = 1 + \frac{\rho_d}{n} K_d \tag{3.36}$$

where ρ_d is the dry bulk density of the soil [M/L^3] and n is porosity. Ahlfeld et al. (1988a) assumed that the free product of contaminants has been removed, and the retardation effect is accounted for by the retardation factor. The resulting transient transport equation,

$$bnR \frac{\partial C}{\partial t} = \nabla \cdot (bn\mathbf{D} \cdot \nabla C) - b\mathbf{u} \cdot \nabla C + fC_f - \sum_{k=1}^{n_p} q_k C_{qk} \delta(x_k,y_k) \tag{3.37}$$

where C, C_f, and C_{qk} are the contaminant concentration in groundwater, leakage fluid, and pumped fluid at location (x_k,y_k), respectively; $\mathbf{u} = -\mathbf{K} \cdot \nabla\phi$ is the Darcy velocity vector; and **D** is the hydrodynamic dispersion tensor and its elements are defined as in equation (3.6). It should be noted that for consistency, the retardation factor is defined in an alternative form from that of Ahlfeld et al. (1988a,b).

Ahlfeld et al. (1988a) reported that the use of steady-state flow condition and time-dependent contaminant transport can be justified because the groundwater remediation usually lasts for years, whereas the groundwater flow can reach its steady state within hours or days. This steady-state assumption, however, usually requires validation through site investigation. In addition, in the formulation above, extraction or injection rates are assumed to be constant during operation. If a time-varying operational schedule is required, a transient flow equation may be necessary. It should be noted that the local equilibrium assumption associated with the linear sorption isotherm is valid when the partitioning is rapid compared with the flow velocity. If the local equilibrium assumption is not valid, a nonlinear sorption isotherm or kinetic (nonequilibrium) sorption model should be used (Fetter, 1993; Haggerty and Gorelick, 1994; Kong and Harmon, 1996).

Optimization Model for Groundwater Remediation Design. Optimization techniques such as linear programming, nonlinear programming, integer programming, and dynamic programming have been widely applied to groundwater management problems (Gorelick, 1983; Willis and Yeh, 1987; Yeh, 1992). A groundwater quality management problem is usually formulated as a nonlinear programming problem due to the existence of the product terms of the unknown flow velocity and unknown concentration. The groundwater flow and contaminant transport equations are part of the constraints of the optimization problem. Additional constraints can be expanded to include budgetary, societal, operational, legislative constraints, and others.

Nonlinear optimization problems are usually solved by a gradient-based algorithm, which solves the nonlinear problem by iterative local linearization using gradient information (Luenburger, 1984). The Gauss–Newton method mentioned previously is one example. Under limited situations, an explicit expression of the gradient vector may be possible. In practice, a simulation model, which is developed to provide necessary information of system response to policy charges in the optimization model, is utilized to generate the gradient vector. The elements of the gradient vector are either partial derivatives of the objective function, or the head, or the concentration with respect to the decision variables.

Pump-and-treat systems usually have migration control of plumes as their primary design premise, or secondary, next to aquifer restoration, in which the aquifer is to be remediated to a concentration specified by health- and environmental-based criteria (U.S. EPA, 1989). These two premises are compatible, since aquifer restoration cannot be accomplished without migration control. Ahlfeld et al. (1988a) demonstrated two different optimization objectives for the design of a pump-and-treat system: Minimization of contaminant residuals in a given operational time frame, and minimization of the operational cost for a given cleanup standard.

The formulation of minimizing contaminant residuals in a given operational time frame can be written as follows (Ahlfeld et al., 1988b):

$$\text{minimize} \sum_{i \in N} V_i R_i C_{i,T}(\mathbf{q}) \tag{3.38a}$$

subject to

$$\sum_{j \in J} q_j^+ \le q_{\max}^+ \tag{3.38b}$$

$$\sum_{j \in J} q_j^- \le q_{\max}^- \tag{3.38c}$$

$$q_j = q_j^+ - q_j^- \qquad j \in J \tag{3.38d}$$

$$0 \le q_j^+ \le r_j \qquad j \in J \tag{3.38e}$$

$$0 \le q_j^- \le s_j \qquad j \in J \tag{3.38f}$$

in which N and J are sets of nodes and potential extraction/injection well locations, respectively; the subscript T indicates the end of the given operational time frame; V_i is the control volume of node i; q_j is divided into a positive component q_j^+ (extraction) and a negative component q_j^- (injection) with upper bounds as r_j and s_j, respectively. Active wells are assumed to be operated at a constant rate either for extraction or injection; thus one additional constraint requires that q_j^+ and q_j^- not both be nonzero at the same time. $\mathbf{q}$ is the vector of extraction/injection rates. The upper limits of total extraction ($q_{\max}^+$) and total injection ($q_{\max}^-$) can be seen as representations of budgetary constraints, safe yield limitation, storage capacity, and even water rights.

The formulation of minimizing operational cost for a given cleanup standard in a given operational period can be written as follows (Ahlfeld et al., 1988b):

$$\text{minimize} \sum_{j \in J} \alpha_i^+ q_j^+ + \alpha_j^- q_j^- \tag{3.39a}$$

subject to

$$C_{i,T}(\mathbf{q}) \le C_i^* \qquad i \in N \tag{3.39b}$$

$$q_j = q_j^+ - q_j^- \qquad j \in J \tag{3.39c}$$

$$0 \le q_j^+ \le r_j \qquad j \in J \tag{3.39d}$$

$$0 \le q_j^- \le s_j \qquad j \in J \tag{3.39e}$$

in which α_j^+ and α_j^- are the unit cost for extraction and injection, respectively, and C_j^* is the given cleanup standard at location i. Cleanup standards such as maximum contaminant level (MCL) are usually defined by laws or regulations based on health or environmental considerations. Less rigorous criteria can be set at observation points to ensure the success of contaminant confinement as set out by Ahlfeld et al. (1988b).

Well installation cost is not incorporated in the formulation above, and the operational cost function is assumed linear. It is very likely that, in reality, the decision maker would prefer using the existing wells for groundwater remediation to avoid the installation cost. Since the pump-and-treat technique is often applied to a local scale problem, the linear operational cost function is generally acceptable. General functions of installation cost and operational cost for groundwater remediation are considered by Lall and Santini (1989) and by McKinney and Lin (1994). However, sometimes the type of cost functions may dictate which optimization method would be preferred in optimization. This issue will be explained further in the application of soil vapor extraction system design.

Ahlfeld et al. (1988b) applied their optimization models to the design of a pump-and-treat system at a federal superfund site, the Woburn aquifer in eastern Massachusetts. Since the 1960s, the Woburn aquifer has been contaminated by two sources of organic compounds, such as trichloroethylene and 1,2-*trans*-dichloroethylene. Figure 3.7a shows the simulated concentration distribution of TCE in 1986. The planning horizon for the pump-and-treat design is 3 years. It was assumed that the contaminant sources cannot be removed from the aquifer over the period. The head contour shown in Figure 3.7b indicates that groundwater generally flows toward the river and exits the aquifer in the direction of river flow. Figure 3.7c shows the distribution of potential well locations and Figure 3.7d the distribution of concentration in 1989 if a "do-nothing" strategy were to be taken.

In the contaminant-minimization problem, Ahlfeld et al. (1988b) assumed that the upper bound of total recharge is 10% of that of total extraction. By varying the value of the extraction upper bound in different trials, the authors reported that the control factor in this optimization is the upper bound of recharge. Figure 3.8a shows one of the optimized system designs for contaminant minimization. It can be seen that the locations of extraction wells generally coincide with the main axis of contaminant distribution, and the recharge wells at the river downstream are meant to create a hydraulic barrier to stop contaminant migrating downstream. Under this optimized operation, the predicted concentration distribution in 1989 is shown in Figure 3.8b.

In the cost-minimization problem, the observation points with concentration requirements are shown in Figure 3.9a. Since the observation points are all located near the exit of the aquifer, it is clear that the purpose of this system design is to confine the contamination. Ahlfeld et al. (1988b) reported a trade-off relationship between the total remediation cost and the maximum contaminant concentration allowed at the observation points; no feasible solution can be found if the contaminant concentration required at the obser-

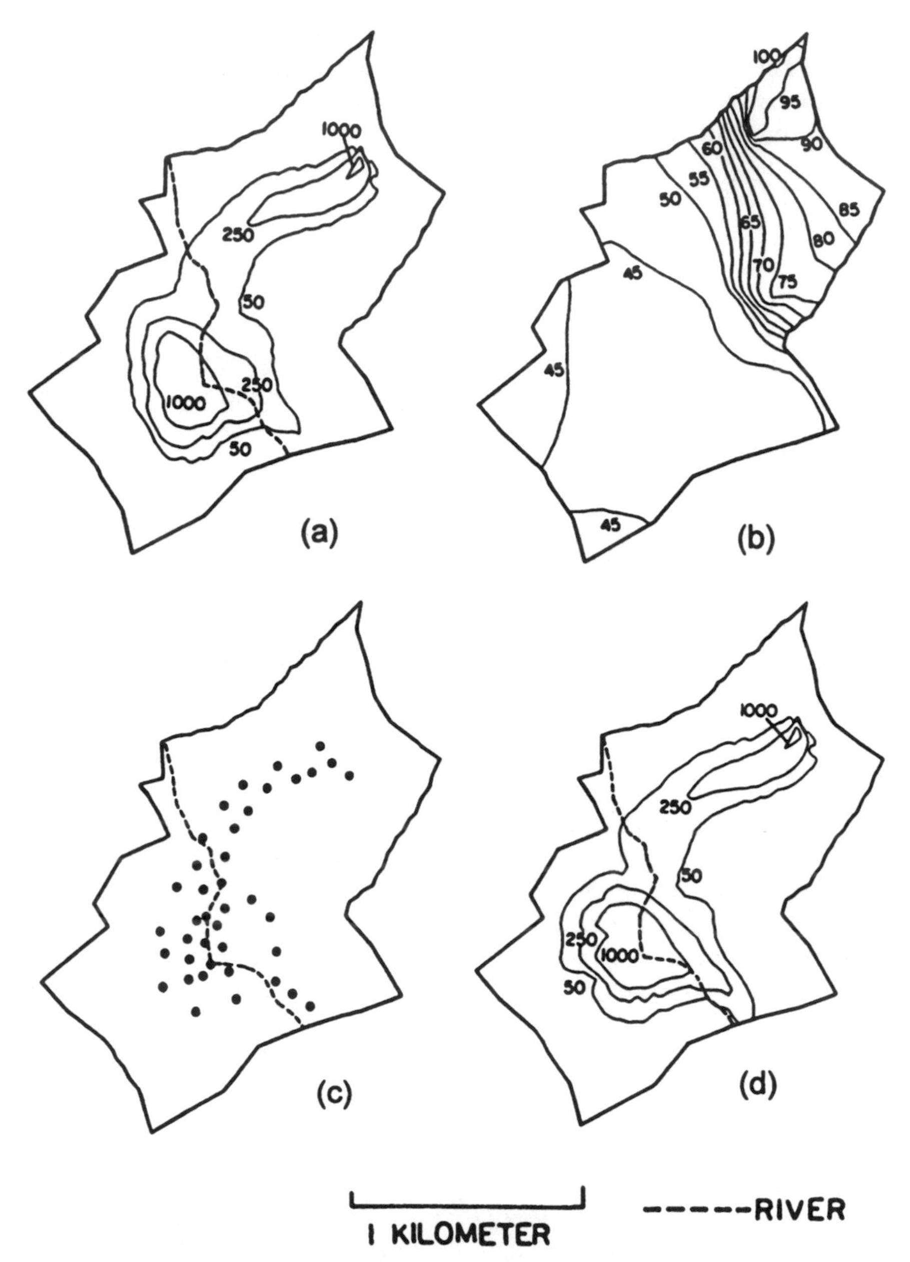

Figure 3.7 (a) Concentration distribution in 1986 (in ppb); (b) head contour in 1986 (in meters); (c) potential well locations; (d) concentration distribution in 1989 with "do-nothing" strategy (in ppb). (After Ahlfeld et al., 1988b.)

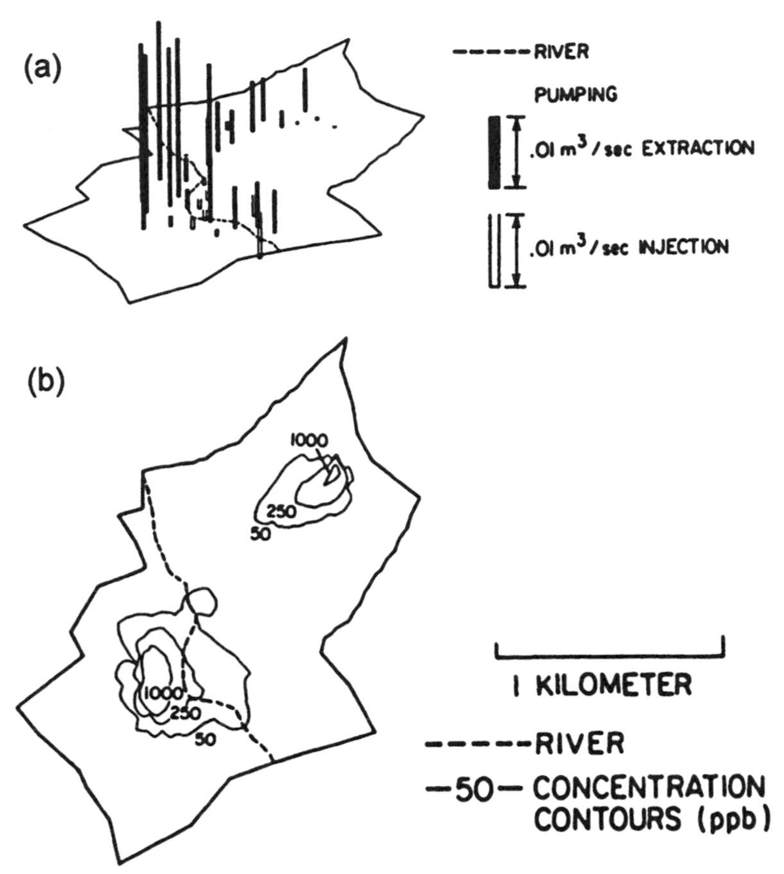

Figure 3.8 (a) Optimized system CM2; (b) concentration distribution in 1989 of CM2. (After Ahlfeld et al., 1988b.)

vation points is 20 ppb (part per billion) or lower. Figure 3.9b shows the optimized design for cost minimization when the contaminant concentration at the observation points is required to be 40 ppb or lower. Figure 3.9c shows the resulting concentration distribution in the study area in 1989. It can be seen that the optimized system has intensive extraction at the upstream side of the observation points with minor recharges. The contamination is confined in the aquifer without migrating into the downstream area, where the decision maker apparently intends to protect by using this optimization model.

The optimization model of remediation design reflects the decision maker's premise of groundwater quality management. While the contaminant sources cannot be removed in the planning period, the decision maker may accept alternatives to confine the contamination. Once the sources are removed, it is possible to use the optimized design for contaminant minimization to achieve a complete aquifer restoration. It should be noted that the optimality of the optimized system design is subject to the problem formulation, and a sound

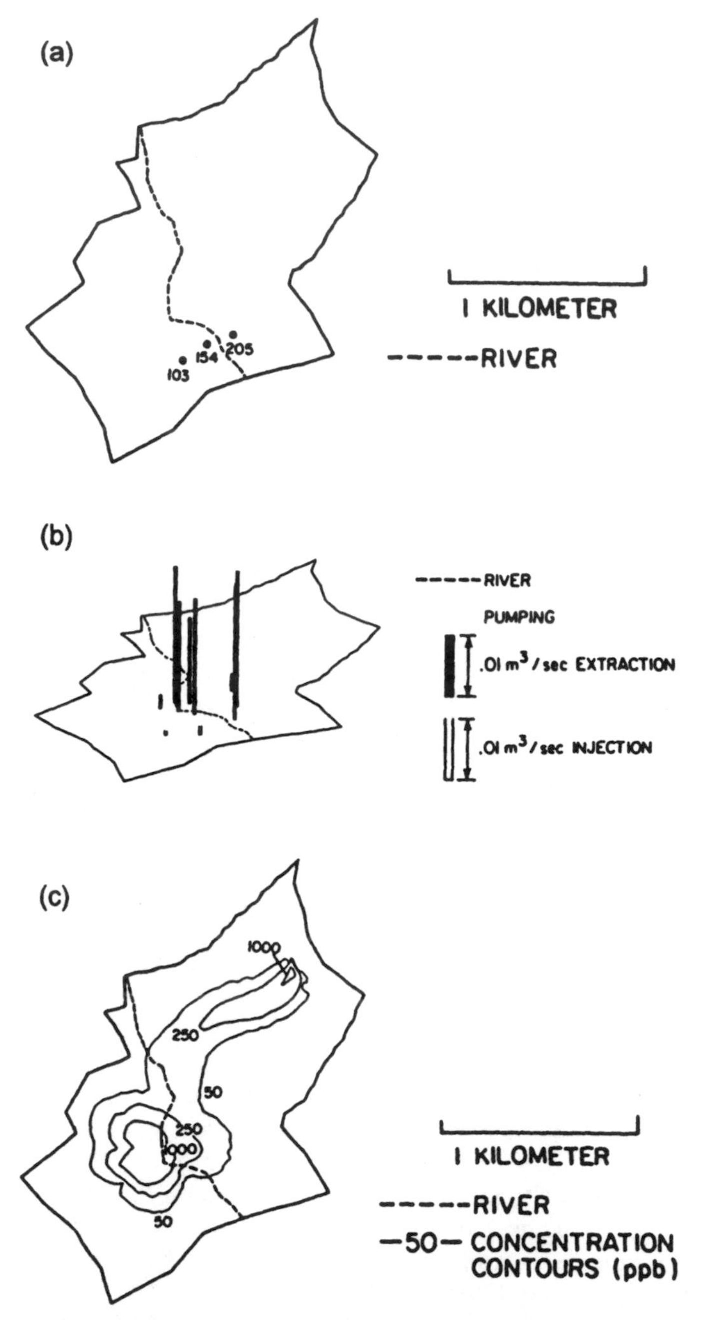

Figure 3.9 (a) Observation locations; (b) optimized system RC2; (c) concentration distribution in 1989 of RC2. (After Ahlfeld et al., 1988b.)

formulation leads to a practical system design for the real-life cleanup problem. In addition, global convergence is not generally guaranteed. This is typical of a gradient-based algorithm in solving nonlinear optimization problems. Multiple initial guesses are often required to improve the likelihood of reaching the global optimum.

Ahlfeld et al. (1988b) solved the nonlinear optimization problem by MINOS (Module Incore Nonlinear Optimization System; Murtagh and Saunders, 1995). MINOS, developed in the Systems Optimization Laboratory at Stanford University, is a commercially available software package for nonlinear optimization. It is worth noting that with many existing optimization packages (More and Wright, 1993), a system analyst can perform his or her duty without developing his or her own optimizer; however, the importance of understanding the solution algorithms and their corresponding limitations cannot be emphasized too strongly.

3.C.3. Groundwater-Related Health Risk Assessment

A contaminated groundwater system may be restored to a health-based standard if the site is favorable and the pump-and-treat system is designed properly, but a significant length of remediation time is generally required (U.S. EPA, 1989). The effect of groundwater contamination on public health needs to be evaluated on a regional scale through long-term monitoring that focuses on identifying the control factors affecting the public health. Cleanup standards must be reevaluated from a health-risk viewpoint.

Although optimization techniques appear to help in assessing groundwater-related health risk problems, such techniques are seldom used because of large problem size and the problem complexity associated with estimating contaminant toxicity, population growth, and contaminant transport through all exposure pathways. There has been much research in the past decade of risk–cost–benefit analysis on the management of contaminated aquifers (Massmann and Freeze, 1987a,b; Freeze et al., 1990, 1992; Massmann et al., 1991). However, since risk in the risk–cost–benefit analysis is defined as the probability of incurring penalties or other financial loss from the contamination, potential human health risks have been overlooked. A general introduction to risk assessment methodology appears in Section 6.C, and an example of exposure calculations to determine risk contours around an industrial site is given in Section 6.C.2.

To relate groundwater management to public health concerns, Pelmulder et al. (1996) developed a simulation framework that includes human health exposure as a criterion in regional-scale aquifer management problems. This simulation framework (see Figure 3.10) includes three models:

1. Groundwater flow and contaminant transport model, which provides the annual time series of contaminant concentration in groundwater supply for irrigation and domestic consumptions.

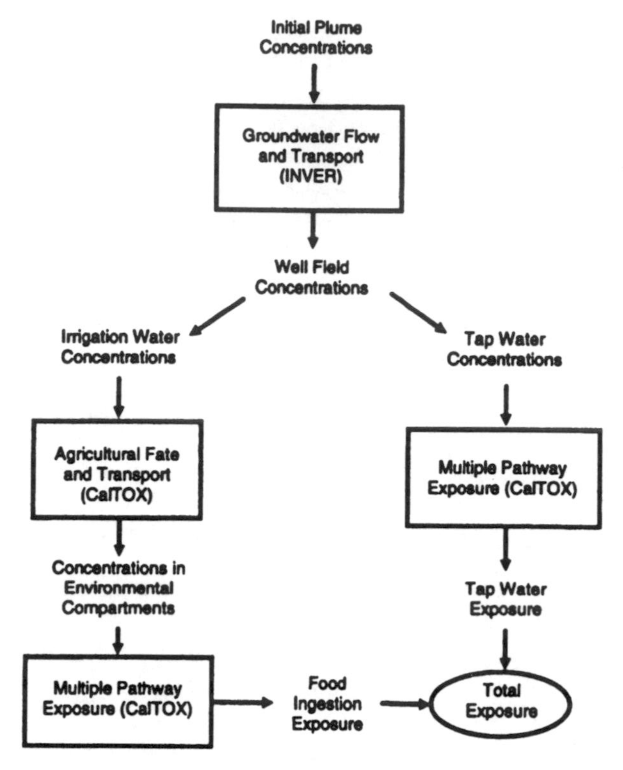

Figure 3.10 Simulation framework of groundwater-related health risk problems. (After Pelmulder et al., 1996.)

2. Environmental compartment model for contaminant transport in agricultural areas, which estimates contaminant concentrations in produce that used in computation of exposure through ingestion.
3. Multiple pathway human exposure model, which determines the overall exposure that can be converted to decease rate or population death.

A controlled scenario used for health risk simulation includes initial conditions of groundwater flow, contaminant distribution, predictions on population growth, and groundwater consumption. Based on the controlled scenario, the framework provides an assessment of health risk in a regional scale. With results of multiple simulations from controlled scenarios, the health risk issues can be investigated and control factors identified for reducing potential health risks. The results of this study framework provide a more reliable base for decision making.

Although the framework of risk assessment defined here is general, the individual models used in the framework can be changed if necessary, since they were developed independently. The criteria for model selection should include applicability to regional scale modeling, ease of modification, and balance between solution accuracy and computation efficiency.

Groundwater Flow and Solute Transport Model. INVER (N.-Z. Sun, 1994a), a model based on the method of multiple-cell balance (MCB; Sun and Yeh, 1983), simulates the water flow and contaminant transport in aquifers. As mentioned previously, MCB preserves superior mass balance in both local and global scales, and the parameters and variables in INVER are such that any consistent set of units can be incorporated. It is assumed that the free product of contaminants has been removed and that chemical reactions and biodegradation are ignored. Sorption process in solute transport is accommodated by the retardation factor. The results of groundwater model simulation will be used as inputs to the environmental compartment model and exposure model.

Environmental Compartment Model. The environmental compartment model monitors the amount of contaminant in irrigation water that is transferred into soil, water, and air and that comes in contact with plants meant for both livestock feed and human consumption. The compartment model used by Pelmulder et al. (1996) is a subset of CalTOX, which is an integrated transport and exposure–risk model developed at Lawrence Livermore National Laboratory for the Department of Toxic Substances Control of the California Environmental Protection Agency (McKone, 1993).

A compartment model divides the physical system into compartments in which the contaminant is assumed to be uniformly distributed. Each compartment may contain more than one phase; for example, a soil compartment may contain solids, water, and air. Mass conservation determines the contaminant mass transferred among compartments, while equilibrium conditions are assumed. Alternative integrated models suitable for this purpose include MMSOILS (Stasko and Fthenakis, 1993), MEPAS (Whelan et al., 1992), and RESRAD (Cheng and Yu, 1993). The CalTOX compartment model was chosen for the ease with which it is modified due to its formulation as a spreadsheet application.

The CalTOX model includes eight compartments: air, ground-surface soil, plants, root-zone soil, vadose-zone soil, surface water, sediments, and groundwater (see Figure 3.11). The mass balance equations in the compartment model are as follows.

Sediment (d): $$L_d N_d = T_{wd} N_w \tag{3.40a}$$

Surface water (w): $$L_w N_w = S_w + T_{aw} N_a + T_{gw} N_g + T_{dw} N_d \tag{3.40b}$$

Air (a): $$L_a N_a = S_a + T_{pa} N_p + T_{ga} N_g + T_{wa} N_w \tag{3.40c}$$

Plants (p): $$L_p N_p = T_{ap} N_a + T_{sp} N_s + T_{gp} N_g \tag{3.40d}$$

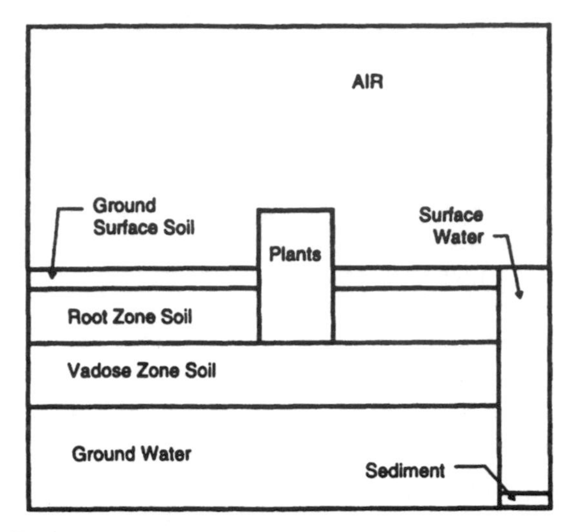

Figure 3.11 Environmental compartment model. (After Pelmulder et al., 1996.)

Ground-surface soil (g): $$L_g N_g = S_g + T_{ag} N_a + T_{sg} N_s + T_{pg} N_p \qquad (3.40e)$$

Root-zone soil (s): $$\frac{dN_s}{dt} = -L_s N_s + T_{gs} N_g + S_s \qquad (3.40f)$$

Vadose-zone soil (v): $$\frac{dN_v}{dt} = -L_v N_v + T_{sv} N_s + S_v \qquad (3.40g)$$

Groundwater (q): $$L_q N_q = T_{vq} N_v \qquad (3.40h)$$

where N's are the number of moles of contaminant in each compartment; T's are transfer coefficients, with the order of subscripts indicating the direction of the transfer [1/T]; L's are total loss rates, including reactions (e.g., biodegradation) and transfers out of the system [1/T]; and S's are constant source terms [1/T]. It should be noted that the groundwater compartment is referred to the shallow irrigation percolation water and is not included as a source in the simulation of INVER. The surface water and the sediment compartments will be minor in this case if the model is being used to represent the aggregate agricultural lands in the region. Reactions such as biodegradation are modeled as a first-order process.

The transfer of contaminant among compartments is determined using the fugacity approach (Mackay, 1979; Mackay and Paterson, 1982, 1991). Fugacity can be regarded as the escaping tendency of a chemical substance from a phase or compartment. It has units of pressure and is assumed to be linearly related to concentration through the fugacity capacity. For the air–water interphase, fugacity is the water vapor pressure, and the fugacity capacity of water is the reciprocal of Henry's constant. At equilibrium, fugacities of the chemical in two neighboring compartments or phases are equal. That is, the concentrations of these two compartments (phases) have a ratio the same as the ratio of their fugacity capacities, which is referred to as the *partition coefficient.* It can be seen that the fugacity approach is consistent with the linear equilibrium assumption for contaminant sorption in INVER. However, to accommodate the contaminant migration through irrigation water, the CalTOX model was modified to include source terms in the root-zone and vadose-zone soil compartments [see equations (3.40f) and (3.40g)].

The irrigation water that percolates through the root zone and into the vadose zone can be thought of as comprising two parts (Bouwer, 1969; Pelmulder et al., 1996): (1) water flows through the root zone in macropores or cracks and contributes directly to the deep percolation; and (2) water flowing through smaller pores replaces the water layer on the soil and forces the previous soil water down to the deep percolation. The former has the concentration of the irrigation water, and the latter has the concentration of the water forced out from the root zone (concentration in the root zone of the previous time step). Thus the source terms can be determined as follows:

$$S_v = D_d C_d = D_d[E_l C_s + (1 - E_l)C_i] \tag{3.40i}$$

$$S_s = D_i C_i - D_d C_d = D_i C_i - S_v \tag{3.40j}$$

where S_v and S_s are the source terms to the vadose- and root-zone compartments (mg/day), respectively; C_s, C_i, and C_d are the contaminant concentrations in the soil water, irrigation water, and deep percolation water (mg/m^3), respectively; D_d is the volume of deep percolation water per day (m^3/day); D_i is the volume of irrigation water applied per day (m^3); and E_l is the leaching efficiency. In stepping through the CalTOX model on a yearly basis, the source terms were calculated using the average well concentration for the coming year, the root-zone soil water concentration of the preceding year, and constant irrigation volumes.

Multiple-Pathway Exposure Model. *Exposure* is a general term for the amount of contaminant with which a person comes in contact, whereas *dose* is the actual amount of contaminant that remains in the body or in a particular target organ. For example, when a person drinks a glass of contaminated water, some of the contaminant will end up in various organs and some will pass through the body. The amount that is absorbed by the body is the dose. The proposed framework is capable of accommodating a special pharmokinetic model for fate and transport within the body; however, Pelmulder et al. (1996) used *potential dose,* which is often used as input in pharmokinetic models, to avoid additional uncertainty associated with the pharmokinetic model and thus to eliminate the model dependence of the final assessment. Potential dose or "exposure" is the amount of contaminant that enters the body. The units of potential dose are typically milligrams of contaminant per kilogram of body weight per day (mg/kg-day). Potential dose can then be aggregated over the duration of the exposure and the number of exposed individuals.

The exposure model of Pelmulder et al. (1996) is also contained in CalTOX (McKone, 1993). Two steps are involved when calculating potential dose. The first step is to determine the concentrations in the exposure media from concentrations in the larger environment. Exposure media are media with which humans come in contact, such as in-house air, tap water, and foods. The second step is to determine the amount of contaminant crossing the human body boundaries with the exposure media. Three major routes for contaminants entering the body are ingestion, inhalation, and dermal contact.

With the focus on the effect of contaminants in groundwater, the only exposure media considered by Pelmulder et al. (1996) were tap water and foods. Foods that were considered included vegetables, milk, eggs, and meat. While the only route considered for food is ingestion, all three major routes are available for tap water. These include ingestion, inhalation of vaporized contaminants, and dermal contact. Potential dose for an individual is determined by summing over all exposure media and all routes. If the risk of added cancers were the final measurement, the various potential doses would be kept separate, as there are different cancer potency factors for ingestion, inhalation, and dermal contact.

Risk Assessment and Sensitivity Analysis. Pelmulder et al. (1996) employed a hypothetical, confined aquifer (see Figure 3.12) to illustrate application of the simulation framework. All model parameters selected represent a typical California situation (Pelmulder, 1994). The region is rectangular, homogeneous, and isotropic, with dimensions of 8000 m in length (general flow direction), 4000 m in width (crossflow direction), and 100 m in thickness. A recharge boundary at the upper edge is maintained at a constant head of 200 m above the base of the aquifer, while the remaining three sides are considered as no-flow boundaries. There are three rows of seven wells, each pumping at the same rate, located at 3600, 4000, and 5500 m from the inflow boundary. The initial contaminant plume is 700 m in length and 350 m in width with a center of mass 2600 m from the first row of wells. The concentration in solution of the initial plume is uniform at 10 ppb of PCE, which is twice the drinking water MCL (maximum contaminant level).

The simulation duration is 200 years. The peak concentration in the first row of wells is 4.7 ppb and occurs at 100 years, while the peak concentration in the second row of wells is 2.1 ppb and occurs at 185 years. Groundwater extraction and plume dispersion contributed to the reduction of peak concentration. It is assumed that 40% of the tap water supply is from groundwater. In addition, it is assumed that 40% of the irrigation water is supplied by groundwater and 60% by PCE-free surface water. Based on the calculation of the environmental compartment model, the highest concentration of PCE is found in the root zone soil, followed by the vadose zone soil, air, and

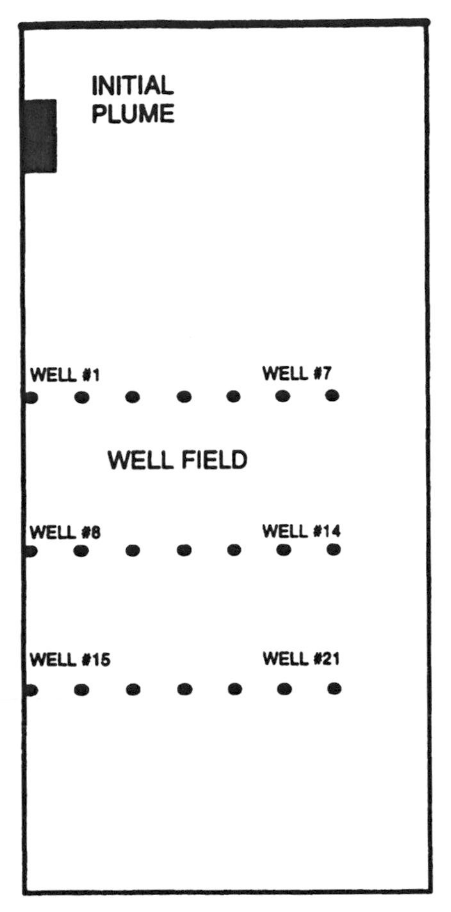

Figure 3.12 Plan view of the hypothetical, confined aquifer. (After Pelmulder et al., 1996.)

surface soil. The results of the exposure model indicate that due to dilution of the contaminant throughout the soil and other compartments, exposure from foods grown with contaminated irrigation water is at least an order of magnitude less than any of the tap water routes when tap water and irrigation water have the same contaminant concentration. Figure 3.13 shows that time series of individual daily exposure in 200 years.

As a first-cut estimate, the societal exposure can be calculated by multiplying the individual exposure time series by the projected population. While individual exposure can be used in assessing cancer risk to a typical individual, exposure of the population is used to compute the total number of illnesses or deaths. These two indices reflect different management objectives.

Sensitivity analysis can be performed to examine the relative sensitivities of parameters to the resulting risk assessment, which are defined similarly to equation (3.14). These parameters include the initial concentration and dis-

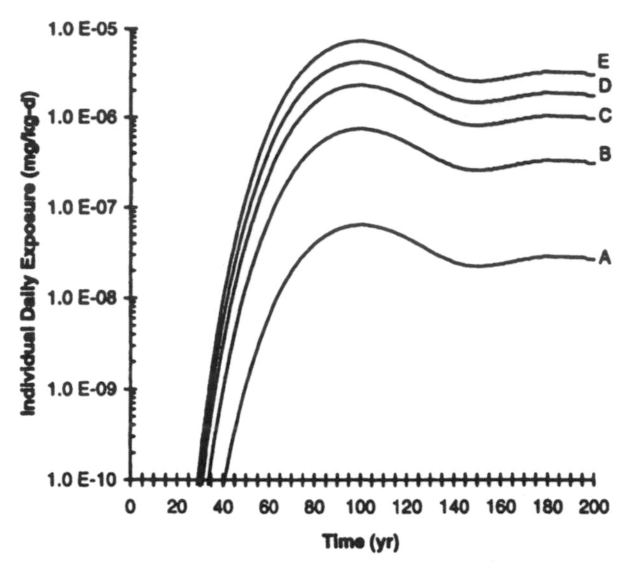

Figure 3.13 Total daily individual exposure. (After Pelmulder et al., 1996.)

tribution of the plume, groundwater fraction in the tap water supply, the distance from the center of the plume to the first row of wells, hydraulic conductivity, retardation factor, and dispersivities (longitudinal and transverse). The duration of exposure for an individual is assumed to be 30 years. Since the simulation time period is much longer than 30 years, a decision must be made regarding which 30-year average exposure is to be used when calculating the sensitivity coefficient. Two different values were selected, reflecting different management approaches. Management decisions based on the average over the 200-year simulation period of the 30-year average exposures allow an individual in one time period to be more exposed if the exposure is balanced by lower exposures in another time period. One characteristic of this approach is that the sensitivity coefficient reflects the total effect over time of the contaminant plume. A more common approach is to take the maximum 30-year average exposure. With the latter approach, the health of one person or group is not sacrificed for another. Table 3.1 shows the relative sensitivities of model parameters and initial conditions, ranking by their absolute value in a descending order.

Exposure is found to be most sensitive to changes in contaminant concentration in the water supply, which can be expected to be a direct cause of health problems. Contaminant concentration in water supply can be controlled by lowering the concentration in the aquifer, reducing the fraction of groundwater in total water supply, or treating all or a portion of the groundwater. Among all hydraulic parameters examined herein, the transverse dispersivity is the least sensitive; however, the conclusion may be site specific. It is important to recognize that the sensitivity rankings of these hydrological

TABLE 3.1 **Rankings of Absolute Value of Sensitivity Parameters**

Parameter	Individual and No Growth		Linear Growth		Exponential Growth	
	200-Year Average	30-Year Average	200-Year Average	30-Year Average	200-Year Average	30-Year Average
Initial concentration	1	1	1	1	1	1
GW fraction	1	1	1	1	1	1
Plume area (×3.8)	2	2	2	3	2	3
Plume area (×2)	3	4	3	4	3	4
Distance to wells	4	3	6	5	6	5
Hydraulic conductivity	5	6	4	7	5	6
Retardation factor	6	7	5	2	5	2
Longitudinal dispersivity	7	5	7	6	7	7
Transverse dispersivity	8	8	8	8	8	8

Source: Adapted from Pelmulder et al. (1996).

parameters depend on the projection of population growth and the definition of yearly average. This is especially obvious for the retardation factor. The existing sorption process causes a slower migration of contaminants. Due to the later population growth, the retardation effect results in delayed impact on the health of a larger population and thus a higher societal exposure.

The interdisciplinary simulation framework proposed by Pelmulder et al. (1996) is helpful to define quantitatively the complex relationship between groundwater management and health risk; this will assist the decision makers in assessing what population is or should be protected, or how decisions of groundwater management today affect the health of the future population. This is especially important for establishing the standards of a protective, but reasonable groundwater remediation program.

3.C.4. Soil Vapor Extraction System Design

Pump-and-treat technique and other groundwater remediation methods focus mainly on the removal of liquid- or aqueous-phase contaminants; however, contaminants leaching through the chemical residuals retained in the soil matrix of the unsaturated zone remain as a potential source of contamination, or even a safety hazard. Crow et al. (1987) reported that vapors of volatile organic compounds (VOCs) migrated through the soil into building basements, sewage systems, and utility vaults and caused incidents that included explosions.

Soil vapor extraction (see Figure 3.14) is a cost-effective way to control VOC contamination in soil and is often used as an enhancement of a pump-and-treat system, or vice versa (U.S. EPA, 1989; Pederson and Curtis, 1991). In this process, VOCs are removed from the unsaturated zone through the

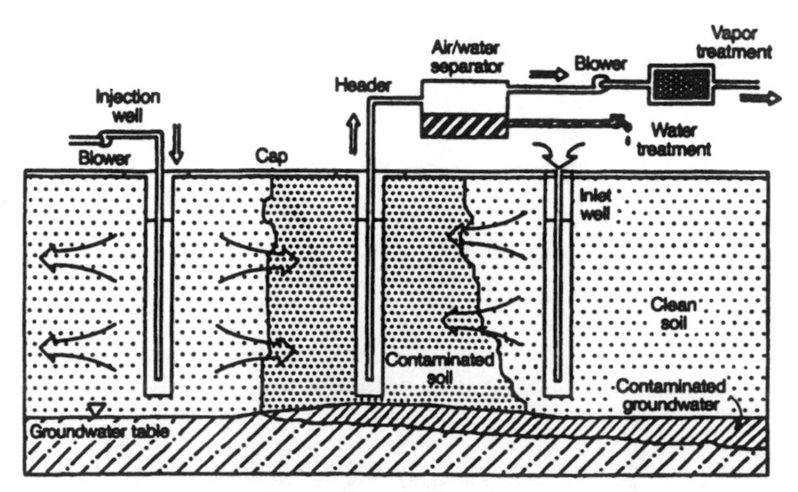

Figure 3.14 Schematic of soil vapor extraction system. (After Hutzler et al., 1990.)

vapor phase by creating advective vapor fluxes throughout the unsaturated soil matrix. These advective fluxes, created by air blowers and/or vacuum pumps, disrupt the vapor–liquid equilibrium of the liquid phase of the VOCs, thereby introducing volatilization and facilitating removal via the vapor phase through extraction wells.

SVES disturbs subsurface soil and surface activities only minimally and is suitable even for an urban environment (Hutzler et al., 1990). The efficiency of SVES is dependent on three principal factors: (1) composition of the contaminant, which must have sufficient volatility; (2) vapor flow rates through the soil matrix; and (3) vapor flow path relative to the contaminant distribution (Johnson et al., 1990a). Vapor flow rates and flow paths depend on the hydrogeological formation of the contaminated site and the configuration and operation of SVES design. Few guidelines for SVES design are available; however, empirical assessments, analytical approximations, and limited trial-and-error simulations are common approaches to the design problem (Massmann, 1989; Johnson et al., 1990b; Pederson and Curtis, 1991). The operation of a soil vapor extraction system introduces highly nonlinear, multiphase contaminant transport in a compressible fluid (vapor). Thus empirical or simplified analytical approaches provide a rough, often overly optimistic evaluation of system performance. The simulation approach improves the system evaluation but is limited to comparing scenarios obtained from empirical and analytical approaches (Massmann, 1989; Sepehr and Samani, 1993; Y.-H. Sun, 1994).

Although soil vapor extraction is conceptually similar to pump and treat, the major difficulty of SVES design comes from the significant well interference during operation. Well interference stems from improper well spacings and extractions. This effect occurs in pump-and-treat systems as well (Haggerty and Gorelick, 1994); however, it is less intensive because water flow rates are usually slower than vapor flow rates. Rosenwald and Green (1974) employed a branch-and-bound method to optimize well installation in a gas reservoir. The response matrix was used as the branching criterion to select the most profitable well location for installation. The design strategy was not successful because high vapor flow rates induced significant pressure drawdowns, amplifying the errors in applying the principle of superposition to gas flow. Iterative solutions improved the results but merely provided a first-order approximation of the optimal design. Empirical and analytical approaches have failed to produce a reliable system design for the same reason. Optimization-based approach has been reported recently (Davert and Yeh, 1994; Sacks et al., 1994; Sun and Yeh, 1995). The optimization-based approach focuses on the development of an adequate design strategy that incorporates the well interference effect during SVES design, usually with a simulation model that evaluates the multiphase, multicomponent contaminant transport under SVES operation. This will be illustrated in this section with the comparison of three design strategies by Sun et al. (1996).

Multiphase, Multicomponent Contaminant Transport Model. The contaminant transport model developed by Rathfelder et al. (1991) was used in Sun et al. (1996) as a simulation module to predict the system response during optimization. The interest in applying soil vapor extraction to remove VOCs in the unsaturated zone has encouraged researchers to develop simulation models for compressible vapor flow and multiphase, multicomponent contaminant transport in the unsaturated zone (Abriola and Pinder, 1985; Corapcioglu and Baehr, 1987; Sleep and Sykes, 1989, 1993a,b; Rathfelder et al., 1991; Culver et al., 1991; Gierke et al., 1992; Armstrong et al., 1994; Ng and Mei, 1996).

The two-dimensional finite difference model developed by Rathfelder et al. (1991) contains four phases: air (a), water (w), NAPL (oil, o), and solid (soil, s). It is further assumed that interphase mass transfer is in equilibrium, air phase the sole mobile phase, and soil particles water-wet. The vapor flow and contaminant transport in the unsaturated zone under SVES operation are characterized by four groups of equations: the density-dependent air-phase flow equation, the multicomponent contaminant transport equation, local equilibrium isotherms for interphase mass transfer, and mass balance totality conditions.

Air-Phase Flow Equation

$$\frac{\partial}{\partial t}\left(nS_a \frac{PM_a}{RT}\right) = \nabla \cdot \left(\frac{PM_a}{RT} \frac{\kappa k_{ra}}{\mu_a} \nabla P\right) + nS_a\rho_a\Gamma \tag{3.41}$$

in which n is the porosity, P the air-phase pressure head, M_a the air-phase molecular weight, R the ideal gas constant, T the system temperature, κ the intrinsic permeability, k_{ra} the air-phase relative permeability, μ_a the air-phase viscosity, Γ a sink–source term, and S_a the air-phase saturation.

Multiphase, Multicomponent Contaminant Transport Equation

$$\frac{\partial}{\partial t}(nS_a\rho_{a\gamma} + nS_o\rho_{o\gamma} + nS_w\rho_{w\gamma} + \rho_s\rho_{s\gamma}) = \nabla \cdot \left(\rho_{a\gamma} \frac{\kappa k_{ra}}{\mu_a} \nabla P + D^h_{a\gamma}\right) + nS_a\rho_{a\gamma}\Gamma \tag{3.42}$$

in which S_o is the oil-phase saturation; S_w the water-phase saturation; ρ_s the bulk soil density; $\rho_{a\gamma}$, $\rho_{o\gamma}$, $\rho_{w\gamma}$, and $\rho_{s\gamma}$ are densities of γ-component in air phase, oil phase, water phase, and soil, respectively; and $D^h_{a\gamma}$ is the hydrodynamic dispersion coefficient of the γ-component in the air phase. It should be noted that since more than one phase participates in mass transfer of contaminants, the retardation factor is not applicable in this case. Thus the third group of equations defines the interphase mass transfer.

Local Equilibrium Isotherms for Interphase Mass Transfer. Local equilibrium state is assumed for all interphase mass transfers. Since the soil particles are assumed water-wet, there is no interphase between oil and soil phases. The evaporation of water can be written by Raoult's law:

$$K_{wa}^{w} = \frac{\rho_{ww}}{\rho_{aw}} = \frac{RT}{P^{w}} \sum_{\gamma=1}^{N} \frac{\rho_{w\gamma}}{M_{r}} \tag{3.43a}$$

in which K_{wa}^{w} is the air–water partition coefficient of water; ρ_{ww} and ρ_{aw} are the densities of water in water and air phases, respectively; P^{w} is the vapor pressure of water; and M_{γ} is the molecular weight of the γ-component.

The interphase mass transfer of contaminants can be written as follows:

$$\text{(air–water)} \qquad K_{wa}^{\gamma} = \frac{\rho_{w\gamma}}{\rho_{a\gamma}} = \frac{M_r K_h^{\gamma}}{RT} \tag{3.43b}$$

$$\text{(air–oil)} \qquad K_{oa}^{\gamma} = \frac{\rho_{o\gamma}}{\rho_{a\gamma}} = \frac{RT}{P^{\gamma}} \sum_{\gamma=1}^{N} \frac{\rho_{o\gamma}}{M_r} \tag{3.43c}$$

$$\text{(water–solid)} \qquad K_{ws}^{\gamma} = \frac{\rho_{w\gamma}}{\rho_{s\gamma}} = K_{oc} f_{oc} \tag{3.43d}$$

in which K_{wa}^{γ}, K_{oa}^{γ}, and K_{ws}^{γ} are the air–water, oil–water, and water–solid partition coefficients of the γ-component; K_h^{γ} and P^{γ} are the Henry's constant and vapor pressure of the γ-component, respectively; K_{oc} is the organic carbon content-normalized partition coefficient, and $\log K_{oc} = -0.54 \log W_{\gamma} + 0.44$; f_{oc} is the soil organic carbon content; and W_{γ} is the solubility of the γ-component in mole fraction. It can be seen that Eq. (3.43b) follows Henry's law, Eqs. (3.43a) and (3.43c) Raoult's law, and Equation (3.43d) linear sorption.

Totality Conditions. The summation of the volumetric fractions should be equal to unity at all times:

$$S_a + S_o + S_w = 1 \tag{3.44a}$$

$$\sum_{\gamma=1}^{N} \rho_{o\gamma} = \rho_o \tag{3.44b}$$

$$\rho_{ww} + \sum_{\gamma=1}^{N} \rho_{w\gamma} = \rho_w \tag{3.44c}$$

$$\rho_{aa} + \rho_{aw} + \sum_{\gamma=1}^{N} \rho_{a\gamma} = \rho_a = \frac{PM_a}{RT} \tag{3.44d}$$

in which ρ_o, ρ_w, and ρ_a are the densities of the oil, water, and air phases, respectively; M_{air} is the molecular weight of pure air; and M_{aq} is the molecular weight of pure water.

Rathfelder et al. (1991) solved the resulting finite difference model by decoupling flow equation and transport equation. At each time step, the flow equation is solved first and the resulting pressure field taken as input to solve the contaminant transport equation. The updating of the other variables, such as the air-phase molecular weight, viscosity, relative permeability, and hydrodynamic dispersion coefficients, is lagged by one time step. The underlying assumption is that the changes of phase saturations and air-phase concentrations within one time step have a negligible effect on the pressure field and the concentration distribution. This assumption is made for computation convenience and is acceptable, since the maximum time step was 0.01 day (14.4 minutes). In addition, the flow equation is solved over a larger domain than the transport equation, to reduce the possible impact from the assumed boundary conditions. The simulation results of this model compared favorably with the one-dimensional analytical solution. Detailed discussion of the simulation model and its numerical approximation can be found in Rathfelder et al. (1991).

Optimization-Based SVES Design. The optimization-based SVES attempts to identify a best alternative, if not the optimum, among competing designs. It should be noted that the optimality of the resulting design is subject to the formulation of the design problem. One of the design problems is the maximization of total contaminant mass removal within a given time frame subject to physical and budgetary constraints:

$$\text{maximize } M = \sum_{l\in\Omega} (C_l(\mathbf{Y},0) - C_l(\mathbf{Y},T)) \cdot V_l \tag{3.45a}$$

subject to

$$\sum_{i\in I} \alpha_i x_i \leq B_c \tag{3.45b}$$

$$\sum_{i\in I} \sum_{j\in J_i} \beta_{i,j} y_{i,j} q_{i,j} \leq B_o \tag{3.45c}$$

$$\sum_{j\in J_i} y_{i,j} = x_i \qquad i \in I \tag{3.45d}$$

$$x_i, y_{i,j} \text{ binary} \qquad j \in J_i, \quad i \in I \tag{3.45e}$$

in which α_i is the capital cost of well installation at well location i; $\beta_{i,j}$ is the operational cost of pumping rate j at well location i within the planning period; Ω is the contaminated area; B_c and B_o are the allowed budgets for well installation and operation, respectively; C_l is the contaminant concentration at location l; V_l is the control volume of location l; I is the set of potential well locations; J_i is the set of possible pumping rates at well location i; $q_{i,j}$ is the extraction rate j at well location i; T is the duration of SVES operation (often imposed by a regulatory agency); x_i is the indicator of well location i; $y_{i,j}$ is the indicator of extraction rate j at well location i; and $\mathbf{Y}$ is the matrix of $y_{i,j}$'s.

The decision variables are the binary indicators of location and pumping rate. Equation (3.45d) restricts the extraction rate of a well to zero unless the well location has been selected. The primary constraints are budgetary. In the formulation above, the total cost of an SVES is separated into capital costs and operational costs. This separation is not mandatory, but can be justified from a practical viewpoint (Davert, 1993; Sun et al., 1996). The separation decouples the optimizations of well locations and extraction rates. The extraction rates are optimized only for active wells, which can be embedded in the location optimization to reduce the computation requirement. Moreover, the separation addresses the difference between location optimization and extraction-rate optimization.

Location optimization is an inherently combinatorial problem due to the binary state of well activeness. Thus a gradient-based algorithm is not applicable, not only because the well interference effect limits application of the principle of superposition, but also because the binary state of well activeness introduces a discontinuous response surface and/or step functions for installation cost if incorporated in the formulation. This is the fundamental reason that Rosenwald and Green (1974) failed in their optimization of the system design for a gas aquifer. By contrast, extraction rate is virtually continuous. Thus, for a given set of active wells, optimization of the corresponding extraction rates can be solved by a gradient-based algorithm, provided that the operational cost function is continuous. It should be noted that the optimization of extraction rates becomes combinatorial only when extraction rates are discretized; however, properties associated with the continuous variables remain.

SVES Design by Combinatorial Optimization. Sun et al. (1996) compared three combinatorial methods in the application of SVES design—implicit enumeration, systematic reduction, and local search—emphasizing the improvement upon location optimization. All three methods converge only to local optima; however, local convergence is common in complex, nonlinear optimization problems. These three methods represent different strategies for SVES design.

Implicit Enumeration. Implicit enumeration is a class of branch-and-bound methods often applied to problems with binary variables (Garfinkel and Nemhauser, 1972). Various tests can be used to determine the most profitable branch and to eliminate many of the total 2^n combinations without explicit enumeration, where n is the total number of binary variables. As demonstrated by Rosenwald and Green (1974), tests based on the response matrix are not applicable to SVES design because the principle of superposition is not valid. Sun et al. (1996) used a modified branching criterion, which is based on the total removal of contaminant mass (the objective function value) instead of its increments. Well interference is reevaluated when a new well location is considered. In other words, extraction rates are reoptimized with the presence of a new well. Therefore, all candidates of system design compete with their maximum contaminant mass removal rate. The procedure repeats until the budgetary constraints are no longer satisfied. Figure 3.15 shows the algorithm of implicit enumeration.

Systematic Reduction. Systematic reduction begins with a system in which all potential wells are active. During optimization, the number of wells is reduced systematically until budgetary constraints are satisfied. Inferior well locations are identified by their zero or near-zero extraction rates and eliminated from the current system configuration. In case every well in the current system is operated at a significant extraction rate, the well that provides the least contribution to the current total contaminant mass removal is the inferior well to be removed.

The contribution of each well is determined by the following procedure. A system candidate is defined by inactivating one well in the current design,

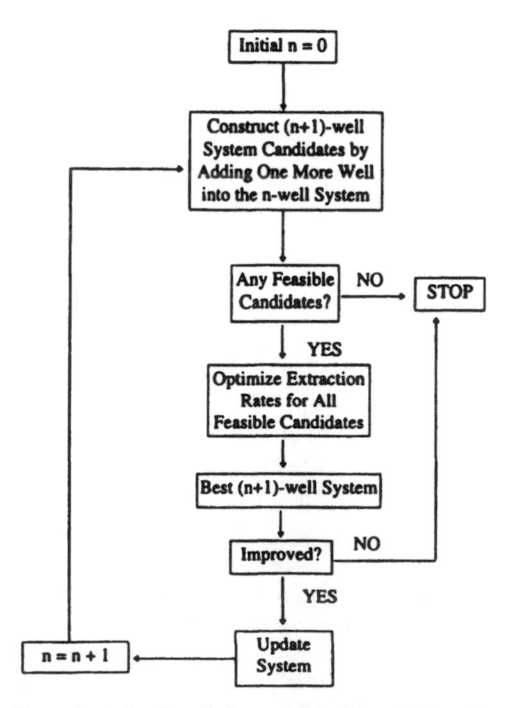

Figure 3.15 Flowchart for implicit enumeration. (After Sun et al., 1996.)

while all other wells operating at the extraction rates are identified for the current design. The contribution of a well is the difference between the total contaminant mass removal rate of the current design and that of the system candidate without the well. It should be noted that the sum of individual contributions is not equal to the total contaminant mass removal because cleanup effect is a nonlinear function. As mentioned previously, the principle of superposition is not applicable to SVES regarding cleanup effect.

The well interference effect is accounted for by reoptimizing the extraction rates for the reduced system in every iteration after the inferior well is removed. The procedure repeats until the resulting system satisfies the budgetary constraints. Figure 3.16 shows the algorithm of systematic reduction.

Local Search. The previous two methods have decisions of well activeness made in sequence. Thus they are considered sequential decision rules. Local search, on the other hand, is a nonsequential decision rule with adaptive features, focusing on system performance rather than individual comparison. Local search requires an initial guess to start optimization. In every iteration, local search restricts its search for improved system design within a user-defined "neighborhood." System candidates in a neighborhood are partially modified from the current system, while retaining the total number of active wells predetermined by the budgetary constraints for installation. The current system is replaced only if a new system candidate in the neighborhood shows improvement in total contaminant mass removal. Consequently, if there is a new neighborhood associated with the updated system, local search will con-

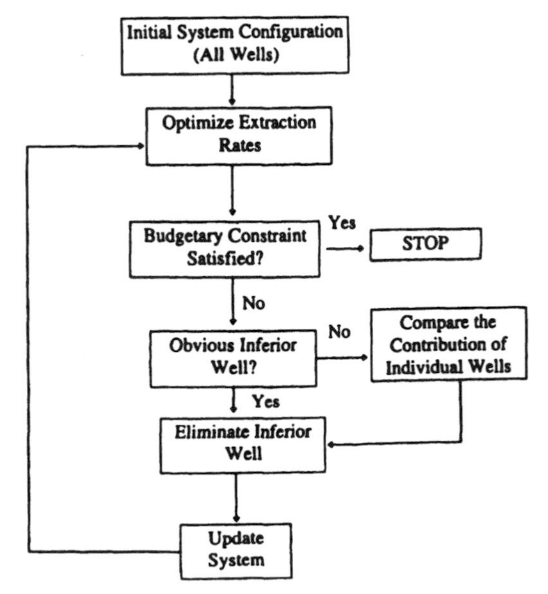

Figure 3.16 Flowchart for systematic reduction. (After Sun et al., 1996.)

tinue its searching based on the new system. The method terminates when no improved design can be found in the neighborhood. The well interference effect is accounted for by optimizing the extraction rates for each system candidate. The system candidates are then compared by their overall contaminant mass removal rate.

Local search can be customized by its three user-defined properties: the definition of a neighborhood, the search pattern within a neighborhood, and the search order among neighborhoods (Papadimitriou and Steiglitz, 1982). A class of neighborhoods often used in local search is the *k*-change neighborhood. A *k*-change neighborhood is defined by a set of feasible solutions in which only *k* decision variables have values that are different from those in the current solution. The search pattern is related to how extensively a neighborhood should be searched, and the search order relates to which *k* variable should be changed first.

Sun et al. (1996) used a two-change neighborhood, which means only one well location currently active is chosen to change. When an active well is determined to change, inactive wells are screened for eligibility of replacement in the order of relative distance to the well currently active. The resulting neighborhood consists of two new system candidates that have not been considered previously. In other words, the relative distance between wells is used as the guideline of neighborhood definition, since proper well spacing is known to prevent well interference effectively. The steepest-descent search pattern was used to perform a complete search in a neighborhood, and a circular search order was used to assign a higher priority of changing to variables that remain unchanged in the previous iteration. Figure 3.17 shows the algorithm of two-change local search.

Methodology Comparison. These three design strategies were compared through a hypothetical benchmark problem, which contains the benzene-contaminated site shown in Figure 3.18 with eight potential extraction well locations (Sun et al., 1996). The design problem is to determine a two-well system that maximizes the total contaminant mass removal within 20 days. Extraction rate for each well was assumed to be constant over the remediation period but could take on values from 0.0 to 8.0 ft^3/min with a 0.1-ft^3/min increment. Installation cost was assumed to be identical for all wells. Thus the budgetary constraint for well installation is expressed explicitly in terms of the number of wells, which is two in this case. In addition, the maximum aggregate extraction rate of active wells was set at 8 ft^3/min for the operational budgetary constraint.

The results and comparison of these three methods are summarized in Table 3.2. To examine the impact of initial guess, three sets of initial extraction rates were used in the optimization of SVES design by systematic reduction, and two initial systems were used in the optimization by local search.

Implicit enumeration and systematic reduction converge to a near-optimum solution, design (3,5), which is ranked second among 28 possible two-well

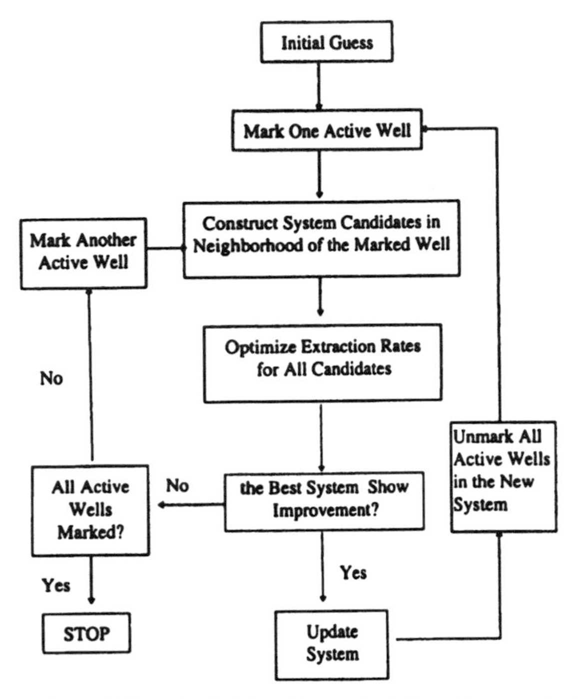

Figure 3.17 Flowchart for local search. (After Sun et al., 1996.)

systems with corresponding optimized extraction rates, while local search reached the global optimum, design (3,7). It should be noted, however, that these two designs do not differ significantly in removing contaminant mass, nor is global convergence guaranteed by any of these three methods. Local search reaches the global optimum because its heuristic of optimality is more appropriate for SVES design.

One reason to have a multiwell system is to improve system performance by the collaborative effort of multiple wells. The heuristics of implicit enumeration and systematic reduction have an inherent problem for this purpose. Implicit enumeration assumes that the best n-well design evolves from the optimal $(n - 1)$-well design, and systematic reduction assumes that the optimal n-well design degenerates from the best $(n + 1)$-well design. In other words, once a well is added into the system configuration by implicit enumeration, it will remain in; once a well is discarded from the system configuration by systematic reduction, it will remain out. These stringent assumptions limit the possibility of developing the collaborative environment for a multiwell system. If the optimization by systematic reduction continued to determine the most effective well location, it would be well 5, as predicted in the first iteration of the optimization by implicit enumeration (Sun et al., 1996). Systematic reduction and implicit enumeration appear to be logically equivalent. By contrast, local search allows a once-removed well location to

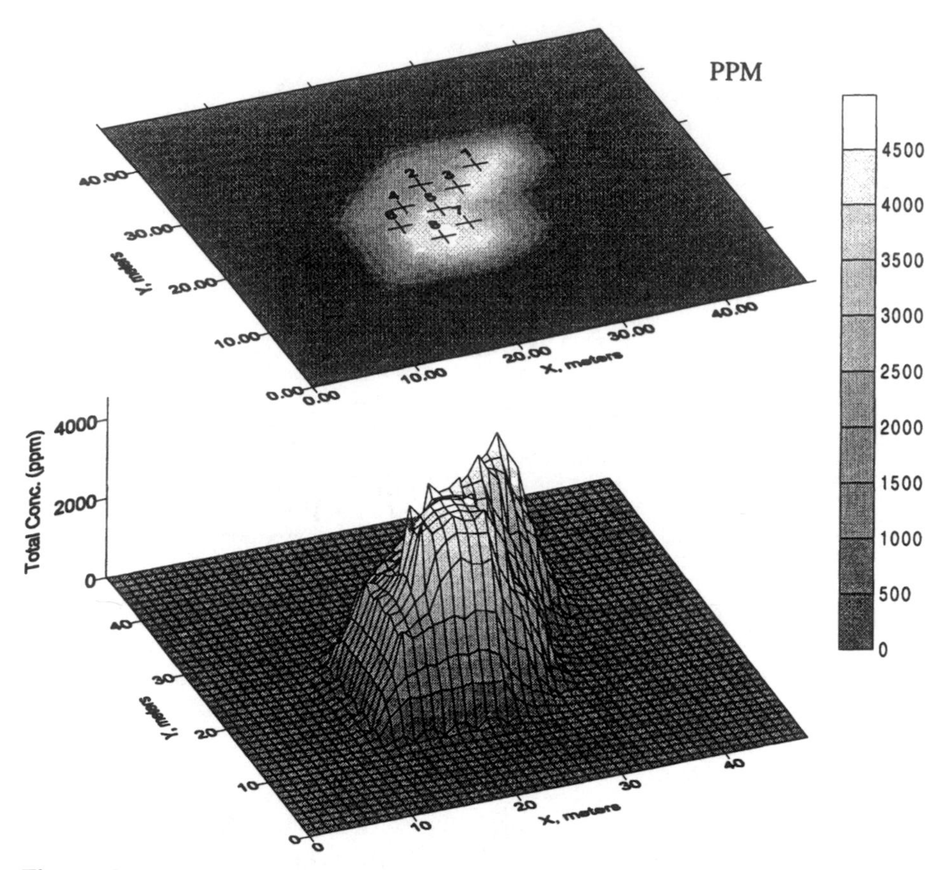

Figure 3.18 Initial distribution of total benzene concentration in the unsaturated zone.

reenter the system design if it assists in improving system performance when working in pair with one of the current active wells. Although the probability of reentrance for a well once removed is subject to the three user-defined properties, it is not totally excluded.

As mentioned previously, local search is categorized as a nonsequential decision rule; implicit enumeration and systematic reduction are considered as sequential decision rules. It is important to recognize that decision rules reflect the premise of decision making. If a sequential decision rule were to be used, SVES design would be separated into partial decisions. The partial decisions are determined in sequence with all previous decisions fully delineated on the current design options. On the other hand, if a nonsequential decision rule is used, decisions would always be made against a complete SVES design; thus this decision premise makes local search more suitable for SVES design.

Methodology comparison is not exclusively for SVES design. It provides a method for seeking a more efficient, more accurate, and more thorough way

TABLE 3.2 Summary of Methodology Comparison

Item	Implicit Enumeration	Systematic Reduction	Local Search
Type of decision rules	Sequential	Sequential	Nonsequential
Heuristic of optimality	The best n-well system evolves from the best $(n - 1)$ system	The best n-well system degenerates from the best $(n + 1)$ system	The best n-well system evolves among n-well systems
Interference among well operations	Incorporated	Incorporated	Incorporated
Convergence	Local convergence	Local convergence	Local convergence
		Benchmark Problem	
System design	(5, 3)	(3, 5)	(3, 7)
Rank among 28 possible designs	2	2	1
Number of simulations	50	151, 208, 229	52, 107
		Applications	
Time-varying extraction rates	Suitable	Suitable	Suitable
Location-dependent well installation cost	Cost constraint may dominate	Cost constraint may dominate	Suitable
Large number of potential well locations	Computationally intensive	Computationally intensive	Suitable; can be stopped when the current design is satisfactory
Parallel computing	Suitable	Not suitable	Suitable

Source: Adapted from Sun et al. (1996).

of problem solving. New methodologies have been introduced in the field of groundwater management. Artificial neural networks (Rogers and Dowla, 1994) and combinatorial methods such as simulated annealing (Dougherty and Marryott, 1991) and genetic algorithm (McKinney and Lin, 1994) have been applied to groundwater remediation and management problems as alternative approaches to the conventional gradient-based algorithms, especially when a discontinuous cost function is used. Due to the similarity in problem characteristics, implicit enumeration, systematic reduction, and local search are also applicable to groundwater management problems. Implicit enumeration and systematic reduction can be seen as improved versions of the conventional gradient-based algorithms, and local search resembles simulated annealing and genetic algorithm but with less complexity. Nonetheless, it is important that a system analyst understand the decision premise that accompanies the selected optimization method or the design strategy selected.

3.D. SUMMARY

The development of groundwater simulation models in the early 1970s provided groundwater system analysts with quantitative techniques for analyzing alternative groundwater management strategies. Optimization techniques have been used to assist further in the management of groundwater resources. Due to the complexity of natural groundwater systems and the various uncertainties involved, system analysis of groundwater resources remains challenging. A useful management strategy or system analysis should be both technically sound and practically feasible. In this chapter the essential concepts of groundwater system analysis are introduced through a demonstration of applications that partially represent the state of the art from a local scale to a regional scale, from water supply to contamination remediation, from a problem of liquid phase to a problem of vapor phase, and even from model authenticity to decision philosophy. While the techniques for groundwater system analysis are improving continuously, the fundamentals remain unchanged. Because groundwater management is such a broad and delicate art, readers are encouraged beyond being skillful in techniques to contemplate the significance and role of system analysis in groundwater management.

EXERCISES

3.1. Use a flowchart to represent the sequential and feedback relationships among the procedures in groundwater system analysis.

3.2. Derive the contaminant transport equation based on the Freundlich sorption isotherm: $C^* = K_F C^N$, where K_F and N are constants.

3.3. Formulate an optimization model for designing a pump-and-treat system which results in the shortest remediation time for a given cleanup standard and an operational budgetary constraint, assuming that contaminant sources and free product have been removed. The required cleanup standard is required for the entire aquifer. Compare this design premise with the contaminant minimization and the cost minimization. Discuss the feasibility of this design premise.

REFERENCES

Abriola, L. M., and G. F. Pinder, 1985. A Multiphase Approach to the Modeling of Porous Media Contamination by Organic Compounds: 1. Equation Development, *Water Resources Research,* 21(1), 11–18.

Ahlfeld, D. P., and M. Heidari, 1994. Applications of Optimal Hydraulic Control to Groundwater Systems, *Journal of Water Resources Planning and Management,* ASCE, 120(3), 350–365.

Ahlfeld, D. P., and C. S. Sawyer, 1990. Well Location in Capture Zone Design Using Simulation and Optimization Techniques, *Ground Water,* 28(4), 507–512.

Ahlfeld, D. P., J. M. Mulvey, and G. F. Pinder, 1986. Designing Optimal Strategies for Contaminated Groundwater Remediation, *Advances in Water Resources,* 9(2), 77–84.

Ahlfeld, D. P., J. M. Mulvey, G. F. Pinder, and E. F. Wood, 1988a. Contaminated Groundwater remediation Design Using Simulation, Optimization, and Sensitivity Theory: 1. Model Development, *Water Resources Research,* 24(3), 431–441.

Ahlfeld, D. P., J. M. Mulvey, and G. F. Pinder, 1988b. Contaminated Groundwater Remediation Design Using Simulation, Optimization, and Sensitivity Theory: 2. Analysis of a Field Site, *Water Resources Research,* 24(3), 443–452.

Ahlfeld, D. P., R. H. Page, and G. F. Pinder, 1995. Optimal Groundwater Remediation Methods Applied to a Superfund Site: From Formulation to Implementation, *Ground Water,* 33(1), 58–70.

Armstrong, J. E., E. O. Frind, and R. D. McClellan, 1994. Nonequilibrium Mass Transfer Between the Vapor, Aqueous, and Solid Phases in Unsaturated Soils During Vapor Extraction, *Water Resources Research,* 30(2), 355–368.

Barcelona, M., J. F. Keely, W. A. Pettyjohn, and A. Wehrmann, 1988. *Handbook of Groundwater Protection,* Hemisphere, New York.

Barlow, P. M., B. J. Wagner, and K. Belitz, 1996. Pumping Strategies for Management of a Shallow Water Table: The Value of the Simulation-Optimization Approach, *Ground Water,* 34(2), 305–317.

Bear, J., 1972. *Dynamics of Fluids in Porous Media,* Elsevier, New York.

Bear, J., 1979. *Hydraulics of Groundwater,* McGraw-Hill, New York.

Becker, L., and W. W.-G. Yeh, 1972. Identification of Parameters in Unsteady Open-Channel Flows, *Water Resources Research,* 8(4), 956–965.

Bedient, P. B., H. S. Rifai, and C. J. Newell, 1994. *Ground Water Contamination,* Prentice Hall, Upper Saddle River, NJ.

Bouwer, H., 1969. Salt Balance, Irrigation Efficiency, and Drainage Design, *Journal of the Irrigation and Drainage Division,* ASCE, 95(1), 153–170.

Cheng, J.-J., and C. Yu, 1993. Using the RESRAD Computer Code to Evaluate Human Health Risks from Radionuclides and Hazardous Chemicals, *Journal of Hazardous Materials,* 35, 353–367.

Connor, J. A., 1994. Hydrogeologic Site Investigation, in P. B. Bedient, H. S. Rifai, and C. J. Newell (eds.), *Ground Water Contamination,* Prentice Hall, Upper Saddle River, NJ, pp. 389–418.

Cooley, R. L., 1982. Incorporation of Prior Information on Parameters into Nonlinear Regression Groundwater Flow Models, 1. Theory, *Water Resources Research,* 13(2), 318–324.

Corapcioglu, M. Y., and A. L. Baehr, 1987. A Compositional Multiphase Model for Groundwater Contamination by Petroleum Products: 1. Theoretical Considerations, *Water Resources Research,* 23(1), 191–200.

Crow, W. L., E. P. Anderson, and E. M. Minugh, 1987. Subsurface Venting of Vapor Emanating from Hydrocarbon Product on Ground Water, *Ground Water Monitoring Review,* 7(1), 51–57.

Culver, T. B., C. A. Shoemaker, and L. W. Lion, 1991. Impact of Vapor Sorption on the Subsurface Transport of Volatile Organic Compound: A Numerical Model and Analysis, *Water Resources Research,* 27(9), 2259–2270.

Davert, M. W., 1993. Optimization of Soil Vapor Extraction Systems, Ph.D. dissertation, UCLA, Los Angeles.

Davert, M. W., and W. W.-G. Yeh, 1994. Optimization of the Design and Operation of Soil Vapor Extraction Systems, *Proceedings of the 21st Annual Conference of the Water Resources Planning and Management Division, ASCE,* Denver, CO, pp. 181–184.

Dougherty, D. E., and R. A. Marryott, 1991. Optimal Groundwater Management: 1. Simulated Annealing, *Water Resources Research,* 27(10), 2493–2508.

Fetter, C. W., 1993. *Contaminant Hydrogeology,* Macmillan, New York.

Freeze, R. A., 1975. A Stochastic-Conceptual Analysis of One-Dimensional Groundwater Flow in Nonhomogeneous Media, *Water Resources Research,* 11(5), 725–741.

Freeze, R. A., and J. A. Cherry, 1979. *Groundwater,* Prentice Hall, Upper Saddle River, NJ.

Freeze, R. A., J. Massmann, L. Smith, T. Sperling, and B. James, 1990. Hydrogeological Decision Analysis: 1. A Framework, *Ground Water,* 28(5), 738–766.

Freeze, R. A., J. Massmann, L. Smith, T. Sperling, and B. James, 1992. Hydrogeological Decision Analysis: 4. The Concept of Data Worth and Its Use in the Development of Site Investigation Strategies, *Ground Water,* 30(4), 574–588.

Garfinkel, R. S., and G. L. Nemhauser, 1972. *Integer Programming,* Wiley, New York.

Gierke, J. S., N. J. Hutzler, and D. B. McKenzie, 1992. Vapor Transport in Unsaturated Soil Column: Implications for Vapor Extraction, *Water Resources Research,* 28(2), 323–335.

Gorelick, S. M., 1983. A Review of Distributed Parameter Groundwater Management Modeling Methods, *Water Resources Research,* 19(2), 305–319.

Gorelick, S. M., C. I. Voss, P. E. Gill, W. Murray, M. A. Saunders, and M. H. Wright, 1984. Aquifer Reclamation Design: The Use of Contaminant Transport Simulation Combined with Nonlinear Programming, *Water Resources Research,* 20(4), 415–427.

Haan, C. T., 1977. *Statistical Methods in Hydrology,* Iowa State University Press, Ames, IA.

Haggerty, R., and S. M. Govelick, 1994. Design of Multiple Contaminant Remediation: Sensitivity to Rate-Limited Mass Transfer, *Water Resources Research,* 30(2), 435–446.

Hoeksema, R. J., and P. K. Kitanidis, 1985. Comparison of Gaussian Conditional Mean and Kriging Estimation in the Geostatistical Solution of the Inverse Problem, *Water Resources Research,* 21(6), 825–836.

Hutzler, N. J., J. S. Gierke, and B. E. Murphy, 1990. Vaporizing VOCs, *Civil Engineering,* 60(4), 57–60.

Johnson, P. C., M. W. Kemblowski, and J. D. Colthart, 1990a. Quantitative Analysis for the Cleanup of Hydrocarbon-Contaminated Soils by In-Situ Soil Venting, *Ground Water,* 28(3), 413–429.

Johnson, P. C., C. C. Stanley, M. W. Kemblowski, D. L. Byers, and J. D. Colthart, 1990b. A Practical Approach to the Design, Operation, and Monitoring of In Situ Soil-Venting Systems, *Ground Water Monitoring Review,* 10(2), 159–178.

Kitanidis, P. K., and E. G. Vomvoris, 1983. A Geostatistical Approach to the Inverse Problem in Groundwater Modeling (Steady State) and One-Dimensional Simulation, *Water Resources Research,* 19(3), 677–690.

Kong, D., and T. C. Harmon, 1996. Using the Multiple Cell Balance Method to Solve the Problem of Two-Dimensional Groundwater Flow and Contaminant Transport with Nonequilibrium Sorption, *Journal of Contaminant Hydrology,* 23, 285–301.

Lall, U., and M. D. Santini, 1989. An Optimization Model for Unconfined Stratified Aquifer Systems, *Journal of Hydrology,* 111, 145–162.

Luenberger, D. G., 1984. *Linear and Nonlinear Programming,* 2nd ed., Addison-Wesley, Reading, MA.

Mackay, D., 1979. Finding Fugacity Feasible, *Environmental Science and Technology,* 13(10), 1218–1223.

Mackay D, and S. Paterson, 1982. Fugacity Revisited, *Environmental Science and Technology,* 16(12), 654A–660A.

Mackay, D., and S. Paterson, 1991. Evaluating the Multimedia Fate of Organic Chemicals: A Level III Fugacity Model, *Environmental Science and Technology,* 25(3), 427–436.

Mantoglou, A., and J. L. Wilson, 1982. The Turning Bands Method for Simulation of Random Fields Using Line Generation by a Spectral Method, *Water Resources Research,* 18(5), 1379–1394.

Massmann, J. W., 1989. Applying Groundwater Flow Models in Vapor Extraction System Design, *Journal of Environmental Engineering,* ASCE, 115(1), 129–149.

Massmann, J., and R. A. Freeze, 1987a. Groundwater Contamination from Waste Management Sites: The Interaction Between Risk-Based Engineering Design and Regulatory Policy, 1. Methodology, *Water Resources Research,* 23(2), 351–367.

Massmann, J., and R. A. Freeze, 1987b. Groundwater Contamination from Waste Management Sites—The Interaction Between Risk-Based Engineering Design and Regulatory Policy: 2. Results, *Water Resources Research,* 23(2), 368–380.

Massmann, J., R. A. Freeze, L. Smith, T. Sperling, and B. James, 1991. Hydrogeological Decision Analysis: 2. Applications to Groundwater Contamination, *Ground Water,* 29(4), 536–548.

McCarthy, J. M., and W. W.-G. Yeh, 1990. Optimal Pumping Test Design for Parameter Estimation and Prediction in Groundwater Hydrology, *Water Resources Research,* 26(4), 779–791.

McKinney, D. C., and M.-D. Lin, 1994. Genetic Algorithm Solution of Groundwater Management Models, *Water Resources Research,* 30(6), 1897–1906.

McKone, T. E., 1993. *CalTOX, A Multimedia Total-Exposure Model for Hazardous-Waste Sites,* UCRL-CR-111456, prepared for the State of California, Department of Toxic Substances Control, Lawrence Livermore National Laboratory, Livermore, CA.

More, J. J., and S. J. Wright, 1993. *Optimization Software Guide,* SIAM, Philadelphia, PA.

Murtagh, B. A., and M. A. Saunders, 1995. *MINOS 5.4 User's Guide,* Technical Report SOL 83-20R, Department of Operations Research, Stanford University, Stanford, CA.

Neuman, S. P., 1980. A Statistical Approach to the Inverse Problem of Aquifer Hydrology: 3. Improved Solution Method and Added Perspective, *Water Resources Research,* 16(2), 331–346.

Ng, C.-O., and C. C. Mei, 1996. Aggregate Diffusion Model Applied to Soil Vapor Extraction in Unidirectional and Radial Flows, *Water Resources Research,* 32(5), 1289–1297.

Papadimitriou, C. H., and K. Steiglitz, 1982. *Combinatorial Optimization: Algorithms and Complexity,* Prentice Hall, Upper Saddle River, NJ.

Pedersen, T. A., and J. T. Curtis, 1991. *Soil Vapor Extraction Technology,* Noyes Data Corporation, Park Ridge, NJ.

Pelmulder, S. D., 1994. A Framework for Regional Scale Modeling of Water Resources Management and Health Risk Problems with a Case Study of Exposure Sensitivity in Aquifer Parameters, Ph.D. dissertation, UCLA, Los Angeles.

Pelmulder, S. D., W. W.-G. Yeh, and W. E. Kastenberg, 1996. Regional Scale Framework for Modeling Water Resources and Health Risk Problems, *Water Resources Research,* 32(6), 1851–1861.

Pettyjohn, W. A. (ed.), 1987. *Protection of Public Water Supplies from Groundwater Contamination,* Noyes Data Corporation, Park Ridge, NJ.

Rathfelder, K., W. W.-G. Yeh, and D. Mackay, 1991. Mathematical Simulation of Soil Vapor Extraction Systems: Model Development and Numerical Examples, *Journal of Contaminant Hydrology,* 8, 263–297.

Rogers, L. L., and F. U. Dowla, 1994. Optimization of Groundwater Remediation Using Artificial Neural Networks with Parallel Solute Transport Modeling, *Water Resources Research,* 30(2), 457–481.

Rosenwald, G. W., and D. W. Green, 1974. A Method for Determining the Optimum Location of Wells in a Reservoir Using Mixed-Integer Programming, *Society of Petroleum Engineers Journal,* 14(1), 44–54.

Sacks, R. L., D. E. Dougherty, and J. F. Guarnaccia, 1994. The Design of Optimal Soil Vapor Extraction Remediation Systems Using Simulated Annealing, *Proceedings of the 1994 Groundwater Modeling Conference,* International Groundwater Modeling Center, Fort Collins, CO, pp. 343–350.

Sepehr, M., and Z. A. Samani, 1993. In Situ Soil Remediation Using Vapor Extraction Wells, Development and Testing of a Three-Dimensional Finite-Difference Model, *Ground Water,* 31(3), 425–436.

Sleep, B. E., and J. F. Sykes, 1989. Modeling the Transport of Volatile Organics in Variably Saturated Media, *Water Resources Research,* 25(1), 81–92.

Sleep, B. E., and J. F. Sykes, 1993a. Compositional Simulation of Groundwater Contamination by Organic Compounds: 1. Model Development and Verification, *Water Resources Research,* 29(6), 1697–1708.

Sleep, B. E., and J. F. Sykes, 1993b. Compositional Simulation of Groundwater Contamination by Organic Compounds: 2. Model Applications, *Water Resources Research,* 29(6), 1709–1718.

Stasko, S., and V. M. Fthenakis, 1993. MMSOILS, Version 2.2, *Risk Analysis,* 13(5), 575–579.

Sun, N.-Z., 1994a. *Inverse Problems in Groundwater Modeling,* Kluwer, Dordrecht, The Netherlands.

Sun, N.-Z., 1994b. *Mathematical Modeling of Groundwater Pollution,* Springer-Verlag, New York.

Sun, Y.-H., 1994. Mixed Nonlinear Optimization of Soil Vapor Extraction Systems, Ph.D. dissertation, UCLA, Los Angeles.

Sun, N.-Z., and W. W.-G. Yeh, 1983. A Proposed Upstream Weight Numerical Method for Simulating Pollutant Transport in Groundwater, *Water Resources Research,* 19(6), 1489–1500.

Sun, N.-Z., and W. W.-G. Yeh, 1985. Identification of Parameter Structure in Groundwater Inverse Problem, *Water Resources Research,* 21(6), 869–883.

Sun, N.-Z., and W. W.-G. Yeh, 1990a. Coupled Inverse Problems in Groundwater Modeling: 1. Sensitivity Analysis and Parameter Identification, *Water Resources Research,* 26(10), 2507–2525.

Sun, N.-Z., W. W.-G. Yeh, 1990b. Coupled Inverse Problems in Groundwater Modeling: 2. Identifiability and Experimental Design, *Water Resources Research,* 26(10), 2527–2540.

Sun, N.-Z., and W. W.-G. Yeh, 1992. A Stochastic Inverse Solution for Transient Groundwater Flow: Parameter Identification and Reliability Analysis, *Water Resources Research,* 28(12), 3269–3280.

Sun, Y.-H., and W. W.-G. Yeh, 1995. Mixed Integer Nonlinear Optimization of Soil Vapor Extraction Systems, *Proceedings of the 22nd Annual Conference of Water Resources Planning and Management Division, ASCE,* Cambridge, MA, pp. 883–886.

Sun, Y.-H., M. W. Davert, and W. W.-G. Yeh, 1996. Soil Vapor Extraction System Design by Combinatorial Optimization, *Water Resources Research,* 32(6), 1963–1873.

TASC (Technical Staff, the Analytic Sciences Corporation), 1974. In A. Gelb (ed.), *Applied Optimal Estimation,* MIT Press, Cambridge, MA.

Todd, D. K., 1980. *Groundwater Hydrology,* Wiley, New York.

Tong, A. Y. S., 1996. Stochastic Parameter Estimation, Reliability Analysis, and Experimental Design in Groundwater Modeling, Ph.D. dissertation, UCLA, Los Angeles.

Tong, A. T. S., and W. W.-G. Yeh, 1995. A Unified Approach for Stochastic Parameter Estimation, Experimental Design, and Reliability Analysis in Groundwater Modeling, *Proceedings of the 22nd Annual Conference of the Water Resources Planning and Management Division, ASCE,* Cambridge, MA, pp. 895–898.

U.S. EPA, 1989. *Evaluation of Ground-water Extraction Remedies,* Vol. 1, *Summary Report,* EPA/540/2-89/054, U.S. Environmental Protection Agency, Washington, DC.

Wang, C., B. Mortazavi, W.-K. Liang, N.-Z. Sun, and W. W.-G. Yeh, 1995. Model Development for Conjunctive Use Study of the San Jacinto Basin, California, *Water Resources Bulletin,* 31(2), 227–241.

Whelan, G. J., W. Buck, D. L. Strenge, J. G. Droppo, B. L. Hoopes, Jr., and R. J. Aiken, 1992. Overview of the Multimedia Environmental Pollutant Assessment (MEPAS), *Hazardous Waste and Hazardous Materials,* 9(2), 191–208.

Willis, R., and W. W.-G. Yeh, 1987. *Groundwater Systems Planning and Management,* Prentice Hall, Upper Saddle River, NJ.

Xiang, Y., J. F. Sykes, and N. R. Thomson, 1996. Optimization of Remedial Pumping Schemes for a Ground-water Site with Multiple Contaminants, *Ground Water,* 34(1), 2–11.

Yeh, W. W.-G., 1986. Review of Parameter Identification Procedures in Groundwater Hydrology: The Inverse Problem, *Water Resources Research,* 22(1), 95–108.

Yeh, W. W.-G., 1992. System Analysis in Ground-water Planning and Management, *Journal of Water Resources, Planning and Management,* ASCE, 118(3), 224–237.

Yeh, W. W.-G., and N.-Z. Sun, 1984. An Extended Identifiability in Aquifer Parameter Identification and Optimal Pumping Test Design, *Water Resources Research,* 20(12), 1837–1847.

Yeh, W. W.-G., and N.-Z. Sun, 1990. Variational Sensitivity Analysis, Data Requirements, and Parameter Identification in a Leaky Aquifer System, *Water Resources Research,* 26(9), 1827–1938.

Yeh, W. W.-G., and Y. S. Yoon, 1976. A Systematic Optimization Procedure for the Identification of Inhomogeneous Aquifer Parameters, *Advances in Groundwater Hydrology,* 72–82.

Yeh, W. W.-G., and Y. S. Yoon, 1981. Aquifer Parameter Identification with Optimum Dimension in Parameterization, *Water Resources Research,* 17(3), 664–672.

Yeh, W. W.-G., W.-K. Liang, and N.-Z. Sun, 1991. *Development of Mathematical Models for the San Jacinto Basin,* Department of Civil Engineering, UCLA, Los Angeles.

Yoon, Y.S., and W. W.-G. Yeh, 1976. Parameter Identification in an Inhomogeneous Medium with the Finite-Element Method, *Society of Petroleum Engineers Journal,* 217–226.

CHAPTER 4

Air Quality Management

J. HUGH ELLIS

4.A. INTRODUCTION

In this chapter three air quality management applications that involve optimization techniques are described. Included as well is an abbreviated taxonomy of applications, categorized according to certain of the mathematical programming attributes associated with the applications. Virtually all of these problems are essentially the same as other optimization-based applications in this book in that they fundamentally involve the cost-effective allocation of limited resources. For example, very little separates stream water quality management problems from air quality management problems, save for the increase in dimensionality (specifically, the spatial dimension) associated with air quality. But for all their similarities, the applications have some pronounced differences, especially in terms of the computational effort needed for their solution. These details and the context within which the models are developed and executed are described next. Included as well is a fairly extensive bibliography of other air pollution–related optimization applications. Notable among those citations are the surveys by Greenberg (1995; a recent paper in *Operations Research* entitled "Mathematical Programming Models for Environmental Quality Controls" and "Survey of Mathematical Programming Models in Air Pollution Management" by Cooper et al., to appear in the *European Journal of Operational Research.*

4.A.1. Policy Relevance

In air quality management as in other systems applications, the key to success lies in formulating problems that have the potential to be policy relevant. That is, the models should have the capability to answer those questions that policymakers deem most important. While policy relevance as a laudable goal is hard to dispute, the best path to achieving that goal is far from obvious as far as model building and analysis are concerned. It can be true that identification of the important questions can be very problematic, even with direct access to policymakers (or whomever is the intended audience or user of the analyses). Without close contact with policymakers, the likelihood of a systems analyst discovering the important questions is far from certain, which leads to the first in a series of modeling and analysis guidelines: Interact early, interact often and repeatedly, and continually pose the question: What is the possible policy relevance of this model and this analysis, stated in terms that policymakers, not just systems analysts, understand. That is a difficult task, but it is central to conducting influential systems analyses. Next is the second guideline that has critical significance in applied systems: Real problems are multiobjective. All decisions involve trade-offs, and explicit consideration of those trade-offs greatly enhances the policy relevance of systems modeling. This issue is explored in somewhat more depth in Section 4.A.2. The third guideline speaks to the challenges posed by the presence of uncertainty in model building, analysis, and communication. The concerns here lay not so much in mathematics or technical sophistication, but rather, in communication. It is very easy to lose one's audience and hence any chance for policy relevance with elegant stochastic analyses that seem to make perfect technical sense—at least to the analyst. With that in mind, some stochastic programming techniques hold much more promise than others in terms of their communicability, about which more is said below.

4.A.2. Mathematical Programming Taxonomy

In this rough taxonomy, we begin with the philosophy that model follows problem. Put another way, the selection of type of model and associated analysis is driven entirely by the characteristics of the problem at hand, unless that model simply cannot be solved, whereupon the central question becomes: Can a related model requiring an altered problem statement still hold promise for policy relevance? There are five categories in the taxonomy described herein, each of which poses more or less difficulty in terms of model execution and communication. Those categories are:

1. Linear versus nonlinear
2. Convex versus nonconvex
3. Single versus multiobjective
4. Deterministic versus stochastic
5. Static versus dynamic

The state of off-the-shelf optimization solvers now is such that there simply is no need to replace a nonlinear problem or model with a linearized approximation, except perhaps for linear programs of truly prodigious size, and even that exception is arguable. A qualifier is needed here, however, involving the nature of the approximation. A convex cost function, for example, may be more realistically modeled as a piecewise linear function if the vertices of the piecewise linearization correspond, say, to discrete control technologies. This situation does not involve approximation in any sense. [Recall that the maximum number of piecewise linear decision variables taking on values not at vertices is limited by the number of binding constraints in the model. That maximum number, depending on the problem, may or may not produce solutions in which most (continuous) decision variables take on vertex values, corresponding, say, to control technology sizes that are actually available.]

Nonconvexity, as distinct from nonlinearity, is always a significant consideration, recalling that many nonlinear optimization problems are indeed convex, and thus are guaranteed to yield global (though not necessarily unique) optima and as well are starting-point independent (to name a few properties of convex programming problems). The presence of nonconvexity most assuredly does not signal the need to formulate and solve a related but different convex programming problem. It does, however, require that additional steps be taken to ensure that good solutions at least are identified, for in most cases this is the best that one can do. The suite of traditional (e.g., reduced gradient) and newer (heuristic) solvers readily available today render the chances of solving even the most recalcitrant nonconvex problems very good.

The need for multiobjective approaches has been cited above and is reaffirmed here with the assertion that every public policy–related management problem, including all air quality management problems, is multiobjective. The nature of the multiple objectives varies greatly from application to application, of course, but typically includes objectives in the categories of cost minimization; achieving environmental standards of many kinds, including standards explicitly linked to human health; a wide variety of stochastic considerations, for example stochasticity manifested in the context of robustness (solutions that perform acceptably well if certain input assumptions prove incorrect); objectives directly or indirectly representing measures of equity and fairness (e.g., allocating emission control costs among polluters); and objectives that reflect aesthetic considerations, to name but a few. The multiobjective approach has with it attributes that realistically speak the language of decision making; that is, all decisions are trade-offs, and the very best information that a systems analyst can provide is information cast explicitly in the form of trade-offs, whereupon decision makers can then apply their own rules, preferences, and value judgments in converging on modeling solutions to support policy.

It is difficult to image a realistic air quality management problem that does not contain important stochastic elements. Mesoscale and long-range pollutant transport and transformation processes are highly stochastic, as are ecological responses and source emission characteristics. Often, the real issue in contemplating a stochastic approach when building models and designing analyses lay in assessing how uncertain information will be presented to decision makers and whether there is policy-relevant *value added* in adopting a stochastic approach. Now it is clear that these are easy questions to pose; their resolution at the onset of an analysis is problematic, especially with regard to the notion of assessing added value. No one approach works for all problems, but a reasonable philosophy to adopt is to attempt to build the (stochastically) simplest model first and then continually subject it to guideline number one described above: Are the analyses possibly policy relevant? Communicating stochastic information is a complex subject unto itself, and if there is a guideline to follow here, it again is: Begin simple, refrain from focusing extensively on those stochastically feasible outcomes that promise little additional insight in terms of policy relevance (admittedly easier said than done), and communicate in terms that decision makers are comfortable with (and to know what those terms are requires adherence to guideline number one again—interact early and often).

Most of the history of environmental systems analysis involves static analyses, not because management problems are static, but for the critically important reason that static problems are typically computationally tractable, at least much more so than their dynamic counterparts. Here is an issue where tractability plays a dominant role, for the transition from static to dynamic models and analyses represents a quantum jump in complexity, analyst effort, model solution, and often, input data requirements. The gradations of dynamic analysis vary greatly from simpler repeated static analysis to explicit modeling efforts involving, say, closed-loop feedback optimal control. But in addition to issues involving optimal control, dynamics also play a critically important role when attempting to model the financial aspects of pollution management realistically and the effects of those aspects on optimal control strategies that may change with time.

4.B. APPLICATIONS

Presented below are three optimization-based air quality management applications. They represent a fairly broad range of mathematical programming techniques, from straightforward linear programs to highly nonlinear, nonconvex programming problems requiring specialized heuristic solvers. The first application involves acid rain control and represents a conventional resource allocation problem. The presentation below derives primarily from Ellis and Bowman (1994). The second application again involves acid rain control but addresses the different problem of estimating the possible ecological consequences of adopting a sulfur dioxide emissions trading program as provided for in the 1990 Clean Air Act Amendments (CAAA; U.S. Congress, 1990). This work was conducted for the Maryland Department of Natural Resources and has not appeared in the published literature. Both of these applications involve multiobjective linear programs and also include pre- and postprocessing routines to automatize operation of the integrated modeling system. The third application concerns air quality monitoring network design and is drawn from Trujillo-Ventura and Ellis (1992, 1991). The associated optimization model is very nonlinear, nonconvex, and multiobjective. The optimization model itself is part of a much larger model that also includes stochastic simulation, optimal spatial interpolation, and adaptive quadrature.

4.B.1. Acid Rain Control

Models designed to identify cost-effective strategies for acid rain control typically include three components: sulfur dioxide (SO_2) and nitrogen oxides (NO_x) emission sources, receptors, and transport/transformation estimates that link source emissions to receptor deposition. Pollutants therefore are emitted from sources, transported, and depending on the pollutant, chemically transformed over considerable distances (Eliassen, 1980; U.S. EPA, 1987) then deposited on terrestrial and aquatic ecosystems, thereby possibly produce an acidic stress. The role of optimization models in this context is to ascertain how that stress can be cost-effectively and equitably reduced to acceptable levels through the imposition of emission controls. Acceptability is defined in this context by the satisfaction of maximum allowable deposition limits—otherwise known as critical loads, defined as the maximum deposition of acidifying compounds that will not cause chemical changes leading to long-term harmful effects on ecosystem structure and function (Nilsson and Grennfelt, 1988).

The critical loads described for this application were generated with the steady-state PROFILE model for 73 sites in Maryland (Sverdrup et al., 1992). Regional MAGIC (Cosby et al., 1985a,b; Hornberger et al., 1986) was used to examine stream chemistry responses to deposition scenarios on a broader spatial scale.

The specific concerns that formed the motivation for this study—that is, our original view of policy relevance—was described by the following questions:

- Can a Maryland control program achieve Maryland critical loads?
- Can a regional approach to emissions reductions achieve Maryland critical loads?
- Given a hypothetical allocation of emission reductions in Maryland (e.g., 100,000 tons SO_2 per year), how can that aggregate emission reduction be allocated to Maryland sources to best achieve Maryland critical loads?
- What control program options present themselves if critical loads are viewed as targets (that can be exceeded) as opposed to inviolate upper limits on deposition? Permitting small critical load exceedences might lead to large savings in pollutant reductions.
- If mitigation at a sensitive site is deemed feasible, might we then permit exceedences of that site's critical load that otherwise would not be tolerated without mitigation efforts? Implicit here is a trade-off between the

costs and benefits of reducing deposition rates through point-source pollutant reductions, versus mitigation efforts applied directly to sensitive receptors (e.g., stream liming).

The optimization models described below were designed to minimize aggregate SO_2 emissions reduction while ensuring that resultant deposition rates do not exceed critical loads at selected sensitive receptor locations. From this basic framework, there emerged variants that reflect different policy objectives. These models are drawn from previous efforts described in Ellis et al. (1985a,b) and Ellis (1988a,b, 1990).

Minimizing Emission Reductions. This model minimized emission reductions such that prescribed upper bounds on net deposition are satisfied. Definitions used in the model follow:

- E_k is the SO_2 emission rate for large controllable point source k. The model has 242 such sources across the United States and Canada.
- R_k represents SO_2 removal level (i.e., the decision variables) with source-specific upper limits R_k^u (e.g., $R_k^u = 0.95$); these are the fractional removal levels that are determined in the optimization procedures.
- The t_{jk} are the transfer coefficients relating emission at source k to deposition at receptor j and are obtained from the Ontario Ministry of the Environment long-range air pollution transport simulation model (Venkatram et al., 1982). The transfer coefficients have units of kg(wet S) deposited per hectare per year per kilotonne of SO_2 emitted.
- B_j is the background deposition rate at receptor j (i.e., deposition not associated with anthropogenic emissions as well as deposition from sources other than those categorized as large in the model).
- C_j is the maximum allowable deposition rate at receptor j (i.e., the critical load).

With these definitions, the net emission from pollutant source k after emission reduction is

$$E_k(1 - R_k) \tag{4.1}$$

The deposition at receptor j resulting from this emission is

$$E_k(1 - R_k)t_{jk} \tag{4.2}$$

The pollutant transport and transformation system is modeled as linear (which is a reasonable assumption, e.g., see generally NAPAP, 1991b, Chap. 3) so that deposition contributions from all sources are simply additive. The deposition at receptor j from all controllable sources j ($j = 1,...,n$) is therefore

$$\sum_{k=1}^{n} E_k(1 - R_k)t_{jk} \tag{4.3}$$

For each receptor j, this deposition rate must be less than the critical load C_j, that is,

$$\sum_{k=1}^{n} E_k(1 - R_k)t_{jk} + B_j \leq C_j \tag{4.4}$$

producing a system of linear inequalities with one constraint per receptor.

The model's objective function serves as a surrogate for cost minimization:

$$\text{minimize} \sum_{k=1}^{n} E_k R_k \tag{4.5}$$

It so happens that for this problem, the deposition constraints cannot be satisfied even with all SO_2 removal levels (R_k) set to their upper limits (R_k^u). In other words, relative to existing deposition rates, the critical loads (C_j) are so small as to be impossible to achieve even with every controllable source removing as much SO_2 as is technologically feasible (via flue gas desulfurization). Feasibility can be obtained (in the model at least) by sequentially removing receptors from the model or, equivalently, arbitrarily increasing maximum allowable deposition rates at those receptor locations where infeasibility occurs (but critical loads are no longer satisfied at those receptors).

Minimizing Average Exceedence. For the next model and its variants, critical loads were viewed as targets rather than inviolate maximum deposition limits. Deposition exceedences are now allowed to occur and in so doing, we can then trade off aggregate measures of exceedences against emissions reduced. The first model minimizes average critical load exceedence subject to a fixed aggregate emission reduction T (e.g., 10 million tons SO_2 per year in the United States). If all violations are considered equal in impact, the model is written

$$\text{Minimize} \sum_{j=1}^{m} V_j \tag{4.6}$$

subject to

$$\sum_{k=1}^{n} E_k(1 - R_k)t_{jk} + B_j + U_j - V_j = C_j \qquad (4.7)$$

$$\sum_{k=1}^{n} E_k R_k = T \qquad (4.8)$$

$$R_k \leq R_k^u \qquad (4.9)$$

For any receptor j, either U_j is positive and V_j equals zero, signifying a resulting deposition rate less than critical load C_j; U_j is zero and V_j is positive, signifying a resulting deposition rate greater than critical load C_j; or both U_j and V_j equal zero, signifying a resulting deposition rate equal to critical load C_j. By inspection, we see that without the constraint that fixes emission reduction [equation (4.8)] the model would set all removal levels R_k to their respective upper bounds, to make the sum of the V_j as close to zero as possible.

Minimizing Maximum Exceedence. This model minimizes the maximum (minmax) exceedence, C^M, and is written

$$\text{Minimize } C^M \qquad (4.10)$$

subject to

$$\sum_{k=1}^{n} E_k(1 - R_k)t_{jk} + B_j + U_j - V_j = C_j \qquad (4.11)$$

$$C^M - V_j \geq 0 \qquad (4.12)$$

$$\sum_{k=1}^{n} E_k R_k = T \qquad (4.13)$$

$$R_k \leq R_k^u \qquad (4.14)$$

Constraints (4.11), (4.13), and (4.14) read as before. Constraints (4.12) serve the purpose of defining the maximum violation (C^M) from among all the possible violations (V_j). When compared to models that minimize average exceedence, minmax solutions generally yield a smaller maximum exceedence but a slightly higher average exceedence, which is the trade-off one would expect.

Model Structure. There exist six categories of SO_2 sources in the models: 184 large out-of-state power plants, 40 out-of-state area (small) power plants, 47 large out-of-state nonpower plants, 50 out-of-state area (small) nonpower plants, 11 large Maryland point sources, and 238 area (grid) Maryland sources. In this inventory "large"' refers to sources with annual SO_2 emissions above 19 kilotonnes; power plants are electric utilities; area or small plants are aggregated into idealized emission-weighted centroids in each state and province; nonpower plants generally are nonferrous smelters and other assorted nonutility sources. The small Maryland area sources refer to a 23 × 41 grid, each grid square 10 km on a side.

Needed as well in the model are the locations and critical loads for the receptors in the model, which are shown in Figure 4.1. Deposition rates at these receptors are linked to source emissions through the Ontario Ministry of the Environment (MOE) long-range air pollution transport simulation model (Venkatram et al., 1982).

With a complete model specified, three steps are executed in an analysis: (1) generate the input for the linear programming solver, (2) solve the linear program using MINOS (Murtagh and Saunders, 1987, and (3) postprocess the linear program output. Steps 1 and 3 are performed with original Fortran routines. MINOS is a commercial Fortran-based general linear and nonlinear optimization solver.

Illustrative Analyses: Sulfur Dioxide Reduction of 10 Million Tons per Year. Described below are some of the results obtained from the modeling effort, intended here to illustrate how the models were used and what was learned from their use. Foremost among the client's interests at the time was whether control efforts would result in satisfaction of Maryland critical loads, and if not, what additional control measures could achieve Maryland critical load satisfaction. The first analysis focused on the aggregate emission level (10 million tons) that most considered would probably be included in the (then not yet released) Clean Air Act Amendments.

A 10 million ton per year emission reduction, optimized for Maryland, was shown to be insufficient to meet critical loads at all 73 modeled receptors using the MOE transfer coefficients. The results show critical load exceed-

Figure 4.1 Receptor map of 73 critical load sites in Maryland.

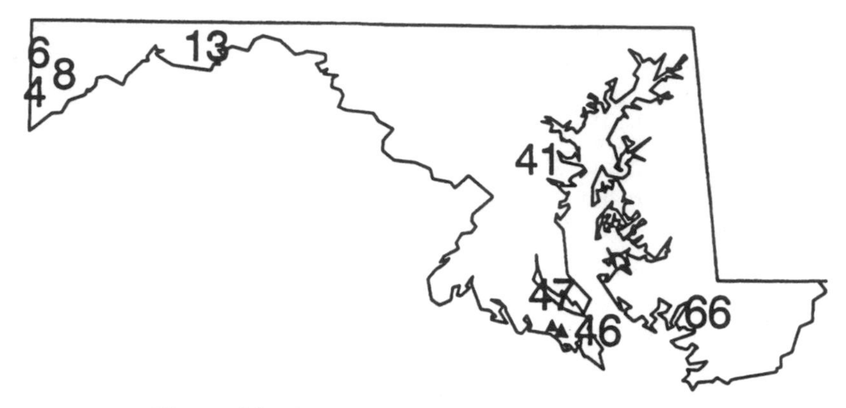

Figure 4.2 Locations of critical load exceedences.

ences at three locations: receptors 4, 41, and 47 shown in Figure 4.2. The numerical results are given in Table 4.1. The emissions reductions that yield these deposition results are shown in Table 4.2. Original emissions refer to the sum of all large and area source annual emissions for both power and nonpower plants in units of kilotonnes SO_2. Emission-weighted reductions are given as percentages. Values with a superscript asterisk denote maximum permissible reductions from the (large) controllable sources in the model. The aggregate annual 10 million ton emission reduction is comprised of non-Maryland power plant reductions totaling 9.076 million tons of SO_2, non-Maryland nonpower plants removing 704,000 tons, and Maryland sources removing 219,430 tons (the maximum attainable for Maryland's 11 large sources).

These results remain unchanged whether the average or maximum critical load exceedence is minimized. There exists little room to move in terms of adjusting removal levels to preferentially reduce either the average critical load exceedence or the maximum exceedence in Maryland. In this case, what is beneficial for reducing average exceedence is also beneficial for reducing the maximum exceedence.

Another way to point up these differences in deposition contributions is brought into focus by observing that complete elimination of all Maryland

TABLE 4.1 Critical Load Exceedences for an Optimized 10 Million Tons per Year SO_2 Reduction[a]

Receptor	Critical Load	Exceedence	% Exceedence
4	4.4	0.2	4.5
41	3.7	0.3	8.1
47	2.5	1.1	44

[a]The deposition units are kg(wet S/ha·yr).

TABLE 4.2 Statewide Emission Reductions for an Optimized 10 Million Tons per Year SO_2 Reduction

State	Original Emission[a]	Emission-Weighted Reduction[b] (%)
Alabama	707	30
Arkansas	92	0
Colorado	122.2	0
Connecticut	65.3	0
Delaware	99	23*
District of Columbia	13.4	0
Florida	993.2	0
Georgia	752.3	10
Illinois	1336.1	58*
Indiana	1818.7	66*
Iowa	299.3	22
Kansas	196.6	0
Kentucky	994.6	76*
Louisiana	276.2	0
Maine	86	0
Maryland	243.8	90*
Massachussetts	312.5	0
Michigan	868.4	47
Minnesota	238.8	0
Mississippi	281.5	0
Missouri	1179.6	62
Montana	153.1	0
Nebraska	68	0
New Hampshire	84.1	0
New Jersey	253.3	21*
New York	864	5
North Carolina	548.4	57*
North Dakota	126.1	0
Ohio	2508.2	76*
Pennsylvania	1812.4	76*
Rhode Island	13.8	0
South Carolina	301.7	0
South Dakota	35.5	0
Tennessee	977.5	74
Texas	1182	0
Vermont	6.1	0
Virginia	370.2	35*
West Virginia	989.4	77*
Wisconsin	544.1	43
Wyoming	168.7	0

[a]The emission units are kilotonnes SO_2 per year.

[b]The asterisks denote states whose large controllable sources are assigned their maximum respective removal levels.

SO_2 emissions still would not result in satisfaction of all 73 critical loads. Therefore, we have a reaffirmation of what is generally expected (e.g., see Figure 4.3, which shows the relative proportion of deposition in Maryland originating at sources outside Maryland) given data currently at our disposal, Maryland cannot unilaterally impose a control strategy that will meet its environmental objectives as represented by attainment of critical loads at 73 sensitive receptor locations.

The conclusion then was that a 10 million ton per year SO_2 emission reduction was not sufficient to achieve all Maryland critical loads, but came close enough to argue that it very well might. Relatively small decreases in transfer coefficient values, and/or increasing the upper SO_2 removal limit beyond 90% at large point sources in the Ohio River Valley, for example, would create a model wherein all 73 critical loads in Maryland would be satisfied.

The notion of a strategy optimized exclusively for Maryland is obviously fiction, but it so happens that in the aggregate, Maryland-targeted strategies look very much like strategies targeted for many other sensitive receptor configurations. The explanation for this behavior is straightforward. Removing 10 million tons of SO_2 per year in the United States requires large reductions where most of the emissions originate: Ohio, Illinois, Indiana, Pennsylvania, West Virginia, and Missouri, no matter where critical loads are assigned. That is, a control strategy targeted to meet critical loads in Maryland will look very much like a strategy targeted to meet critical loads in New York.

Relaxation Analyses. Previous results show that attainment of critical loads is highly site specific. To see in more detail where the hot-spot receptors are, we successively remove receptors at which exceedences occur (these analyses serve to identify a core set of receptors and associated sources that dominate

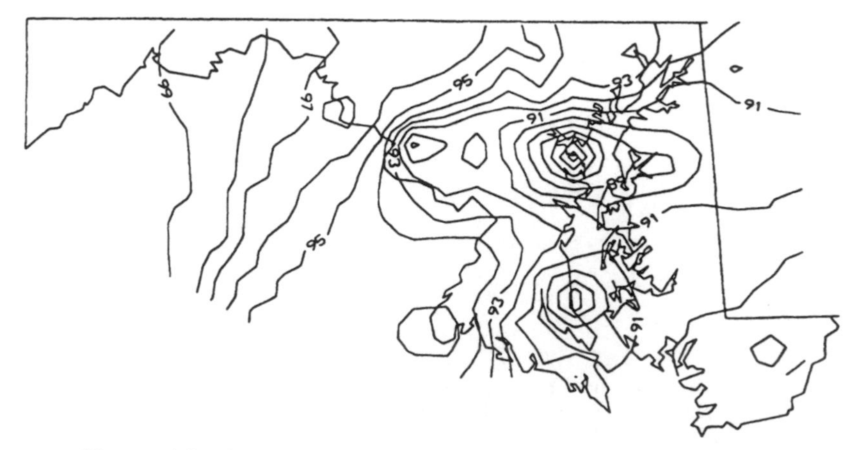

Figure 4.3 Deposition in Maryland from non-Maryland sources.

the emissions allocation process). In these analyses, this core group is studied in somewhat greater detail by performing a succession of runs in which receptors are removed from the model and the model rerun to estimate the incremental effect of having one less critical load to achieve. These incremental effects can vary considerably and are strongly a function of which receptors are relaxed (and where they are located). All of these runs use as a framework the basic minimize average population-weighted exceedence model, with aggregate emission reduction held to 10 million tons SO_2 per year. The analyses yielded the results shown in Table 4.3. Note that number of exceedences includes exceedences at receptors already relaxed (when they are relaxed, their critical load no longer need be satisfied in the model). Average exceedence can change in unexpected ways because we calculate average exceedence based on the sum of all exceedences divided by the number of exceedences that occur—instead of dividing by the total number of receptors. Finally, once having relaxed eight receptors, no more exceedences occur for an aggregate emission reduction of 10 million tons of SO_2 per year.

In general, then, these results reaffirm the special nature of receptors 4, 41, and 47; they stand out in terms of the relative difficulty in meeting their critical loads. These three receptors might be good candidates for stream liming in that liming could obviate some of the need for large deposition reductions at these sites, reductions that we have shown to be relatively expensive compared to other critical load locations in Maryland. For that matter, this argument could be extended to include all eight receptors listed above (which are shown on the map presented in Figure 4.2).

Ecological Loss Functions. One interpretation of a critical load is that it represents a point on an environmental loss function—more specifically, the point where the loss function becomes zero. In that sense, therefore, the use of critical loads can be construed as a surrogate for loss functions. Such interpretations and associated analyses have usefulness in their own right, but

TABLE 4.3 Deposition Results for the Constraint Relaxation Analyses

Run	Relax	Number of Exceedences	Average	Maximum	Relax Next
Origin	—	3	0.5	1.1	4
R1	4	3	0.5	1.1	41
R2	4,41	3	0.6	1.1	47
R3	4,41,47	4	1.1	1.6	46
R4	4,41,47,46	5	0.9	1.6	66
R5	4,41,47,46,66	6	0.8	1.7	13
R6	4,41,47,46,66,13	7	1.2	1.9	8
R7	4,41,47,46,66,13,8	8	0.9	1.9	6
R8	4,41,47,46,66,13,8,6	8	0.9	2.2	—

inescapably, in the final instance they depend on our ability to attach relative losses or penalties to exceedences of critical loads. In most formulations, linear loss functions were implied; that is, we presume that a larger exceedence yields proportionately more loss than a smaller exceedence, but we know nothing more. Even if that presupposition is true, there remains the problem of aggregating exceedences. For example, and all else equal, is a control strategy that yields one large exceedence better or worse than another strategy that yields five small exceedences? In other words, how do we aggregate exceedences across space? With loss functions approximated by critical loads, these issues of aggregation can be addressed only in ad hoc procedures that are something less than completely satisfactory (e.g., minimizing average or maximum exceedence). The point here is that loss functions represent a natural and very satisfying means of aggregation.

4.B.2. Acid Rain Emissions Trading

One of the most interesting and contentious aspects of the 1990 Clean Air Act Amendments (CAAA; U.S. Congress, 1990) was the provision for a market-based emissions trading program. The general notion here is that both economic efficiency and equity might be served through relatively unconstrained market interaction. For example, sources faced with relatively high marginal costs of SO_2 control might find it advantageous to buy the right to pollute rather than to install pollution control equipment on site. Alternatively, sources that already have large installed pollution control capabilities, and hence relatively low marginal costs of control, might benefit from overcontrolling—beyond some nominal level specified in law.

One piece missing from this admittedly oversimplistic view of trading is that relating to the environmental consequences of trading. That is, trading begs the obvious question relating to the concern that in the quest to achieve economic efficiency, trading should result in potentially severe local degradations in environmental quality. An assessment of the environmental consequences of trading requires, in part, estimates of SO_2 emissions across eastern North America that are consistent with trading.

Described next is a simple model for assessing the potential environmental consequences of trading *without modeling the dynamics of a market-based emissions trading program.* Rather, our approach is to estimate bounds on the deposition changes that trading *could* produce. The analyses are presented below, beginning with a description of the mathematical models used to produce the above-mentioned bounds.

Optimization Models. The first formulation—labeled the *disaggregated model*—is

$$\text{Minimize} \sum_i M_i + \epsilon \sum_j \sum_k x_{jk} \tag{4.15}$$

subject to

$$\sum_j \left(P_j - \sum_k x_{jk} + \sum_k x_{kj} \right) t_{ji} \le M_i \qquad i = 1,...,m \qquad (4.16)$$

$$P_j - \sum_k x_{jk} + \sum_k x_{kj} \le P_j^u \qquad j = 1,...,n \qquad (4.17)$$

$$\sum_k x_{jk} \le P_j - P_j^l \qquad j = 1,...,n \qquad (4.18)$$

$$\sum_j \sum_k x_{jk} = T \qquad (4.19)$$

where, x_{jk} are the decision variables that represent the amount of SO_2 emission traded from j to k; P_j is the original (before trading) SO_2 emission level at source j; t_{ji} are the transfer coefficients relating emission at source j to wet sulfate deposition at receptor i; P_j^u and P_j^l are, respectively, the upper and lower emission limits for source j; T is a stipulated total amount of emissions to be traded; ϵ is a prespecified small number; and M_i represents wet sulfate deposition rate for receptor i.

For a given large point source j, the quantity

$$P_j - \sum_k x_{jk} + \sum_k x_{kj} \qquad (4.20)$$

represents the net emission rate after trading has taken place, that is, the original emission rate P_j minus emissions exported to all other sources k ($\Sigma_k x_{jk}$), plus emissions imported from all other sources k ($\Sigma_k x_{kj}$). This net emission, when multiplied by the source-receptor transfer coefficient t_{ji}, yields the resulting (wet sulfate) deposition at receptor i. That is, the deposition at receptor i resulting from the net (after trading) emission at source j is

$$\sum_j \left(P_j - \sum_k x_{jk} + \sum_k x_{kj} \right) t_{ji} \qquad (4.21)$$

When summed over all sources j, this incremental deposition contribution yields the total deposition at receptor i from all sources. In turn, this quantity must be less than or equal to M_i, which we attempt to minimize. Constraints (4.3) state that the original (before trading) emission level at source j, minus exports, plus imports must be less than or equal to a prespecified upper bound P_j^u. This upper bound could be the 1980 emission level for that source. The implication here is that emission rate at the onset of trading (P_j, in 1995) would generally be somewhat smaller than the baseline 1980 level.

Each source has a maximum tradable amount of emissions ($P_j - P_j^l$); thus total exports ($\Sigma_k x_{jk}$) must be less than or equal to this amount, as reflected in constraints (4.4). Constraint (4.5) fixes the total amount of traded emissions (exports and imports) across all sources to be T million tons of SO_2 per year. Finally, the objective function (4.1) acts to minimize average deposition rate over all receptors (which is equivalent to minimizing the sum of all depositions, $\Sigma_i M_i$) and also includes the term (i.e., a second objective)

$$\epsilon \sum_j \sum_k x_{jk} \tag{4.22}$$

which is simply a programming artifice which precludes simultaneous imports and exports from any pair of sources. Its operation is simple and innocuous. To see why, compare two cases, both of which involve a net export of 50,000 tons of SO_2 per year, from source *a* to *b*: in the first, source *a* exports 100,000 tons to source *b* and imports 50,000 tons from *b*; the net export is 50,000 tons. The value of the second objective (which is to be minimized) is $\epsilon \cdot$ 150,000. Now consider the second and more realistic case, in which source *a* exports 50,000 tons to *b* and imports nothing. The value of the second objective in this case is $\epsilon \cdot$ 50,000, which represents the same net trading result but is less than and hence preferable to the first case, and would be selected in the optimization procedures.

This model represents only one of two required models in the analyses. Its closely related counterpart acts to *maximize* the average deposition rate across the eastern United States. For a given traded amount *T*, therefore, the minimizing model shown above and its counterpart will yield an estimate of the deposition *differences* that trading could produce. It follows that these differences will vary in magnitude and location as functions of *T*. Furthermore, these differences can be treated in an absolute sense, as is the case above, or alternatively, we can express deposition rates relative to site-specific critical loads [Sverdrup et al. (1992) document the study that developed critical load estimates for Maryland; for more general critical loads background, see Nilsson and Grennfelt (1988), Cosby et al. (1985a,b), and Hornberger et al. (1986)]. In the latter case we would minimize (then maximize) critical load violations, where violations represent net deposition rates in excess of critical loads.

Aggregated Model. As noted earlier, the disaggregated model was not implemented, owing to its very large size and the fact that for these analyses, knowledge of source-specific trading transactions is neither necessary nor warranted. We did, however, implement a variant on the disaggregated formulation, called here the aggregated model. This simplified and much smaller version replaces every summation of the decision variables x_{jk} and x_{kj} with single decision variables. Total export from source *j* to all other sources is y_j. Total import from all other sources to source *j* is z_j. The mathematical model therefore becomes

$$\text{Minimize} \sum_i M_i + \epsilon \sum_j (y_j + z_j) \tag{4.23}$$

subject to

$$\sum_j (P_j - y_j + z_j)t_{ji} \leq M_i \qquad i = 1,...,m \tag{4.24}$$

$$P_j - y_j + z_j \leq P_j^u \qquad j = 1,...,n \tag{4.25}$$

$$z_j \leq P_j - P_j^l \qquad j = 1,...,n \tag{4.26}$$

$$\sum_j (y_j + z_j) = T \tag{4.27}$$

Again, the advantage here is a dramatic reduction in problem size. In the aggregated model, therefore, it is not explicitly known who trades with whom. We know only the net traded emission amount either exported out of or imported into a point-source or gridded region.

Implementation. The implementation of the aggregated model involves a process of several related steps, and recall that two related models (deposition minimized and deposition maximized) must be solved for each scenario analyzed:

1. Create the input data file for the linear programming solver. This input data file itself requires several input data files:
   a. Point- and area-source emission inventories (reflecting the emissions anticipated after the CAAA phase 1 reductions).
   b. Source-receptor transfer coefficient matrices (obtained from the Ontario Ministry of the Environment statistical model [Venkatram et al. (1982); this is one of several long-range transport models extant, evaluations of which can be found in ISDME (U.S. EPA, 1987); of the three performance groupings (A, B, C) used in the ISDME study, the MOE model placed in the A or best category].
   c. Critical load estimates for 73 sites in Maryland (Sverdrup et al., 1992); a map of these sites is shown in Figure 4.1.
   d. User-prescribed trading target (aggregate amount of traded emissions in kilotonnes of SO_2 per year).
2. Solve the deposition-minimizing model variant.
3. Solve the deposition-maximizing model variant.
4. For each of the two model outputs generated above, perform the first stage of output postprocessing. That is, extract decision-variable values and solution status information from the linear programming solver solution files.
5. Create two (companion) reports, processed in LATEX, that summarize the results of the analyses.

The linear programming solver used for this work is MINOS (Murtagh and Saunders, 1987). The input (MPS) files to MINOS are created using originally coded Fortran routines. The output file postprocessing routines that extract specific information from the MINOS output files and create LATEX reports are written in portable Fortran. A master coordinating program was developed (written in QuickBasic) that calls and executes all the programs noted above.

Analyses. There are two basic groups of analyses, the first of which uses the aggregated trading models and assesses the effects on changes in deposition as aggregate traded amount is varied (and recall that aggregate traded amount is user prescribed). The second group of analyses fixes the aggregate traded amount of SO_2 emissions at 1 million tons per year and places addition restrictions on allowable trades. Based on results from the first group of runs (discussed next), these restrictions targeted certain key states and precluded their participation in trading, thus forcing the trading models to find other allocations of emission reductions, traded amounts, and trading partners.

Initial Scenario. The initial scenario to be discussed uses the basic aggregated trading models (deposition minimizing and deposition maximizing) and prescribes a relatively modest traded emission amount of 100,000 tons of SO_2 per year. The pretrading emissions inventory corresponds to 1980 emissions reduced by amounts given in the CAAA phase 1 reductions (U.S. Congress, 1990, Sec. 404, pp. 2597–2601). The deposition units given below are kg(wet S) per hectare per year. All emission units are kilotonnes of SO_2 per year. We first discuss the model that minimizes deposition across 73 sites in Maryland. Selected deposition statistics for this scenario are shown in Table 4.4.

The emissions results show only Florida imports emissions in this scenario, while both Maryland and West Virginia export emissions in the amounts shown. Note that this type of trading movement is intuitively to be expected; that is, reducing deposition in Maryland necessitates reducing precursor emissions at source locations that contribute significantly to deposition in Maryland (West Virginia and Maryland itself). It therefore follows that the model seeks to place these reduced emissions where they would have least impact on deposition in Maryland, which in this case is Florida.

Proceeding now to the companion trading model that maximizes deposition in Maryland (again with a total traded emission amount of 100,000 tons of SO_2 per year), we show the associated deposition statistics in Table 4.5. Comparing with the previous results for the deposition minimizing model, the

TABLE 4.4 Deposition Statistics: Initial Minimizing Scenario

Statistic	Value	Location
Average deposition rate	9.49	—
Minimum deposition rate	7.06	Receptor 71
Maximum deposition rate	13.20	Receptor 7
Maximum CL violation	7.76	Receptor 4

TABLE 4.5 Deposition Statistics: Initial Maximizing Scenario

Statistic	Value	Location
Average deposition rate	9.89	—
Minimum deposition rate	7.22	Receptor 71
Maximum deposition rate	13.66	Receptor 7
Maximum CL violation	9.20	Receptor 4

average, minimum, and maximum rates shown above are larger (which must be the case) but not much larger, owing to the modest level of emissions trading prescribed for this scenario. We do, however, see a relatively larger difference in critical load violations (7.76 versus 9.20) between the deposition minimizing and maximizing results. The location of maximum violation is receptor 4 (in far western Maryland) and the behavior we see here reflects the large effect that emission changes in West Virginia have on deposition in that area.

The emission results for this case reflect behavior opposite to that described for the minimizing case. Increasing deposition in Maryland is best accomplished by increasing emissions in West Virginia, which is shown in import 42.26 kilotonnes of SO_2 per year. Maryland imports the rest (the difference between 50 and 42.26), and these emissions originate from emission reductions in Florida. Note that these results and those for the minimizing model are not symmetric—nor should they be. To understand this behavior, recall first that the maximum tradable amount for any point source is bounded: from above by the 1980 emission rate and from below by the emission rate associated with maximum technologically feasible SO_2 reduction (i.e., 90%). Here we see that it is possible in the maximizing case to import relatively more emissions into West Virginia (than Maryland) because West Virginia's phase 1 reductions are collectively much larger than those for Maryland. But that is only part of the story. These relatively larger phase 1 reductions in West Virginia would be exploited in the model only if it were desirable to do so in terms of increasing deposition in Maryland—and that is indeed the case. Put another way, the deposition maximizing model identifies a set of point-source emissions, different from the original post-phase 1 reductions, that both increases deposition in Maryland and stays within the trading limits prescribed for this scenario. We can do worse (i.e., greater deposition rates in Maryland) than the situation corresponding to the post-phase 1 emission levels with no trading.

Sensitivity to the Traded Amount. The minimizing and maximizing models were next run for different levels of total traded amount, beginning with 200,000 tons of SO_2 per year, and then increasing to 500,000, then 1,000,000, and thereafter up to 5,000,000 tons per year in increments of 500,000. For

TABLE 4.6 Deposition Changes with Traded Amount

	Minimize Deposition		Maximize Deposition	
Traded Amount	Average	Maximum	Average	Maximum
100,000	9.49	13.20	9.89	13.66
200,000	9.28	13.18	10.01	14.00
500,000	8.79	11.95	10.30	14.92
1,000,000	8.15	10.91	10.65	15.77
1,500,000	7.71	9.51	10.92	16.25
2,000,000	7.44	8.90	11.16	16.59
2,500,000	7.22	8.50	11.30	16.80
3,000,000	7.02	8.15	11.39	16.94
3,500,000	6.85	7.96	11.46	17.04
4,000,000	6.72	7.90	11.52	17.13
4,500,000	6.62	7.79	11.56	17.19
5,000,000	6.59	7.75	11.59	17.23

this suite of runs, the deposition results are summarized in Table 4.6 (and include the above-noted 100,000-ton results as well for reference).

We see, therefore, that as traded amount increases, the differences between minimum and maximum deposition metrics increases, which is the trade-off expected. The greatest differences in either average or maximum deposition occur, of course, when the traded amount is greatest—in this case, the relatively large value of 5,000,000 tons of SO_2 per year. These differences are 5 kg(wet S)/ha per year for the average deposition rate and 9.48 kg(wet S)/ha per year for the maximum deposition rate. For smaller and perhaps more likely traded amounts, say, 1,000,000 tons per year, deposition differences between the maximizing and minimizing scenarios are commensurately smaller: 2.5 kg(wet S)/ha per year for average deposition; 4.86 for maximum deposition rate.

The next discussion takes these same numerical results but now looks at deposition rates expressed relative to critical load values. Moreover, we focus on traded amounts of 1,000,000 tons of SO_2 per year and less. The entries shown in Table 4.7 represent the difference in critical load violations between

TABLE 4.7 Critical Load Violations: Sensitivity to Trading Level

	Receptor				
Traded Amount	4 (4.37)	41 (3.73)	47 (2.45)	66 (4.05)	67 (4.59)
100,000	1.44	0.44	0.23	0.22	0.27
200,000	1.84	0.69	0.54	0.57	0.68
500,000	3.05	1.24	2.06	1.20	1.26
1,000,000	5.37	2.13	2.54	1.70	1.85

the minimizing and maximizing models for selected receptors (where the largest violations occur). The numbers shown in parentheses under the receptor number headings are the critical loads for those respective receptors.

In terms of absolute values, the violations shown in Table 4.7 are generally small, but the question of their potential significance is decidedly difficult to answer. Issues of significance must necessarily involve other considerations, notably, the accuracy of long-range transport estimates and the establishment of a realistic context within which such violations can be interpreted. While the transport estimates that we used are reasonable and have been subject to scientific peer review, they nonetheless are uncertain, and must be, for the meteorologic processes that drive long-range transport are themselves highly variable. One can, for example, modify transfer coefficients only moderately and produce deposition changes commensurate with the number shown in Table 4.7. The related issue of context, as interpreted here, speaks to the problem of deciding when a violation is large enough to raise serious concern, transport uncertainties notwithstanding. But for that we need deposition-based ecological damage functions—and they do not yet exist. It seems reasonable to interpret a critical load as one (admittedly important) point on a loss or damage function, but how then does one interpret violations other than by assuming that greater violations probably cause more damage? Set against these rather harsh but realistic measures of assessment, calling a violation of 5 kg(wet S)/ha per year at receptor 4 large because it equals 110% of the critical load value is a statement more of environmental predisposition than scientific fact. So then, what *is* large? Without damage functions, the answer is elusive.

4.B.3. Monitoring Network Design

Objectives for air pollution monitoring networks have been reported frequently in the literature (see, e.g., Leavitt et al., 1963; Seinfeld, 1972; Darby et al., 1973; Ludwig et al., 1976; Liu et al., 1977; Ott, 1977; Munn, 1981), and generally fall into one of the following categories: objectives related directly to air pollution legislation (long-term land use planning and the declaration of alert and emergency situations); the evaluation of the exposure of population and other vulnerable receptors (e.g., fragile exosystems, historic and/or artistic valuable property); the control of emissions from singularly important sources (e.g., thermal or nuclear power plants); and the creation of databases containing air pollution data to permit the analysis of long-term trends in air pollution or for other research purposes (one of the most important of which is the validation of mathematical models for air pollutant diffusion, transport, and transformation). Generally subsumed within these objectives is the minimization of the cost of the network.

Optimal Spatial Interpolation. The first objective function in the model describes the network's ability to estimate seasonal or annual average air pollution concentrations. This requires, in turn, estimation of the spatially averaged error associated with interpolating concentrations (or depositions) from data collected at measurement stations. Real errors cannot be known, but an estimate (obtained through calibrated and verified simulation) is typically available. An efficient method for incorporating the spatial structure of pollutant concentrations into the model is through optimal spatial interpolation (Fedorov, 1989). Described next is a popular technique—simple punctual kriging—the description of which follows Matheron (1965).

Let $\mathbf{u}_i$ be denoted by $Q_i = Q(\mathbf{u}_i)$. Kriging serves to interpolate values for Q at user-specified locations $\mathbf{x}$, $\hat{Q}(\mathbf{x})$. The covariance structure of Q is assumed to be

$$E[Q(\mathbf{x} + \mathbf{h}) - Q(\mathbf{x}) = m(\mathbf{h}) \tag{4.28}$$

$$\mathrm{Var}[Q(\mathbf{x} + \mathbf{h}) - Q(\mathbf{x})] = 2\gamma(\mathbf{h}) \tag{4.29}$$

$\gamma(\mathbf{h})$ is the semivariogram function, related to the covariance between points $\mathbf{x}$ and $\mathbf{x} + \mathbf{h}$ by

$$\gamma(\mathbf{h}) = \mathrm{Var}[Q(\mathbf{x})] - \mathrm{Cov}[Q(\mathbf{x} + \mathbf{h}), Q(\mathbf{x})] \tag{4.30}$$

Interpolated values are (by assumption) written as a linear combination of known values:

$$\hat{Q}(\mathbf{x}) = \sum_{i=1}^{N} \lambda_i Q_i \tag{4.31}$$

The condition of unbiasedness is imposed:

$$E[\hat{Q}(\mathbf{x}) - Q(\mathbf{x})] = 0 \tag{4.32}$$

with the result that the sum of the kriging weights equals unity:

$$\sum_{i=1}^{N} \lambda_i = 1 \tag{4.33}$$

By definition, the expected value of $\hat{Q}(\mathbf{x}) - Q(\mathbf{x})$ is zero, so that the expected value of the mean-square interpolation error is

$$\hat{S}_Q^2(\mathbf{x}) = E[(\hat{Q}(\mathbf{x}) - Q(\mathbf{x}))^2] \tag{4.34}$$

With some algebraic manipulations, the mean-square error of the interpolation can be expressed as

$$\hat{S}_Q^2(\mathbf{x}) = -\sum_{i=1}^{N}\sum_{j=1}^{N} \lambda_i\lambda_j\gamma(\mathbf{u}_i - \mathbf{u}_j) + 2\sum_{i=1}^{N} \lambda_{i\gamma}(\mathbf{u}_i - \mathbf{x}) \tag{4.35}$$

The weights λ_i are given by the optimal solution of the following problem:

Given

$$\gamma(\mathbf{h}):\mathbb{R}^2 \rightarrow \mathbb{R} \tag{4.36}$$

$$N \in \mathbb{N} \qquad N > 0 \tag{4.37}$$

$$\mathbf{u}_i \in \mathbb{R}^2 \qquad i = 1,\ldots,N \tag{4.38}$$

find

$$\lambda_i \in \mathbb{R} \qquad i = 1,\ldots,N \tag{4.39}$$

so as to

$$\text{Minimize } \hat{S}_Q(\mathbf{x}) = -\sum_{i=1}^{N}\sum_{j=1}^{N} \lambda_i\lambda_j\gamma(\mathbf{u}_i - \mathbf{u}_j) + 2\sum_{i=1}^{N} \lambda_i\gamma(\mathbf{u}_i - \mathbf{x}) \tag{4.40}$$

subject to

$$\sum_{i=1}^{N} \lambda_i = 1 \tag{4.41}$$

This problem is readily solved using the Lagrange multiplier technique. If μ is the Lagrange multiplier for the unbiasedness constraint [equation (4.41)] in the previous expression, the problem is equivalent to solving the following set of linear kriging equations:

$$\sum_{j=1}^{N} \lambda_j\gamma(\mathbf{u}_i - \mathbf{u}_j) + \mu = \gamma(\mathbf{u}_i - \mathbf{x}) \qquad i = 1,\ldots,N \tag{4.42}$$

$$\sum_{i=1}^{N} \lambda_i = 1 \tag{4.43}$$

which in more compact form is

$$\mathbf{G}\Lambda = \Gamma \tag{4.44}$$

G denotes the kriging matrix and is given by

$$\mathbf{G} = \begin{pmatrix} \gamma_{11} & \gamma_{12} & \gamma_{13} & \cdots & \gamma_{1N} & 1 \\ \gamma_{21} & \gamma_{22} & \gamma_{23} & \cdots & \gamma_{2N} & 1 \\ \vdots & \vdots & \vdots & \ddots & \vdots & \vdots \\ \gamma_{N1} & \gamma_{N2} & \gamma_{N3} & \cdots & \gamma_{NN} & 1 \\ 1 & 1 & 1 & 1 & 1 & 0 \end{pmatrix} \tag{4.45}$$

with

$$\gamma_{ij} = \gamma(\mathbf{u}_i - \mathbf{u}_j) \tag{4.46}$$

The goal of the optimal interpolation (like any interpolation scheme) is to find weights to be applied to existing data points. Λ is the vector of those (unknown) weights and is given by

$$\Lambda = \begin{pmatrix} \lambda_1 \\ \lambda_2 \\ \vdots \\ \lambda_N \\ \mu \end{pmatrix} \tag{4.47}$$

Γ is the right-hand side of the system of equations and is denoted

$$\Gamma = \begin{pmatrix} \gamma_1 \\ \gamma_2 \\ \vdots \\ \gamma_N \\ 1 \end{pmatrix} \tag{4.48}$$

where

$$\gamma_i = \gamma(\mathbf{u}_i - \mathbf{x}) \tag{4.49}$$

The mean-square error of the interpolated estimates is calculated from

$$\hat{S}^2_Q(\mathbf{x}) = \Gamma\Lambda - \gamma(\mathbf{O}) \tag{4.50}$$

Estimation Error Minimization. $\hat{Q}_i(\mathbf{x};\mathbf{u})$ is the kriging estimate at point $\mathbf{x}$ of, in this case, the yearly averaged concentration of pollutant i ($i = 1$ to $\mathcal{Q}$), given that measurements at locations $\mathbf{u}$; $\hat{S}_{Q_i}(\mathbf{x};\mathbf{u})$ is an estimate of the error of the interpolation; and Ω denotes the spatial domain. The coverage objective function for pollutant i (conditioned on the stations located at $\mathbf{u}$) is

$$Z_{1,i}(\mathbf{u}) = \frac{\int_{\Omega} \hat{S}_{Q_i}(\mathbf{x};\mathbf{u})\, d\mathbf{x}}{\int_{\Omega} d\mathbf{x}} \tag{4.51}$$

We next modify the spatial coverage objective to place added emphasis on estimation errors associated with relatively large concentrations. Areas with relatively low concentration are attached relatively less weight. The spatial coverage objective function therefore becomes

$$Z_{1,i}(\mathbf{u}) = \frac{\int_{\Omega} \hat{Q}_i(\mathbf{x};\mathbf{u})\, \hat{S}_{Q_i}(\mathbf{x};\mathbf{u})\, d\mathbf{x}}{\int_{\Omega} \hat{Q}_i(\mathbf{x};\mathbf{u})\, d\mathbf{x}} \tag{4.52}$$

To use kriging to determine $\hat{Q}_i(\mathbf{x};\mathbf{u})$ and $\hat{S}_{Q_i}(\mathbf{x};\mathbf{u})$ requires that the values of the pollutant concentrations at the measurement locations be known. These can be obtained from either historical data or from a simulation model for air pollution transport and transformation.

Detecting Violations. The concentrations of any given pollutant is a random field whose distribution can be estimated through simulation. This distribution is characterized by its moments at every point in space, which in the case of an unshifted log-normal distribution (the standard model for concentration of air pollutants) are the median and geometric standard deviation. The probability that a standard is violated can then be calculated. When historical observations are available, the distribution of the field of concentration of pollutant can also be estimated with an interpolation scheme. The distribution so derived is then used to determine the aforementioned probabilities of violation. It is assumed that a violation at a specific location will always be detected if a measurement station exists there (measurement error is disregarded). An aggregate measure of the ability of the network to detect violations is then calculated by summing the violation probabilities for each of its stations. If $P_j(\mathbf{u}_i)$ is the probability that standard j (j ranges from 1 to $\mathfrak{S}$) is violated at point $\mathbf{u}_i$, the violation detection objective (to be maximized) for standard j, evaluated with N_j stations located at $\mathbf{u}$ can be written as

$$Z_{2,j}(\mathbf{u}) = \sum_{i=1}^{N_j} P_j(\mathbf{u}_i) \tag{4.53}$$

This objective does not represent an overall probability of violation detection because it does not account for double (or triple, etc.) counting when the same violation of a standard is detected by several stations. Clustering of the stations is therefore not prevented, as evidenced by the fact that the global optimum of the objective for standard j is achieved when all the stations are located at the (single) point where the maximum of $P_j(\mathbf{u})$ occurs.

Uniform Spatial Coverage. Faced with the task of manually designing a monitoring network, an intuitive approach is to distribute stations evenly in space. An objective function that accomplished even distribution is

$$Z_3(\mathbf{u}) = \sum_{i=1}^{N} \min_{j \neq i} |\mathbf{u}_i - \mathbf{u}_j|^{1/2} \qquad (4.54)$$

Maximizing this (deceptively simple-looking) objective acts to push all stations toward a regular triangular distribution, which has been reported (Dalenius et al., 1961; Olea, 1984) as the optimum sampling pattern for an isotropic plane stochastic process. Furthermore, a regular triangular distribution has been shown (McBratney et al., 1981) to be superior to the square grid if, as is the case here, the spatial covariance structure varies with direction. This objective function grows unboundedly as the stations are separated. Therefore, their locations must be constrained to lie in a given domain. The bounds of this domain introduce perturbations in the regular distribution of the stations. The amount of the perturbation at a given point is a function of the relation between the distance of this point to the boundary and the distance between stations.

From a mathematical programming perspective, an important characteristic of this objective function is that its gradient is discontinuous precisely at the optimum, due to the nondifferentiability of the minimization operator included in the summation. This discontinuity can present a serious problem for the optimization procedure used to solve the model, but as will be described later, other objectives also present discontinuities, and therefore an algorithm that can deal with them must be used in any event.

Computational Implementation. The complete statement of the model is:

Given

$$\Omega \in \mathbb{R}^2 \qquad \Omega \text{ compact} \qquad (4.55)$$

$$N \in \mathbb{N} \qquad N > 0 \qquad (4.56)$$

$$N_j \in \mathbb{N}, \quad N_j > 0, \quad N_j \leq N \qquad j = 1,\ldots,\mathcal{S} \tag{4.57}$$

$$N_i \in \mathbb{N}, \quad N_i > 0, \quad N_i \leq N \qquad i = 1,\ldots,\mathcal{Q} \tag{4.58}$$

$$\hat{Q}_i(\mathbf{x};\mathbf{u}):\Omega \times \Omega^{2N_i} \rightarrow \mathbb{R} \; i = 1,\ldots,\mathcal{Q} \tag{4.59}$$

$$\hat{S}_{Qi}(\mathbf{x};\mathbf{u}):\Omega \times \Omega^{2N_i} \rightarrow \mathbb{R} \; i = 1,\ldots,\mathcal{Q} \tag{4.60}$$

$$P_j(\mathbf{x}):\Omega \rightarrow [0,1] \; j = 1,\ldots,\mathcal{S} \tag{4.61}$$

find $\mathbf{u} \in \Omega^N$ so as to

$$\text{minimize } Z_{1,i}(\mathbf{u}) = \frac{\int_\Omega \hat{Q}_i(\mathbf{x};\mathbf{u})\hat{S}_{Qi}(\mathbf{x};\mathbf{u})\,\mathbf{dx}}{\int_\Omega \hat{Q}_i(\mathbf{x};\mathbf{u})\,\mathbf{dx}} \qquad i = 1,\ldots,\mathcal{Q} \tag{4.62}$$

$$\text{maximize } Z_{2,j}(\mathbf{u}) = \sum_{i=1}^{N} P_j(\mathbf{u}_i) \; j = 1,\ldots,\mathcal{S} \tag{4.63}$$

$$\text{maximize } Z_3(\mathbf{u}) = \sum_{i=1}^{N} \min_{j \neq i} |\mathbf{u}_i - \mathbf{u}_j|^{1/2} \tag{4.64}$$

The objectives [equations (4.62)–(4.64)] are aggregated into a weighted objective. For the fourth objective, the number of stations to be located is fixed in a given run and thus is equivalent to using the constrained method for the cost objective function (Cohon, 1978). The approximation of the noninferior set is therefore obtained using a hybrid (weighting plus constraint) method. The conceptual process used to evaluate the objective functions involves four steps:

1. *Computation of the necessary moments of the pollutant concentration distributions at the current location of the stations* (median and geometric standard deviation). This is performed using a model for the transport and transformation of air pollutants. The input to this step is the current location of the stations and the information needed by the air pollution model (source locations, emission rates, meteorology, etc.). The output consists of the first and second moments of the air quality indices at the locations of the stations.
2. *Interpolation of air quality indices.* This step is accomplished using kriging. Input consists of the current location of the stations and the expected values of the air quality indices at those locations. Output includes the expected values of the air quality indices and associated interpolation errors at all points specified in the domain.

3. *Integration of air quality indices.* This step is accomplished via numerical integration. The inputs to this step are the expected values of air quality indices and interpolation errors at selected points in the domain. The outputs are, for each air quality index, the numerators and denominators of the spatial coverage objective functions [equation (4.62)].
4. *Final assembly of objectives.* This step includes the evaluation of the standard violation control objectives and data validity objective, as well as weighting (i.e., aggregating) the different objectives.

The first step (modeling the transport and transformation of air pollution) will depend on the particular model used for each case. In the example presented, an enhanced Gaussian model was especially developed and calibrated for the region studied (BRAIN Ingenieros, 1988). This model is based on a continuous-plume Gaussian model, and includes corrections to account for plume rise, aerodynamic effects of stacks and other structures, and topography.

The second step involves the interpolation of the expected values of pollutant concentrations by kriging. As described earlier, the principal task in this process is the resolution of a set of linear equations for each pollutant. Notice, however, that the coefficient matrix of this set of equations depends only on the location of the stations and not on the location of the point where the interpolation is desired. Therefore, only one factorization of the matrix is needed each time the objectives are evaluated, regardless of the number of points at which the integration routine requires interpolation. Only the factorized matrix and the pivot information must be saved to perform the back-substitution step and hence the interpolation. Furthermore, the kriging matrix is symmetric and positive definite, making possible the use of efficient factorization techniques (see, e.g., Dongarra et al., 1979, or Golub and van Loan, 1983).

The integration step is the most delicate of the entire process. While integration rules for functions of one variable have been in use for many years, efficient quadrature rules for multivariate functions are still the subject of research, especially for nonproduct rules. The particular rule used here is based in two different Lyness–Jespersen integration formulas (Lynnes and Jespersen, 1975) defined over triangular domains, embedded in an adaptive quadrature routine using a global partitioning strategy.

The use of an adaptive quadrature rule brings with it the complication that the values obtained for the objectives are discontinuous numerical approximations to the exact objective functions. This is due to switching between coarser and finer discretizations of the integration domain as a function of the estimate of the integration error (a more detailed explanation of this phenomenon is provided by Lynnes, 1983). Clearly, this is inconvenient for the optimization procedure, but it cannot be avoided. The use of a nonadaptive integration routine would require a very fine mesh everywhere in the domain to capture abrupt variations of the functions in relatively small regions surrounding the stations. The requisite computing effort would be prohibitive.

To enforce the constraints, a combination of both barrier and penalty functions is used, specifically a generalized Heaviside barrier function plus a quadratic penalty function written as

$$G(\mathbf{x}) = k_1 H_\Omega(\mathbf{x}) + k_2 \, [d_\Omega(\mathbf{x})]^2 \tag{4.65}$$

where k_1 and k_2 are constants, $H_\Omega(\mathbf{x})$ is the generalized Heaviside function (1 if $\mathbf{x}$ belongs to Ω and 0 otherwise), and $d_\Omega(\mathbf{x})$ is the distance between point $\mathbf{x}$ and the domain Ω, defined as

$$d_\Omega(\mathbf{x}) \equiv \min_{\mathbf{u}\in\Omega} |\mathbf{u} - \mathbf{x}|^{1/2} \tag{4.66}$$

The barrier function acts as an infinite penalty on stations that violate the constraint on location. In practice, the constant k_1 is selected to be large enough (with respect to the weighted objective) that the constraint on the location of the stations is strictly enforced. The additional quadratic penalty is used to guide the optimization routine back to the domain should a station be located in an inadmissible region.

Optimization Method. The choice of optimization routine is a major issue here. Consider first the reduced gradient-based approaches such as methods of feasible directions or augmented and projected Lagrangian methods (see, e.g., Gill et al., 1981). The use of coded analytical gradients is possible for this model and generally would reduce the number of evaluations of the objective function needed to solve the problem. It would be necessary, however, to determine the derivatives of concentration with respect to position in the simulation model. Another, more important limitation of this approach involves the calculation of two sets of integrals [one for the derivatives of the numerator and another for those of the denominator of the fraction in equation (4.62)], all of them similar to those of the spatial coverage objective, for each direction of the gradient (twice the number of stations) and for each objective. The amount of work added by these additional integrations does not compensate the reduction in the number of evaluations of the objective function required by a routine that uses gradient information.

The use of finite difference schemes to estimate the gradients is impossible, given the discontinuities of the objective functions described above. These will greatly amplify the intrinsic instability of finite difference approximation to derivatives. Alternatively, there are procedures that do not require gradient information (direct search methods, methods of steepest descent, polytope methods). A series of tests were conducted to assess the ability of each of those routines to cope with the particular difficulties posed by the model proposed. As a consequence of this series of tests, Hooke and Jeeves' method with discrete steps (Hooke and Jeeves, 1961) was chosen as the optimization routine.

Application. The model was applied to design an air quality monitoring network for a region surrounding the city of Tarragona, Spain. This case is particularly interesting as a test for the model proposed: four pollutants are involved (sulfur dioxide, nitrogen oxides, particulate matter, and unsaturated hydrocarbons), and heavy industry coexists with densely populated areas and holiday resorts, resulting in a conflict between the objectives of the network.

The application of the model to this case example permits the designer of the network to generate trade-offs typified by that shown in Figure 4.4, between the coverage objectives (i.e., spatial coverage and violation detection), denoted by $\bar{Z}_1$, and the number of stations. The aggregate coverage objective is built by adding together (with equal weights) spatial coverage and violation detection objectives for all pollutants. It represents a measure of the overall

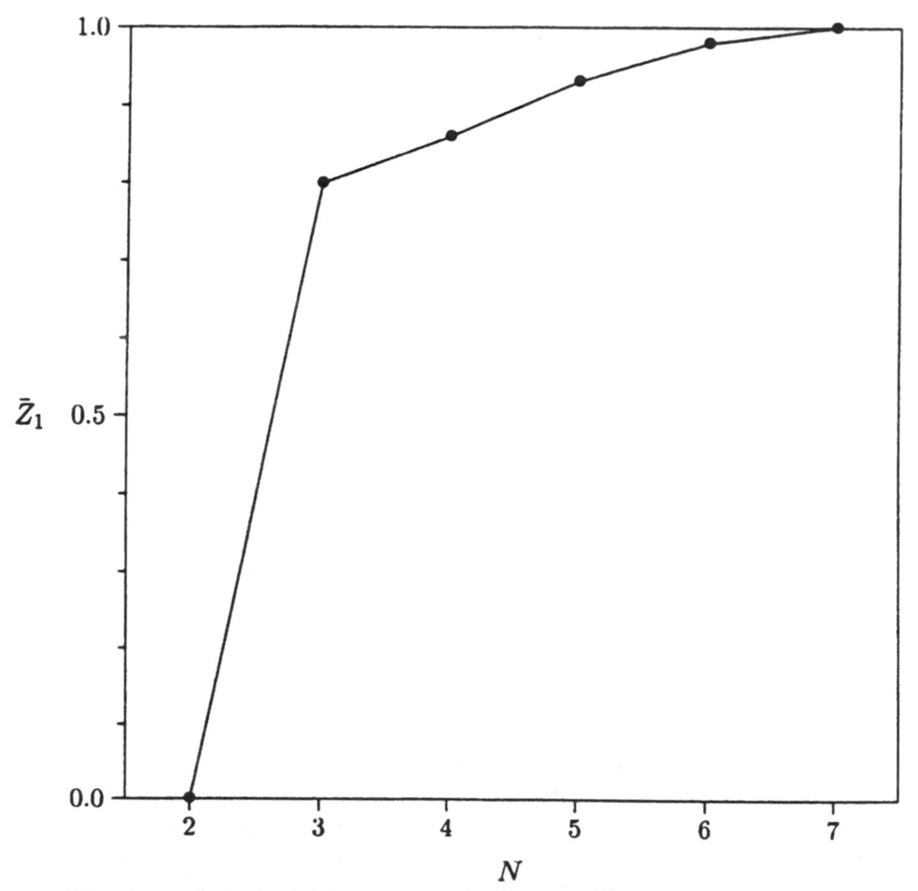

Figure 4.4 Trade-off between coverage and number of monitoring stations.

quality of the network under the assumption that the diffusion model is accurate. The problem is set up so as to maximize this measure, (i.e., to obtain a solution that has a good overall performance with the given number of stations). The coverage aggregate is normalized, so that its values range from 0 to 1.

This trade-off can be used, for instance, to determine the number of stations required in the network. A first conclusion is that little additional performance is gained by adding stations beyond six. Therefore, this should be an upper bound for the size of the network. Furthermore, increasing from two to three stations captures about 80% of the overall performance of the network. The marginal gain in performance in this section of the trade-off is very high; thus the addition of a third station is clearly advisable. From this point onward, the marginal increases in performance continually decrease to the point of insignificance when six stations comprise the network.

Another trade-off is shown in Figure 4.5, relating the aggregate coverage objectives (i.e., spatial coverage and violation detection), denoted by $\overline{Z}_1$, and the data validity objective, denoted as $\overline{Z}_2$, for a network of five stations. The problem is set up to maximize both objectives, that is, to obtain a solution

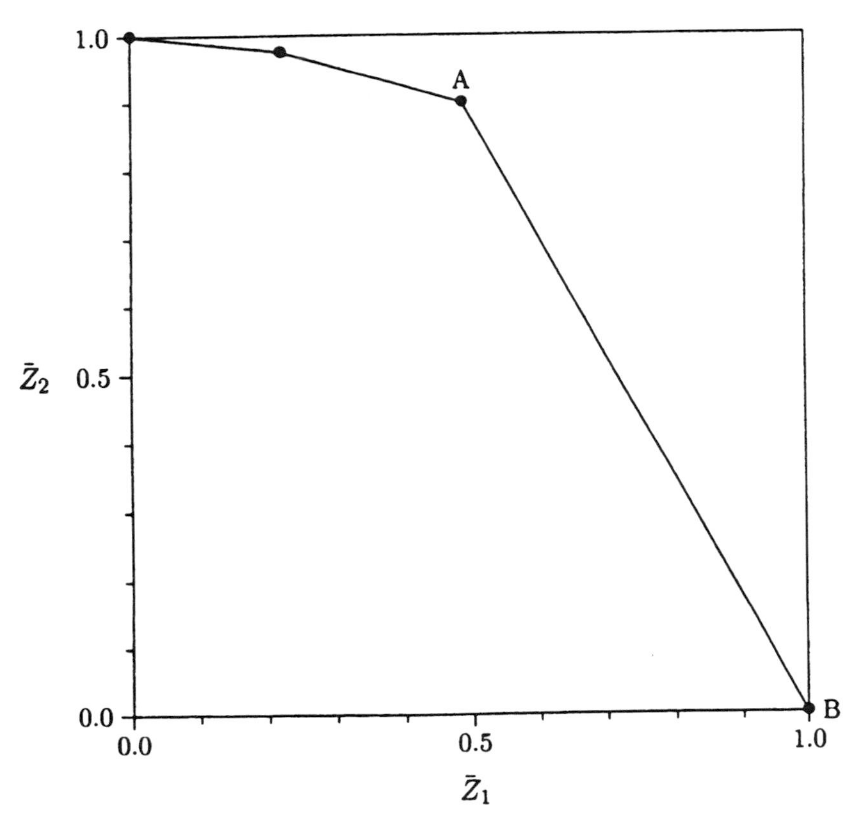

Figure 4.5 Trade-off between coverage and uniformity objectives.

that exhibits broad spatial coverage and the ability to detect violations, yet is also relatively less dependent on simulation model outputs.

To interpret this result, consider the solutions denoted A ($\bar{Z}_1 = 0.487$ and $\bar{Z}_2 = 0.899$) and B ($\bar{Z}_1 = 1.0$ and $\bar{Z}_2 = 0.0$). Moving from A to B places more reliance upon simulation model outputs, but significantly increases the spatial coverage of the network and its ability to detect violations. Alternatively, that portion of the trade-off to the left of A indicates large reductions in coverage and violation detection as reliance upon simulation model outputs decreases. While perhaps intuitive from the outset, this result is nonetheless very informative. Given spatial coverage and violation detections as performance attributes, a network designed through uniform placement of monitoring stations throughout the domain is decidedly inferior to a network designed with explicit consideration of the spatial covariance structure of air pollutant concentrations.

4.C. CONCLUSIONS

The three applications presented above give a reasonably representative picture of some of the useful ways in which operations research tools can aid in the development of air pollution management decisions. A common theme throughout these applications is the critically important task of making models that speak to the questions relevant to policymakers. Satisfaction of that task demands close interaction with policymakers and in turn places exceptional demands on systems modelers; demands that have, in fact, rather little to do with the mathematics of optimization and simulation, but concern instead the challenge of communicating said mathematics and models in ways that policymakers can understand. Another common methodologic theme in this work is that the vast majority of the programming effort for these applications lay in pre- and postprocessing information: that is, the steps before and after the optimization and simulation occurs. People interested in building applications would be well advised to become as conversant as possible with the many different ways that pre- and postprocessing can be accomplished, from the traditional (and very limited) approaches involving, for example, Fortran-coded routines described above, to much more powerful and convenient approaches using object-oriented programming languages of varying kinds, along with their attendant interactive graphical capabilities. Another very powerful (and more recent) approach abandons the tradition of stand-alone dedicated optimization solvers in favor of spreadsheet-based routines. The state of hardware and software today has created a situation wherein almost any problem of most any size can be solved efficiently in a multiplicity of ways, platforms, programming environments, and the like. These methodologic advances augur well for an even more influential role for environmental systems modelers and operations research–based approaches for policy support.

REFERENCES AND BIBLIOGRAPHY

Abilock, H., and L. B. Fishborne, 1979. *Users Guide for MARKAL (BNL Version)*, Brookhaven National Laboratory, Upton, NY.

Adar, Z., and J. M. Griffen, 1976. Uncertainty in the Choice of Pollution Control Instruments, *Journal of Environmental Economics and Management*, 3(3), 178–188.

Ahn, S. J., 1992. MARKAL-MACRO/2: An Energy Environmental Modeling System, Engineer's thesis, Stanford University, Stanford, CA.

Alcamo, J. R., R. Show, and L. Hordijk (eds.), 1990. *The RAINS Models of Acidification*, Kluwer, Amsterdam, The Netherlands.

Anadalingam, G., 1987. A Multiple Criteria Decision Analytic Approach for Evaluating Acid Rain Policy Choices, *European Journal of Operational Research*, 29, 336–352.

Anderson, R. J., Jr., 1982. Using Mathematical Programming Models for Cost-Effective Management of Air Quality, in G. Fronza and P. Melli (eds.), *Mathematical Models for Planning and Controlling Air Quality*, pp. 59–74.

Atkinson, S. E., and D. H. Lewis, 1974. A Cost-Effective Analysis of Alternative Air Quality Control Strategies, *Journal of Environmental Economics and Management*, 1, 237–250.

Atkinson, S. E., and D. H. Lewis, 1976. Determination and Implementation of Optimal Air Quality Standards, *Journal of Environmental Economics and Management*, 3, 363–380.

Batterman, S. A., 1989. Selection of Receptor Sites for Optimized Acid Rain Control Strategies, *ASCE Journal of Environmental Engineering*, 115(5), 1046–1058.

Batterman, S., 1992. Optimized Acid Rain Abatement Strategies Using Ecological Goals, *Environmental Management*, 16(1), 133–141.

Bawa, V. S., 1975. On Optimal Pollution Control Policies, *Management Science*, 21, 1397–1404.

Blumstein, A., R. G. Cassidy, W. L. Gorr, and A. S. Walters, 1972. Optional Specifications of Air Pollution Emission Regulations Including Reliability Requirements, *Operations Research*, 20, 752–763.

BRAIN Ingenieros, S. A., 1988. *Projecte de Xarxa de Vigilància i Previsió de la Contaminació Atmosférica al Camp de Tarragona.* Delegació Territorial a Tarragona de la Conselleria de Sanitat i Seguretat Social de la Generalitat de Catalynya, Tarragona, Spain.

Brown, S. L., 1985. Quantitative Risk Assessment of environmental Hazards, in *Annual Review of Public Health*, Annual Reviews, Palo Alto, CA, pp. 247–267.

Burton, E. S., E. H. Pechan, and W. Sanjour, 1973. A Survey of Air Pollution Control Models, in R. A. Deininger (ed.), *Models for Environmental Pollution Control*, pp. 219–235.

Charnes, A., and W. W. Cooper, 1963. Deterministic Equivalents for Optimizing and Satisficing in Chance Constrained Programming, *Operations Research*, 18–39.

Charnes, A., and W. W. Cooper, 1977. Goal Programming and Multiple Objective Optimization, *European Journal of Operational Research*, 39–54.

Charnes, A., W. W. Cooper, and K. O. Kortanek, 1963. Duality in Semi-infinite Programs and Some Works of Haar and Caratheodory, *Management Science,* 9, 209–228.

Chen, W. H., and J. H. Seinfeld, 1975. Optimal Location of Process Measurements, *International Journal of Control,* 21(6), 1003–1014.

Chilton, C. H., J. H. Broehl, R. W. Sullivan, and A. W. Lemmon, Jr., 1972. *Task Report on EPA Energy Quality Model Exercises for 1975,* Battelle, Columbus Laboratories, Columbus, OH.

Cohon, J. L., 1978. *Multiobjective Programming and Planning,* Academic Press, San Diego, CA.

Cooper, W. W., H. Hemphill, V. Lelas, and D. Sullivan, Survey of Mathematical Programming Models in Air Pollution Management, to appear in *European Journal of Operational Research.*

Cosby, B. J., G. M. Hornberger, J. N. Galloway, and R. F. Wright, 1985a. Modeling the Effects of Acid Deposition: Assessments of a Lumped Parameter Model of Soil Water and Stream Water Chemistry, *Water Resources Research,* 21, 51–63.

Cosby, B. J., R. F. Wright, G. M. Hornberger, and J. N. Galloway, 1985b. Modeling the Effects of Acid Deposition: Estimation of Long-Term Water Quality Responses in a Small Forested Catchment, *Water Resources Research,* 21, 1591–1601.

Dalenius, T., J. Hájek, and S. Zubrzycki, 1961. On Plane Sampling and Related Geometrical Problems, in J. Neyman (ed.), *Proceedings of the 4th Berkeley Symposium on Mathematical Statistics and Probability,* Statistical Laboratory. University of California, June 20–July 30, 1960, University of California Press, Berkeley, CA.

Darby, W. P., P. J. Ossenbruggen, and C. J. Gregory, 1973. Optimization of Urban Air Monitoring Networks, *ASCE Journal of the Environmental Engineering Division,* 100(3), 577–591.

de Marsily, G., 1986. *Quantitative Hydrogeology: Groundwater Hydrology for Engineers,* Academic Press, San Diego, CA.

Dongarra, J. J., C. B. Moler, J. R. Bunch, and G. W. Stewart, 1979. *LINPACK Users' Guide,* Philadelphia, PA.

Eliassen, A., 1980. A Review of Long-Range Transport Modeling, *Journal of Applied Meteorology,* 19(3), 231–240.

Ellis, J. H., 1988a. Acid Rain Control Strategies: Options Exist Despite Scientific Uncertainties, *Environmental Science and Technology,* 22(11), 1248–1255.

Ellis, J. H., 1988b. Multiobjective Mathematical Programming Models for Acid Rain Control, *European Journal of Operations Research,* 35, 365–377.

Ellis, J. H., 1990. Integrating Multiple Long-Range Transport Models into Optimization Methodology for Acid Rain Policy Analysis, *European Journal of Operational Research,* 46, 313–321.

Ellis, H., and M. L. Bownan, 1994. Critical Loads and Development of Acid Rain Control Options, *ASCE Journal of Environmental Engineering,* 120(2), 273–290.

Ellis, J. H., E. A. McBean, and G. J. Farquhar, 1985a. Deterministic Linear Programming Model for Acid Rain Abatement, *ASCE Journal of Environmental Engineering,* 111(2), 119–140.

Ellis, J. H., E. A. McBean, and G. J. Farquhar, 1985b. Chance Constrained/Stochastic Linear Programming Model for Acid Rain Abatement: I. Complete Colinearity and Noncolinearity, *Atmospheric Environment,* 19(6), 925–937.

Ellis, J. H., E. A. McBean, and G. J. Farquhar, 1986. Chance Constrained/Stochastic Linear Programming Model for Acid Rain Abatement: II. Limited Colinearity, *Atmospheric Environment,* 20(3), 501–511.

Ellis, J. H., P. R. Ringold, and R. Holdren, 1996. Emission Reductions and Ecological Response: Management Models for Acid Rain Control, *Socio-Economic Planning Sciences,* 30(1), 15–26.

Escuddero, L. F., 1975. The Air Pollution Abatement MASC-AP Model, in C. Brebbia (ed.), *Mathematical Models for Environmental Control,* pp. 173–181.

Federal Energy Administration, 1976. *The National Coal Model: Description and Documentation,* Report FEAtB-771047, ICF, Inc., Washington, DC.

Fedorov, V. V., 1989. Kriging and Other Estimators of Spatial Field Characteristics (with Special Reference to Environmental Studies), *Atmospheric Environment,* 23, 175–184.

Fishborne, L. G., and H. Abilock, 1981. MARKAL, A Linear Programming Model for Energy System Analysis: Technical Description of the BNL Version, *Energy Research,* 5, 353–375.

Fishburn, P. C., and R. K. Sarin, 1991. Dispersive Equity and Social Risk, *Management Science,* 37, 751–769.

Fortin, M., and E. A. McBean, 1983. A Management Model for Acid Rain Abatement, *Atmospheric Environment,* 17(11), 2331–2336.

Fronza, G., and P. Melli, 1984. Assignment of Emission Abatement Levels by Stochastic Programming, *Atmospheric Environment,* 18(3), 531–535.

Fuessle, R. W., E. D. Brill, Jr., and J. C. Liebman, 1987. Air Quality Planning: A General Chance Constrained Model, *ASCE Journal of Environmental Engineering,* 113(1), 106–123.

Gill, P. E., W. Murray, and M. H. Wright, 1981. *Practical Optimization,* Academic Press, San Diego, CA.

Golub, G. H., and C. F. van Loan, 1983. *Matrix Computations,* Johns Hopkins University Press, Baltimore, MD.

Gopalan, R., K. S. Kolluri, R. Batta, and M. K. Karwan, 1990. Modeling Equity of Risk in the Transportation of Hazardous Materials, *Operations Research,* 38(6), 961–973.

Gordon, S. I., 1985. *Computer Models in Environmental Planning,* Van Nostrand Reinhold, New York.

Gorr, W. L., S. A. Gustafson, and K. O. Kortanek, 1972. Optimal Control Strategies for Air Pollution Standards and Regulatory Policy, *Environment and Planning,* 4(2), 183–192.

Greenberg, H. J., 1995. Mathematical Programming Models for Environmental Quality Control, *Operations Research,* 43(4), 578–622.

Guariso, G., and H. Werthner, 1989. *Environmental Decision Support Systems,* Ellis Horwood, Chichester, West Sussex, England.

Guldman, J.-M., 1978. Industrial Location, Air Pollution Control, and Meteorological Variability: A Dynamic Optimization Approach, *Socio-Economic Planning Sciences,* 12(4), 197–214.

Guldman, J.-M., 1986. Interactions Between Weather, Stochasticity, and the Locations of Pollution Sources and Receptors in Air Quality Planning, *Geographical Analysis,* 18(3), 198–214.

Guldman, J.-M., 1988. Chance Constrained Dynamic Model of Air Quality Management, *ASCE Journal of Environmental Engineering,* 114(5), 1116–1135.

Guldman, J.-M., and D. Shefer, 1977. Optimal Plant Location and Air Quality Management Under Indivisibilities and Economics of Scale, *Socio-Economic Planning Sciences,* 11, 77–93.

Guldman, J.-M., and D. Shefer, 1980. Air Quality Control, Industrial Siting, and Fuel Substitution: An Optimization Approach, in J. N. Pitts, Jr., R. L. Metcalf, and D. Grosjean (eds.), *Advances in Environmental Science and Technology,* Vol. 10, Wiley, New York, pp. 301–367.

Gustafson, S. A., and K. O. Kortanek, 1976. On the Calculation of Optimal Long-Term Air Pollution Abatement Strategies for Multiple Source Areas, in Brebbia (ed.), *Mathematical Models for Environmental Control,* pp. 161–171.

Gustafson, S. A., and K. O. Kortanek, 1982a. Semi-infinite Programming and Applications, in Bachem et al. (eds.), *Mathematical Programming: The State of the Art,*

Gustafson, S. A., and K. O. Kortanek, 1982b. A Comprehensive Approach to Air Quality Planning: Abatement, Monitoring Networks, and Real-Time Interpolation, in G. Fronza and P. Melli (eds.), *Mathematical Models for Planning and Controlling Air Quality,* pp. 75–89.

Haith, D. A., 1982. *Environmental Systems Optimization,* Wiley, New York.

Hickey, H. R., W. D. Rowe, and F. Skinner, 1971. A Cost Model for Air Quality Monitoring Systems, *Journal of the Air Pollution Control Association,* 21(11), 689–693.

Hooke, R., and T. A. Jeeves, 1961. Direct Search Solution of Numerical and Statistical Problems, *Journal of the Association for Computing Machinery,* 8, 212–229.

Hornberger, G. M., B. J. Cosby, and J. N. Galloway, 1986. Modeling the Effects of Acid Deposition: Uncertainty and Spatial Variability in Estimation of Long-Term Sulfate Dynamics in a Region, *Water Resources Research,* 22(8), 1293–1302.

Hougland, E. S., and N. T. Stephens, 1976. Air Pollutant Monitor Siting by Analytical Techniques, 26(1), 51–53.

Jagannathan, R., 1974. Chance Constrained Programming with Joint Constraints, *Operations Research,* 22(2), 358–372.

Jones, C. V., 1994. Visualization in Mathematical Programming, *ORSA Journal of Computing,* 6(3), 221–257.

Keeney, R. L., 1980. Equity and Public Risk, *Operations Research,* 28(3), 527–534.

Kortanek, K. O., and W. L. Gorr, 1972. Numerical Aspects of Pollution Abatement Problems: Optimal Control Strategies for Air Quality Standards, in *Proceedings in Operations Research,*

Kessler, C. J., T. Sager, H. Hemphill, and T. Porter, 1990. Factor Analysis of Texas Trends in Acidic Deposition, *Atmospheric Environment,*

Kilaru, V. A., 1994. Screening Analysis of Ambient Monitoring Data Sets in Support of the Urban Area, *Source Program,* U.S. Environmental Protection Agency, Washington, DC.

Kohn, R. E., 1971a. Optimal Air Quality Standards, *Econometrica,* 39(6), 983–995.

Kohn, R. E., 1971b. Application of Linear Programming to a Controversy of Air Pollution Control, *Management Science,* 17, B609–B621.

Kohn, R. E., 1973. Labor Displacement and Air Pollution Control, *Operations Research,* 21, 1063–1070.

Kohn, R. E., and D. E. Burlingame, 1971. Air Quality Control Model Combining Data on Morbidity and Pollution Abatement, *Decision Science,* 2, 300–310.

Langstaff, J., C. Seigneur, and M.-K. Liu, 1987. Design of an Optimal Air Monitoring Network for Exposure Assessment, *Atmospheric Environment,* 21(6), 1393–1410.

Leavitt, J. M., F. Pooler, Jr., and R. C. Wanta, 1963. Design and Interim Meteorological Evaluation of a Community Network for Meteorological and Air Quality Measurements, *Journal of the Air Pollution Control Association,* 7(3), 211–215.

Lehmann, R., 1991. Uncertainty Analysis for a Linear Programming Model for Acid Rain Abatement, *Atmospheric Environment,* 25(2), 231–240.

Lehmann, R., 1992. On Properties of Linear Programming Models for Acid Rain Abatement, *Atmospheric Environment,* 26A(7), 1347–1359.

Liu, M. K., J. Meyer, R. Pollack, P. M. Roth, J. H. Seinfeld, J. V. Behar, L. M. Dunn, J. L. McElroy, P. N. Lem, A. M. Pitchford, and N. T. Fischer, 1977. *Development of a Methodology for the Design of a Carbon Monoxide Monitoring Network,* EPA-600/4-77-019. U.S. Environmental Protection Agency, Las Vegas, NV.

Ludwig, F. L., N. J. Berg, and A. J. Hoffman, 1976. The Selection of Sites for Air Pollutant Monitoring, 69th Annual Meeting of the Air Pollution Control Association, Portland, OR.

Lynnes, J. N., 1983. When Not to Use an Automatic Quadrature Routine, *SIAM Review,* 25(1), 63–87.

Lynnes, J. N., and D. Jespersen, 1975. Moderate Degree Symmetric Quadrature Rules for the Triangle, *Journal of the Institute of Mathematics and Its Applications,* 15, 19–32.

MacKenzie, J. J., 1989. *Breathing Easier: Taking Action on Climate Change, Air Pollution, and Energy Insecurity,* World Resources Institute, New York.

Manne, A. S., and C.-O. Wene, 1992. *MARKAL-MACRO: A Linked Model for Energy Economy Analysis,* Informal Report BNL-47161, Brookhaven National Laboratory, Upton, NY.

Matheron, G., 1965. *Les Variables régionalisées et leur estimation: une application de la théorie des fonctions aléatoires aux sciences de la nature,* Masson, Paris.

McBean, E. A., J. H. Ellis, and M. Fortin, 1985. A Screening Model for the Development and Evaluation of Acid Rain Abatement Strategies, *Journal of Environmental Management,* 21, 287–299.

McBratney, A. B., R. Webster, and T. M. Burgess, 1981. The Design of Optimal Sampling Schemes for Local Estimation and Mapping of Regionalized Variables: 1. Theory and Method, *Computers and Geosciences,* 7, 335–365.

Meetham, A. R., et al., 1981. *Atmospheric Pollution: Its History, Origins, and Prevention,* Pergamon Press, Oxford.

Miller, B. L., and H. M. Wagner, 1965. Chance Constrained Programming with Joint Constraints, *Operations Research,* 13(6), 930–945.

Misch, A., 1994. Assessing Environmental Health Risks, in *State of the World 1994,* Worldwatch Institute, New York.

Morrison, J. L., 1970. A Link Between Cartographic Theory and Mapping Practice: The Nearest Neighbor Statistic, *Geographical Review,* 60(4), 494–510.

Morrison, M. B., and E. S. Rubin, 1985. A Linear Programming Model for Acid Rain Policy Analysis, *Journal of the Air Pollution Control Association,* 35(11), 1137–1148.

Munn, R. E., 1981. *The Design of Air Quality Monitoring Networks,* Macmillan, London.

Murtagh, B. A., and M. A. Saunders, 1987. *MINOS 5.1 User's Guide,* Technical Report SOL 83-20R, Systems Optimization Laboratory, Stanford University, Stanford, CA.

NAS, 1991. *National Academy of Sciences on Troposphere Ozone Formation and Measurement: Rethinking the Ozone Problem in Urban and Regional Air Pollution,* National Research Council, Washington, DC.

NAPAP (National Acid Precipitation Assessment Program), 1991a. *The Experience and Legacy of NAPAP, Report to the Joint Chairs Council of the Interagency Task Force on Acidic Deposition,* NAPAP Office of the Director, Washington, DC.

NAPAP (National Acid Precipitation Assessment Program), 1991b. *Acidic Deposition: State of Science and Technology,* NAPAP Office of the Director, Washington, DC.

Nilsson, J., and P. Grennfelt (eds.), 1988. Critical Loads for Sulfur and Nitrogen, *Report from a Workshop,* Skokloster, Sweden, March 19–24, UN/ECE and Nordic Council of Ministers, NORD 1988:15.

Olea, R. A., 1984. *Systematic Sampling of Spatial Functions,* Series on Spatial Analysis, Kansas Geological Survey, University of Kansas, Lawrence, KS.

Ott, W. J., 1977. Development of Criteria for Siting Air Monitoring Stations, *Journal of the Air Pollution Control Association,* 27(6), 543–547.

Peck, S. C., and T. J. Teisberg, 1992. CETA: A Model for Carbon Emissions Trajectory Assessment, *Energy Journal,* 13(1), 55–77.

Pinter J., 1991. Stochastic Modeling and Optimization for Environmental Management, *Annals of Operations Research,* 31, 527–544.

Pinter J., J. W. Meeuwig, D. J. Meeuwig, M. Fels, and D. S. Lycon, 1993. ESIS: An Intelligent Decision Support System for Assisting Industrial Waste Water Management, *Annals of Operations Research,*

Purcel, R. Y., and G. S. Shareef, 1988. *Handbook of Control Technologies for Hazardous Air Pollutants,* Hemisphere, New York.

Querner, I., 1993. *An Economics Analysis of Severe Industrial Hazards,* Physica-Verlag, Heidelberg, Germany.

Renka, R. J., 1984. Algorithm 624: Triangulation and Interpolation at Arbitrarily Distributed Points in the Plane, *ACM Transactions on Mathematical Software,* 10(4), 440–442.

Roberts, S. V., 1990. The Clean Air Sweepstakes, *U.S. News & World Report,* April 16, pp. 22–24.

Ruckelshaus, W., 1984. Risk in a Free Society, *Risk Analysis,* 157.

Samson, J., 1988. Uncertainties in Source–Receptor Relationships, in White (ed.), *Acid Rain: The Relationships Between Sources and Receptor,*

Seinfeld, J. H., 1972. Optimal Location of Pollutant Monitoring Stations in an Airshed, *Atmospheric Environment,* 6, 847–858.

Seinfeld, J. H., and C. P. Kyan, 1971. Determination of Optimal Air Pollution Control Strategies, *Socio-Economic Planning Sciences,* 5, 173–190.

Smith, D. G., and B. A. Egan, 1979. Design of Monitor Networks to Meet Multiple Criteria, *Journal of the Air Pollution Control Association,* 29(7), 710–714.

Sverdrup, H., P. Warfvinge, M. Rabenhorst, A. Janicki, R. Morgan, and M. Bowman, 1992. Critical Loads and Steady-State Chemistry for Streams in the State of Maryland, *Environmental Pollution,* 77, 195–203.

Teller, A., 1968. The Use of Linear Programming to Estimate the Cost of Some Alternative Air Pollution Abatement Policies, Proc., IBM Scientific Computing Symposium on Water and Air Resource Management, 345–353.

Trijonis, J. C., 1974. Economic Air Pollution Control Model for Los Angeles County in 1975, *Environmental Science and Technology,* 8(9), 811–826.

Trujillo-Ventura, A., and J. H. Ellis, 1991. Multiobjective Air Pollution Monitoring Network Design, *Atmospheric Environment,* 25A(2), 469–479.

Trujillo-Ventura, A., and J. H. Ellis, 1992. Nonlinear Optimization of Air Pollution Monitoring Networks: Algorithmic Considerations and Computational Results, *Engineering Optimization,* 19, 287–308.

U.S. Congress, 1990. Public Law 101-549, November 15.

U.S. EPA, 1987. *International Sulfur Deposition Model Evaluation (ISDME),* EPA-600/3-87-008, U.S. Environmental Protection Agency, Washington, DC.

Venkatram, A., B. E. Ley, and S. Y. Wong, 1982. A Statistical Model to Estimate Long-Term Concentrations of Pollutants Associated with Long-Range Transport, *Atmospheric Environment,* 16, 249–257.

Wareham, D. G., E. A. McBean, and J. M. Byrne, 1988. Linear Programming for Abatement of Nitrogen Oxides Acid Rain Deposition, *Water, Air, and Soil Pollution,* 40, 157–175.

Watanabe, T., and H. Ellis, 1993a. Robustness in Stochastic Programming, *Applied Mathematical Modeling,* 17, 547–554.

Watanabe, T., and H. Ellis, 1993b. Stochastic Programming Models for Air Quality Management, *Computers and Operations Research,* 20(6), 651–663.

Wendling, R. M., and R. H. Bezdek, 1989. Acid Rain Abatement Legislation: Costs and Benefits, *OMEGA: International Journal of Management Science,* 17(3), 251–261.

Zannetti, 1990. *Air Pollution Modeling,* Van Nostrand Reinhold, New York.

CHAPTER 5

Solid Waste Management

JON C. LIEBMAN

5.A. INTRODUCTION

A solid waste "system" is so diffuse as to be barely definable as a system. There are many actors, responsible in varying degrees for different (and often overlapping) segments of the system and with a wide variety of managerial roles. Partly for this reason, systems management models are fragmented, dealing with individual components or subcomponents more frequently than with the overall system or even large components of the system. This fragmentation is also a result of the fact that more comprehensive models are very difficult to formulate and nearly impossible to solve.

It is also the case that solid waste systems vary widely from place to place. In some locations, the entire system is public and under single management. Elsewhere, collection is by private collectors, who may contract with government for exclusive collection rights in the entire area or in subdistricts, or who may compete with one another for individual household collection. Transfer and disposal may be by private agencies as well. Often, many different political entities are included in an area that, for various reasons, should be considered as a single solid waste management system.

With this wide variety of different configurations, models of entire systems have very limited applicability since each system is likely to be unique in some aspect. On the other hand, model applicability tends to be greater when the model is of system components that are often the same or that are common between many systems.

Solid waste is, of course, not a single material, nor is it even a single stream of materials that can be modeled homogeneously. Ordinary household waste is the material that most people think of in first considering solid waste management, but there are many other streams of materials that require different handling and therefore different modeling. These include, for example, larger and less common household waste (refrigerators, furniture), dead animals, abandoned automobiles, construction waste, and demolition rubble. Commercial and industrial wastes are even more complex.

Since solid waste collection and disposal is generally regarded as a public good, the overall objective should be the maximization of social welfare. This implies some consideration of benefits minus costs. In most published solid waste management models, however, the objective function is to minimize cost. Thus the level of benefits is implicitly taken as constant across the entire feasible region of the model, so that cost minimization is equivalent to welfare maximization. This, of course, begs the question of the desirable level of benefits, which is often the most important and most difficult question to deal with. In many cases, some light can be shed on these issues by incorporating level of service (or other benefit surrogate) into the model as a constraint and then solving with different right-hand-side values, thus providing optimal cost information for varying benefits. Such an approach is simply one way of looking at a multiobjective problem.

The bulk of solid waste management modeling is based on one form or another of a network flow model. Such models attempt to determine how materials should be routed through a collection–separation–processing–recycling–disposal system that may include many alternative steps. Thus these models are not only useful for determining allocation of flows to various processes but may also be used to select among alternative locations of facilities. Network flow models are discussed in Section 5.B, and facility location models are covered in Section 5.C.

A second important model form in solid waste management deals with the routing of individual collection vehicles. Here the attention is on the selection and sequencing of links from a network, in such a way that a closed tour is formed that satisfies various other criteria. Vehicle routing models are discussed in Section 5.D.

This chapter is not intended to be exhaustive or even to cover all types of problems and models that have been considered in the solid waste system. Most of the fundamental work was done in the 1960s and 1970s, and for this reason most of the references are from that time. More recent work frequently extends the earlier models. Some other modeling applications are mentioned briefly in Sections 5.E and 5.F, and some concluding remarks are given in Section 5.G.

5.B. NETWORK FLOW MODELS

5.B.1. Hitchcock Transportation Model

The simplest network flow model is the widely known Hitchcock transportation model, which is reviewed here for completeness. In this case, there are two distinct sets of nodes: a set of m sources which represent collection areas and a set of n destinations which represent disposal or treatment sites (Figure 5.1). At the sources, the sum of the outflows must be equal to the amounts generated. At the destinations, the amounts arriving cannot exceed capacities.

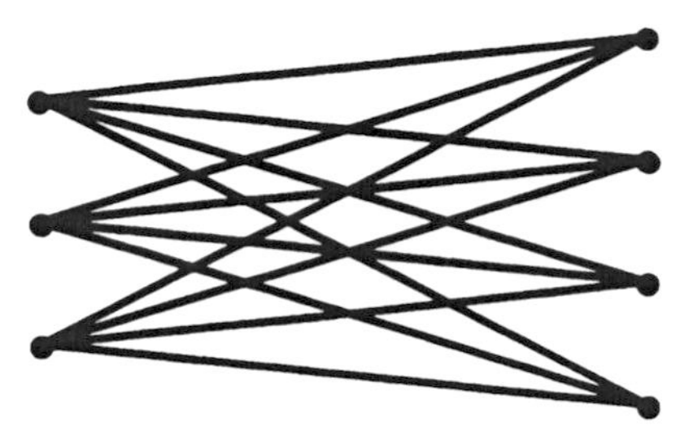

Figure 5.1

For each link between a source and a destination, a unit cost of transportation is known. The model determines the amounts to ship from each source to each destination to minimize the total cost:

$$\text{minimize} \sum_{i=1}^{m} \sum_{j=1}^{n} a_{i,j} x_{i,j} \tag{5.1}$$

subject to

$$\sum_{j=1}^{n} x_{i,j} = g_i \qquad i = 1,2,\ldots,m \tag{5.2}$$

$$\sum_{i=1}^{m} x_{i,j} \leq h_j \qquad j = 1,2,\ldots,n \tag{5.3}$$

Here g_i is the known amount generated or collected in neighborhood i, and h_j is the known capacity of facility j. The unknown, $x_{i,j}$, is the amount shipped from source i to destination j. The objective function (5.1) minimizes the total cost of transportation. Equation (5.2) requires the amount leaving neighborhood i to be equal to that generated, and equation (5.3) requires that the amount going into facility j be less than or equal to its capacity. This very simple linear programming model can be used in planning to allocate the material collected in neighborhoods or collection areas to disposal sites when there are no intermediate transfer or processing stations (Figure 5.2). In Chapter 7 (Section 7.C.3) the Hitchcock problem is discussed in the context of urban transportation planning.

5.B.2. Transshipment Model

A more general version, called the transshipment model, includes intermediate nodes inserted between the sources and the destinations. These nodes represent intermediate facilities such as transfer stations. Added mass balance equations are included for each intermediate node:

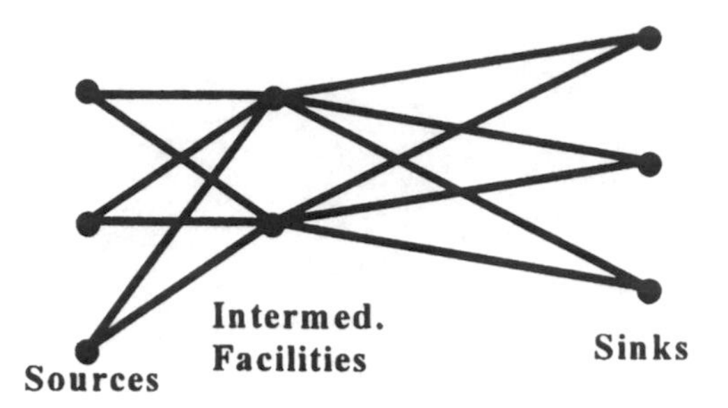

Figure 5.2

$$\text{minimize} \sum_{i=1}^{m} \sum_{k=1}^{p} a_{i,k} x_{i,k} + \sum_{k=1}^{p} \sum_{j=1}^{n} b_{k,j} w_{k,j} \tag{5.4}$$

subject to

$$\sum_{k=1}^{p} x_{i,k} = g_i \qquad i = 1,2,\ldots,m \tag{5.5}$$

$$\sum_{i=1}^{m} x_{i,k} - \sum_{j=1}^{n} w_{k,j} = 0 \qquad k = 1,2,\ldots,p \tag{5.6}$$

$$\sum_{k=1}^{p} w_{k,j} \leq h_j \qquad j = 1,2,\ldots,n \tag{5.7}$$

Here $x_{i,k}$ is the amount shipped from source i to intermediate facility k, and $w_{k,j}$ is the amount shipped from intermediate facility k to destination j.

5.B.3. Minimum-Cost Circulation Model

An even more general network flow model that is sometimes useful in modeling selections among facilities and processes is the minimum-cost circulation model. This model envisions a network with a single source and a single sink and directed links on which there are both capacities and lower bounds on flows. Unlike the transportation and transshipment models, there are not "ranks" of facilities that have no interconnections; instead, any node may be connected to any other node (Figure 5.3). In its simplest form, the model assumes the existence of an arc (the *reverse arc*) from the sink returning to the source, so that the network flow forms a complete circulation. Multiple sources and sinks can be represented simply by including a single *supersource* with arcs leading out of it to all the sources, and a single *supersink* with arcs leading into it from all the sinks. The reverse arc then flows from supersink to supersource.

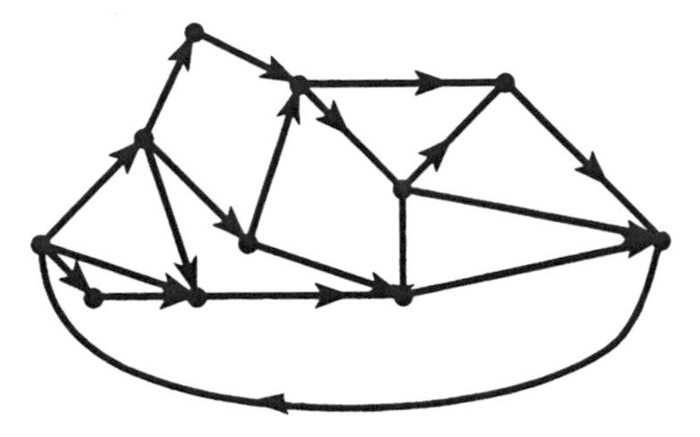

Figure 5.3

In this model, a set of constraints enforces mass balance at each node, including the source and the sink. Unit costs are given for flows in each arc, and the model attempts to find a minimum-cost flow pattern that satisfies all the upper and lower bound constraints:

$$\text{minimize} \sum_{i=1}^{n} \sum_{j=1}^{n} a_{i,j} x_{i,j} \tag{5.8}$$

subject to

$$x_{i,j} \geq s_{i,j} \qquad i = 1,2,\ldots,n, \quad j = 1,2,\ldots,n \tag{5.9}$$

$$x_{i,j} \leq c_{i,j} \qquad i = 1,2,\ldots,n, \quad j = 1,2,\ldots,n \tag{5.10}$$

$$\sum_{i=1}^{n} x_{i,k} - \sum_{j=1}^{n} x_{k,j} = 0 \qquad k = 1,2,\ldots,n \tag{5.11}$$

where $a_{i,j}$ is the unit cost of shipment on the arc from node i to node j, $s_{i,j}$ and $c_{i,j}$ are the lower bound and upper bound on flow in that arc, respectively, and $x_{i,j}$ is the flow in the arc. Equation (5.8), the objective function, minimizes the cost of flow in all arcs. Equations (5.9) and (5.10) require that the flow in each arc be greater than the lower bound and less than the upper bound. Equation (5.11) is the conservation of mass equation, requiring that the inflow be equal to the outflow at every node.

If costs are positive, there must be a nonzero lower bound on flow in at least one arc in order to have a nonzero solution. Frequently, this bound appears on the reverse arc and specifies the required flow through the network; that is, the total amount of refuse that must be picked up from the neighborhoods and transferred to ultimate disposal. Of course, the bound in this case can actually appear as an equality rather than a lower bound.

It should be clear that the reverse arc is present only for modeling convenience. It provides an easy place to specify the required flow in the network, and it permits writing all node constraints in the same "sum of inflow arcs − sum of outflow arcs = 0" form. The same problem may be represented without the reverse arc, by specifying that the sum of the outflows at the starting node be equal to the required flow and that the sum of the inflows at the finish node be equal to the same required flow. This is simply algebraic elimination of the equality constraint for flow in the reverse arc by substitution. It should also be noted that the mass balance constraint at any one node can be eliminated as redundant.

This linear program can be used to determine flows of material from any set of sources to any set of destinations. For example, sources could be neighborhoods in which refuse is collected, intermediate nodes could be transfer stations and recycling separation, and final destinations could be various disposal sites, including incinerators and landfills. Flow costs on the arcs leading into facilities (transfer stations, incinerators, landfills) could include the unit cost of processing at these facilities.

5.B.4. Capacitated Facilities

The flow model given above does not include capacities on intermediate facilities, since only arcs carry upper and lower bounds. A simple modification, however, will permit including facility capacities (as well as required minimum flows, if necessary) and also a more straightforward representation of processing cost. Any node, representing a facility, can be split into two nodes (which may be thought of as the separate receiving and shipping docks at the facility) with a single arc joining them. All incoming arcs enter the receiving node, and all outgoing arcs leaving from the shipping node. The arc joining these two nodes then represents the processing in the facility; it can have an upper bound representing capacity, a lower bound representing a required minimum flow, and the processing cost can be more naturally placed on this arc instead of on all the incoming arcs.

5.B.5. Representation of Material Removal

The facility nodes in the network representation may be processes that remove a fraction of the material while passing the remaining material on. For example, a node might represent a recycling separation facility, where a known (estimated) percentage of the refuse is removed for recycling.

It is relatively easy to modify the minimum-cost circulation model to represent this sort of processing. We begin with the model form without the reverse arc, with the sum of the outflows from the starting node set equal to the required (collected) flow. We also drop the mass balance requirement at the finish node (since, as observed above, it is redundant). Then the mass balance constraint at any processing node can be modified to reflect the change in material amount due to the processing. For example, if a particular node represents processing that removes 20% of the material, the constraint is modified to "0.8 (sum of inflows) − sum of outflows = 0."

This simple modification can represent any process that removes material as long as the fractional removal is independent of previous processing of the flow stream. Unfortunately, in many cases in solid waste, that assumption is not true. For example, suppose it is desired to model an incinerator as one node of the network and to consider the material flow of the ash and continuing to ultimate disposal. In this case, the fractional reduction of mass at the incinerator will certainly depend on whether paper has been removed from the flow stream for recycling at an earlier node.

Golueke and McGauhey (1970) used a model similar to the one described to consider the collection, treatment, and disposal of solid waste from a materials flow standpoint. They constructed a multicommodity transshipment network in which source nodes represented origin points of various types of waste, sink nodes represented existing or potential disposal sites, and intermediate nodes represented transfer and treatment facilities. At nodes representing treatment, flows were separated in fixed ratios to one another. Although the model is in linear programming form, the solution method used was a modified out-of-kilter algorithm, which presumably was more efficient. This model and similar ones are particularly useful in the preliminary planning of a new system or in evaluating proposed additions to an existing system. Because of the rather rigid and simplified problem structure, it is unlikely that such a model can be used for detailed operational issues.

5.C. FACILITY SITING MODELS

Models for the selection of facilities or facility location may be viewed as extensions of the flow models described above. In fact, the flow models are, in a sense, facility selection models in that any node through which there is zero flow represents a facility that should not be constructed. The trouble, of course, is that the linear cost structure of the flow models does not reflect the usual costs for constructing a facility, and construction decisions made without considering these costs will not be optimal. Construction costs are highly nonlinear, commonly consisting of a very high unit cost for the first capacity increments, and decreasing unit costs as capacity is made greater. This relationship is often approximated as a fixed-charge problem, in which there is a large initial cost for building any facility at all (which may be thought of as the cost of the land on which the facility is built), followed by a linear cost as capacity is increased. An even simpler assumption is that a facility is either not built or is built to a prespecified capacity. Then the facility cost is either zero or it is equal to the known cost of construction to that capacity. In any of these cases, the form of the function prevents the use of piecewise linearization as a solution technique, since the later segments of cost are more attractive than the early ones. Modeling of such problems frequently involves the use of logical variables. In Chapter 8 (Section 8.B) model formulations are presented for solving capacitated plant location models in the context of freight transportation.

5.C.1. Simple Terminal Facility Selection Model

The Hitchcock transportation model may be modified so that it selects terminal facilities with prespecified capacity as well as allocating collection amounts to these facilities. In addition to the flow variables, introduce a 0–1 valued variable for each prospective facility site. This variable is to be 0 if the facility is not built, and 1 if the facility is built to the known capacity. The model is then

$$\text{minimize} \sum_{i=1}^{m} \sum_{j=1}^{n} a_{i,j}x_{i,j} + \sum_{j=1}^{m} q_j y_j \tag{5.12}$$

subject to

$$\sum_{j=1}^{n} x_{i,j} = g_i \qquad i = 1,2,\ldots,m \tag{5.13}$$

$$\sum_{i=1}^{m} x_{i,j} - h_j y_j = 0 \qquad j = 1,2,\ldots,n \tag{5.14}$$

and, of course, $y_j = \{0,1\}$. The only change in the objective function (5.12) is the addition of the fixed cost of construction for each facility, which, because it is multiplied by a 0–1 variable, is charged only if the facility is built. The source mass balance constraint (5.13) is unchanged. The destination mass balance constraint (5.14) has the capacity multiplied by the logical variable, so that the capacity exists only if the facility is built.

This model closely resembles one initially proposed by Efroymson and Ray (1966). Because of the 0–1 variables, it must be solved by some form of integer linear programming, and branch and bound is the most common approach.

If it is desired to consider several different capacities for the same facility, each different capacity can be treated as a different facility. A constraint that forces the sum of the alternative y_j's to be no greater than 1 will prevent the selection of more than one of the alternatives; depending on the form of the cost function as capacity increases, this constraint may not be necessary.

5.C.2. Intermediate Facility Location Model

Although the foregoing model permits siting of terminal facilities, it does not consider the location of intermediate facilities such as transfer stations. A similar modification of the transshipment model provides that capability, at the cost of introducing logical variables and requiring solution by branch and bound:

$$\text{minimize} \sum_{i=1}^{m} \sum_{k=1}^{p} a_{i,k} x_{i,k} + \sum_{k=1}^{p} \sum_{j=1}^{n} b_{k,j} w_{k,j} + \sum_{k=1}^{p} q_k y_k \tag{5.15}$$

subject to

$$\sum_{k=1}^{p} x_{i,k} = g_i \qquad i = 1,2,...,m \tag{5.16}$$

$$\sum_{k=1}^{p} w_{k,j} \leq h_j \qquad j = 1,2,...,n \tag{5.17}$$

$$\sum_{i=1}^{m} x_{i,k} - \sum_{j=1}^{n} w_{k,j} = 0 \qquad k = 1,2,...,p \tag{5.18}$$

$$\sum_{i=1}^{m} x_{i,k} - q_k y_k = 0 \qquad k = 1,2,...,p \tag{5.19}$$

As with the previous location model, the objective function (5.15) is modified from that of the transshipment model only by including the cost of the facility (q_k), if constructed. Constraints (5.16)–(5.18) are not modified at all. The added constraint (5.19) limits inflow at intermediate facilities to their capacity if constructed, and zero otherwise. This model, originally due to Marks (1969), is particularly useful in siting transfer stations in an existing collection system. A similar model is extended and applied to the city of Istanbul, Turkey, by Kirca and Erkip (1988). Gottinger (1988) has incorporated the representation of material removal discussed in Section 5.B.5 into a similar model, and applied it to an area near Munich. Uncertainty is considered in a related model by Huang et al. (1992) demonstrating the approach of "gray linear programming."

5.4. ROUTING OF VEHICLES

Once the facilities of the collection and disposal system are in place, there remains the issue of transporting the refuse. Optimization has been used in several different aspects of the transport system, with the most common applications being in routing and scheduling of the collection vehicles themselves.

The objective in collection is to provide the required level of service with minimum cost. Cost consists of two components: the costs associated with the vehicle and those associated with labor. Vehicle costs are both capital and operating costs. It is common to view the variable component of most of these costs as being proportional to distance traveled and/or time. This assumes, for example, that the labor cost involved with collection at an individual pickup is essentially invariant (or at least not controlled by the system designer), so that only the labor cost utilized in travel is controllable. Taking this point of view permits modeling the collection routing problem as one of minimizing either distance or time of travel or some function of both. Thus we can take the collection region as a network of links, on each of which is a travel cost, with an objective of minimizing the sum of the travel costs to perform the collection. In what follows these costs will be referred to as *distances*.

The overall vehicle routing problem requires dividing up the entire collection region into individual areas that each generate a single truckload of waste (called *districting*); and then determining, for each truckload a route from the depot to the collection area, through the collection area, and back to the depot (called *routing*). These two tasks should be accomplished simultaneously if true optimality is to be obtained, but optimization of that overall system has yet to be done successfully, and the two steps are commonly examined independently.

There are two fundamental models of vehicle collection. One views collection as occurring at discrete points, which are explicitly represented on the network. The problem then becomes one of traveling through the network so as to visit every point at least once, while minimizing the total distance (cost). This model most nearly represents rural refuse collection and some forms of industrial or commercial collection, such as collection from dumpsters. Its simplest form is the well-known traveling salesman problem, which is probably the most widely researched integer optimization problem in mathematics. No efficient algorithm for solving this problem has been found, and it belongs to a class of integer problems called NP-complete, for which truly efficient algorithms are believed not to be possible. Because this problem and its variants have been so widely studied, they will not be discussed further herein. A good starting reference is Lawler et al. (1985) and a more recent survey paper is that of LaPorte (1992). Applications to refuse collection are discussed in detail in Beltrami and Bodin (1974).

The second model treats collection as occurring continuously along the links of the network. This most closely represents the situation in urban areas where collection is from closely adjacent individual buildings. The problem is then to traverse every link of the network at least once, while minimizing the total distance. This problem, called the *Chinese postman's problem* because its first solution was published in the *Chinese Journal of Mathematics* with application to postal routes, has been solved efficiently for a number of cases. These solutions are the subject of the following sections. An excellent reference for these sections is Eiselt et al. (1995). In Chapter 8 (Section 8.C) model formulations are presented for vehicle routing problems that arise in the area of freight transportation.

5.D.1. Individual Vehicle Routing: Chinese Postman's Problem

An optimal solution to the Chinese postman's problem requires the vehicle to traverse every link at least once and to minimize the distance of additional traverses, if any. Thus the sum of the lengths of all the links is an absolute lower bound on the required travel. In some cases, no additional traverses will be necessary. Such a network is called *unicursal.*

The Chinese postman's problem may be formulated as an integer linear program in which there is a variable $x_{i,j}$ that represents the number of times the link between nodes i and j is traversed in the direction from i to j. Such a formulation is:

$$\text{minimize} \sum_{i=1}^{N} \sum_{j=1}^{N} c_{i,j} x_{i,j} \tag{5.20}$$

subject to:

$$\sum_{k=1}^{N} x_{k,i} - \sum_{k=1}^{N} x_{i,k} = 0 \qquad i = 1,2,\ldots,N \tag{5.21}$$

$$x_{i,j} + x_{j,i} \geq 1 \qquad \text{for all links } (i,j) \tag{5.22}$$

and $x_{i,j} \geq 0$ and integer. N is the number of nodes in the network. The objective function (5.20) minimizes the total length of link traversal. Constraint (5.21) requires that the number of entries into node i be equal to the number of exits from node i, and constraint (5.22) requires that every link be traversed at least once, in one direction or the other.

Although this formulation describes the problem, it is not particularly practical, because of the integer requirement. Branch and bound, or other integer techniques, are not able to solve problems on networks of moderate size. A more practical solution of such a problem was developed by Edmonds and Johnson (1973), based on an algorithm by Edmonds (1965). Edmonds' technique requires some additional background.

A fundamental theorem that leads to the solution method is that of Leonhard Euler (1736), who demonstrated that an undirected network is unicursal if and only if the network is connected[1] and the number of streets incident on each intersection is even.[2] This result is fairly intuitive: For every entry into some intersection there must also be an untraversed exit street, so that a vehicle uses the streets incident on the intersection in pairs. If the degree of the intersection is odd, there will be a street remaining untraversed after the last exit from the intersection. Thus when there are intersections in the network whose degree is not even, it will be necessary to retrace some streets in order to traverse all of them at least once. In refuse collection, of course, this retracing is *deadheading,* traversing a street without collecting along it,

[1]A connected network is one in which there is some way to get from every node to every other node.

[2]In graph theory, the number of edges incident on a node is called the *degree* of the node. Euler's requirement, then, is that every node in the network be of even degree.

either because it has already been collected, will be collected in some future pass, or requires no collection (and therefore need not be traversed at all).

This result leads to a closely related observation. In a network in which some nodes are of odd degree, any tour that does traverse every link at least once will contain retracings that, if added to the network as links, convert every node to be of even degree. That is, there must be an even number of entries and exits from every node, considering both original tracings of the links and retracings.

Now, if there is an odd-degree node in the network, it is clearly necessary to retrace at least one link touching that node, so as to make it even. But if the opposite end of that link is already even, the retracing will serve to make it odd, and so an additional link out of that node must also be retraced. This reasoning leads to the conclusion that the retracings of links necessary to permit a complete closed tour to be made must serve to connect together pairs of originally odd-degree nodes.[3] In collection systems, this means that deadheading will take place between intersections that are of odd degree. It is this deadheading whose length or cost is to be minimized.

In a connected network with only a few odd-degree nodes, it may be obvious which links to retrace to connect the odd-degree nodes in pairs with minimum total distance, but as soon as there are more than just a few widely scattered odd-degree nodes, there are many different solutions. Figure 5.4 shows two of the many possible pairings of the 12 odd-degree nodes, with the retracings shown as added arcs. Of these two, the first is slightly better than the second (assuming the networks are drawn to scale).

We can narrow our attention to the heart of the problem by converting it to a network in which only nodes that must be paired are shown. This is done by using a shortest-path algorithm to find the shortest paths between all the odd-degree nodes and then drawing a new network consisting only of the odd-degree nodes connected by edges whose length is that of the shortest path between them. For example, the seven-node network shown in Figure 5.5 has four odd-degree nodes (shown as rectangles). Numbers marked on the edges are their lengths. Using the shortest-path algorithm on this network to find the shortest paths between all the odd-degree nodes results in the four-node network shown in Figure 5.6. On the edges, the numbers in parentheses indicate the intermediate (even-degree) nodes through which the shortest path goes, and the other numbers are the lengths of the shortest paths. On this network, which must always have an even number of nodes, the equivalent of our "pairing-up" problem is to select links joining the nodes in such a way that every node is touched by exactly one selected link and that the sum

[3]This observation naturally leads to a further question: What happens when a network has an odd number of odd-degree nodes? The answer is simple: Such a condition cannot occur. Since every link touches two nodes, the sum of the degrees of all nodes in the network must be even. The sum of the degrees of the even-degree nodes is obviously even; and in order for the sum of the degrees of the odd-degree nodes to be even, there must be an even number of them.

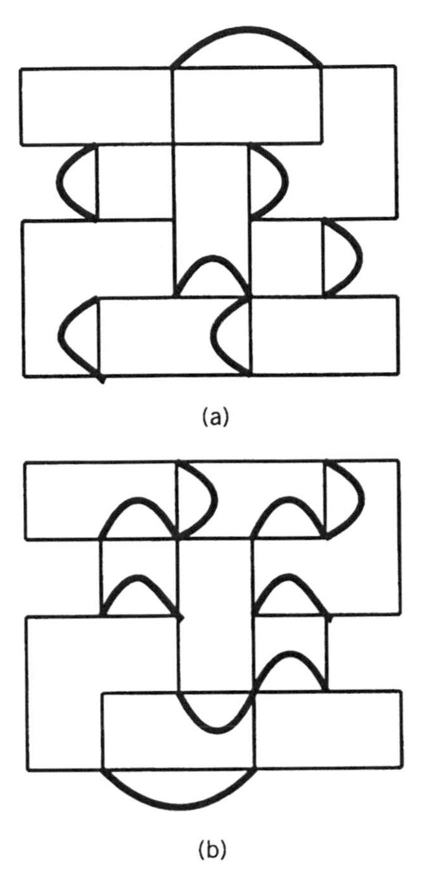

Figure 5.4

of the lengths of the selected links is minimum. Inspection of the sample problem indicates that the solution is to pair up nodes 2 and 3 (by means of the shortest path, of length 5, passing through node 4 on the original network) and nodes 6 and 7 (by means of the shortest path, of length 5, directly between the two nodes). Thus in the original network we should retrace edges 2–4, 4–3, and 6–7, for a total length of retracing of 10.

In a larger problem, of course, finding the pairings on the constructed network of shortest paths between odd nodes cannot be done easily by inspection. If we can solve this problem, known as the *matching problem,* we can translate back to our original "pairing-up" problem by deadheading on the shortest paths between odd-degree nodes represented by the selected links in the matching.

It is the matching problem to which Edmonds (1965) provides an elegant solution method. Although the method is beyond the scope of this book, the formulation used by Edmonds is fairly easy to describe. For each link j in the matching network there is a boolean (0–1) variable x_j. This variable will

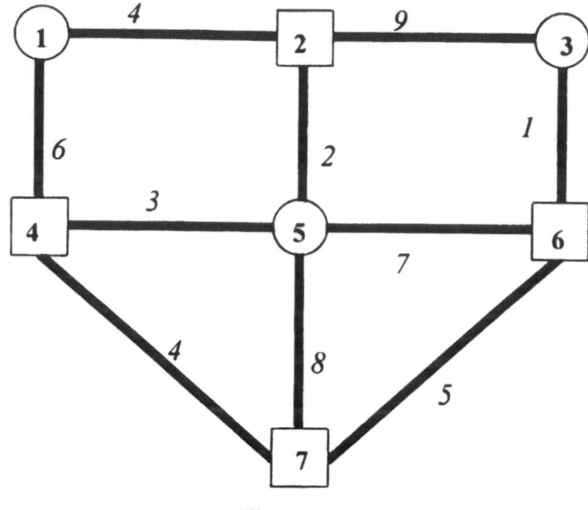

Figure 5.5

be 1 if the link is selected to be part of the matching, and 0 otherwise. If link j is of length c_j, the objective function is simply

$$\text{minimize} \sum_{j=1}^{N} c_j x_j \tag{5.23}$$

The requirement that there be exactly one link in the matching touching each

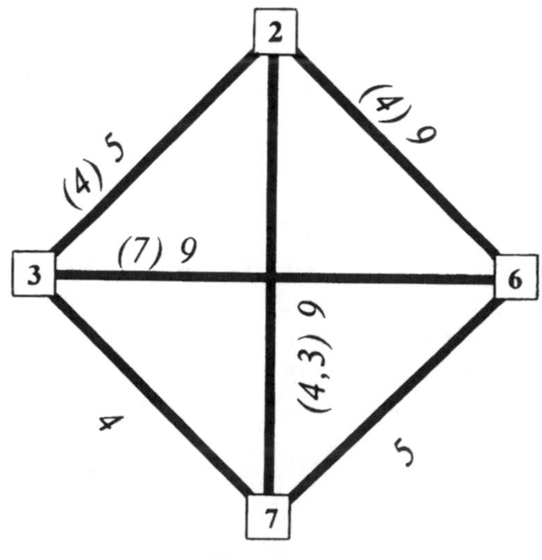

Figure 5.6

node is obtained by writing the sum of all links touching each node i and setting it to 1:

$$\sum_{\text{all links incident on node } i} x_j = 1 \tag{5.24}$$

This formulation alone serves to define the matching problem, but it, too, is very difficult to solve. Although the number of variables is smaller than that of the original Chinese postman's model, and the variables are 0–1 instead of integer, solution by ordinary integer programming methods is not practical for moderate-to-large problems.

There is, however, a set of added constraints that makes the resulting formulation unimodular.[4] If we consider any set of three nodes in the network and all of the links that have both ends incident in this set of nodes (there are at most three such links), the matching solution cannot contain more than one of these edges. This must be true since if two of these edges are in the matching, one of the three nodes is touched by more than one matching edge. By similar reasoning, in any set of five nodes in the network, among all the network edges that have both ends incident in this set of five nodes, no more than two of these edges can be in the matching. And the same argument can be extended to sets of seven nodes (where no more than three edges can be in the matching), nine nodes, and so on. If we write out *all* these constraints, for every subset containing an odd number of nodes up to one less than the number of nodes in the matching problem, the resulting linear program is unimodular. Thus if we could solve it by ordinary linear programming, the variables would all be $\{0\text{–}1\}$. Unfortunately, of course, the problem is far too big to be solved in that way. However, Edmonds developed an elegant special solution algorithm, based on the dual of this problem (in which there is a relatively small number of constraints but a huge number of variables, most of which are zero at any extreme point). Efficient computer implementations of Edmonds' matching algorithm are available and can be used to solve fairly large matching problems.

It should be noted that narrowing attention to the deadheading links, as the matching approach does, also simplifies the problem of estimating costs for the formulation. The cost of travel on collection links or passes is assumed to be invariant (at least with respect to routing), and therefore need not be estimated. Only the cost of deadheading along the links is needed, since it is that cost that appears in the matching formulation. This cost is much easier to estimate, as a function of distance and average travel speed.

With the solution of the matching problem, it is easy to determine which paths between odd-degree nodes to retrace in the routing network. These

[4]A unimodular formulation is one for which every extreme point has all variables $\{0\text{–}1\}$. Thus no matter what the objective function, solution by ordinary linear programming will find an optimum in which all variables have values of 0 or 1.

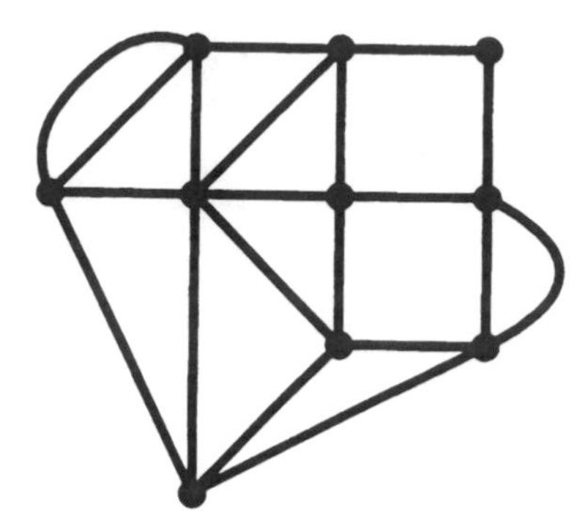

Figure 5.7

retracings make all nodes even-degreed, so that there does exist a tour that starts and finishes at the depot and traverses every link exactly once (considering the retracings as duplicate links).[5]

There remains the issue of how to construct such a tour. In fact, there are many different tours that can be built on an even-degree network, and their construction is based on the observation that a unicursal network can be decomposed (in many different ways) into a set of independent cycles with every link contained in exactly one loop. Figures 5.7 and 5.8 demonstrate this decomposition (with the dashed lines in Figure 5.8 joining replications of the same node). The ability to always decompose a unicursal network in this way can be proven by the following procedure. In a unicursal network, identify any cycle and remove all its links from the network. Since a cycle touches

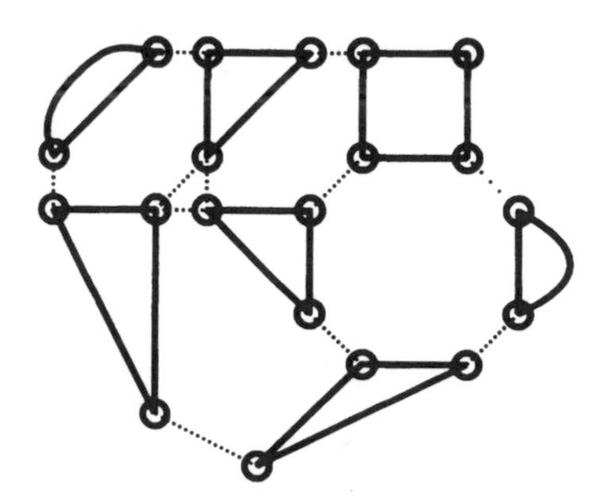

Figure 5.8

[5]Edmonds and Johnson (1973) provide an algorithm for determining the deadheading links directly on the collection network, without the intermediate step of finding all shortest paths between odd-degree nodes.

those nodes it contains an even number of times, its removal from the network leaves the network unicursal (although it may no longer be connected). Obviously, it is possible to continue in this way, removing cycles and leaving a unicursal network, until all links have been removed.

A tour in a unicursal network may be constructed by reversing this process:

1. Start at the depot and construct a tour traversing any closed loop, over any number of the links.
2. Find any node in the constructed tour that is touched by a link not yet in the tour. Since the node's degree is even, there are at least two links incident on this node that are not yet in the tour.
3. Starting with one of these links, construct a new closed loop consisting entirely of links that have not previously been traversed, returning to the starting node.
4. Break the tour at this node, and insert the new closed loop into the tour.
5. Using this new tour, repeat steps 2, 3, and 4. Continue until there is no node in the tour that has untraversed links incident upon it; this tour must then traverse all links in the network.

Thus we have a way of determining a tour on a single-vehicle collection region that covers the region in minimum total distance (or cost). There are, however, a number of practical difficulties with such solutions. These will be discussed later.

5.D.2. Chinese Postman's Problem with One-Way Streets

If the connected network over which collection is to take place consists entirely of one-way streets, the general approach is the same but the actual solution method is somewhat simpler. It has been shown (Ford and Fulkerson, 1962) that the equivalent of Euler's even-degree requirement for unicursality in a directed network is a requirement that the number of incoming arcs at each node be exactly equal to the number of outgoing arcs; that is, the nodes must be balanced. When this is not the case and there are unbalanced nodes, some paths will have to be retraced, and the retraced paths will have to originate at a node that has too few outgoing arcs and terminate at a node that has too few incoming arcs. The difference between this and the undirected case is that now the unbalanced nodes are in two disjoint groups: those with an excess of incoming arcs and those with an excess of outgoing arcs. Thus, instead of solving a matching problem on odd-degree nodes without having any indication of which nodes should be matched with which, it is only necessary to solve a similar problem on a bipartite network. In addition, of course, the imbalance may be greater than 1, so nodes may have to be paired up additional times.

This problem turns out to be considerably easier to model and to solve. As in the undirected case, a new network is constructed consisting only of the unbalanced nodes and shortest paths between them. But only the paths from nodes with a shortage of outgoing arcs (sources) to nodes with a shortage of incoming arcs (sinks) are needed. Then, to determine retracings, it is not necessary to solve a matching problem. Instead, an ordinary Hitchcock transportation problem may be solved, in which each source is considered to have available a number of units of flow equal to its imbalance, each sink is considered to require a number of units of flow equal to its imbalance, and the cost of shipping a unit of flow from source to sink is equal to the length of the shortest path. The solution of this problem specifies how many times various routes from source nodes to destination nodes are to be retraced. Given these retracings, tours may be constructed by the same loop-insertion algorithm, modified only in that the appropriate street direction must always be observed.

Figure 5.9 shows a 12-node network of directed arcs. It can be seen that nodes 3, 4, 5, 6, 7, 9, and 11 are unbalanced, with nodes 3, 5, 7, and 9 having an excess of incoming arcs, and nodes 4, 6, and 11 having an excess of outgoing arcs. At nodes 4, 5, and 11, the excesses are of magnitude 2, while all others are 1. Thus arcs must be retraced leading out of nodes 3, 5, 7, and 9, and leading into nodes 4, 6, and 11 in order to balance these nodes.

Shortest-path problems may be solved on this network to find all the shortest paths from nodes 3, 5, 7, and 9 to nodes 4, 6, and 11. These paths and their lengths are shown in Figure 5.10, and Table 5.1 indicates the intermediate nodes involved in each path.

An ordinary Hitchcock transportation problem may be solved on this transformed network, with the amounts available on the left being equal to the excess of incoming arcs at these nodes, and the amounts required at the nodes

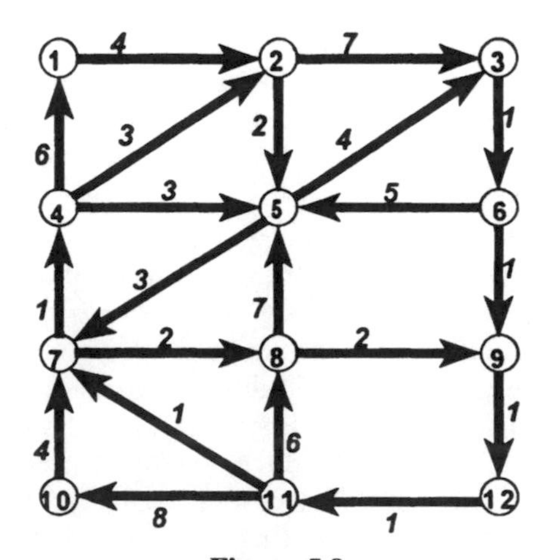

Figure 5.9

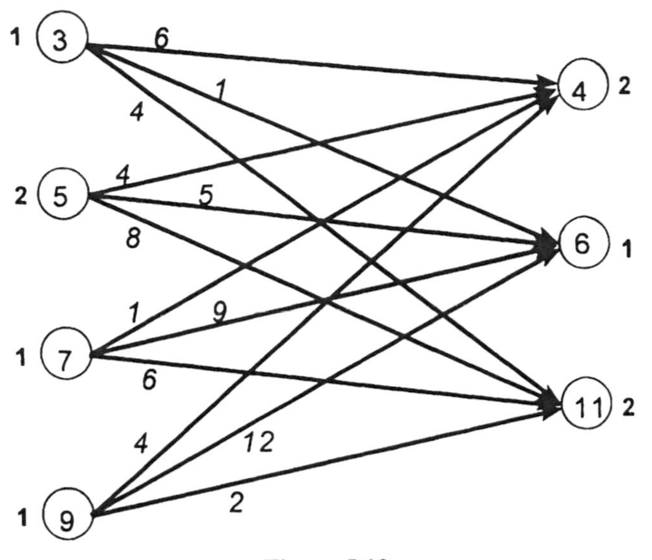

Figure 5.10

on the right being equal to the excess of outgoing arcs. Shipping costs are the lengths of the shortest paths. The solution of this small problem indicates retracing arcs 3–11, 5–4, 5–6, 7–4, and 9–11. The retracings of these shortest path are shown in Figure 5.11, which indicates that all nodes are now balanced. Total length of retracings is 16.

5.D.3 Chinese Postman's Problem with Mixed One- and Two-Way Streets

When there are some one-way streets and some two-way streets in the network, the problem is of significantly greater difficulty. Part of the difficulty arises from the fact that the necessary and sufficient conditions for unicursality on the network are much more complex. These conditions are as follows (Ford and Fulkerson, 1962):

TABLE 5.1 Intermediate Nodes in Each Path

	To		
From	4	6	11
3	6,9,12,11,7		6,9,12
5	7	3	3,6,9,12
7		4,5,3	8,9,12
9	12,11,7	12,11,7,4,5,3	12

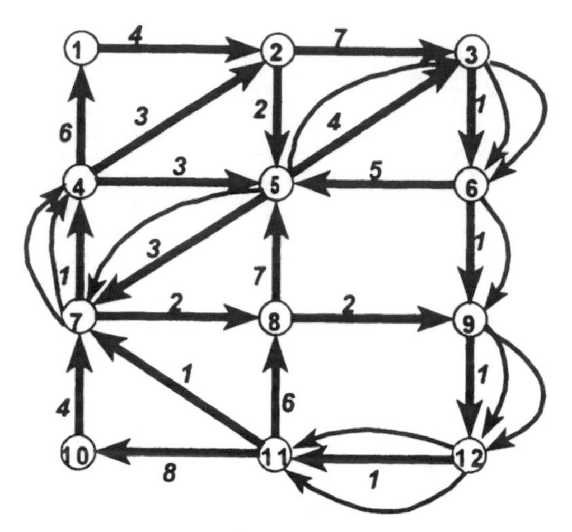

Figure 5.11

a. The network must be connected.
b. Every node must be of even degree.
c. The number of undirected links incident on each node must be at least as great as the imbalance in directed links on the node.
d. The number of undirected links incident on each subset of nodes must be at least as great as the imbalance in directed links on the subset.

Condition (c) may be seen intuitively as allowing sufficient undirected links to be used in the appropriate direction to make up the imbalance in directed links. Condition (d) is a generalization of condition (c), necessary because the satisfaction of condition (c) does not guarantee that (d) will also be satisfied. As shown in Figure 5.12, although every node satisfies requirement (c), the three-node subset in the upper left corner does not satisfy (d). There are three directed links into this region, and only one (undirected) link that can be used to leave the region. A tour is not possible on this network without retracing that undirected link twice.

When a mixed network is not unicursal, there two distinct cases for solution. If every node has even degree, Edmonds and Johnson (1973) provide a flow solution similar to that given above for the all one-way case, but with a number of links added to the network before the problem is solved. However, in the more general case in which some nodes are of odd degree, Papadimitriou (1976) has shown that the problem is NP-complete, and is therefore probably not solvable for large cases. Minieka (1979) presents a flow-with-gains model, which must then be solved by some integer programming technique and thus is usable only for small problems. The same model can be used for instances in which the cost of travel on some links is different in opposite directions.

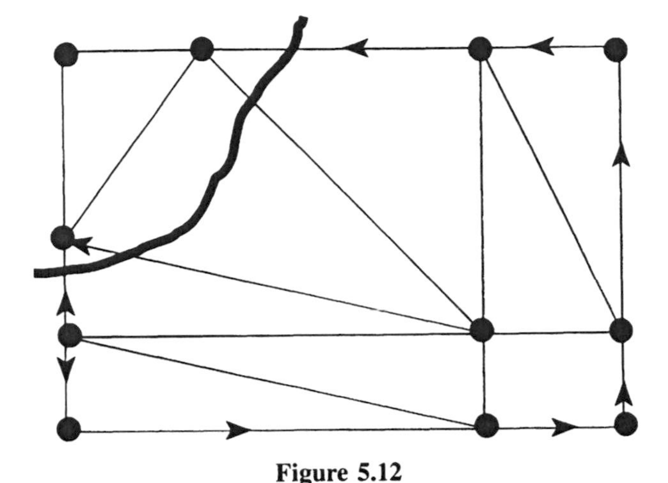

Figure 5.12

5.D.4. Chinese Postman's Problem with One-Side Collection

In some collection systems, collectors are not permitted to cross the streets; that is, collection may only take place on the side of the street that the vehicle is traversing. In the all-two-way-street case this restriction makes the problem particularly easy to solve. If every two-way street is represented as two oppositely directed one-way streets, we now have an all-one-way-street case. Further, since each original street is now both an incoming and an outgoing link at both incident nodes, all nodes are balanced. That is, a tour is always possible on a set of two-way streets by traversing each street once in each direction, without additional retracing.

If a one-side requirement is imposed on a system with all one-way streets, the solution method is not changed. Each one-way link is represented as two links, both in the same direction. A transportation problem is then solved with the unbalanced nodes as sources and sinks, and the results indicate additional traverses of the links beyond the two necessary to cover both sides of the street.

Similarly, in a mixed one- and two-way system the one-side collection requirement does not change the solvability or the approach. Each street is represented as two links, either in opposite directions or in the same direction, depending on whether the street is two-way or one-way. After these links are added, if all nodes are of even degree, there is a solution method given by Edmonds and Johnson (1973); otherwise, the problem is very difficult and Minieka's (1979) formulation may be used for small cases.

In the most general case, the one-side collection requirement exists only for some streets. There is no difficulty in modeling this problem by representing streets that require one-side collection with two arcs in the appropriate direction, while other streets are represented as a single link. The solution method then depends on whether one has a mixed or directed case after the extra arcs are added.

5.D.5 Chinese Postman's Problem with Optional Streets

All the models of the Chinese postman's problem are based on the assumption that travel (and therefore collection) must take place on every link in the network. In a collection system, however, it is quite likely that there will exist streets along which there is no collection. However, it is not appropriate to omit these streets entirely, since they may provide shorter travel distances while deadheading.

Since the algorithms given above concentrate on determining the deadheading links (retracings) between odd-degree nodes, it is easy to incorporate noncollection streets into the model. While the degree (in the undirected case) or the imbalance (in the directed case) of each node is being determined, the noncollection links are ignored, since they need not be traversed and they do not contribute to the required travel. However, when shortest paths between odd-degree nodes or from source nodes to destination nodes are being determined, the noncollection links are included for consideration as part of any shortest path. Thus, if they provide a shorter deadheading distance, they will be used, and otherwise, they will not be traversed at all.

5.D.6. Districting or Partitioning

The preceding sections describe algorithms for determining optimal routing of a single vehicle in a predetermined district that is known to constitute a load for the vehicle. If true minimum cost is to be obtained, of course, the entire collection region must be considered, and minimization must be done to include the determination of districts and with consideration for the vehicles' trips between the districts and the depot. This problem is known as the *m-postmen's problem* in the edge-routing case, and as the *m-salesmen's problem* when the model requires a visit to each node. Both of these problems are difficult and are generally solved by heuristics. The m-salesmen's problem is not addressed in this chapter.

In a large collection area, the minimum travel distance is achieved by a single collection vehicle traversing a tour found by solving the Chinese postman's problem as described in previous sections. If the area is too large for a single vehicle, it must be subdivided into multiple collection districts. Then it is likely that to reach their own districts, some vehicles will have to deadhead over streets that are part of the collection district of other vehicles, increasing the total travel distance beyond that of a single vehicle. That is, a lower bound on the total travel distance in the m-postmen's problem is given by the optimal solution of the single-postman's problem on the collection area, no matter how large.

In the m-postmen's problem, there are two additional criteria for good routing in addition to the minimization of total distance. The first is load balancing: Each collection district must generate no more load than can be carried by a collection vehicle in one pass, and districts should not generate amounts that are far less than a vehicle load. A second desirable feature is that all the streets being serviced by a single vehicle should be geographically contiguous or nearly so.

A simple way of determining contiguous, balanced districts is given by Liebman et al. (1975). The method is first to solve the matching problem on the odd-degree nodes and add the duplicated links to the network. Then simply draw district boundaries through the network, passing through each node in such a way as to leave it of even degree on both sides of the boundary. Figure 5.13 shows one way of doing this. Since each node is of even degree *within each district,* the individual districts are still unicursal, and a tour can therefore be constructed inside each district. Before constructing the tours, however, it is possible to balance the loads heuristically. Any cycle of streets can be moved from one district to an adjacent one without changing the unicursality of the districts. It is a fairly simple matter to locate districts with light loads and to move the boundaries slightly to include one or more cycles from adjacent districts with heavy loads, continuing until the districts are closely balanced. Then the individual tours can be constructed in each district.

Of course, the method given above does not account for the distances traveled by the collection vehicles between the depot and their collection areas. A heuristic algorithm that attempts to minimize these added distances was provided by Male and Liebman (1978). The algorithm utilizes five steps:

1. The Chinese postman's problem is solved on the entire region. The duplicated links found in this step are considered part of the network in the remaining steps.
2. The resulting network is decomposed into a set of cycles. The cycles should be small, elementary loops, such that every link is in one and only one cycle. These cycles will be the building blocks for the districts.
3. A districting network is created in which each of the cycles is represented by a node. Adjacent cycles are joined in the districting network by links between their respective nodes. These links are considered to

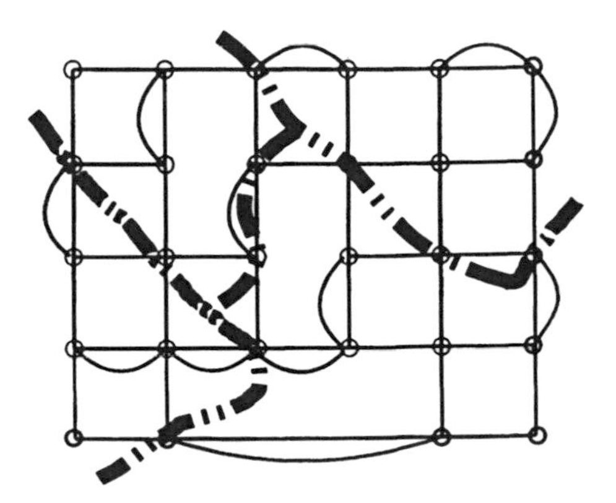

Figure 5.13

have zero length. In addition, a node representing the depot is added to the districting network. A link is added from the depot to each node, with length equal to twice the length of the shortest path from the depot to the closest node in the cycle on the original network.

4. On the districting network, any subtree rooted at the depot, with only one link touching the depot, represents a unicursal tour on the set of cycles represented by the nodes of the subtree, including the travel from the depot and return. The *added* length of the tour is the length of the tree. Of course, since all links have zero length except the link touching the depot, so the length of the tree is simply the length of that link, which is the distance from the depot to the tour and back. If there are to be m collection tours, the algorithm first selects the m shortest links touching the root and attempts to build balanced subtrees connected to these links. If it is unable to do so, it successively selects longer and longer starting links until it is able to build the balanced subtrees.
5. The m subtrees are decoded into the corresponding tours.

This algorithm is heuristic because it considers only a subset of all possible m-postman tours. First, only tours that contain all of the duplicate links of the optimal Chinese postman's solution are considered, but the optimal m-postman solution may not contain some of those links. Second, the arbitrary selection of the individual cycles to be represented as nodes in the districting tree eliminates a large number of possible tours. Finally, the search for a balanced set of subtrees based on a particular set of m starting links is not exhaustive and may therefore miss the optimal set. Improvement in the solution may be obtained after the districting has been completed by considering each district independently and looking for changes that reduce the length of the tour. Among the possible improvements is elimination of one of the deadheading links in the district (which leaves the district nonunicursal), and routing the collection vehicle to enter the district via one node and leaving the district from a different node.

5.D.7. Manual Routing and Districting

The exact algorithms for single-vehicle routing and the heuristic methods solving multivehicle problems are complex and data intensive. Further, they do not allow full representation of the many other considerations that fully describe a refuse collection route problem. Among these, for example, are the different traffic conditions encountered at different times of day (which make some link costs time dependent) and the additional costs of left turns as compared with right turns. For these reasons it is unlikely that simple algorithmic computer codes will provide solutions that can be implemented directly. A number of proprietary codes are used by routing consultants, but it is clear that these codes contain many heuristic procedures developed through experience for obtaining good routes, and often even their results are fine-tuned manually.

An understanding of the theory as described in previous sections provides a great deal of aid in attempting to determine routes without resorting to computer codes. For example, much better routes can be developed by attempting to find a good solution to the matching problem on odd-degree nodes first rather than simply beginning to draw tours.

After a first-cut matching solution has been found, a result due to Kwan (1962) can be used to search for improvement. Kwan, seeking a general algorithm for the Chinese postman's problem (prior to the work of Edmonds), noted that in any cycle of the street network, if the sum of the lengths of streets that are deadheaded is greater than the sum of the lengths of streets that are not deadheaded, deadheading can be reduced by reversing the deadheading and nondeadheading streets in the cycle. This reversal, of course, keeps all nodes of even degree, so the network remains unicursal. Clearly, in an optimal solution no such reversal is possible, and Kwan proved that in any nonoptimal solution there exists at least one cycle in which such a reversal can be done. Kwan's proposed solution algorithm, which required checking every cycle in the network, is not practical for computer implementation for large networks. However, it does provide a heuristic method for manual improvement, simply by inspecting the network for cycles that can be reversed to reduce the length of deadheading.

A general manual procedure was proposed by Marks and Stricker (1971). In their procedure, districting is done first, by manually determining good districts, keeping an even number of odd-degree nodes in each. Then a matching solution in each district is estimated by making a first pairing and then seeking improvement as described above. Finally, routes are drawn using the matching solution to determine deadheading links. An excellent set of heuristics for constructing the routes is given by Shuster and Schur (1974).

5.E. SIMULATION MODELS

Simulation has been used in a number of instances to explore specific issues in solid waste collection systems. Queuing issues at the disposal point were studied by Quon et al. (1965). The effect of changes in work rules and collection policies (including frequency) were considered by Quon et al. (1969). Truitt et al. (1969) constructed a detailed simulation of the Baltimore collection system and used it to investigate the effect of adding transfer stations and the cost of increased collection frequency.

Simulation models can be very useful in investigating specific issues in specific systems. However, it must be recognized that because of the wide variation in collection systems, simulation programs are unlikely to be transportable to other systems, and their results are unlikely to be applicable in other situations.

5.F. OTHER MODELS

This brief chapter is not intended to be exhaustive in its discussion of models applied and applicable to solid waste management. Modeling has been used to approach a number of other issues, including the prediction of refuse amounts (Hudson et al., 1973), the scheduling of recycling measures and landfill replacement (Lund, 1990), crew scheduling (Bodin et al., 1983), and the analysis of hazardous waste transport and treatment (Jennings and Sholar, 1984).

5.G. CONCLUSIONS

Solid waste management is an excellent example of the often-repeated comment that mathematical modeling is a decision-making *aid* rather than a decision-making *procedure*. The observations made in Section 5.D.7 on manual routing apply equally well to other modeling issues. Like many other public systems, solid waste systems are highly individualistic. They are also highly interrelated with many other municipal systems, and decisions made in one have major impacts on others. Further, there are social, political, and aesthetic issues that are by no means fully understood, with outcomes that are best characterized as "unpredictable." Thus it is unlikely that the full complexity of these interactive decisions can even be adequately modeled (at least in a *solvable* model) for automatic decision selection.

Under such circumstances, the appropriate use of optimization modeling is to provide guidance to those involved in making decisions, not to provide immediately implementable solutions. Outcomes of individual models must be recognized as suboptimal at best, and used to suggest families of solutions that are reasonable and should be explored further. Of equal or perhaps greater importance is the system insight and understanding that can be gained by the analyst who exercises models appropriately. Merely solving a model once does not do the job; but doing extensive sensitivity analysis, making model modifications, and asking a range of what-if questions will develop in the analyst a comprehensive grasp of the system behavior that may in the end be the best decision-making tool of all.

EXERCISES

5.1. In Sections 5.C.1 and 5.C.2, the Hitchcock and transshipment models of Sections 5.B.1 and 5.B.2 were extended to be facility siting models by the addition of boolean site variables. Use the same technique to extend the minimum-cost circulation model of Section 5.B.3.

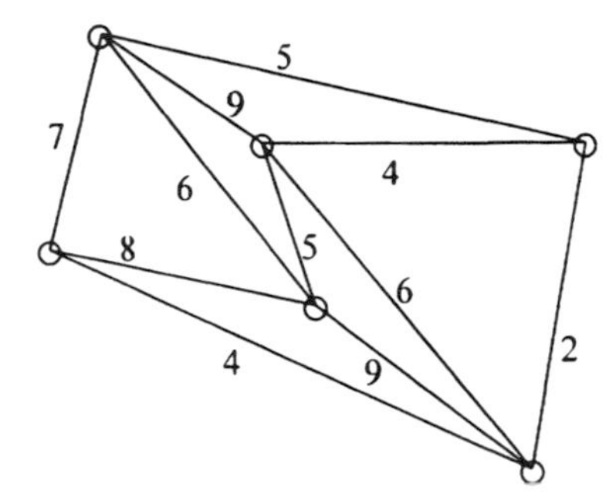

Figure 5.14

5.2. Use any linear programming routing to solve the minimum matching problem on the network shown in Figure 5.14, by adding the full set of required constraints to make the problem unimodular.

REFERENCES

Beltrami, E. J., and L. D. Bodin, 1974. Networks and Vehicle Routing for Municipal Waste Collection, *Networks* 4(1), 65–94.

Bodin, L. D., B. L. Golden, A. A. Assad, and M. O. Ball, 1983. Routing and Scheduling of Vehicles and Crews: The State of the Art, *Computing and Operations Research,* 10, 63–211.

Edmonds, J., 1965. Maximum Matching and a Polyhedron with 0,1 Vertices, *Journal of Research, National Bureau of Standards,* 69B, January–June.

Edmonds, J., and E. Johnson, 1973. Matching, Euler Tours and the Chinese Postman, *Mathematical Programming,* 5(1), 88–124.

Efroymson, M., and T. Ray, 1966. A Branch-and-Bound Algorithm for Plant Location, *Operations Research,* 14, 361.

Eiselt, H. A., M. Gendreau, and G. Laporte, 1995. Arc Routing Problems: I. The Chinese Postman Problem, *Operations Research,* 43(2), 231–242.

Euler, L. 1736. Solutio Problematis ad Geometriam Situa Pertinentis, *Comment. Academiae Sci. 1 Petropolitanae,* 8, 128–140. See also The Konigsberg Bridges, *Scientific American,* 189, 1953, 66–70.

Ford, L. R., Jr., and D. R. Fulkerson, 1962. *Flows in Networks,* Princeton University Press, Princeton, NJ.

Golueke, C. G., and P. H. McGauhey, 1970. *Comprehensive Studies of Solid Waste Management* (First and Second Annual Reports), Public Health Service Publication 2039, U.S. Government Printing Office, Washington, DC.

Gottinger, H. W., 1988. A Computational Model for Solid Waste Management with Application, *European Journal of Operational Research,* 35, 350–364.

Huang, G., B. W. Baetz, and G. G. Patry, 1992. A Grey Linear Programming Approach for Municipal Solid Waste Management Planning Under Uncertainty, *Civil Engineering Systems,* 9, 319–335.

Hudson, J. F., D. S. Grossman, and D. H. Marks, 1973, *Analysis Models for Solid Waste Collection,* Department of Civil Engineering, Massachusetts Institute of Technology, Cambridge, MA.

Jennings, A. A., and R. L. Sholar, 1984. Hazardous Waste Disposal Network Analysis, *Journal of Environmental Engineering, ASCE* 110, 325–342.

Kirca, O., and N. Erkip, 1988. Selecting Transfer Station Locations for Large Solid Waste Systems, *European Journal of Operational Research,* 38, 339–349.

Kwan, M.-K., 1962. Graphic Programming Using Odd or Even Points, *Chinese Mathematics,* 1, 273–277.

LaPorte, G., 1992. The Traveling Salesman Problem: An Overview of Exact and Approximate Algorithms, *European Journal of Operational Research,* 59, 231–247.

Lawler, E. L., J. K. Lenstra, A. H. G. Rinnooy Kan, and D. B. Shmoys (eds.), 1985. *The Traveling Salesman Problem: A Guided Tour of Combinatorial Optimization,* Wiley, Chichester, West Sussex, England.

Liebman, J. C., J. W. Male, and M. Wathne, 1975. Minimum Cost in Residential Refuse Vehicle Routes, *Journal of the Environmental Engineering division, ASCE,* 101 EE3, 399–412.

Lund, J. R., 1990. Least-Cost Scheduling of Solid Waste Recycling, *Journal of the Environmental Engineering Division, ASCE,* 116, 182–197.

Male, J. W., and J. C. Liebman, 1978. Districting and Routing for Solid Waste Collection, *Journal of the Environmental Engineering Division, ASCE,* 104 EE1, 1–14.

Marks, D. H., 1969. Facility Location and Routing Models in Solid Waste Collection Systems, Ph.D. dissertation, The Johns Hopkins University, Baltimore, MD.

Marks, D. H., and R. Stricker, 1971. Routing for Public Service Vehicles, *Journal of the Urban Planning and Development Division, ASCE,* 97 UP2, 165–178

Minieka, E., 1979. The Chinese Postman Problem for Mixed Networks, *Management Science,* 25(7), 643–648.

Papadimitriou, C., 1976. On the Complexity of Edge Traversing, *Journal of the Association for Computing Machinery,* 23, 544–554.

Quon, J. E., A. Charnes, and S. J. Wersan, 1965. Simulation and Analysis of a Refuse Collection System, *Journal of the Sanitary Engineering Division, ASCE,* 91 SA5, 17–36.

Quon, J. E., M. Tanaka, and S. G. Wersan, 1969, Simulation Model of Refuse Collection Policies, *Journal of the Sanitary Engineering Division, ASCE,* 95 SA3, 575–592.

Shuster, K. A., and D. A. Schur, 1974. *Heuristic Routing for Solid Waste Collection Vehicles,* U.S. Environmental Protection Agency Publication SW-113, U.S. Government Printing Office, Washington, DC.

Truitt, M. M., J. C. Liebman, and C. W. Krusé, 1969. Simulation Model of Urban Refuse Collection, *Journal of the Sanitary Engineering Division, ASCE,* 95 SA2, 289–298.

CHAPTER 6

Hazardous Waste Management

MARK A. TURNQUIST and LINDA K. NOZICK

6.A. INTRODUCTION

The term *hazardous waste* covers many different types of waste materials. At home, you may have old paint cans or used flashlight batteries. These are considered hazardous wastes and generally cannot be disposed of in municipal landfills. However, you probably do not consider handling these items in your home as a risky activity. The dry cleaner in your community must dispose of used dry cleaning fluid, which typically contains tetrachloroethylene. If breathed or ingested in sufficient quantity, tetrachloroethylene can cause central nervous system damage and is a suspected human carcinogen. Thus this waste must be handled more carefully, but the total volume of waste generated by a single business is relatively small, and it can be packaged and handled with relatively low risk. A local manufacturing facility may produce a variety of solid and liquid wastes that must be treated as hazardous, and these wastes may be produced in relatively large quantities. Bulk shipment of these wastes to a storage or treatment facility may create risks in the transportation system, and air emissions or other discharges from the treatment facilities can create additional risks. The electric utility that produces the power you use in your home may produce some of that power using nuclear plants. The spent fuel assemblies from these plants contain several highly radioactive elements, and these high-level radioactive wastes require very careful handling and disposal.

Because the character of these hazardous wastes varies so widely, there are many types of treatment and disposal technologies that have application to some type or other of hazardous waste. This is a text on civil and environmental engineering systems, and thus this chapter focuses primarily on planning and management activities related to hazardous wastes rather than on the physical, chemical, or biological aspects of their treatment. We want to describe how elements of the waste management system are connected and how models can be used to assist major decisions regarding that system. We focus on models of risk assessment; models for locating storage, treatment, and disposal facilities; and models to aid decisions about how to transport the wastes.

The use of systems models and extensive analyses is most effective when we are concerned with wastes that pose a substantial hazard. Quite clearly, the level of hazard (and corresponding level of care in handling, transportation, treatment and disposal) for different types of wastes spans an enormous range, as indicated by the small set of examples in the first paragraph of this chapter. We focus our attention on management of wastes that are at the high-hazard end of the spectrum, where substantial attention to careful planning is required of both public- and private-sector decision makers.

We begin with a discussion of what constitutes a hazardous waste, including a brief synopsis of the legislative and regulatory framework for managing hazardous wastes in the United States. We then discuss modeling approaches to understanding and analyzing the level of risk involved in particular hazardous waste management activities. The analysis of risks forms the basis for models of various types of decisions that must be made in managing these wastes, including siting of facilities and transporting the wastes. The models (and the data to support them) are inherently spatial in structure, so we emphasize the use of geographical information systems (GIS) technology as a means of supporting and implementing the models.

6.B. CHARACTER OF HAZARDOUS WASTES

During the 1960s public concern for the most noticeable effects of pollution in the environment led to congressional action in the Clean Air Act, the Clean Water Act, and the Solid Waste Disposal Act. By the 1970s, increasing attention to a less obvious form of pollution—contamination of groundwater used for drinking water—led to additional federal action. In 1976, the Resource Conservation and Recovery Act (RCRA) and the Toxic Substances Control Act (TSCA) were enacted to control the generation, transportation, and management of hazardous wastes and the use of toxic substances.

In 1980, largely in response to the Love Canal incident, Congress passed the Comprehensive Environmental Response, Compensation, and Liability Act (CERCLA), which has become known as Superfund. This legislation provided a framework for cleaning up past hazardous waste disposal sites that presented a dangerous threat to public health and the environment.

The Hazardous and Solid Waste Act (HSWA) of 1984 substantially amended RCRA, to place additional restrictions on land disposal of wastes and further define what wastes are to be considered hazardous. The reauthorization of CERCLA in 1986 resulted in the Superfund Amendments and Reauthorization Act (SARA), which mandates increased activity to clean up inactive waste sites. This array of legislation constitutes the basis for the nation's hazardous waste regulatory system.

RCRA and amendments added in HSWA provide the basic framework for classifying wastes as hazardous and for specifiying requirements on their handling, storage, treatment, and disposal. The Hazardous Materials Transportation Uniform Safety Act (HMTUSA) of 1990 establishes policies and standards that apply specifically to transporting hazardous materials, regardless of whether or not they are wastes.

RCRA defines four characteristic properties that qualify a waste as being hazardous (Griffin, 1988): (1) ignitability, (2) corrosivity, (3) reactivity, and (4) toxicity. Any one of these characteristics can cause a waste to be considered hazardous.

Ignitability refers to the characteristic of being able to sustain combustion. It is measured by flashpoint (the temperature at which the substance will ignite). RCRA defines any waste with a flashpoint below 140°F as hazardous.

Corrosive wastes are those with extreme values of pH. They may destroy containers, contaminate soil and groundwater, or react with other materials to cause toxic gas emissions, and they present a very specific hazard to human and animal tissue. RCRA defines any waste substance whose pH is below 2, or above 12.5, as a corrosive waste.

Reactive wastes may be unstable or explosive during handling. Substances whose reactions with other materials generate pressure are also included in this category, as are substances that react with water.

Toxicity refers to potential poisonous effects of specified concentrations of the substance on humans, fish, or wildlife. Toxicity is often expressed in terms of the dose that is received, usually measured in milligrams of the substance per kilogram of weight of the recipient (mg/kg). Toxicity is species specific as well as chemical specific (Kamrin, 1988). Toxicologists frequently use the concept of a dosage of a specific toxin that is fatal to 50% of an exposed population of a particular species as a measure of relative toxicity. This measure of acute toxicity is called the *LD_{50} level.* Most of the testing to establish acute toxicity levels has been done using rats. For example, the LD_{50} value in rats of arsenic is 48 mg/kg, and that of dioxin is 0.001 mg/kg (Kamrin, 1988), so dioxin is considered much more toxic than arsenic.

Radioactive wastes constitute a category that is different from typical RCRA-defined hazardous wastes. The exposure hazard for radioactive wastes arises from high-energy radiation rather than from fire or explosion, or from breathing, touching, or ingesting the material. From a regulatory standpoint, the movement, storage, and disposal of radioactive wastes is covered by separate, specific regulations and standards. *Low-level radioactive wastes* (characterized by low levels of alpha, beta, and gamma emissions) are generally solids that can be packaged in drums and disposed of in specially designated landfills. *Intermediate-level radioactive wastes* have somewhat higher levels of emissive activity, and may be either liquids or solids but usually can also be packaged in specially designed drums for transportation or disposal. *High-level radioactive wastes* are typically spent-fuel assemblies from nuclear reactors or highly radioactive materials resulting from defense activities. These wastes require very special consideration in transportation and disposal.

6.C. RISKS ASSOCIATED WITH HAZARDOUS WASTES

The risks associated with hazardous wastes may be either acute or chronic. An *acute risk* is the potential for danger that is the result of a specific event: for example, a transportation accident that results in release of a cloud of toxic gas from a ruptured tank truck. A *chronic risk* is associated with long-term exposure to low levels of some substance: for example, as a result of chemicals leaching from a landfill into groundwater that is used for a community's water supply. In either case the overall structure we can construct for analyzing risks is similar, but the relative emphasis on different parts of the analysis changes depending on the nature of the risks that are deemed most important.

Risk analysis is centered around five basic steps, as shown in Figure 6.1 (Rowe, 1977). Initially, there is some *causal event*. This may be a dramatic event, as in a transportation accident, or may be completely unnoticed, as in the leaching of chemicals from a landfill. The first step of a risk analysis is concerned with defining the causal event(s). In the case of analyzing acute risks, there is typically considerable effort on estimating a probability of occurrence for each such event.

When a causal event occurs, a variety of *outcomes* may be defined, each with associated probabilities. Defining these outcomes, and their probabilities, is the second step of the process defined in Figure 6.1. For example, in the case of a tank truck accident, the tank may rupture or not, valves may break open or not, and so on. In such a case, the definition of the outcomes, along with their probabilities of occurrence, could be used to define a probability distribution for the magnitude of released material.

The third step involves *exposure* of people, property, or the environment to the results of the event outcome. The nature of exposure needs to be defined, and the pathways of possible exposure must be identified. We also

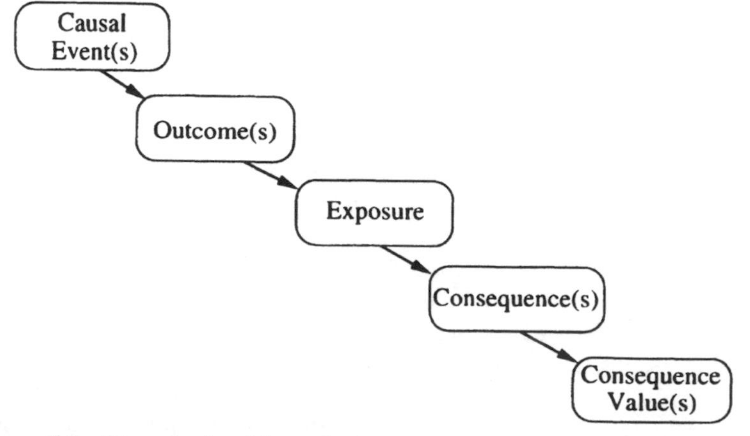

Figure 6.1 Steps in the risk analysis process. (Adapted from Rowe, 1977.)

need to estimate the probability of exposure along each pathway. For example, Figure 6.2 illustrates the multiple pathways through which constitutents of a hazardous waste might move from sources to recipients. For any specific hazardous waste, or any specific causal event and outcome, some of these pathways may be irrelevant. However, to analyze specific risks, we must identify which pathways are relevant and how likely exposure is along each.

For the people, animals, or pieces of property that are exposed, the fourth step involves estimation of *consequences*. The total level of exposure along all pathways must be summed and levels of consequence defined. For example, if a tank truck is involved in an accident and a valve breaks open allowing a toxic gas to be released, a person near the accident may be exposed to the gas by inhalation as well as contact with the skin, eyes, and so on. The consequences of this exposure may range from temporary discomfort through various degrees of illness to death. A key element in assessment of consequences is often a *dose–response relationship*, which attempts to measure the probable level of consequences associated with differing degrees of exposure. There is typically great uncertainty in such relationships, especially when the response may not appear until months or years after the exposure.

Finally, the fifth step in the risk analysis process focuses on *valuation* of the consequences of exposure. For damage to property, this can be done on a monetary basis (using repair or replacement costs) with modest difficulty. However, when the consequence is the death of a person (even if a "statistical" person), this becomes much more difficult.

In the analysis of acute risks, it is common to focus primary attention on the first three steps of the risk analysis process: events, outcomes, and ex-

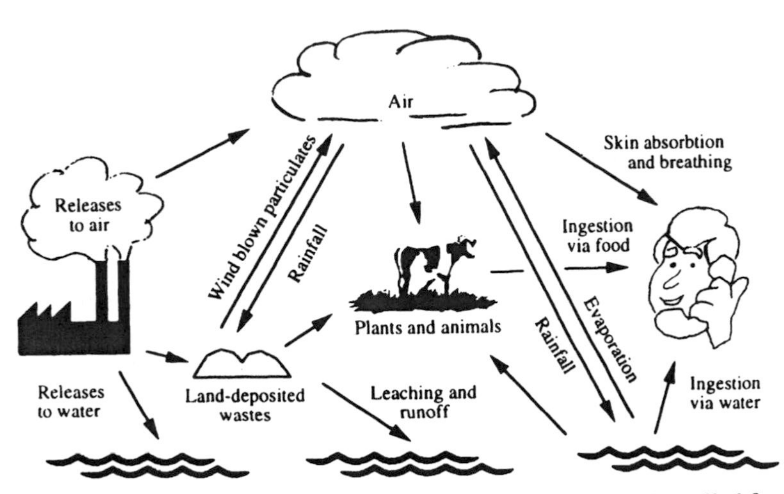

Figure 6.2 Potential pathways of exposure to hazardous wastes. Compiled from Figure 6.1 from "Controlling Cross-Media Pollutants: An Issue Report," published by The Conservation Foundation 1984. Used with permission of the World Wildlife Fund.

posure. The risk arises from the occurrence of very specific (and rare) events, such as the sudden release of material from an industrial or transportation accident. For those people, animals, or property exposed to the material, the level of exposure is likely to be quite high and the consequences quite severe. However, the events themselves are very uncommon, so the analysis focuses on determining *how likely* certain events (or event sequences) are, *how much material* might be released, and what size and shape a *region of exposure* might be.

By contrast, the analysis of chronic risks focuses primarily on the last three steps of the risk analysis process: exposure, consequences, and valuation. The risk arises from low levels of exposure over long periods of time, so the major questions of concern are: Who is exposed? At what levels? What is the dose–response relationship? What are the potential consequences? What is an acceptable level of risk for the general public, or for workers exposed while on the job?

A detailed description of the methods of risk analysis is well beyond the scope of what we can cover in this chapter. Some useful additional sources on the general methods and issues in risk analysis are the works by Rowe (1977), Henley and Kumamoto (1981), Fischoff et al. (1981), and Glickman and Gough (1990). Because our specific focus in this chapter is on site location for storage, treatment, and disposal facilities, and on transportation of wastes to those facilities, we concentrate on one element of risk analysis that is very closely associated with both site location and transportation decisions, the determination of exposure areas around facilities and transportation routes.

6.C.1. Exposure Areas

Much of the concern in siting facilities for storage, treatment, and disposal of hazardous wastes is based on the desire of people not to be near such a facility. The result is the so-called NIMBY ("not in my backyard") syndrome, which arises from concern about exposure to the effects of normal facility operation (e.g., air emissions or leaching of chemicals into groundwater), as well as concern about potential accidents resulting in release of hazardous materials in the vicinity of the facility. Thus the *radius of potential exposure* around facilities is an important input into site location decisions (and the models used to assist in making those decisions). Additional details on exposure to hazardous waste in groundwater are discussed in Chapter 3 (Section 3.C.3).

Transportation risks are associated with accidents that result in release of hazardous materials to the area surrounding the accident location. Thus one criterion for transportation routing decisions may be the population in exposure areas around transportation facilities (e.g., highways, rail lines, terminal facilities, etc.).

Population in exposure areas is unlikely to be the sole criterion in either siting a facility for treatment, storage, or disposal of hazardous wastes, or for making decisions on transportation options for moving such wastes. Many other criteria are also likely to be considered in the decisions. However, exposure area provides a good example of how risk analysis, facility siting, and transportation routing are all linked together. Thus we will use this example criterion as the basis for further discussion. In Section 6.C.2, we describe an example analysis of risk exposure for a chemical processing facility in the Netherlands. In the following sections we show how exposure area calculations are used in facility siting models and transportation routing models.

6.C.2. Example of Exposure Calculation

In Limburg, in the southeastern part of the Netherlands, the Dutch State Mines (DSM) site is the main facility for one of the major chemical companies in the country. The company employs about 10,000 people at the site and produces ammonia, acrylonitrile, and other intermediate chemical products that are used in fertilizer, polymers, and many other end products. Until about 1960, the site contained a large coal mine, and housing developments around the site are still influenced by the previous planning of housing for the miners, which means that residential areas are in close proximity to the site and the current chemical installations.

In 1975, a huge explosion occurred on this site, killing 14 people and injuring 104. That event, as well as subsequent industrial accidents in various parts of the world, led the government of the Netherlands to initiate a comprehensive environmental review for the area, including a risk analysis. The results of that analysis are described by van Kuijen (1989).

Of the 35 separate facilities on the overall site, an initial screening showed that 10 posed a risk outside the site boundary. Subsequent analysis focused on these 10 facilities. For each, a set of potential failure scenarios was created, and probabilities for each of these accident types were estimated using models based on an underlying database of failure frequencies for various components in a specific processing facility. The outcomes from these failure events were then estimated, based on underlying properties of the chemical substances involved. The possible outcomes included fires, explosions, and toxic vapor clouds. These outcomes were then combined with data on local weather conditions, population distribution, and so on, to develop estimated risk contours around the plant site.

The end result is depicted in Figure 6.3, which shows a risk contour plot for the site. The contour lines plotted are lines of equal risk, with the risk level labeled. A value of 10^{-6}, for example, means that the added risk to the population along this contour line is approximately 1 chance in 1,000,000 of death per year of exposure. The shaded areas in Figure 6.3 indicate villages in the vicinity of the site. The region shown in Figure 6.3 is 6 km (east–west) by 7.5 km (north–south). The tick marks along the edges of the figure are 1 km apart. Thus the area in Figure 6.3 that has a risk above 10^{-6} (the area enclosed by the 10^{-6} risk contour) is approximately 8 km^2.

The development of exposure areas like the example shown in Figure 6.3 plays an important role in modeling location decisions for hazardous waste

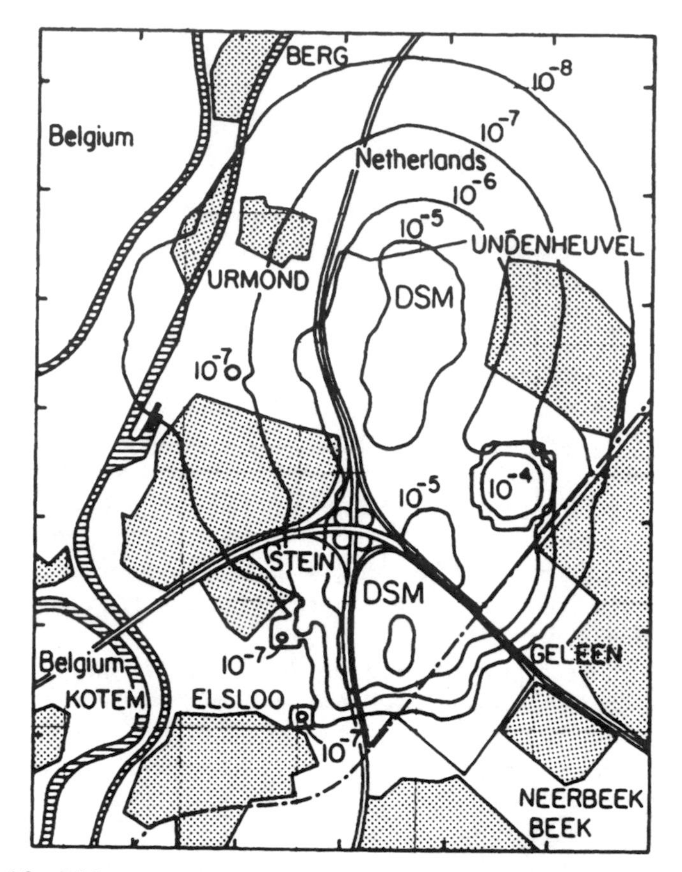

Figure 6.3 Risk contours around the DSM site. (From Maltezou et al., 1989.)

treatment, storage, and disposal facilities. In this example from the Netherlands, the analysis was performed on an existing facility, in an effort to determine policy on changes in land use in the area surrounding the facility. However, the same type of analysis could be performed on a proposed facility and used in making location decisions for the facility itself. This is the type of analysis described in Section 6.D.

6.D. LOCATING STORAGE, TREATMENT, AND DISPOSAL FACILITIES

Facilities for storing, treating, or disposal of hazardous wastes are usually not viewed by the public as desirable neighbors. As a result, finding sites for such facilities is often a difficult task. The problem of locating an undesirable facility (sometimes called a noxious facility) can be formulated in several different ways, depending on what criteria are to be applied. One simple form of such a model might be based on exposure contours such as the ones shown in Figure 6.3 for the DSM site.

An approximate representation might be to say that any point within a distance d_r of the facility will experience a risk level greater than r. For example, in Figure 6.3, any point within approximately 1 km of the facility will experience a risk level greater than 10^{-6}. Then, within some specified geographic area where the facility is to be located, we could try to identify the location that would minimize the population within a distance d_r of the facility. This is what is sometimes called an *anticovering problem* (Ratick and White, 1988) because the objective is to minimize the population "covered" by (i.e., within distance d_r of) the facility.

6.D.1. Anticovering Formulation

Consider the example situation shown in Figure 6.4. A hazardous waste treatment, storage, or disposal (HWTSD) facility is to be sited within the region defined by the outer boundary (perhaps a county). Five potential sites have been identified, designated by the letters *A* through *E*. Within this region there are six political subdivisions (perhaps towns), designated by the numbers 1 to 6, whose boundaries are as shown. Within each town, the value in parentheses is the population. It might be of interest to identify which of the five potential sites "covers" the minimum population.

One approach to this problem is illustrated in Figure 6.5. Each town has been replaced by a single point, and the town's population is all assumed to be at that point. Circles of radius d_r have been drawn around each of the five potential facility locations, and if the point representing each town lies within a circle, the entire population of that town is assumed to be covered (or affected) by the potential HWTSD facility site at the center of the circle.

Inspection of Figure 6.5 allows construction of the estimates for population coverage shown in Table 6.1. Based on the values in Table 6.1, location *E* clearly appears to have the minimum coverage. However, it is also nearly certain that 0 is not the real population within a distance d_r of location *E*. By aggregating the population of areas into a series of points, errors have been introduced. In the example shown in Figures 6.4 and 6.5, it is likely that the

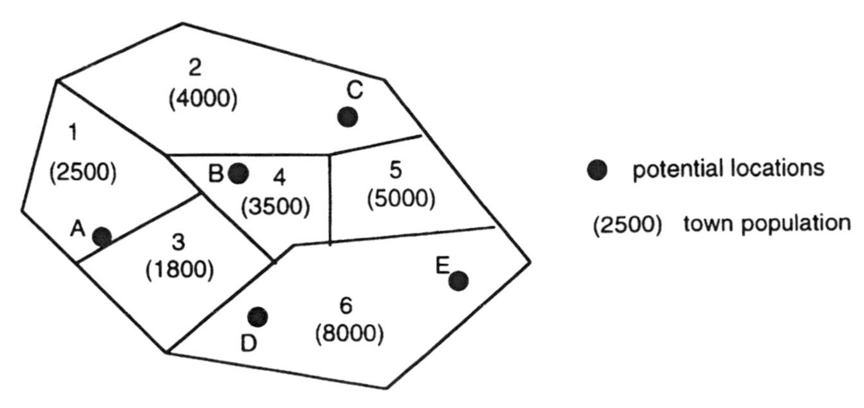

Figure 6.4 Facility locations problems.

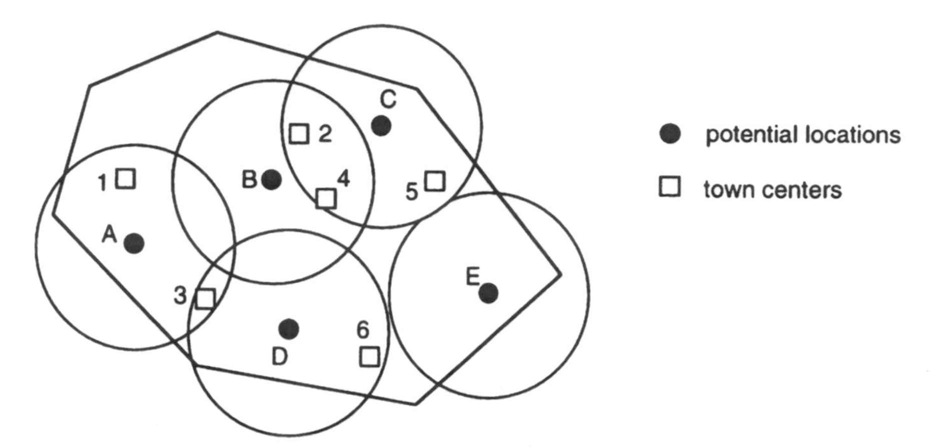

Figure 6.5 Coverage based on point locations of population.

errors introduced by aggregation are large enough to undermine the usefulness of the analysis.

Aggregation of population into a set of discrete points is a convenience in location analysis. This aggregation simplifies the problem of determining what populations are to be considered covered, because it is easy to determine whether a single point is within distance d_r of a potential facility location. The magnitude of the errors introduced depends on the sizes of the geographic areas being aggregated and the distance d_r. In Figures 6.4 and 6.5, the areas being aggregated are of approximately the same size as the area of coverage around a facility location, and the aggregation introduces substantial errors.

For situations like the one illustrated in Figures 6.4 and 6.5, a better way is needed to estimate the population covered by each potential facility location. One way would be to break each of the towns down into smaller geographic areas, and then treat each of these areas as a single point. These areas might be either politically based divisions or divisions created for other specific purposes (e.g., census tracts, traffic analysis zones, zipcodes, etc.). Since the objective in this analysis is population coverage, census tracts might be especially useful.

Geographic information systems (GIS) tools are extremely helpful in manipulating spatial information such as population in census tracts or other

TABLE 6.1 Estimates of Population Coverage for the Five Potential Facility Locations

Location	Towns Covered	Population Covered
A	1, 3	4,300
B	2, 4	7,500
C	2, 4, 5	12,500
D	3, 6	9,800
E	None	0

geographically defined regions. A GIS can also be used very effectively to identify which geographic regions are within a specified distance from some point and do operations on attributes of those regions (e.g., sum their populations). A more complete discussion of using GIS tools for hazardous waste management problems is given in Section 6.F.

Suppose, for example, that we used a GIS to sum the populations that are inside both the regional (county) boundary and the circular exposure areas around the potential HWTSD facility locations. The areas are indicated by the shaded portions of Figure 6.6. Summing the population in census tracts within the shaded portions of Figure 6.6 would probably produce quite different results from the estimates shown in Table 6.1, and this could easily lead to a different conclusion about which potential HWTSD facility location minimizes the population exposure.

The estimates of population exposure (either those shown in Table 6.1 or those implied by Figure 6.6) explicitly exclude population residing outside the specified regional boundary but inside the exposure areas of possible HWTSD facility locations. This has been done intentionally in this example to illustrate how political considerations often influence, and sometimes conflict with, the objectives of a thorough analysis. If a political decision has been made that a HWTSD facility is to be located within this region and the siting decision is controlled by governmental entities within that region, population exposure outside the region may frequently be excluded from the analysis. Such exclusion may not be analytically defensible, but decisions on siting of hazardous waste facilities have a substantial political component, and it may be politically expedient to focus entirely on population within the governmental region where the HWTSD facility is to be located. If the site selected is very near the regional boundary, this approach may, of course, lead to conflicts with adjacent regions.

In the treatment of this example HWTSD facility location problem (regardless of how we estimate the population within the exposure areas), there has been an assumption that a discrete set of possible locations has been identified, and the location problem is to choose from among those discrete locations. If only one facility is to be located, the problem can be solved by

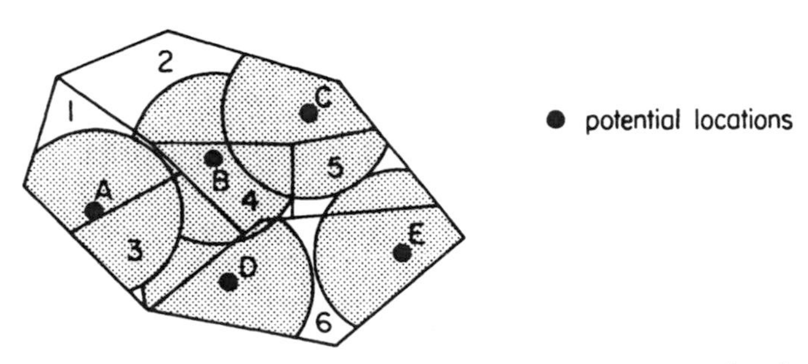

Figure 6.6 Selection of areas for population exposure estimation.

a simple procedure of estimating the population exposure for each potential location and then choosing the location with the minimum value.

Because the exposure areas of the potential locations can overlap (as shown in Figure 6.6), this simple procedure does not generalize to situations in which more than one location is to be chosen. That is, the best choice of two facility locations (i.e., the two locations that minimize the total population covered by both) is not necessarily obtained by choosing the two sites with the smallest values of individual population coverage. In small problems (like the one we're dealing with here), it is not too difficult to compute the population coverage for all combinations of two chosen sites (A and B, A and C, etc.). However, as the number of potential sites increases, or as the number of sites to be chosen increases, this explicit enumeration of possible solutions becomes intractable, and more sophisticated methods of exploring the set of feasible solutions must be developed.

One such method involves using integer programming. The particular formulation we describe here uses discrete nodes to represent the aggregated population in potential covered areas. Suppose there are N locations (nodes) that serve as the centroids of population areas for which exposure is to be measured (and minimized), and that these locations are indexed by $i = 1,..,N$. Also suppose that we have M possible facility sites (indexed by $j = 1,..,M$) and that we want to select P of them in such a way as to minimize total population exposed (covered). We define two sets of input variables: h_i is the population in the area represented by node i, and $a_{ij} = 1$ if a facility at site j will expose (cover) node i and 0 otherwise. The optimization formulation is then

$$\text{minimize} \sum_{i=1}^{N} h_i z_i \tag{6.1}$$

subject to

$$z_i - a_{ij}x_j \geq 0 \qquad \forall i,j \tag{6.2}$$

$$\sum_{j=1}^{M} x_j = P \tag{6.3}$$

$$x_j = 0,1 \qquad \forall j \tag{6.4}$$

$$z_i = 0,1 \qquad \forall i \tag{6.5}$$

where z_i is 1 if node i is covered by one or more of the selected locations and 0 otherwise, and x_j is 1 if potential site j is selected and 0 otherwise.

The objective function (6.1) represents the minimization of covered population. Constraint (6.2) ensures that z_i will be 1 if any selected site covers it (if $x_j = 1$ for any j for which $a_{ij} = 1$). Constraint (6.3) forces selection of P sites for facilities. For small instances of such a problem, a general method for solving integer programming problems can be used, but for problems of realistic size, specialized techniques are important to find solutions in reasonable amounts of computational time. Description of such solution methods is not included in this chapter, but Erkut and Neuman (1989) and Daskin (1995) provide more details on solving multiple-facility models, as well as references to other articles in the location theory literature.

We do not want to leave the impression that the anticovering approach based on a discrete set of potential locations is the only way to approach HWTSD facility siting problems, nor that minimizing population exposure is the only criterion to use in selecting sites. The exposure minimization problem can be formulated in several different ways, and the siting of HWTSD facilities almost always involves multiple criteria.

6.D.2. Minimax Formulation

As an example of an alternative formulation of the HWTSD facility siting problem, consider again the region (county) and points 1 through 6 (town centers) shown in Figure 6.4. We will define some coordinate system (e.g., latitude and longitude) so that any location in the region can be designated by a set of coordinates (x,y). The town centers can be denoted by the locations (a_i,b_i), for $i = 1,\ldots,6$, and we will associated with each of these points a weight, w_i. This weight may represent population, some measure of sensitivity to exposure, or another factor. For purposes of this example, suppose that w_i is the population of town i. The whole region (county) will be denoted by S.

Then a formulation of the problem of locating the site for a HWTSD facility might be to find the location (x,y) that minimizes the maximum effect on any town, as measured by the inverse square distance to the town center (Melachrinoudis and Cullinane, 1986):

$$\min_{(x,y) \text{ in } S} \left[\max_{1 \le i \le n} \left(\frac{w_i}{d_i^2} \right) \right] \tag{6.6}$$

where $d_i^2 = (x - a_i)^2 + (y - b_i)^2$.

The assumption that the impact of the HWTSD facility on a point, i, is proportional to the inverse square of the distance between that point and the facility location is consistent with the physical laws of mass or energy transfer for many different types of effects (e.g., blast energy from an explosion, radiation density, noise, etc.). This minimax formulation, like the first anticovering formulation described in Section 6.D.1, assumes that the impacts of the HWTSD facility location on town i are reflected accurately by a measurement taken at the town center. It also assumes that the relative impacts on the various towns in the region are weighted by their populations.

Figure 6.7 shows a plot of

$$z = \max \left(\frac{w_i}{d_i^2} \right) \tag{6.7}$$

for a simple case where there are $n = 2$ points (P_1 and P_2) in a four-sided region S defined by the corners labeled A, B, C, and D. We want to located the HWTSD facility where the value of z is minimized. In the example shown

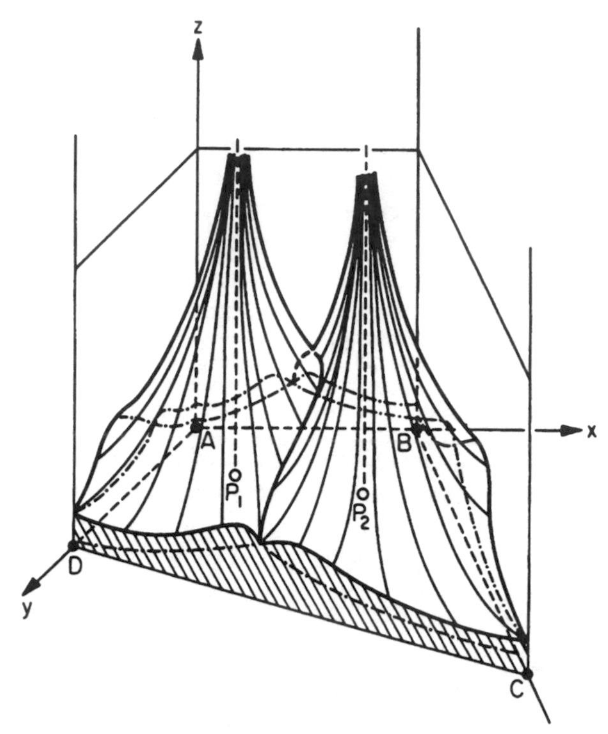

Figure 6.7 Impact measure z in a case with $n = 2$ and a four-sided region $S = ABCD$. (From Melachrinoudis and Cullinane, 1986.)

in this figure, it is clear that the lowest value of z will somewhere on the boundary of S. By looking at Figure 6.7, we can also imagine that if points P_1 and P_2 were moved toward opposite corners of the region S, the minimum value of z might occur along the line between P_1 and P_2.

The observation from Figure 6.7 that there are two sets of places to look for the optimal location (along the boundary of S and along the line between P_1 and P_2) is important, because is greatly simplifies the search for the solution. In general, when the number of points, n, is greater than two, it can be shown (Melachrinoudis and Cullinane, 1986) that the solution must lie either on the boundary of S or within the convex hull of the points P_i, $i = 1,...,n$. Figure 6.8 shows the convex hull (the shaded area denoted as H) of the points P_i, $i = 1,...,6$, for the example from Figure 6.4.

Figure 6.8 provides an important insight about the nature of optimal solutions to the HWTSD facility location problem formulated in equation (6.6). If all the points, P_i, are near the boundary of region S, area H will be large, and it is possible that the best location is in the interior of the region, quite far from all the points of interest. However, if there are points near the center of the region (like point 4 in this case), interior to the convex hull, H, it is quite likely that the optimal solution will be along the boundary of S. That

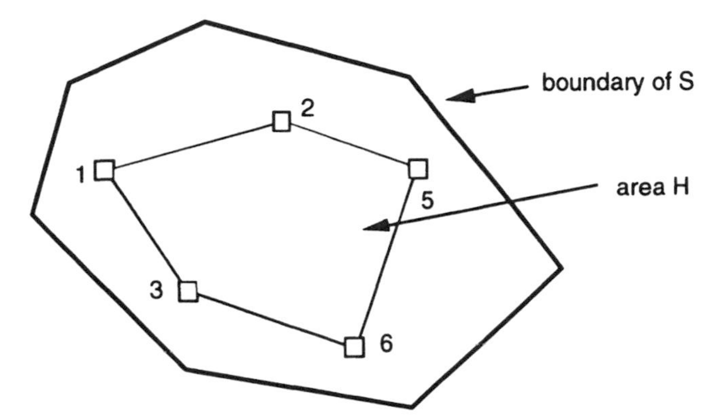

Figure 6.8 Possible solution space, including the boundary of S and the convex hull (H) of the points P_i, $i = 1,\ldots,n$.

is, the best location is going to be right next to an adjacent region. This is because the consideration of exposure impact is limited to points within S, and the model has an incentive to push the facility as far away as possible without leaving the region S. Thus solutions to the problem formulation expressed in equation (6.6) may lead to the same kind of political conflicts described in the discussion of the anticovering formulation.

In the examples shown in Figures 6.7 and 6.8, the region S is a convex polygon. As long as S is a convex polygon, we can guarantee that the area H is entirely within S, so we do not have to worry about possible solutions within H that are infeasible because they are outside S. In fact, this property holds as long as S is a convex region, even if some of the sides are nonlinear. A formal solution algorithm for the problem in (6.6) can be created to search for a solution within a region S', a convex polygon that is an approximation to S (if the original region S is not). Constructing S' may involve approximating nonlinear sides of S by a sequence of linear segments, or enclosing some areas that are not inside S (to make the search region convex) if S is not a convex region. Melachrinoudis and Cullinane (1986) describe in detail a solution algorithm to search for the optimal location in a convex polygon S'.

Alternatively, and perhaps more usefully, a graphical approach can be developed for finding a solution in a region of any shape. If we specify some relatively large value of z, the contours of the objective function (6.6) will be circles centered on the points P_i, $i = 1,\ldots,n$. The radius of the circle around P_i will be

$$r_i = \sqrt{\frac{w_i}{z}} \tag{6.8}$$

Figure 6.9 illustrates this idea using the example data from Figure 6.5. In this case the values of w_i are the town populations. Note that some of the circles

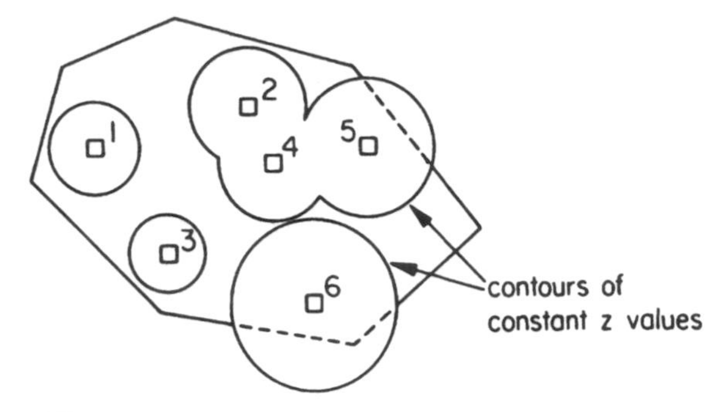

Figure 6.9 Contours of the objective function, z.

may intersect (e.g., the circles around points 2, 4, and 5 in this example), and in this case the contour lines follow the "outer" parts of the intersecting circles.

It is clear that points inside the circles have higher values of z, and points outside the circles have lower values of z (see Figure 6.7). Since our objective is to minimize z, the possible solution points are those inside S but outside all the circles. This area of search for possible solutions is shown as the shaded area in Figure 6.10. The graphical presentations in Figures 6.6 and 6.10 appear relatively similar, but they represent quite different views of the problem. In Figure 6.6 the circles represent areas of exposure centered on the possible HWTSD facility locations, and in Figure 6.10 the circles represent areas of impact around the town center locations. In the problem formulation represented by Figure 6.6, we want to choose the facility location that has the minimum population within its circle of exposure. In the problem formulation represented by Figure 6.10, we want to find a potential location that is outside all the "impact" circles.

If we decrease the specified value of z, each of the circles in Figure 6.10 gets larger, and the sets of points with smaller z values is reduced, as shown in Figure 6.11. Using interactive computer graphics, we can make adjustments in z, redraw the region and the circles around the points P_i, and find the

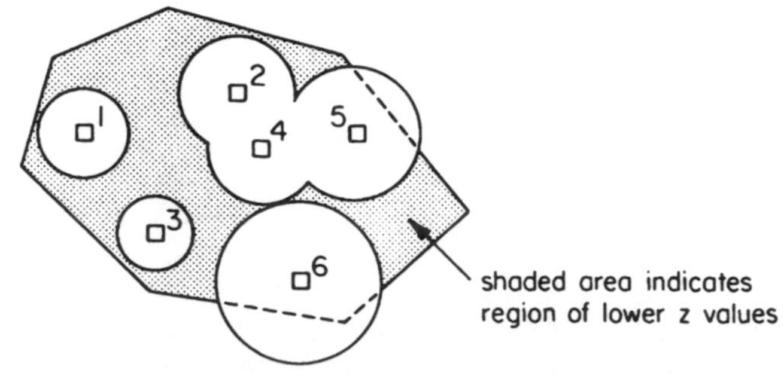

Figure 6.10 Shaded areas indicate possible locations with lower z value.

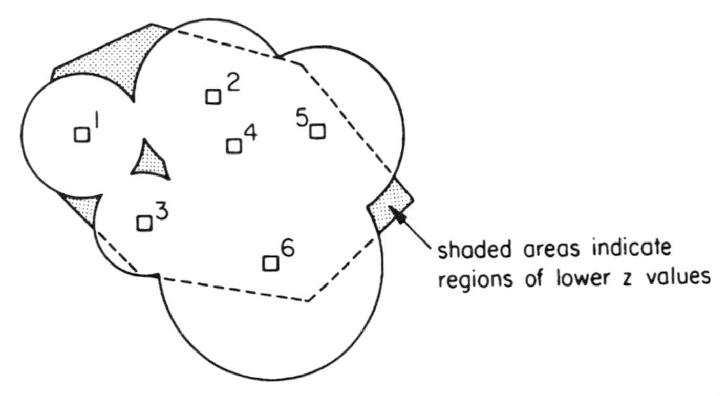

Figure 6.11 Reducing the set of possible solution by decreasing the value of z.

solution graphically by reducing the set of points that are inside S but outside all the circles of constant z value. Figure 6.12 shows that the solution obtained for this example problem will lie along the "northwest" boundary of the region S. This is quite a different solution than the one obtained for the anticovering formulation, because none of the candidate sites (A through E) considered in that formulation was in the area selected as the best by this minimax formulation. This illustrates the value of looking at HWTSD facility siting problems from multiple perspectives, because different perspectives may yield different types of solutions.

6.D.3. Using Multiple Criteria for Location Analysis

The anticovering and minimax formulations described in the preceding two sections represent different criteria for determining an optimal location, and these two criteria led us to different solutions for the example problem. Both

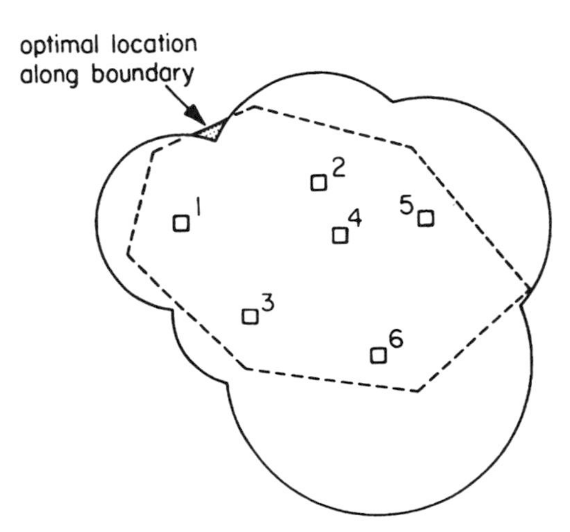

Figure 6.12 Identifying the location with the lowest z value.

criteria are likely to be valid measures of the desirability of potential HWTSD facility locations, so we are left with something of a quandary. Which solution is really "best?" The answer, unfortunately, is that there usually is no single "best" solution, and the more criteria we use to evaluate possible solutions, the more likely it is that no single solution will be preferred under all criteria.

This places us squarely in the domain of multiobjective optimization, a topic described in much more detail in Chapter 12. We will not venture very far into the study of multiobjective location models for HWTSD facilities here, but the work of Ratick and White (1988) is an important example of such a formulation. It should also be noted that in their review of location modeling for undesirable facilities, Erkut and Neuman (1989) state that they believe "multiobjective decision making tools are underutilized in the undesirable facility location literature. The multiple constituency, multiobjective nature of the problem severely limits the usefulness of single objective models." Thus further development of multiobjective location models for HWTSD facilities is a fertile area for further research.

6.E. TRANSPORTATION OF HAZARDOUS WASTES

Management of hazardous wastes involves transporting those wastes from points at which they are generated to the locations chosen for treatment, disposal, or storage. Hazardous wastes are part of a broader category of hazardous materials that are transported in substantial volumes. It is estimated that about 500,000 shipments of hazardous materials occur every day in the United States (Dungan, 1991). Most of these shipments are transported without incident, but in 1993 there were nearly 13,000 transportation accidents involving hazardous materials reported to the U.S. Department of Transportation (U.S. DOT, 1994). Most of these accidents were minor, but in total there were 15 deaths and 626 injuries reported that year. Approximately 86% of the incidents, all of the fatalities, and 82% of the injuries in 1993 were associated with highway transportation.

Despite a very good overall safety record, there is considerable public concern about the risks involved in transporting hazardous wastes and other hazardous materials. In Section 6.C we have looked at ways of analyzing risks, including transportation risks. In this section we focus on how transportation decisions can improve management of those risks, with particular attention to routing decisions. The discussion of routing algorithms presumes a basic familiarity with methods for finding a shortest path between two points in a network. Readers who need additional background reading on shortest-path algorithms can consult any of several excellent texts (e.g., Bertsekas, 1991; or Evans and Minieka, 1992).

The principal federal legislation governing hazardous materials transportation (the Hazardous Materials Transportation Uniform Safety Act of 1990) identifies route control as a significant policy option and mentions several potential factors that could influence routing decisions, including:

1. Population density
2. Type of highway
3. Type and quantity of hazardous material to be shipped
4. Emergency response capabilities
5. Exposure and other risk factors
6. Delays in transportation

The challenges for analysis of hazardous materials routing decisions are to (1) convert the general "factors" described in legislation and regulations into specific, measurable criteria; (2) develop reasonable numerical estimates of those criteria as they apply to specific links in the transportation network; and (3) develop algorithms that can operate on the estimated values to identify desirable routes for hazardous materials shipments. In the remainder of this section we treat these three elements of routing analyses. Because much of the hazmat movement, and most of the reported incidents, are in highway shipments, we focus attention on movements by truck. However, the same basic principles can be applied to movements by other transportation modes.

6.E.1. Identifying Routing Criteria

The routing factors *population density* and *exposure* both address concerns with the exposure of population along the route to the possible effects of an accidental release of material. This illustrates the connection between risk analysis and transportation routing decisions that we have emphasized in this chapter. A typical approach is to define one routing criterion as the population within a band of specified width along each link in the network. The bandwidth may be a function of the type and quantity of material being transported.

A variation on the use of bandwidth along a link to reflect population exposure is the use of traffic volumes on the link itself. Concern with exposure of other travelers on the same link as the hazmat-carrying truck is appropriate for many types of chemicals for which the primary risk is flammability (so the hazard area is quite limited), and when the movements are on grade-separated highways (such as interstates) where the traffic stream is well separated from adjacent businesses and residential areas. Two examples of using on-link exposure measures are the work of Glickman and Macouley (1991), focusing on the Capital Beltway around Washington, DC, and the routing/scheduling analysis by Nozick et al. (1995).

The term used in HMTUSA, *exposure and other factors,* can also imply specific concern with exposure of people or facilities that are particularly vulnerable, such as children in schools or patients in hospitals. This can lead to an additional exposure criterion over and above general population exposure. For example, Turnquist and List (1993) used the number of schools within $\frac{1}{2}$ mile of the route as a criterion in addition to total population within the exposure area.

The type of highway is a factor, in part because it may determine whether the primary risks are to other travelers or to population along the roadway, but mostly because different types of highways have different accident rates. Differences in accident probability constitute a significant basis for routing decisions, and thus accident probability is the primary criterion reflecting highway type.

Emergency response capability is a factor that can play a relatively complex role in routing analyses. On the one hand, it is desirable for a community's fire department, emergency medical personnel, and so on, to have the capability to respond to hazardous material spills in an effective manner. On the other hand, if it is desirable to route hazmat shipments along routes where there is better emergency response capability, the act of increasing that capability is likely to draw more shipments, which may not be the community's objective. Most previous routing analyses that have incorporated emergency response capability have done so in a qualitative way, but it is also possible to construct quantitative measures, based on either levels of responder training, or response time to a link along the route, or a combination of both.

The factor *delays to transportation* clearly reflects economic concerns in the routing decisions. These concerns may be reflected in measures of time, distance, or cost associated with a route. One of the main purposes of routing analysis is often to illuminate trade-offs between risk measures (exposure, accident probability, etc.) and economic measures (time, distance, or cost), and thus we want to have both types of measures included in the analysis.

The foregoing discussion illustrates that a variety of measures or routing criteria can be constructed to represent the range of general factors identified in legislation and regulations. These criteria represent different objectives to be pursued in the determination of transportation routes for hazardous wastes and other hazardous materials. However, to implement these objectives in routing algorithms, we must be able to estimate values for the criteria of interest, and attach those values to links in a network.

6.E.2. Estimating Values for Routing Criteria

Estimating exposure along a route through a network is similar to estimating exposure around a treatment, storage, or disposal facility, except that the vehicle carrying the hazardous material is moving. This leads to an exposure "band" along the links in the network rather than a circular area around a fixed facility. A typical means of creating a criterion for routing analysis is to estimate the population within a specified distance of each link and attach that population exposure to the link as an attribute. GIS tools are an effective means of doing this estimation. In the GIS context, the process of creating a band around a geographical entity (such as a highway) is called *buffering*. This process is discussed more thoroughly, with an example, in Section 6.F.

Most analyses of population exposure are based on census or employment figures, and thus count residential population or numbers of employees within specific areas. Such values do not reflect variations that are likely during the course of the day. For example, the actual number of people present in a residential area is likely to be much higher at night than during the middle of the day, whereas the number of people present in an office building will be much higher during the day than at night. We may be concerned about routes that pass close to schools during normal school hours, but much less concerned if the shipment passes by the school during the night. If time-varying attributes are used in routing analyses, there is an explicit incorporation of scheduling of the trip, as well as choosing the route to be followed. Some work has been done on simultaneous routing/scheduling models for hazardous materials shipments (Turnquist, 1987, 1991; Nozick et al., 1995), but considerable opportunities exist for improvements in this area.

In addition to various exposure measures, a second category of criteria that is of interest in routing hazardous materials is accident rates. There is a long history of research on estimating accident rates for various types of vehicles on various types of highways, as part of general efforts to improve highway safety. The most widely quoted estimates of accident rates for hazardous materials shipments are those developed by Harwood et al. (1993). Their results are summarized in Table 6.2.

The estimates in Table 6.2 distinguish between links in rural and urban areas and among highways of several different types. Note also that most accidents do not result in release of hazardous material. The likelihood of releasing any material, given that an accident occurs, varies between 0.055 and 0.09, depending on the area type and roadway type. Since risk analyses are usually predicated on the release of material, the release accident rates in

TABLE 6.2 Default Truck Accident Rates and Release Probabilities for Use in Hazardous Materials Routing Analyses

Area type	Roadway Type	Truck Accident Rate (accidents per million vehicle-km)	Probability of Release Given an Accident	Releasing Accident Rate (releases per million vehicle-km)
Rural	Two-lane	1.36	0.086	0.12
	Multilane undivided	2.79	0.081	0.22
	Multilane divided	1.34	0.082	0.11
	Freeway	0.40	0.090	0.04
Urban	Two-lane	5.38	0.069	0.37
	Multilane undivided	8.65	0.055	0.48
	Multilane divided	7.75	0.062	0.48
	One-way street	6.03	0.056	0.34
	Freeway	1.35	0.062	0.09

Source: Harwood et al. (1993).

the last column of Table 6.2 are usually the most appropriate values for use in routing studies. The values in Table 6.2 represent reasonable default accident and release rates, to be used in the absence of more specific local data. However, if better local data are available, they should be used instead.

Once measures of exposure, accident probability, emergency response, travel time or distance, cost, and so on, have been estimated for each link in a highway network of interest, they can be used for routing analyses. We can think of these values as a vector of attributes attached to each network link. Most routing analyses are based on minimizing the sum of values (e.g., population exposure, time, etc.) along all the links in a route, but it is also possible to develop routes on the basis of minimizing the maximum value (e.g., emergency response time) for any link in the route. If a single criterion is selected for the routing, the problem is generally a simple shortest-path problem, where *shortest* is defined to be the minimization of the selected criterion. However, selection of routes for hazardous material shipments almost always involves multiple criteria, so multiobjective methods are very important tools. The following section illustrates both single and multiobjective path selection using an example based on a truck shipment over the interstate highway system in the northeastern United States.

6.E.3. Routing Methods for Hazardous Materials and Wastes

As an example of routing based on a single criterion, suppose that we wanted to find the minimum on-link exposure path (or route) through the interstate highway network from a shipment origin point in Wilmington, Delaware, to a destination in Portland, Maine. Figure 6.13 shows this minimum exposure route, which zigzags its way through the network, in an effort to avoid urban areas. The length of this path is 720 miles (1160 km), 285 miles (462 km) longer than the shortest path from Wilmington to Portland.

Some hazardous materials, including combustibles and explosives, are prohibited from using certain tunnels, bridges, and highways. To ensure that these facilities are not part of a selected route, the links that represent these facilities can be removed prior to running the shortest-path algorithm. A GIS is useful for organizing information pertaining to specific facilities, such as prohibitions on use by hazmat trucks, and integration of routing analysis tools with GIS tools leads to more effective data organization, visualization, and solution display.

In the example shown in Figure 6.13, the only criterion considered is exposure. Often there are other criteria that are important in the final section of the route. In general, importance of multiple criteria results from the presence of multiple interested parties, or stakeholders, each of whom may bring a different set of values and/or criteria to bear on assessing any particular solution. Thus multiple objectives must be incorporated into the analysis and decision making. The existence of multiple criteria means that it is not usually possible to identify a single "best" route between a given origin and desti-

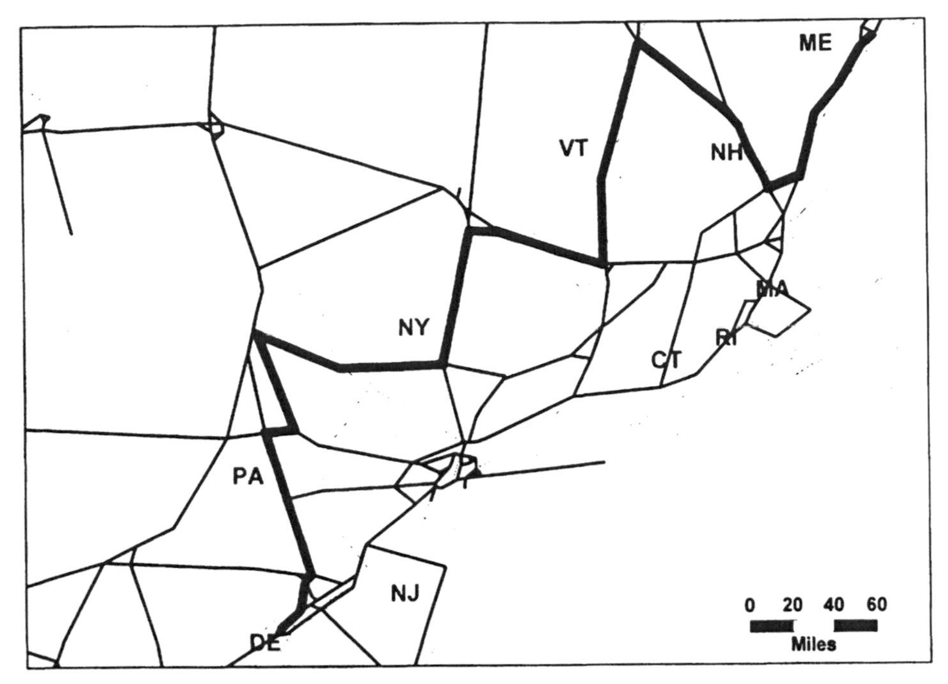

Figure 6.13 Path from Wilmington to Portland that minimizes exposure.

nation. One route may minimize the number of people at risk, for example, while a second minimizes the probability of an accident, and a third minimizes distance traveled. In this context, attention should be focused on finding a set of *nondominated routes* that represents the available trade-offs explicitly. One route dominates another if there is at least one measure, or criterion, on which it is preferred and there is no measure on which the other route is preferred. The nondominated set of routes is that which remains after all dominated routes have been removed.

Algorithms for determining the nondominated set of routes using several criteria are generalizations of shortest-path algorithms, but they are more complex because the underlying problem to be solved is much more complicated. We will describe one method for finding multiobjective nondominated routes using an example problem with two criteria. The method itself is quite general and can operate with more than two criteria, but it is easier to understand how it works using a relatively simple example.

The method operates on a *directed* network like the one shown in Figure 6.14. A directed network contains arcs that allow passage in only one direction. When travel is permitted in both directions, a pair of directional arcs are defined (like the example between nodes 4 and 5). Each arc in Figure 6.14 is labeled with a two-element vector indicating routing criteria of interest (e.g., exposure, accident rate, etc.), and we will assume that both of these criteria are additive for arcs in a path. For example, if we are interested in

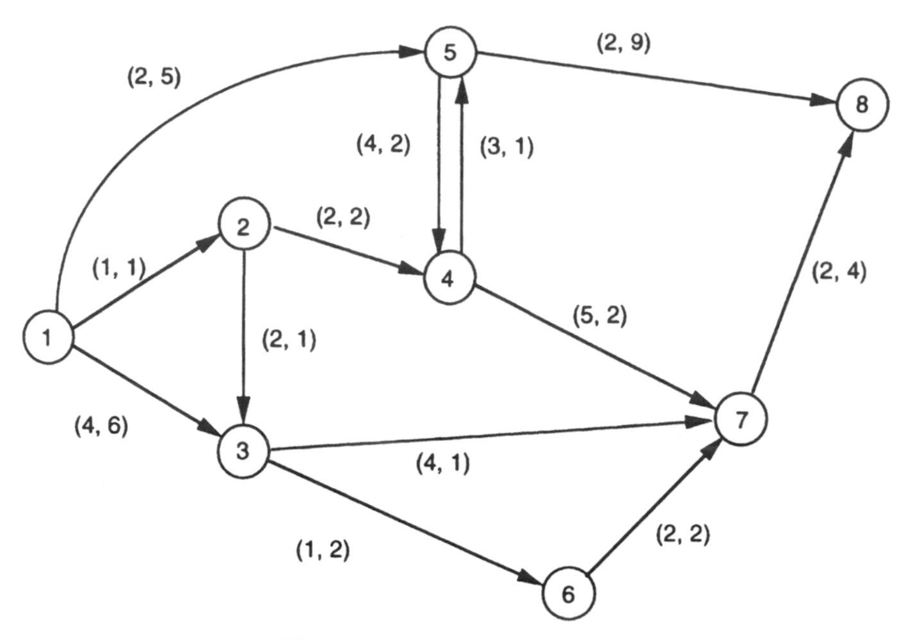

Figure 6.14 Directed network.

finding paths from node 1 to node 8, the particular path 1–5–8 would have a value [4,14], because both criteria are additive along the two arcs in the path. We will further assume that both criteria are to be minimized.

The algorithm to identify the set of nondominated routes is very similar in principle and operation to a class of shortest-path algorithms called *label-correcting algorithms.* The algorithm begins by associating with each node in the network a vector of dimension r, where r is the number of attributes (in this example, $r = 2$). That vector represents our current understanding of the cost to reach that node with respect to each attribute or criterion. Initially, the vector for the origin node (node 1 in this example) is a vector of zeros, and for all other nodes all values are infinite. Then we consider all nodes that can be reached by crossing a single arc from the origin. For each of these nodes a temporary label is created, where the elements are equal to the attribute measures for the arc traversed to reach that node. This temporary label replaces the now-dominated initial label vector.

Figure 6.15 illustrates the status after the first iteration, where new temporary labels have been created for nodes 2, 3, and 5. The notation used for the labels includes the partial path information, so that the paths can be traced. For example, the label at node 5 ([2,5]:1) indicates that the partial path has accumulated values of 2 and 5 for the two attributes and that the path came directly from node 1. At the conclusion of this iteration, node 2 can be permanently labeled because any path other than the direct path would have to pass through either node 5 or 3, and the partial paths from 1 through both these nodes are dominated by the partial path to node 2 already identified.

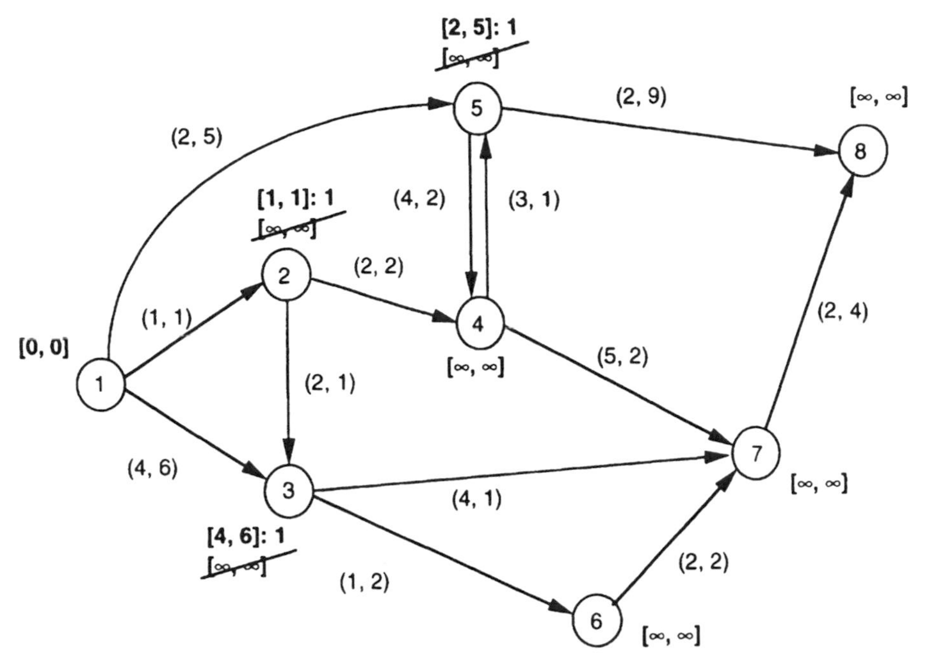

Figure 6.15 Network after first iteration.

The algorithm operates recursively. At each iteration we branch off all nodes that received new temporary labels in the previous iteration. Thus at iteration j, node labels are generated corresponding to paths composed of j arcs starting from the origin. For each node there is a set of temporary labels corresponding to tentatively identified nondominated paths from the origin to that node. When we attempt to add a temporary label to a node, we first check that it is not dominated by any existing temporary label. If it is dominated by an already existing temporary label, the new temporary label is not added to the set. If it dominates any preexisting temporary labels in the set, the dominated temporary label is removed. A node is considered permanently labeled when all nondominated paths have been found to that node from the origin. All nondominated paths to a node k have been found if at the conclusion of an iteration j, all paths of j or fewer arcs that terminate at nodes other than k are dominated by some path of j or fewer arcs that terminates at node k. The algorithm terminates either when the destination node is permanently labeled, or at the conclusion of interation j if no acyclic paths of $j + 1$ arcs exist. If no acyclic paths of $j + 1$ arcs exist, all paths already identified are nondominated.

Figures 6.16 through 6.19 illustrate iterations of the algorithm necessary to solve the example problem. In Figure 6.16 (iteration 2), notice that a new path to node 3 (1–2–3) is identified that dominates the path found at the first iteration (because both criteria are improved simultaneously). Thus we discard the first path and its label, indicated by crossing out the label. During the

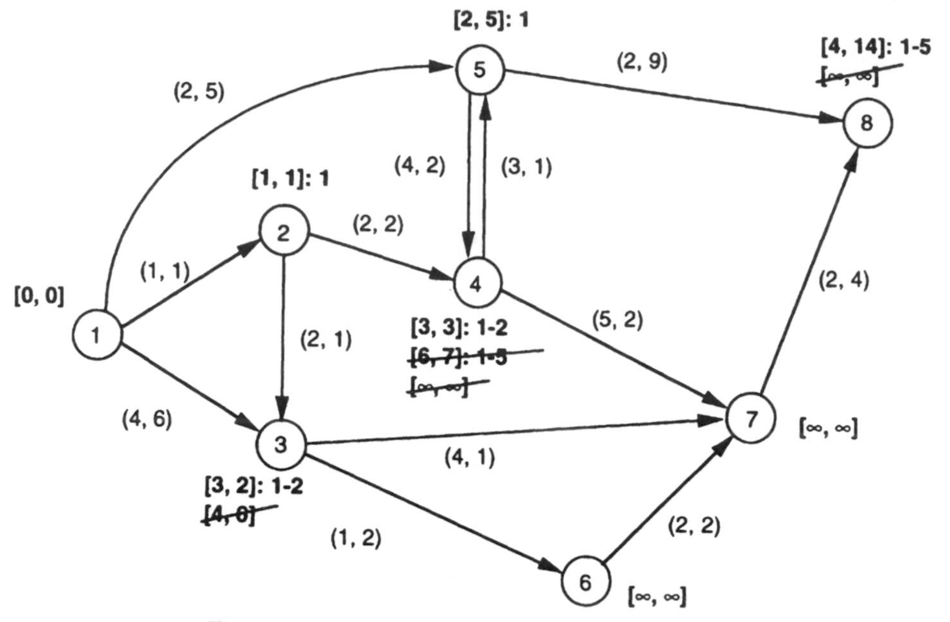

Figure 6.16 Network after the second iteration.

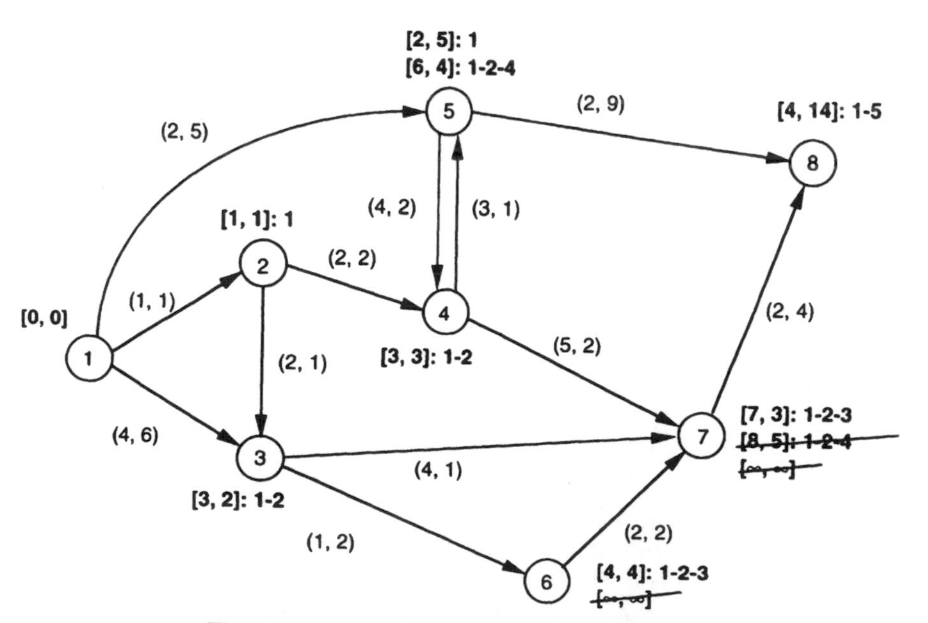

Figure 6.17 Network after the third iteration.

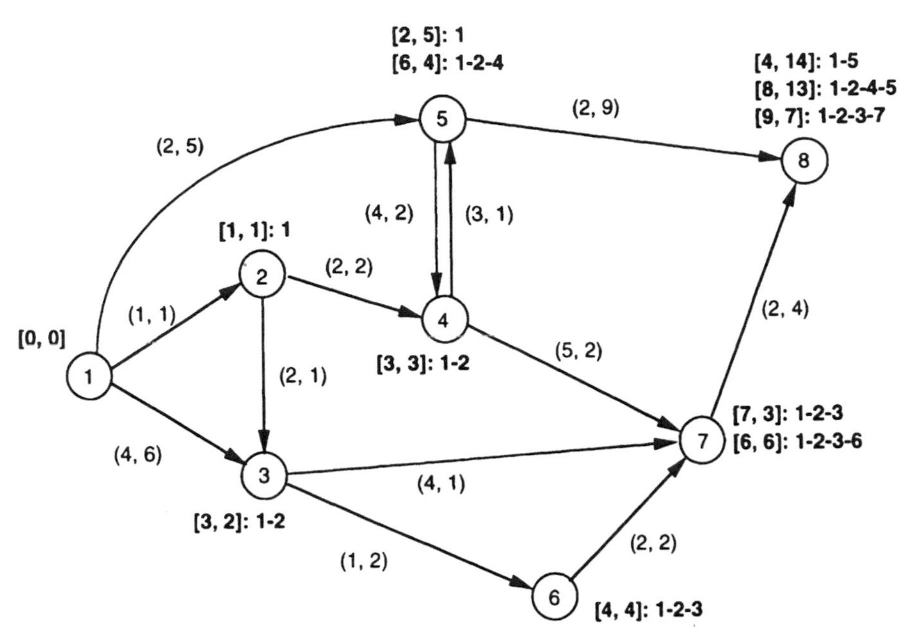

Figure 6.18 Network after the fourth iteration.

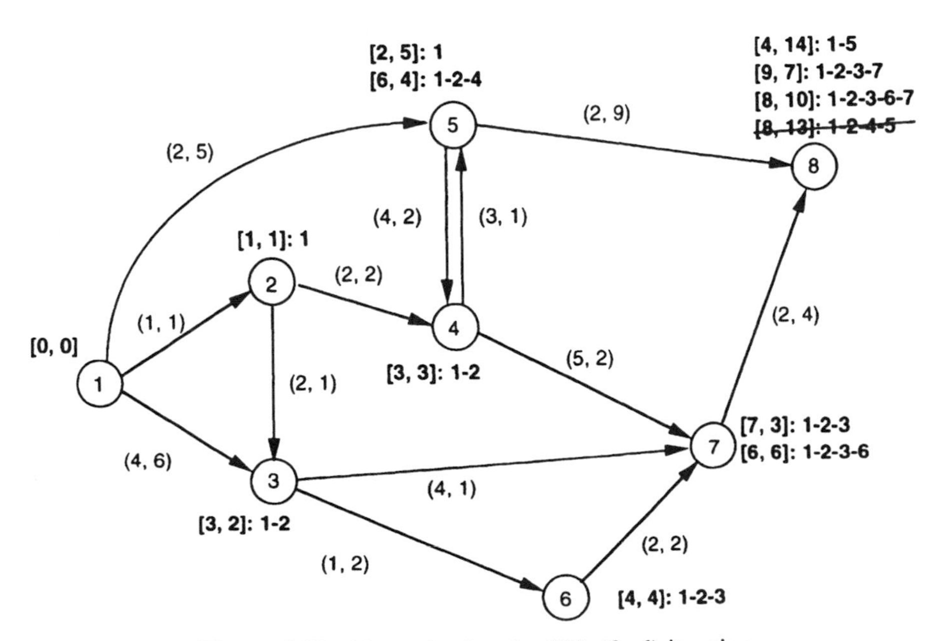

Figure 6.19 Network after the fifth (final) iteration.

second iteration, two different paths are identified to node 4 (1–2–4 and 1–5–4). The path via node 2 dominates the path via node 5, so the trial label for the path via node 5 is crossed out and that path is discarded.

At the conclusion of the algorithm (see Figure 6.19) there are three nondominated paths from node 1 to node 8. Path 1–5–8 has the lowest value for the first criterion, and path 1–2–3–7–8 has the lowest value for the second criterion. Path 1–2–3–6–7–8 is a trade-off solution, with intermediate values for both objectives.

Consider again the example of the chemical shipment from Wilmington, Delaware, to Portland, Maine. In Figure 6.13 we showed the minimum exposure route. Now imagine that the set of routing criteria is expanded to include accident rate and route length. Using these three criteria, there are 20 nondominated routes. However, we can gain a general understanding of the range of solutions generated by examining six of the routes. The values of the three criteria for these six routes are as shown in Table 6.3. Note that the ranges among the solutions are quite different for the three criteria. The route lengths range from 435 miles (705 km) to 720 miles (1166 km). The longest nondominated route is thus 66% longer than the shortest. The exposure values range from 9859 vehicle-minutes to 13,077 vehicle-minutes (33% above the minimum value). The release accident rates range from 0.0412 accident per 1000 trips to 0.0461 accident per 1000 trips (12% above the minimum value), and five of the six routes have accident rates that are within 4% of the minimum.

The six routes summarized in Table 6.3 are illustrated in Figure 6.20. Route *A* is the shortest, but because it passes directly through several cities (including New York City), it has the highest exposure value. Route *B* has the smallest exposure. It zigzags its way from Wilmington to Portland in an effort to avoid urban areas. This route is the longest and also has the highest accident rate (0.0461 release accident per 1000 trips). Route *C* has the minimum accident rate. It is 570 miles long (31% longer than the minimum) and has an exposure of 10,211 (4% above the minimum). Routes *A*, *B*, and *C* are the

TABLE 6.3 Summary of Criteria for Six Nondominated Routes[a]

Route	Length (miles)	Accident Rate (per 1000 trips)	Exposure (vehicle-min)
A	435 [—]	0.0427 [4%]	13,077 [33%]
B	720 [66%]	0.0461 [12%]	9,859 [—]
C	571 [31%]	0.0412 [—]	10,211 [4%]
D	504 [16%]	0.0412 [—]	11,025 [12%]
E	454 [4%]	0.0417 [1%]	12,243 [24%]
F	480 [11%]	0.0416 [1%]	11,658 [18%]

[a]Values in brackets indicate % above the minimum value for that attribute.

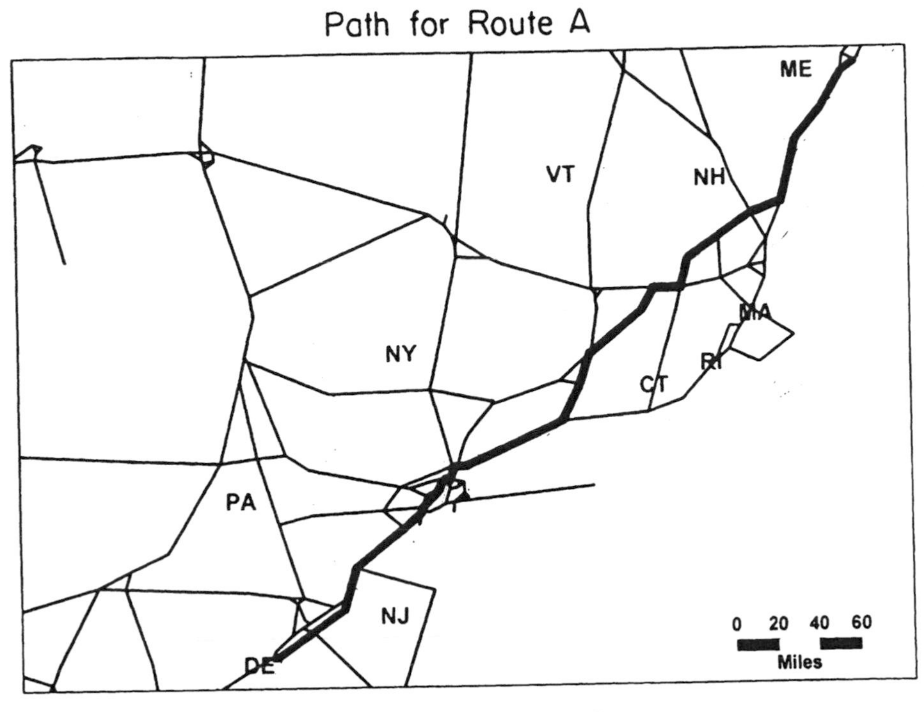

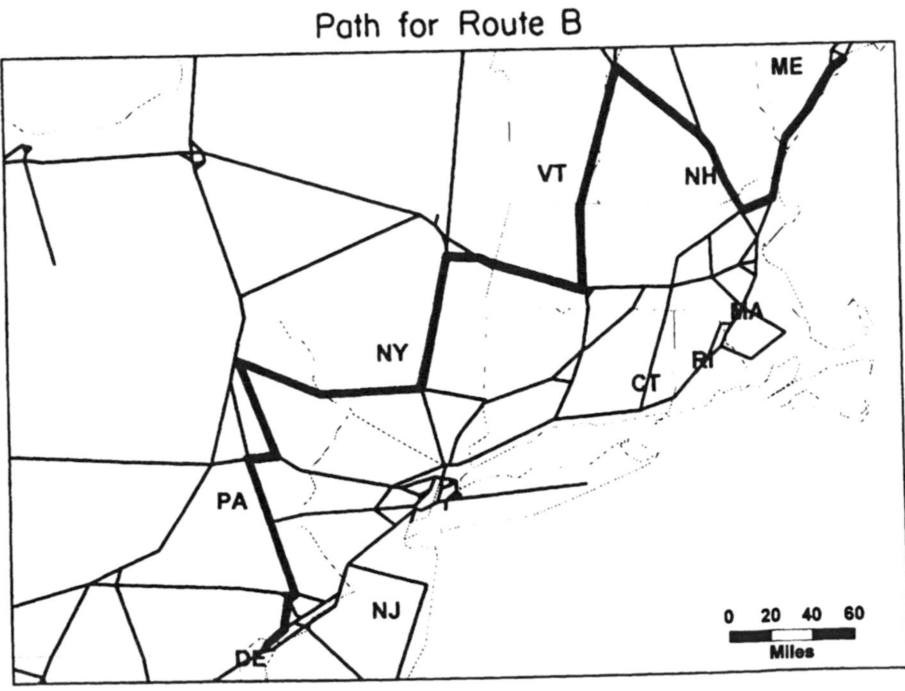

Figure 6.20 Six nondominated routes.

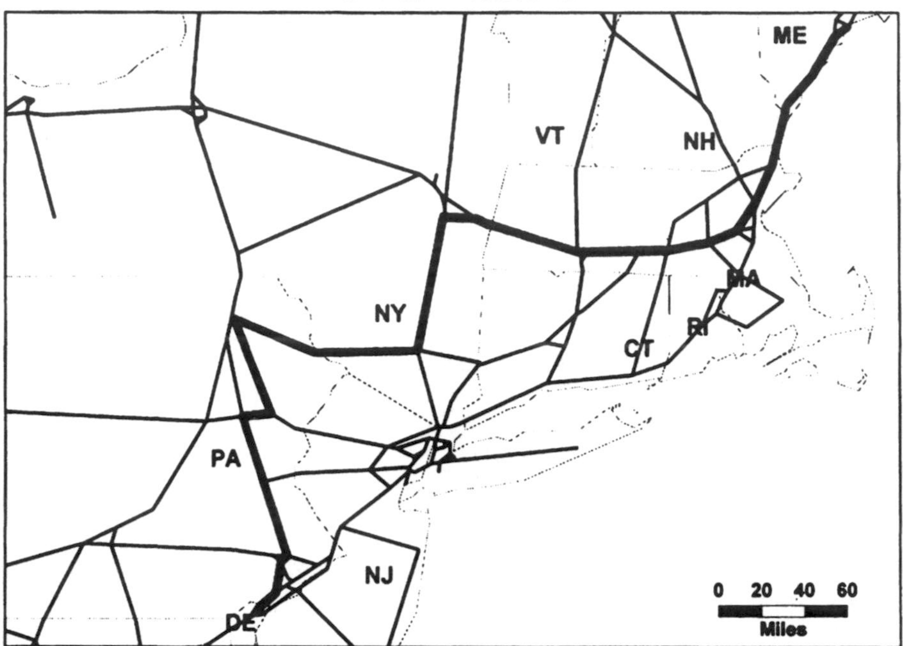

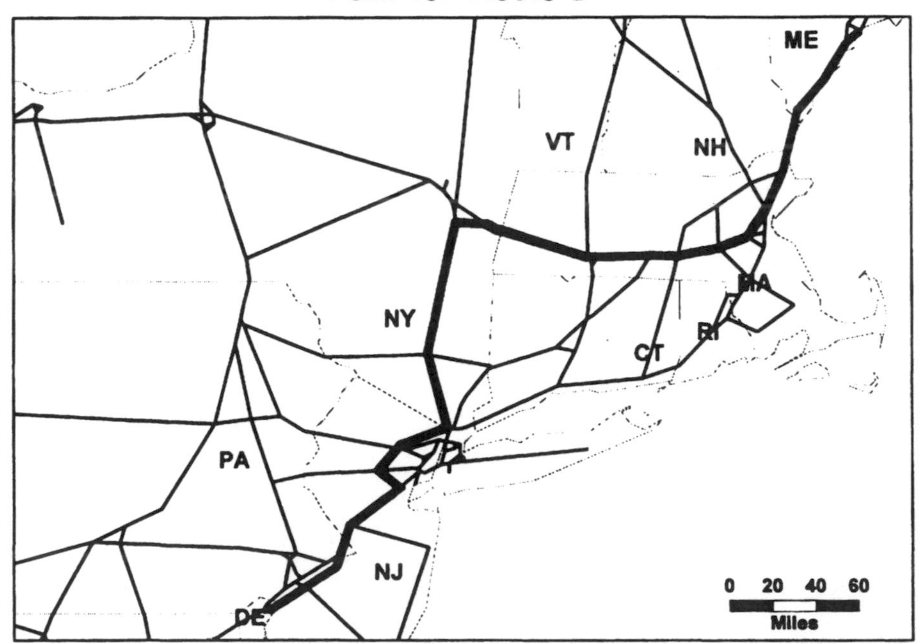

Figure 6.20 (*Continued*)

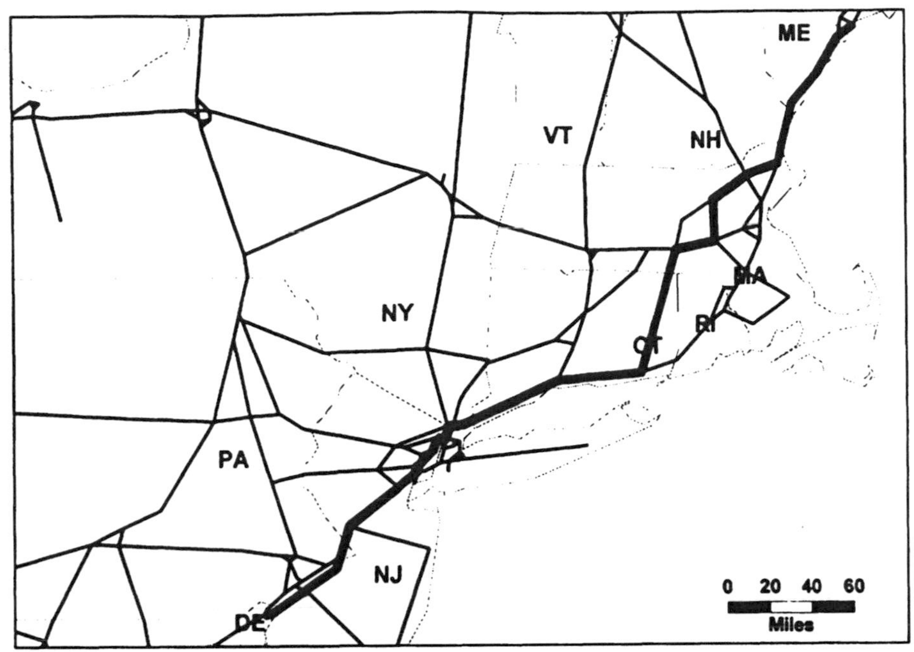

Path for Route F

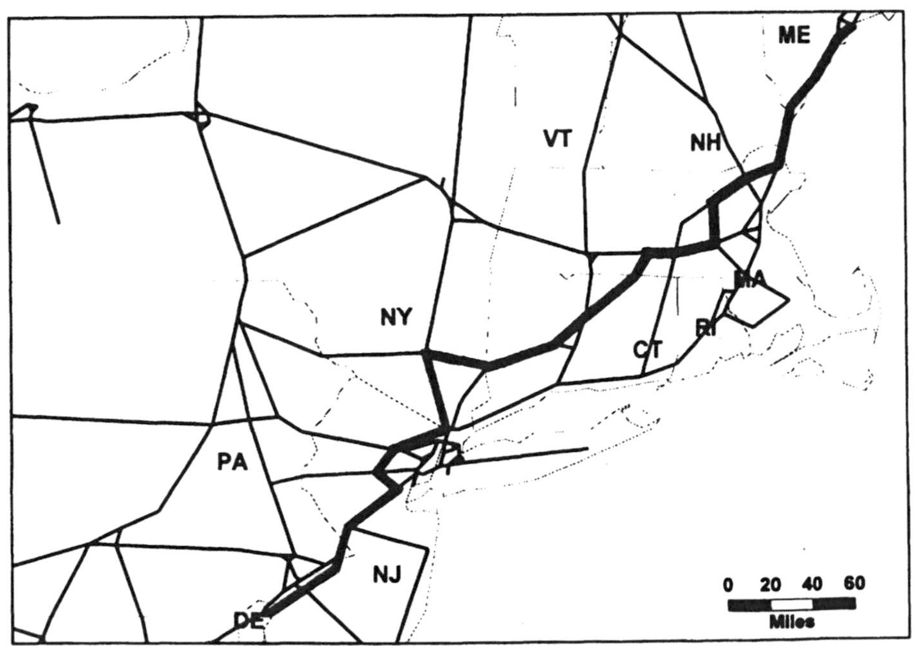

Figure 6.20 (*Continued*)

"extreme" solutions, those that individually minimize each of the three criteria. However, each of these three solutions rates very badly for one or both of the other two criteria.

Routes *D*, *E*, and *F* illustrate different types of "compromise" solutions. Route *D* has the same low accident rate as route *C*, but if offers a different trade-off of the other two criteria (shorter, but higher exposure). Route *E* is only slightly longer than route *A* but offers improvements in both of the other criteria as a trade-off. Finally, route *F* results in intermediate values of all three criteria (11% above the minimum distance, 18% above the minimum exposure, and 1% above the minimum accident rate).

Adding other links to the analysis network (noninterstate federal and state highways) would, of course, change some of the routing solutions. In particular, a minimum exposure solution that zigzags less than route *B* could probably be found by adding in some noninterstate routes in New England. However, the purpose of this example is not to create detailed arguments for or against some specific route, but to illustrate an analysis methodology.

It is important to notice the difference in the results obtained through multiobjective analysis and the results of a single-objective analysis. The single-objective analysis results in one route being identified; in the multiobjective case we generate far more routes. Among those routes is the best with respect to each of the objectives in isolation as well as a series of compromise solutions. In most cases it is these trade-off solutions that are more likely to be implemented.

6.F. USING GEOGRAPHIC INFORMATION SYSTEMS FOR HAZARDOUS WASTE MANAGEMENT

A geographic information system (GIS) is a computer information system designed to work with spatially referenced data (Star and Estes, 1990). As discussed in previous sections, hazardous waste management problems involve locating treatment, storage, and disposal facilities, transporting wastes to those sites, and analyzing risks to populations and the environment in the vicinity of facilities and transportation activities. GISs are very effective at two main tasks that are central to hazardous waste management. The first task is managing, and helping the decision maker(s) visualize, relevant data. The second task is linking different types of data by geographic location. In this section we describe some of the basic elements of a GIS and illustrate how GIS tools can be used to support decision making in hazardous waste management.

We begin by discussing the concept of a spatial object and how this concept allows representation of the important elements in waste management problems. A spatial object is a delimited geographic area defined to exist at a location or cover a finite area, to which we can assign descriptive information (or attributes). The three main types of spatial objects are points, arcs, and polygons (Star and Estes, 1990). A *point* is a location that has no area. An *arc* is a connected sequence of points and, as such, also has no area. Typically, the endpoints of an arc or the intersections between arcs are called *nodes,* and the points along arcs are referred to as *shape points.* A *polygon* is a enclosed area. With these three spatial objects we can represent a variety of locations and facilities, including roadways, railways, hospitals, and schools.

Figure 6.21 illustrates the location of schools and hospitals in several cities in the vicinity of Boston, Massachusetts. Each school and hospital is represented by a point and is referenced to a common location system so that the geographic relationship between these locations is explicit. Associated with each point location is a variety of nonspatial information. For instance, the geographic location of the Newton–Wellesley Hospital is indicated on the map by the symbol for hospital, and there is a record in a database table for that location which gives its name and an identification code, lists the facility as a hospital, and may contain other information about the facility.

Figure 6.22 illustrates the highway network in the same region as covered by Figure 6.21. The heavier arcs represent interstate highways. Attached to each arc is a list of attributes which includes length, number of lanes, and so on. The list of attributes could also include special items of information useful in hazardous waste management, such as accident rate, average daily traffic, or population within some distance of the arc. In the example shown in Figure 6.22, the length value of 1.02 miles refers to an arc defined as the portion of I-95 that is between state route 9 and state route 16.

Figure 6.23 illustrates the boundaries of several cities in the Boston area. Each city is represented by a polygon that encloses an area proportional in shape to the real city. Within the boundaries of each city there are small polygons. These smaller polygons represent ponds and other bodies of water.

In the illustration each city has been labeled by the value in the name field in the appropriate database record. The ability to label spatial objects based

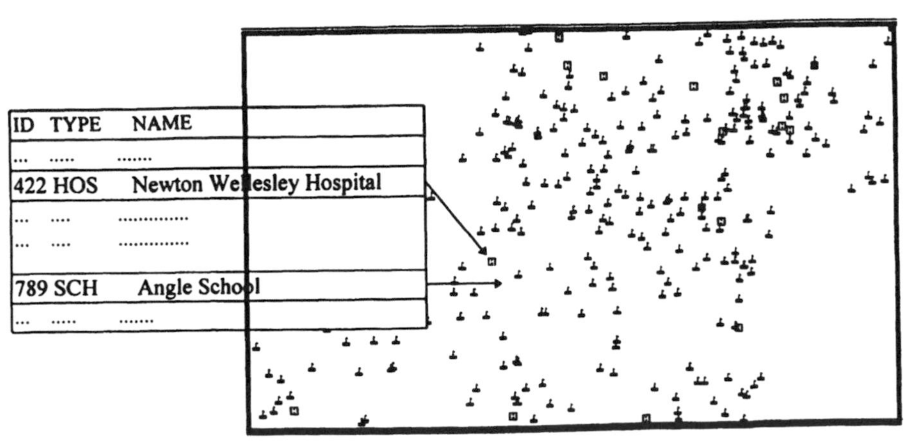

Figure 6.21 Locations of schools and hospitals in several cities near Boston.

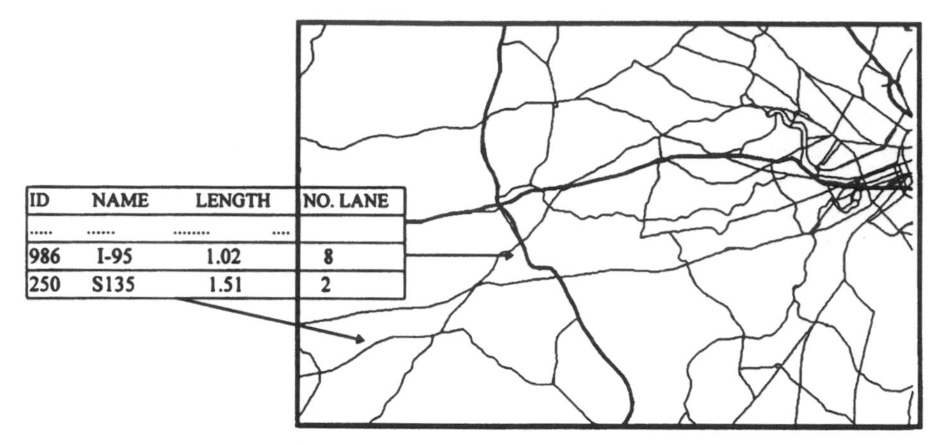

Figure 6.22 Highway facilities.

on an entry in a database record is a common feature in all GIS systems, and improves the readability of the displays dramatically when used properly.

Figure 6.24 is a thematic map of the population of several cities. Thematic maps are created by developing a color scheme or some type of graphical indicator for the value of an attribute or a function of attributes for each spatial object. In Figure 6.24 it is clear that Boston has a larger population then the other cities illustrated. A thematic map of population density could have been created by coloring the spatial objects based on the ratio of population to area. By indicating a value in this very visual manner, important relationships in the data can be more easily understood and communicated.

In contrast to points and lines, polygons have areas. Arcs were chosen as the representation for the highway database illustrated in Figure 6.22, implying an area equal to zero. This representation for roads is called a centerline

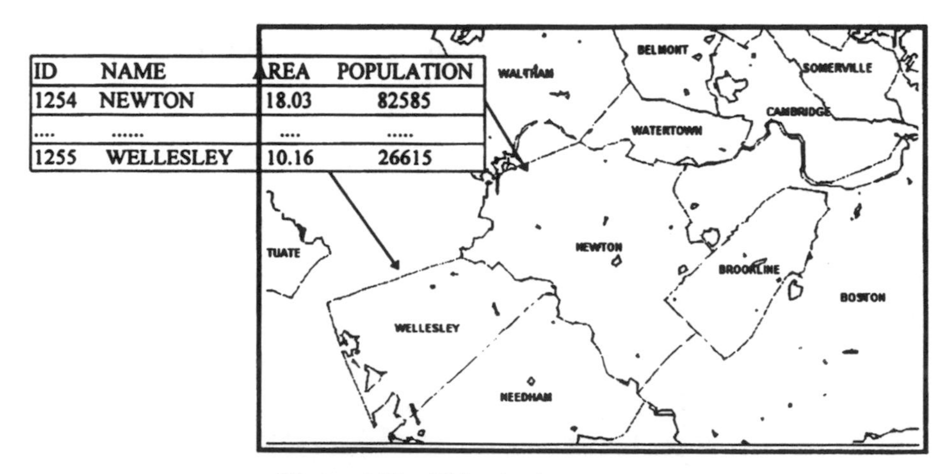

Figure 6.23 Cities in the same area.

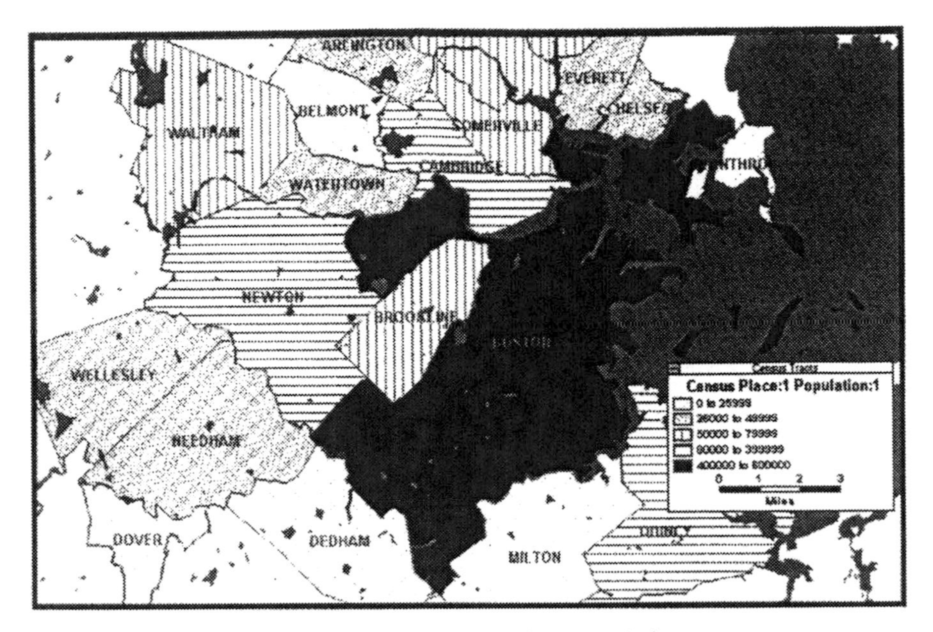

Figure 6.24 Cities coded by population.

representation because the geographic locations associated with the arcs are based on the geographic center of the facility. Highways clearly have area; however, for many purposes (including hazardous material routing), the arc representation is sufficient and it is unnecessary to represent highways in a more detailed manner (using polygons).

Figures 6.21 through 6.23 illustrate how a GIS can be used to organize both spatial and nonspatial information on facilities and on natural and administrative boundaries. They also illustrate that the geographic relationship between these elements is always explicit and maintained. Generally, information on each type of feature (e.g., roads, schools, city or county boundaries, etc.) is organized into a separate layer in the database. Each layer consists of a set of geographic features and the associated attribute data.

One of the major strengths of a GIS is its ability to relate information in one layer to information in another through the location and distance between specific features. One way is simply to display the layers (using the same geographic locational system) one on the top of the other, to more easily understand the geographic relationships between the features stored across all the layers. Figure 6.25 illustrates this type of display with the city layer, the school and hospital layer, and the water layer. Based on this display, it is clear what schools and hospitals are located in which cities.

For layers that share the same geographic locational system, queries based on geography can be written. A geographic query asks for information from the database layers, such as "How many hospitals are within a given fixed distance of this potential location for a hazardous waste facility?" or "What

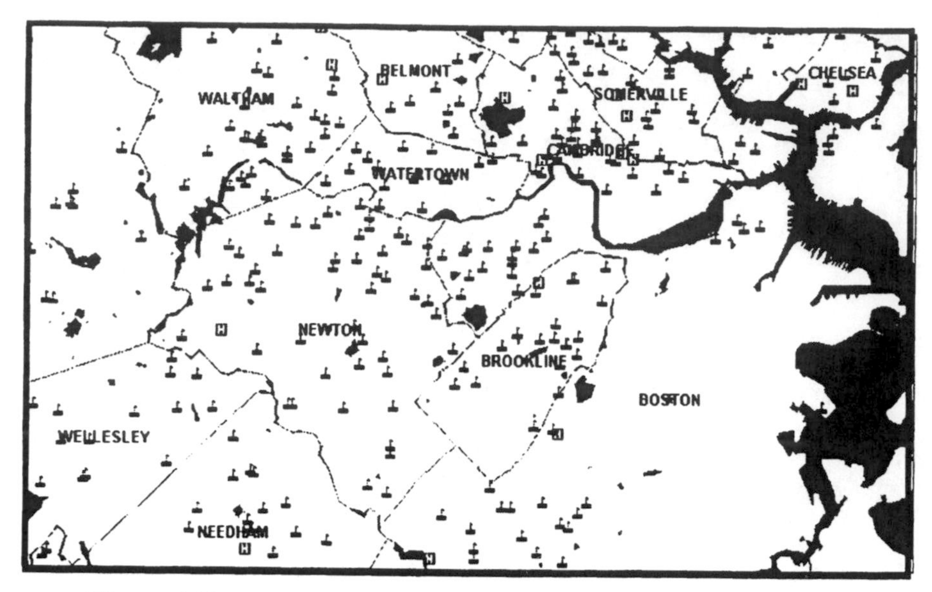

Figure 6.25 Overlay of cities, schools, hospitals, and bodies of water.

is the total population within half a mile of a selected set of arcs?" Through geographic queries a potential location for a hazardous waste repository can be evaluated or the impact of a proposed route for a specific hazardous materials shipment can be analyzed.

Buffering is a geographic operation that identifies an area within a specified distance of a map feature (Caliper, 1996). Figure 6.26 illustrates the use of buffers to compute the number schools that are within 0.25 mile and 0.5 mile of route I-90 through the cities of Newton, Watertown, Brookline, and Boston. The first ring indicates that 11 schools are within 0.25 mile, and the second ring indicates that 16 schools are between 0.25 and 0.5 mile of the highway.

A polygon overlay is a geographic operation that merges the attributes of polygons from two layers into a third layer (ESRI, 1990). For instance, if we consider the buffers illustrated in Figure 6.26, we might want to estimate the population within each of the buffer rings. This could be done by adding a census tract layer with population statistics. Then if we assume that the population is distributed uniformly within a census tract, the buffer rings can be overlaid on the census tracts and the population estimated. Figure 6.27 illustrates the necessary map layers: the regions in which we want the estimate (buffer rings) and the regions in which the data of interest is located (census tracts). In this example about 4410 people live within 0.25 mile of the selected section of I-90, and about 4379 people live between 0.25 and 0.5 mile from the highway.

The results of geographic queries and overlays can be saved as attributes of map features. For instance, a geographic query could be used to estimate

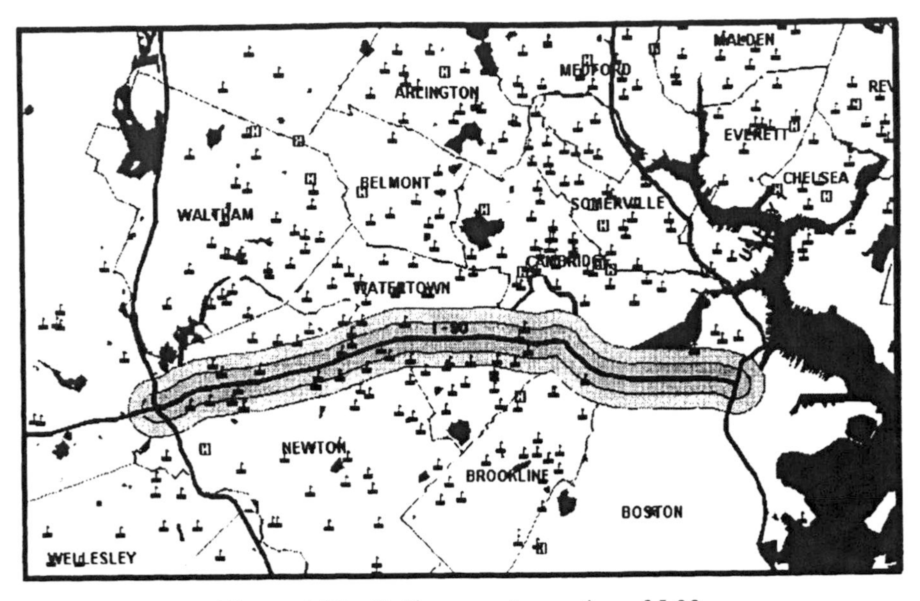

Figure 6.26 Buffer around a portion of I-90.

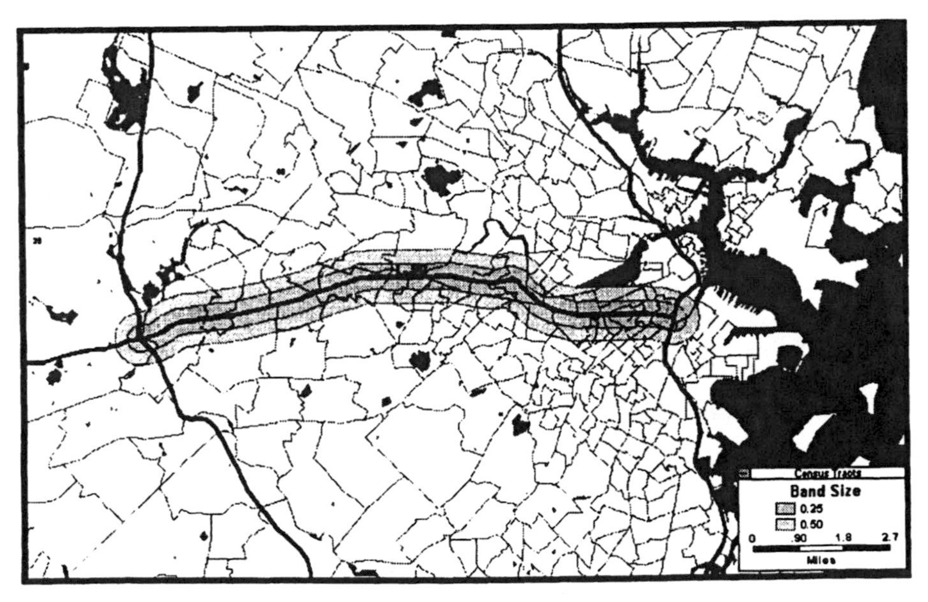

Figure 6.27 Census tracts and interstate highways.

the population within a specific distance of an arc, and that population estimate could then be added to the database record for that arc. Once in the database, the results can be displayed or used as input for a routing analysis, for example.

All of the GIS operations discussed thus far are based on a model for storing the geographic information which is called the *vector model.* A vector model or a vector data structure is based on the notion that every geographic object can be described by its elemental points and that the locations of these points are known. A vector is described by an elemental point, a direction, and magnitude. If we consider the three spatial objects described above, a point is just one elemental point. An arc is composed of a sequence of connected elemental points, or alternatively by a series of vectors. A polygon is really a closed arc in which the interior is considered to be part of the spatial object. A vector model is very efficient at representing elements for which there are definite boundaries, such as roads, buildings, or governmental units.

The *raster data model* is an alternative data model for use when the boundaries of the elements are not well defined. An example of a case in which a raster representation might be more appropriate is in representing elevation of a geographic area. In a raster representation, the geographic area is divided into cells and the value(s) of the attribute(s) of interest are stored for each cell. Raster and vector layers can be displayed together and geographic queries and operations can be used between layers using different data models. Therefore, the decision of which representation to adopt for a particular layer should be determined by which one yields the most efficient representation given the analysis to be conducted.

A GIS can be used to store locational information for many important elements of hazardous waste management problems, including the transportation infrastructure, topography, the locations of environmentally sensitive areas such as wetlands and aquifers, the spatial distribution of people, and the location of emergency service facilities. Through ordinary database operations such as relates and joins, descriptive (attribute) data from a variety of sources can be brought into the GIS database layers and linked to the appropriate spatial objects or raster cells. The geographic queries and thematic displays can be used to gain a clear understanding of the data and the options available. Also, the data files necessary for an optimization or evaluation routine can be constructed from the data in the layers as well as the results of geographic queries and operations.

Once the optimization or evaluation routines have been executed, results can be brought back into the GIS for display. By using a GIS to display the results of the analysis, a clearer understanding of the solution can be achieved. This is illustrated in Figure 6.20, where several different routing solutions have been displayed on maps, using GIS software. This form of display is much more effective than looking at lists of links in an output table.

In addition to the general GIS capabilities discussed here, there are specialized types of GIS software designed particularly for facilities management, and these systems can also be of significant use in managing hazardous wastes. Facility management GIS (FM-GIS) is the integration of computerized maps of facilities and attribute data about those facilities (Douglas, 1994). FM-GIS is a natural environment in which to conduct risk exposure analyses for input into facility location decisions. However, a full discussion of FM-GIS is beyond the scope of this chapter.

In the course of this chapter GIS has been discussed exclusively for planning analysis, in contrast to real-time decision making. GIS can also be a useful tool in deciding what actions to take during the course of an emergency. Although this is primarily a promising area for future research, Lepofsky et al. (1993) present an interesting case study.

6.G. AREAS FOR CONTINUING STUDY

In the course of this chapter we have defined hazardous waste and touched on several important elements of hazardous waste management, including risk analysis, location modeling, transportation routing, and the use of GIS tools. The material in this chapter is intended to provide a general overview of the subject and indicate how several of the major ideas are related to one another. Each of the previous sections contains references that can be consulted for more complete treatments of specific subproblems.

We want to close by highlighting a few problems that are particularly worthy of continued study, as an indication of research directions that are likely to prove useful over the next several years. The first of these is development of multiobjective location analysis methods for siting treatment, storage, and disposal facilities. These location problems invariably involve multiple stakeholders and multiple objectives, yet our analytic methods are mostly single-objective models. Development of better multiobjective models for locating facilities is clearly an important undertaking.

A second important area is further development of multiobjective transportation models that incorporate time-of-day variation in the routing criteria. Some initial work in this area has been done, as noted in Section 6.E, but these efforts have only scratched the surface of very substantive issues. It is quite clear that many of the important routing criteria do have identifiable diurnal patterns, and there is a significant opportunity for managing hazardous materials transportation to take better advantage of these patterns. Rapidly evolving technology for vehicle location monitoring and real-time rerouting may make it possible to implement transportation strategies that would have been impossible only a few years ago.

Third, there is significant opportunity for integrating models of facility location and transportation options. There has been some initial work in this area as well (Zografos and Samara, 1990; List and Mirchandani, 1991a,b; ReVelle et al., 1991), but many important elements of this type of model integration remain unstudied.

Fourth, more work is needed in the area of integration of GIS technology with the analysis tools used in location modeling and transportation routing/scheduling. The ideas described in this chapter illustrate how effectively these various tools can support each other. However, most decisions made in practice at this time do not yet take advantage of all these tools, at least in part because it is still difficult to get all the pieces to work together. Additional efforts in this direction are likely to pay substantial dividends.

EXERCISES

6.1. Consider an incinerator that destroys liquid hazardous wastes by high-temperature combustion. A "causal event" for risks (in the sense of Figure 6.1) might be a clogged feed line which reduces the available oxygen in the incineration chamber and results in incomplete combustion of the wastes. The stack gases then contain the unburned waste chemicals. Follow the sequence of steps in Figure 6.1 and identify short lists of outcomes and exposure pathways (as in Figure 6.2) for a population in the vicinity of the incinerator.

6.2. The risk contour map in Figure 6.3 shows that part of a major highway is inside the 10^{-5} risk contour from the DSM site. It also shows that parts of the villages of Stein and Undenheuvel are inside the 10^{-6} contour. Which of these would you consider more serious? Why?

6.3. The village of Smallville contains a chemical plant that produces chlorine gas. One of this plant's major customers is an industrial plant in Middletown, 20 miles away. Every workday (250 days per year), three tank cars of chlorine are shipped by rail from the Smallville plant to the plant in Middletown. For simplicity, assume that every day the train contains 25 cars and that the three chlorine cars are together somewhere in the train. We are concerned with analyzing the risks due to possible derailment and rupture of the chlorine cars, which would result in release of chlorine gas, a very poisonous substance.

(a) Is the possible derailment and release of chlorine an acute risk or a chronic risk? Why?

(b) Suppose that each time a derailment occurs on this track section, exactly five cars derail (obviously a simplification). Assume that every car in the train is equally likely to initiate the derailment and that the five cars which derail must be adjacent to one another in the train. What is the probability that two of the five cars that derail carry chlorine?

(c) Suppose that of the number of chlorine cars that derail, the number that release their contents can be modeled using a binomial distribution:

$$\text{Prob}(r|d) = \binom{d}{r} p^r(1 - p)^{d-r}$$

where p is the probability that a derailed car will release its contents, r the number of chlorine cars that release their contents, and d the number of chlorine cars that derail. Some analysis of rail accidents in general has indicated that p is a function of train speed, v, measured in miles per hour:

$$p = 0.013\sqrt{v}$$

Suppose that the train speed on the rail line between Smallville and Middletown is 45 mph. For the situation specified in part (b), what is the probability that of the two chlorine cars that derail, one releases its contents?

(d) Suppose that the expected frequency of derailment accidents is 1 per 50,000 train-miles. Assuming, as before, that all derailments involve five cars, what is the expected frequency of accidents in which two cars carrying chlorine release their contents?

(e) A risk profile is a graph that illustrates the frequency of occurrence of specific outcomes, or the probability of those outcomes occurring over a specified time period. Assuming that the outcomes are the numbers of chlorine cars that release their contents in a given year, and that no more than one accident occurs in a given year, create a risk profile for zero, one, two, and three cars of released chlorine over one year. [You should have the answer for two cars from part (d).]

6.4. For the example problem illustrated in Figures 6.4 and 6.5 and Table 6.1, it is clear by inspection that the optimal (i.e., minimum population coverage) two-facility solution is A and E. Using the formulation shown in equations (6.1)–(6.5), construct the integer programming version of this problem and solve it (using some available computer package) to confirm that this is the optimal solution.

6.5. The graph in Figure 6.28 shows two attributes along each directed arc. Consider them to be a cost (in hundreds of dollars) and a measure of population exposure (in thousands of people) for a truckload of hazardous wastes to cross the arc. Thus for the arc from node 1 to node 2, the cost is \$300 and the exposure is 6000 people. Suppose we are interested in finding paths from node 1 to node 10 that minimize both cost and population exposure.

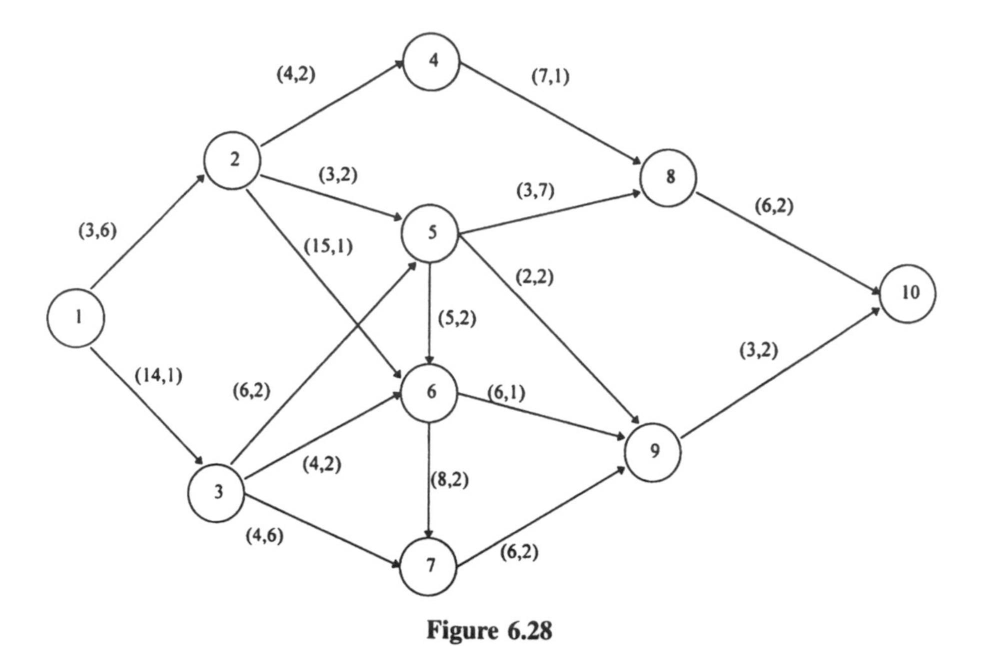

Figure 6.28

(a) Using the algorithm described in Section 6.E.3, find the set of all nondominated paths from 1 to 10.

(b) Consider each of the paths you have found to be a point on a graph like the one shown in Figure 6.29.

Plot the points corresponding to your nondominated paths to create a picture of the efficient frontier for this network.

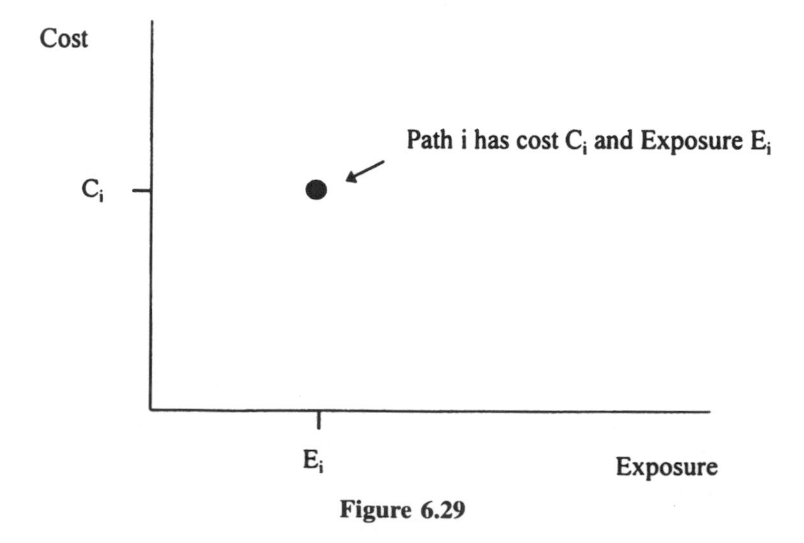

Figure 6.29

(c) Describe the evaluation and trade-off process you might go through to choose one of these alternative routes for the truck to follow.

REFERENCES

Bertsekas, D. P., 1991. *Linear Network Optimization: Algorithms and Codes,* MIT Press, Cambridge, MA.

Caliper, 1996. *TransCAD: Transportation GIS Software,* Caliper Corporation, Newton, MA.

Daskin, M. S., 1995. *Network and Discrete Location: Models, Algorithms, and Applications,* Wiley, New York.

Douglas, W. J., 1994. *Environmental GIS Applications to Industrial Facilities,* Lewis Publishers, Boca Raton, FL.

Dungan, T. P., 1991. The Hazardous Materials Transportation Uniform Safety Act of 1990: The U.S. Department of Transportation Perspective, *Hazmat Transport '91, Proceedings of a National Conference on Transportation of Hazardous Materials and Wastes,* Northwestern University, Evanston, IL, pp. 1–5 to 1–12.

Erkut, E. and S. Neuman, 1989. Analytical Models for Locating Undesirable Facilities, *European Journal of Operational Research,* 40, 275–291.

ESRI, 1990. *Understanding GIS: The Arc/Info Method,* Environmental Systems Research Institute, Inc. Redlands, CA.

Evans, J. R., and E. Minieka, 1992. *Optimization Algorithms for Networks and Graphs,* 2nd ed., Marcel Dekker, New York.

Fischoff, B., S. Lichtenstein, P. Slovic, S. L. Derby, and R. L. Keeney, 1981. *Acceptable Risk,* Cambridge University Press, Cambridge.

Glickman, T. S., and M. Gough (eds.), 1990. *Readings in Risk,* Resources for the Future, Washington, DC.

Glickman, T. S., and M. K. Macouley, 1991. Highway Robbery: The Social Costs of Hazardous Materials Accidents on the Capital Beltway, in M. D. Abkowitz and K. G. Zografos (eds.), *State and Local Issues in Transportation of Hazardous Waste Materials: Towards a National Strategy,* American Society of Civil Engineers, New York, pp. 148–162.

Griffin, R. D., 1988. *Principles of Hazardous Materials Management,* Lewis Publishers, Boca Raton, FL.

Harwood, D. W., J. G. Viner, and E. R. Russell, 1993. Procedure for Developing Truck Accident and Release Rates for Hazmat Routing, *Journal of Transportation Engineering,* 119(2), 189–199.

Henley, E. J., and H. Kumamoto, 1981. *Reliability Engineering and Risk Assessment,* Prentice Hall, Upper Saddle River, NJ.

Kamrin, M. A., 1988. *Toxicology: A Primer on Toxicology Principles and Applications,* Lewis Publishers, Boca Raton, FL.

Lepofsky, M., M. Abkowitz, and P. Cheng, 1993. Transportation Hazard Analysis in Integrated GIS Environment, *Journal of Transportation Engineering,* 119(2), 239–254.

List, G. F., and P. B. Mirchandani, 1991a. A Community-Focused Routing and Siting Model for Hazardous Materials and Wastes, in M. D. Abkowitz and K. G. Zografos (eds.), *State and Local Issues in Transportation of Hazardous Waste Materials: Towards a National Strategy,* American Society of Civil Engineers, New York, pp. 163–176.

List, G. F., and P. B. Mirchandani, 1991b. An Integrated Network/Planar Multiobjective Model for Routing and Siting for Hazardous Materials and Wastes, *Transportation Science,* 25(2), 146–156.

Maltezou, S. P., A. K. Biswas, and H. Sutter (eds.), 1989. *Hazardous Waste Management,* Tycooly Publishing, London.

Melachrinoudis, E., and T. P. Cullinane, 1986. Locating an Undesirable Facility with a Minimax Criterion, *European Journal of Operational Research,* 24, 239–246.

Nozick, L. K., G. F. List, and M. A. Turnquist, 1995. Integrated Routing and Scheduling in Hazardous Materials Transportation, working paper, School of Civil and Environmental Engineering, Cornell University, Ithaca, NY.

Ratick, S. J., and A. L. White, 1988. A Risk-Sharing Model for Locating Noxious Facilities, *Environment and Planning B,* 15, 165–179.

ReVelle, C., J. Cohon, and D. Shobrys, 1991. Simultaneous Siting and Routing in the Disposal of Hazardous Wastes, *Transportation Science,* 25(2), 138–145.

Rowe, W. D., 1977. *An Anatomy of Risk,* Wiley, New York.

Star, J., and J. Estes, 1990. *Geographic Information Systems: An Introduction,* Prentice Hall, Upper Saddle River, NJ.

Turnquist, M. A., 1987. Routes, Schedules and Risks in Transporting Hazardous Materials," in B. Lev et al. (eds.), *Strategic Planning in Energy and Natural Resources,* North-Holland, Amsterdam, pp. 289–302.

Turnquist, M. A., 1991. Using Real-Time Location Information for Hazardous Materials Shipments, in Y. Stephanedes and K. C. Sinha (eds.), *Applications of Advanced Technologies in Transportation Engineering,* American Society of Civil Engineers, New York, pp. 152–156.

Turnquist, M. A., and G. F. List, 1993. Multiobjective Policy Analysis of Hazardous Materials Routing, in L. N. Moses and D. Lindstrom (eds.), *Transportation of Hazardous Materials,* Kluwer, Boston, pp. 103–116.

U.S. Dot, 1994. *Biennial Report on Hazardous Materials Transportation: Calendar Years 1992–1993,* U.S. Department of Transportation, Washington, DC.

van Kuijen, C. J., 1989. Risk Management in the Netherlands: A Quantitative Approach, in S. P. Maltezou, A. K. Biswas, and H. Sutter (eds.), *Hazardous Waste Management,* Tycooly, London, pp. 200–212.

Zografos, K. G., and S. S. Samara, 1990. A Combined Location-Routing Model for Hazardous Waste Transportation and Disposal, *Transportation Research Record 1245,* pp. 52–59.

CHAPTER 7

Urban Transportation

DAVID E. BOYCE and MARK S. DASKIN

7.A. INTRODUCTION

7.A.1. Statement of the Problem

Designing urban transportation systems requires a capability to predict the future travel that will use these systems. Since the future demand for travel will respond to the characteristics and travel costs on the system supplied, this capability must consider future system attributes as well as future travel requirements that are generally independent of the transportation system, such as the size and extent of the urban area.

These design requirements have resulted in the extensive development of urban travel forecasting models and methods, beginning in the 1950s and continuing actively today. Optimization models have proven to be effective tools both in forecasting flows on urban transportation networks and in designing such networks. Indeed, one of the first formulations of an urban travel model (Beckmann et al., 1956) was based directly on the seminal paper of Kuhn and Tucker (1951) on nonlinear programming. Similarly, these models make extensive use of network concepts and methods, which are also an integral part of mathematical programming methods (Potts and Oliver, 1972; Shapiro, 1979; Ahuja et al., 1993; Bazaraa et al., 1993).

As in any field, the conceptualization of the problem solved by urban travel forecasting models evolved in a particular way that reflects the thinking and contributions of engineers and planners at the outset. For our purposes it is sufficient to understand that system design requires a forecast of travel activity for a typical weekday in the future. How far in the future depends on the facility in question. For large-scale facilities such as urban freeways or rail transit lines, 20 to 25 years in the future has been regarded as appropriate, even though the physical and economic life of such a facility may be longer. Equipment investments or operational plans may require forecasts for shorter time horizons, such as five or 10 years or even less.

For system planning, or even for a single large facility serving an urban corridor, the travel forecast is prepared for the entire urban region. This approach recognizes that urban travel choices involve alternative destinations, modes (e.g., auto drive alone, shared ride, bus and rail transit) and routes over a broad geographic region. For the purpose of preparing a forecast, this region is divided into many subareas, or zones, which typically range from 0.2 to 4 square miles in area, depending on the density of development. External zones represent trips originating and/or terminating outside the region. Similarly, the road and transit systems are represented by detailed networks comprised of nodes and links. The nodes correspond to roadway intersections and transit stations or boarding locations, as well as other points where supply characteristics change. Special nodes called *zone centroids* represent the points at which trips leave (origins) and enter (destinations) each zone. Directed links represent travel over roads from node to node in automobiles, or a route segment of a transit service.

Based on these concepts, the urban travel forecasting problem can be stated as follows: Given a description of the urban area in terms of zonal activities, and the specification of the transportation systems in terms of road and transit networks, determine the pattern of origin–destination flows by mode and route for a typical future weekday and peak travel period within that day, and determine the corresponding vehicle and person flows on the links of the networks for those time periods. The following comments may be helpful in understanding and limiting this definition.

First, travel occurring over the peak period, or even the 24-hour day, is traditionally considered to be in the form of constant or static flows. Therefore, queuing of traffic and severe congestion resulting from incidents are not considered in this problem statement. These conditions, however, are addressed in dynamic transportation network models, an active research area during the past 10 years that resulted in part from the emerging field of intelligent transportation systems (Ran and Boyce, 1996).

Second, the travel choices are considered to represent a long-term equilibrium condition with regard to where to travel and by which mode and route. Clearly, some choices, such as where to travel, are related to long-term decisions about where to live and work that involve factors other than transportation systems. Other choices, such as which mode to use, involve decisions concerning auto availability and the number of autos owned by the household, decisions that are typically made only once every few years. In contrast, choices of route may be made daily in response to specific needs, travel conditions on the road, and transit networks, weather, and other factors. Nevertheless, in the conventional travel forecasting model, this mix of long- and short-term factors is collapsed into a forecast for only one typical day and peak period in the near or distant future. This approach suggests that many factors are represented in the models in a rather aggregate manner, and that the detailed behavior of individuals and households is not considered explicitly. Similarly, it means that mechanisms are needed to account for these unrepresented aspects.

Third, the important issue of urban goods movement, or freight traffic, is not considered in the problem statement. This omission does not mean that trucks and delivery vehicles are totally ignored in the roadway traffic flows; in fact, they are typically represented as fixed interzonal flows. However, the forecasting of these flows is not considered in this chapter. Insufficient attention and effort are typically devoted by urban transportation planners to this important source of economic activity and congestion.

Fourth, an alternative view of the travel forecasting problem, the estimation of origin–destination (OD) flows from observed link flows, is not considered in this chapter. Models for this *OD estimation* problem have been studied extensively and occasionally are applied in practice. Some of these models are closely related to the travel choice models considered in Section 7.C. A literature review and comparison of two main approaches to solving this problem have been provided by Abrahamsson (1996).

Finally, as suggested by the way the problem has been framed, the difficult problem of transportation network design, including the choice of which links to build or improve and the phasing of such network improvements, is not generally regarded as part of the scope of urban transportation planning. In professional transportation planning practice, urban travel forecasts are typically used to evaluate, and eventually to design, road and transit systems. Despite a number of efforts to optimize network designs (e.g., Ridley, 1968; Scott, 1969; Boyce et al., 1973; Steenbrink, 1974; LeBlanc, 1975; Rothengatter, 1979; Poorzahedy, 1980), systems planning typically proceeds on a trial-and-error basis, largely based on conceptual schemes and practical issues. Indeed, many of these problems are NP-hard even when congestion effects are ignored (Magnanti and Wong, 1984). Thus in this chapter we are concerned primarily with the problem of forecasting flows on a network *given* the topology of the network. In other words, we take as given the links in the network and their capacities (e.g., the number of lanes on each link).

The remainder of the chapter is organized as follows. First, we summarize the *basic notation, variables, and constraints* that are common to all travel forecasting problems. These equations and constraints are not sufficient to allow us to forecast flows. Additional assumptions about the behavior of travelers and the nature of the information available to travelers are needed to forecast flows. Then in Section 7.B we outline several concepts of *route choice,* based on alternative behavioral assumptions, show how these concepts can be converted into optimization models, and illustrate the impact of these different models on the flows that result in a small network.

Section 7.C, on *origin–destination and mode choice,* extends the route choice problem to the choice of travel modes and destinations. Then we turn our attention to *solution algorithms* for these problems in Section 7.D and outline two different approaches: the first is appropriate for the pure route choice problem, while the second is needed for the more general travel choice models. We summarize some of our experience with the *application* of these algorithms in Section 7.E, followed in Section 7.F by a discussion of some of the issues related to implementing these algorithms in *professional practice.* Finally, *extensions* of the models are considered briefly in Section 7.G.

7.A.2. Basic Notation, Variables, and Constraints

In this subsection the notation, variables, and key constraints necessary to solve the urban travel forecasting problem are defined. First, we define the basic variables and notation. Then we define an optimization framework for the design of transportation systems in which travel is endogenous, or internal to the design method, and consistent with transportation supply.

As already noted, an urban transportation system can be represented by nodes and links. Let i represent a node and N the set of nodes in the network. Let a represent a directed link from node i to node j and A the set of links in the network. Hence the network is defined by G(N,A).

Associated with link a are travel time and monetary cost functions which depend on f_a, the flow of vehicles per hour on link a. In urban transportation planning practice, functions of the following form are typically used:

$$t_a(f_a) = t_a^0 \left[1 + \alpha \left(\frac{f_a}{C_a} \right)^{\beta} \right]$$

where $t_a(f_a)$ is the travel time on link a at flow f_a, t_a^0 the travel time on link a at zero flow, known as the free-flow travel time, and C_a the maximum flow on link a at a given level of service, often referred to as the *capacity* of link a. The parameters α and β are generally taken to be 0.15 and 4, respectively, but other values are sometimes used. Note that this function is strictly increasing with flow for $\alpha > 0$ and $\beta > 0$. Preferably, a function based on traffic flow theory should be used, such as the class of time-dependent functions proposed by Akcelik (1988); such functions are turning-movement specific. Another alternative function proposed by Davidson (1966) has some attractive theoretical properties but results in challenging numerical problems when incorporated into the solution algorithms. For purposes of this exposition, the familiar function above, known as the *BPR function* for its originators at the U.S. Bureau of Public Roads, is adopted as an example.

Monetary costs may also be an important determinant of travel choice. For road costs, the following functional form incorporates the effects of travel distance and congestion, through a travel time term:

$$k_a(f_a) = r_1 d_a + r_2 t_a(f_a)$$

where $k_a(f_a)$ is the monetary cost at flow f_a, d_a is the length of link a, and parameters r_1 and r_2 reflect the operating costs per unit distance and per unit time, respectively, of the average auto. Fixed costs at destinations, such as parking fees, can also be represented as links. By defining weights representing the relative importance of time and money, γ_1 and γ_2, these two functions can be combined into a generalized cost function,

$$c_a(f_a) = \gamma_1 t_a(f_a) + \gamma_2 k_a(f_a)$$

Note that the ratio $0.6\gamma_1/\gamma_2$ is an implied value of time in dollars per hour if the units of t_a and k_a are minutes and cents, respectively. For the 1990s, the value of time is typically in the range \$5 to \$15 per hour.

Next, define $\delta_{ra} = 1$, if route r includes link a, and 0 otherwise. The generalized cost c_r of travel on route r is assumed to be the sum of the costs of the links comprising route r, and can be expressed as follows:

$$c_r = \sum_{a \in A} \delta_{ra} c_a(f_a)$$

If h_r is the flow in vehicles per hour on route r, then

$$f_a = \sum_{r \in R} \delta_{ra} h_r$$

where R is the set of all routes in the network. This definition need not imply that all the routes are enumerated. Note that c_r depends on all the route flows in the network through the link flows (f_a); this fact can be made explicit by writing the route costs as $c_r(h)$, where h is the vector of route flows.

The two equations above define the interrelationships between link costs and route costs and route flows and link flows. The following equations complete the definition of the flows on a network:

$$\sum_{r \in R_{pq}} h_r = d_{pq} \qquad p \in P, \quad q \in Q$$

Here d_{pq} is the flow from origin zone p to destination zone q, also known as the travel demand from p to q. R_{pq} is the set of routes from p to q. This definition states that the sum of the flows on the individual routes connecting p and q must equal the total flow d_{pq}. This statement must hold for all zone pairs with positive flow. These definitions could also hold for separate trip purposes, such as the journey to work or shopping trips. These equations are called the *conservation of flow constraints*.

In contrast to the link–route network representation described above, a link–node representation is also widely used, as in some other civil engineering systems. In this case, a conservation of flow equation is defined for each node. Fur further details, see Patriksson (1994, pp. 36–39).

Using the definitions above, the following objective function can be stated for *designing* an urban transportation system:

$$\min \sum_{r\in R} c_r(h)h_r \equiv \min \sum_{r\in R} h_r \left[\sum_{a\in A} \delta_{ra}c_a(f_a)\right] \equiv \min \sum_{a\in A} c_a(f_a)f_a$$

This objective minimizes the total generalized travel cost in the network. The equivalence of these three expressions for the total network travel cost may be demonstrated by reversing the order of the summation in the middle expression. This definition of the function presumes that the benefits of travel are fixed; a more general definition would include a benefit function also.

The design variables for this objective function are the capacities (C_a) embedded in the travel cost functions $c_a(f_a)$, including the possibility of adding new links to the network, whose current capacities may be regarded as zero. To solve this problem, forecasts of the link flows f and the OD flows d are needed. The constraints outlined above, however, are not sufficient to determine these flows; additional behavioral assumptions are needed. In the following sections we show that under common behavioral assumptions the problem of forecasting the flows can be formulated as one of minimizing some objective function $z(f)$ subject to the constraints outlined above and possibly some additional constraints specific to the behavioral model being postulated. With this background we obtain the following general framework for the *design* of a transportation network:

$$\text{U:} \qquad \underset{C\in B}{\text{minimize}} \sum_{a\in A} c_a(f_a, C_a)f_a$$

where f_a solves

$$\text{L:} \qquad \underset{f\in F}{\text{minimize}}\ z(f, C)$$

The objective of the upper problem U is to minimize the total travel cost in the network. The decision variables are the link capacities (C_a) in the network G(N,A) which are feasible in terms of a set of budget constraints B. The lower problem L represents the user behavior (route, mode, and origin–destination choice) subject to the conservation of flow constraints F and the link capacities C determined by the upper problem. This formulation has the structure of a bilevel programming problem. In general, this problem is nonconvex and therefore may have multiple local optima. Nevertheless, the formulation captures abstractly the essence of the system design problem faced by urban transportation planners. By elaborating on the lower problem, we seek to elucidate the properties and challenges of this class of models.

7.B. ROUTE CHOICE

7.B.1. System-Optimal Auto Route Choice

Perhaps the most intuitively obvious approach to forecasting or estimating the flows on a network would be to determine the link flows so that (1) all demands are satisfied, and (2) the total cost of the flows, as represented by the objective function of the upper problem, is minimized. In this subsection we formulate such a model and outline its properties. In addressing these questions, we assume the origin-to-destination (OD) flows are auto trips in vehicles per hour.

This flow optimization problem may be stated as follows:

$$\underset{h_r}{\text{minimize}}\ c(h) = \sum_{r \in R} c_r(h) h_r \equiv \sum_{a \in A} c_a(f_a) f_a$$

subject to

$$\sum_{r \in R_{pq}} h_r = d_{pq} \qquad p \in P, \quad q \in Q$$

$$h_r \geq 0 \qquad r \in R$$

where

$$f_a = \sum_{r \in R} \delta_{ra} h_r \qquad a \in A$$

To analyze this problem, we form the Lagrangian equation and take partial derivatives with respect to the unknown variables (h_r):

$$\mathcal{L}(h, \pi) = \sum_{a \in A} c_a(f_a) f_a + \sum_{p \in P} \sum_{q \in Q} \pi_{pq} \left(d_{pq} - \sum_{r \in R_{pq}} h_r \right)$$

To take the partial derivatives of $\mathcal{L}(h, \pi)$ with respect to h_r, apply the chain rule to the first term as follows:

$$\frac{\partial \Sigma_{a \in A}\, c_a(f_a) f_a}{\partial h_r} = \frac{\partial \Sigma_{a \in A}\, c_a(f_a) f_a}{\partial f_a} \frac{\partial f_a}{\partial h_r}$$

$$= \sum_{a \in A} \left[c_a(f_a) + \frac{dc_a(f_a)}{df_a} f_a \right] \delta_{ra} \equiv \sum_{a \in A} m_a(f_a) \delta_{ra}$$

Note that the term δ_{ra} results from taking the partial derivative of the definition of f_a with respect to h_r.

The partial derivative of the second term of $\mathcal{L}$ with respect to h_r is simply $(-\pi_{pq})$. Therefore, the optimality or Karush–Kuhn–Tucker conditions for the nonlinear programming problem above may be stated as follows:

$$h_r \left[\sum_{a \in A} m_a(f_a)\delta_{ra} - \pi_{pq} \right] = 0 \qquad r \in R_{pq}, \quad p \in P, \quad q \in Q$$

$$\sum_{a \in A} m_a(f_a)\delta_{ra} - \pi_{pq} \geq 0 \qquad r \in R_{pq}, \quad p \in P, \quad q \in Q$$

$$\sum_{r \in R_{pq}} h_r = d_{pq} \qquad p \in P, \quad q \in Q$$

$$h_r \geq 0 \qquad r \in R$$

where $m_a(f_a) \equiv c_a(f_a) + f_a(dc_a(f_a)/df_a)$. The procedure for constructing the first-order optimality conditions from the partial derivatives of the Lagrangian function and the constraints, as well as their properties, may be found in any textbook on mathematical programming, such as that of Bazaraa et al. (1993). The application of these concepts directly to the problem of forecasting transportation network flows is also reviewed by Sheffi (1985).

Let us restate in words what we have derived so far and then interpret the optimality conditions. We seek the route flows that minimize the total network travel costs subject to the constraints that the sum of the route flows between each OD pair equals the fixed OD flow and the logical constraints that the route flows are nonnegative. The optimality conditions stated above may be interpreted as follows:

1. For OD pair pq, if $h_r > 0$, then $\Sigma_{a \in A}\, m_a(f_a)\delta_{ra} = \pi_{pq}$, $r \in R_{pq}$.
2. For OD pair pq, if $h_r = 0$, then $\Sigma_{a \in A}\, m_a(f_a)\delta_{ra} \geq \pi_{pq}$, $r \in R_{pq}$.

That is, if there is a flow on route r from zone p to zone q, the route "cost" equals π_{pq}, which is constant for all routes with positive flow from p to q. In other words, routes from p to q with positive flow have equal values of π_{pq}. Moreover, if there is no flow, the *cost* from p to q is at least π_{pq}.

Next, we need to interpret the nature of these *costs* obtained by minimizing the total network cost. Define

$$m_r(h) \equiv \sum_{a \in A} m_a(f_a)\delta_{ra} \equiv \sum_{a \in A} \left[c_a(f_a) + f_a \frac{dc_a(f_a)}{df_a} \right] \delta_{ra}$$

Note that $c_a(f_a)$, the travel cost actually experienced by each vehicle, is the first term of $m_a(f_a)$. The second term, $f_a(dc_a(f_a)/df_a)$, is positive if $c_a(f_a)$ is assumed to be a strictly increasing function of f_a and f_a is positive. The derivative of $c_a(f_a)$ with respect to f_a is the change in the actual travel cost per unit of flow, or the incremental cost experienced by each vehicle already on link a for one additional unit of flow. (The term *actual travel cost* denotes the cost experienced by each traveler.) Since this unit incremental cost is multiplied by the total link flow, the product is the total incremental cost to the existing users of link a of a unit increase in flow, f_a. The sum of this incremental cost and the actual travel cost $c_a(f_a)$ represents the total cost of adding one unit of flow to the link, which is known in economic analysis as the *marginal cost.* The definition of $m_r(h)$ indicates that the marginal route cost is the sum of the marginal link costs.

Therefore, the optimality conditions of the route (and link) flows that minimize total network cost may be stated as follows:

1. All used routes between each OD pair have equal marginal costs.
2. No unused route between that OD pair has a lower marginal cost than the used routes.

These interesting conditions, known as the *system-optimal conditions,* indicate how the optimal route flows on an urban road system may be achieved: The flows should be adjusted among alternative routes for each OD pair until the marginal route costs are equal, that is, until the decrease in total costs from a unit reduction in flow on route 1 just equals the increase in total costs from a unit addition on flow on route 2. Such system-optimal flows might be achieved by an intelligent transportation system through the introduction of link tolls.

To illustrate these concepts, consider the network shown in Figure 7.1. In this example the BPR function introduced in Section 7.A.2 is used with $\alpha = 0.15$ and $\beta = 4$. Table 7.1 gives the free-flow travel times and nominal link capacities for the five links. In this example, route choice is assumed to depend only on travel time and not on operating cost. We assume that 5000 vehicles per hour travel from node A to node D, and 5000 vehicles per hour travel from node B to node D. Node C is neither an origin nor a destination.

Table 7.2 shows the link flows, marginal link travel times, and link travel times actually experienced by the drivers on the five links. Note that the

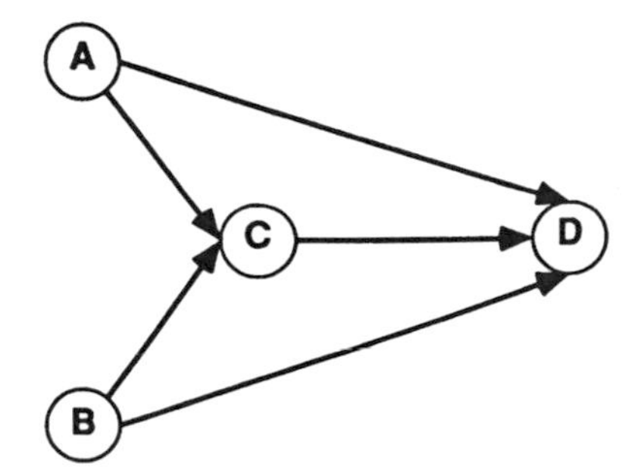

Figure 7.1 Example four-node network.

TABLE 7.1 Characteristics of the Links of the Network of Figure 7.1

Origin Node	Destination Node	Free-Flow Time (min)	Capacity (vehicles/hr)
A	*D*	20.0	1500
A	*C*	8.0	1000
C	*D*	10.0	2000
B	*C*	16.0	1000
B	*D*	24.0	1500

marginal travel time on the direct route (single link *AD*) from *A* to *D* is 245.008 min, which is identical to the sum of the marginal travel times on links *AC* and *CD*, which comprise the indirect route from *A* to *D*. Similarly, the marginal travel time on the direct route from *B* to *D* (link *BD*) is 326.037 min, which is identical to the sum of the marginal times on links *BC* and *CD*, which constitute the indirect route from *B* to *D*. The total time in the network is 746,910 vehicle-minutes per hour, or a mean travel time of 74.691 min. Note that this mean time is substantially less than the node-to-node marginal travel times.

7.B.2. User-Optimal Auto Route Choice

A little reflection should indicate that individual drivers left to their own discretion do not choose system-optimal routes if each individual's objective is to minimize his or her own travel costs. For example, in Figure 7.1, the actual travel time from *A* to *D* using the direct route (link *AD*) is 65.0 min. The actual travel time using the indirect route (links *AC* and *AD*) is only 63.4 min. Eventually, some drivers using link *AD* may realize that they can reduce their travel time by switching to the indirect route. Similarly, the actual travel time on the direct route from *B* to *D* is 84.41 min, whereas the actual time on the indirect route (links *BC* and *CD*) is 86.01 min. Again, eventually, some

TABLE 7.2 System Optimal Flows and Marginal and Actual Travel Times on Links of the Network of Figure 7.1

Origin Node	Destination Node	Link Flow (vehicles/hr)	Marginal Time (min)	Actual Time (min)
A	*D*	2952.01	245.008	65.00
A	*C*	2047.99	113.551	29.11
C	*D*	4012.09	131.457	34.29
B	*C*	1964.10	194.580	51.72
B	*D*	3035.90	326.037	84.41

drivers on the indirect route may realize that they can reduce their travel time by switching to the direct route from B to D. Thus individuals prefer the routes given by the following conditions:

$$h_r \left[\sum_{a \in A} c_a(f_a)\delta_{ra} - \pi_{pq} \right] = 0 \qquad r \in R_{pq}, \quad p \in P, \quad q \in Q$$

$$\sum_{a \in A} c_a(f_a)\delta_{ra} - \pi_{pq} \geq 0 \qquad r \in R_{pq}, \quad p \in P, \quad q \in Q$$

$$\sum_{r \in R_{pq}} h_r = d_{pq} \qquad p \in P, \quad q \in Q$$

$$h_r \geq 0 \qquad r \in R$$

These *user-optimal* route choice conditions have the interpretation that:

1. For OD pair pq, if $h_r > 0$, then $\Sigma_{a \in A} c_a(f_a)\delta_{ra} = \pi_{pq}$, $r \in R_{pq}$.
2. For OD pair pq, if $h_{r'} = 0$, then $\Sigma_{a \in A} c_a(f_a)\delta_{r'a} \geq \pi_{pq}$, $r' \in R_{pq}$.

Equivalent conditions were stated by Wardrop (1952) as his first principle, which pertains to a single OD pair: "The journey times on all the routes actually used are equal, and less than those which would be experienced by a single vehicle on any unused route." Wardrop also stated a second principle, which corresponds to *system-optimal* flows, as follows: "The average journey time is a minimum." What remains is to define the objective function that results in the user-optimal flow conditions. This function was identified by Beckmann et al. (1956) in their seminal book, and is defined as follows:

$$\underset{h}{\text{minimize}}\ c(f) = \sum_{a \in A} \int_0^{f_a} c_a(x)\, dx$$

This function has no interpretation in the formulation considered here; however, see Oppenheim (1995) for its interpretation in a random utility framework. Minimizing this function with respect to the route flows, given the same conservation of flow and nonnegativity constraints, results in the user-optimal conditions stated above, since

$$\frac{\partial c(f)}{\partial h_r} = \sum_{a \in A} c_a(f_a)\delta_{ra} \equiv c_r(h_{UO})$$

Hence $c_r(h_{UO})$ replaces $m_r(h_{SO})$ in the system-optimal analysis, where h_{UO} is a vector of user-optimal route flows and h_{SO} is a vector of system-optimal route flows. Thus we can state the following user-optimal route choice conditions:

1. All used routes between each OD pair have equal costs.
2. No unused route between that OD pair has a lower cost than those of the used routes.

Recall that the term *cost,* as used here, means the actual cost experienced by an individual driver, as contrasted with the marginal cost in the system-optimal case. Table 7.3 gives the user-optimal flows and travel times for the network shown in Figure 7.1. Notice that now the actual travel times between nodes A and D are the same (64.11 min) for the two routes. Also, the actual travel times between B and D are the same (85.30 min) for the two routes. Now, however, the marginal travel times are different for the routes between each OD pair. For example, the marginal travel time on the direct route from A to D is 240.53 min, while the marginal travel time on the indirect route is 248.53 min, suggesting that vehicles should be diverted from the indirect route to the direct route to reduce total travel time. This diversion is just what occurred in the system-optimal solution shown in Table 7.2.

We conclude this introduction to user-optimal and system-optimal route choice with a discussion of the basic properties of this class of models.

1. Implicit in both models is the assumption that drivers have perfect information concerning link and route costs. As a result, the solutions are deterministic. However, no mechanism is specified concerning how they obtain this information or how these user-optimal choices could be achieved in reality. The extent to which user-optimal conditions do prevail in actual road systems is unknown, there being few data on such conditions; however, it is generally believed that the deterministic user-optimal model is a valid first approximation of real-world conditions.
2. As already noted, both models are static, or have no time-of-day dimension. Flows are assumed to occur at constant levels for the period of the analysis, such as the peak hour or a longer peak period. There are no departure or arrival times in such a static model; in its dynamic

TABLE 7.3 User-Optimal Flows and Marginal and Actual Travel Times on Links of the Network of Figure 7.1

Origin Node	Destination Node	Link Flow (vehicles/hr)	Marginal Time (min)	Actual Time (min)
A	D	2937.23	240.53	64.11
A	C	2062.77	116.53	29.73
C	D	4015.77	131.90	34.38
B	C	1952.99	190.58	50.92
B	D	3047.01	330.48	85.30

or time-dependent analog, flows are departure-time (or arrival-time) specific.

3. Assuming that the link cost functions are strictly increasing with flow and that each link's cost depends only on the link's own flow, the solutions of both route-choice problems exist and are unique in the link flows. However, they are not unique in the route flows for a given OD pair or across OD pairs (see Exercise 7.6).
4. For a given OD flow vector d, the total travel time for the system-optimal flows is less than or equal to the total travel time for the user-optimal flows. This is intuitive since the system-optimal flows minimize the total, or mean, travel time. For example, the mean travel time of the system-optimal flows for the network shown in Figure 7.1 is 74.691 min. For the user-optimal flows, the mean travel time is 74.701 min. The solutions are identical if $c_a(f_a) = \bar{c}_a$, $a \in A$, where $\bar{c}_a$ is independent of flow.
5. Similarly, the *artificial* user-optimal objective function is smaller when evaluated for the user-optimal flows than for the system-optimal flows, since the user-optimal flows minimize this objective function. Thus the user-optimal objective function for this network evaluated for the user-optimal flows is 325,227.11, while the system-optimal objective function evaluated for the user-optimal flows is 325,247.82.
6. What is much less intuitive is the result that $c(h_{UO})$ can increase following the addition of one or more links to the network. This condition, known as *Braess's paradox*, is shown in Exercise 7.1; see also Murchland (1970).
7. It is possible to find a set of link tolls that alters the user-optimal route choices so as to achieve the system-optimal solution. One set of such tolls is found by setting the toll on each link equal to the difference between the marginal cost in the system-optimal solution and the user cost in the same solution. The toll makes users behave as if they are basing their choices on the marginal costs (Dafermos and Sparrow, 1971). Many different sets of tolls may produce the same system-optimal flows as discussed by Bergendorff et al. (1996) (see Exercise 7.2).

A detailed mathematical treatment of this class of models, including notes on the history of these developments, is given by Patriksson (1994, pp. 29–51).

7.B.3. Stochastic Auto Route Choice

The assumption that drivers have perfect information about link and route travel costs is clearly strong and can be justified primarily as providing a first approximation of their route choice behavior. Desirably, the assumption should be relaxed by introducing link perception errors into the model, by representing link travel costs as random variables, or both. A comprehensive approach to formulating this problem for flow-dependent models, as well as for the flow-independent case, is provided by Mirchandani and Soroush (1987). In this section we consider the perception error approach with deterministic link travel costs.

There are several ways of accomplishing this objective. One approach in the context of the optimization framework described above is to introduce a constraint representing the composite effect of suboptimal route choices resulting from perception errors. We shall follow this approach and then point out the relationship of the result to an unconstrained optimization formulation.

The solution to the user-optimal route choice model places each OD flow on routes of minimal and equal cost, implicitly assuming that drivers have perfect knowledge of these routes. If drivers do not possess this knowledge, their perception errors necessarily result in the choice of higher-cost routes; that is, there is a dispersion of choices to higher-cost alternatives. If this dispersion can be represented as a function of the route choices, the user-optimal formulation can be constrained to a certain suboptimal solution representing these perception errors.

An appropriate measure of the dispersion or spread of a discrete probability distribution is the entropy function (Theil, 1967):

$$S(I) = -\sum_{i \in I} P_i \ln P_i$$

where $S(I)$ is the *entropy* of the distribution defined on the set I and P_i is the probability of event i, $i \in I$, the set of alternative events. The value of $S(I)$ ranges from 0 if only one alternative is chosen, to $\ln n$ if all n alternatives have equal probability; that is, $P_i = 1/n$. Hence the entropy achieves its maximum when the probabilities are uniformly distributed. Note that the value of S depends only on the probabilities of the events, not on a variable such as travel cost as would be the case with the variance, another measure of the dispersion of a distribution.

In the case of auto route choice, the entropy function may be stated as

$$S(PQR) = -\sum_{p \in P} \sum_{q \in Q} \sum_{r \in R_{pq}} P_{pqr} \ln P_{pqr}$$

This definition is based on the joint probability P_{pqr} of traveling from zone p to zone q by route r. Since we presently assume that the interzonal flow d_{pq} is fixed, the use of the conditional probability is more appropriate:

$$P_{r|pq} = \frac{P_{pqr}}{P_{pq}}$$

where

$$P_{pq} = \frac{d_{pq}}{T} \quad \text{and} \quad T = \sum_{p \in P} \sum_{q \in Q} d_{pq}$$

Then

$$S(R \mid PQ) = -\sum_{p \in P} \sum_{q \in Q} P_{pq} \sum_{r \in R_{pq}} P_{r|pq} \ln P_{r|pq}$$

The reader may wish to verify that $S(R \mid PQ) = S(PQR) - S(PQ)$, where $S(PQ) = -\Sigma_{p \in P} \Sigma_{q \in Q} P_{pq} \ln P_{pq}$.

In the user-optimal solution, if only one route is used between each OD pair, $S^{UO}(R \mid PQ) = 0$. If a few routes are used between each OD pair, $S^{UO}(R \mid PQ)$ is somewhat greater than 0. If many routes are used as a result of perception errors, $S(R \mid PQ) > S^{UO}(R \mid PQ)$.

Suppose that we wish to model a pattern of route choices whose dispersion is characterized by $\bar{S}(R \mid PQ) > S^{UO}(R \mid PQ)$, where the bar denotes an observed or measured value of S. Then we can formulate the following stochastic route choice model:

$$\underset{P_{r|pq}}{\text{minimize}} \frac{1}{T} \sum_{a \in A} \int_0^{f_a} c_a(x)\, dx$$

subject to

$$-\sum_{p \in P} \sum_{q \in Q} P_{pq} \sum_{r \in R_{pq}} P_{r|pq} \ln P_{r|pq} \geq \bar{S}(R \mid PQ)$$

$$\sum_{r \in R_{pq}} P_{r|pq} P_{pq} T = d_{pq} \qquad p \in P, \quad q \in Q$$

$$P_{r|pq} \geq 0 \qquad r \in R_{pq}, \quad p \in P, \quad q \in Q$$

where

$$f_a = \sum_{r \in R} P_{r|pq} P_{pq} T \delta_{ra} \qquad a \in A$$

Forming the Lagrangian, we obtain

$$\mathfrak{L}(P, \pi) = \frac{1}{T} \sum_{a \in A} \int_0^{f_a} c_a(x)\, dx + \frac{1}{\theta} \left[\bar{S}(R \mid PQ) + \sum_{p \in P} \sum_{q \in Q} P_{pq} \sum_{r \in R_{pq}} P_{r|pq} \ln P_{r|pq} \right]$$

$$+ \sum_{p \in P} \sum_{q \in Q} \pi_{qp} \left(d_{pq} - \sum_{r \in R_{pq}} P_{r|pq} P_{pq} T \right)$$

Taking partial derivatives with respect to $P_{r|pq}$, we obtain

$$\sum_{a\in A} c_a(f_a)\delta_{ra} + \frac{1}{\theta}(\ln P_{r|pq} + 1) + \pi_{pq}(-T)$$

Letting $c_r(h) = \Sigma_{a\in A} c_a(f_a)\delta_{ra}$, $r \in R_{pq}$, the first-order optimality conditions are as follows:

$$P_{r|pq}\left[c_r(h) + \frac{1}{\theta}(\ln P_{r|pq} + 1) - \pi_{pq}T\right] = 0 \qquad r \in R_{pq}, \quad p \in P, \quad q \in Q$$

$$\left[c_r(h) + \frac{1}{\theta}(\ln P_{r|pq} + 1) - \pi_{pq}T\right] \geq 0 \qquad r \in R_{pq}, \quad p \in P, \quad q \in Q$$

$$-\sum_{p\in P}\sum_{q\in Q} P_{pq} \sum_{r\in R_{pq}} P_{r|pq} \ln P_{r|pq} \geq \bar{S}(R \mid PQ)$$

$$\sum_{r\in R_{pq}} P_{r|pq}P_{pq}T = d_{pq} \qquad p \in P, \quad q \in Q$$

$$P_{r|pq} \geq 0 \qquad r \in R_{pq}, \qquad p \in P, \quad q \in Q$$

$$\frac{1}{\theta} \geq 0$$

Since $\ln 0 \rightarrow -\infty$, intuitively we can expect that $P_{r|pq} > 0$, ensuring that the second condition is not violated. Solving for $P_{r|pq}$ from the first and third conditions, we obtain

$$\ln P_{r|pq} = \theta[\pi_{pq}T - c_r(h)] - 1 \quad \text{or} \quad P_{r|pq} = \exp\,[\theta(\pi_{pq}T - c_r(h)) - 1]$$

Since $\Sigma_{r\in R_{pq}} P_{r|pq} = 1 = \exp(\theta\pi_{pqr}T - 1)\Sigma_{r\in R_{pq}} \exp(-\theta c_r(h))$,

$$P_{r|pq} = \frac{\exp(-\theta c_r(h))}{\Sigma_{s\in R_{pq}} \exp(-\theta c_s(h))}$$

or

$$h_r = d_{pq}\frac{\exp(-\theta c_r(h))}{\Sigma_{s\in R_{pq}} \exp(-\theta c_s(h))} \qquad r \in R_{pq}, \quad p \in P, \quad q \in Q$$

The properties of this stochastic route choice model may be summarized as follows:

1. The functional form of the model is the well-known logit or logistic function. The parameter θ of the model may be seen to be related to

the dispersion of route choice and hence is referred to as the *dispersion parameter.*

2. The formulation implicitly assumes that all routes in R_{pq} receive a portion of the flow; these flows decrease monotonically in relation to the increasing route travel costs $c_r(h)$. Although this property is a serious drawback, solution methods based on sampling of routes can generate solutions that are reasonable in terms of the number of routes used.
3. As $\theta \to +\infty$, the solution of the stochastic route choice model tends to the solution of the deterministic user-optimal model. Similarly, as $\theta \to 0$, all routes R_{pq} receive an equal share of the flow d_{pq}. This property is illustrated in Table 7.4, which gives the link flows for the network of Figure 7.1 as a function of the value of θ. As the value of θ increases, the link flows approach the value obtained for the user-optimal case shown in the final column of the table.
4. A characteristic of the logit function is the independence of irrelevant alternatives (IIA) property. In this context it means that overlapping routes with several links in common are treated as if they are independent, resulting in these routes receiving an overallocation of flow.
5. The addition of the nonlinear entropy function constraint makes the formulation strictly convex, with the result that the route flows are unique.

TABLE 7.4 Link Flows for the Network of Figure 7.1 as a Function of the Value of the Dispersion Parameter θ

(a) Original Link Capacities

Link	$\theta = 0.00$	$\theta = 0.01$	$\theta = 0.10$	$\theta = 0.50$	$\theta = 1.0$	User Opt.
AD	2500	2790	2915	2933	2935	2937
AC	2500	2210	2085	2067	2065	2063
CD	5000	4316	4059	4025	4020	4016
BC	2500	2106	1974	1957	1955	1953
BD	2500	2894	3026	3043	3045	3047
Mean time	99.18	76.74	74.74	74.71	74.70	74.70

(b) Link Capacities Equal to Twice the Original Capacities

Link	$\theta = 0.00$	$\theta = 0.01$	$\theta = 0.10$	$\theta = 0.50$	$\theta = 1.0$	$\theta = 10.0$	User Opt.
AD	2500	2532	2618	2653	2658	2662	2663
AC	2500	2468	2382	2347	2342	2338	2337
CD	5000	4866	4437	4187	4137	4086	4080
BC	2500	2398	2055	1839	1955	1795	1748
BC	2500	2602	2945	3161	3205	3252	3258
Mean time	26.82	26.38	25.51	25.39	25.40	25.43	25.43

6. A solution of the model with flows slightly more dispersed that the user-optimal solution can have a somewhat lower total travel cost than the user-optimal total travel cost. (See Table 7.4b in which the stochastic route choice model is again applied to the network of Figure 7.1, but now with all link capacities doubled.) This solution is similar to the system-optimal solution in that somewhat more circuitous routes are used. However, as the solution becomes more dispersed, the total travel cost increases as the route flows become more and more circuitous.

As an alternative to, and generalization of, the stochastic route choice model above, an unconstrained optimization problem with the same optimality conditions was formulated by Sheffi and Powell (1982); see also Sheffi (1985, pp. 312–322).

$$\underset{h}{\text{minimize}}\ z(f,h) = -\sum_{p \in P} \sum_{q \in Q} d_{pq}\, \mathrm{E}\left(\min_{r \in R_{pq}} C_r \mid c(h) \right) + \sum_{a \in A} c_a(f_a) f_a - \sum_{a \in A} \int_0^{f_a} c_a(x)\, dx$$

where $\mathrm{E}(\min_{r \in R_{pq}} C_r \mid c(h))$ is the expectation (mean) of the minimum of the perceived route travel costs C_r, $r \in R_{pq}$, given the vector of network travel costs $c(h) = (c_r(h),\ r \in R)$, and the vector of current route flows h. The perceived travel cost C_r is assumed to equal the sum of the actual route travel cost $c_r(h)$ and a random error term ε_r that varies from traveler to traveler:

$$C_r = c_r(h) + \varepsilon_r \qquad r \in R_{pq},\quad p \in P,\quad q \in Q$$

Since $\mathrm{E}(\varepsilon_r) = 0$, $\mathrm{E}(C_r) = c_r(h)$. Also note that the conservation of flow constraints are satisfied automatically by the definition that $h_r = d_{pq} P_{r|pq}$, $r \in R_{pq}$. If the random errors (ε) are identically and independently distributed Weibull–Gumbel variates, the resulting model is the logit function given above with $C_r = c_r(h) - \varepsilon/\theta$, where ε is the Gumbel variate and θ is related to the variance of ε.

Alternatively, if the perception errors are distributed according to the multivariate normal distribution, the resulting stochastic choice model is the probit model. In this case, one possible model has

$$\mathrm{Var}(C_r) = \beta c_r^o, \quad r \in R \quad \text{and} \quad \mathrm{Cov}(C_r, C_s) = \beta \sum_{a \in A} \delta_{ra} \delta_{sa} c_a^o, \quad r,s \in R$$

where β is a proportionality constant. Here the variance of the perceived cost of route r is linearly proportional to its free-flow cost c_r^o. The covariance of the perceived cost of routes r and s is linearly related to the total free-flow costs of the links shared by the two routes. Hence overlapping routes are correlated, a major improvement over the logit model. The probit model, however, cannot be expressed analytically. See Patriksson (1994, pp. 62–65) and Daganzo (1979) for further details.

7.B.4. Transit Route Choice

Thus far our analysis of route choice has focused entirely on the auto mode. Link travel times, as well as operating costs, have formed the basis for both deterministic and stochastic models. For transit networks, similar variables have been employed in the past in formulating simple models of transit route choice. Since passenger-related delays in transit systems are related primarily to boarding times, link travel times and fares may be regarded as fixed. If boarding delays are also assumed fixed, the best route for each OD pair may readily be identified for a transit network and the total OD flow allocated to this route. Such a model is termed an *all-or-nothing assignment,* since the entire flow is allocated to the minimum-cost route and no flow to higher-cost routes. Since all times and costs are fixed, the route choice is deterministic and straightforward. Even so, the result is not very satisfying, since transit choices do tend to be dispersed among alternative routes if such choices are available.

This treatment of the transit route choice problem also ignores an important difference between auto and transit systems. Whereas auto travel generally can begin on demand, travel by transit begins when the transit vehicle (bus, train) arrives at the point of service or service node (bus stop, station). Hence transit riders incur waiting and transfer times that depend on the service timetable and the arrival times of passengers at the service node. These waiting times contribute to a different type of stochastic attribute than the one found in auto route choice.

In this section a simple version of the stochastic transit route choice model of Spiess and Florian (1989, pp. 92–93) is examined. We show that the model can be formulated as a separate linear programming problem for each destination zone, since the link travel times are fixed and the waiting and transfer times depend on the service provided and the passenger arrival pattern. The model is especially appropriate for complex transit systems that offer more than one service alternative between many OD pairs.

We begin by assuming that the objective of each traveler is to minimize his or her generalized travel cost, a weighted linear function of in-vehicle travel time, waiting and transfer times, and fare. We assume that the weights are known for a homogeneous group of travelers. Since the waiting and transfer times are not deterministic but depend on the service frequency and passenger arrival pattern, each passenger's objective is to minimize the expected generalized travel cost; since these individual costs do not depend on flow, this is equivalent to minimizing the total expected travel cost.

If transit services are frequent, it is reasonable to assume that travelers arrive at their service nodes at random with respect to the service schedule (uniformly distributed arrivals) and take the first vehicle belonging to the set of transit routes serving their single destination q, including combinations of routes that involve transferring between routes. This set of routes, termed a *strategy,* is denoted by a partial network, $G_q(N,\bar{A})$. For a strategy to be feasible it must contain at least one route from the node to the destination.

The total waiting time at node i depends on the flow of travelers arriving at node i from other links $\Sigma_{a\in A_i^-} f_a$, where A_i^- denotes the subset of links arriving at node i. If d_i is the flow of travelers entering the network at node i, the total flow requiring service at node i is

$$F_i = \sum_{a\in A_i^-} f_a + d_i \qquad i \in I$$

The total waiting time also depends on the frequency of service at node i on links that are in the feasible strategy set $\bar{A}$. For reasonable assumptions concerning the distribution of arrival times, the total waiting time at node i may be defined as

$$\omega_i = \frac{F_i}{2\,\Sigma_{a\in A_i^+} q_a x_a} \qquad i \in I$$

where $x_a = 1$ if $a \in \bar{A}$, and 0 otherwise; q_a is the frequency of service on link a; and A_i^+ is the set of links leaving node i. That ω_i is the total waiting time may be demonstrated as follows: $q_q x_a$ is the frequency of service on link a, a member of the strategy set for reaching the destination, and $\Sigma_{a\in A_i^+} q_a x_a$ is the sum of these frequencies. The reciprocal of the aggregate frequency of service is the mean time-headway; therefore,

$$\bar{h}_i = \frac{1}{\Sigma_{a\in A_i^+} q_a x_a} \qquad i \in I$$

Assuming random arrivals of passengers and deterministic headways, the mean waiting time is one-half of the headway. Hence we obtain the expression shown above. Given the total waiting time, the transit route choice model may be stated as the following linear programming problem.

$$\underset{f,w}{\text{minimize}} \sum_{a\in A} c_a f_a + \sum_{i\in I} \omega_i$$

subject to

$$\sum_{a\in A_i^-} f_a + d_i = \sum_{a\in A_i^+} f_a \qquad i \in I$$

$$f_a \le q_a \omega_i \qquad a \in A_i^+, \quad i \in I$$

$$f_a \ge 0 \qquad a \in A$$

Note that the terms in the denominator of ω_i are inputs to the flow estimation problem. Therefore, ω_i is a linear function of the decision variables (f_a). In the problem statement above, the first constraint is the link-node form of the conservation of flow conditions, which state that the flow into node i plus the flow entering the network at node i equals the flow out of node i. The second constraint states that the passenger flow on link a is no more than the frequency of service on link a times the total waiting time, which is equivalent to the total waiting time divided by the mean headway, which gives the maximum flow on link a.

The following comments are offered concerning the linear program above.

1. This formulation is a "relaxed" version of a formulation involving the (0,1) variables $x_a, a \in A$. In Spiess and Florian (1989, p. 92), the equivalence of the two formulations is demonstrated.
2. Transit fares can be introduced into the formulation in several ways if different fares are charged for different routes; if a flat fare is charged for the entire trip (free transfers), fares need not be explicitly included. For example, fares can be represented as an additional, equivalent boarding time.
3. The formulation above represents only one destination. Since travel times are flow independent, a sequence of such problems can be solved, one for each destination.
4. A nonlinear flow-dependent extension of the formulation above is also given by Spiess and Florian (1989).

Given a transit route choice model with properties comparable to the auto route choice models introduced previously, we are now ready to consider models of choices of mode and origin–destination pairs, given the rates of originations and terminations.

7.C. ORIGIN–DESTINATION AND MODE CHOICE

7.C.1. Underlying Concepts

Urban travel choices involve more dimensions than which routes to take through the road or transit networks. In large cities, at least, many travelers make a choice of mode, either daily or less frequently, such as in deciding whether to purchase another auto for their household. Modal alternatives may also be a factor in deciding where to work or shop with respect to one's household's residence, or conversely where to reside, given one's place of work. These choices are clearly made less frequently than mode or route choices.

Moreover, travel choices pertaining to the frequency of travel, whether to share an auto for a certain trip, when to purchase another auto, or when to replace an existing one are made periodically. These choices are especially intertwined with the household, cultural, and socioeconomic circumstances of the individual traveler. Accordingly, their prediction is even more difficult.

Modeling of this array of choices has evolved over the past 40 years in several competing styles. Of these, the traditional four-step forecasting procedure remains the prevailing practitioner paradigm: how often to travel, where to travel, by what mode, and by which route. Within this four-step framework, advances over the past 25 years have provided a more satisfactory foundation for some individual choice functions. For example, random utility theory, based on microeconomic theory, places the four principal travel choices on a sound utility-maximizing basis; see Ben-Akiva and Lerman (1985) and Oppenheim (1995). Activity analysis seeks to account for the role and effect of daily activities, as described by their time and place, in shaping travel choices; see Kitamura et al. (1996).

In this section we extend the modeling framework stated above to mode and origin–destination choices. Working within the optimization framework defined by the route choice models, assumptions about what is known or given are relaxed so as to make the model more comprehensive. Unlike the four-step approach, a single model incorporating several choices in a consistent manner is derived by this process.

In Section 7.C.2 the auto and transit route choice formulations are combined into a model of mode and route choice. Since transit route choices are generally flow independent, we summarize the expected transit generalized travel cost from origin p to destination q as c_{pqt}. For auto with its flow-dependent costs, we assume as before that the generalized cost consists of travel time and operating cost components. Moreover, since we are now concerned with travel choices of persons rather than drivers of vehicles, the role of vehicle occupancy needs to be made explicit. One definition of person travel cost by auto for link a is

$$c_a(f_a) \equiv \gamma_1 t_a(f_a) + \frac{\gamma_2}{\eta} k_a(f_a) \tag{7.1}$$

where (γ_1, γ_2) are weights representing the relative importance of travel time and operating cost on an individual's choice, η is the average occupancy (persons/auto), and t_a and k_a are the travel time and operating cost functions, respectively. This definition supposes that the operating costs are divided equally among the occupants of the vehicle. As before, f_a is the link flow in vehicles per hour. In addition, fixed times and costs at either end of the trip need to be considered.

Until now, the total OD flow has been denoted by $d_{pq} = P_{pq}T$. From now on this flow is understood to include both auto and transit. As is conventional in practice, nonmotorized modes are ignored in this treatment, but they could be treated in a similar way.

7.C.2. Mode Choice

We begin by extending the route choice models stated in Section 7.B to include the choice of mode. Of several possible formulations, the ones given below illustrate the modeling procedure. Initially, we assume that individual travelers have perfect information for making route choices based on travel times and costs of the auto and transit modes. Other modal variables, such as comfort, convenience, privacy, and safety, are not observable and result in some travelers not choosing the strictly low-cost and mode–route combination. As in the stochastic auto route choice case, we represent this dispersion of choices to the higher-cost mode with a conditional entropy function:

$$S(M \mid PQ) = -\sum_{p \in P} \sum_{q \in Q} P_{pq} \sum_{m \in M} P_{m|pq} \ln P_{m|pq} \geq \bar{S}(M \mid PQ)$$

where $\bar{S}(M \mid PQ)$ is the observed or measured conditional dispersion of mode choice, and $M \equiv \{a,t\}$ is the set of modes consisting of auto and transit. For our purposes here, all transit submodes (bus, rapid rail, commuter rail, etc.) are considered to be represented as a composite transit mode. Also note that P_{pq} is an input and not a decision variable to this mode-choice model.

In addition to this representation of unobserved modal characteristics by the entropy function, we also require that the predicted modal share for the entire region equal the observed modal share. As shown below, this constraint is equivalent to determining the value of a term γ_0 in the generalized cost function corresponding to the constraint

$$\sum_{p \in P} \sum_{q \in Q} P_{pqt} = \bar{P}_t$$

where $\overline{P}_t$ is the observed transit share for the entire region. Since we assume only two modes, a similar constraint for the auto mode is redundant.

Now we are ready to state the mode and route choice model, as follows:

$$\underset{P,h}{\text{minimize}}\ Z(P,h) = \frac{1}{T}\sum_{a\in A}\left[\eta\gamma_i\int_0^{f_a} t_a(x)\,dx + \gamma_2\int_0^{f_a} k_a(x)\,dx\right]$$

$$+ \sum_{q\in Q}\left(\sum_{p\in P} P_{pqa}\right)\gamma_2 c_{qa} + \sum_{p\in P}\sum_{q\in Q} P_{pqt}c_{pqt}$$

subject to

$$-\sum_{p\in P}\sum_{q\in Q} P_{pq}\sum_{m\in M} P_{m|pq}\ln P_{m|pq} \geq \overline{S}(M \mid PQ)$$

$$\sum_{m\in M} P_{pqm} = P_{pq} \qquad p\in P,\quad q\in Q$$

$$\sum_{p\in P}\sum_{q\in Q} P_{pqt} = \overline{P}_t$$

$$\sum_{r\in R_{pq}} P_{pqar} = P_{pqa} \qquad p\in P,\quad q\in Q$$

$$P_{pqar} \geq 0 \qquad r\in R_{pq},\quad p\in P,\quad q\in Q$$

$$P_{pqm} \geq 0 \qquad p\in P,\quad q\in Q,\quad m\in M$$

where

$$P_{m|pq} \equiv \frac{P_{pqm}}{P_{pq}}, \quad \text{and} \quad f_a \equiv \sum_{p\in P}\sum_{q\in Q}\sum_{r\in R_{pq}} \frac{\delta^a_{pqr}P_{pqar}T}{\eta} \qquad a\in A$$

Here c_{qa} is a destination-specific charge for the auto mode, which might represent a parking fee at destination q. Recall that c_{pqt} is the fixed (flow-independent) cost of traveling from origin p to destination q by transit. This cost includes all out-of-pocket costs as well as the value of the travel time associated with the transit mode. It does not include unobservable costs, such as comfort, convenience, privacy, and safety, which are represented by the entropy constraint and the modal bias term γ_0, introduced below as the Lagrange multiplier of the regional modal share constraint. Finally, P_{pqar} is the probability of traveling from origin p to destination q by the auto mode a on route r.

In the following Lagrangian analysis, we restate the entropy constraints in terms of the joint probabilities P_{pqm} to facilitate the solution process.

$$
\begin{aligned}
\mathcal{L}(P,\mu^{-1},\alpha,\pi,\gamma_0) = {} & Z(P,h) + \frac{1}{\mu}\left[\bar{S}(M \mid PQ) + \sum_{p\in P}\sum_{q\in Q}\sum_{m\in M} P_{pqm} \ln \frac{P_{pqm}}{P_{pq}}\right] \\
& + \sum_{p\in P}\sum_{q\in Q} \alpha_{pq}\left(P_{pq} - \sum_{m\in M} P_{pqm}\right) \\
& + \gamma_0\left(\bar{P}_t - \sum_{p\in P}\sum_{q\in Q} P_{pqt}\right) \\
& + \sum_{p\in P}\sum_{q\in Q} \pi_{pq}\left(P_{pqa} - \sum_{r\in R_{pq}} P_{pqar}\right)
\end{aligned}
\tag{7.2}
$$

From the partial derivatives of $\mathcal{L}$ with respect to P_{pqar} and the constraints, we obtain the following optimality conditions for auto route choice:

$$
\begin{aligned}
P_{pqar}\left(c_{pqar} - \frac{\pi_{pq}T}{\eta}\right) &= 0 \qquad r \in R_{pq},\ \ p \in P,\ \ q \in Q \\
c_{pqar} - \frac{\pi_{pq}T}{\eta} &\geq 0 \qquad r \in R_{pq},\ \ p \in P,\ \ q \in Q \\
\sum_{r\in R_{pq}} P_{pqar} &= P_{pqa} \quad p \in P,\ \ q \in Q \\
P_{pqar} &\geq 0 \qquad r \in R_{pq},\ \ p \in P,\ \ q \in Q
\end{aligned}
$$

where

$$
c_{pqar} \equiv \sum_{a\in A} \delta^a_{pqr}\left[\gamma_1 t_a(f_a) + \frac{\gamma_2 k_i(f_a)}{\eta}\right]
$$

Note that the auto occupancy η now appears in the denominator of the auto operating cost term, as assumed at the outset. These conditions are essentially the same as those for the user-optimal route choice model and therefore require no further discussion.

Rather than stating the equivalent optimality conditions on the modal choice proportions, we solve the partial derivatives for P_{pqm} in terms of the generalized travel costs.

$$\frac{\partial \mathcal{L}}{\partial P_{pqa}} = \gamma_2 c_{qa} + \frac{1}{\mu}\left(\ln \frac{P_{pqa}}{P_{pq}} + 1\right) - \alpha_{pq} + \pi_{pq}$$

$$\frac{\partial \mathcal{L}}{\partial P_{pqt}} = c_{pqt} + \frac{1}{\mu}\left(\ln \frac{P_{pqt}}{P_{pq}} + 1\right) - \alpha_{pq} - \gamma_0$$

For $P_{pqm} > 0$, and $P_{pqar} > 0$, $r \in R_{pq}$, we obtain

$$P_{pqm} = P_{pq} \exp(\mu\alpha_{pq} - 1) \exp(-\mu c'_{pqm})$$

where

$$c'_{pqa} = \pi_{pq} + \gamma_2 c_{qa} = c_{pqa} + \gamma_2 c_{qa}$$

$$c_{pqa} = c_{pqar} \qquad \text{if } P_{pqar} > 0$$

$$c'_{pqt} = c_{pqt} - \gamma_0$$

Applying the modal conservation of flow constraints yields

$$\exp(\mu\alpha_{pq} - 1) = \left[\sum_{m \in M} \exp(-\mu c'_{pqm})\right]^{-1} \tag{7.3}$$

Therefore,

$$P_{pqm} = P_{pq} \frac{\exp(-\mu c'_{pqm})}{\sum_{m \in M} \exp(-\mu c'_{pqm})} \tag{7.4}$$

or

$$d_{pqm} = d_{pq} \frac{\exp(-\mu c'_{pqm})}{\sum_{m \in M} \exp(-\mu c'_{pqm})} = \frac{d_{pq}}{1 + \exp(-\mu(c'_{pqm'} - c'_{pqm})}$$

where m' is the alternative to mode m. We observe the following points about this result.

1. The mode choice function has the form of the logit function, which stems from the use of the entropy function $S(M \mid PQ)$ to describe the dispersion of choices to the higher-cost mode.
2. The constraint that the predicted regional modal share equals the observed share results in a modal bias coefficient γ_0 in the transit generalized cost.
3. Zone-related origin or destination costs, such as parking fees, can readily be incorporated into the model.

TABLE 7.5 Origin–Destination Demand and Transit Characteristics for a Mode and Route Choice Model for the Network of Figure 7.1

OD Pair	Total OD Flow (persons)	Transit Fare (cents)	Transit Time (min)
AD	5460	85	74
BD	5716	100	90

4. The term π_{pq} linking the auto route choice model and the mode choice model may be interpreted as the generalized travel cost for used routes from zone p to zone q. The correspondence of these results to the model assumptions confirms that the equivalent objective function is correctly stated.

To illustrate these concepts, consider once again the network shown in Figure 7.1. Table 7.5 gives origin–destination demand, now including transit demand, as well as the transit fare in cents and transit travel time in minutes. Assume that there is a $1.00 parking fee at destination *D*, of which one-half ($0.50) is charged to the one-way trip from the origins (*A* or *B*) to destination *D*. The other half would be assessed to the return trip. Finally, we assume that the value of time is $0.10 per minute or $6.00 per hour. With these values, the generalized transit costs c_{pqt} are $8.25 and $10.00 per trip for OD pairs *AD* and *BD*, respectively.

Table 7.6 shows the results of solving this problem if the modal bias $\gamma_0 = 65$, the parameter $\mu = 0.012$, and $\gamma_2 = 1$. Note that the overall transit share is 10.52% of the total flow of 11,176 trips per hour, assuming one person per auto. Also note that the auto flows in this case turn out to be 5000 for each OD pair and thus correspond (miraculously) to the flows used in Table 7.3, in which we solved the route choice problem for these same auto OD flows. Refer to Table 7.3 for the user-optimal link flows and travel times.

The mode and route choice model above can be further generalized by relaxing the assumption that auto route choices are user optimal. To this end we add a conditional entropy constraint on route choice to the formulation above:

TABLE 7.6 Mode Flows and Travel Times and Generalized Costs[a]

OD Pair	Transit Flow (persons)	Transit Share (%)	Transit Cost	Auto Flow (vehicles)	Auto Time (min)	Auto Cost
AD	460	8.43	$ 8.25	5000	64.11	$6.91
BD	716	12.53	10.00	5000	85.30	9.03

[a]Auto occupancy is one person per vehicle.

$$-\sum_{p\in P}\sum_{q\in Q} P_{pqa} \sum_{r\in R_{pq}} P_{r|pqa} \ln P_{r|pqa} \geq \bar{S}(R \mid PQa)$$

Setting $P_{r|pqa} \equiv P_{pqar}/P_{pqa}$ and taking derivatives with respect to P_{pqar}, we obtain

$$P_{pqar} = P_{pqa} \frac{\exp(-\theta c_r)}{\sum_{s\in R_{pq}} \exp(-\theta c_s)} \qquad r \in R_{pq}, \quad p \in P, \quad q \in Q$$

where

$$c_r \equiv \sum_{a\in A} \delta^a_{pqr} \left[\gamma_1 t_a(f_a) + \frac{\gamma_2}{\eta} k_a(f_a) \right] \qquad r \in R_{pq}$$

Substituting for P_{pqa}, we obtain

$$P_{pqar} = P_{pq} \frac{\exp(-\mu c'_{pqa})}{\sum_{m\in M} \exp(-\mu c'_{pqm})} \frac{\exp(-\theta c_r)}{\sum_{r\in R_{pq}} \exp(-\theta c_r)}$$

$$r \in R_{pq}, \quad p \in P, \quad q \in Q$$

where

$$c'_{pqa} \equiv \pi_{pq} + \frac{1}{\theta} + \gamma_2 c_{qa}$$

From the derivation of P_{pqar}, we know that

$$\pi_{pq} = -\frac{1}{\theta} \ln \left[\sum_{r\in R_{pq}} \exp(-\theta c_r) \right] \equiv \tilde{c}_{pqa}$$

which is known as the composite cost of travel from P to Q. Therefore,

$$c'_{pqa} = \tilde{c}_{pqa} + \frac{1}{\theta} + \gamma_2 c_{qa}$$

7.C.3. Origin–Destination Choice

Next, we relax the assumption that the OD flows (d_{pq}) are known and replace the corresponding constraints with constraints on the zonal departures (origins) and terminations (destinations). In transportation planning practice, the prediction of these quantities is known as trip generation modeling, which largely consists of using statistical models based on zonal activity variables such as the number of households, autos, and jobs, and the floor area devoted to employment, shopping, and other economic activities. Note that transportation facilities and services are generally not included as explanatory variables. Because the estimation of these relationships and their use in predicting the frequency of travel are unrelated to our optimization framework, we do not consider this problem. Refer to Oppenheim (1995) for a treatment of trip frequency within the travel choice model framework.

Origin–destination flows can also be modeled directly using demand functions for each OD pair. Such functions were the basis for the original formulation of Beckmann et al. (1956) and are sometimes called elastic or variable demand functions. Such models tend to be more abstract and less used in professional practice than the approach described here.

Empirical examination of regional travel patterns, as represented by the matrix d_{pq}, indicates a strong inverse relationship between OD flows and generalized travel cost, as well as a direct relationship with the level of zonal activities. Such an empirical relationship has historically been described as a *gravity* model, because of its resemblance to the physical law of gravity governing the attractive force between two bodies:

$$d_{pq} = O_p D_q g(c_{pq})$$

where O_p is the flow of persons departing from zone p and D_q is the flow terminating at zone q; the cost function $g(c_{pq})$ is observed to be inversely related to c_{pq}.

The representation of this problem as a two-dimensional matrix also reminds us of the classical *transportation problem of linear programming,* which originated in applications of linear programming methods to commodity shipment problems in the early years of the field of operations research (Hitchcock, 1941; Koopmans, 1947). This problem is stated as follows:

$$\underset{d}{\text{minimize}}\ c(d) = \sum_{p \in P} \sum_{q \in Q} d_{pq} c_{pq}$$

subject to

$$\sum_{q \in Q} d_{pq} = O_p \qquad p \in P$$

$$\sum_{p \in P} d_{pq} = D_q \qquad q \in Q$$

$$d_{pq} \geq 0 \qquad p \in P, \quad q \in Q$$

where $\Sigma_{p \in P} O_p = \Sigma_{q \in Q} D_q = T$, and the travel costs (c_{pq}) are fixed. Note that d_{pq} is a decision variable in this model and not an input. The solution of this linear program results in at most $n(P) + n(Q) - 1$ positive flows, where $n(P)$ and $n(Q)$ are the number of origin and destination zones, respectively.

If the problem is well conditioned or *nondegenerate,* the solution has exactly $n(P) + n(Q) - 1$ positive flows. In Chapter 5 (Section 5.B) the Hitchcock problem is discussed in the context of solid waste management.

As compared with empirical OD flows, such a flow pattern is extremely sparse. For example, if the zone system has 1000 zones, a typical size for a large metropolitan region, the number of cells in the flow matrix is 1 million, whereas the optimal solution of the transportation problem has at most 1999 positive flows. By comparison, in an OD matrix for auto and transit trips based on a sample of household of perhaps 10%, typically about one-half of all cells are positive; for the remainder, no trips are observed. For a larger sample, or a complete census of travel, the proportion of cells with positive flow would undoubtedly rise. Hence the transportation problem does not correspond well to our OD choice problem, despite the similarity of the formulation.

By applying the elegant device of the entropy constraint to the transportation problem, however, we can obtain a gravity type function, as first suggested by Erlander (1977, 1980). Constraining $c(d)$ to a suboptimal solution by imposing an entropy constraint yields a function in which each zone pair has a positive flow. The entropy function for the problem above is defined as

$$-\sum_{p \in P} \sum_{q \in Q} d_{pq} \ln d_{pq} \geq \bar{S}(PQ)$$

where $\bar{S}(PQ)$ represents an empirically observed level of entropy. Forming the Lagrangian function and taking derivatives with respect to d_{pq}, we obtain

$$\ln d_{pq} + 1 = \beta(\sigma_p + \tau_q - c_{pq})$$

where β is the reciprocal of the Lagrange multiplier associated with the entropy constraint and σ_p and τ_q are Lagrange multipliers associated with the origin and destination constraints, respectively. Solving for d_{pq} yields

$$d_{pq} = \exp(\beta\sigma_p - 1) \exp(\beta\tau_q) \exp(-\beta c_{pq})$$

Applying the origin and destination constraints leads to

$$d_{pq} = A_p O_p B_q D_q \exp(-\beta c_{pq})$$

where

$$A_p = \left[\sum_{q \in Q} B_q D_q \exp(-\beta c_{pq}) \right]^{-1} \qquad p \in P$$

$$B_q = \left[\sum_{p \in P} A_p O_p \exp(-\beta c_{pq}) \right]^{-1} \qquad q \in Q$$

These sets of balancing factors ensure that the origin and destination constraints are satisfied. They can readily be computed by applying a simple iterative procedure: Set $B_q^1 = 1$; solve for A_p^1; then solve for $B_q^2, A_p^2, \ldots$. This procedure converges to any reasonable level of accuracy (e.g., 10^{-4}) in 30 to 60 iterations, assuming that the travel costs are positive and finite and the origin and destination totals are positive.

The function for d_{pq} is known as the doubly constrained negative-exponential gravity model and was derived by Wilson (1967), among others, using the related entropy-maximizing technique. The parameter β is inversely related to the total travel cost, or its mean, if all the constants are standardized by the total flow T. Similarly, it is inversely related to the right-hand side of the entropy constraint (see Figure 7.2).

To illustrate these concepts, consider a three-origin, four-destination transportation problem with costs, supplies, and demands, as shown in Table 7.7. Figure 7.2a plots the total cost versus the entropy function S, defined as

$$S = -\sum_{p \in P} \sum_{q \in Q} P_{pq} \ln P_{pq}$$

where $P_{pq} = d_{pq}/T$ and $0 \equiv 0 \ln 0$. Note that at relatively low values of the entropy function, the total cost is minimized and the flows equal those of the *transportation problem,* as shown in Table 7.8. At these low values of the entropy function, the value of β is quite large. As the entropy increases and β decreases, the total cost increases and the flows become more dispersed (see Table 7.8).

This entropy-constrained model has the following properties (cf. Evans, 1973):

1. All OD flows are positive if $O_p > 0$, $D_q > 0$, and $c_{pq} < +\infty$.
2. As $\beta \to \infty$, the values of d_{pq} approach the solution of the transportation problem as shown in Figure 7.2b.
3. As $\beta \to 0$, $d_{pq} = O_p D_q / T$.

Having established these basic results, we now apply them to the origin–destination mode and route choice model. We replace the constraint in the mode and route choice model of Section 7.C.2,

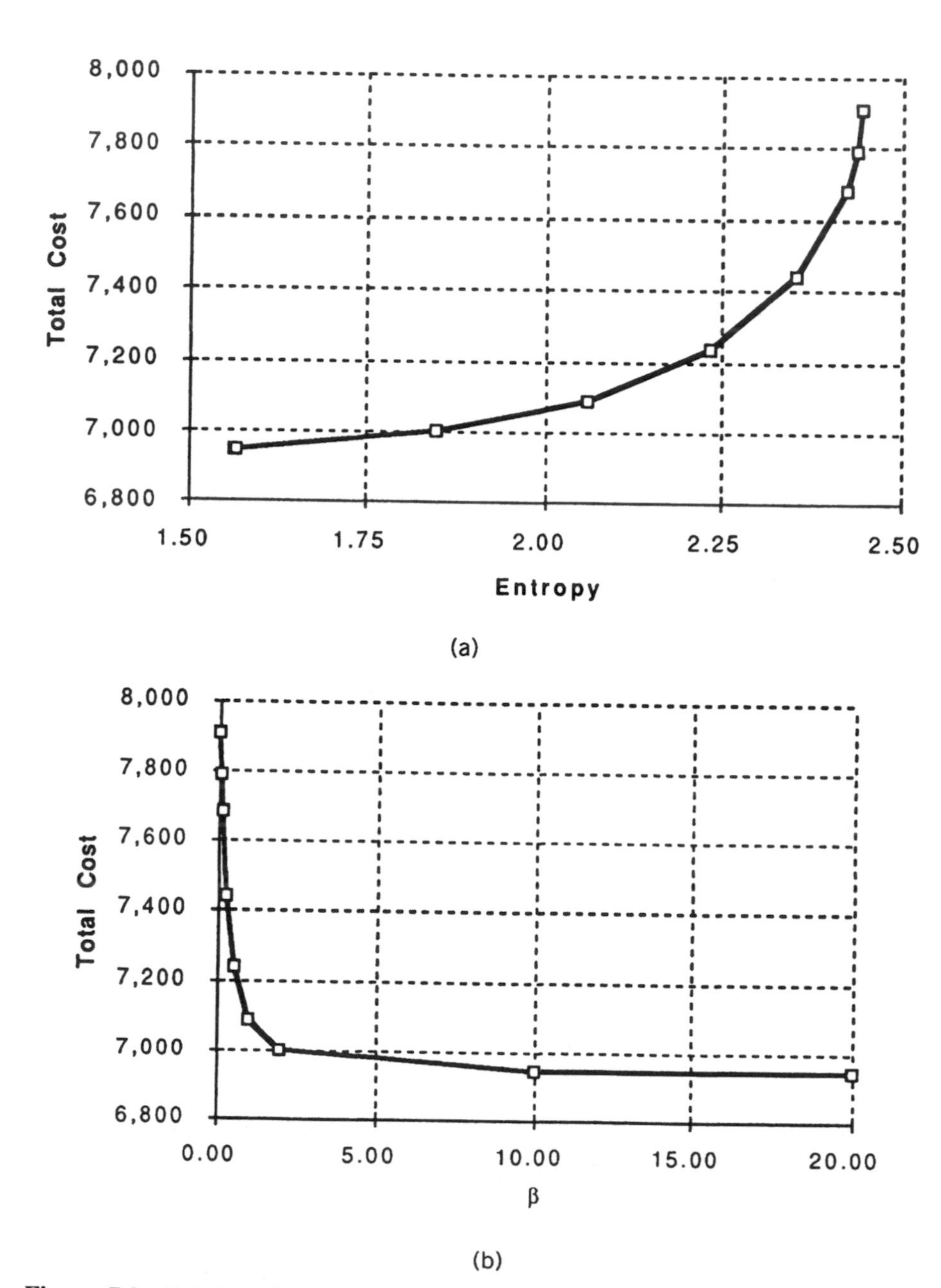

Figure 7.2 Relationships between entropy, total cost and β: (a) entropy versus total cost; (b) β versus total cost; (c) entropy versus β.

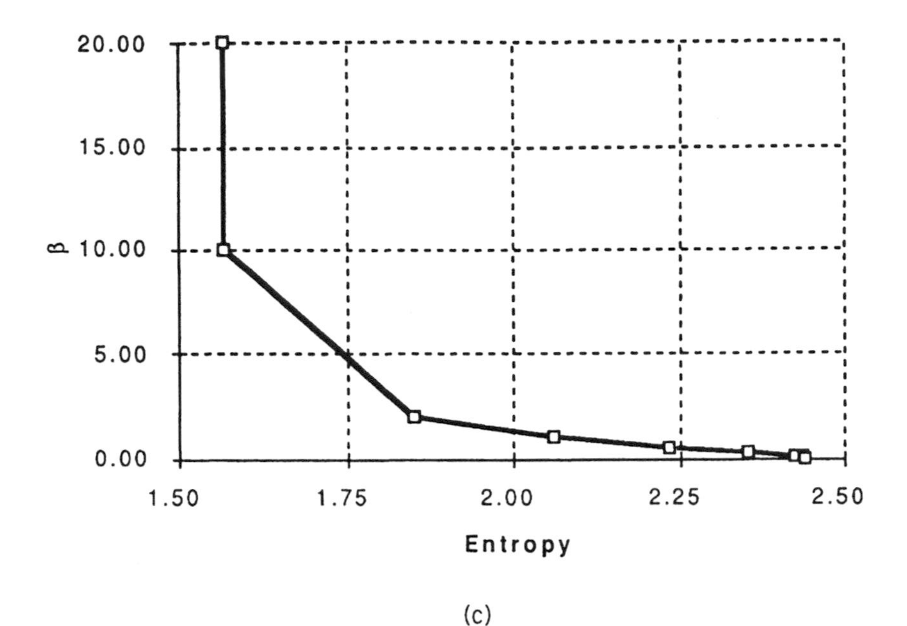

(c)

Figure 7.2 *(Continued)*

TABLE 7.7 Unit Shipment Costs, Supplies, and Demands for an Entropy-Constrained Transportation Problem

(a) Unit Link Cost

	Destination			
Origin	*D*	*E*	*F*	*G*
A	19	16	14	13
B	12	16	14	15
C	5	8	10	13

(b) Supplies at the Origin Nodes

Origin	Total Supply
A	175
B	265
C	165
All	595

(c) Demands at the Destination Nodes

Destination				
D	*E*	*F*	*G*	All
115	150	130	200	595

TABLE 7.8 OD Flows, Total Cost, and Entropy as a Function of β for the Entropy-Constrained Transportation Problem

		OD Flows for Values of β				
Origin	Destination	0.0	0.1	0.5	2.0	20.0
A	*D*	33.824	19.034	0.855	0.000	0.000
A	*E*	44.118	41.452	23.480	2.165	0.000
A	*F*	38.235	40.370	38.144	9.726	0.000
A	*G*	58.824	74.143	112.521	163.110	175.000
B	*D*	49.286	54.044	54.967	71.289	100.000
B	*E*	64.286	58.445	45.592	26.730	0.000
B	*F*	55.714	56.921	74.064	120.092	130.000
B	*G*	85.714	85.589	80.376	36.889	25.000
C	*D*	31.891	41.921	59.178	43.711	15.000
C	*E*	41.597	50.103	80.927	121.105	150.000
C	*F*	36.050	32.709	17.792	0.183	0.000
C	*G*	55.462	40.267	7.103	0.001	0.000
Total cost		7908.0	7679.3	7233.9	6998.2	6945.0
Entropy		2.44259	2.42402	2.23340	1.84978	1.56533

$$\sum_{m\in M} P_{pqm} = P_{pq} \qquad p \in P, \quad q \in q$$

with the following constraint set:

$$-\sum_{p\in P}\sum_{q\in Q} P_{pq} \ln P_{pq} \geq \bar{S}(PQ)$$

$$\sum_{q\in Q} P_{pq} = o_p \equiv \frac{O_p}{T} \qquad p \in P$$

$$\sum_{p\in P} P_{pq} = d_q \equiv \frac{D_q}{T} \qquad q \in Q$$

Augmenting the Lagrangian stated as equation (2), we obtain

$$\begin{aligned}\mathcal{L}(P,\mu^{-1},\pi,\gamma_0,\beta^{-1},\sigma,\tau) &= \mathcal{L}(P,\mu^{-1},\pi,\gamma_0) \\ &+ \frac{1}{\beta}\left[\bar{S}(PQ) + \sum_{p\in P}\sum_{q\in Q} P_{pq} \ln P_{pq}\right] \\ &+ \sum_{p\in P} \sigma_p \left(o_p - \sum_{q\in Q} P_{pq}\right) + \sum_{q\in Q} \tau_q \left(d_q - \sum_{p\in P} P_{pq}\right)\end{aligned}$$

Differentiating with respect to P_{pq}, the following additional conditions are obtained:

$$\frac{\partial \mathcal{L}}{\partial P_{pq}} = \frac{1}{\beta}(\ln P_{pq} + 1) - \sigma_p - \tau_q + \alpha_{pq} - \frac{1}{\mu}$$

For $P_{pq} > 0$, $P_{pqm} > 0$, and $P_{pqar} > 0$,

$$P_{pq} = \exp\left[\beta\left(\sigma_p + \tau_q + \frac{1}{\mu} - 1\right)\right]\exp(-\beta\alpha_{pq}) \qquad p \in P, \quad q \in Q$$

Applying the origin and destination constraints, we have

$$P_{pq} = a_p o_p b_q d_q \exp(-\beta\alpha_{pq}) \qquad p \in P, \quad q \in Q$$

where, as above,

$$a_p = \left[\sum_{q \in Q} b_q d_q \exp(-\beta\alpha_{pq})\right]^{-1} \qquad p \in P$$

$$b_q = \left[\sum_{p \in P} a_p o_p \exp(-\beta\alpha_{pq})\right]^{-1} \qquad q \in Q$$

From equation (3), we know that

$$\exp(-\mu\alpha_{pq} - 1) = \sum_{m \in M} \exp(-\mu c'_{pqm})$$

Therefore,

$$\alpha_{pq} = -\frac{1}{\mu}\left[\ln \sum_{m \in M} \exp(-\mu c'_{pqm}) + 1\right] \equiv \tilde{c}_{pq}$$

which leads to

$$P_{pq} = a_p o_p b_q d_q \exp(-\beta\tilde{c}_{pq})$$

Substituting into the mode choice function, equation (7.4), we obtain

$$P_{pqm} = a_p o_p b_q d_q \exp(-\beta\tilde{c}_{pq}) \frac{\exp(-\mu c'_{pqm})}{\Sigma_{m \in M} \exp(-\mu c'_{pqm})}$$

This origin–destination mode choice model is another example of a hierarchical travel choice function. An alternative form of this model may be obtained by replacing the two entropy constraints with a single constraint,

$$\sum_{p \in P} \sum_{q \in Q} \sum_{m \in M} P_{pqm} \ln P_{pqm} \geq \bar{S}(PQ)$$

Then the corresponding choice function is

$$P_{pqm} = a_p o_p b_q d_q \exp(-\beta c'_{pqm})$$

This last model result represents the simplest version of the origin–destination mode choice function that can be derived in this framework. The choice among alternative functional forms should be based on the difficulty of solution of the model and the goodness of fit of the model to the data. These topics are considered next.

7.D. SOLUTION ALGORITHMS

Although the models described in Sections 7.B and 7.C have interesting properties in their own right, their practical value for transportation network analysis depends on our ability to solve them for large-scale networks. We review the status of solution algorithms suited for such network analyses in this section. Then in Section 7.E we consider the implementation of such models for the Chicago region.

Our consideration of solution algorithms follows the dichotomy of Sections 7.B and 7.C. First, we consider algorithms for fixed-demand route choice models, including both the deterministic and stochastic cases. Then we introduce a generalization of the fixed demand case to solve the OD mode and route choice model. For a rigorous mathematical statement of these results and a review of the history of their development, see Patriksson (1994, Chap. 4).

7.D.1. Algorithms for Fixed-Demand Route Choice Models

The principal algorithm used in practice to solve the fixed-demand deterministic route choice model is the method of Frank and Wolfe (1956), also known as the linearization or convex combinations method for minimizing a convex function $g(x)$ subject to linear constraints, where x is a vector. This nonlinear program is called the *main problem* in the following description of the algorithm. Intuitively, the concept of the algorithm is straightforward:

1. *Search direction.* Given a feasible solution x to the main problem, find a linear approximation $\underline{g}(y)$ to the objective function and minimize it subject to the same constraints as the main problem, obtaining the vector y. This linearized problem is known as the *subproblem.*
2. *Line search.* Find a convex combination of x and y [i.e., a weighted average with weights $(1 - \lambda)$ and λ], respectively, that minimizes the objective function of the main problem.

Steps 1 and 2 are repeated until a specified termination criterion is satisfied.

A property of the method is that a lower bound on the objective function is available based on x and y. Hence the algorithm can be applied repeatedly until $g(x)$ is as close as desired to $g(x^*)$, as estimated by the lower bound, where x^* denotes the optimal solution. Having stated the general principle of the method, we now apply it to the deterministic route choice problem.

The linear (or tangential) approximation of $g(x)$ is defined for the $n \times 1$ vector x as

$$\underline{g}(y) \equiv g(x) + \nabla g(x)'(y - x)$$

where $\nabla g(x) = (\partial g(x)/\partial x_i,\ i = 1,\ldots,n)$ and $'$ denotes transpose. In the deterministic route choice model,

$$g(f) \equiv \frac{1}{T} \sum_{a \in A} \int_0^{f_a} c_a(x)\, dx$$

Therefore, the linear approximation of $g(f)$ at f^k, the flow vector obtained at iteration k of the algorithm, is

$$\underline{g}(y) = \frac{1}{T} \sum_{a \in A} \int_0^{f_a^k} c_a(x)\, dx + \frac{1}{T} \sum_{a \in A} c_a(f_a^k)(y_a - f_a^k)$$

$$= \frac{1}{T} \sum_{a \in A} c_a(f_a^k) y_a + \text{constant} = \tilde{\underline{g}}(y) + \text{constant}$$

Minimizing $\underline{g}(y)$ subject to the linear constraints of the deterministic route choice model, we have

$$\underset{y}{\text{minimize}}\ \tilde{\underline{g}}(y) = \frac{1}{T} \sum_{a \in A} c_a(f_a^k) y_a$$

subject to

$$\sum_{r \in R_{pq}} h_{pqr} = d_{pq} \qquad p \in P,\ q \in Q$$

$$h_{pqr} \geq 0 \qquad r \in R_{pq},\ p \in P,\ q \in Q$$

where

$$y_a = \sum_{p \in P} \sum_{q \in Q} \sum_{r \in R_{pq}} \delta_{pqr}^a h_{pqr} \qquad a \in A$$

Note that the linear function $\tilde{\underline{g}}(y)$ is the mean travel cost of the network flows given the constant link costs $\overline{c}_a(f^k)$. Since this linear program decomposes by

OD pair, $\tilde{g}(y)$ is minimized by "assigning" the entire OD flow d_{pq} to the shortest path (minimum-cost route) from p to q. In the discussion of this algorithm and the algorithm of Section 7.D.2, we refer to the *shortest-path* problem, to be consistent with the algorithmic literature. For our purposes, the terms *path* and *route* may be used interchangeably. The shortest-path problem is readily solved by applying one of several algorithms for finding the shortest-path tree from one node representing origin zone p to all nodes in the network, including the nodes corresponding to the destination nodes q. See Gallo and Pallottino (1984, 1988) for a comprehensive review of shortest-path algorithms as applied to transportation network modeling.

Let y^k be the solution to the subproblem at iteration k and f^k be the corresponding solution to the main problem. Since $g(f)$ is convex, it follows that

$$\underline{g}(y^k) \equiv g(f^k) + \nabla g(f^k)'(y^k - f^k) \leq g(f^k) + \nabla g(f^k)'(f^* - f^k) \leq g(f^*)$$

where f^* denotes the optimal solution to the main problem. Note that the term

$$\nabla g(f^k)'(y^k - f^k) \leq 0$$

has the interpretation of the difference between (1) the mean travel cost of the subproblem solution, evaluated at the current main problem link flows, and (2) the mean travel cost of the main problem solution. This condition must hold since the subproblem solution represents the lowest-cost assignment of the OD flows to the network at the prevailing main problem costs. Also, at the user-optimal solution, these two costs must be equal, according to Wardrop's principle.

Hence $\underline{g}(y^k)$ is a lower bound on $g(f^*)$, the optimal value of the main problem objective function. Therefore, at each iteration k we can evaluate how far our solution is from the optimum. Since the optimal solution is never reached exactly for a problem of realistic size, this lower bound is very useful in judging when to stop the solution process. Use of the lower-bound stopping criterion is especially important in comparing solutions of the model for different networks.

Although the values of the objective function for successive solutions, $g(f^k)$, $g(f^{k+1})$,..., are strictly decreasing due to the line search procedure, the values of the lower bound are not strictly increasing from below. Hence we require a definition of the best lower bound, as follows:

$$\text{BLB}^{k+1} = \max(\text{BLB}^k, \underline{g}(y^k))$$

Now we are ready to state the Frank–Wolfe (F-W) algorithm in detail, following Patriksson (1994, pp. 96–97).

Step 0. Initialization. Let f^1 be a feasible solution to the fixed-demand route choice model. Set $\text{BLB}^1 = -10^{+6}$, $\epsilon > 0$ (e.g., 0.01 and $k = 1$).

Step 1. Search direction generation. Solve the linearized subproblem stated above, obtaining y^k. The search direction is $d^k = y^k - f^k$. For each origin zone p:

(a) Apply a shortest-path (route) algorithm to the origin zone centroid node.

(b) Assign (load) the fixed demand (d_{pq}, $q \in Q$) originating at zone p to the link flows of the shortest routes to each zone q; the accumulated flows for all origins result in (y_a^k).

(c) Evaluate $\underline{g}(y^k)$, and set $\text{BLB}^{k+1} = \max(\text{BLB}^k, \underline{g}(y^k))$.

Step 2. Line search. Find a step length λ^k that solves

$$\underset{0 \le \lambda^k \le 1}{\text{minimize}} \; \frac{1}{T} \sum_{a \in A} \int_0^{f_a^{k+1}} c_a(x)\, dx$$

where $f_a^{k+1} = (1 - \lambda^k) f_a^k + \lambda^k y_a^k$, $a \in A$.

Step 3. Update. Set f^{k+1} to the values defined above. Increment k.

Step 4. Convergence check. If $|\underline{g}(f^k) - \text{BLB}^k| \le \epsilon$, stop; otherwise, go to step 1.

Note that the convergence check could be performed after step 1; as a practical matter, it is worthwhile to update f^k once y^k is available. The convergence criterion above can also be computed as a relative quantity by dividing by BLB^k. In our experience the definition above is preferred, since it does not depend on the magnitude of BLB^k, which may include arbitrary constants.

The following comments apply to the algorithm.

1. The algorithm can be proven to converge to the user-optimal solution of the route choice problem, or alternatively, to the system-optimal solution if the link travel time functions are modified appropriately (see Exercise 7.5).
2. A good initial solution can be obtained by solving step 1 for $f = 0$ (i.e., with zero link flows); an improved initial solution may be obtained by using the converged solution to a related problem or by using an incremental heuristic (see Sheffi, 1985, pp. 111–116).
3. The number of iterations required to achieve a given level of convergence depends on the level of congestion in the network: less-congested networks require fewer iterations than do more-congested networks. For a moderately congested large network, satisfactory convergence can typically be achieved in 15 to 25 iterations. A relative convergence measure

can also be computed by dividing the measure in step 4 by BLB. Both the absolute and relative measures are useful in monitoring the convergence of the solution.

4. The Frank–Wolfe algorithm converges rapidly in the first few iterations and then exhibits a characteristic tailing off. Described in terms of the route choice problem, this slow convergence occurs for at least two reasons: (a) a nonoptimal route introduced in early iterations is never completely removed, and its influence decreases very slowly; (b) although the optimal solution cannot contain cycles (sequences of links that form a circular route), such cycles may be introduced during the iterative process or by a heuristical initial solution, since the algorithm does not check for their presence (see Newell, 1980; Janson and Zozaya-Gorostiza, 1987).
5. Although various improvements to the F-W algorithm have been proposed (Lawphongpanich and Hearn, 1984; Florian et al., 1987), they are often more computationally intensive than F-W itself for a moderate level of convergence. Unless a very high level of convergence is required, they have not proven to be very effective.
6. An alternative stopping criterion involves checking whether the maximum change in any link flow is less than some value (i.e. $\max_a | f_a^{k+1} - f_a^k| < \epsilon$, where ϵ is specified exogenously). Rose et al. (1988) showed that attaining a tight convergence using this measure may require far more iterations than suggested in comment 3 above. One reason for this phenomenon is that the objective function being minimized is quite flat. In other words, rather different flow patterns may produce objective functions with valves that are close to the optimal value.
7. The deterministic route choice problem is solved in terms of link flows only, where the number of links is typically on the order of 10^4. In contrast, the number of routes is typically on the order of 10^7; moreover, as noted in Section 7.B, the route flows are not unique. This feature of the algorithm is attractive in terms of its memory requirements, even for today's computing environment. It was a critical property for former generations of computers with memory limitations of 512 kilobytes.
8. As noted in Section 7.B.2, the route flows of deterministic route choice models are not unique. Janson (1993) suggests a method for finding the most likely set of OD-specific link flows for either the UO or SO objective function. He also shows that the F-W algorithm converges to a very likely set of OD-specific link flows.

These last two comments lead to the issue of stochastic route algorithms, to which we turn next; recall from Section 7.B.3 that the solution to the stochastic route choice model is unique in the route flows. In general, an algorithm that requires storing of routes is not practical for solving the stochastic route choice model for networks of realistic size. Moreover, such algorithms require either route enumeration or prespecification of routes. Route enumeration is intractable for problems of realistic size, and prespecified routes may be highly arbitrary.

Two practical algorithms for solving the stochastic route choice model are available, however, Both utilize the *method of successive averages* (MSA) in place of the line search step. The simplest version of MSA sets the step length λ^k, or averaging weight, equal to $1/k$, where k is the iteration number; therefore, $(1 - \lambda^k) = (k - 1)/k$. The sequence of weights generated by this procedure typically resembles fairly closely the values that result from the line search, especially for $k > 5$. MSA can be shown to converge to the optimal solution, although more iterations are required. Note, however, that use of MSA destroys the property that the sequence of objective function values is strictly decreasing. Hence an additional convergence check should be added to determine whether the last iteration actually yielded the best solution found so far. More general predetermined step-length formulas may improve convergence somewhat (see Patriksson, 1994, p. 142).

Replacing the line search step with MSA obviates the need to evaluate the objective function at each iteration, which is not possible if route flows are unavailable. Of course, this means that the convergence of the algorithm is more difficult to monitor; simpler convergence measures such as the proportion of links with a change in flow of more than 5 or 10% can be utilized.

What remains, then, is to specify a procedure for finding the search direction. If the logit route choice model is being solved, one method is the STOCH algorithm of Dial (1971). This clever procedure succeeds in allocating flows to links without actually enumerating the routes. Moreover, the links considered can be restricted to *efficient* links (i.e., links whose head node is closer to the destination and farther from the origin than the tail node). Although intuitively appealing, the use of efficient links makes the approach excessively sensitive to the coding of the network. In particular, the addition of a single link may alter the pattern of flows significantly (see Exercises 7.3 and 7.4).

The STOCH algorithm also may allocate flows to an excessive number of routes compared with the size of the flow. For example, for one suburban arterial network, the algorithm identified 10^4 routes for a zone pair about 10 miles apart. Since the OD flow might only be on the order of 10^2 or even less, the assignment of this flow to so many routes seems unacceptable. In addition, recall the undesirable IIA properties of the logit model.

A second approach that can be applied to solving a stochastic route choice model with any assumed distribution of perception errors (including the logit-based Gumbel distribution) is Monte Carlo sampling. In this approach, link perception errors are sampled from the assumed error distribution for each link. The shortest routes are found and the OD flows assigned to them. For each execution of the search direction step, this procedure is repeated many times (10 to 50) and the resulting link flows are averaged. In this manner a reasonable number of routes are identified for each OD pair. For many zone pairs, a few routes may be found repeatedly, whereas for more widely separated zone pairs, a larger number of different routes is likely to result.

The Monte Carlo sampling approach was initially introduced by Powell and Sheffi (1982) in conjunction with an application of the MSA method. It has recently been applied to large-scale networks by Hicks et al. (1992) and Tatineni et al. (1996).

7.D.2. Algorithm for OD Mode and Route Choice Models

The convex combination or F-W method can also be applied to solve the variable-demand models described in Section 7.C.3, as proposed by Florian et al. (1975) and Florian and Nguyen (1978). As shown initially by Frank (1978) as well as LeBlanc and Farhangian (1981), this idea does not lead to an efficient algorithm. The reason can be readily demonstrated.

The form of the F-W subproblem of the OD mode and route choice model is equivalent to the *transportation problem of linear programming*. As already noted, the optimal solution to this problem results in a highly sparse matrix. In contrast, the optimal solution to the travel choice model results in a positive matrix of OD mode flows. Hence an unacceptably large number of iterations is required to generate this positive matrix by averaging a sequence of highly sparse matrices.

Fortunately, Evans (1976) recognized that the linearization algorithm can be generalized to a partial linearization algorithm. She also formulated the model of OD mode and user-optimal route choice, proved that its solution exists and is unique, proved that her proposed algorithm converges to the optimal solution, and stated criteria for monitoring its convergence. In recognition of this path-breaking contribution, her partial linearization algorithm was termed the Evans algorithm by Boyce et al. (1983, 1985, 1988, 1992).

The crucial insight of Evans was that any of the variables for which a closed-form function is available from the optimality conditions can be solved for directly without linearization of the corresponding objective function terms. Hence the OD mode flows can be solved for directly from the doubly constrained negative exponential function. Therefore, only the terms related to link flows need to be linearized.

Because of its potential significance, the Evans algorithm is stated in full for the simplest origin–destination mode and user-optimal route choice model,

$$P_{pqm} = a_p o_p b_q d_q \exp(-\beta c_{pqm}) \tag{7.5}$$

where

$$a_p = \left[\sum_{q \in Q} b_q d_q \exp(-\beta c_{pqm}) \right]^{-1} \quad \text{and} \quad b_q = \left[\sum_{p \in P} a_p o_p \exp(-\beta c_{pqm}) \right]^{-1}$$

The corresponding form of the objective function is

$$g(f,p) = \frac{1}{T} \sum_{a \in A} \int_0^{f_a} c_a(x\, dx + \sum_{p \in P} \sum_{q \in Q} P_{pqt} c_{pqt}$$

$$+ \frac{1}{\beta} \sum_{p \in P} \sum_{q \in Q} \sum_{m \in M} P_{pqm} \ln P_{pqm}$$

For solving the subproblem and for the convergence test, we require its partial linear approximation, which is

$$\underline{g}(y^k, Q^k) \equiv \frac{1}{T} \sum_{a \in A} \left[\int_0^{f_a^k} c_a(x)\, dx + c_a(f_a^k)(y_a^k - f_a^k) \right]$$

$$+ \sum_{p \in P} \sum_{q \in Q} Q_{pqt}^k c_{pqt} + \frac{1}{\beta} \sum_{p \in P} \sum_{q \in Q} \sum_{m \in M} Q_{pqm}^k \ln Q_{pqm}^k$$

Q_{pqm}^k is the subproblem variable for OD mode flows given by equation (7.5). The algorithm is stated as follows.

Step 0. Initialization. Let (f^1, P^1) be a feasible solution to the variable-demand route choice model. Set $\text{BLB}^1 = 10^{-6}$; $\epsilon > 0$; $k = 1$.

Step 1. Search direction generation. Solve the partially linearized subproblem, obtaining (y^k, Q^k). The search direction is $d^k = (y^k - f^k, Q^k - P^k)$. For all origin–destination pairs:

(a) Find the shortest path (route) on the road network based on link flows f^k.
(b) Solve for Q_{pqm}^k, using the choice function in equation (7.5).
(c) Assign the flow TQ_{pqm}^k to the shortest route on the road network from p to q.
(d) Evaluate $\underline{g}(y^k, Q^k)$ and set $\text{BLB}^{k+1} = \max(\text{BLB}^k, \underline{g}(y^k, Q^k))$.

Step 2. Line search. Find a step length λ^k that solves

$$\underset{0 \le \lambda^k \le 1}{\text{minimize}}\ \frac{1}{T} \sum_{a \in A} \int_0^{f_a^{k+1}} c_a(x)\, dx + \sum_{p \in P} \sum_{q \in Q} c_{pqt} P_{pqt}^{k+1}$$

$$+ \frac{1}{\beta} \sum_{p \in P} \sum_{q \in Q} \sum_{m \in M} P_{pqm}^{k+1} \ln P_{pqm}^{k+1}$$

where $f_a^{k+1} = (1 - \lambda^k) f_a^k + \lambda^k y_a^k$ and $P_{pqm}^{k+1} = (1 - \lambda^k) P_{pqm}^k + \lambda^k Q_{pqm}^k$.

Step 3. Update. Set (f^{k+1}, P^{k+1}) to the values shown above. Increment k.

Step 4. Convergence check. If $|g(f^k, P^k) - \text{BLB}^k| < \epsilon$, stop; otherwise, go to step 1.

The following comments pertain to the Evans algorithm.

1. As with the FW algorithm, an initial solution may be found by solving step 1 with $f^o = \mathbf{0}$.
2. Because Q^k and P^k are based on the choice function from the optimality conditions, these variables tend to converge to near-optimal values much faster than f^k. In fact, the Evans algorithm may converge in fewer iterations than a fixed-demand route choice problem, in effect because the OD mode flows are adjusting to the road network conditions. Under some conditions, Q^k does not need to be computed each at iteration, thereby saving some computational effort (See Huang and Lam, 1992).
3. The model can generally be solved to a satisfactory level of convergence in 10 to 20 iterations.
4. The line search step is quite computationally intensive, especially if the number of zones is large. If computer time is limited, MSA can be substituted for the line search; however, additional iterations will then be required to achieve the same level of convergence.

7.E. APPLICATION OF A TRAVEL CHOICE MODEL

Travel choice models of the type presented in Sections 7.B and 7.C have been implemented by Boyce et al. (1985, 1992), Safwat and Magnanti (1988), Lam and Huang (1992), Abrahamsson and Lundqvist (1997) and Arezki et al. (1996). A variant of this class of model (Florian, 1977) is available in the EMME/2 System (INRO, 1996).

In this section we summarize the findings of Boyce et al. for an implementation for the Chicago region. This synthesis focuses on three aspects of their research:

1. Convergence of the Frank–Wolfe and Evans algorithms
2. Form of the OD mode choice function
3. Estimation and empirical evaluation of the results

7.E.1. Model Implementation

An aggregate zone system and network representation was selected for the initial model implementation for the Chicago region in 1980. The motivation for this choice was the desire to solve a problem of more manageable size than in existing models, given the available memory on mainframe computers of that generation (512 kilobytes) and the recognition that the research would require solving the models hundreds of times. Although computing capabilities have undergone several generations of improvement since, the original scale of implementation has been retained, with the result that the model can now be solved on a midsized personal computer (8 megabytes) instead of a relatively large mainframe machine.

The Chicago region aggregate, or sketch planning, model consists of 317 zones of 9 and 36 square miles each in the inner and outer parts of the region, respectively, and an aggregated road network representation of 1060 nodes and 2902 links. By comparison, the Chicago regional model of the Chicago Area Transportation Study (CATS) had 1800 zones, 12,000 nodes, and 37,000 links in 1980. In the aggregated road network, expressways and tollways are represented explicitly; however, arterial roads are aggregated into a square grid defined on the zone centroids, as shown in Figure 7.3. Origin–destination mode matrices were provided by CATS corresponding to forecasts prepared for 1990 using a conventional travel forecasting procedure. The traditional BPR travel-time function was applied to links of the aggregated road network using zero-flow travel times between adjacent zones and aggregated capacities. An operating cost function based on travel distance and time was employed to capture the effect of congestion on auto operating costs. Transit

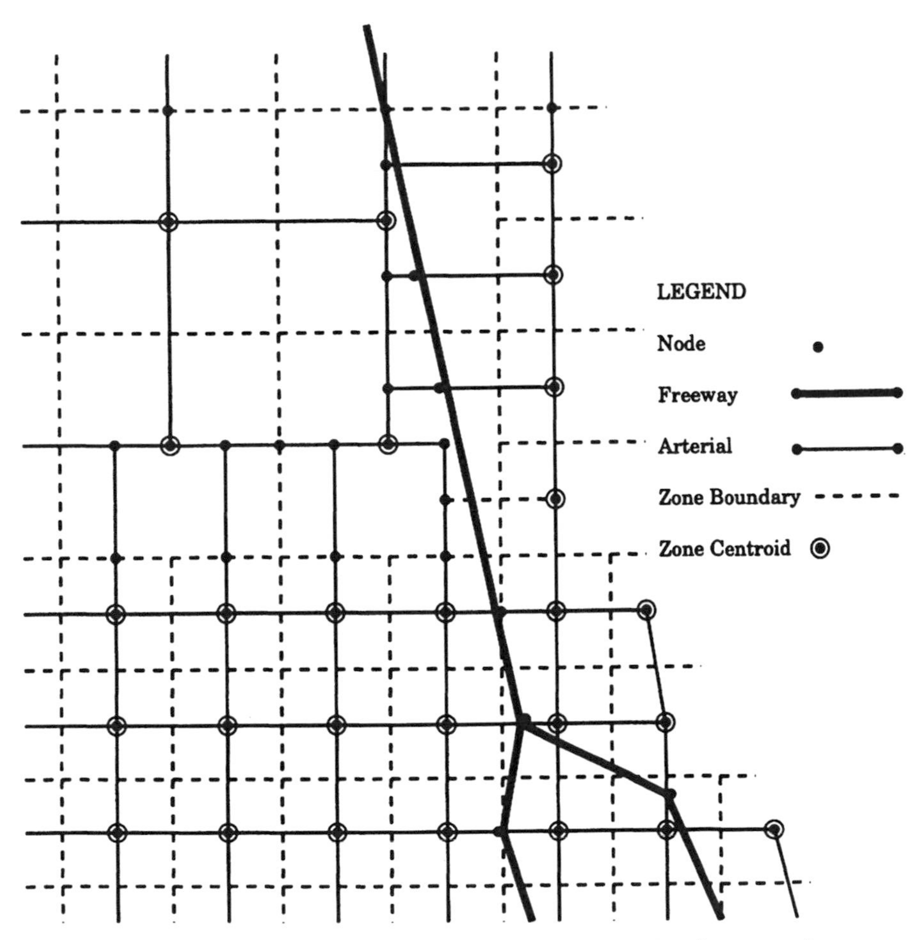

Figure 7.3 Representative portion of the sketch planning network.

travel times and fares were aggregated from the Chicago regional transit network.

7.E.2. Model Solution and Convergence

Solution of the auto route choice model and the OD mode and route choice model using the Frank–Wolfe and Evans algorithms is now considered. Solution of the auto route choice model requires the full linearization of the objective function. Figure 7.4 shows the objective function and lower-bound values found by the F-W algorithm for 20 iterations. Note that the objective function is measured in terms of generalized cost, not travel time; see equation (7.1). To emphasize the convergence properties, the initial values of the objective function and lower bound are omitted. The convergence criterion ϵ reaches a value of less than 0.01 in 21 iterations, which may be regarded as quite well converged.

Figure 7.5 shows similar results of the OD mode and route choice model solved with the Evans algorithm. Because the objective function stated in equation (7.5) includes terms related to the entropy function as well as transit times and costs, its value is lower than for the auto route choice case. This partial linearization algorithm reaches convergence ($\epsilon \leq 0.01$) after 17 iterations. Note that Figures 7.4 and 7.5 are only roughly comparable because the models being solved are quite different.

Figure 7.6 shows the values of the step length λ for both models. For comparison, the corresponding values of $1/k$ for the method of successive averages (MSA) are also shown. As can be seen, the optimal step lengths for the F-W and Evans algorithms are roughly similar to the MSA step lengths but deviate substantially at some iterations.

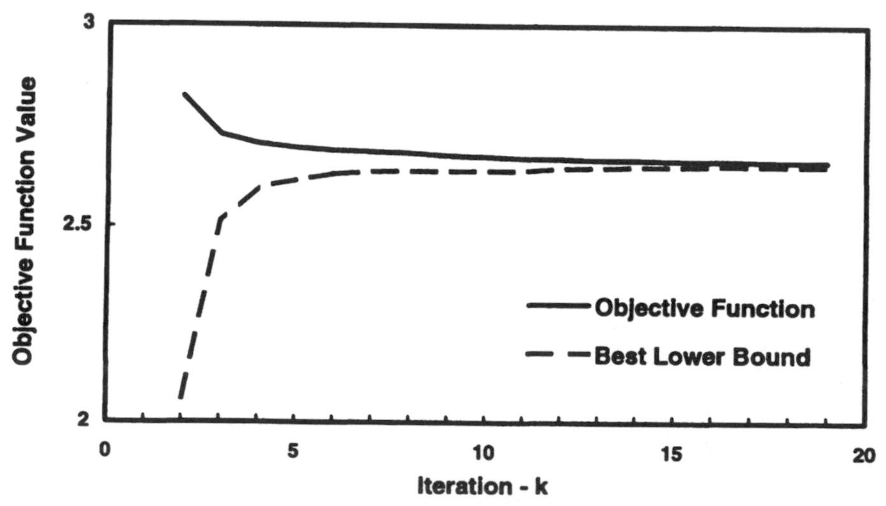

Figure 7.4 Convergence of the Frank–Wolfe algorithm.

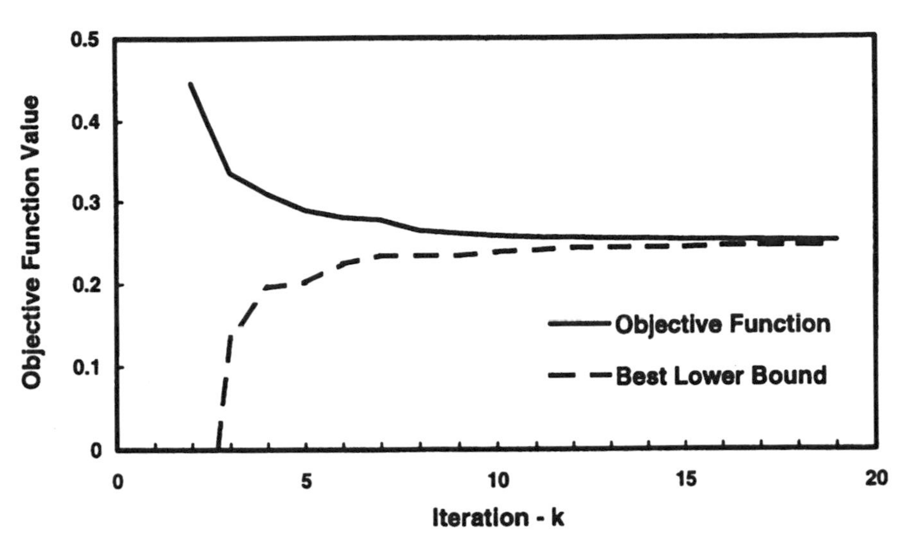

Figure 7.5 Convergence of the Evans algorithm.

7.E.3. Model Estimation

Having formulated a model and identified an effective solution procedure, the next step of model implementation is to determine the appropriate values of the parameters in applying the model. For the models presented in this chapter, two types of parameters can be distinguished: exogenous values such as the weights γ in the generalized cost function; and Lagrange multipliers,

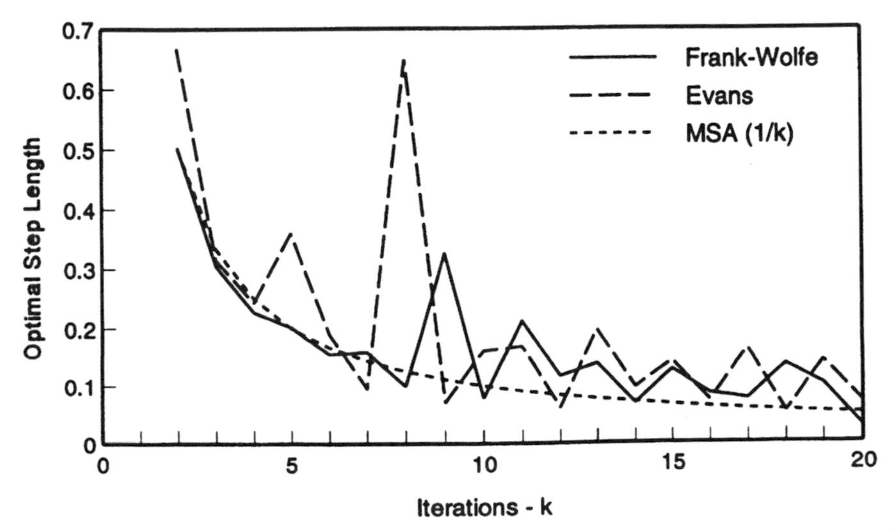

Figure 7.6 Comparison of the optimal step length for Frank–Wolfe, Evans, and MSA.

which may also be regarded as exogenous once the functional form has been identified.

Three approaches to identifying appropriate values for parameters are the following:

1. Borrow values from other studies.
2. Solve for the values of Lagrange multipliers in terms of their associated constraints.
3. Apply a statistical estimation procedure.

We discuss these approaches briefly and then give an example of the third approach.

Borrowing coefficient values from other studies in a reasonable method for a quick analysis. For example, a mode-choice analysis may have been conducted yielding relative values of travel time and out-of-pocket costs. Appropriate weights for a generalized cost function can be inferred from the results of such a study. Even when more sophisticated optimization-based or statistical estimation approaches are used, the results should be checked against those found in other studies. Thus an estimate of the value of time of $100 per hour is likely to indicate an error of some form, such as with the data or estimation technique used.

If the right-hand-side values of constraints are available from data, the corresponding values of the Lagrange multipliers can be computed, typically on a trial-and-error basis. An example is the value of the regional transit bias coefficient necessary to predict correctly the regional transit share. A problem with this approach is that the right-hand-side values may not be available for the desired application. Or as in the case of the entropy constraint, the value may not be available at all for the actual population of travelers. In this case, a sample-based OD mode matrix, with its substantial number of empty cells, is not an adequate basis for computing the entropy value.

Another problem with these two approaches is that the coefficient values may not be consistent with each other, especially if values are drawn from other studies. In this case, a statistical estimation approach is a better choice. For the travel choice functions formulated above, parameters μ and γ may be estimated using the following maximum likelihood function:

$$\underset{\mu,\gamma}{\text{maximize}} \sum_{p\in P} \sum_{q\in Q} \sum_{m\in M} \overline{P}_{pqm} \ln P_{pqm}(\mu,\gamma)$$

where

$$P_{pqm}(\mu,\gamma) = a_p o_p b_q d_q \exp[-\mu(\gamma_0 + \gamma_1 t_{pqm} + \gamma_2 k_{pqm} + \gamma_3 w_{pqm})]$$

and $\overline{P}_{pqm}$ are the observed travel choice proportions. The variables t, k, and

w are in-vehicle travel time, operating cost/fare, and out-of-vehicle travel time, respectively. The parameters $(\gamma_1, \gamma_2, \gamma_3)$ are generalized cost weights; γ_0 is the transit bias, which was derived as a Lagrange multiplier related to regional transit share, and μ is the reciprocal of the Lagrange multiplier related to the entropy constraint. Since μ and γ_n always appear as a product, it is not possible to estimate them separately. Hence we may set $\mu = 1.0$ and estimate (γ).

The maximum likelihood method is a standard approach for estimating the values of parameters of logit choice functions (Ben-Akiva and Lerman, 1985). In a typical application, the values of in-vehicle and out-of-vehicle travel times, as well as out-of-pocket travel costs, are available from survey data. These data may be observations on individual travelers or aggregations to zone pairs. In the latter case, if there is no observation for a specific OD mode combination, this term is simply omitted from the likelihood function.

In urban transportation planning applications, however, travel choice data are often drawn from the decennial census as well as household surveys, and travel cost data are compiled from networks. In this case, road network travel times and operating costs are not fixed but rather, depend on link flows. The travel choice function is the solution to one of the choice models already described above. The structure of the resulting problem is a bilevel programming problem (see Section 7.A) where the likelihood problem is the upper problem and the travel choice problem is the lower problem. This problem was studied extensively by Zhang (1995). The heuristic solution procedure applied in her research does converge to stable and reasonable values of the model parameters, with a high value of the standard goodness-of-fit measure,

$$\rho^2 = \frac{L_M - L_N}{L_P - L_N}$$

where L_M is the likelihood value of the model, L_P the likelihood value for the data $(\bar{P}_{pqm})$, which may be regarded as the value for a perfect fit of the model to the data, and L_N the likelihood value of the null hypothesis given by

$$P^N_{pqm} = o_p d_q p_m$$

where p_m is the observed regional modal share and o_p and d_q are the observed origin and destination shares. Note that the model cannot be estimated for route choice parameters because route choice data are unavailable.

As an example of the results of this procedure, we consider three alternative choice functions. Function 1 is the function stated in equation (7.5). Function 2 is the same as function 1 except that the values of (γ) are estimated separately by mode. The reason for this mode-specific version is that the modal times and distances predicted by function 1 are not satisfactory.

The third function is a nested logit function of the form

$$P_{pqm} = a_p o_p b_q d_q \exp(-\beta \tilde{c}_{pq}) \frac{\exp(-\mu c_{pqm})}{\sum_{m \in M} \exp(-\mu c_{pqm})}$$

where

$$\tilde{c}_{pq} \equiv -\frac{1}{\mu} \ln \left[\sum_{m \in M} \exp(-\mu c_{pqm}) \right]$$

It was hypothesized in the research that this nested function would somewhat improve the goodness of fit of the estimated model to the data. The results of the estimation of these three functions are shown in Table 7.9. Also shown in the table are several regional evaluation measures: (1) regional transit share, (2) regional transit share for trips to CBD destinations, (3) mean in-vehicle travel time, (4) mean travel distance, and (5) goodness-of-fit measure ρ^2

TABLE 7.9 Comparison of Parameter Estimates and Evaluation Measures

Variable	Observed Value	Choice Function 1	Choice Function 2	Choice Function 3
Estimated Parameter Values				
Transit bias, γ_0		0.7756	4.9015	4.4818
In-vehicle time, γ_1		0.0680	—	—
Auto		—	0.0900	0.0940
Transit		—	0.0308	0.0328
Operating cost/fare, γ_2		0.0203	—	—
Auto		—	0.0142	0.0127
Transit		—	0.0081	0.0085
Out-of-vehicle time, γ_3		0.0119	—	—
Auto		—	0.2530	0.2278
Transit		—	0.0052	0.0083
Modal dispersion, μ		1.0	1.0	1.0
OD dispersion, β		—	—	0.9942
Evaluation Measures				
Regional transit share (%)	15.97	15.96	15.99	16.06
Transit share to CBD (%)	63.20	65.62	61.69	61.15
Mean in-vehicle time (min)				
Auto	17.20	21.50	17.18	17.00
Transit	33.54	24.82	33.54	33.38
Mean travel distance (miles)				
Auto	8.12	9.02	8.29	8.25
Transit	13.00	9.83	14.03	13.96
Goodness of fit, ρ^2	—	85.74	89.37	89.31

$(0 \leq \rho^2 \leq 1)$. Some of these evaluation measures have a direct correspondence to the parameter values.

1. The transit bias γ_o was derived as the Lagrange multiplier of the regional transit share constraint; its maximum likelihood estimate seeks to equate the predicted and observed transit share.
2. The in-vehicle and out-of-vehicle travel time parameters (γ_1, γ_3) are estimated by the maximum likelihood method to equate the predicted and observed mean travel times by mode.
3. In contrast, there is no direct correspondence between operating cost/fare (γ_2) and mean travel distance; however, they are related empirically.

As noted above, the value of μ is set to 1.0. The parameter values for function 1 yield a reasonably high value of the goodness-of-fit measure, but the model predicts auto and transit travel times and distances poorly and slightly overpredicts the transit share of trips to the CBD, the primary destination served by the regional transit system.

Function 2 has a slightly higher goodness of fit, predicts modal times and distances rather well, and predicts transit's CBD share slightly better. The parameter values are rather different from those of function 1, possibly indicating substantial multicollinearity between the variables corresponding to the transit bias and the generalized cost weights. This result suggests the values of some parameters need to be tested to determine if they are statistically different from zero.

The results for function 3 are quite similar to those for function 2. The value of β is very close to 1.0, in which case function 3 simplifies to function 2. The parameter values and evaluation measures are only slightly different. These differences may indicate that the estimates are not as tightly converged as might be desired. A problem in this regard is that the lower-level network problem cannot be solved exactly, so the upper-level estimation problem tends to be somewhat unstable in the region of the optimal solution. Nevertheless, the values shown are regarded as acceptable. The number of significant digits shown is intended to facilitate comparison of the results rather than to claim such a high level of accuracy.

The combined model incorporating functions 1 and 2 has been applied extensively in scenario analyses of the Chicago region. Descriptions of the results are found in Boyce et al. (1992) and Tatineni et al. (1994). Subsequently, a version of the model has been implemented in the EMME/2 system by Metaxatos et al. (1995). CATS has also updated the network representation and zone system and is applying the model in a land use–transportation planning exercise. Finally, a two-period model was implemented by Compere (1996) and used to investigate the effects of congestion pricing on the region's road network and travel patterns.

7.F. TRANSPORTATION PLANNING PRACTICE

The perspective on urban travel choice models offered in this chapter is definitely at odds with the traditional travel forecasting approach, the four-step sequential procedure. The reasons for this dichotomy are twofold. First, in the context of this book we present an optimization-based approach to formulating and solving travel choice models. Second, we seek to provide a framework in which the four-step procedure itself can be more fully understood and evaluated. In this section we address the second point.

The four-step sequential procedure consists of separate models for OD mode and route choice, preceded by a method for estimating the frequency of travel. In Figure 7.7 the traditional names of these models are shown together with their sequential relationship. Several variants of the procedure have been applied, including ones in which the order of the OD and mode choice steps are reversed. Descriptions of these variants may be found in the textbook by Ortúzar and Willumsen (1990).

The venerable sequential travel forecasting procedure has been roundly criticized elsewhere; it is not our intention to repeat these criticisms here. Rather, we note that it was the intention of the originators of such schemes that the procedure be iterative in nature. Such iterations were regarded as

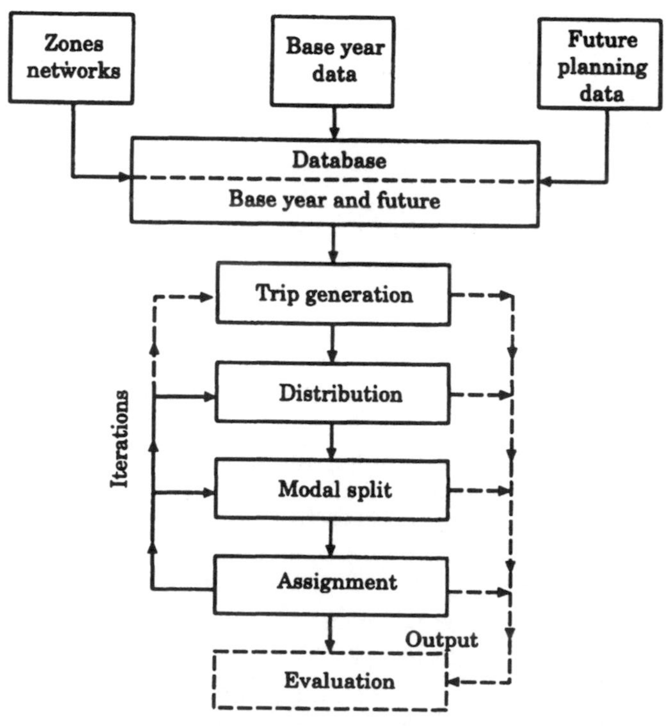

Figure 7.7 Four-step sequential procedure. (From Ortúzar and Willumsen, 1990.)

necessary because the procedure was always seen as forecasting an equilibrium condition, however ill-defined. In particular, the equilibrium travel times and costs needed to solve the trip distribution (OD choice) model are available only from the solution of the trip assignment (route choice) model. Hence initial values of interzonal travel times and costs must be assumed at the outset and replaced by improved values as the procedure iterates to the desired equilibrium solution. This iterative procedure is denoted by an arrow in Figure 7.7.

In recent years, the iterative solution of the four-step procedure, as contrasted with the once-through approach, has come to be known as *solving the four-step procedure with feedback.* Indeed, such a solution is now mandated by legislation in the United States, and the Federal Highway Administration has issued a handbook on incorporating feedback in travel forecasting (Comsis, 1996a; see Boyce et al., 1994, for an earlier contribution to this issue).

What should be apparent by now is that solving the four-step model with feedback is effectively equivalent to applying the class of solution algorithms proposed by Evans (1976). Since the iterative travel forecasting procedure, as originally defined, had neither an objective function nor an averaging step (line search and update), there is no mechanism by which the procedure could or should converge to the desired equilibrium solution. Recently, practitioners have begun to utilize the method of successive averages in their iterative schemes, thereby remedying this omission.

Viewed from this perspective, the four-step sequential procedure may be seen as simply a heuristic, albeit a rather primitive one, for solving an unstated network equilibrium problem belonging to the class of travel choice models derived in this chapter. By making the statement of the problem explicit, one obtains the advantages of an algorithm that can be proven to converge to the equilibrium solution, well-defined convergence criteria, and a framework in which to state behavioral assumptions about urban travel.

In the title of her seminal paper, Evans refers to "models for combining trip distribution and assignment." Accordingly, several authors began to refer to these formulations as *combined models.* For many reasons the replacement of the four-step procedure with combined models has not occurred, and it remains largely a research curiosity. As an appreciation and understanding of optimization-based models continues to grow, however, acceptance of this approach may also improve.

Concurrent with the development of combined models, the era of the microprocessor generation of computers has convincingly demonstrated that methods are adopted in practice for which microcomputer software is available. This experience explains very well the course of development of travel forecasting procedures over the past 15 years.

By 1980, development of the Urban Transportation Planning System (UTPS) of the U.S. Department of Transportation had effectively reached its culmination. This mainframe-based system, which was under development in various forms since the late 1950s, was deemed inappropriate for further enhancement and offered to the private sector for implementation on the newly available generation of personal computers. This privatization effort led to several software systems being marketed to planning agencies, consultants, and universities: MINUTP, TModel 2, and TRANPLAN (Comsis, 1996b; TModel, 1995; UAG, 1995). These and other software systems have been reviewed by Ferguson et al. (1992). In parallel, several independent developers began devising transportation planning software from scratch. Prominent examples are EMME/2, QRS II, SYSTEM II, TransCAD, TRIPS, and VISEM/VISUM (INRO, 1996; AJH, 1995; JHK, 1996; Caliper, 1996; MVA, 1996; PTV, 1996). While offering the basic capabilities of the four-step procedure, some of the latter systems also adopt some of the conceptual viewpoints presented in this chapter and include capabilities for solving network equilibrium models with variable demand, or combined models.

As these concepts become more widely accepted, and as requirements for solving the traditional models to an equilibrium are further emphasized, we are confident that more software developers will add this capability to their systems. In the final analysis, then, the professional practitioner community will determine the pace and extent of the success of the application of these research innovations.

7.G. EXTENSIONS

We conclude this chapter with a brief examination of extensions of the models considered here and prospective future research. First, we consider extensions of static models and then turn to their dynamic counterparts. We conclude where we began in Section 7.A with the question of more integrated approaches focused on design methods.

A principal extension of static models concerns the generalization of constrained optimization formulations to an equilibrium framework known as the *variational inequality* (VI) *problem.* As an example, consider the following formulation of the auto route choice model:

$$\sum_{a \in A} c_a(f^*)(f_a - f_a^*) \geq 0 \qquad f \in F$$

where $f = (f_a, a \in A)$ is the vector of link flows; $c_a(f)$ a link travel cost function that depends on the flow on link a *and* flows on other selected links, such as opposing and conflicting flows at an intersection; and F represents the conservation of flow constraints. An asterisk denotes the equilibrium solution. If $c_a(f) = c_a(f_a)$, the VI problem above is equivalent to the optimization formulation in Section 7.B.2. If $c_a(f) \neq c_a(f_a)$, a more general model results that in appropriate for detailed modeling of intersection flows; see Berka and Boyce (1996). If travel costs depend on the vector of link flows, the solution of the integral in the user-optimal objective function may not be independent of the path of integration. If so, its solution is not unique, and no equivalent optimization problem exists. See Patriksson (1994, pp. 51–54) for details.

The utility of variational inequality models has been demonstrated in improving and generalizing the formulation of travel choice models. Similarly, some progress has occurred in improving solution methods. The requirements for the existence and uniqueness of solutions to VI models remain quite strong (restricted) and many realistic traffic engineering-based cost functions do not belong to the class of functions for which unique solutions can be guaranteed. Nevertheless, one study concluded that solutions to such models are locally unique (Meneguzzer, 1995). Another study with two classes of vehicles found a bifurcation of solutions (Wynter, 1996). For additional developments, see, Friesz (1985), Nagurney (1993), Florian and Hearn (1995), and Nagurney and Zhang (1996).

The other principal new thrust pertains to dynamic travel choice models, in the sense of time-of-day dynamics. Some observers would claim that without a dynamic dimension, travel choice models are too simplistic to be useful for applications. Although this statement is quite harsh in view of the demonstrated utility of static models for strategic transportation planning, it is true that the instantaneous propagation of flows over a network is a difficult notion to accept. Moreover, dynamic models are clearly needed for real-time control problems.

Despite remarkable progress in the formulation of dynamic models in less than 10 years, a deep understanding of these models is an elusive objective. Moreover, solution algorithms remain at an early stage of development, despite some recent success for large networks based on a variational inequality formulation (Lee et al., 1996). Simulation approaches combined with dynamic route choice concepts have also resulted in meaningful solutions for small test networks (Mahmassani and Peeta, 1993).

Finally, we return to the hierarchical optimization problem introduced at the end of Section 7.A.2. Although appealing conceptually for those working in the optimization paradigm, the capability to solve bilevel programs is still at an early stage. Some progress is noted for specific models, often with linear objective functions (Ben-Ayed et al., 1993; Gendreau et al., 1992). In general, bilevel programs do not possess unique solutions, since their single-level equivalent program is nonconvex. Nevertheless, we remain optimistic that this conceptual framework will provide needed algorithms for the design of transportation networks, as well as the analysis and prediction of travel choices for more complex sets of behavioral assumptions.

ACKNOWLEDGMENTS

The authors thank Dr. Bruce Janson and Dr. Michael Patriksson for helpful suggestions and critical comments, and Dr. Der-Horng Lee and Ms. Xin Tian for their assistance with text editing and preparing the examples of the algorithms. David Boyce acknowledges the support of the National Science Foundation (Grant DMS 9313113) through the National Institute of Statistical Sciences, Research Triangle Park, NC, and the National Center for Supercomputing Applications, University of Illinois at Urbana–Champaign, Urbana, IL.

EXERCISES

7.1. Consider network E-1 (Figure 7.8), where travel times on the links are as follows:

$$t_{12} = 10.0 + 0.01 f_{12}$$

$$t_{13} = 30.0 + 0.002 f_{13}$$

$$t_{34} = 10.0 + 0.01 f_{34}$$

$$t_{24} = 30.0 + 0.002 f_{24}$$

The demand from origin 1 to destination 4 is 1000 vehicles per hour. The user-optimal solution is

$$f_{12} = f_{13} = f_{24} = f_{34} = 500$$

$$t^{14} = 46.0$$

where origin–destination values are given with superscripts and link values are given with subscripts. We now consider adding a one-way link

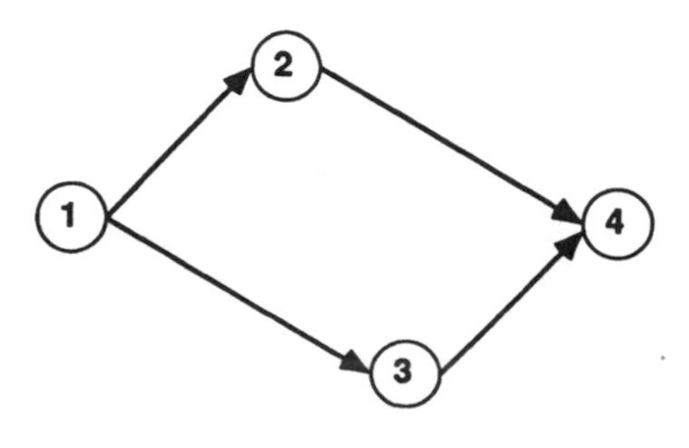

Network E-1

Figure 7.8

from 2 to 3 to define network E-2 (Figure 7.9), where the travel-time function on link (2,3) is

$$t_{23} = 15.0 + 0.01 f_{23}$$

The new user-optimal solution is given by

$$f_{12} = f_{34} = 531.25$$

$$f_{13} = f_{24} = 468.75$$

$$f_{23} = 62.5$$

$$t^{14} = 46.25$$

Let's not worry too much about fractional people. Assume that they are schizophrenic! Note, however, that the new travel time exceeds the old time!

(a) One might argue that the new solution found for network E2 is not a valid equilibrium, that eventually everyone will revert to the old system, and that link (2,3) will not be used. Do you agree or not? Why?

(b) Formulate the system-optimal problem for network E-1. Solve for the system-optimal flows. Does the solution differ at all from the user-optimal solution?

(c) Solve for the system-optimal flows for network E-2.

(d) Compute the user-optimal objective function and the system-optimal objective function for each network using both the user and system flows. That is, complete the following table:

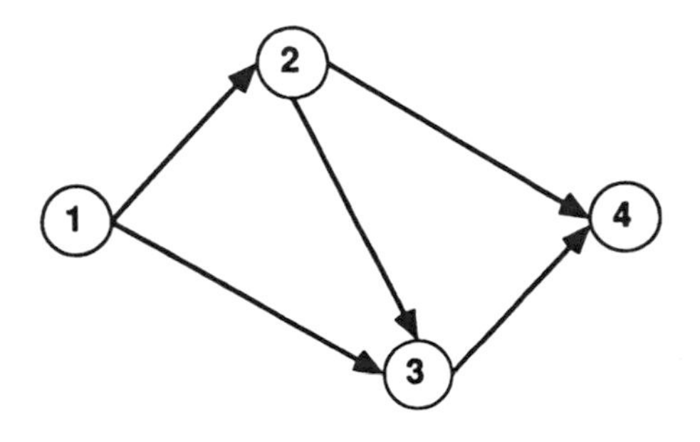

Network E-2

Figure 7.9

Flow Pattern	User-Optimal Objective Function	System-Optimal Objective Function
User-optimal		
System-optimal		

(e) This problem is known as Braess's paradox. What do you think is the resolution of the paradox?

7.2. For the network shown in Figure 7.1, assuming a value of time of $9.00 per hour and assuming that the times are given in minutes, identify a set of tolls on the links that causes the flows to correspond to the system-optimal flows if all users minimize their own total out-of-pocket (toll) and travel-time cost.

7.3. As indicated in Section 7.D.1, the STOCH algorithm proposed by Dial (1971) defines an *efficient* route to be one such that traversing each link takes the traveler farther from the origin and closer to the destination. In network E-3a (Figure 7.10), both routes from origin A to destination B would be used, although the top route, with a travel time of 10 min, would be used by fewer people than the lower route, with a travel time of only 7 min. Discuss why only the bottom route with a travel time of 7 min would be used if we add a link from A to C as shown in network

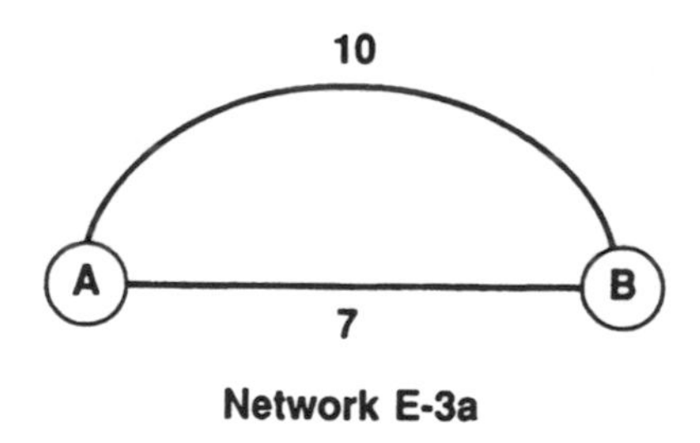

Network E-3a

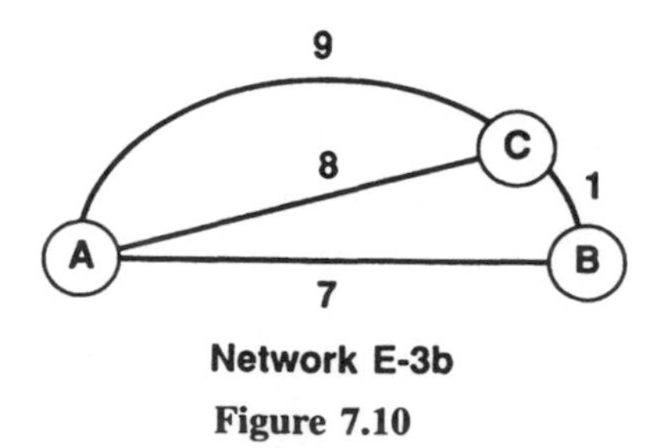

Network E-3b

Figure 7.10

E-3b (Figure 7.10). What does this imply about the sensitivity of the algorithm to changes in the coding of the network?

7.4. The fraction of vehicles using route r in the STOCH algorithm is given by

$$P_r = \frac{\exp(-\theta c_r)}{\Sigma_{s\in R} \exp(-\theta c_s)}$$

where θ is a parameter of the model ($\theta \geq 0$); c_r, c_s the costs (or travel times) on routes r and s, respectively; R the set of efficient routes from the origin to the destination (routes that take the traveler farther from the origin and closer to the destination for each link traversed); and P_r the probability that route r is chosen.

(a) For network E-4a (Figure 7.11), compute the probability of a traveler using each of the two routes for $\theta = 0.2$.

(b) Network E-4b (Figure 7.11) is identical to network E-4a except that we have now included slightly more detail about the last small part of the trip from A to B when using the bottom route. This creates two routes where there had been only one. In effect, there are now

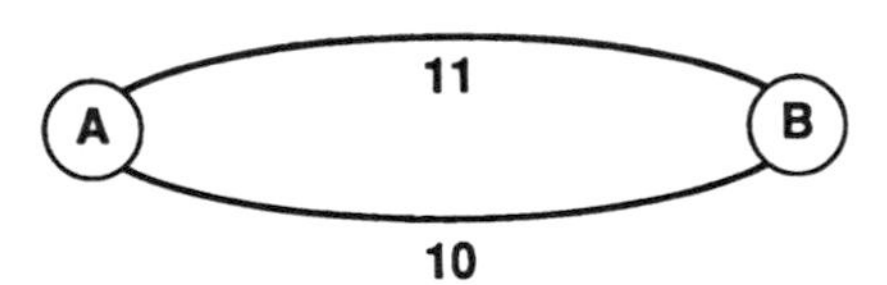

Network E-4a

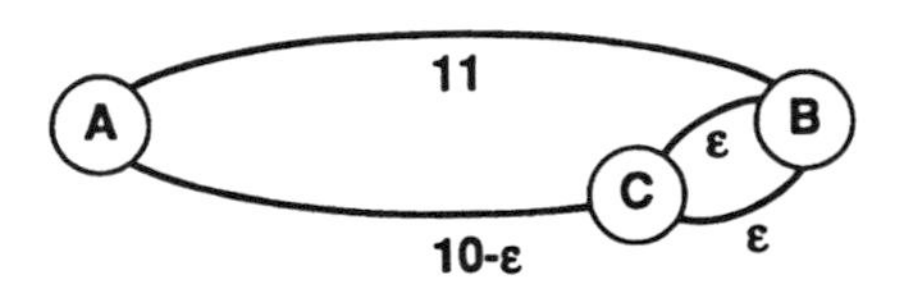

Network E-4b

Figure 7.11

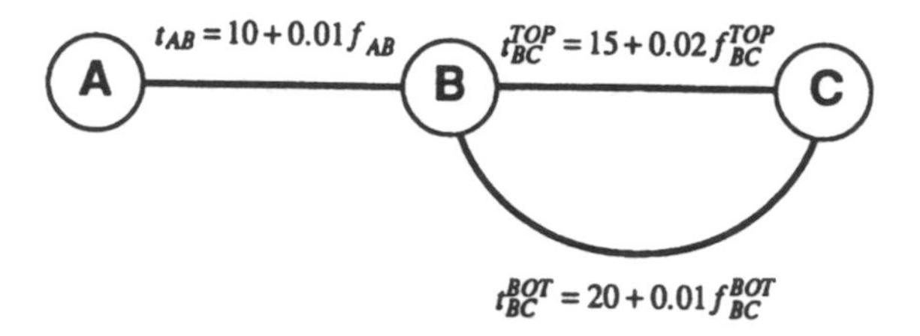

Network E-6

Figure 7.12

three routes from A to B. Compute the probability of a traveler choosing each of the routes in this case when $\theta = 0.2$.

(c) Briefly discuss the implications that changes in the coding of the network will have on the results when using this model.

7.5. Show that you can modify the *inputs* to a network equilibrium problem in such a way that the Frank–Wolfe algorithm (also known as the convex combinations algorithm) can be used to solve for the *system-optimal* flows when the travel-time functions are BPR functions. State clearly how this can be done and why the result is valid.

7.6. We can show that the link flows that result from solving the user-optimal problem are unique. However, the route flows are not. Find at least two sets of route flows that satisfy the user-optimal conditions for network E-6 (Figure 7.12), when the demand from A to C is 600 vehicles per hour and the demand from B to C is 400 vehicles per hour. The linear travel time functions are shown next to the links.

REFERENCES

Abrahamsson, T., 1996. *Network Equilibrium Approaches to Urban Transportation Markets: Combined Models and Efficient Matrix Estimation,* Department of Infrastructure and Regional Planning, Royal Institute of Technology, Stockholm, Sweden.

Abrahamsson, T., and L. Lundqvist, 1997. Formulation and Estimation of Combined Network Equilibrium Models with Applications to Stockholm, *Transportation Science,* forthcoming.

Ahuja, R. K., T. L. Magnanti, and J. B. Orlin, 1993. *Network Flows: Theory, Algorithms, and Applications,* Prentice Hall, Upper Saddle River, NJ.

AJH, 1995. *Quick Response System II for Windows Reference Manual, Version 4.0,* and *Advanced General Network Reference Manual, Version 5.1,* AJH Associates, Milwaukee, WI.

Akcelik, R., 1988. The Highway Capacity Manual Formula for Shared Intersections, *ITE Journal,* 58(3), 23–27.

Arezki, Y., M. D. Hall, and G. Hyman, 1996. Simultaneous Assignment, Distribution and Mode Choice in NAOMI, *Transportation Planning Methods,* 24th European Transport Forum, PTRC Education and Research Services, London.

Bazaraa, M. S., H. D. Sherali, and C. M. Shetty, 1993. *Nonlinear Programming: Theory and Algorithms,* 2nd ed., Wiley, New York.

Beckmann, M., C. B. McGuire, and C. B. Winsten, 1956. *Studies in the Economics of Transportation,* Yale University Press, New Haven, CT.

Ben-Akiva, M., and S. Lerman, 1985. *Discrete Choice Analysis: Theory and Application to Travel Demand,* MIT Press, Cambridge, MA.

Ben-Ayed, O., C. E. Blair, D. E. Boyce, and L. J. LeBlanc, 1992. Construction of a Real-World Bilevel Linear Programming Model of the Highway Network Design Problem, *Annals of Operations Research,* 34, 219–254.

Bergendorff, P., D. W. Hearn, and M. V. Ramana, 1996. Congestion Toll Pricing of Traffic Networks, in P. Pardalos, D. W. Hearn, and W. W. Hager (eds.), *Network Optimization,* Springer-Verlag, Berlin.

Berka, S., and D. E. Boyce, 1996. Generating Highway Travel Times with a Large-Scale Asymmetric User Equilibrium Assignment Model, in L. Bianco and P. Toth (eds.), *Advanced Methods in Transportation Analysis,* Springer-Verlag, Berlin.

Boyce, D. E., A. Farhi, and R. Weischedel, 1973. Optimal Network Problem: A Branch-and-Bound Algorithm, *Environment and Planning A,* 5, 519–533.

Boyce, D. E., K. S. Chon, Y. J. Lee, K. T. Lin, and L. J. LeBlanc, 1983. Implementation and Computational Issues for Combined Models of Location, Destination, Mode and Route Choice, *Environment and Planning A,* 15, 1219–1230.

Boyce, D. E., K. S. Chon, M. E. Ferris, Y. J. Lee, K. T. Lin, and R. W. Eash, 1985. *Implementation and Evaluation of Combined Models of Urban Travel and Location on a Sketch Planning Network,* Final Report to Chicago Area Transportation Study, Department of Civil Engineering, University of Illinois, Urbana, IL.

Boyce, D. E., L. J. LeBlanc, and K. S. Chon, 1988. Network Equilibrium Models of Urban Location and Travel Choices: A Retrospective Survey, *Journal of Regional Science,* 28, 159–183.

Boyce, D. E., M. Tatineni, and Y.-F. Zhang, 1992. *Scenario Analyses for the Chicago Region with a Sketch Planning Model of Origin–Destination, Mode and Route Choice,* Final Report to the Illinois Department of Transportation, Urban Transportation Center, University of Illinois, Chicago.

Boyce, D. E., Y.-F. Zhang, and M. R. Lupa, 1994. Introducing Feedback into Four-Step Travel Forecasting Procedure Versus Equilibrium Solution of Combined Model, *Transportation Research Record,* 1443, 65–74.

Caliper, 1996. *TransCAD User's Guide, Version 3.0,* Caliper Corporation, Newton, MA.

Compere, K. J., 1996. Impact of Tolls on Departure Time, Destination, Mode and Route: A Combined Model Implemented in EMME/2, M.S. thesis, University of Illinois, Chicago.

Comsis, 1996a. *Incorporating Feedback in Travel Forecasting: Methods, Pitfalls and Common Concerns,* Final Report to the Federal Highway Administration, DOT-T-96-14, Washington, DC.

Comsis, 1996b. *MINUTP Technical Reference Manual, Version 96A,* Comsis Corporation, Silver Spring, MD.

Dafermos, S., and F. T. Sparrow, 1971. Optimal Resource Allocation and Toll Patterns in User-Optimised Transport Networks, *Journal of Transport Economics and Policy,* 5, 184–200.

Daganzo, C. F., 1979. *Multinomial Probit: The Theory and Its Application to Demand Forecasting,* Academic Press, San Diego, CA.

Davidson, K. B., 1966. A Flow–Travel Time Relationship for Use in Transportation Planning, *Proceedings, Australian Road Research Board,* 3, 183–194.

Dial, R. B., 1971. A Probabilistic Multipath Traffic Assignment Model Which Obviates Path Enumeration, *Transportation Research,* 5, 83–111.

Erlander, S., 1977. Accessibility, Entropy and the Distribution and Assignment of Traffic, *Transportation Research,* 11, 149–153.

Erlander, S., 1980. *Optimal Spatial Interaction and the Gravity Model,* Lecture Notes in Economics and Mathematical Systems, Vol. 173, Springer-Verlag, Berlin.

Evans, S. P., 1973. A Relationship Between the Gravity Model for Trip Distribution and the Transportation Problem in Linear Programming, *Transportation Research,* 7, 39–61.

Evans, S. P., 1976. Derivation and Analysis of Some Models for Combining Trip Distribution and Assignment, *Transportation Research,* 10, 37–57.

Ferguson, E., C. Ross, and M. Meyer, 1992. PC Software for Urban Transportation Planning, *Journal of the American Planning Association,* 58, 238-243.

Florian, M., 1977. A Traffic Equilibrium Model of Travel by Car and Public Transit Modes, *Transportation Science,* 11, 166–179.

Florian, M., and D. Hearn, 1995. Network Equilibrium Models and Algorithms, in M. O. Ball (ed.), *Handbooks in Operations Research and Management Science, Network Routing,* Vol. 8: Elsevier Science, Amsterdam, Chapter 6.

Florian, M., and S. Nguyen, 1978. A Combined Trip Distribution Modal Split and Trip Assignment Model, *Transportation Research,* 12, 241–246.

Florian, M., S. Nguyen, and J. Ferland, 1975. On the Combined Distribution-Assignment of Traffic, *Transportation Science,* 9, 43–53.

Florian, M., J. Guélat, and H. Spiess, 1987. An Efficient Implementation of the "PARTAN" Variant of the Linear Approximation Method for the Network Equilibrium Problem, *Networks,* 17, 319–339.

Frank, C., 1978. A Study of Alternative Approaches to Combined Trip Distribution–Assignment Modeling, Ph.D. thesis, University of Pennsylvania, Philadelphia, PA.

Frank, M., and P. Wolfe, 1956. An Algorithm for Quadratic Programming, *Naval Research Logistics Quarterly,* 3, 95–110.

Friesz, T. L., 1985. Transportation Network Equilibrium, Design and Aggregation, *Transportation Research,* 19A, 413–427.

Gallo, G., and S. Pallottino, 1984. Shortest Path Methods in Transportation Models in M. Florian (ed.), *Transportation Planning Models,* North-Holland, Amsterdam, 227–256.

Gallo, G., and S. Pallottino, 1988. Shortest Path Algorithms, *Annals of Operations Research,* 13, 3–79.

Gendreau, M., P. Marcotte, and G. Savard, 1992. A Hybrid Tabu-Ascent Algorithm for the Linear Bilevel Programming Problem, *Journal of Global Optimization,* 9, 1–14.

Hicks, J. E., D. E. Boyce, and A. Sen, 1992. *Static Network Equilibrium Models and Analyses for the Design of Dynamic Route Guidance Systems,* Report to the Illinois Department of Transportation, Urban Transportation Center, University of Illinois, Chicago.

Hitchcock, F. L., 1941. The Distribution of a Product from Several Sources to Numerous Localities, *Journal of Mathematics and Physics,* 20, 224–230.

Huang, H.-J., and W. H. K. Lam, 1992. Modified Evans' Algorithms for Solving the Combined Trip Distribution and Assignment Problem, *Transportation Research,* 26B, 325–337.

INRO, 1996. *EMME/2 User's Manual, Release 8,* INRO Consultants, Montreal, Quebec, Canada.

Janson, B. N., 1993. Most Likely O–D Link Uses from Equilibrium Assignment, *Transportation Research,* 28B, 333–350.

Janson, B. N., and C. Zozaya-Gorostiza, 1987. The Problem of Cyclic Flows in Traffic Assignment, *Transportation Research,* 21B, 299–310.

JHK, 1996. *SYSTEM II Reference Manual,* JHK and Associates, Alexandria, VA.

Kitamura, R., E. I. Pas, C. V. Lula, T. K. Lawton, and P. Benson, 1996. The Sequenced Activity-Mobility Simulator (SAMS): A New Approach to Transportation, Land-Use and Air Quality Modeling, *Transportation,* 23, 267–291.

Koopmans, T. C., 1947. Optimum Utilization of the Transportation System, *Proceedings of the International Statistical Conferences,* 5, 136–146.

Kuhn, H. W., and A. W. Tucker, 1951. Nonlinear Programming, in J. Neyman (ed.), *Proceedings of the 2nd Berkeley Symposium on Mathematical Statistics and Probability,* University of California Press, Berkeley, CA, pp. 481–492.

Lam, W. H. K., and H.-J. Huang, 1992. A Combined Trip Distribution and Assignment Model for Multiple User Classes, *Transportation Research,* 26B, 275–287.

Lawphongpanich, S., and D. W. Hearn, 1984. Simplicial Decomposition of the Asymmetric Traffic Assignment Problem, *Transportation Research,* 18B, 123–133.

LeBlanc, L. J., 1975. An Algorithm for the Discrete Network Design Problem, *Transportation Science,* 9, 183–199.

LeBlanc, L. J., and K. Farhangian, 1981. Efficient Algorithms for Solving Elastic Demand Traffic Assignment and Mode-Split Assignment Problems, *Transportation Science,* 15, 306–317.

Lee, D.-H., D. E. Boyce, and B. N. Janson, 1996. *Formulation and Solution of a Dynamic User-Optimal Route Choice Model on a Large-Scale Traffic Network,* Report to the National Institute of Statistical Sciences, Urban Transportation Center, University of Illinois, Chicago.

Magnanti, T. L., and R. T. Wong, 1984. Network Design and Transportation Planning: Models and Algorithms, *Transportation Science,* 18, 1–55.

Mahmassani, H. S., and S. Peeta, 1993. Network Performance Under System Optimal and User Equilibrium Dynamic Assignments: Implications for ATIS, *Transportation Research Record,* 1408, 83–93.

Meneguzzer, C., 1995. An Equilibrium Route Choice Model with Explicit Treatment of the Effect of Intersections, *Transportation Research,* 29B, 329–356.

Metaxatos, P., D. E. Boyce, M. Florian, and I. Constantine, 1995. Introducing "Feedback" Among the Origin–Destination, Mode and Route Choice Steps of the Urban

Travel Forecasting Procedure in the EMME/2 System, *Transportation Planning Methods Applications, 5th National Conference,* I, Transportation Research Board, Washington, DC, pp. 11–17.

Mirchandani, P., and H. Soroush, 1987. Generalized Traffic Equilibrium with Probabilistic Travel Times and Perceptions, *Transportation Science,* 21, 133–152.

Murchland, J. D., 1970. Braess's Paradox of Traffic Flow, *Transportation Research,* 4, 391–394.

MVA, 1996. *TRIPS Manuals, Version 7,* MVA Systematica, Woking, Surrey, England.

Nagurney, A., 1993. *Network Economics: A Variational Inequality Approach,* Kluwer, Norwell, MA.

Nagurney, A., and D. Zhang, 1996. *Projected Dynamical Systems and Variational Inequalities with Applications,* Kluwer, Norwell, NA.

Newell, G. F., 1980. *Traffic Flow on Transportation Networks,* MIT Press, Cambridge, MA.

Oppenheim, N., 1995, *Urban Travel Demand Modeling,* Wiley, New York.

Ortúzar, J. de D., and L. G. Willumsen, 1990. *Modelling Transport,* Wiley, Chichester, West Surrey, England.

Patriksson, M., 1994. *The Traffic Assignment Problem, Models and Methods,* VSP, Utrecht, The Netherlands.

Poorzahedy, H., 1980. Efficient Algorithms for Solving Network Design Problems, Ph.D. thesis, Northwestern University, Evanston, IL.

Potts, R. B., and R. M. Oliver, 1972. *Flows in Transportation Networks,* Academic Press, San Diego, CA.

Powell, W. B., and Y. Sheffi, 1982. The Convergence of Equilibrium Algorithms with Predetermined Step Sizes, *Transportation Science,* 16, 45–55.

PTV, 1996. *VISION System Manual,* PTV System—Software and Consulting GmbH, Karlsruhe, Germany.

Ran, B., and D. Boyce, 1996. *Modeling Dynamic Transportation Networks,* Springer-Verlag, Berlin.

Ridley, T. M., 1968. An Investment Policy to Reduce the Travel Time in a Transportation Network, *Transportation Research,* 2, 409–424.

Rose, G., M. S. Daskin, and F. S. Koppelman, 1988. An Examination of Convergence Error in Equilibrium Traffic Assignment Models, *Transportation Research,* 22B, 261–274.

Rothengatter, W., 1979. Application of Optimal Network Selection to Problems of Design and Scheduling in Urban Transportation Networks, *Transportation Research,* 13B, 49–63.

Safwat, K. N. A., and T. L. Magnanti, 1988. A Combined Trip Generation, Trip Distribution, Modal Split, and Trip Assignment Model, *Transportation Science,* 18, 14–30.

Scott, A. J., 1969. The Optimal Network Problem: Some Computational Procedures, *Transportation Research,* 3, 201–210.

Shapiro, J. F., 1979. *Mathematical Programming: Structures and Algorithms,* Wiley, New York.

Sheffi, Y., 1985. *Urban Transportation Networks: Equilibrium with Mathematical Programming Methods,* Prentice Hall, Upper Saddle River, NJ.

Sheffi, Y., and W. B. Powell, 1982. An Algorithm for the Equilibrium Assignment Problem with Random Link Times, *Networks,* 12, 191–207.

Spiess, H., and M. Florian, 1989. Optimal Strategies: A New Assignment Model for Transit Networks, *Transportation Research,* 23B, 83–102.

Steenbrink, P. A., 1974. Transport Network Optimization in the Dutch Integral Transportation Study, *Transportation Research,* 8, 11–27.

Tatineni, M. R., M. R. Lupa, D. B. Englund, and D. E. Boyce, 1994. Transportation Policy Analysis Using a Combined Model of Travel Choice, *Transportation Research Record,* 1452, 10–17.

Tatineni, M., D. E. Boyce, and P. Mirchandani, 1996. *Solution Properties of Stochastic Route Choice Models,* Report to the National Institute of Statistical Sciences, Urban Transportation Center, University of Illinois, Chicago.

Theil, H., 1967. *Economics and Information Theory,* North-Holland, Amsterdam.

TModel, 1995. *TModel 2 User's Manual, Version 2.49,* TModel Corporation, Vashon, WA.

UAG, 1995. *URBAN/SYS Version 8.0, User Manuals for TRANPLAN,* Highway NIS and Transit NIS, The Urban Analysis Group, Danville, CA.

Wardrop, J. G., 1952. Some Theoretical Aspects of Road Traffic Research, *Proceedings of the Institution of Civil Engineers, Part II,* 1, 325–378.

Wilson, A. G., 1967. A Statistical Theory of Spatial Distribution Models, *Transportation Research,* 1, 253–269.

Wynter, L., 1996. *Solving the Asymmetric Traffic Assignment Problem with a Nonmonotone Operator,* Report PRISM 96/026, Département d'Informatique, Université de Versailles, Versailles, France.

Zhang, Y.-F., 1995. Parameter Estimation for Combined Models of Urban Travel Choices Consistent with Equilibrium Travel Costs, Ph.D. thesis, University of Illinois, Chicago.

CHAPTER 8

Planning Models for Freight Transportation

TEODOR GABRIEL CRAINIC and GILBERT LAPORTE

8.A. INTRODUCTION

Transportation is an important domain of human activity. It supports and makes possible most other social and economic activities and exchanges. Freight transportation, in particular, is one of today's most important activities, not only as measured by the yardstick of its own share of a nation's gross national product (GNP), but also by the increasing influence that the transportation and distribution of goods have on the performance of virtually all other economic sectors.

A few figures illustrate these assertions (Cordeau and Laporte, 1996). It has been estimated (Taff, 1978) that transportation accounts for approximately 10% of the U.S. GNP, and current figures could very well be significantly larger. In the United Kingdom, for example, transportation represents some 15% of national expenditures (Button, 1993). These figures are similar to those observed for Canada (some 16%; Zalatan, 1993) and France (around 9%; Quinet, 1990). Furthermore, transportation represents a significant part of the cost of a product. In Canada, for example, it has been estimated that this part may reach 13% for the primary industrial sector and 11% for the transformation and production industry (Owoc and Sargious, 1992).

Transportation is also a complex domain, with several players and levels of decision, where investments are capital intensive and usually require long implementation delays. Furthermore, freight transportation has to adapt to rapidly changing political, social, and economic conditions and trends. It is thus a domain where accurate and efficient methods and tools are required to assist and enhance the analysis of planning and decision-making processes.

The objective of this chapter is to identify some of the main issues in freight transportation planning and operations and to present appropriate operations research models, methods, and computer-based planning tools. We start this presentation by recalling the contemporary environment of the freight transportation sector, its main players, and a general decision-making framework.

As all other economic sectors, the freight transportation industry has to achieve high performance levels in terms of both economic efficiency and service quality: economic efficiency because a transportation firm has to make profits while competing in an increasingly open and competitive market (Ash, 1993, reports an average profit rate of only 2% for the Canadian motor carrier industry in 1992) where cost (or cost for a given quality level) is still the major decision factor in selecting a carrier or distribution firm; yet one also observes an increasing emphasis on the quality of the service offered. Indeed, the new paradigms of production and management, such as small or no inventories associated with just-in-time procurement, production, and distribution, quality control of the entire logistics chain driven by customer demand and requirements, and so on, impose high service standards on the transportation industry. This applies, in particular, to total delivery time (be there fast) and service reliability (be there within specified limits and be consistent in your performance).

The political evolution of the world also has an impact on the transportation sector. The emergence of free trade zones, in Europe and on the American continent in particular, has tremendous consequences for the evolution of freight transportation systems, not all of which are yet apparent or well understood. For example, open borders generally mean that firms are no longer under the obligation to maintain a major distribution center in each country. Then, distribution systems are reorganized, and this often results in fewer warehouses and transportation over longer distances (which still have to perform according to low cost–high service standards). A significant increase in road traffic is a normal consequence of this process, as may be observed in Europe. A study conducted for the European Parliament forecasts a 34% increase in land-based transport for the countries of the European Economic Community between 1988 and the year 2000 (Button, 1993).

Additional factors that impact on the organization, operation policies, and competitiveness conditions in the transportation industry are the internationalization of the economy and the opening of new markets due to political changes, mainly in central Europe and Asia, and the evolution of the regulatory environment. The first two imply larger economic spaces and transportation networks. Thus, from 1971 to 1988, the total volume of goods moved by ship has doubled while the total number of kilometers covered by air cargo has quadrupled (Button, 1993). Changes to the regulatory environment of transportation, particularly significant in North America and starting to gather momentum in Europe and elsewhere, also has a powerful impact on the operation and competitive environment of transportation firms. The deregulation drive of the 1980s has seen governments remove numerous rules and restrictions, especially with regard to the entry of new firms in the market and the fixing of tariffs and routes, resulting in a more competitive industry and in changes in the number and characteristics of transportation firms. At the same time, more stringent safety regulations have been imposed, resulting in more complex planning and operating procedures.

There are several different types of players in the transportation field, each with its own set of objectives and means. Drawing a complete picture is well beyond the scope of this chapter. Here, we only identify some of the most important players. Producers of goods require transportation services to move raw materials and intermediate products and to distribute final goods to meet demands. Hence they determine the demand for transportation and are often called *shippers.* (Other players, such as brokers, may also fall in this category.) Transportation is usually performed by *carriers,* such as railways, shipping lines, motor carriers, and so on. Thus one may describe an intermodal container service or a port facility as a carrier. Governments constitutes another important group of players. First, they still regulate several aspects of freight transportation (dangerous and obnoxious goods transportation, for example). They also provide a large part of the transportation infrastructure: roads and highways, and often a significant portion of the port, internal navigation, and rail facilities.

Transportation systems are rather complex organizations that involve a great deal of human and material resources and display intricate relationships and trade-offs among the various decisions and management policies affecting their components. It is then convenient to classify these policies according to the following three *planning levels:*

1. *Strategic* (long-term) planning at the firm level typically involves the highest level of management and requires large capital investments over long time horizons. Strategic decisions determine general development policies and broadly shape the operating strategies of the system. Prime examples of decisions at this planning level are the design of the physical network and its evolution (upgrading or resizing), the location of main facilities (rail yards, multimodal platforms, etc.), resource acquisition (motive power units, rolling stock, etc.), and the definition of broad service and tariff policies, among others. Strategic planning also takes place at the international, national, and regional levels, where the transportation networks or services of several carriers are considered simultaneously. State transportation departments, consultants, international shippers, and so on, engage in this type of activity.
2. *Tactical* (medium-term) planning aims to ensure, over a medium-term horizon, an efficient and rational allocation of existing resources to improve the performance of the whole system. At this level, data are aggregated, policies are somewhat abstracted, and decisions are sensitive only to broad variations in data and system parameters (such as the seasonal changes in traffic demand) without incorporating the day-to-day information. Tactical decisions need to be made primarily concern-

ing the design of the service network: route choice and type of service to operate, general operating rules for each terminal and work allocation among terminals, traffic routing using the available services and terminals, and repositioning of resources (e.g., empty vehicles) for use in the next planning period.

3. *Operational* (short-term) planning is performed by local management (yardmasters and dispatchers, for example) in a highly dynamic environment where the time factor plays an important role and detailed representations of vehicles, facilities, and activities are essential. Scheduling of services, maintenance activities, crews, and so on, routing and dispatching of vehicles and crews, and resource allocation are important operational decisions.

This classification highlights how the data flow among the decision-making levels and how policy guidelines are set. The strategic level sets the general policies and guidelines for the decisions taken at the tactical level, which determines goals, rules, and limits for the operational decision level regulating the transportation system. The data flow follows the reverse route, each level of planning supplying information essential for the decision making process at a higher level. This hierarchical relationship prevents the formulation of a unique model for the planning of freight transportation systems and calls for different model formulations addressing specific problems at specific levels of decision making.

The remainder of the chapter is dedicated to presentation of important issues in planning and managing freight transportation systems. The presentation is organized according to the three planning levels just described. In each case we identify the major players, describe the problem and the main related questions, and review how the questions have been addressed through the development of operations research models, methods, and computer systems.

8.B. STRATEGIC PLANNING MODELS

For a firm, strategic decisions determine general development policies and broadly shape the operating strategies of the system over relatively long time horizons. Several such decisions affect the design of the physical infrastructure network: where to locate facilities (loading and unloading terminals, consolidation centers, etc.) and of what type (capacity, for example), on what lines to add capacity or which ones to abandon, and so on. These issues, which may be collectively identified as *logistics system design*, are the subject of the first two sections where location and network design models are considered. In the third, the focus is broadened and we examine strategic planning issues and models aimed at the international, national, and regional levels, where the transportation networks or services of several carriers are considered simultaneously.

8.B.1. Location Models

Location problems involve the siting of one or several facilities, usually at vertices of a network, to facilitate the movement of goods or the provision of services along the network. The main location models are often classified as follows:

1. *Covering models.* Locate facilities at the vertices of a network so that the remaining vertices are covered by a facility (i.e., they lie within a given distance of a facility). The problem can be to minimize the cost of locating facilities, subject to a constraint stating that all remaining vertices are covered. If one operates within a fixed budget, an objective can be to maximize the demand covered by the facilities.
2. *Center models.* Locate p facilities at vertices of a network to minimize the maximal distance between a vertex and a facility.
3. *Median models.* Locate p facilities at vertices on the network and allocate demands to these facilities to minimize the total weighted distance between facilities and demand points. If facilities are uncapacitated and p is fixed, one obtains the *p-median problem.* In such a case, each vertex is associated to its closest facility. If p is a variable and facilities are uncapacitated, this defines the *uncapacitated plant location problem* (UPLP). If p is a decision variable and facilities are capacitated, one obtains the *capacitated plant location problem* (CPLP).

Covering problems are typically associated with the location of public facilities such as health clinics, post offices, libraries, and schools. Center problems often arise in the location of emergency facilities such as fire or ambulance stations. Median problems are directly relevant to freight distribution. Here we describe the CPLP. For a review of these and other models, see Daskin (1995) or Labbé et al. (1995).

The CPLP can be formulated as follows. Assume that there are n points in the plane called *vertices* and define

f_j = cost of locating a facility at vertex v_j

d_i = demand at vertex v_i

c_{ij} = travel cost per unit of demand between vertices v_i and v_j

u_j = capacity of a facility located at vertex v_j

y_j = a 0, 1 variable equal to 1 if and only if a facility is located at vertex v_j

x_{ij} = fraction of the demand of vertex v_i served by a facility located at a vertex v_j

The model is then

$$\text{minimize} \sum_j f_i y_j + \sum_i \sum_j d_i c_{ij} x_{ij} \tag{8.1}$$

subject to

$$x_{ij} \leq y_j \qquad \text{for all } i \text{ and } j \tag{8.2}$$

$$\sum_j x_{ij} = 1 \qquad \text{for all } i \tag{8.3}$$

$$\sum_i d_i x_{ij} \leq u_j y_j \qquad \text{for all } j \tag{8.4}$$

$$y_j = 0 \text{ or } 1 \qquad \text{for all } i \tag{8.5}$$

$$x_{ij} \geq 0 \qquad \text{for all } i \text{ and } j \tag{8.6}$$

In this model the objective function represents the sum of fixed facility costs and transportation costs. It is assumed that these costs are scaled over the same planning horizon. Constraints (8.2) express the condition that vertex v_i can be served by vertex v_j only if a facility is located at v_j. Constraints (8.3) state that the entire demand of each vertex must be allocated to facilities. Constraint (8.4) ensures that the capacity of a facility is never exceeded by its assigned demand. In Chapter 5 (Section 5.C) a similar model is presented for locating terminal facilities in solid waste management problems.

The standard methodology for this model is to relax capacity constraints (8.4) in a Lagrangian fashion to obtain an objective of the form

$$\text{minimize} \sum_j \alpha_j y_j + \sum_i \sum_j \beta_{ij} x_{ij} \tag{8.7}$$

where $\alpha_j = f_j - \mu_j u_j$, $\beta_{ij} = d_i c_{ij} + \mu_j d_j$, and μ_j are positive Lagrangian multipliers. The resulting problem is an uncapacitated plant location problem that can be solved by the DUALOC procedure devised by Erlenkotter (1978). For a detailed description of this approach, the reader is again referred to Daskin (1995).

More complex location problems arise in production and transportation planning. As an example, Crainic et al. (1989) present a multicommodity location allocation with balancing requirements model. The formulation aims to determine the best logistics structure of the land distribution and transportation component of an international container shipping company, including the selection of inland depots, the assignment of customers to depots for each container type and direction of operations (allocation of empty containers from depots to customers and return of empties from customers to depots), and the determination of the main repositioning flows of empty containers to counter the regional differences between supplies and demands. The problem presents characteristics similar to a multicommodity simple plant location problem (the selection of the depots as discrete choice variables) with a complete multicommodity network flow structure (the allocation and repositioning of containers as continuous decision variables). Dual-ascent methods, of DUALOC type, have been proposed (Delorme and Crainic, 1993), as well as branch-and-bound (Gendron and Crainic, 1995) and tabu search procedures (Crainic et al., 1993b).

8.B.2. Network Design Models

Network design problems are a generalization of location formulations. They are defined on graphs containing *nodes* or *vertices,* connected by *links.* Typically, links are directed and are represented by *arcs* in a network. When it is not necessary to specify a direction, they are represented by *edges.* Some of the vertices represent *origins* of some transportation demand for one or several commodities or products, while others (possibly the same) stand for the *destinations* of this traffic. Links may have various characteristics, such as length, capacity, and cost. In particular, *fixed costs* may be associated with some or all links, signaling that as soon as one chooses to use that particular arc, one has to incur the fixed cost, in excess of the utilization cost, which is in most cases related to the volume of traffic on the link. Note that when the fixed costs are associated to nodes, one obtains the formulations of a location problem. Such representations are generally used to model the cost of constructing new facilities, or of offering new transport services, or of adding capacity to existing facilities. In network design problems, the aim is to choose links in a network, along with capacities, eventually, to enable goods to flow between their origin and destination at the lowest possible system cost (i.e., the total fixed cost of selecting the links plus the total variables cost of using the network).

The simplest version of this problem is probably the *shortest spanning tree problem* (SSTP), which consists of determining a minimal-length tree linking all vertices of an undirected graph $G = (V,E)$, where V is a vertex set and E is an edge set. To formulate the SSTP, define c_{ij} to be the length of edge (v_i,v_j) and x_{ij} a 0–1 variable equal to 1 if and only if edge (v_i,v_j) belongs to the SSTP. The problem is then

$$\text{minimize} \sum_{i<j} c_{ij} x_{ij} \tag{8.8}$$

subject to

$$\sum_{\substack{v_i \in S, v_j \in V \setminus S \\ \text{or} \\ v_i \in V \setminus S, v_j \in S}} x_{ij} \geq 1 \qquad S \subset V, S \neq \emptyset \tag{8.9}$$

$$x_{ij} = 0, 1 \qquad \text{for all } (v_i, v_j) \in E \tag{8.10}$$

Here the objective function represents the total length of the tree, while constraints (8.9) state that every subset S of nodes must be linked to its complement. Since it is uneconomical to overconnect the graph, there will be no cycles and the solution will be a tree.

The SSTP can be solved to optimality by applying one of several simple greedy algorithms (see Ahuja et al., 1993). Here we describe Kruskal's algorithm (Kruskal, 1956).

Step 1. Sort the edges in nondecreasing order of their lengths. Let T be the set of edges included in the tree. Set $T := \emptyset$.

Step 2. Consider the next edge in the list. If it does not form a cycle with the edges of T, include it in T. If $|T| = |V| - 1$, stop. Otherwise, repeat this step.

More complex network design problems arise in transportation planning and operations (see also Section 8.D). In their most general formulation (see Magnanti and Wong, 1984; Minoux, 1986; Ahuja et al., 1993; and Magnanti and Wolsey, 1995), the models contain the same two types of variables defined in the previous formulations: y_{ij}: 0,1 variables modeling discrete choice design decisions; for a link (v_i,v_j), y_{ij} will equal to 1 if and only if link (v_i,v_j) is "opened"; and x_{ij}^p: continuous flow decision variables indicating the amount of flow of commodity p using link (v_i,v_j). The model then becomes

$$\text{minimize} \sum_{ij} f_{ji} y_{ji} = \sum_{ij} \sum_{p} c_{ij}^p x_{ij}^p \tag{8.11}$$

subject to

$$\sum_{j \in N} x_{ji}^p - \sum_{j \in N} x_{ij}^p = d_i^p \qquad \text{for all } p \text{ and } i \tag{8.12}$$

$$\sum_{p} x_{ij}^p \leq u_{ij} y_{ij} \qquad \text{for all links } (v_i,v_j) \tag{8.13}$$

$$(y_{ij}, x_{ij}^p) \in S \qquad \text{for all links } (v_i,v_j) \text{ and all } p \tag{8.14}$$

$$y_{ij} = 0 \text{ or } 1 \qquad \text{for all links } (v_i,v_j) \tag{8.15}$$

$$x_{ij}^p \geq 0 \qquad \text{for all links } (v_i,v_j) \text{ and all } p \tag{8.16}$$

where

f_{ij} = fixed cost of "opening" link (v_i,v_j)

c^p_{ij} = travel cost per unit of flow of product p on link (v_i,v_j)

w^p = total demand of product p

d^p_i = demand at vertex v_i

$$= \begin{cases} -w^p & \text{if vertex } v_i \text{ is the origin of commodity } p \\ w^p & \text{if } v_i \text{ is the destination of commodity } p \\ 0 & \text{otherwise} \end{cases} \tag{8.17}$$

u_{ij} = capacity of link (v_i,v_j)

This is the linear-cost, multicommodity version of the formulation. There exist important applications (e.g., variants of the service network design problem; see Section 8.C.1) that require nonlinear formulations, but this subject is beyond the scope of this chapter. Similarly, we focus on multicommodity formulations, since they represent the vast majority of applications in transportation and in other areas such as telecommunications and production.

In the network design formulation, the objective function (8.11) measures the total cost of the system. An interesting point of view is to consider this objective as also capturing the trade-offs between the costs of offering transportation infrastructure or services and those of operating the system. Equations (8.12), together with the demand definition (8.17), express the usual flow conservation and demand satisfaction restrictions. (Here, each commodity is associated to one origin–destination pair, but this assumption may easily be relaxed.) Constraints (8.13), often identified as *bundle* or *forcing* constraints, state that the total flow on link (v_i,v_j) cannot exceed its capacity u_{ij} if it is chosen in the design of the network (i.e., if $y_{ij} = 1$) and must be 0 if (v_i,v_j) is not part of the selected network (i.e., if $y_{ij} = 0$). Relations (8.15) and 8.16) specify the range of admissible values for each set of decision variables.

Equations (8.14) capture additional constraints related to the design of the network or relationships among the flow variables. Together, they may be used to model a wide variety of practical situations, and this is what makes network design problems so interesting. For example, the set S may represent topological restrictions imposed on the design of the network, such as precedence constraints [choose link (v_i,v_j) only if link (v_p,v_q) is chosen] or multiple-choice constraints (select at most or exactly a given number of arcs out of a specified subset). An important type of additional constraint reflects the usually limited nature of available resources:

$$\sum_{v_i,v_j} f_{ji} y_{ji} \leq B \tag{8.18}$$

These *budget* constraints illustrate a relatively general class of restrictions imposed on resources shared by several (or all) links. Note that quite often, budget constraints replace the fixed-cost term in the objective function [equation (8.11)]. *Partial capacity* constraints also belong to this group:

$$x_{ij}^p \leq b_{ij}^p \qquad \text{for all links } (v_i,v_j) \text{ and all } p \tag{8.19}$$

and reflect restrictions imposed on the use of some facilities by individual commodities. Such conditions may be used to model, for example, the quantity of some hazardous goods moved by a train or a ship.

Much effort has been dedicated to uncapacitated versions of the problem and significant results have been obtained. The references already indicated provide an extensive review of the most significant results and real-life applications. In particular, Balakrishnan et al. (1989) present a dual-ascent procedure which, in conjunction with an add-drop heuristic, is capable of efficiently solving realistically sized instances of less-than-truckload consolidation problems (see Section 8.C.1). Less effort has been directed toward capacitated problems, which are more difficult to solve and pose considerable algorithmic challenges. See Gendron and Crainic (1994, 1996) and Crainic et al. (1996) for a review of research in this area and for a presentation of bounding and tabu search procedures associated with this formulation.

8.B.3. Regional Multimodal Planning

Strategic planning activities must also be able to perform on a wider—regional, national, or even international—scale. The issues considered at this planning level usually concern the entire transportation system or a significant part of it, the products that use it, and the interaction between passenger travel and freight flows.

The main questions addressed relate to the evolution of a given transportation system and its response to various modifications in its environment. Here we describe three such issues.

1. What would be the impact on the system's performance of infrastructure modifications? Several types of modifications may affect the utilization and performance of a transportation system. One may build new facilities or improve existing ones through modernization or capacity expansion, for example. One may also deconstruct the transportation network: the abandonment of underused, unprofitable rail lines is a typical example. These questions are often part of cost–benefit analyzes and comparative studies of investment alternatives, especially when the available monetary resources are very limited, and are asked by regional or national planning authorities as well as by international financial institutions such as the World Bank.

2. How would the evolution of demand affect utilization of the system? Several changes may affect demand.
   a. *Volume.* The quantities that have to be moved for each product, or product group, may increase or decrease in future years.
   b. *Spatial distribution.* New communities appear and grow, new economic areas are developed, resources dry up in certain regions, and the economic profile of a country or region evolves; these are only a few reasons why the spatial distribution of trade and freight flows may change in the future.
   c. *Composition.* The relative importance of each product group in the trade exchanges between two zones varies according to the economic and social evolution (e.g., from rural to urban, from exporter of raw materials to a high-technology industrial zone, etc.) of each of the two zones.

   Generally, one notices that several factors change simultaneously and one aims to evaluate the performance of the system under forecast demand conditions.
3. What would be the impact of government or industry policies? Government policies may significantly affect the distribution of traffic and thus the utilization and performance of a transportation system. Common examples include energy pricing and taxation, mode imposition (in some energy-importing countries, legislation specifies that bulk goods have to be transported by energy-efficient modes such as rail), infrastructure utilization pricing (such as the new trends in pricing highway utilization according to weight and distance), and so on. Decisions made by firms also affect the performance of the transportation systems: rail line abandonments, which may change the ability of the transportation system to serve all regions of a country; mergers of carriers (the current trend of mergers and partnerships among carriers of various modes involved in container transportation is profoundly changing the rules by which intermodal transportation is operated); the introduction of new services (e.g., transcontinental double stack container trains); and so on.

Planning and regulatory agencies at various levels of government are particularly interested in such issues (in the United States, in particular, the new intermodal legislation seriously challenges the state and regional transportation departments in this respect), as are international agencies involved in financing major projects in developing countries. Private firms are also interested in these questions: for example, companies involved in the financing of transportation infrastructures, or firms that plan and operate the distribution of goods using several transportation modes. In all cases, the focus is on the specific representation of several transportation modes, the corresponding intermodal transfer operations, the various criteria used to determine the movement of freight, and the associated estimation of the traffic distribution over the transportation system considered.

The prediction of multicommodity freight flows over a multimode network is therefore an important component of transportation science and has attracted significant interest in recent years. One notes, however, that perhaps due to the inherent difficulties and complexities of such problems, the study of freight flows at the national or regional level has not yet achieved full maturity, in contrast to passenger transportation, where the prediction of car and transit flows over multimode networks has been studied extensively and several of the research results have been transferred to practice (see, e.g., Florian, 1984, 1986; Sheffi, 1985; or Florian and Hearn, 1995).

A class of models well studied in the past for the prediction of interregional commodity flows is the *spatial price equilibrium model* and its variants. This class of models determines simultaneously the flows between *producing* and *consuming* regions, as well as the *selling* and *buying* prices. The transportation network is usually modeled in a simplistic way (bipartite networks), and these models rely to a large extent on the *supply* and *demand* functions of the producers and consumers, respectively. The calibration of these functions is essential to the application of these models, and the transportation costs are unit costs or may be functions of the flow on the network. There have so far been few multicommodity applications of this class of models, with the majority of applications having been carried out in the agricultural and energy sectors in an international or interregional setting. See the review by Florian and Hearn (1995) or the book by Nagurney (1993).

The class of *network models* is generally considered to be more appropriate for the type of planning issues considered here. These formulations enable the prediction of multicommodity flows over a multimode network, where the physical network is modeled at a level of detail appropriate for a nation or a large region and represents the physical facilities with relatively little abstraction. The demand for transportation services is exogenous and may originate from an input–output model if one is available, or from other sources, such as observed demand or scaling of observed past demand. The choice of mode, or subsets of modes, used is exogenous and intermodal shipments are permitted. In this sense these models may be integrated with econometric demand models as well. The emphasis is on the network representation and on the proper representation of congestion effects in a static model to be used for comparative studies or for discrete-time multiperiod analyses.

Several studies in the 1970s used rather simple network representations. Guélat et al. (1990) and Crainic et al. (1990a) review and discuss these efforts. Several studies were also aimed at extending spatial equilibrium models to include more refined network representations and to consider congestion effects and shipper–carrier interactions (see, e.g., Friesz and Harker, 1985; Harker and Friesz, 1986a,b). This line of research has not, however, yet yielded practical planning models and tools. This is not the case of the model presented by Friesz et al. (1986). This is a sequential model that uses two network representations: an aggregate, user-perceived network which serves to determine, by using traffic equilibrium principles, the carriers chosen by the shippers, and detailed separate networks for each carrier, where commodities are transported at least total cost. This approach has proven quite successful for studying the logistics of products (e.g., coal), where a limited number of "shippers" (e.g., the electric utilities in the United States and their suppliers in the exporting countries) strongly determine the behavior of the system.

The modeling framework we present is based on the work of Guélat et al. (1990). The formulation does not consider shippers and carriers as distinct actors in the decisions made in shipping freight. The level of aggregation appropriate for strategic planning of freight flows results in origins and destinations that correspond to relatively large geographical areas and leads to the specification of supplies and demands representing, for each of the products considered, the total volumes generated by all the individual shippers. Furthermore, demand for strategic freight analyses are often determined from data sources (national freight flow statistics, economic input/output models) which enable the identification of the mode used but do not contain information on individual shippers. It is thus assumed that shippers' behavior is reflected in the origin-to-destination product matrices and in the specification of the corresponding mode choice, as indicated in the following.

In this modeling framework a mode is a means of transportation having its own characteristics, such as vehicle type and capacity, as well as specific cost measures. Depending on the scope and level of detail of the contemplated strategic study, a mode may represent a carrier or part of its network representing a particular transportation service, an aggregation of several carrier networks, or specific transportation infrastructures such as highway networks or ports.

The base network is the network that consists of the nodes, links, and modes that represent all possible physical movements on the available infrastructure. To capture the modal characteristics of transportation, a link a is defined as a triplet (i,j,m), where i is the origin node, j is the destination node, and m is the mode allowed on the arc. Parallel links are used to represent situations where more than one mode is available for transporting goods between two adjacent nodes. This network representation is compact and enables easy identification of the flow of goods by mode, as well as cost and delay functions by product and mode. Furthermore, the network model resembles the physical network, since, for example, the rail and road infrastructures are physically different. Also, when on a physical link there are two different types of services, such as diesel and electric train services on rail lines, a separate link may be assigned to each service to capture the fact that they have different cost and delay functions.

Once the network representation is chosen, it is necessary to model intermodal shipments, and indeed, to allow for mode transfers at certain nodes of the network and to compete the associate costs and delays. Intermodal transfers t at a node of the network are modeled as link to link, hence mode to mode, allowed movements and are represented as (i,m_1,j,m_2,k), where (i,m_i,j)

stands for the incoming link, while (j,m_2,k) indicates the outgoing link of the transfer.

A path in this network consists of a sequence of directed links of a mode, a possible transfer to another mode, a sequence of directed links of the second mode, and so on. Thus a mode change is possible only at a transfer node. This representation also allows for the restriction of flows of certain commodities to subsets of modes (e.g., iron ore may be shipped only by rail and ship) to capture the mode restrictions that occur in the operation of freight networks and transshipment facilities.

A product is any commodity (collection of similar products), good, or passenger that generates a link flow. Each product p transported over the multimode network is shipped from some origins to some destinations of the network. The demand for each product for all origin–destination (OD) pairs is exogenous and is specified by a set of OD matrices. The mode choice for each product is also exogenous and is indicated by defining for each of these OD matrices a subset of modes allowed for transporting the corresponding demand. For example, one may indicate that the traffic out of certain regions has to use rail, while in other regions there is a choice between rail and barges. Let $g^{m(p)}$ be a demand matrix associated with product p, where $m(p)$ is the subset of modes that may be used to move this particular part of product p.

In the context of strategic planning of freight flows on a national or regional scale, the most efficient use of the transportation infrastructure is to carry the freight at the least total cost. Even though it is reasonable to assume that even in countries where a central authority controls and regulates the shipment of goods, a variety of circumstances in fact prevent the precise achievement of the goal of shipping at least cost, the model we present is based on the objective of minimizing total costs. The notion of cost is central to the model and is interpreted in the most general way, in the sense that it may have different components, such as monetary cost, delay (in terminals and on the lines of the network), energy consumption, noise and pollution level, risk (in case of incidents and accidents involving hazardous goods), and so on.

The flows on product p on the multimode network are the decision variables of the model. Flows on links a are denoted by v_a^p and flows on transfers t are denoted by v_t^p. Cost functions, which depend on the volume of goods, are associated to the links and transfers of the network. For product p, the average cost functions $s_a^p(v)$, on links, and $s_t^p(v)$, on transfers, correspond to a given flow vector v. Then the total cost of the flow on arc a for the product p is the product $s_a^p(v)v_a^p$; and the total cost of the flow on transfer t is $s_t^p(v)v_t^p$. The total cost of flows for all products over the multimode network is the function F that is to be minimized over the set of flows that satisfy the conservation of the flow and nonnegativity constraints:

$$F = \sum_{p \in P} \left(\sum_{a \in A} s_a^p(v)v_a^p + \sum_{t \in T} s_t^p(v)v_t^p \right) \tag{8.20}$$

To write these constraints, let $K_{od}^{m(p)}$ denote the set of paths that for product p lead from origin o to destination d using only modes in $m(p)$. The flow conservation equations are then

$$\sum_{k \in K_{od}^{m(p)}} h_k = g_{od}^{m(p)} \qquad \text{for all } o,d,m(p),p \tag{8.21}$$

where h_k is the flow on path k. These constraints specify that the total flow moved over all the paths that may be used to transport product p must be equal to the demand for that product. The nonnegativity constraints are

$$h_k \geq 0 \qquad \text{for all } k \in K_{od}^{m(p)}, o,d,m(p),p \tag{8.22}$$

Equations (8.21) stand for the same type of constraints as equations (8.12) in the network design formulations but are written in terms of path flow variables. The relation between arc flows and path flows is

$$v_a^p = \sum_{k \in K^p} \delta_{ak} h_k \qquad \text{for all } a,p \tag{8.23}$$

where K^p is the set of all paths that may be used by product p, and

$$\delta_{ak} = \begin{cases} 1 & \text{if } a \in k \\ 0 & \text{otherwise} \end{cases}$$

is the indicator function which identifies the arcs of a particular path. Similarly, the flows on transfers are

$$v_t^p = \sum_{k \in K^p} \delta_{tk} h_k \qquad \text{for all } t,p \tag{8.24}$$

where

$$\delta_{ak} = \begin{cases} 1 & \text{if } t \in k \\ 0 & \text{otherwise} \end{cases}$$

Transfer t belongs to path k if the two arcs that define the transfer belong to k. Then the system optimal multiproduct, multimode assignment model consists of minimizing (8.20), subject to restrictions (8.21) and (8.22) with the definitional constraints (8.23) and (8.24). The algorithm developed for this problem exploits the natural decomposition by product and results in a Gauss–Seidel-like procedure which allows the solution of large problems in reasonable computational times (see Guélat et al., 1990).

This network model allows for a detailed representation of the transportation infrastructure, facilities, and services as well as the simultaneous assignment of multiple products on multiple modes. It captures the competition of products for the service capacity available, a feature of particular relevance when alternative scenarios of network capacity expansion are considered. On the other hand, the model is sufficiently flexible to represent the transport infrastructure of one carrier only.

The model, embedded in the STAN interactive-graphic system (Crainic et al., 1990), is used by several agencies and organizations in a number of countries and has been applied successfully in practice. Crainic et al. (1990) present the application of this methodology to the study of freight rail transportation, while several other applications are discussed in Guélat et al. (1990), Crainic et al. (1990), and others.

8.C. TACTICAL PLANNING PROBLEM

When examining freight transportation, we distinguish between *producers* of goods who own or operate their own transportation fleet, and *carriers,* who perform transportation services for various shippers. From a planning point of view, a more interesting classification differentiates between transportation operations that are concerned primarily with long-distance movements of goods, such as rail transportation and less-than-truckload (LTL) trucking and those that perform several pickup and delivery operations, mainly by truck, over relatively short distances. The first case is often referred to as the *service network design* problem, while the second types of operations are usually identified as *vehicle routing problems.*

In all cases we are considering transportation systems where one vehicle (e.g., truck, railway wagon, ship) or convoy (e.g., rail, truck, or barge "trains") may serve to move freight of different customers with possibly different initial origins and final destinations. *Consolidation*-type operations are thus central to such systems and one of the main issues in tactical model development. This is in contrast to the "door-to-door" transportation operations performed, for example, by truckload motor carriers or by intermodal container transportation firms, and addressed in the next section.

When demands of several customers are served simultaneously by using the same "vehicle," one cannot tailor a service for each customer individually. Carriers have to establish regular services (e.g., a container ship from Seattle to Singapore) and adjust their characteristics (route, intermediary stops, frequency, vehicle and convoy type, capacity, speed, etc.) to satisfy the expectations of the largest number of customers possible. Externally, the carrier then proposes a series of services (or *routes* grouped into a *network*), each with its operational characteristics. Internally, the carrier builds a series of rules and policies that affect the entire system and are often collected in an *operational* (or *load* or *transportation*) plan. The aim is to ensure that the proposed services are performed as stated (or as close as possible) while operating in a rational and efficient way.

Customers' expectations have traditionally been expressed in terms of "going there" at the lowest cost possible. This, combined with the usual cost consciousness of any firm, has implied that the primary objective of a freight carrier was, and still is for many carriers, to operate at the lowest possible cost. Increasingly, however, customers not only expect low tariffs but also require a high-quality service, mostly in terms of speed, flexibility, and reliability. The significant increase in the market share achieved by motor carriers, mainly at the expense of railway transportation, is due to a large extent to this phenomenon. Consequently, tactical planning not only aims at an adequate allocation and utilization of existing resources but also strives to achieve the best trade-off between operating costs and service performance.

Tactical planning thus appears as a vital link in the planning process of a freight transportation carrier. Its output, the *transportation plan,* is used to determine the day-to-day policies that guide the operations of the system and is also a privileged evaluation tool for "what-if" questions raised during strategic planning. In the following, we examine models and methods that may be used to build the transportation plan in each of the two cases identified at the beginning of the section.

8.C.1. Service Network Design for Intermodal Transportation

In intercity freight transportation, service design is particularly relevant to firms and organizations which both supply or regulate transportation services and control, at least partially, the routing of goods through the service network. The presence of terminals where cargo and vehicles are consolidated, grouped, or simply moved from one service to another further characterizes this type of transportation. Main examples of such transportation systems are:

1. Transportation by rail where various train services (e.g., normal, rapid, direct, unit, etc.) correspond to various "modes."
2. Less-than-truckload trucking, eventually incorporating multitrailer assemblies and use of rail transportation.
3. Container transportation through a combination of air, sea, road, and rail modes.
4. Freight transportation in developing countries where a central authority more or less controls a large part of the transportation system.

The underlying structure of any large freight transportation system consists of a rather complex network of terminals connected by physical links (e.g., rail tracks) or conceptual links (e.g., sea or truck lines). On this network, freight demand is specified by commodity class according to its origin and destination, in addition to physical and service characteristics, and is moved by carrier services performed by a large number of vehicles (e.g., railcars, trailers, etc.). Vehicles move, usually on specified routes and sometimes following a given schedule, either individually or grouped in convoys such as trains or multitrailer assemblies. Convoys are formed and dismantled in terminals. Also in terminals, freight may be consolidated and loaded in and unloaded from vehicles, vehicles may be changed from one convoy to another, and so on. Consequently, terminals come in several types and sizes. For railways, for example, one identifies large and small classification *yards* where railcars are consolidated into blocks and trains are formed, pickup and delivery stations, junction points, and so on. Similarly, an LTL network may encompass *end-of-line* terminals, where the local traffic is delivered (by smaller pickup trucks; see Section 8.C.2) and consolidated into larger shipments while loads from other parts of the network are unloaded and moved into smaller delivery trucks (see Section 8.C.2), and *breakbulk* terminals, where traffic from many end-of-line terminals is unloaded, sorted, and consolidated for the next portion of the journey. (Railyards and breakbulks may be classified further according to their importance and role: regional, national, specialized services, etc.) The traffic on the links of an intermodal transportation network thus represents vehicle and, eventually, convoy movements, while operations are performed at terminal nodes on traffic, vehicles, and convoys.

To clarify these notions further, consider the case of railway transportation. Here everything begins when an order for a number of empty vehicles is used by a customer or, alternatively, when freight is brought into the loading facility following a pickup operation. At the appropriate yard, railcars are selected, inspected, and then delivered to the loading point. Once loaded, cars are moved to the origin yard (possibly the same), where they are sorted, or *classified,* and assembled into *blocks.* A block is a group of cars, with possibly different final destinations, considered arbitrarily as a single unit for handling purposes from the yard where it is made up, to its destination yard, where its component cars are to be resorted. Rail companies use blocks as a means to take advantage of some of the economies of scale related to full trainloads and to the handling of longer car strings in yards.

The block is eventually put on a *train,* and this signals the beginning of the journey. During the long-haul part of this journey, the train may overtake other trains or be overtaken by trains with different speeds and priorities. When the train travels on single-track lines, it may also meet trains traveling in the opposite direction. Then the train with the lowest priority has to give way and wait on a side line for the train with the higher priority to pass by. At the yards where the train stops, cars and engines are regularly inspected. Also, blocks of cars may be *transferred* (i.e., taken off one train and put into another). When a block finally arrives at destination, it is taken off the train, its cars are sorted, and those cars that have reached their final destination are directed to the unloading station. Once empty, the cars are prepared for a new assignment, which may be either a loaded trip or an empty movement.

One source of complication in rail freight transportation is the complex nature of yard activities, in particular its main operations: the classification of cars and the composition of trains. The modeling of yard operations and of their interactions with the remainder of the system is a critical component of any comprehensive rail model. It is interesting to note that, traditionally, in most rail systems cars spend most of their lifetime in yards: being loaded and unloaded, being classified, waiting for an operation to be performed or for a train to come, or simply sitting idle on a side track. Also of interest is the fact that most rail companies have dedicated yards to intermodal services in an attempt to cut on the delays associated with yard operations.

Less-than-truckload motor carrier transportation follows the same basic operational structure but on a simpler scale and with significantly more flexibility due to the fundamental difference in infrastructure: while rail transportation is "captive" of the rail tracks, trucks may use any of the existing links of the road and highway network as long as they comply with the weight regulations. Furthermore, a truck is only formed of a tractor and one or several trailers (when more than one trailer is used, these are smaller and are called "pups"). Consequently, the terminal operations are also significantly simpler: Only freight is handled to consolidate outbound movements. Transfer operations may still take place, however, since trailers may be exchanged between convoys. Actually, this operation is performed quite regularly since not only does it not require any specialized equipment (most switches take place at truck stops along highways) but it also facilitates the building of efficient driver schedules.

To clarify the preceding description further and to illustrate the types of decisions and trade-offs characteristic of tactical planning, we refer to Figure 8.1, which displays in a very simplified form part of an intermodal service network (rail or LTL, for example). Five terminals make up this network and

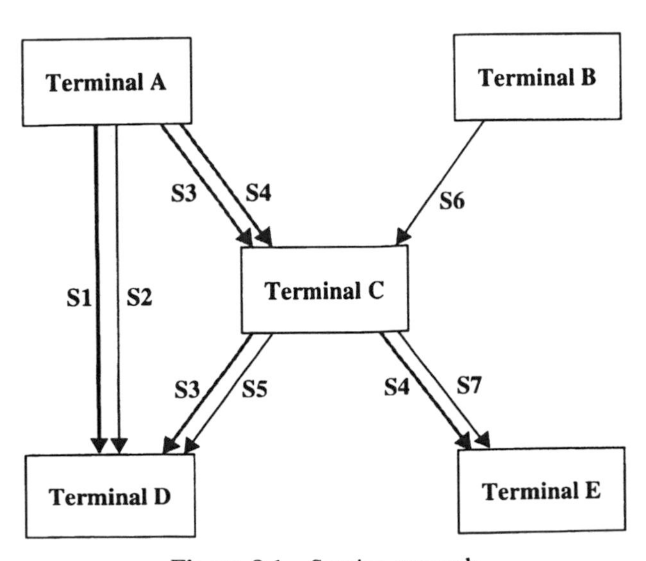

Figure 8.1 Service network.

seven *services* can be offered: one express direct service (S1), four normal direct services (S2, S5, S6, and S7) and two services with intermediary stops (S3 from terminal A to D with a stop at terminal C and S4 from A to E with a stop at C). Not indicated on the figure is the physical route followed by each service. Thus it may happen that services S1, S2, and S3 follow exactly the same route, their differences being only at the level of their respective speeds (and eventually other characteristics, such as capacity, preferred traffic, and priority) and intermediary stops. Let us examine the transportation options for a shipment of a given commodity from terminal A to terminal D. The shipment is classified at terminal A and it may be:

1. Consolidated with other shipments going directly to D and put on a direct service, S1 or S2, according to its characteristics.
2. Consolidated into a load for D but put into a vehicle on service S3, which stops at terminal C to drop and pick up traffic.
3. Consolidated into a load for terminal C, put on service S4, unloaded at terminal C, reclassified and consolidated together with traffic originating in B and C into a load for terminal D, which will be put on service S5.

Which alternative is "best"? Each has its own cost and delay characteristics, which are a direct consequence of the service characteristics of each terminal and service. Thus, if the direct services from A to D are offered only rarely (due to generally low level of traffic demand, for example), it may be more efficient to consolidate the A-to-D traffic together with the A-to-C and A-to-D volumes and ship through terminal C. This would probably result in a higher utilization of the equipment and in a decrease of the waiting time at the original terminal, hence in more rapid service for the customer. However, the same decision would also result in additional unloading, consolidation, and loading operations, creating additional delays and a higher congestion level at terminal C, as well as a decrease in the total reliability of the shipment. On the other hand, to increase the frequency of a direct service from A to D would imply faster and more reliable service for the corresponding traffic as well as a decrease in the level of congestion at terminal C, but would also require additional resources, which increases direct costs. Thus, to select the "best" solution for the customer and the company, one has to consider simultaneously the routing of all traffic demands, as well as the costs and service characteristics of each terminal and line operation.

More formally, the main decisions made at the tactical level concern the following issues:

1. *Service network design.* Selection of the routes (origin and destination terminal, physical route, and intermediate stops) on which services will be offered and determination of the characteristics of each service, particularly their *frequency.*

2. *Traffic distribution.* Routing specification for the traffic of each origin–destination pair (usually according to the specific commodity): services used, terminals passes through, and operations performed at the terminals.
3. *Terminal policies.* General rules specifying for each terminal the type of consolidation to perform (for a rail application, these rules would specify the blocks into which cars should be classified, the trains that are to be formed and the blocks that should be put on each train, etc.). An efficient allocation of work among terminals is an important policy objective.
4. *Empty balancing.* How to reposition empty vehicles to meet the forecast needs of the next planning period.
5. *Crew and motive power scheduling.* How to allocate and reposition in preparation for the next planning period the resources required by the selected transportation plan.

These problems and decisions have network-wide impacts and are strongly and complexly interconnected in both their economic aspects and their space–time dimensions. Thus the strategies developed for different planning problems interact and mutually influence each other. Furthermore, trade-offs have to be made between operating costs resulting from a given policy and service quality, as measured, in most cases, by delays incurred by freight and rolling stock, or with respect to predefined performance targets. These trade-offs are important when a single problem and decision are considered and even more so when the relationships among decisions for different problems are contemplated. Therefore, decision should be made globally, network-wide, in an integrated manner (Crainic and Roy, 1988).

Several efforts have been directed toward the formulation of tactical planning models. Network models that take advantage of the structure of the system and integrate policies affecting several terminal and line operations are the most widely developed. These formulations may be classified into two main groups: network simulation and optimization models.

Rail companies have used simulation models for quite a long time (see Assad 1980, and references therein). Models of this type simulate the movements of trains and cars through the rail network, given a set of operating policies for the yards and lines of the system and a set of train schedules. The user has to input, among other data, the classification and train formation policies at each yard, a complete set of train schedules and traffic demand information. The results are detailed cost information, an evaluation of the occupancy of each facility in the network (yards and lines) and sometimes an estimation of transit traffic times. Based on these results, the user may evaluate a given operating policy in detail. The main limitation of this approach is that simulation models are not able to generate new operating strategies that would incorporate a network-wide analysis of their impact and the evaluation of a number of apparently conflicting objectives. Also, simulation models usually require prohibitive data input and running times, which makes their repeated use with different sets of operating policies impractical.

Network optimization models, on the other hand, are less detailed but offer the advantage of fast generation, evaluation, and selection of integrated, network-wide operating strategies with respect to some objective function, usually involving both operating costs and service criteria. See Crainic (1988) for a comprehensive review of network, yard, and line models in rail tactical planning, and Delorme et al. (1988) for a similar review of models and methods for LTL transportation. Interesting recent contributions to the field are found in Keaton (1989, 1991, 1992), Haghani (1989), Roy and Delorme (1989), Jovanović and Harker (1991), Powell and Sheffi (1989), Braklow et al. (1992), Farvolden and Powell (1994), and Barnhart and Kim (1995).

The network optimization model that follows is based on the work of Crainic (1982) and Crainic et al. (1984). It integrates the service network design and traffic multimode routing problems with general terminal policies. Its goal is the generation of global strategies to improve the cost and service performance of the system. It is a modeling framework for the tactical planning problem of intermodal freight transportation systems in the sense that while it may represent a large variety of real situations, it has to be adapted to each application. In fact, it has been applied to problems from the Canadian (Crainic, 1984) and French (Crainic and Nicolle, 1986) railways. It has also been adapted with considerable success to the multimode LTL trucking problem (Roy, 1984; Roy and Delorme, 1989). In the following we present a simplified model, to emphasize the main issues and challenges of formulating tactical planning models and tools.

Let $G = (V,E)$ represent the physical network. Vertices of V represent the terminals selected for the particular application (terminals A, B, C, D, and E of Figure 8.1). For simplicity, assume that all terminals can perform all operations. The set E is the set of links representing the connections between terminals. The *transportation demand* is defined in terms of volume (e.g., number of vehicles) of a certain commodity c to be moved from an origin node $o \in V$ to a destination $d \in V$. To simplify, we refer to the market or *traffic class* $m = (o,d,c)$, with a positive transportation demand d^m.

The *service network* specifies the transportation services that could be offered to satisfy this demand. Each *service* s in this structure is defined by the route it follows through the physical network from its origin to its destination, by the sequence of intermediate terminals, and by its service characteristics: mode, speed, capacity, and so on. For example, in Figure 8.1 service S3 may represent a normal rail train service from terminal A to terminal D with an intermediate stop at terminal C. Similarly, for an LTL application, service S2 could represent the operation of a direct truck between terminals A and D, while service S1 could stand for the option to move the trailer TOFC, taking advantage of existing long-distance railway connections. The *frequency* F_s is the other important service characteristic. It defines the level of service offered on the route: how often the service is run during the planning period. A tactical model determines the (integer) values of the frequencies of the selected services, sometimes subject to lower (more rarely, upper) limits. Note that Figure 8.1 actually illustrates a service network since the physical routes are not displayed.

Traffic moves following predefined *itineraries* that specify the paths of services to use and the sequence of intermediate terminals on this path where operations are to be performed. For a given traffic class, one or several such itineraries may be used, according to the level of congestion in the system and the service and cost criteria of the particular application. The *routing* of freight (also called the distribution of freight), as given by the amount of flow X_k^m of traffic class m moved by using itinerary k, must also be determined by the model.

Frequencies and itineraries are the central elements of the model. Fixing frequency values determine the design of the service network, the level of service, and the feasible domain for the traffic distribution problem. On the other hand, selection of the best itineraries for each traffic class solves the traffic distribution problem and determines the workloads and the general consolidation strategies for each terminal.

The model states that the total system cost has to be minimized while satisfying the demand for transportation and the service standards. It also contains the usual nonnegativity and integrality constraints:

$$\text{minimize } \Psi(X_k^m, F_s)$$

subject to

$$\sum_k X_k^m = d^m \qquad \text{for all } m$$

$$X_k^m \geq 0 \qquad \text{for all } k,m$$

$$F_s \geq 0 \text{ and integer} \qquad \text{for all } s$$

The objective function $\Psi(X_k^m, F_s)$ defines the total system cost and includes (1) the total cost of operating a given service network at level F_s; this is the "fixed' cost of the transportation system; and (2) the total cost of moving freight by using the selected itineraries for each traffic class (X_k^m); this is the "variable" cost of satisfying demand by a service network operated at level F_s.

What the objective function represents is a generalized cost, in the sense that both operating and service costs are included. It is at this level that the relationships and trade-offs among the various system and policy components are considered. Indeed, the objective combines the real costs of handling and moving freight and vehicles and penalties related to service reliability (generally based on the mean and variance of transportation delays). These delays are incurred by vehicles, convoys, and freight due to congestion and operational policies in terminals and on the road. The resulting model has the structure of a nonlinear, mixed-integer, multimode, multicommodity flow problem and may be solved by an efficient heuristic algorithm (Crainic and Rousseau, 1986) based on decomposition, column generation, and descent techniques.

Equation (8.25) illustrates one approach for integrating service considerations into the total generalized system cost:

$$\begin{aligned}\Psi(X_k^m, F_s) = & \sum_s C_s^t F_s + \sum_{m,k} C_{mk}^t X_k^m \\ & + \sum_s C_s^d E(T_s) F_s \\ & + \sum_{m,k} C_{mk}^d E(T_k^m) X_k^m + \sum_s C_s^p (\min\{0, \alpha_s F_s - X_s\})^2 \qquad (8.25)\end{aligned}$$

Here delays are converted into delay costs, compatible with operating costs (C_s^t and C_{mk}^t for each service and itinerary, respectively), via user-defined unit time costs for each traffic class (C_{mk}^d) and type of service (C_s^d). These costs are usually based on equipment depreciation values, goods inventory costs, and time-related characteristics such as priorities or different degrees of time sensitivity for specific traffic classes. Equation (8.25) also illustrates the use of penalty terms to capture various restrictions and conditions. Thus the service capacity restrictions are considered as utilization targets, and overassignment of traffic is permitted at the expense of additional costs and delays. Then, while solving the resulting mathematical programming problem, tradeoffs between the cost of increasing the level of service and the extra costs of insufficient capacity may be addressed.

The terms most likely to appear in such a function, especially in the computation of the mean delays $E(T_s)$ and $E(T_k^m)$ (for services and itineraries, respectively), are application specific. Generally, one attempts to include those that reflect the cost and delay characteristics of the terminal and line operations most significant for the system. Typically, for a rail application one may have:

1. Handling costs associated with car classification operations at yards (sorting, blocking, etc.).
2. The cost of transferring cars or blocks among trains at terminals.
3. The costs of breaking down and of making up trains at yard terminals.
4. Hauling cost (by service class) for trains and cars over the lines of the network; these costs may relate to energy, motive power, and crews.
5. The average (and possibly the variance) of the delay due to yard operations: car and train inspection, car classification and blocking, con-

nection delays (the waiting time for the designated service to be available), train formation, and so on.

6. The mean delays incurred by trains on the lines of the network due to congestion conditions and meet/overtake operations. Stopping times at intermediate yards may be included.

The same main elements are found in LTL applications, with the normal adjustment for the specificity of trucking operations:

1. Costs and durations related to loading and unloading freight at terminals.
2. Costs and delays due to transdock operations and consolidation functions at terminals.
3. Waiting for departure and, sometimes, for an unloading gate to become available.
4. Cost and time required to move vehicles and cargo from one terminal to another (note that congestion might appear even in this context when terminals are located within the congested part of urban highway and road networks).
5. The associated labor costs in terminals and for driving the vehicles.

Queueing models are generally used to derive classification, consolidation, connection, and other terminal delay functions, as well as over-the-road delays which reflect the congestion and physical characteristics of the system. For a review of such models, the interested reader may consult Crainic (1988) as well as the more recent contributions of Chen and Harker (1990), Harker and Hong (1990), and references quoted by these authors.

A second approach to integrating service considerations into the total generalized system cost is illustrated by the equation (Roy, 1984)

$$\begin{aligned}\Psi(X_k^m, F_s) = \sum_s C_s^t F_s + \sum_{m,k} C_{mk}^t X_k^m \\ + \sum_{m,k} C_{mk}^S(\min\{0, S_m - E(T_k^m) - n\sigma(T_k^m)\}) X_k^m \\ + \sum_s C_s^p(\min\{0, \alpha_s F_s - X_s\})^2 \qquad (8.26)\end{aligned}$$

In this case the operating cost coefficients (C_s^t and C_{mk}^t) for vehicle (service) and freight handling and transport have the same general meaning as previously, as does the second penalty term of the equation. The first penalty term, appearing on the second line of equation (8.26), represents the case when service quality targets are announced. Here, for example, each traffic class has an associated delivery objective (24 hours for 90% of deliveries, for example) and a penalty for not achieving it. Then the service target may be compared with the mean expected delay of the entire journey. Trade-offs may be achieved during optimization between the costs of improving operations and those of achieving the promised service quality. Furthermore, scenario analyses may be conducted following the optimization phase to examine, for example, the effect of the relative importance of each factor on the global performance of the system. For further examples of such postoptimal analyses, see Crainic and Roy (1988) and Roy and Crainic (1992).

An additional planning issue particularly important and challenging for freight carriers is the need to move empty vehicles. Indeed, the geographic differences in demands and supplies for each commodity type often result in an accumulation of empty vehicles in a region where they are not needed, or in a deficit of vehicles in other regions that require them. Freight carriers must therefore reposition vehicles for use in the following planning periods. This is a complicated issue, and numerous studies reflect the significant research and development effort that has been dedicated to it. *Empty balancing* models are often associated with the operational level of planning. Interested readers may start exploring this field with the review of Dejax and Crainic (1987). This issue must also be considered at the tactical level. In rail transportation, for example, empty rail cars are put on the same trains as loaded ones and thus contribute to an increase in the number of trains, in the volume of vehicles handled in terminals, and ultimately, in system costs and delays. For planning purposes, the demand for empty cars may be approximated and introduced in the tactical model by viewing empties as another commodity to be transported. The issue is also relevant in LTL trucking, where empty balancing is an integral part of a transportation plan. In this case a transportation plan is first obtained for the actual traffic demands, and an empty balancing model is then solved to reposition the empties (see, e.g., Roy and Delorme, 1989; Braklow et al., 1992).

8.C.2. Vehicle Routing Problems

The models just described typically apply to long-distance freight transportation, where sorting and consolidation occurs at freight terminals. The next planning level takes place in more restricted geographical areas. It involves the distribution of goods at the local or regional level and comprises activities such as pickup, delivery, or a combination of both. In some industries, such as in the food and drink business, distribution costs at this level can account for up to 70% of the value-added costs of goods (Golden and Wasil, 1987).

Distribution management problems arising at this level have been studied extensively by operations researchers over the last 40 years. Interesting references are those of Eilon et al. (1971), Bodin et al. (1983), Golden and Assad (1988), and Daganzo (1991). Such problems are generally referred to as *vehicle routing problems* (VRPs), but this designation covers a wide range of setups rather than a specific problem. VRPs involve the design of pickup or delivery routes from one or more central depots to a set of geographically scattered customers. Several versions of the problem can be defined, depending on a number of factors, constraints, and objectives.

1. Does the problem involve deliveries, collections, or a combination of both? Are there precedence relations between deliveries or collections?
2. Does distribution take place from a single depot or from several centers?
3. How may vehicles are involved? Is the number fixed or does it constitute a decision variable? Is the vehicle fleet homogeneous or heterogeneous? What are the capacity, speed, and operating costs of these vehicles?
4. What are the work conditions of the drivers? What is the pay structure? What is the length of a normal workday? What are the conditions on overtime? Are multiple same-day trips allowed?
5. Is the demand known in advance, or is it revealed in a dynamic fashion during the course of operations?
6. How often or when must each customer be visited during the planning period (one week, say)? On a given day, must customers be visited within specific time windows?

These are just some of the questions that must be addressed when solving a VRP. For a broader coverage of these issues, see Assad (1988). It is therefore important to have a clear understanding of the situation at hand and of the rules of the game.

How does one solve vehicle routing problems? Unfortunately, even for the most basic version of the problem, exact methods can handle only relatively small instances. To this day, no optimization algorithm can *consistently* solve instances involving more than 50 customers. In practice (and in all known VRP software), the only option is to use heuristics. There exists an abundant literature on exact and approximate algorithms for the VRP. For recent surveys, see Laporte (1992b), Fisher (1995), and Desrosiers et al. (1995).

To gain more insight into the nature of the problem and into some of the better known algorithms, we focus on a very basic, but commonly studied version of the problem: the single-depot capacitated VRP or CVRP. [A different class of vehicle routing problems, where customers are on links rather than nodes, known as the Chinese postman problem, is presented in Chapter 5 (Section 5.D) relative to the issue of collection of solid wastes.] The CVRP is usually defined on an undirected graph $G = (V,E)$, where $V = \{v_0,v_1,\ldots,v_n\}$ is the vertex set, and $E = \{(v_i,v_j) \mid v_i,v_j \in E, i < j\}$ is the edge set. Vertex v_0 represents a depot at which are based m identical vehicles of capacity Q, and the remaining vertices represent customers. The demand of customer v_i is equal to a positive integer q_i. With each edge (v_i,v_j) is associated a cost c_{ij}, typically proportional to distance or travel time. The CVRP consists of designing m vehicle routes starting and ending at the depot, in such a way that each customer is visited exactly once by only one vehicle, the demand on any vehicle route does not exceed Q, and the total cost is minimized. Several formulations have been proposed for this problem (see Laporte and Nobert, 1987, for a survey). Here we describe a simple and commonly employed *two-index vehicle flow formulation.* Let x_{ij} be an integer variable equal to 2 if a vehicle makes a return trip between v_i and v_j, equal to 1 if it makes a single trip between v_i and v_j, and equal to 0 otherwise. The case x_{ij} can occur only if one of the two vertices is the depot. In what follows, x_{ij} must be interpreted as x_{ji} whenever $i > j$. The CVRP is then formulated as follows:

$$\text{minimize} \sum_{v_i, v_j \in V} c_{ij} x_{ij} \tag{8.27}$$

subject to

$$\sum_{j=1}^{n} x_{0j} = 2m \tag{8.28}$$

$$\sum_{i<j} x_{ik} + \sum_{j>k} x_{kj} = 2 \qquad v_k \in V \setminus \{v_0\} \tag{8.29}$$

$$\sum_{\substack{v_i \in S, v_j \in V \setminus S \\ \text{or} \\ v_j \in S, v_i \in V \setminus S}} x_{ij} \geq 2 \left(\frac{\Sigma_{v_i \in S}\, q_i}{Q} \right) \qquad S \subseteq V \setminus \{v_0\}, \quad 2 \leq |V| \leq n-2 \tag{8.30}$$

$$x_{0j} = 0, 1, \text{ or } 2 \qquad v_j \in V \setminus \{v_0\} \tag{8.31}$$

$$x_{ij} = 0 \text{ or } 1 \qquad v_i, v_j \in V \setminus \{v_0\} \tag{8.32}$$

In this formulation the objective function states that the total travel cost must be minimized. Constraints (8.28) express the requirement that each of the m vehicle trips must start and end at the depot, while constraints (8.29) state that the degree of every customer vertex must be equal to 2. Constraints (8.30) are connectivity constraints. They can be interpreted as follows. Given as set $S \subseteq V \setminus \{v_0\}$ of customers and their total demand $\Sigma_{v_i \in S} q_i$, in any feasible solution there must be at least $\lceil \Sigma_{v_i \in S} q_i / Q \rceil$ vehicles servicing S and twice that number connecting S to its complement $V \setminus S$. These constraints play a double role: They ensure that all customers are connected to the depot (since the right-hand side is never less than 2) and that the total demand of any route never exceeds the vehicle capacity. In constraints (8.31), the case $x_{0j} = 2$ corresponds to a return trip between the depot and customer v_j. In all other cases, x_{ij} is equal to 0 or 1. One interest of this formulation is that m, the number of vehicles, can be regarded as a constant or as a variable. In the latter case one can add to the objective function an extra term fm, where f is the fixed cost of a vehicle.

The main drawback of this formulation is that the number of integer variables is relatively large in most practical situations and the number of connectivity constraints is exponential in n'. Therefore, this model is never solved directly. It is usually tackled by means of a *branch-and-cut algorithm*. Initially, a simplified version of the problem without integrality requirements and without connectivity constraints is solved. A heuristic is applied to determine whether any of the connectivity constraints are violated. If this is the case, several of the most violated constraints are introduced into the model, which is then reoptimized. If no violated connectivity constraint can be identified, a check for integrality is made. If the solution is feasible, it is then optimal. Otherwise, two subproblems are created by branching on a fractional variable x_{ij}. A search tree is thus created and the same process is reapplied at each node of the search tree by following the usual branch-and-bound rules. This method was first applied by Laporte et al. (1985), who solved to optimality some loosely constrained 60-customer problems (with vehicles filled at about 70% of their capacity). The method does not perform so well on tightly constrained problems because of the very large number of violated connectivity constraints that have to be generated. Recently, a number of other valid, but more complicated constraints have been identified for the CVRP. Some of these results are presented in Cornuéjols and Harche (1993).

Countless heuristics have been proposed for the CVRP. We first sketch two classical methods: the savings method (Clarke and Wright, 1964) and the sweep heuristic (Gillett and Miller, 1974). These two algorithms are based on intuitive principles and have gained a fair degree of acceptance over the years. We then briefly describe a more recent and more powerful algorithm based on tabu search (Glover, 1989, 1990).

The classical *savings algorithm* applies to problems for which the number of vehicles is not determined a priori. It starts with n vehicle routes, each containing the depot and a single customer. At each step, two routes are merged according to the largest achievable saving. The method can be outlined as follows.

Step 1. Compute the savings $s_{ij} = c_{i0} + c_{0j} - c_{ij}$, for $i,j = 1,2,...,n$. Create n vehicle routes (v_0,v_i,v_0) for $i = 1,2,...,n$.

Step 2. Order the savings in a nonincreasing fashion.

Step 3. Consider two vehicle routes containing edges (v_0,v_i) and (v_0,v_j), respectively. If $s_{ij} > 0$, tentatively merge these routes by introducing edge (v_i,v_j) and by deleting edges (v_0,v_i) and (v_0,v_j). Implement the merge of the resulting route if feasible. Repeat this step until no improvement is possible (see Figure 8.2).

The myopic aspect of this procedure and the fact that it tends to produce a number of circumferential routes make it a relatively uninteresting alternative. However, several improvements suggested over the years tend to re-

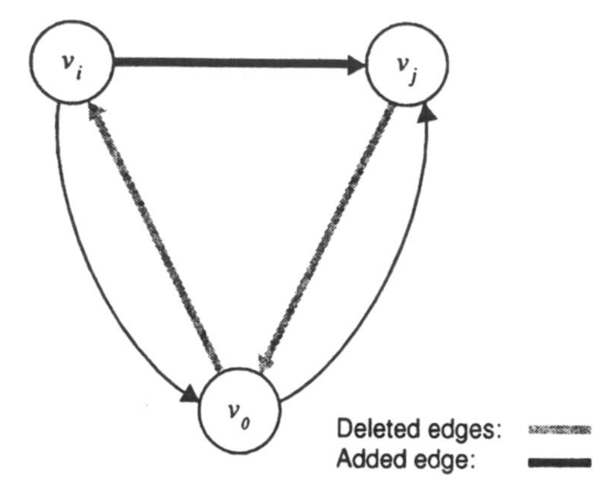

Figure 8.2 Merging two routes in the savings algorithm.

duce these effects. One is to construct several routes in parallel (Altinkemer and Gavish, 1991), another is to weigh the term c_{ij} by a user-controlled positive parameter α (Yellow, 1970); others are the use of sophisticated data structures to speed up the computations (Nelson et al., 1985).

The sweep algorithm works on planar problems for which the number of vehicles is not fixed. The method is easily implemented using an arbitrary ray first drawn from the depot. Customers v_i are represented by their polar coordinates (θ_i,ρ_i), where θ_i is the angle v_i makes with the depot and the ray, and ρ_i is the length of the segment $v_0 - v_i$. Assume that all customers are ranked in increasing order of their polar angle.

Step 1. Select an unused vehicle k.

Step 2. Starting from the unrouted vertex with the smallest angle, assign the next customer (with the next largest angle) to vehicle k, as long as vehicle capacity is not exceeded. If unrouted vertices remain, go to step 1.

Step 3. Optimize each route separately by means of a heuristic for the traveling salesman problem (TSP) (see Laporte, 1992a).

To illustrate how the algorithm works, consider the following 10-customer example. The vehicle capacity is 10. Customers and their demands are given in Table 8.1. In Figure 8.3 these customers are shown in a plane. Each customer v_i with demand q_i is represented by a pair (v_i,q_i). The initial ray is given by the full line; each dashed line corresponds to a vehicle. In this example the algorithm would create four vehicle routes with customer sets $\{4,5,7\}$, $\{10\}$, $\{2,6\}$, $\{1,3,8,9\}$.

In recent years, several *metaheuristics* have been proposed for approximate resolution of a number of combinatorial optimization problems, including the VRP. Metaheuristics are general methods that guide subordinate heuristics to perform a clever search of the solution space. These include simulated an-

TABLE 8.1 Customers and Demands

Customer	Demand
1	2
2	4
3	3
4	1
5	5
6	6
7	3
8	3
9	2
10	7

nealing, tabu search, and genetic algorithms (see Pirlot, 1992; Reeves, 1993; and Osman and Kelly, 1996, for recent surveys). Tabu search (TS) was introduced in the mid-1980s by Glover (1986) and Hansen (1986). It has been applied successfully to the VRP by a number of authors (see, e.g., Taillard, 1993; Gendreau et al. 1994; Rochat and Taillard, 1995). Typically, the method starts from an initial solution obtained by means of any heuristic. At a given iteration it explores neighbor solutions (i.e., solutions that can be reached from the current solution by performing a given type of operation). It then moves to the best neighbor and the process is repeated until a termination criterion is met. Since successive solutions do not necessarily improve upon one another (as is the case in classical local search methods), solutions that

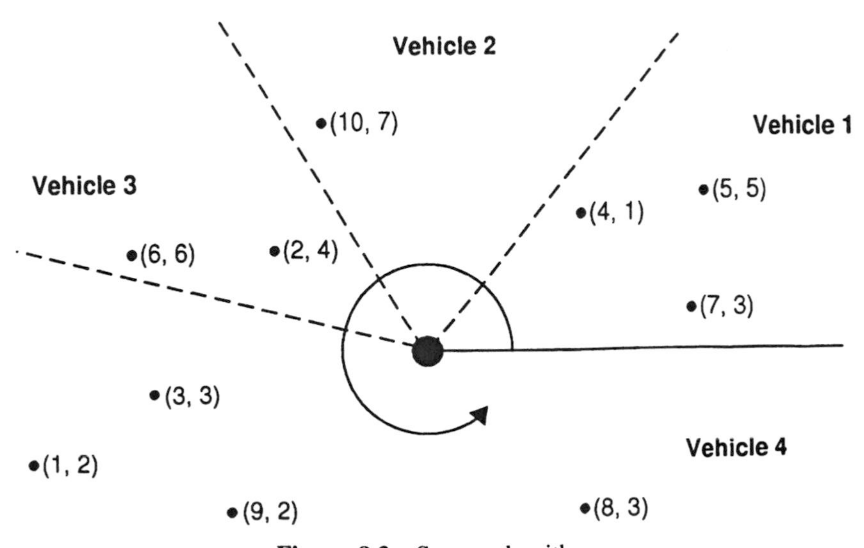

Figure 8.3 Sweep algorithm.

were recently examined are inserted in a constantly updated *tabu list* (i.e., a list of forbidden solutions). Several implementation devices, such as diversification and intensification, are now commonly employed (see Glover and Laguna, 1993, for an overview). Here follows a short and simplified description of TABUROUTE, a tabu search algorithm designed by Gendreau et al. (1994) for the VRP with capacity and maximal route length restrictions.

Step 1. Construct a giant tour over all vertices. (Here the GENIUS TSP heuristic due to Gendreau et al. 1992, is used). Then break the tour into feasible VRP routes. Denote this solution by x and its costs by $F(x)$. Set the tabu list $T := \emptyset$.

Step 2. Attempt to improve the solution by applying the following search procedure to a subset W of $V \setminus \{v_0\}$ and by then selecting the best solution. Define $N(x)$, the neighborhood of x, as the set of all solutions that can be reached by inserting a vertex of W into a route containing one of its closest neighbors, using the GENI procedure (Gendreau et al., 1992). If $N(x) \setminus T = \emptyset$, go to step 3. Otherwise, identify a least-cost solution y in $N(x) \setminus T$ and set $x := y$. Update the best known solution.

Step 3. If a preset number of iterations has been reached, go to step 4. Otherwise, update T and go to step 2.

Step 4. Attempt to improve the current solution by applying the search procedure to a larger set W and allowing fewer iterations. If $N(x) \setminus T = \emptyset$, go to step 5. Otherwise, identify a least-cost solution y in $N(x) \setminus T$ and set $x := y$. Update the best known solution.

Step 5. If the maximum number of iterations has been reached, stop. Otherwise, update T and go to step 4.

TABUROUTE has been applied to 14 benchmark problems with $50 \leq n \leq 199$, taken from Christofides et al. (1979). It always produces better solutions than simpler heuristics, at the expense of increased computation time. The more recent TS implementation by Rochat and Taillard (1995) yields best known solutions on all of the 14 test problem.

Whether one should use a rather involved heuristic such as TABUROUTE, as opposed to easy-to-program heuristics such as the savings method or the sweep algorithm, depends on the context. If quick and dirty solutions are required, the latter methods are probably appropriate. But if large sums of money are at stake, it may be worth investing in a more sophisticated heuristic to obtain a more economical solution.

8.D. OPERATIONAL ISSUES AND MODELS

The ultimate goal of any transportation firm is to make profits and improve, or at least maintain, its competitive position. To this end, strategic and tactic plans can be drawn up to guide operations, but the operational capabilities of the firm will ultimately determine its performance. Hence models and tools aimed at assisting decision making at an operational level are an important component of a comprehensive decision support system.

Many different issues have to be addressed at the operational level to ensure that demand is satisfied within the required service criteria and that the resources of the carrier are used efficiently. Most of these issues have to consider the *time* factor: A response may be required in real or near-real time (e.g., assigning an empty container to a customer request), today's decisions may have a significant impact on future decisions and performances (e.g., if empty rail cars are not repositioned, in a few days equipment will be idle in some yards while it will not be possible to satisfy demands at some other terminals), schedules may be needed for vehicles and crews, time conditions are imposed on operations (e.g., a container has to arrive in time to be loaded on the departing ship or a truck has to pick up a load within a specified time window), and so on. In other type of operation the very notion of a planned solution does not make sense. Consider, for example, courier services operating in urban contexts. Here, drivers often start their working day with a short list of collection or delivery requests. As the day progresses, new requests will be communicated to the drivers and will have to be incorporated into their routes.

Most models traditionally used in transportation planning use as their input known static data. For examples, VRP models operate with given customer demands and travel times, tactical planning formulations consider aggregated forecast demand data as "known," and so on. However, the real world in which these models are implemented is in a constant state of change and solutions cannot always be implemented as planned. If traffic is slower than predicted, vehicles may arrive late at the customers' locations and at the depot. If supplies are larger than forecast, vehicles may become full prematurely and it may become necessary to pay penalties (e.g., pay drivers at overtime rate for part of the day) or to implement recourse actions (e.g., when a delivery vehicle becomes empty, it could go back to the depot to replenish and resume deliveries starting at the next point in its planned route).

Consequently, the *dynamic* aspect of operations is further compounded by the *stochasticity* inherent in the system, that is, by the set of uncertainties that are characteristic of real-life management and operations. Increasingly, these characteristics are reflected into the models and methods aimed at operational planning and management issues, as illustrated in the following.

8.D.1. Dynamic Modeling to Support Carrier Operations

The problems most often encountered in the operational planning and management of transportation carriers include the following (see also Powell et al., 1995b, and references therein):

1. *Scheduling of services.* A tactical load plan will often indicate which service to offer and, eventually, how often to run it over the planning period.

Often, however, service will be offered according to a *schedule* that indicates the time of departure at the origin and the time of arrival at the destination, as well as the time and length of stops at intermediate terminals, where appropriate. The schedule may be fixed (e.g., airlines) or it may indicate a departure interval (e.g., a truck of a LTL motor carrier leaves the terminal between 7 and 9 P.M.), in which case customers and other operating units of the company will be given a cutoff time for their loads to make the service in time. Passenger transportation usually takes place according to fixed schedules and so usually is air cargo. LTL trucking is generally operated according to interval schedules. Traditionally, North American railways did not operate scheduled services, or the schedule represented only a goal or a general idea of how the trains were run. Currently, however, there is a trend toward totally or partially scheduled services to increase service levels and the competitiveness of the firm. In Europe, there is a longer tradition of operating scheduled rail services (with varying degrees of success), even of including booking schemes for freight trains (see, e.g., Joborn, 1995). In this book, schedule generation is also treated in the context of public-sector urban transportation [Chapter 9 (Section 9.D)] and the commercial airline industry [Chapter 10 (Section 10.B.1)].

2. *Empty vehicle distribution or repositioning.* The imbalance between freight demand and supply is reflected in the fact that at any point in time there are terminals with a surplus number of vehicles of a certain type while some other terminals show a shortage. Then vehicles have to be moved empty (or additional loads have to be found) to bring them where they will be needed to satisfy known and forecast demand. Moving vehicles empty does not contribute directly to the profit of the firm but is essential to its continuing operations. Consequently, one attempts to minimize empty movements within the limits imposed by the demand and service requirements. Empty vehicle distribution is a central component of planning and operations of many transportation firms, especially in the rail, container, and LTL motor carrier industries. There is, therefore, a very rich literature addressing these issues. The Dejax and Crainic (1987) survey reviews contributions going back to the 1960s and spans the entire spectrum of modeling approaches from simple static transport models to formulations that integrate the dynamic and stochastic characteristics of the problem (e.g., the work of Jordan and Turnquist, 1983). Significant research efforts continue to be directed toward this class of problems, as illustrated by the recent contributions of Chih (1986), Adamidou et al. (1993), Turnquist (1994), Joborn (1995), and others.

3. *Crew scheduling.* Crews have to be assigned to vehicles and convoys in order to support the planned operations. There are also numerous other issues related to labor management, such as the scheduling of reserve crews, terminal employees, maintenance crews, and so on. A significant body of methodological and technological knowledge has been developed to deal with these issues, especially in the context of transit (bus and passenger rail) and

airline transportation (see, e.g., Desrosiers et al., 1995; Barnhart and Talluri, 1997). These methodologies were developed for applications where detailed schedules are known and adhered to. Consequently, although some efforts have been aimed in this direction (see, e.g., Crainic and Roy, 1990, 1992), currently it appears that better results can be achieved by applying the class of methodologies used to dynamically allocate resources to tasks. In this book, methodologies for crew scheduling are also presented for problems in the areas of public-sector urban transportation [Chapter 9 (Section 9.F)] and the commercial airline industry [Chapter 10 (Section 10.B.4)].

4. *Allocation of resources.* Many operational problems may be viewed as dynamically allocating resources to tasks (Powell, 1995). For example, one has to allocate empty vehicles (trailers, railcars, etc.) to the appropriate terminals, motive power (tractors, locomotives, etc.) to services, crews to movements or services, customer loads to driver–truck combinations, and empty containers from depots to customers and returning containers from customers to depots. This is an extremely rich field both for research and development and for applications. Dynamic and stochastic network formulations have been and continue to be studied extensively. This has resulted in important modeling and algorithmic results; see, for example, Powell (1988), Frantzeskakis (1990) and Frantzeskakis and Powell (1990), Crainic et al. (1993a), Powell and Cheung (1994a,b) and Cheung and Powell (1996). A number of these results have been transferred to industry (Powell et al., 1992, for example). The interested reader should consult the excellent synthesis and review by Powell et al. (1995b) and the numerous references quoted in that work. The recently proposed logistics queueing networks methodology (Powell et al., 1994, 1995a; Carvalho, 1996) appears, however, to offer an even more interesting framework for a wide variety of real situations which may be efficiently represented and solved. In this book, vehicle assignment and scheduling are also treated in the contexts of public-sector urban transportation [Chapter 9 (Section 9.E)] and the commercial airline industry [Chapter 10 (Section 10.B.2)].

It is not possible in the present chapter to provide an exhaustive review of the models and methods developed for the issues just described. We encourage the interested reader to consult the documents mentioned. In the following we briefly illustrate two such modeling approaches.

The first example concerns the issue of determining the dynamic service network of a carrier that operates according to more or less fixed schedules. Rail companies and less-than-truckload motor carriers are examples of such firms. The main issues are similar to those of the tactical planning problem described in Section 8.C.1, with one major difference. While in tactical planning one is concerned with the *where* and *how* issues (selecting services of given types and traffic routes between spatial locations), here one is interested above all in *when* issues: when to start a given service, when the vehicle arrives at destination or at an intermediary terminal, when is the traffic delivered, and so on.

A very simple instance of such a problem is illustrated in Figure 8.4. For the sake of simplicity, assume that for the service network illustrated in Figure 8.1, one has already decided that out of terminal A only services S1, S2, and S3 will be offered. One must then determine the departure schedule for these services over the next periods, as well as the corresponding traffic routing. Part of the dynamic time–space network that may be used to support these decisions is shown in Figure 8.4. Here each terminal is drawn for each of the

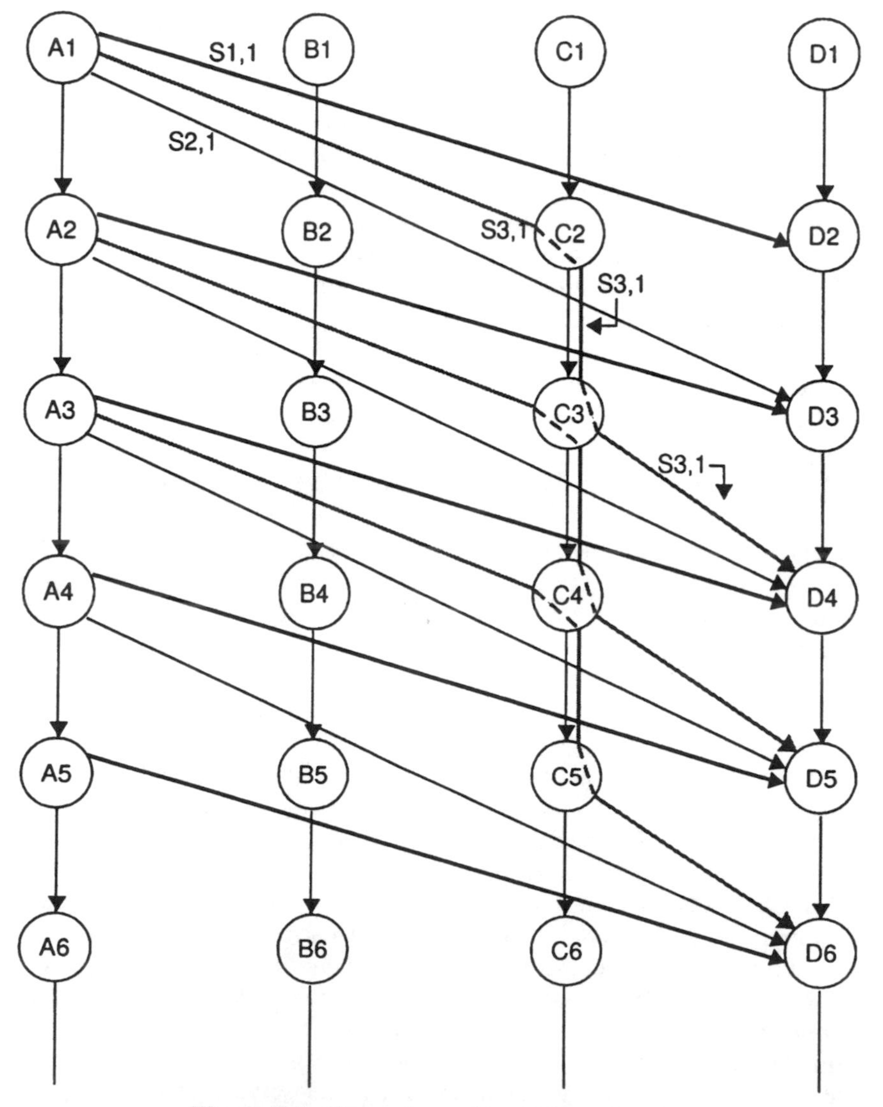

Figure 8.4 Dynamic service design network.

periods considered (six, in this example). Each service is then drawn at each of its possible departure periods and its route in space and time is shown. Hence, service S1 departing in period 1 arrives at terminal D one period later, while service S2 takes two periods to reach the same destination. A more complex route is shown for service S2, which, departing at the same period as the other two, arrives at terminal C one period later, waits there for one period for loading, unloading, and consolidation operations, to finally arrive at terminal D in period 4. The same *dynamic service arcs* are also drawn starting at the other periods. Other dynamic arcs are the *holding* links, which represent waiting for traffic in terminals; the arc from node A1 to node A2 is such a holding arc. Different holding arcs could be used to represent various terminal activities. Note that in order not to overload the picture, we have not shown all the dynamic arcs, nor all terminals, nor all the other arcs required to model the entry and exit of demand.

Models for this class of problems may be written similarly to the network design formulations of Section 8.B [equations (8.11) to (8.16)]. Here one associates to each dynamic service arc an integer variable equal to 1 if the corresponding departure is chosen, and to zero otherwise. Other variables capture the flow of traffic on the services and through terminals. Additional constraints, such as terminal and service capacities or the time windows imposed on the delivery of goods at destination, further complicate the problem (see Farvolden and Powell, 1991, 1994, for a complete formulation).

As already indicated, the problem of allocating limited resources to requests and tasks is one of the most typical and critical issues relative to operating a transportation system (as well as most industrial and service systems, for that matter). Problems involving repositioning equipment, engines, tractors, rail cars, trailers, ships, and so on, to respond to sure and forecast demands, and allocating crews to transportation services or trucks and containers to loads, all have several common characteristics:

1. Some future demands are known, but most can only be forecast, and requests that may be made are, of course, not predictable.
2. Many requests materialize in real or quasi real time and have to be acted upon in a relatively short time.
3. Once a resource is allocated to an activity it is no longer available for a certain duration (whose length may be subject to variations as well).
4. Once a resource again becomes available, it is often in a different location than its initial one.
5. The value of an additional unit of a given resource at a location depends greatly on the total quantity of resources available (which are determined by previous decisions at potentially all terminals in previous periods) the current demand, and other factors.

One may represent such issues by an activity graph similar to the one displayed in Figure 8.5. Here the operations of a simple four-terminal system

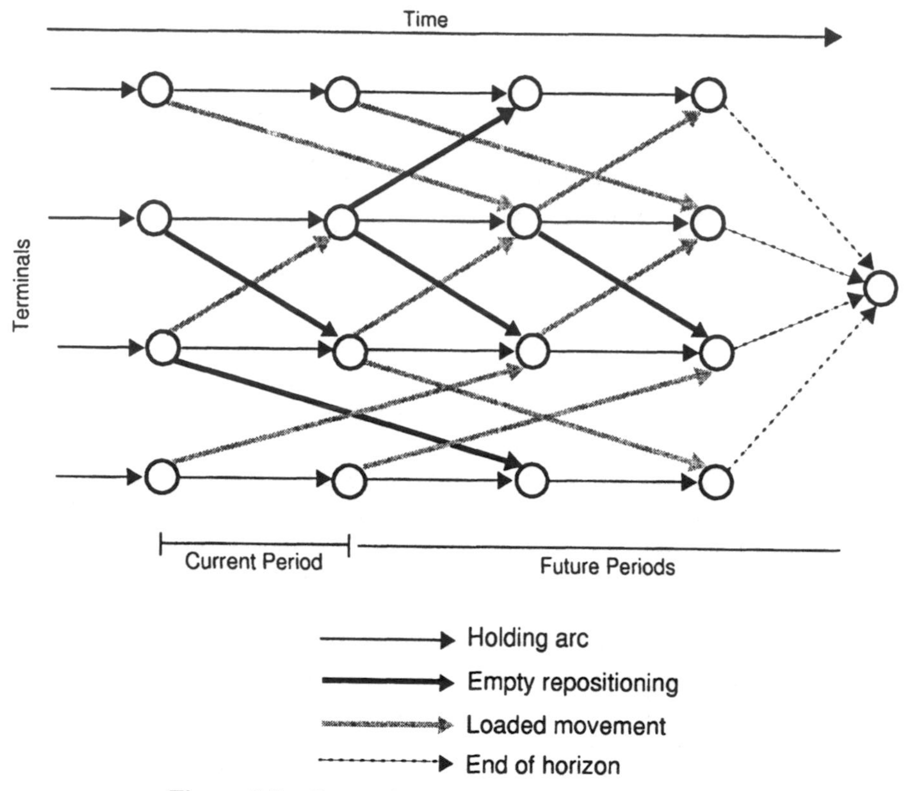

Figure 8.5 Space–time diagram resource allocation.

are drawn schematically for a certain length of time (one week, for example), arbitrarily divided into three periods (half-days, for example). At each terminal there are a number of vehicles (containers, for example) which are currently available to satisfy customer requests during the current period and in future ones. Customer demands have precise characteristics, such as origin and destination of movement, pickup and delivery dates (with time windows, eventually), and so on. At any period, a vehicle may be assigned to a customer demand at the current period and location, or it may be moved to another location to satisfy a known future request, or it may be held at the current location or moved empty to another location in preparation for future, forecast demands. The network of Figure 8.5 illustrates these possibilities and decisions.

Accepting requests and performing the corresponding movements implies expenses and generates revenues. Some models address the issue of whether the request is profitable with respect to operation of the system and should therefore be accepted (Powell et al., 1992, for example). Repositioning empty vehicles does not generate immediate revenues. One may be ready to incur these expenses, however, in the hope that, as a consequence, vehicles will be adequately posted to take advantage of future (known, forecast, or only guessed at) opportunities. Refused requests represent lost business opportunities, while accepted but unsatisfied ones generally result in penalties.

A classical modeling approach for this class of problems is to consider the entire planning horizon, with the objective of maximizing the *total system profit,* computed as the sum of the profit resulting from decisions taken for the current period, plus the expected profit over future periods. The usual constraints (e.g., satisfy demands, do not use more than the number of available vehicles, etc.) apply. When the state of the system and its environment in future periods is known, or assumed to be known, the resulting formulation is deterministic and is often written as a network optimization model with additional constraints.

The major difficulty with this approach becomes apparent when the uncertainties in future demands (as well as, eventually, uncertainties related to performing the operations) are explicitly considered. In this case, decisions taken "now" for future periods cannot be based on "sure" data, only on estimations of how the system will evolve, which demand will materialize, and so on. From a mathematical programming point of view, random variables are used to represent the stochastic elements and decisions in future periods. Consequently, the expectation of future profits that appears in the objective function of the model becomes a very complex recursive stochastic equation where the statistical expectation of the total profit has to be computed over all possible realizations of all random variables.

To address this complex issue, the model generally takes the form of a recourse formulation. Such formulations are based on the idea that today's decisions are taken within today's deterministic context but using an "estimation" of the variability of the random factors, and that their consequences are reflected in later decisions. The recourse represents the later decisions that have to be taken to adjust the initial policies once the actual realization of the random variables is observed. In the simplest possible recourse formulation, called *simple recourse,* it is assumed that one does not attempt to change the decisions but pays a penalty when the observed value of a random variable is different from the estimation. More complex formulations, such as nodal and network recourse (see Powell and Frantzeskakis, 1994, and Powell and Cheung, 1994, and references therein), attempt to evaluate the possible modifications to the initial decisions and the impact on the total expected profit. An excellent analysis of the application of these approaches to the dynamic fleet management problems for truckload motor carriers, as well as a discussion of the merits and difficulties of stochastic formulations, may be found in Powell (1996).

These formulations, which are generally difficult to solve, also make use of various criteria to discretize aggregate and end time. For example, in Figure 8.5, the theoretically infinite future planning and operations horizon has been reduced to three periods. When the recourse formulation is solved, the periods could be aggregated further, all future ones being considered as one; this corresponds to a two-period formulation, as opposed to an n-period formulation otherwise. Then, in actual application, the models are used in a rolling horizon environment where, as time advances, a new period is added at the end of the horizon. An important issue is then how to approximate what happens in all the periods beyond the artificially fixed end of the horizon and how to integrate this approximation into the recourse function. Powell et al. (1995b) present an excellent review of this class of formulations.

A different approach recently championed by Powell (1995; see also Powell et al., 1995a; Carvalho, 1996) addresses resource allocation problems as *logistic queueing networks*. In this case, at each node of the time–space diagram there are two queues: one of resources and one of tasks requesting resources. Figure 8.6 illustrates a possible configuration for the first period of a four-terminal case. The basic idea is that to evaluate the worth of allocating a vehicle to a loaded movement, one has to know, or evaluate, not only the operating costs and the price of the load, but also the value of the vehicle to the destination terminal. Then at each terminal and for each possible destination, these values have to be computed to decide on the allocations. Similar values have to be computed for empty movements. Various other considerations, such as time windows, labor restrictions, substitutions, and so on, may be included in the model. The general solution approach makes a series of forward (to allocate vehicles) and backward (to evaluate vehicle

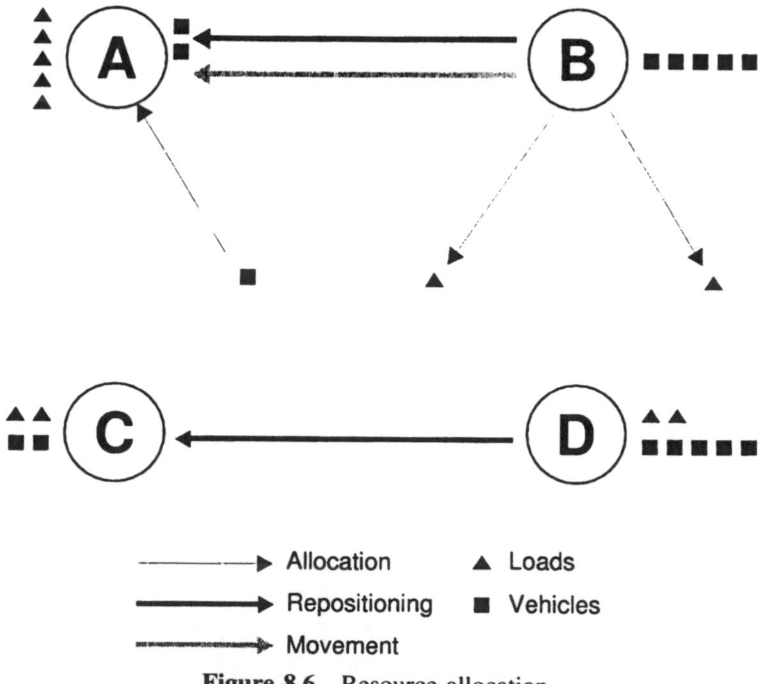

Figure 8.6 Resource allocation.

values at nodes) passes, until convergence is ensured. Several issues have to be addressed before implementing such an approach, but the initial results appear very promising.

8.D.2. Capacitated Routing with Uncertainties

We now describe the two approaches commonly used for dealing with stochasticity in transportation planning, in general, and vehicle routing, in particular: a priori (Bertsimas et al., 1990) optimization and real-time optimization.

In a priori optimization, problems are formulated within the framework of stochastic programming and are modeled in two stages. In a first stage, a planned or a priori solution is designed. The values taken by some random variables are then disclosed, and in a second stage, a recourse action is taken, such as paying drivers overtime or returning to the depot to replenish. A stochastic program is usually modeled as a chance constrained program (CCP) or as a stochastic program with recourse (SPR). In CCP, the planned solution is designed such that its probability of failure lies below a certain acceptable threshold. In SPR, the first-stage solution must be such that the expected cost of the second-stage solution is minimized. SPR is more realistic than CCP in that it works on the true expected cost of a solution, but it is also more difficult to develop solution methodologies for this case.

Before proceeding we illustrate these two solution concepts in the context of the CVRP. Consider the simple three-customer problem depicted in Figure 8.7. Customers 1 and 2 have a deterministic demand of 4, and the demand of customer 4 is 2 with probability 0.9 and 4 with probability 0.1. Let the vehicle capacity be equal to 10. For the CCP model, assume that the probability of route failure is $\alpha = 0.05$. Then an optimal solution is to use two vehicles and thus ensure that route failure will never occur. To implement an SPR solution, one must first define an appropriate recourse policy. Here the

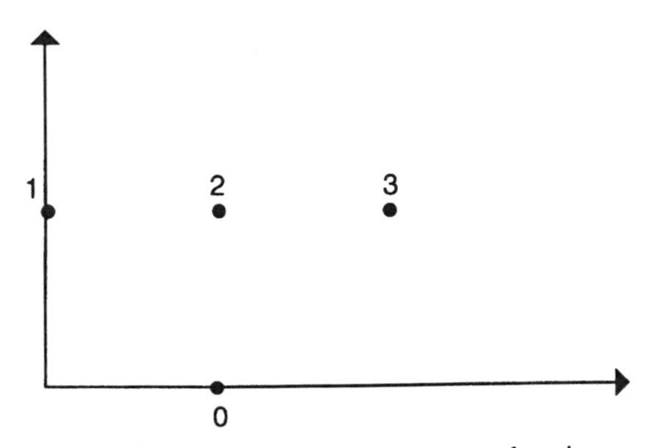

Figure 8.7 Stochastic VRP: customer locations.

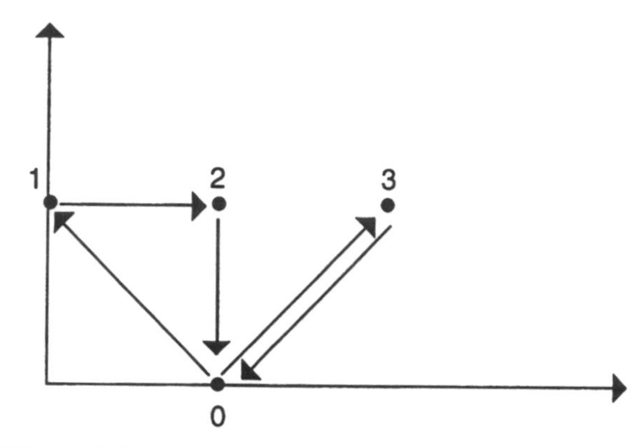

Figure 8.8 Optimal CCP solution using two vehicles.

policy is for the delivery vehicle to follow its planned route until its capacity becomes empty. In such a case it returns to the depot to replenish and resumes its deliveries along the planned route. The optimal planned route is (0,1,2,3,0) and has a cost of $2 + 2\sqrt{2}$. If the demand of customer 3 is 2, this is also the cost of the second-stage solution. If the demand of customer 3 is 4, then upon arriving at that customer, the vehicle must return to the depot to replenish and go back to customer 3 to carry out its last delivery. Thus the expected cost of that solution is $2 + 2\sqrt{2} + 0.1\ (2\sqrt{2}) = 2 + 2.2\sqrt{2}$. Note that in this case the reverse first-stage solution (0,3,2,1,0) yields the same expected cost, but as shown by Dror and Trudeau (1986), this is not always the case. Figures 8.8 and 8.9 illustrate the solutions of the two models.

How does one solve stochastic VRPs in practice? Due to the extreme complexity of such problems, most authors resort to heuristics, typically adaptations of known heuristics for the deterministic VRP. Thus Dror and Trudeau (1986) use a modification of the Clarke and Wright (1964) savings

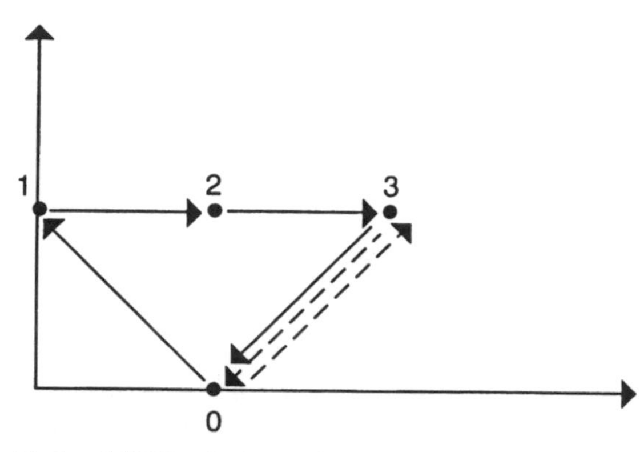

Figure 8.9 Optimal SPR solution using one vehicle and a recourse strategy.

algorithm. More recently, Gendreau et al. (1996a) have devised a tabu search heuristic for a more general version of the stochastic VRP in which customers are present with some probability (and are skipped if they are not present) and have stochastic demands. The method uses a proxy objective function to avoid computing at each candidate solution the true expected solution cost. Until quite recently it was difficult to assess the quality of solutions found by heuristics since there were no good lower bounds and no exact solution available. Some progress is, however, being made in these areas. A new type of exact algorithm, called the *integer L-shaped method,* was recently proposed by Laporte and Louveaux (1993) for a wide class of stochastic integer programs with recourse. Briefly, the method solves in a branch-and-cut fashion the first-stage program containing a lower bound on the cost of recourse. At any feasible solution, the true cost of recourse is computed. If the solution is dominated, fathoming occurs; otherwise, the branching process is reinitiated from that solution until dominance occurs or until the cost of the relaxed problem coincides with the cost of the true solution. This method was applied successfully to a family of SVRPs by Gendreau et al. (1995). For some values of the parameters it was able to solve to optimality VRPs with stochastic demands involving up to 70 customers. Smaller sizes were attained when both the customer presence and their demands were stochastic. Using these results the authors were able to demonstrate that when an optimal solution was available, their tabu search algorithm yielded solutions that were optimal more than 89% of the time, and the average deviation from optimality was only 0.38%. For a survey on stochastic VRP, interested readers are referred to Gendreau et al. (1996b).

The type of stochastic VRP just described is relatively basic in that information on demands arrives only when drivers reach the customers' locations, and then a very simple type of recourse action is applied: Return to the depot or carry on along the planned route. More sophisticated types of recourse are available. For example, as preventive return trips to the depot could be made even if route failure has not occurred, to avoid a potentially longer penalty later. Such preventive breaks could take place, for example, when the vehicle is near the depot. A more sophisticated type of recourse action would be to obtain information on the demand of the following customer whenever the vehicle arrives at a location and to decide at that stage whether a prevention break is necessary. At the other end of the spectrum, one could envisage a situation where information on customers' demands is transmitted to the driver continuously as he follows his route, and where the remaining portion of the route is reoptimized in real time. This requires, of course, the availability of efficient and reliable optimization algorithms.

Some of these scenarios are now a step closer to reality with the dissemination of new information technologies (see, e.g., Psaraftis, 1995; Powell et al., 1995b), which brings us to the problem of real-time vehicle dispatching.

As mentioned earlier, it is now common for courier firms to provide instructions in real time to their drivers upon reception of new requests. What distinguishes these problems from the SVRP is that the set of potential customers is to all practical purposes infinite. In such contexts one can only hope to devise good operating policies. Here simulation can be used to generate requests and to compare alternative policies.

In a recent paper, Shen et al. (1995) describe a dispatching system for a courier service company. Throughout the day, customers who have a letter to deliver phone the dispatching office and state their preferences (a time window) with regards to pickup and delivery (e.g., the letter should be picked up within the next 3 hours). The dispatcher must assign that request to one of the available vehicles, taking into account the current location and current planned route of each driver, locations of pickup and delivery points of the new request, the distances and travel points between points, and other details. The authors have developed a learning system, based on neural networks, to assign new requests to drivers. Decisions taken by the system are then compared to those made by professional dispatchers, which enables the system to readjust its decision rules. The authors found a good fit between the decisions made by the system and those made by the dispatchers. They conclude that neural network methodology can be applied successfully to such a system provided that sufficient information can be obtained in real time on drivers' locations.

The development of real-time routing systems is still in its early stages but is quickly emerging as a rich research area. There is no doubt that its growth is closely linked to that of new information technologies. However, technology alone is not sufficient to ensure success. The need remains to develop quick and efficient algorithms capable of providing good solutions in real time.

8.E. CONCLUSIONS

Freight transportation lies at the heart of our economic life. In industrialized countries, it accounts for a significant share of the gross national product. In developing countries it is the essential ingredient of sustainable development. With free trade zones emerging in several parts of the world and with the globalization of the economic system, transportation will in all likelihood play an even more major role in years to come.

The trend toward larger, more integrated, and more efficient transportation systems is likely to remain and should create the need for better planning at the strategic, tactical, and operational levels. Transportation planning is undoubtedly one of the great success stories of operations research. Classical OR models and algorithms have consistently proved highly suitable for the solution of complex transportation problems at all planning levels and involving just about any mode. Several planning tools developed by operations researchers are now widely available and are used routinely by transportation planners.

The rapid growth in the development of computerized planning systems witnessed over the last 15 to 20 years is due primarily to three factors: the design of more realistic models, the development of more powerful algorithms, and the availability of more performing computers and of friendlier user interfaces. These advances have had a direct impact on the realism, complexity, and size of planning models in the field of freight transportation and on their level of acceptance by planners and operators.

The growth of transportation planning methods will probably continue to be driven by the same factors in years to come. For example, major research efforts are being devoted to the design of models for dynamic and stochastic problems, bilevel programming enables the representation of competition between players, and so on. Major developments are also taking place in the artificial intelligence-related area of metaheuristics, such as tabu search, genetic algorithms, and neural networks. These have already given a new impetus to the area of global optimization and have led to a rethinking of the entire field of heuristics. These developments, coupled with the growth of parallel methods, mean that in the near future larger and more complex problems should be amenable to analysis and optimization. In particular, significant advances should be expected in the areas of dynamic, stochastic, and real-time programming, central to so many transportation systems.

ACKNOWLEDGMENTS

Funding for this project has been provided by the Natural Sciences and Engineering Council of Canada and by the Fonds F.C.A.R. of the Province of Québec. We would like to thank Lucie Larin for her assistance with the editing of the manuscript and the professional drawing of the figures.

REFERENCES

Adamidou, E. A., A. L. Kornhouser, and Y. A. Koskosidis, 1993. A Game Theoretic/Network Equilibrium Solution Approach for the Railroad Care Management Problem, *Transportation Research B,* 27B, 237–252.

Ahuja, R. K., T. L. Magnanti, and J. B. Orlin, 1993. *Network Flows: Theory, Algorithms, and Applications.* Prentice Hall, Upper Saddle River, NJ.

Altinkemer, K., and B. Gavish, 1991. Parallel Savings Based Heuristic for the Delivery Problem, *Operations Research,* 39, 456–469.

Ash, L. L., 1993. Leadership and the North American Trucking Industry, in B. G. Bison (ed.), *A Look Back from the Year 2000,* 28th Annual Meeting of the Canadian Transportation Research Forum, pp. 30–40.

Assad, A. A., 1980. Models for Rail Transportation, *Transportation Research A,* 14A, 205–220.

Asad, A. A., 1988. Modeling and Implementation Issues in Vehicle Routing, in B. L. Golden and A. A. Assad (eds.), *Vehicle Routing: Methods and Studies,* North-Holland, Amsterdam, pp. 7–45.

Balakrishnan, A., T. L. Magnanti, and R. T. Wong, 1989. A Dual-Ascent Procedure for Large-Scale Uncapacitated Network Design, *Operations Research,* 37(5), 716–740.

Barnhart, C., and D. Kim, 1995. Routing Models and Solution Procedures for Regional Less-Than-Truckload Operations, *Annals of Operations Research,* 61, 67–90.

Barnhart, C., and K. Talluri, 1996. Airlines Operations Research, in A. E. McGarity and C. ReVelle (eds.), *Civil and Environmental Engineering Systems: An Advanced Applications Text,* Wiley, New York.

Bertsimas, D. J., P. Jaillet, and A. R. Odoni, 1990. A Priori Optimization, *Operations Research,* 38(1), 1019–1033.

Bodin, L. D., B. L. Golden, A. A. Assad, and M. O. Ball, 1983. Routing and Scheduling of Vehicles and Crews: The State of the Art, *Computers and Operations Research,* 10(2), 63–211.

Braklow, J. W., W. W. Graham, S. M. Hassler, K. E. Peck, and W. B. Powell, 1992. Interactive Optimization Improves Service and Performance for Yellow Freight System, *Interfaces,* 22(1), 147–172.

Button, K. J., 1993. *Transport Economics,* Edward Elgar, Aldershot, Hants, England.

de Carvalho, T. A., 1996. *A New Approach to Solving Dynamic Resource Allocation Problems,* Ph.D. thesis, Princeton University, Princeton, NJ.

Chen, B., and P. T. Harker, 1990. Two Moments Estimation of the Delay on a Single-Track Rail Line with Scheduled Traffic, *Transportation Science,* 24(4), 261–275.

Cheung, R. K.-M., and W. B. Powell, 1996. An Algorithm for Multistage Dynamic Networks with Random Arc Capacities, with an Application to Dynamic Fleet Management, *Operations Research,* 44(6), 951–963.

Chih, K. C.-K., 1986. A Real Time Dynamic Optimal Freight Car Management Simulation Model of the Multiple Railroad, Multicommodity Temporal Spatial Network Flow Problem, Ph.D. thesis, Princeton University, Princeton, NJ.

Christofides, N., A. Mingozzi, and P. Toth, 1979. The Vehicle Routing Problem, in N. Christofides, A. Mingozzi, P. Toth, and C. Sandi (eds.), *Combinatorial Optimization,* Wiley, New York, pp. 315–338.

Clarke, G., and J. W. Wright, 1964. Scheduling of Vehicles from a Central Depot to a Number of Delivery Points, *Operations Research,* 12, 568–581.

Cordeau, J.-F., and G. Laporte, 1996. Le Transport des marchandises en quelques chiffres, *Routes et Transports,* 25, 43–44.

Cornuéjols, G., and F. Harche, 1993. Polyhedral Study of the Capacitated Vehicle Routing Problem, *Mathematical Programming,* 60, 21–52.

Crainic, T. G., 1982. *Un Modèle de plannification tactique pour le transport ferroviaire des marchandises,* Ph.D. thesis, Université de Montréal, Montreal, Quebec, Canada.

Crainic, T. G., 1984. A Comparison of Two Methods for Tactical Planning in Rail Freight Transportation, in J. P. Brans (ed.), *Operational Research '84,* North-Holland, Amsterdam, pp. 707–720.

Crainic, T. G., 1988. Rail Tactical Planning: Issues, Models and Tools, in L. Bianco and A. La Bella (eds.), *Freight Transport Planning and Logistics,* Springer-Verlag, Berlin, pp. 463–509.

Crainic, T. G., and M.-C. Nicolle, 1986. Planification tactique du transport ferroviaire des marchandises: quelques aspects de modélisation, in *Actes du Premier Congrès International en France de Génie Industriel, CEFI-AFCET-GGI, Paris,* Vol. 1, pp. 161–174.

Crainic, T. G., and J.-M. Rousseau, 1986. Multicommodity, Multimode Freight Transportation: A General Modeling and Algorithmic Framework for the Service Network Design Problem, *Transportation Research B,* 20B, 225–242.

Crainic, T. G., and J. Roy, 1988. O.R. Tools for Tactical Freight Transportation Planning, *European Journal of Operational Research,* 33(3), 290–297.

Crainic, T. G., and J. Roy, 1990. Une Approche de recouvrement d'ensembles pour l'établissement d'horaires des chauffeurs dans le transport routier de charges partielles, *R.A.I.R.O.—Recherche Opérationnelle,* 24(2), 123–158.

Crainic, T. G., and J. Roy, 1992. Design of Regular Intercity Driver Routes for the LTL Motor Carrier Industry, *Transportation Science,* 26(4), 280–295.

Crainic, T. G., J.-A. Ferland, and J.-M. Rousseau, 1984. A Tactical Planning Model for Rail Freight Transportation, *Transportation Science,* 18(2), 165–184.

Crainic, T. G., P. J. Dejax, and L. Delorme, 1989. Models for Multimode Multicommodity Location Problems with Interdepot Balancing Requirements, *Annals of Operations Research,* 18, 279–302.

Crainic, T. G., M. Florian, J. Guélat, and H. Spiess, 1990a. Strategic Planning of Freight Transportation: STAN, An Interactive-Graphic System, *Transportation Research Record,* 1283, 97–124.

Crainic, T. G., M. Florian, and J.-E. Léal, 1990b. A Model for the Strategic Planning of National Freight Transportation by Rail, *Transportation Science,* 24(1), 1–24, 1990b.

Crainic, T. G., M. Gendreau, and P. J. Dejax, 1993a. Dynamic Stochastic Models for the Allocation of Empty Containers, *Operations Research,* 43, 102–302.

Crainic, T. G., M. Gendreau, P. Soriano, and M. Toulouse, 1993b. A Tabu Search Procedure for Multicommodity Location/Allocation with Balancing Requirements, *Annals of Operations Research,* 41, 359–383.

Crainic, T. G., M. Gendreau, and J. M. Farvolden, 1996. *Simplex-Based Tabu Search for the Multicommodity Capacitated Fixed Charge Network Design Problem,* Publication CRT-96-07, Centre de Recherche sur les Transports, Université de Montréal, Montreal, Quebec, Canada.

Daganzo, C. F., 1991. *Logistics Systems Analysis,* Lecture Notes in Economics and Mathematical Systems 361, Springer-Verlag, Berlin.

Daskin, M. S., 1995. *Networks and Discrete Location,* Wiley, New York.

Dejax, P. J., and T. G. Crainic, 1987. A Review of Empty Flows and Fleet Management Models in Freight Transportation, *Transportation Science,* 21(4), 227–247.

Delorme, L., and T. G. Crainic, 1993. Dual-Ascent Procedures for Multicommodity Location–Allocation Problems with Balancing Requirements, *Transportation Science,* 27(2), 90–101.

Delorme, L., J. Roy, and J.-M. Rousseau, 1988. Motor-Carrier Operation Planning Models: A State of the Art, in L. Bianco and A. La Bella (eds.), *Freight Transport Planning and Logistics,* Springer-Verlag, Berlin, pp. 510–545.

Desrosiers, J., Y. Dumas, M. M. Solomon, and F. Soumis, 1995. Time Constrained Routing and Scheduling, in M. Ball, T. L. Magnanti, C. L. Monma, and G. L. Nemhauser (eds.), *Handbooks in Operations Research and Management Science: Network Routing,* Vol. 8, North-Holland, Amsterdam, pp. 35–139.

Dror, M., and P. Trudeau, 1986. Stochastic Vehicle Routing Problem, *Journal of Operational Research,* 23, 282–235.

Eilon, S., C. D. T. Watson-Gandy, and N. Christofides, 1971. *Distribution Management: Mathematical Modelling and Practical Analysis,* Griffin, London.

Erlenkotter, D., 1978. A Dual-Based Procedure for Uncapacitated Facility Location, *Operations Research,* 26, 992–1009.

Farvolden, J. M., and W. B. Powell, 1991. *A Dynamic Network Model for Less-Than-Truckload Motor Carrier Operations,* Working Paper 90-05, Department of Industrial Engineering, University of Toronto, Toronto, Ontario, Canada.

Farvolden, J. M., and W. B. Powell, 1994. Subgradient Methods for the Service Network Design Problem, *Transportation Science,* 28(3), 256–272.

Fisher, M. L., 1995. Vehicle Routing, in M. Ball, T. L. Magnanti, C. L. Monma, and G. L. Nemhauser (eds.), *Handbooks in Operations Research and Management Science: Network Routing,* Vol. 8, North-Holland, Amsterdam, pp. 1–33.

Florian, M., 1984. An Introduction to Network Models Used in Transportation Planning, in M. Florian (ed.), *Transportation Planning Models,* North-Holland, Amsterdam, pp. 137–152.

Florian, M., 1986. Nonlinear Cost Network Models in Transportation Analysis, *Mathematical Programming Study,* 26, 167–196.

Florian, M., and D. Hearn, 1995. Networks Equilibrium Models and Algorithms, in M. Ball, T. L. Magnanti, C. L. Monma, and G. L. Nemhauser (eds), *Handbooks in Operations Research and Management Science: Network Routing,* Vol. 8, North-Holland, Amsterdam, pp. 485–550.

Frantzeskakis, L., 1990. Dynamic Networks with Random Arc Capacities, with Application to the Stochastic Dynamic Vehicle Allocation Problem, Ph.D. thesis, Princeton University, Princeton, NJ.

Frantzeskakis, L., and W. B. Powell, 1990. A Successive Linear Approximation Procedure for Stochastic Dynamic Vehicle Allocation Problems, *Transportation Science,* 24(1), 40–57.

Friesz, T. L., and P. T. Harker, 1985. Freight Network Equilibrium: A Review of the State of the Art, in A. F. Daughety (ed.), *Analytical Studies in Transport Economics,* Cambridge University Press, Cambridge, Chapter 7.

Friesz, T. L., J. A. Gottfried, and E. K. Morlok, 1986. A Sequential Shipper–Carrier Network Model for Predicting Freight Flows, *Transportation Science,* 20, 80–91.

Gendreau, M., A. Hertz, and G. Laporte, 1992. New Insertion and Postoptimization Procedures for the Traveling Salesman Problem, *Operations Research,* 40(6), 1086–1094.

Gendreau, M., A. Hertz, and G. Laporte, 1994. A Tabu Search Heuristic for the Vehicle Routing Problem, *Management Science,* 40, 1276–1290.

Gendreau, M., G. Laporte, and R. Séguin, 1995. An Exact Algorithm for the Vehicle Routing Problem with Stochastic Demands and Customers, *Transportation Science,* 29, 143–155.

Gendreau, M., G. Laporte, and R. Séguin, 1996a. A Tabu Search Algorithm for the Vehicle Routing Problem with Stochastic Demands and Customers, *Operations Research,* 44(3), 469–477.

Gendreau, M., G. Laporte, and R. Séguin, 1996b. Stochastic Vehicle Routing, *European Journal of Operational Research,* 88, 3–12.

Gendron, B., and T. G. Crainic, 1994. *Relaxations for Multicommodity Network Design Problems,* Publication CRT-965, Centre de Recherche sur les Transports, Université de Montréal, Montreal, Quebec, Canada.

Gendron, B., and T. G. Crainic, 1995. A Branch-and-Bound Algorithm for Depot Location and Container Fleet Management, *Location Science,* 3(1), 39–53.

Gendron, B., and T. G. Crainic, 1996. Bounding Procedures for Multicommodity Capacitated Network Design Problems, Publication CRT-96-06, Centre de Recherche sur les Transports, Université de Montréal, Montreal, Quebec, Canada.

Gillett, B., and L. Miller, 1974. A Heuristic Algorithm for the Vehicle Dispatch Problem, *Operations Research,* 22, 340–349.

Glover, F., 1986. Future Paths for Integer Programming and Links to Artificial Intelligence, *Computers and Operations Research,* 1(3), 533–549.

Glover, F., 1989. Tabu Search, Part I, *ORSA Journal on Computing,* 1(3), 190–206.

Glover, F., 1990. Tabu Search, Part II, *ORSA Journal on Computing,* 2(1), 4–32.

Glover, F., and M. Laguna, 1993. Tabu Search, in C. R. Reeves (ed.), *Modern Heuristic Techniques for Combinatorial Problems,* Blackwell Scientific Publications, Oxford, pp. 70–150.

Golden, B. L., and A. A. Assad, (eds.), 1988. *Vehicle Routing: Methods and Studies,* North-Holland, Amsterdam.

Golden, B. L., and E. A. Wasil, 1987. Computerized Vehicle Routing in the Soft Drink Industry, *Operations Research,* 35, 6–17.

Guélat, J., M. Florian, and T. G. Crainic, 1990. A Multimode Multiproduct Network Assignment Model for Strategic Planning of Freight Flows, *Transportation Science,* 24(1), 25–39.

Haghani, A. E., 1989. Formulation and Solution of Combined Train Routing and Makeup, and Empty Car Distribution Model, *Transportation Research B,* 23B(6), 433–431.

Hansen, P., 1986. The Steepest Ascent Mildest Descent Heuristic for Combinatorial Optimization, in *Congress on Numerical Methods in Combinatorial Optimization,* Capri, Italy.

Harker, P. T., and T. L. Friesz, 1986a. Prediction of Intercity Freight Flows: I. Theory, *Transportation Research,* 20B(2), 139–153.

Harker, P. T., and T. L. Friesz, 1986b. Prediction of Intercity Freight Flows: II. Mathematical Formulations, *Transportation Research,* 20B(2), 155–174.

Harker, P. T., and S. Hong, 1990. Two Moments Estimation of the Delay on a Partially Double-Track Rail Line with Scheduled Traffic, *Journal of the Transportation Research Forum,* 31(1), 38–49.

Joborn, M., 1995. Empty Freight Car Distribution at Swedish Railways: Analysis and Optimization Modeling, Ph.D. thesis, University of Linköping, Linköping, Sweden.

Jordan, W. C., and M. A. Turnquist, 1983. A Stochastic Dynamic Network Model for Railroad Car Distribution, *Transportation Science*, 17, 123–145.

Jovanović, D., and P. T. Harker, 1991. Tactical Scheduling of Rail Operations: The SCAN I Decision Support System, *Transportation Science*, 25(1), 46–64.

Keaton, M. H., 1989. Designing Optimal Railroad Operating Plans: Lagrangian Relaxation and Heuristic Approaches, *Transportation Research B*, 23B, 415–431.

Keaton, M. H., 1991. Service–Cost Tradeoffs for Carload Freight Traffic in the U.S. Rail Industry, *Transportation Research A*, 25A(6), 363–374.

Keaton, M. H., 1992. Designing Optimal Railroad Operating Plans: A Dual Adjustment Method for Implementing Lagrangian Relaxation, *Transportation Science*, 26(3), 263–279.

Kruskal, J. B., 1956. On the Shortest Spanning Tree of Graphs and the Travelling Salesman Problem, *Proceedings of the American Mathematical Society*, 7, 48–50.

Labbé M., D. Peeters, and J.-F. Thisse, 1995. Location on Networks, in M. Ball, T. L. Magnanti, C. L. Monma, and G. L. Nemhauser (eds.), *Handbooks in Operations Research and Management Science: Network Routing*, Vol. 8, North-Holland, Amsterdam, pp. 551–624.

Laporte, G., 1992a. The Traveling Salesman Problem: An Overview of Exact and Approximate Algorithms, *European Journal of Operational Research*, 59, 231–247.

Laporte, G., 1992b. The Vehicle Routing Problem: An Overview of Exact and Approximate Algorithms, *European Journal of Operational Research*, 59, 345–358.

Laporte, G., and F. V. Louveaux, 1993. The Integer L-Shaped Method for Stochastic Integer Programs with Complete Recourse, *Operations Research Letters*, 13, 133–142.

Laporte, G., and Y. Nobert, 1987. Exact Algorithm for the Vehicle Routing Problem, *Annals of Discrete Mathematics*, 31, 147–184.

Laporte, G., Y. Nobert, and M. Desrochers, 1985. Optimal Routing Under Capacity and Distance Restrictions, *Operations Research*, 33, 1050–1073.

Magnanti, T. L., and L. A. Wolsey, 1995. Optimal Trees, in M. Ball, T. L. Magnanti, C. L. Monma, and G. L. Nemhauser (eds.), *Handbooks in Operations Research and Management Science: Network Models*, Vol. 6, North-Holland, Amsterdam, pp. 503–615.

Magnanti, T. L., and R. T. Wong, 1984. Network Design and Transportation Planning: Models and Algorithms, *Transportation Science*, 18(1), 1–55.

Minoux, M., 1986. Network Synthesis and Optimum Network Design Problems: Models, Solution Methods and Applications, *Networks*, 19, 313–360.

Nagurney, A., 1993. *Network Economics: A Variational Inequality Approach*, Kluwer, Dordrecht, The Netherlands.

Nelson, M. D., K. E. Nygard, J. H. Griffin, and W. E. Shreve, 1985. Implementation Techniques for the Vehicle Routing Problem, *Computers and Operations Research*, 12(3), 273–283.

Osman, I. H., and J. P. Kelly (eds.), 1996. *Meta-Heuristics: Theory and Applications*, Blackwell Scientific Publications, Oxford.

Owoc, M., and M. A. Sargious, 1992. The Role of Transportation in Free Trade Competition, in N. Waters (ed.), *Canadian Transportation: Competing in a Global Context,* pp. 23–32.

Pirlot, M., 1992. General Local Search Heuristics in Combinatorial Optimization: A Tutorial, *Belgian Journal of Operations Research, Statistics and Computer Science,* 32(2), 8–67.

Powell, W. B., 1988. A Comparative Review of Alternative Algorithms for the Dynamic Vehicle Allocation Problem, in B. L. Golden and A. A. Assad (eds.), *Vehicle Routing: Methods and Studies,* North-Holland, Amsterdam, pp. 249–292.

Powell, W. B., 1995. Toward a Unified Modeling Framework for Real-Time Control of Logistics, *Military Journal of Operations Research.*

Powell, W. B., 1996. A Stochastic Formulation of the Dynamic Assignment Problem, with an Application to Truckload Motor Carriers, *Transportation Science,* 30(3), 195–219.

Powell, W. B., and R. K.-M. Cheung, 1994a. A Network Recourse Decomposition Method for Dynamic Networks with Random Arc Capacities, *Networks,* 24, 369–384.

Powell, W. B., and R. K.-M. Cheung, 1994b. Stochastic Programs over Trees with Random Arc Capacities, *Networks,* 24, 161–175.

Powell, W. B., and L. Frantzeskakis, 1994. Restricted Recourse Strategies for Dynamic Networks with Random Arc Capacities, *Transportation Science,* 28(1), 3–23.

Powell, W. B., and Y. Sheffi, 1989. Design and Implementation of an Interactive Optimization System for the Network Design in the Motor Carrier Industry, *Operations Research,* 37(1), 12–29.

Powell, W. B., Y. Sheffi, K. S. Nickerson, K. Butterbaugh, and S. Atherton, 1992. Maximizing Profits for North American Van Lines' Truckload Division: A New Framework for Pricing and Operations, *Interfaces,* 18(1), 21–41.

Powell, W. B., T. A. Carvalho, G. A. Godfrey, and H. P. Simaõ, 1994. Optimal Control Methods for Logistics Queueing Networks, in *TRISTAN II, TRIennial Symposium on Transportation ANalysis, Preprints,* pp. 185–199.

Powell, W. B., T. A. Carvalho, G. A. Godfrey, and H. P. Simaõ, 1995a. Dynamic Fleet Management as a Logistics Queueing Network, *Annals of Operations Research,* 61, 165–188.

Powell, W. B., P. Jaillet, and A. Odoni, 1995b. Stochastic and Dynamic Networks and Routing, in M. Ball, T. L. Magnanti, C. L. Monma, and G. L. Nemhauser (eds.), *Handbooks in Operations Research and Management Science: Network Routing,* Vol. 8, North-Holland, Amsterdam, pp. 141–295.

Psaraftis, H. N., 1995. Dynamic Vehicle Routing: Status and Prospects, *Annals of Operations Research,* 61, 143–164.

Quinet, E., 1990. *Analyse économique des transports,* Presses Universitaires de France, Paris.

Reeves, C. R. (ed.), 1993. *Modern Heuristic Techniques for Combinatorial Problems,* Kluwer, Norwell, MA.

Rochat, Y., and E. Taillard, 1995. Probabilistic Diversification and Intensification in Local Search for Vehicle Routing, *Journal of Heuristics,* 1(1), 147–167.

Roy, J., 1984. *Un Modèle de planification globale pour le transport routier des marchandises.* Ph.D. thesis, Université de Montréal, Montreal, Quebec, Canada.

Roy, J., and T. G. Crainic, 1992. Improving Intercity Freight Routing with a Tactical Planning Model, *Interfaces,* 22(3), 31–44.

Roy, J., and L. Delorme, 1989. NETPLAN: A Network Optimization Model for Tactical Planning in the Less-Than-Truckload Motor-Carrier Industry, *INFOR,* 27(1), 22–35.

Sheffi, Y., 1985. *Urban Transportation Networks: Equilibrium Analysis with Mathematical Programming,* Prentice Hall, Upper Saddle River, NJ.

Shen, S., J.-Y. Potvin, J.-M. Rousseau, and S. Roy, 1995. A Computer Assistant for Vehicle Dispatching with Learning Capabilities, *Annals of Operations Research,* 61, 189–211.

Taff, C. A., 1978. *Management of Physical Distribution and Transportation,* Irwin, Homewood, IL.

Taillard, E., 1993. Parallel Iterative Search Methods for Vehicle Routing Problems, *Networks,* 23, 661–673.

Turnquist, M. A., 1994. Economies of Scale and Network Optimization for Empty Car Distribution, presented at the TIMS/ORSA National Meeting, Detroit, MI.

Yellow, P., 1970. A Combinatorial Modification to the Savings Method of Vehicle Scheduling, *Operational Research Quarterly,* 21, 281–283.

Zalatan, P., 1993. Economic Cycles, Structural Change and the Transportation Sector, in B. G. Bison (ed.), *A Look Back from the Year 2000,* 28th Annual Meeting of the Canadian Transportation Research Forum, pp. 111–121.

CHAPTER 9

Public Transport Operations Planning[1]

AVISHAI CEDER and NIGEL H. M. WILSON

9.A. INTRODUCTION

In this chapter we describe how systems methods can be used to solve planning and operational problems in public transport authorities. The emphasis is on techniques that have been applied or that have potential for application. Public transport problems are examined in isolation from two larger problems in which they are embedded. First, public transport is just one element of the larger urban transport problem. Street congestion, availability of reserved lanes (road space reserved for public transport), parking policy and pricing, intermodal services, and limits on private car utilization can all have a significant impact on public transport planning and operations. Second, urban development plans may directly influence decisions concerning the location of major public transport infrastructure (e.g., rail or other fixed facilities) so as to encourage desired urban development. These important interrelations are not considered here.

9.B. COMPONENTS OF PUBLIC TRANSPORT PLANNING

Public transport operations planning includes four basic components performed in sequence: (1) network and route design, (2) setting timetables, (3) scheduling vehicles to trips, and (4) scheduling crews. Clearly, it would be desirable for all four components to be planned simultaneously so as to exploit the interactions and interdependencies and to maximize the system's productivity and efficiency. However, this planning process is extremely complex, and mathematically it is not yet feasible to consider all these components simultaneously. Typically, the full problem is decomposed so that each com-

[1]Parts of this chapter are similar to parts of Chapter 5, "Models in Urban and Air Transportation," by Odoni, Rousseau, and Wilson in Pollock et al. (1994).

ponent is treated separately, with the outcome of one becoming an input to the next component.

9.B.1. Planning Process

The overview of this planning process is shown in Figure 9.1, with an emphasis on the scheduling components. The first component in Figure 9.1 represents the more strategic planning activity of designing new routes or modifying existing routes given the existing infrastructure, travel demand, and operating policies and objectives. The second component is aimed at meeting the general demand for public transportation travel described both spatially and temporarily. Demand varies across hours of the day, days of the week, and seasons of the year, as well as changing from year to year. This demand reflects the business, industrial, cultural, educational, social, and recreational transportation needs of the community. It is the purpose of this component of the planning process to set appropriate frequencies and hence timetables for each transit route, to meet the variation in the demand across routes and across time. These decisions are based on passenger counts, either actual or forecast, and policies on passenger crowding and minimal acceptable service frequencies.

The third component in Figure 9.1 is scheduling vehicles to trips so as to satisfy the given timetables. A transit trip can be either planned to transport passengers along a route or to operate out of service (a deadhead trip) to connect two service trips or to move the vehicle between the depot and the start or end locations of service trips. The vehicle scheduling task is to develop all daily chains (or blocks) of trips (sequences of service and deadhead trips) for each vehicle, so as to satisfy the timetable requirements and other operational requirements (e.g., vehicle maintenance and refueling). The fourth component of Figure 9.1 is scheduling crew duties (also referred to as *runs*) so as to cover all the vehicle blocks. These schedules must comply with labor contract terms and provisions governing issues such as working hours, run length, and number of pieces of work as well as organization policies and priorities. Any public transport operator that wishes to utilize its resources efficiently has to deal with problems resulting from various wage scales (e.g., regular, overtime) and preferences for different types of duties. Components 3 and 4 in Figure 9.1 are very sensitive to both internal and external constraints, which can easily lead to inefficient solutions if not treated carefully.

9.B.2. Problem Decomposition

An important first step in solving many large and complex problems is to break them down into pieces that are significant, yet tractable. The decomposition represented in Figure 9.1 is certainly not the only one possible, but it is a good reflection of the approach taken in most public transport authorities as well as most analytical treatments of the public transport planning

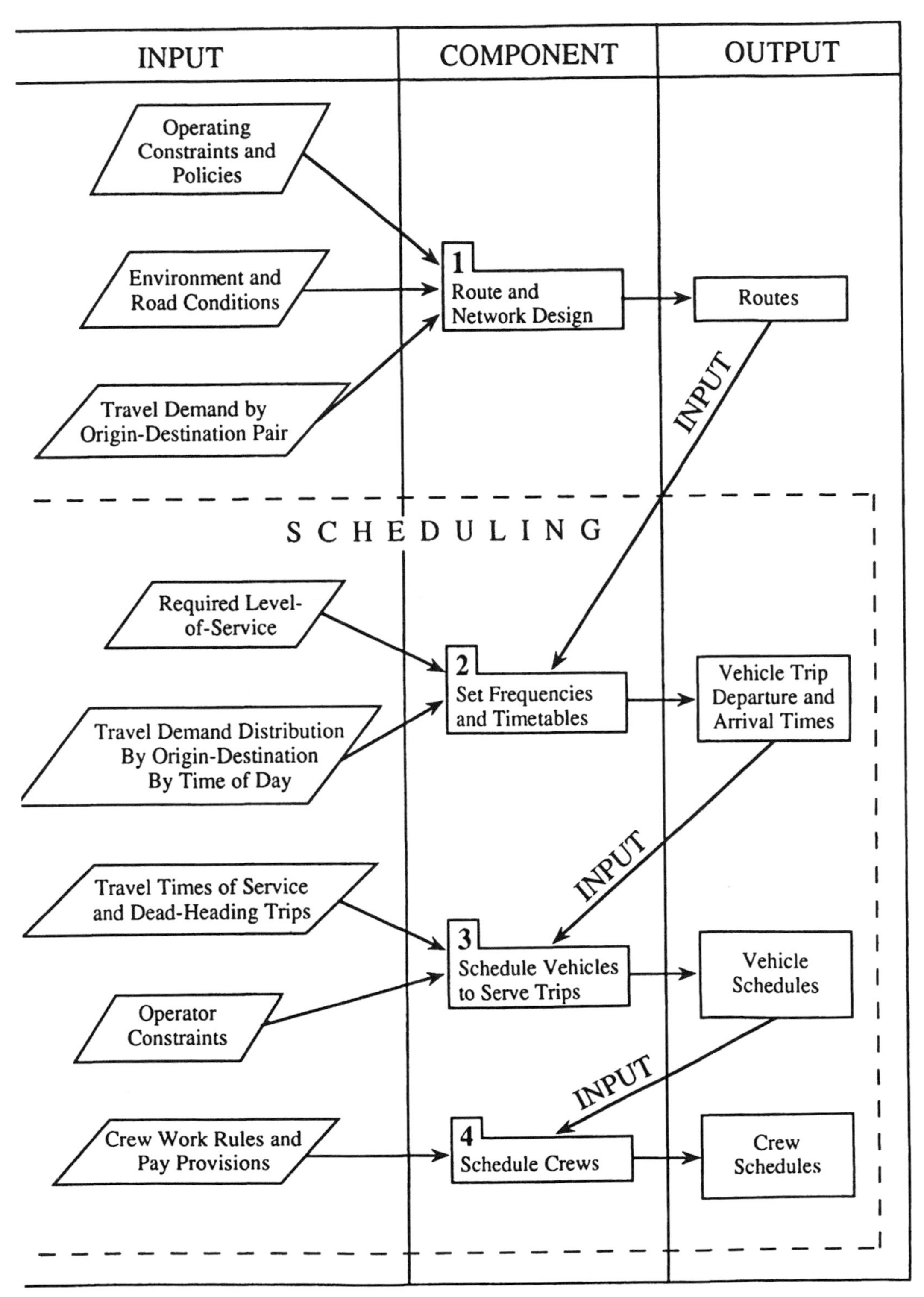

Figure 9.1 Components of the public transport operational planning process.

problem. Generally, strategic planning activities such as designing new routes or constructing new infrastructure are separated from the timetabling and vehicle and crew scheduling components, which are in turn separated from the day-to-day management of operations. We describe these four components of the problems in more detail in the following sections of this chapter, together with some of the approaches used to tackle each component.

We can also classify these components of the public transport planning process into those at the strategic, tactical, and operations levels. Strategic planning is concerned with long-term development of the system, including fixed infrastructure investments and major modifications to the route network. Since it is difficult for the public to adapt to major revisions in the network, most changes of this type are the result of comprehensive studies and must stay in place for some time to achieve their full potential impact. As such, there is a stability to the bus network structure, making it appropriately part of the strategic planning problem.

The tactical planning problem includes minor route revisions, the structuring of routes in high-ridership corridors, and the assignment of service frequencies to routes. These aspects of the overall problem often do not receive the attention they deserve, with some of the attendant strategies being implicitly part of the service planning function while others are included in the operational planning function. Operational planning encompasses all scheduling-related activity, including the design of timetables, the scheduling of vehicles, and the generation of crew duties for drivers together with their days off.

9.C. DESIGN OF PUBLIC TRANSPORT ROUTES AND NETWORKS

9.C.1. Problem Description

Conflicting Goals. The public transport network design problem is one of a class of very difficult network design problems covering most scheduled carrier systems, including airlines and rail passenger systems. There are several reasons for this complexity. First is the existence of the route as an intermediate structure between the link and the full network. Second is the strong interdependence between the design of routes by the operator and the choice of travel itinerary carried out independently by users. We are in a user optimum context rather than a system optimum context because the network planning agency does not decide the actual routing (or routing strategies) of the "traffic" on its network (which is the case for companies in industries such as telephone, pipeline, trucking). In addition, as in many design problems, the objective function is ill defined. The planner certainly wants to minimize the total travel time of the population, but also wants to give "equitable" access to the system to every taxpayer, design "attractive routes" that will be "easy" to use, and perhaps "favor" the use of heavy infrastructure such as subways to minimize operating cost, while taking into account "acquired rights" to certain services and dealing with political pressures for better services in individual political jurisdictions.

9.C.2. Alternative Approaches

Interactive Methods. For all these reasons, it may be better to take an approach in which the design of new infrastructure or new routes is undertaken interactively, with the planner choosing scenarios (routes and network structures) and the computer evaluating these scenarios and providing the planner with information to help in the design of new scenarios. Indeed, while many attempts have been made to design routes heuristically by computer (see, e.g., Lampkin and Saalmans, 1967; Last and Leak, 1976; Mandl, 1980; Hasselström, 1981) all transit network models that have received widespread application are of an interactive type.

VIPS-II, developed jointly by VTS and Stockholm Transport, is the only one of these models that allows the automatic generation of transit routes, as described in Hasselström (1981). It is first assumed that there is a direct transportation link between each pair of zones, and the frequency of service is calculated using a simple model of the type described in Section 9.D. The algorithm assigns passengers to paths between each origin–destination pair so as to minimize their travel time. At each iteration it sequentially drops from the network the least productive routes, reallocating the affected passengers to the remaining routes and recalculating the frequencies of services until a satisfactory solution is reached (see Figure 9.2). In addition to this automatic generation of routes, routes can be specified by the planner, with the model then being used for network evaluation as in the approaches to be described below.

Network Simulation. Several computer packages have been designed for transit network simulation and evaluation and so can be used in the interactive design of bus routes and transit infrastructure. Among these are Transcom (Chapleau, 1974), Transept (Last and Leak, 1976), NOPTS (Rapp et al., 1976), VIPS (Andreasson, 1976) and its successor VIPS-II, U load (UMTA/FHWA, 1977), TRANSPLAN (Osleeb and Moellering, 1976), IGTDS (General Motors, 1980), Madituc (Chapleau et al., 1982) and EMME/2 (Babin et al., 1982; Florian and Spiess, 1986). Of these, VIPS-II and EMME/2 appear to be among the most sophisticated evaluation models. Both provide extensive graphical displays that assist the user in the evaluation of a transit network. In each, the planner can define scenarios interactively and the program then forecasts how the passengers will use the system, providing answers through a series of graphical and numerical outputs. Whereas VIPS-II is strictly a public transport planning tool, EMME/2 provides a multimodal urban transportation planning environment, letting the user define up to 10 modes of transportation. Both systems have been used extensively around the world.

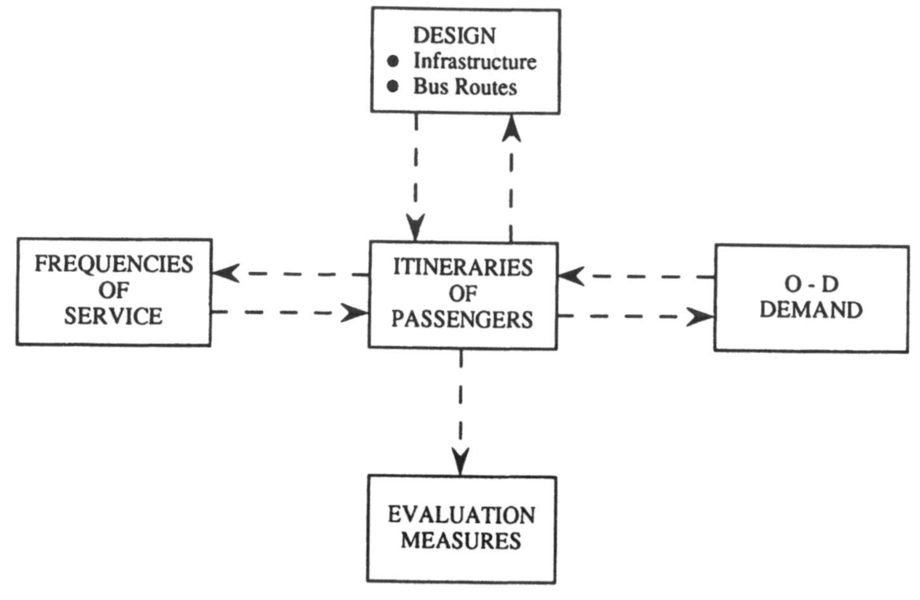

Figure 9.2 Network design problem.

Any transit network evaluation tool supposes that the transit network is given, including the frequency of service on each route, and that total travel demand by transit between all zones is known (in some cases a modal split function is used to extract the transit demand from a total origin–destination travel matrix). From this information it calculates the itinerary of passengers and the transit flow on each line together with average trip duration between each origin–destination zone pair and other local and global measures. In this context one possible feedback loop is the reevaluation of the matrix by using a mode choice function. The process may be repeated until a stable solution is obtained.

Passenger Route Selection. A key problem in any public transport network evaluation model, as well as in any heuristic approach to network design, is to determine the passenger assignment to routes in the network. The key behavioral principle is well accepted; the passenger wants to travel at minimum "inconvenience," this being a weighted combination of access time, waiting time, travel time, travel cost, and transfer inconvenience. Any generalized cost functions can be calibrated and tested in each context assigning different weights to the various factors comprising it.

However, even given this objective function, the solution of the passenger assignment problem is not as simple as it might at first sight appear. First, the origin–destination matrix is a zonal aggregate; it will not always be true that persons living in the same zone and traveling to the same destination zone will choose the same route. One reason for this is that their exact origins and destinations in either zone may dictate different choices, given the importance of access time. Second, several transit routes may serve a given street segment or a given intersection. The optimal choice of itinerary for a transit user may very well be dictated by the arrival of the first bus on any of these routes. In the same way, the traveler may well change itinerary given the arrival of a particular bus at a certain street intersection, and so on.

The transit passenger assignment problem has been studied by many authors, either as a separate problem (see, e.g., Dial, 1967; LeClerq, 1972; Chriqui, 1974; Chriqui and Robillard, 1975; Andreasson, 1976; Rapp et al., 1976; Spiess, 1983; Marguier and Ceder, 1984) or as a subproblem of more complex models, such as transit network design (see, e.g., Lampkin and Saalmans, 1967; Schéele, 1977; Mandl, 1980; Hasselström, 1981; Jansson and Ridderstolpe, 1992) or multimodal network equilibrium (Florian, 1977; Florian and Spiess, 1983).

Following Dial (1967), many of these algorithms are based on the assumptions of deterministic running times and exponentially distributed headways on all routes, in which case the market share for a particular route is simply its frequency share. Jansson and Ridderstolpe (1992) and Israeli and Ceder (1996) show that when headways are deterministic rather than exponential, the frequency share model does not always hold, with the result depending on the degree of schedule coordination. Chriqui and Robillard (1975) and Marguier and Ceder (1984) both examine choice among routes on a common path. Chriqui and Robillard proposed a heuristic for selecting a set of acceptable paths among which the passenger would select the first bus to arrive so as to minimize the expected total travel time. They showed that their heuristic is optimal when all routes have exponentially distributed headways and identical running times, but Marguier and Ceder showed that this heuristic is not optimal for all headway distributions. Jansson and Ridderstolpe present an alternative heuristic for this problem for the deterministic headway case.

Spiess (1983), De Cea et al. (1988), and Spiess and Florian (1989) present linear programming solutions to the transit path assignment problem as a relaxation of a mixed-integer program under the assumption of exponentially distributed headways. There are two main differences between these models and other approaches. First and foremost, in these models the traveler does not simply choose a path, but rather, selects a strategy. While in earlier methods the transit traveler's route choice is limited to one path, defined as a sequence of route segments, links, or transfer points, the strategies considered in Spiess's approach allow the transit rider to select any subset of paths leading to the destination with the first vehicle to arrive determining which of the alternative routes is actually taken on an individual trip. This concept of optimal strategies allows more realistic modeling of the traveler's behavior. Second, the resulting transit assignment problem, formulated as a mathematical programming problem, may be solved directly without recourse to heuristic approaches.

The principal difference between the approaches of Speiss and De Cea is in the representation of the problem in network form. Speiss's algorithm is in the form of a shortest-path problem, whereas De Cea's approach results in network search to define feasible paths, followed by Chriqui's heuristic to determine optimal paths. Speiss's algorithm solves the problem in polynomial time and so may be applied even to very large transit networks. This method is used in EMME/2. A method based on similar ideas has also been developed and implemented in Torino by Nguyen and Pallottino (1988).

9.C.3. Mathematical Programming Approach

Problem Formulation. In this section a mathematical programming approach is presented which aims to be practical, given typical data availability, and is less complex than other models.[2] This should increase the chances of acceptance within most public transport agencies. This approach combines the philosophy of the mathematical programming approaches with multiobjective decision-making techniques, so as to allow the user to select from a number of alternatives.

The transit route design problem is based on two objective functions, Z_1 and Z_2:

$$\text{minimize } Z_1 = \alpha_1 \sum_{i,j \in N} \text{PH}(i,j) + \alpha_2 \sum_{i,j \in N} \text{WH}(i,j) + \alpha_3 \sum_{r} \text{EH}_r \quad (9.1)$$

$$\text{minimize } Z_2 = \text{FS} \quad (9.2)$$

where PH(i,j) represents the passenger hours between nodes i and j (defined as passengers' bus riding time between nodes i and j), WH(i,j) the waiting time between nodes i and j (defined as the amount of waiting time between nodes i and j), EH_r the empty space-hours on route r, FS the fleet size, and α_1, α_2, and α_3 are coefficients reflecting the relative importance of passenger riding time, passenger waiting time, and unused capacity.

The nature of the overall formulation is nonlinear, mixed-integer programming and it is an analog problem to the generalized network design problem described by Magnanti and Wong (1984) with NP-hard computational complexity. Thus conventional approaches are incapable of providing an exact solution even with significant simplification.

In the first stage, the problem dimension is reduced through the construction of a skeleton route network that satisfies a maximum travel time constraint. The skeleton network is the basis for an optimization routine to determine the shortest direct and indirect (including transfers) paths between each pair of nodes. The second stage relies on a procedure that incorporates optimization and enumeration processes to derive the minimal Z_1 objective function. This procedure, while searching for minimum Z_1, creates various Z_2 solutions, each associated with a different Z_1 solution. Finally, the most desirable set of (Z_1, Z_2) is derived through multiobjective programming tech-

[2]This approach is based heavily on the method developed by Ceder and Israeli (1992).

niques (Cohon, 1978; Goicoechea et al., 1982). Further details of the theoretical dimension of this methodology can be found in Israeli (1992). A thorough treatment of multiobjective programming methods is presented in Chapter 12.

A simple eight-node example is used throughout this section to demonstrate the procedures used. The basic network with two terminals (the only nodes from which routes can be initiated) is shown in Figure 9.3 with the input demand presented in Table 9.1.

Solution Algorithm. The overall methodology comprises seven elements, as shown in Figure 9.4. In the first element, the system generates every possible route, from all terminals, which satisfies the route-length constraints. In other words, the algorithm screens out potential routes based on a given upper bound on route length. In addition, there is a limit on the travel time between each origin–destination pair. That is, a given demand cannot be assigned to a bus route if its travel time would exceed the minimum travel time on the network by more than a given percentage. The outcome of this part is a set of routes directly connecting only some of the origin–destination pairs. In this component the user can also introduce additional routes (which may be

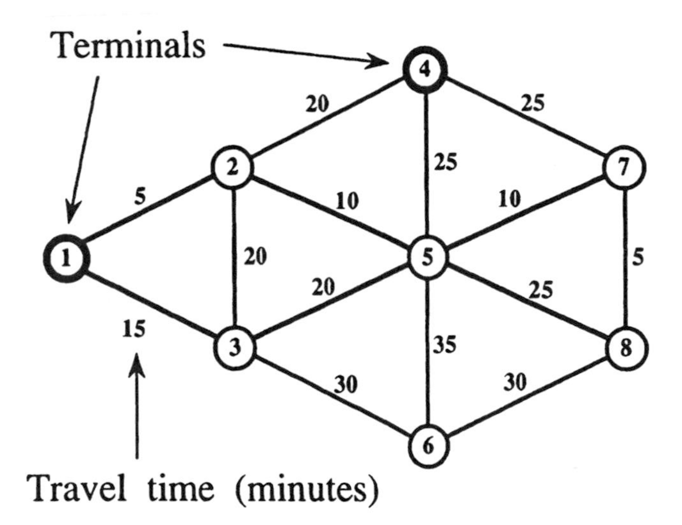

Maximum route circuity factor (α)	= 0.40
Maximum number of transfers	= 2
Bus capacity	= 50

Figure 9.3 Eight-node network with basic input.

TABLE 9.1 Origin–Destination Demand in the Example

	To							
From	1	2	3	4	5	6	7	8
1	0	80	70	160	50	200	120	60
2	80	0	120	90	100	70	250	70
3	70	120	0	180	150	120	30	250
4	160	90	180	0	80	210	170	230
5	50	100	150	80	0	250	40	130
6	200	70	120	210	250	0	130	120
7	120	250	30	170	40	130	0	70
8	60	70	250	230	130	120	120	0

either existing routes or manually designed routes). The outcome of the first component is presented in Table 9.2, while using $\alpha = 0.4$ (no portion of a route can exceed its shortest travel time by more than 40%).

The second element is intended to make sure that a feasible path exists for every origin–destination pair. The first step of the algorithm is to establish additional direct routes between high-demand origin–destination pairs (a predetermined origin–destination demand threshold). Since these direct routes were not generated in the first stage, they must originate and/or terminate at nonterminal nodes, and consequently, deadheading is required to connect them to designated terminal nodes. Low-demand origin–destination pairs without a direct route are not considered for additional service.

Next, feasible paths requiring transfers are identified for all origin–destination pairs, recognizing a maximum number of transfers on any path within the bus network (typically, one or two). The number of transfers is simply the number of routes on any origin–destination path minus one. An origin–destination transfer path is created subject to an additional limit on its travel time: It cannot be greater than a certain percentage over the shortest travel time. If for an origin–destination pair, none of the transfer paths are feasible, the one with the least violation is chosen. The algorithm is based on a recurrence relationship so that initially it searches for one transfer path, with travel time less than the given limit, and then continues to the two transfer paths. The output of this element is a route, or set of routes, for each origin–destination pair that defines acceptable travel paths, including transfer nodes.

At this stage there is a large set of routes, which provide connectivity throughout the network and satisfy the travel-time and route-length constraints. However, the large set of routes is likely to contain many overlapping segments, defined as segments that serve origin–destination pairs that are completely served by other routes. An overlapped route is one in which all segments are overlapping; this case will be treated in the next element of the methodology.

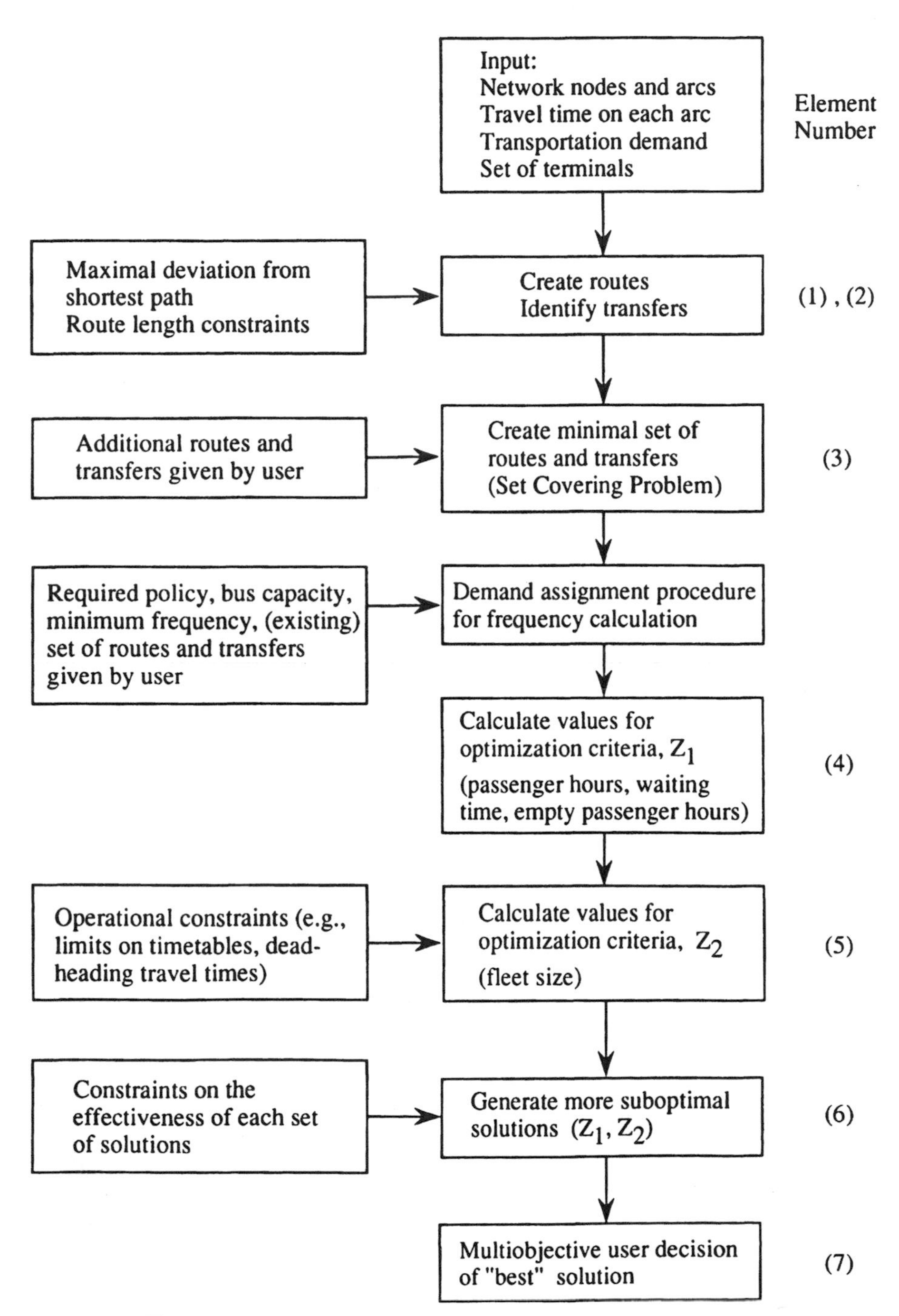

Figure 9.4 Methodology for designing routes and networks.

TABLE 9.2 All Routes Generated in First Element

Route	Description
1	1 → 2
2	1 → 2 → 4
3	1 → 2 → 5
4	1 → 2 → 5 → 6
5	1 → 2 → 5 → 7
6	1 → 2 → 5 → 7 → 8
7	1 → 2 → 5 → 7 → 8 → 6
8	1 → 3
9	1 → 3 → 6
10	4 → 2
11	4 → 2 → 1 → 3
12	4 → 2 → 1 → 3 → 6
13	4 → 2 → 3
14	4 → 2 → 3 → 6
15	4 → 2 → 5
16	4 → 2 → 5 → 6
17	4 → 5
18	4 → 5 → 3
19	4 → 5 → 6
20	4 → 5 → 7
21	4 → 5 → 7 → 8
22	4 → 5 → 7 → 8 → 6
23	4 → 7
24	4 → 7 → 5
25	4 → 7 → 5 → 3
26	4 → 7 → 5 → 6
27	4 → 7 → 8
28	4 → 7 → 8 → 6

In the example there are 18 transfer paths that meet the criteria as shown in Table 9.3, where the numbers in parentheses are the route numbers (see Table 9.2) that comprise segments of the path. In the transfer path description, the numbers outside the parentheses represent (transfer) nodes, while those inside the parentheses represent routes.

In the third element, the system creates minimal set(s) of routes and their related transfers, such that connectivity between nodes is maintained and their total deviation from the shortest path is minimized. This problem can be formulated as a set covering problem (SCP), (see Minieka, 1978; Syslo et al., 1983) in which each column represents either a feasible route or a transfer path and each row represents an origin–destination pair. A "1" in the matrix means that the corresponding origin–destination demand can be served by that route or transfer path; otherwise, the coefficient is 0. The word *covering*

TABLE 9.3 Transfer Paths Generated in Second Element

Transfer Number	Description
29 (5,18,27)	3(18) → 5(5) → 7(27) → 8
30 (5,18,28)	3(18) → 5(5) → 7(28) → 8
31 (6,18)	3(18) → 5(6) → 7(6) → 8
32 (6,25)	3(25) → 5(25,6) → 7(6) → 8
33 (7,18)	3(18) → 5(7) → 7(7) → 8
34 (7,25)	3(25) → 5(25,7) → 7(7) → 8
35 (18,20,27)	3(18) → 5(20) → 7(27) → 8
36 (18,20,28)	3(18) → 5(20) → 7(28) → 8
37 (18,21)	3(18) → 5(21) → 7(21) → 8
38 (18,22)	3(18) → 5(22) → 7(22) → 8
39 (18,24,27)	3(18) → 5(24) → 7(27) → 8
40 (18,24,28)	3(18) → 5(24) → 7(28) → 8
41 (18,26,27)	3(18) → 5(26) → 7(27) → 8
42 (18,26,28)	3(18) → 5(26) → 7(28) → 8
43 (21,25)	3(25) → 5(25,21) → 7(21) → 8
44 (22,25)	3(25) → 5(25,22) → 7(22) → 8
45 (25,27)	3(25) → 5(25) → 7(27) → 8
46 (25,28)	3(25) → 5(25) → 7(28) → 8

refers here to every row having at least one column with coefficient 1. The transfer paths are combined columns in the SCP matrix, which increases the complexity of the problem. No exact solution algorithm is known for this problem, but a heuristic algorithm has been developed and tested with a random network.

The outcome of this element is a set of the minimum number of routes that cover all the origin–destination pairs in the network. Israeli (1992) provides a heuristic search algorithm which includes three covering variations:

1. Covering only the shortest paths between all origin–destination pairs
2. Covering all origin–destination paths while minimizing the number of paths that are not the shortest
3. Covering all origin–destination paths while minimizing the combined paths' travel time

Nine sets of routes are generated in this example as presented in Table 9.4, with routes and transfer paths as defined in Tables 9.2 and 9.3.

In the fourth element the entire origin–destination demand is assigned to the chosen set of routes. The assignment algorithm (see Israeli and Ceder, 1996) is based on a probabilistic function for passengers who either take the bus that arrives first, or alternatively, wait for a faster bus. The passengers' assumed strategy is to minimize total travel time. The methodology used is

TABLE 9.4 All Route Subsets Generated in Third Element

Set	Description
1	{4,6,9,11,25,28,32,46}
2	{7,9,11,19,25,27,34,35}
3	{7,9,11,25,28,34,46}
4	{6,9,11,16,25,28,32,46}
5	{6,12,19,25,28,32,46}
6	{7,12,25,27,34,45}
7	{7,12,25,28,34,46}
8	{4,6,12,25,28,32,46}
9	{6,12,16,25,28,32,46}

similar to that developed by Marguier and Ceder (1984) but with a different probabilistic function. This function considers the route length and the length of a transfer path between each origin–destination pair, where both the routes and transfer paths are divided into "slow" and "fast" categories. The bus frequency is used as a variable in the algorithm and is derived from the passenger load profile of each route. The heart of the methodology is a set of equations (of degree 3), which are solved iteratively. The outcome of this element is: bus frequencies for the set of routes, passenger load profiles, demand assignment across the set of routes, and the optimization parameters PH, WH, and EH for computing Z_1.

The fifth element represents the operator's perspective by finding the minimum fleet size required to meet the demand as well as to satisfy the determined frequency on each route. This objective is designated Z_2, in the formulation, and it complements the operator's term in Z_1, which is to minimize empty space-hours. The method used for evaluating the fleet size is based on the deficit function theory (see Section 9.E) proposed by Ceder and Stern (1981). For this case, the lower-bound calculation described in Stern and Ceder (1983) is used, based on the frequencies determined in the fourth element of the model. The final outcome of this step is the value Z_2, which represents the minimum required fleet size.

The sixth element is concerned with constructing alternative sets of routes in order to search locally for additional (Z_1,Z_2) values in the vicinity of the current best solution. The procedure for this search is based on incremental changes in the set of routes, much like the reduced gradient methods. Given the set of routes associated with the minimum Z_1 value, the single route that contributes least to Z_1 is deleted and then the SCP (element three) is resolved, followed by the execution of the fourth and fifth elements. This process could continue, but there is no guarantee that a previous alternative will not be repeated. To avoid this problem, a new matrix is constructed with the idea of finding the minimal set of candidate routes for possible deletion in each iteration [i.e., a new SCP matrix is constructed in which the candidate routes are the columns and each row represents a previous set of routes that was already identified in the vicinity of the optimal (Z_1,Z_2) setting]. The solution to this new SCP matrix is a set of rejected routes selected so as to prevent a previous alternative solution being repeated. During this process, a number of unique sets of routes are termed *prohibited columns,* as they are the only ones that can transport certain demand. These prohibited columns are assigned an artificially high cost, to ensure their exclusion from the solution. This process also involves some bounds to ensure convergence in a desired number of iterations, or number of (Z_1,Z_2) solutions.

In the example problem we set $\alpha_1 = \alpha_2 = \alpha_3 = 1$ in the objective function, equation (9.1), for the weighted sum of the optimization parameters and the lower bound on the fleet size in equation (9.2). Nine "good" sets were produced by the sixth element, as shown in Table 9.4, and their (Z_1,Z_2) values appear in Table 9.5.

The seventh and final element of the methodology involves multiobjective programming of the two objective functions Z_1 and Z_2. Given the alternative sets of routes served in the sixth element, the purpose is to select the best overall (Z_1,Z_2) solution. The method selected in this element, the *compromise set method,* is based on Zeleny (1973,1974). It fits linear objective functions from which Duckstein and Opricovic (1980) derived solutions for discrete variables. The outcome of this method is the theoretical points at which (Z_1,Z_2) can attain their minimal combined value. The results can be presented as a table or a two-dimensional graph that shows the trade-offs between Z_1 and Z_2. The decision maker can then decide which of the possible solutions to accept based on their relative values of Z_1 and Z_2. The trade-off situation for the example problem is shown in Figure 9.5. The lower left corner of the envelope contour of the nine solutions represents the best alternatives. The user is then able to choose the desired solution. In the example, the choice is between $(Z_1 = 788, Z_2 = 106)$ and $(Z_1 = 866, Z_2 = 102)$.

9.D. FREQUENCY DETERMINATION AND TIMETABLE SETTING

Tactical Planning. The tactical planning problem for public transport has received relatively little serious attention. Figure 9.6 summarizes the various elements of this problem. The transit network is assumed known and the objective is to allocate the available resources (buses and drivers) to provide the best possible service at the least cost. Several problems can be studied at this level, including the transit assignment problem already discussed at length in Section 9.C. Another problem that has received some attention (Fearnside and Draper, 1971; Last and Leak, 1976; Schéele, 1977) is that of choosing the optimal frequencies for bus routes, given a fixed fleet of vehicles, to minimize total travel time. The frequency of service influences the itinerary and duration of travel, which in turn may influence travel demand. Schedule generation in private-sector transportation planning is discussed in this book in the contexts of freight transportation [Chapter 8 (Section 8.D.1)] and airline transportation [Chapter 10 (Section 10.B.1)].

TABLE 9.5 Alternative Sets of Routes in the Example

Set	Z_1	Z_2 = FS
1	788	106
2	900	109
3	1105	117
4	866	102
5	937	103
6	997	101
7	1213	113
8	869	103
9	961	105

There are two fundamentally different ways of defining the frequency of service in a public transport network. Most operators set frequencies, or the number of vehicle trips required per hour, so as to obtain a given mean number of passengers per vehicle at the peak loading point on each route, subject to a minimum frequency constraint. Typically, both acceptable peak loads and minimum frequencies vary by time of day and perhaps by route type. This type of approach has been formulated (with extensions) by Ceder

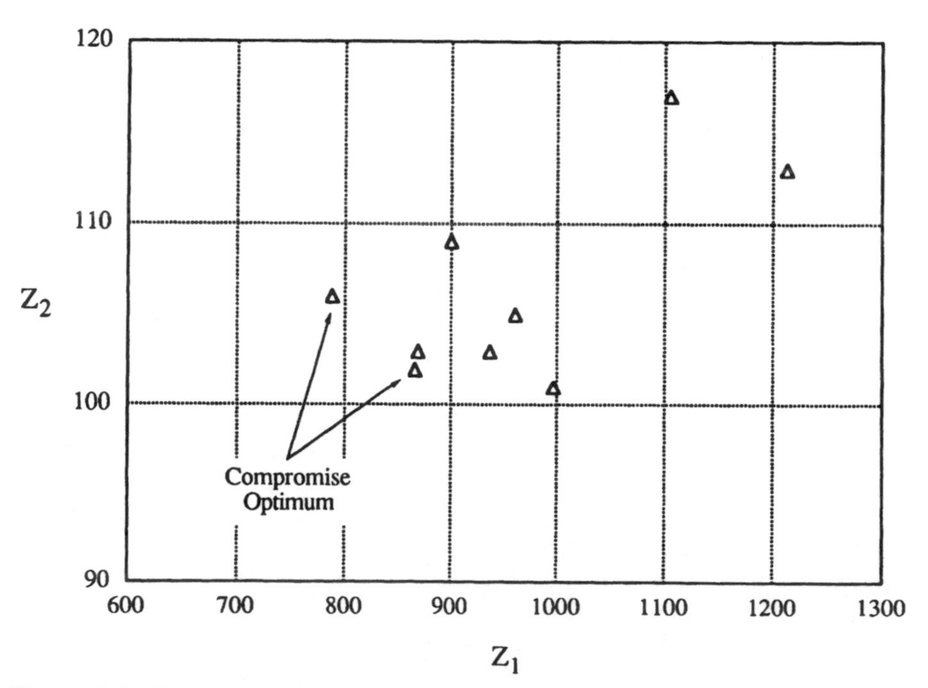

Figure 9.5 Trade-off between Z_1 and the minimal fleet size, Z_2 in the example problem.

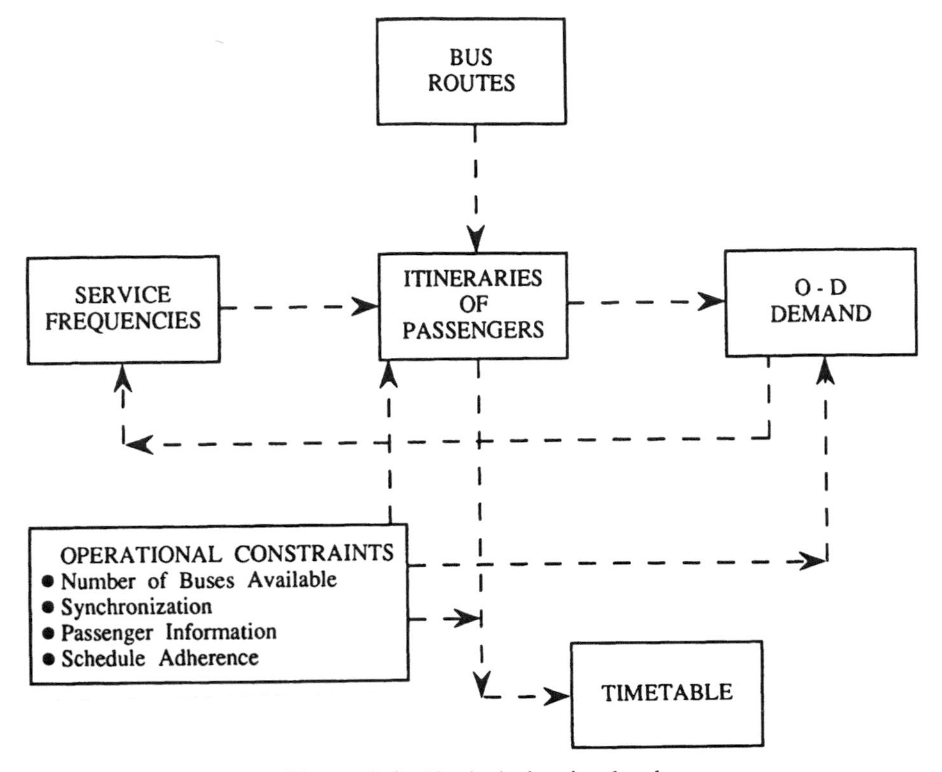

Figure 9.6 Tactical planning level.

(1984). An alternative approach treats the problem as one of resource allocation, with the resources, for example bus-hours, being allocated across routes and across time periods so as to maximize some objective subject to passenger loading constraints. The theories of determining optimal frequencies are described in the next section.

9.D.1. Optimization Approaches

Square Root Rule. The best known simple theory for setting frequencies on bus routes is the square root rule developed by Mohring (1972), which is derived below in terms of the headway, defined as the scheduled time between consecutive buses on a route. Headway is simply the inverse of frequency. The total cost can be defined as the sum of the operator and passenger time cost:

$$C = \sum_{i=1}^{N} \left(\frac{ct_i}{h_i} + \frac{bh_i r_i}{2} \right) \tag{9.3}$$

where

C = total cost per hour of service

i = route number, $i = 1,...,N$

c = operating cost per bus hour

b = value of passenger waiting time

t_i = round trip travel time for route i

h_i = headway on route i

r_i = ridership per hour on route i

The total cost can be minimized by setting the headway on each route to

$$h_i = \left(\frac{2ct_i}{br_i}\right)^{1/2} \tag{9.4}$$

This relationship implies that the ridership per bus trip will vary from route to route, being highest on the longest, heaviest ridership routes, quite different from the standard industry practice of having approximately equal passenger loads across all routes in the same time period.

There are several assumptions behind the square root rule, which help explain its total lack of acceptance by the industry:

1. It does not consider the effects of bus capacity constraints.
2. It assumes that ridership is fixed and independent of the service frequency provided.
3. It does not allow for minimum service level constraints.
4. It does not allow for a budget constraint.
5. It treats each route as being independent and thus ignores any complementary or route competition effects.

Optimal Resource Allocation. Optimization methods to determine bus frequency have been proposed by several researchers (e.g., Furth and Wilson, 1981; Wilson and Banasiak, 1984; Ceder and Stern, 1984; Koutsopoulos et al., 1985). To illustrate these methods we present the formulation and solution approach proposed by Furth and Wilson (1981).

Problem Formulation. The problem can be stated as follows: Find the frequencies on each of a number of routes so as to maximize net social benefit subject to constraints on total subsidy, fleet size, and maximum headways. In the following formulation, headway is used as the basic decision variable.

Maximize:

$$Z = \sum_{j=1}^{P} D_j \sum_{i=1}^{N_j} \left[\frac{b}{2} \int_{h_{ij}}^{\infty} r_{ij}(u)\, du + a_{ij} r_{ij}(h_{ij}) \right] \tag{9.5}$$

subject to the following constraints:

Subsidy:

$$\sum_{j=1}^{P} D_j \sum_{i=1}^{N_j} \left[\frac{k_{ij}}{h_{ij}} - F_{ij} r_{ij}(h_{ij}) \right] = S_0 \tag{9.6}$$

Fleet size:

$$\sum_{i=1}^{N_j} \frac{t_{ij}}{h_{ij}} \leq M_j \qquad j = 1,\ldots,P \tag{9.7}$$

Headway:

$$h_{ij} \leq x_{ij} \qquad j = 1,\ldots,P, \quad i = 1,\ldots,N_j \tag{9.8}$$

where

P = number of time periods

N_j = number of routes operated during time period j

D_j = duration of period j

b = value of passenger wait time

h_{ij} = headway on route i during period j

a_{ij} = surplus marginal ridership benefit on route i during period j (marginal ridership benefit minus fare)

r_{ij} = ridership on route i during period j (a function of h_{ij})

k_{ij} = operating cost per run on route i during period j

F_{ij} = fare on route i during period j

S_0 = subsidy available

t_{ij} = run time (round trip) on route i during period j

M_j = fleet size during period j

x_{ij} = maximum headway for route i during period j

The objective function, equation (9.5), can easily be shown to be equivalent to maximizing the wait-time savings plus the social ridership benefit minus fare; this is the net social benefit. Equation (9.6) simply states that the operating cost minus the revenue must be equal to the known subsidy. Equation (9.7) is the fleet-size constraint, and equation (9.8) constrains the headway to be less than the policy headway and the headway at which the loading constraint is binding.

This general formulation could be simplified or made more complex (e.g., by defining classes of riders, each of which has a separate marginal benefit) in specific applications, but all important facets of the problem are included. Before the method developed to solve this mathematical program is presented, it is necessary to recognize and discuss perhaps the most important limitation of the model, the assumption of the independence of all routes in the system.

The assumption that both costs and benefits due to a headway on a specific route are independent of headways on other routes is not always true, at least on the benefit side. In this model, ridership on a route depends on the headway of only that route, whereas, in general, ridership will also depend on the headways on competing and complementary routes.

When passengers have a choice among several routes, an improvement in service on one of those routes will divert riders from the other routes. Such route competition is less common in North America than in most other parts of the world, in which an approach that considers route competition directly is called for. An improvement in service on one route can also increase the demand on another route when there is a large transfer volume between the two routes. Care must be taken, therefore, both in applying the model and in interpreting its results in situations in which strong route competition or complementarity exists.

Solution Algorithm. Optimality (Kuhn–Tucker) conditions can be derived as a set of equations that relate headways to other variables in the model; these equations then become the optimal decision rules for the operator. These optimality conditions are applied in the following step-by-step algorithm to determine the optimal set of headways by route and by period.

Step 1. Relax the fleet-size and maximum-headway constraints on all routes and for all time periods not yet constrained and solve the following set of equations for the headways, h_{ij}:

$$\frac{b}{\lambda}\left(\frac{r_{ij}}{2}\right) h_{ij}^2 - \left[F_{ij} + \frac{a_{ij}}{\lambda}\left(\frac{dr_{ij}}{dh_{ij}}\right) h_{ij}^2 - k_{ij}\right] = 0 \qquad (9.9)$$

where λ is determined so as to exhaust the available subsidy. If no routes violate their maximum-headway constraint, go to step 3.

Step 2. For routes and periods for which the maximum headway constraints, equation (9.8), are violated, set $h_{ij} = x_{ij}$. Compute the deficit incurred on those routes and time periods and reduce the available subsidy by that amount. Go to step 1.

Step 3. Identify time periods in which the fleet-size constraint, equation (9.7), is binding. For each of these time periods solve the following set of equations:

$$\frac{b}{\lambda}\left(\frac{r_{ij}}{2}\right) h_{ij}^2 - \left[F_{ij} + \frac{a_{ij}}{\lambda}\left(\frac{dr_{ij}}{dh_{ij}}\right) h_{ij}^2 - (k_{ij} + w_j t_{ij})\right] = 0 \qquad (9.10)$$

where w_j is the shadow price of run time during period j and is determined to use all available buses.

Step 4. If no routes violate their maximum-headway constraint, equation (9.8), go to step 5. Otherwise, for every route that violates its maximum-headway constraint, set $h_{ij} = x_{ij}$, compute the number of buses required by all such routes in each period j, and reduce the number of available busses in period j by this amount. Go to step 3.

Step 5. Compute the deficit incurred by the fleet-constrained time periods and reduce the available subsidy by this amount. Let $\lambda_c = \lambda$.

Step 6. Repeat steps 1 and 2 for the unconstrained time periods to find a new value of λ, which is λ_u.

Step 7. If $\lambda_u = \lambda_c$, stop; otherwise, set $\lambda = \lambda_u$ and return to step 3.

The theory behind this algorithm will not be presented in detail here. However, the computational burden of the algorithm is very small, since it consists basically of a sequence of one-dimensional searches that are performed very rapidly. Equations (9.9) and (9.10) can be solved very efficiently by using Newton's method (provided that the demand function has continuous second derivatives), and values of λ and w_j can be found by making successive linear approximations.

9.D.2. High-Ridership-Corridor Strategies

Another topic that falls within the tactical planning area is design of operating strategies in high ridership corridors, which might be amenable to market segmentation techniques designed either to increase service quality or to increase productivity (or both). Turnquist (1979) formulated the design of zonal express service as a dynamic programming problem, and this work was extended by Furth (1986) to zonal design for bidirectional local service, including light direction deadheading, and to branching as well as linear corridors. These models have received little application by transit planners, largely because most real transit corridor planning situations have such limited design options that they can typically be solved by enumeration.

9.D.3. Practice in Timetable Preparation

Given a frequency that has been determined for a route in a time period, the basic steps in developing a timetable are as follows:

1. Running times are established for each route segment by time of day (using the most recent running time data obtained either manually or automatically).
2. Headways are determined at the peak flow point (usually, this is the time point at which the maximum passenger flow is observed; a time point is generally a bus stop at a major intersection which appears on the public timetable), based on the specified frequencies.
3. Departure (passage) times are set at the peak flow point.
4. Corresponding departure times are set at all time points on the route, including the terminals by using the established running times and the times at the peak point.

This completes the basic route timetable, although it may be modified subsequently in the vehicle scheduling step (see Section 9.E). While all four steps can be quite straightforward, steps 2 and 3 are worth discussing in somewhat greater detail. First, we can think of alternative types of headways (Ceder, 1984, 1986). An *equal headway* simply means constant time intervals between consecutive departures in each time period, or evenly spaced departures. A *balanced headway* refers to headways which are set so that the expected passenger loads on all buses are similar. Finally, a *smoother headway* is simply a compromise between the equal and the balanced headways. It is an option in cases where the data are not sufficient for concrete conclusions about balanced headways, but at the same time, the scheduler believes that equal headways will result in significant variation in loads. Such situations may occur around start and end times for work and school.

Equal headways are most common since it can be shown that under certain conditions, equal headways will minimize expected passenger waiting times. This result is shown below under the following assumptions:

1. The passenger arrival rate at any stop is independent of the vehicle departure process and is constant over the period.
2. Passengers can always board the first vehicle to depart.

Expected passenger wait time, E(WT)

$$= \frac{\text{expected total passenger wait time/run}}{\text{expected passengers/bus}}$$

$$= \frac{\int_0^\infty r(h)\text{E}(\text{WT}(h))g(h)\,dh}{\int_0^\infty r(h)g(h)\,dh} \tag{9.11}$$

where $r(h)$ represents the passengers arriving in headway h, $g(h)$ the probability density function for headway h, and $\text{E}(\text{WT}(h))$ the expected passenger wait time for passengers arriving in headway h. Given assumption 1 (above),

$$r(h) = kh$$

where k = passenger arrival rate and $\text{E}(\text{WT}(h)) = \frac{h}{2}$. Therefore,

$$\text{E(WT)} = \frac{\int_0^\infty (kh^2)g(h)\,dh}{2\int_0^\infty khg(h)\,dh} = \frac{\text{E}(h^2)}{2\text{E}(h)} = \frac{\text{E}(h)}{2}\left[1 + \frac{\text{Var}(h)}{(\text{E}(h))^2}\right] \tag{9.12}$$

where $\text{Var}(h)$ = variance of the headway. Thus the expected passenger wait time is minimized, at half the headway, when the variance of the headway is zero.

In practice, assumptions 1 and 2 may not always hold, which may argue in favor of unequal headways. In particular, if there are surges in the passenger arrival rate at predictable times, for example at the start and end of a school day or a factory shift, it will be most appropriate to add bus trips at times to accommodate these heavy loads. If one follows this argument to the extreme, a headway sequence for a route could be derived to provide approximately equal passenger loads on each trip: This is what is referred to as *balanced headways.* This will only make sense, however, when vehicles are operating close to their passenger capacity and there is a good deal of certainty about the arrival pattern of passengers wishing to travel on the route. In practice, such conditions are rare.

The second issue with respect to the timetable is dealing with special scheduling requests. One characteristic of many transit timetables is the repetition of departure times, every hour. These easy-to-memorize departure times are "clockface headways" and include headways of 6, 7.5, 10, 12, 15, 20, 30, and 60 minutes. Headways of less than 6 minutes are also acceptable since at such short headways passenger arrivals at bus stops are not generally influenced by the timetable.

9.E. VEHICLE SCHEDULING

Vehicle scheduling refers to the class of problems in which vehicles are assigned to timetabled trips in such a way that each trip is carried out by one vehicle, a given set of constraints is satisfied, and cost is minimized. For simplicity here, the term *trip* will be used to denote all vehicle tasks defined in the timetable (whether trips or blocks of trips). Very often, the cost structure reflects the fact that the number of vehicles must be minimized. We assume that for each trip, the departure time, trip duration, and terminals are known. In addition to regular trips, we assume that deadheading trips are allowed between terminals. A survey paper and several papers on specific vehicle scheduling problems appear in Daduna (1995). Vehicle assignment and scheduling problems in private-sector transportation planning are discussed in this book in the contexts of freight transportation [Chapter 8 (Section 8.D.1)] and airline transportation [Chapter 10 (Section 10.B.2)].

9.E.1. Objectives

The main objective is to construct vehicle blocks (simply chains of vehicle trips) while either:

1. Minimizing total deadheading kilometers and changes to the timetable while using the existing number of vehicles, or
2. Minimizing the number of vehicles required to operate timetable.

These objectives can be pursued through the following functions:

1. Deadhead trip construction
2. Intertrip deadheading trip insertions
3. Timetable shifting
4. Vehicle trip chaining

In the literature edited by Daduna and Wren (1988), Desrochers and Rousseau (1992), and Daduna (1995), one can find various approaches to the vehicle scheduling function. One of these approaches, based on an interactive graphical method, is described below.

9.E.2. Deficit Function Method

The deficit function method is a special graphical representation of the trip timetable for improving or generating new vehicle schedules. It is based on the deficit function theory developed by Ceder and Stern (1981).

A deficit function is simply a step function that increases by one at the time of each trip departure and decreases by one at the time of each trip arrival. Such a function may be constructed for each terminal in a multiterminal transit system based only on the transit timetable. The main advantage of the deficit function is its visual and intuitive nature.

Let $d(k,t)$ denote the deficit for a point k at time t. This point k is typically a terminal or any other point at which some trips are initiated and/or terminated. The value of $d(k,t)$ represents the total number of departures less the total number of arrivals up to and including time t. The maximal value of $d(k,t)$ over the schedule horizon is designated $D(k)$. It is possible to partition the schedule horizon of $d(k,t)$ into a sequence of alternating hollow and maximal intervals. The maximal intervals define the interval of time over which $d(k,t)$ takes on its maximum value. A *hollow interval* is defined as the interval between two maximal intervals.

If we denote the set of all route terminals and other points of interest as E, the sum of $D(k)$ (for all $k \in E$) is equal to the minimum number of vehicles required to service the set E. This is known as the *fleet size formula*, derived independently by Bartlett (1957), Salzborn (1972), and Gertsbach and Gurevich (1977). Mathematically, for a given fixed schedule,

$$N = \sum_{k \in E} D(k) = \sum_{k \in E} \max_{t} d(k,t) \tag{9.13}$$

where N is the minimum number of vehicles required to service the set E.

When deadheading (DH) trips are allowed, the fleet size may be reduced below the level defined in equation (9.13). Ceder and Stern (1981) describe this procedure based on the construction of a unit reduction deadheading chain (URDHC). Such a chain is comprised of a set of nonoverlapping DH trips which, when inserted into the schedule, reduces the fleet size by one. The procedure continues inserting URDHCs until either no more can be inserted or a lower bound on the minimum fleet is reached. Determination of the lower bound is detailed in Stern and Ceder (1983). The deficit function theory for transit scheduling has been extended by Ceder and Stern (1982, 1985) to include possible shifting of departure times within bounded tolerances.

Example Using Shifting Departure Time. A simple example of shifting a trip departure time to affect the fleet size at a given terminal is shown in Figure 9.7, comprising 10 trips between three terminals, m, u, and v. Figure 9.7 shows the trips and the deficit function for terminal m with the trip num-

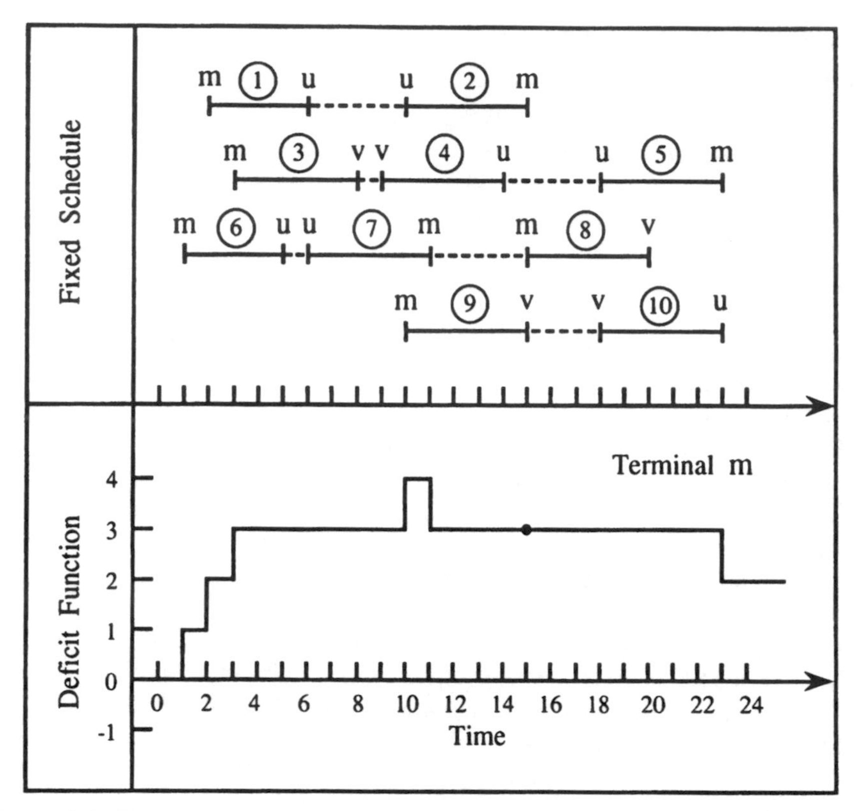

Figure 9.7 Ten trips arranged in four chains according to the maximum deficit function value.

bers circled and the departure and arrival terminals as indicated. According to the fleet size formula, a minimum of four buses are required at *m*. The four trip chains (or blocks) are shown in Figure 9.7 with the dashed lines representing waiting time at a terminal between trips.

Even if we assume that the number and duration of all trips are fixed, the maximum value of the deficit function can still be reduced by one, and only three buses will be required to satisfy the schedule if either trip 7 can be moved 1 time unit earlier or trip 9 can be moved 1 time unit later. In either case the revised schedule can be satisfied by the following three trip chains: (1–2–8), (3–4–5), and (6–7–9–10). This example demonstrates how the deficit function can serve as a guide for examining possible changes in times of particular trips.

Example of the Lower Bound on the Fleet Size. The fleet size formula states that the sum of the maximum values of each of the deficit functions is equal to the minimum fleet size. This value is normally greater than the maximum number of trips in simultaneous operation. A function representing the number of trips in simultaneous operation can be constructed by summing the individual deficit functions for each terminal. The resultant function is referred to as the *overall deficit function* $g(t)$:

$$g(t) = \sum_{k \in E} d(k,t) \tag{9.14}$$

and let

$$G(t) = \max_{t} g(t)$$

where $G(t)$ is the maximum number of trips in simultaneous operation and was used as a lower bound on the fleet size by Ceder and Stern (1981).

Stern and Ceder (1983) improved the lower bound by construction of $g'(t) \geq g(t)$ when $g'(t)$ is the overall deficit function of an extended schedule. The extended schedule is based on a method to extend each trip's arrival time to its nearest feasible trip's departure time in the schedule (taking into account the deadheading trip time from the trip's arrival terminal to the terminal of the nearest trip's departure time, if both do not coincide). By creating these extensions, one obtains $G' \geq G(t)$, where

$$G'(t) = \max_{t} g'(t)$$

and hence an improved lower bound on the fleet size. The following example demonstrates these findings.

Consider the seven-trip schedule S, shown in Figure 9.8a. From the graph of the overall deficit function, the lower bound is $G(S) = 4$. Let the deadheading times be $t_{ba} = 3$, $t_{bc} = 4$, $t_{ac} = 5$ between terminals (b,a), (b,c), and (a,c), respectively. Construct the extended schedule, S', which is the same as S except that the arrival times of trips 4 and 6 (trips starting at a and ending at b) are extended to time 9. The maximum value of the overall deficit function for the extended schedule S' is $G(S') = 6$ (see Figure 9.8b). This gives a stronger lower bound on the optimal deadheading solution than $G(S)$. The optimal solution is also 6, as can be seen by examining the chain possibilities for trips 4 and 6:

- *Case 1.* Trips 4 and 6 are left as the ends of chains. This would give a solution of seven chains, since trip 2 cannot follow 1.
- *Case 2.* Either trip 4 or 6 is left as the end of a chain, and the other is followed by trip 2. This results in a solution of six chains.
- *Case 3.* Trips 4 and 6 are each followed by separate trips in separate chains. This case is impossible.

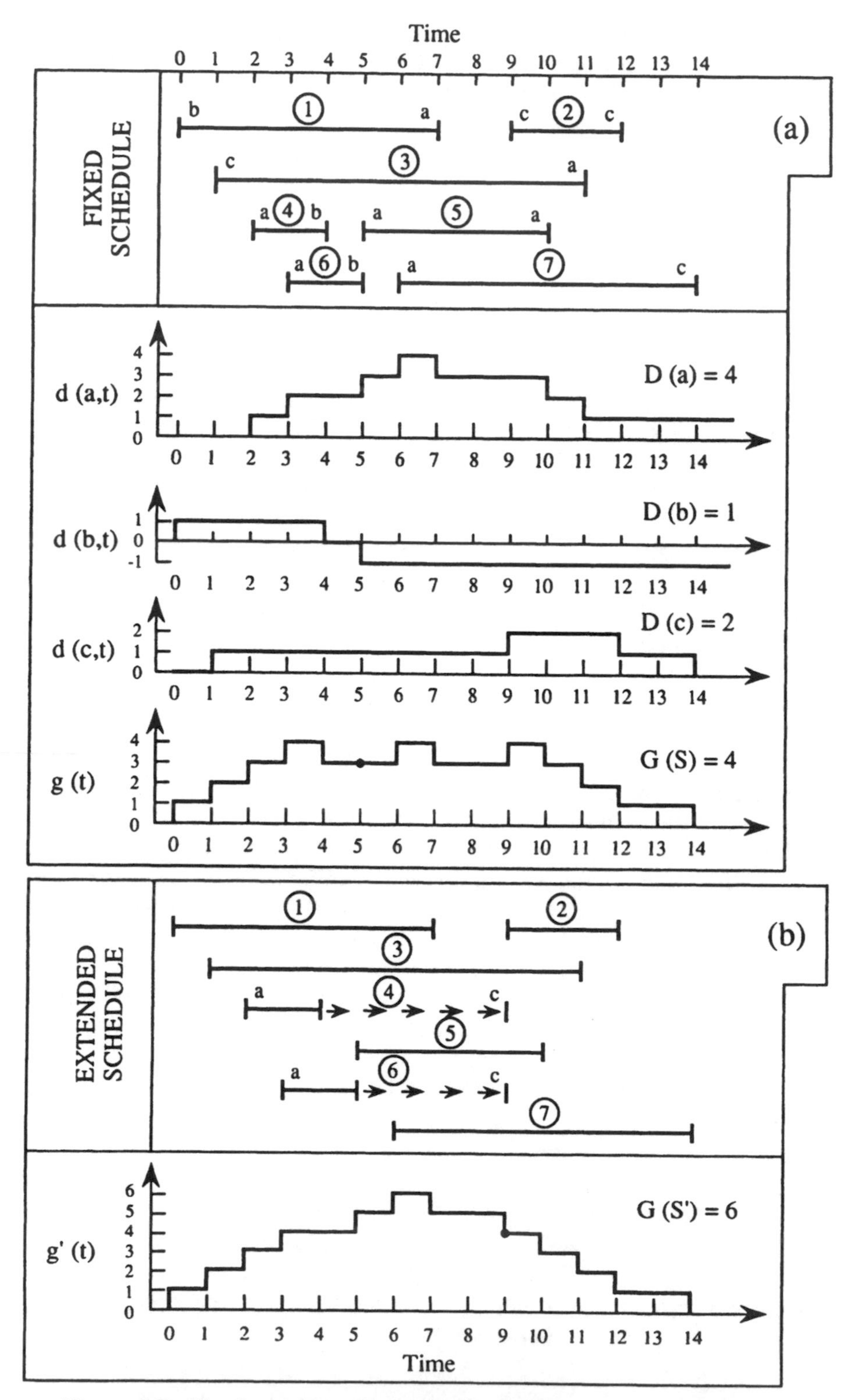

Figure 9.8 Fixed schedule with the improved lower-bound determination.

Therefore, the optimal solution is case 2 with a value of six vehicles, which the improved lower bound attains. It should be noted that the improved lower bound will not always be the same as the optimal fleet size.

Example of Minimum Fleet Size for a Complete Timetable. A simple example is used to illustrate the deficit function approach to minimizing the fleet size. This example (see Figure 9.9) shows a timetable for two routes that covers about a 2-hour period with the times shown referring to the departure times at the maximum load points. The route is comprised of three timepoints, A, B, and C, with travel times for service and deadheading trips as given in Figure 9.9.

Based on the deficit function approach the minimum number of vehicles required without deadheading trips is $D(A) + D(C) = 11$. However, a DH trip can be inserted from A to C—departing after the last maximal interval of $d(A,t)$ and arriving just before the start of the first maximal interval of $d(C,t)$. Both $d(A,t)$ and $d(C,t)$ are then changed according to the dashed line in Figure 9.9. It results in reducing $D(C)$ from 6 to 5 and the overall fleet size from 11 to 10. After that, it is impossible to reduce the fleet size further through DH trip insertions.

The latter observation can also be detected automatically by the lower bound test. Even for the nonimproved lower-bound test and following the DH trip insertion procedure, the maximum of the combined functions is 10, and therefore the fleet size is at its lower bound.

9.F. CREW SCHEDULING

Crew scheduling in a transit company involves the generation of the drivers' daily duties, and in some cases, the assignment of a set of daily duties (called a *roster*) to each driver. We consider only the first of these problems in this chapter. The proceedings of the most recent workshops on computer scheduling of public transport constitute an excellent set of references on the subject (Daduna and Wren, 1988; Desrochers and Rousseau, 1992; Daduna, 1995). Wren and Rousseau (1994) also provide a survey of this problem. Crew scheduling problems in private-sector transportation planning are discussed in this book in the contexts of freight transportation [Chapter 8 (Section 8.D.1)] and airline transportation [Chapter 10 (Section 10.B.4)].

9.F.1. Set Covering Formulation

The problem can be expressed simply as the creation of a set of duties of minimum cost, ensuring that each bus trip is covered by a driver (or crew) and that all union contract rules are respected. Several authors favor an approach based on the set-covering formulation of the problem. Theoretically, in this approach all feasible duties are generated based on the contract rules,

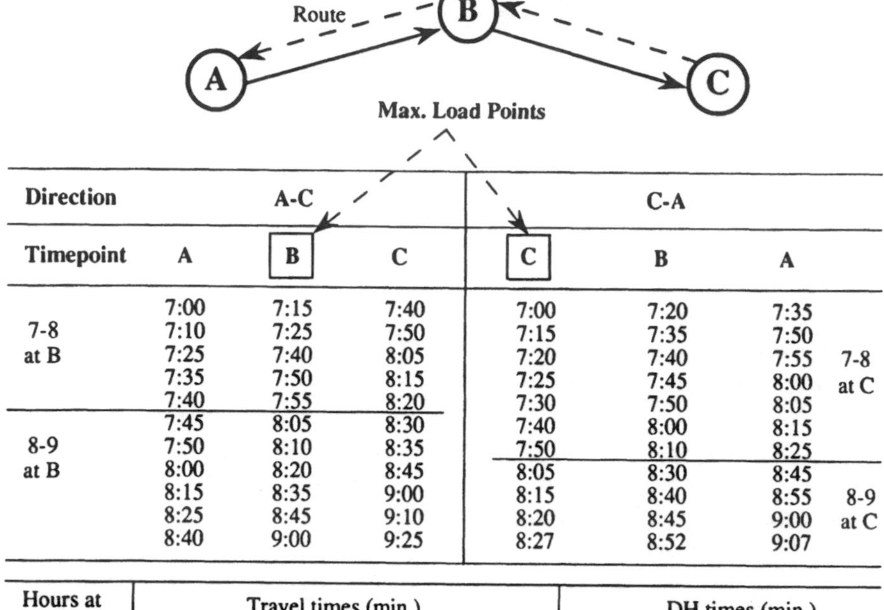

Direction	A-C			C-A			
Timepoint	A	[B]	C	[C]	B	A	
7-8 at B	7:00	7:15	7:40	7:00	7:20	7:35	7-8 at C
	7:10	7:25	7:50	7:15	7:35	7:50	
	7:25	7:40	8:05	7:20	7:40	7:55	
	7:35	7:50	8:15	7:25	7:45	8:00	
	7:40	7:55	8:20	7:30	7:50	8:05	
8-9 at B	7:45	8:05	8:30	7:40	8:00	8:15	
	7:50	8:10	8:35	7:50	8:10	8:25	
	8:00	8:20	8:45	8:05	8:30	8:45	8-9 at C
	8:15	8:35	9:00	8:15	8:40	8:55	
	8:25	8:45	9:10	8:20	8:45	9:00	
	8:40	9:00	9:25	8:27	8:52	9:07	

Hours at max. load point	Travel times (min.) A-B	B-C	C-B	B-A	DH times (min.) A-C	A-B	B-C
7-8	15	25	20	15	25	15	5
8-9	20	25	25	15			

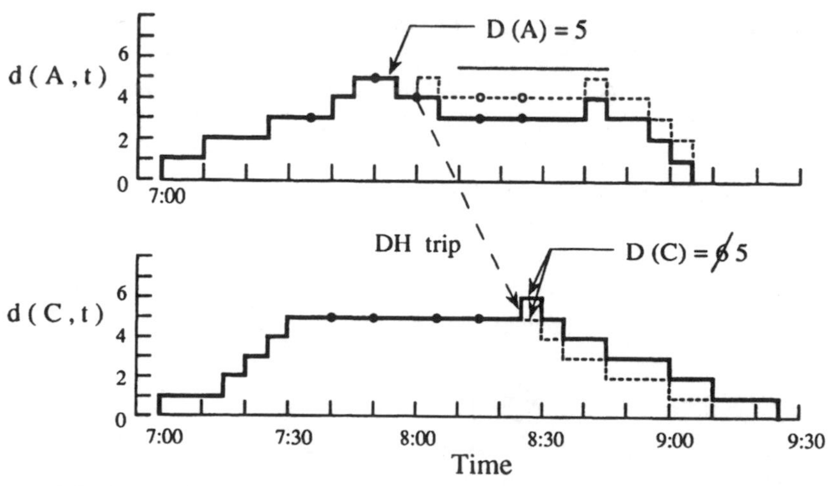

Figure 9.9 Determination of the minimum fleet size using deadheading trip insertions.

and the least-cost subset of duties that covers all bus trips is selected. The mathematical formulation is as follows:

$$\text{minimize} \sum_j c_j x_j$$

subject to

$$\sum_j a_{ij} x_j = 1 \qquad \text{for all trips } i = 1,\ldots,N \tag{9.15}$$

$$\sum_j b_{ij} x_j \geq d_i \qquad \text{for duty types } i = 1,\ldots,M \tag{9.16}$$

where

$$c_j = \text{cost of duty}$$

$$x_j = \begin{cases} 1 & \text{if duty } j \text{ is included in the solution} \\ 0 & \text{otherwise} \end{cases}$$

$$a_{ij} = \begin{cases} 1 & \text{if duty } j \text{ includes trip } i \\ 0 & \text{otherwise} \end{cases}$$

$$b_{ij} = \begin{cases} 1 & \text{if duty } j \text{ is of type } i \\ 0 & \text{otherwise} \end{cases}$$

Constraint set (9.15) ensures that each trip is covered by a duty, while constraint set (9.16) covers other constraints on the solution, such as requiring a minimum number of duties of defined types.

If rules restrict a driver to working on only one bus route, and if strict work rules limit the number of feasible duties, this approach may be feasible. Otherwise if, for example, full interlining is permitted (meaning that a driver is allowed to work on more than one route during a duty), the number of feasible duties can be extremely large, making it impossible to generate them all. Advocates of the set-covering approach then either have to split the problem into several subproblems—evening problem, morning problem, etc. (see Ward et al. 1981)—or heavily restrict the generation of duties according to heuristic rules (e.g., by eliminating relief points that are unlikely to be useful). This is the case for Busman, Ramcutter, and Crewsched, among other software packages (see Daduna, 1995 for several papers based on this approach).

The main difficulty with this approach is that there does not exist a good set of rules for identifying what are likely to be good duties. A typical duty may be very efficient under certain rules while being very inefficient under another set of rules. Desrochers (1986) and Desrochers and Soumis (1989) used the set-covering approach to the problem, but sequentially generated only the duties useful to improve the solution at each step of the algorithm. Thus only a small subset of the duties had to be generated. It seems that this technique is likely to apply only to small to medium-sized problems, but it could be quite useful, for example, where drivers can work on only one bus route. Rousseau et al. (1995) have reported on the results obtained with this approach.

9.F.2. Manual and Computer-Based Methods

The most common approach to the transit crew scheduling problem, in both manual and computer-based methods, proceeds as follows: First, the vehicle blocks are split into pieces of work (the period of time during which a crew works continuously); second, the pieces of work are combined to form driver duties; and finally, the solution is improved heuristically and/or manually. A matching or an assignment problem formulation for the construction of duties is generally used, such as

$$\text{minimize} \sum_i \sum_j c_{ij} x_{ij} \tag{9.17}$$

subject to

$$\sum_i x_{ij} = 1 \qquad \text{for all pieces } j = 1,\ldots,k \tag{9.18}$$

$$\sum_j x_{ij} = 1 \qquad \text{for all pieces } i = 1,\ldots,k \tag{9.19}$$

$$x_{ij} = 0 \text{ or } 1$$

where x_{ij} is the duty that combines pieces i and j at a cost c_{ij}. The constraints guarantee that each piece of work will be covered once and the minimum cost set of duties is selected. This formulation is equivalent to the Hitchcock transportation problem with all supplies and demands set equal to unity. The Hitchcock formulation is presented in this book in Sections 5.B.1 and 7.C.3.

Various strategies are possible for splitting the blocks into pieces of work. Most approaches use heuristics adapted to the type of problem considered. In RUCUS (Wilhelm, 1975), the first general package developed for transit crew scheduling, a set of parameters enables the user partially to determine the method by which the pieces will be defined. In other packages, such as RUCUS II (Luedtke, 1985), an interactive system for transit crew scheduling, the user has direct control over the ways the pieces are defined.

A particularly interesting decomposition approach is taken in HASTUS, which uses a mathematical model to split the blocks into pieces of work. A full description of the algorithm is that of Rousseau et al. (1985). The authors solve a relaxation of the crew scheduling problem first, retaining the main features of the problem while relaxing many of the details. The idea is to retain the essence of the bus schedule and the contract rules and cost provisions while greatly simplifying the precise bus schedule.

First, the user must define periods, normally about half an hour in length. The schedules of all bus blocks are approximated so that all both start and end at the beginning (end) of a period. Driver reliefs can also occur only at the beginning of a period. Moreover, the problem is further relaxed by requiring that the duties selected be sufficient to cover the total requirement of drivers in each period instead of requiring that they exactly cover all the blocks individually. Two additional characteristics of the real problem are also relaxed. First, all spatial aspects of the problem beyond those already incorporated into the vehicle blocks are ignored. Second, the integer restriction on duties is relaxed. Collectively, the relaxation results in a much simpler problem which can be formulated as a linear program.

All feasible (in light of the work rules) duties covering an integer number of periods are generated and costed out. The generating rules eliminate, under instructions from the user, duties which are of no interest or which do not correspond to unwritten rules or common practice (e.g., a duty with a long break during a peak hour). A linear program then searches for the set of duties that will provide at least the number of drivers required during each period and satisfy additional constraints at minimum total cost.

Additional constraints are often related to the contract. All the constraints related to the total number of duties of a certain type are considered. These include constraints on the minimum (or maximum) number of duties without a break and on the maximum number of duties with certain undesirable characteristics such as long spreads (the time between driver sign on and sign off for the day) or short breaks.

The relaxation is called HASTUS-macro and can also be used as a planning tool for evaluation of the economic impact of changes in the contract or in the service characteristics. By changing the work rules, parameters, or the service profile, a new solution can be obtained and compared with the current solution. The paper by Blais and Rousseau (1982) describes in detail the HASTUS-macro system and its use in that context.

The HASTUS-macro solution suggests the number of duties of each type that should be included in a final solution. The next step uses this information to split the blocks into pieces of work that correspond as closely as possible to the ones generated in the HASTUS-macro relaxation.

Each block is split into pieces of work by solving a shortest-path problem. It is interesting to note that the problem of splitting a block into pieces can be expressed as the problem of finding a path from the starting time of the block to its ending time using arcs that correspond to pieces of work. Figure 9.10 illustrates this concept with the feasible pieces of work (we are assuming

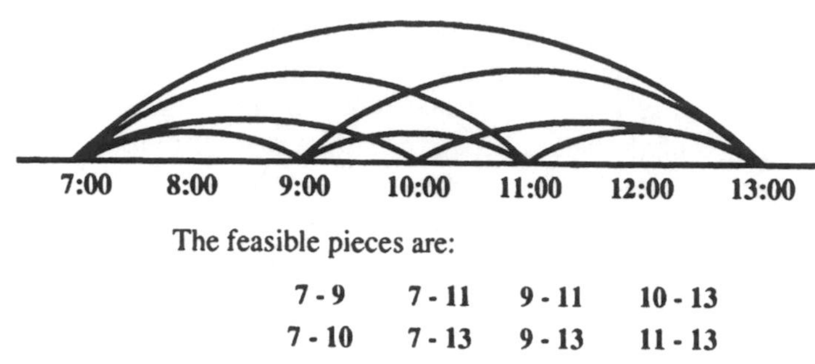

Figure 9.10 Flow formulation for the partition of a block.

here a minimum length of 2 hours) represented by arcs. We have assumed for simplicity that a relief opportunity exists at every hour in this example but there is a 2-hour minimum for every piece of work.

The cost associated with each piece of work in the flow formulation corresponds to a penalty indicating whether this given piece will increase or decrease the difference between the desired number of pieces of work of that type and the number of pieces of work of that type already present in the current solution. The algorithm cycles through all blocks until no improvement can be achieved toward matching the number of desired pieces of each type indicated by HASTUS-macro.

Once the blocks have been split into pieces of work, a matching algorithm is used to generate duties (Rousseau et al., 1985; Lessard et al., 1989) and several heuristics are used to improve the solution. This application has been described in *Interfaces* (Blais et al., 1990), where the mathematical model is also provided.

9.G. CONCLUSIONS

This chapter has been brief, with very little detail. We have demonstrated, however, that in the transit area, operations research techniques are being used widely and effectively to assist in the planning of many transit systems around the world. Some of the problems have not been solved, and even for the ones that have been solved, better techniques could be found. New problems are arising from the application of new information technology in transit. It is clear that there will be plenty of challenging operations research problems in the transit area for many years to come.

EXERCISES

9.1. Given a bus network comprised of four nodes and five links (arcs) (Figure 9.11). The numbers on the links represent the total travel time, *in*

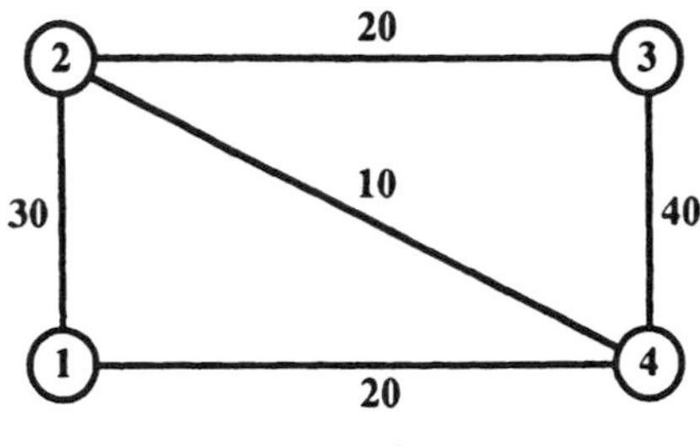

Figure 9.11

minutes, required to travel from one node to the other. Following is an origin–destination or $\{q_{ij}\}$ demand matrix for a *2-hour* peak period.

	To			
From	1	2	3	4
1	0	960	380	160
2	960	0	220	200
3	380	220	0	240
4	160	200	240	0

The arrival pattern of passengers is assumed to be uniformly distributed for each cell q_{ij}, where q_{ij} represents the demand (number of passengers) from origin i to destination j.

The desired bus occupancy is 60 passengers per bus. Bus routes can be generated (start and end) *only at node 1,* and deadheading trips *are not allowed.* Two routes serve the given demand:

Route A: node 1 ⟶ node 2 ⟶ node 3 ⟶ node 2 ⟶ node 1

Route B: node 1 ⟶ node 2 ⟶ node 4 ⟶ node 2 ⟶ node 1

Boarding, alighting, and *transfer times* can be neglected. Answer the following questions:

(a) Derive, for the existing two routes, the required headways based on the expected maximum load not exceeding the desired bus occupancy. If you derive more than one alternative, select the most appropriate one and explain your decision.

(b) What is the minimum fleet size required for routes A and B based your solution to part (a).

(c) Your task now is to evaluate all possible routes in the given network that visit all nodes during the 2-hour peak period. Note that trips may end after the peak period.

(1) Define all possible routes.
(2) Select an appropriate criterion (or criteria) for your evaluation.
(3) Construct the load profile for each possible route as a function of travel time.
(4) What is the minimum fleet size required for each route?
(5) Select the best route and compare it with the two existing routes.

9.2. Given a bus system comprised of two cyclic routes departing and arriving at the same terminal. The mean trip for each route is 3 hours. Based on observed passenger demand, the headways for each route are determined by hour of day as follows:

Route 1		Route 2	
Hour of Day	Headway (min)	Hour of Day	Headway (min)
6 A.M.–9 A.M.	7.5	6 A.M.–8 A.M.	6
9 A.M.–12 noon	12	8 A.M.–12 noon	7.5
12 noon–2 P.M.	10	12 noon–3 P.M.	10
2 P.M.–4 P.M.	15	3 P.M.–6 P.M.	7.5
4 P.M.–6 P.M.	6	3 P.M.–6 P.M.	7.5
6 P.M.–11 P.M.	10	6 P.M.–11 P.M.	10

What is the *minimum fleet size* required for this bus system? Comment on the generalization of your approach. (*Note:* Buses may be switched from one route to another if necessary.)

9.3. Given a gus route in which the mean wait time of passengers, E(WT), can be estimated by $E(WT) = E(h^2)/2E(h)$, where h is the bus headway.

(a) Using the wait-time formula, explain and show (mathematically) how E(WT) can be reduced for a given $E(h)$. Consider headway distributions ranging from regular (deterministic) to exponential (random).

(b) Outline the feasible ways to reduce E(WT) given that the passenger demand is fixed.

9.4. With high volumes of bus service on major arterial roads, it becomes very expensive, and often ineffective, for bus operators to use curbside inspectors and mobile supervisors. The inspector's or supervisor's knowledge of the situation is limited to the immediate vicinity, with no indication of conditions elsewhere. This can result in bus drivers being instructed to act in a fashion that is counterproductive from the system point of view. This represents only one group of problems that can be resolved by introducing an automatic vehicle monitoring (AVM) system. In general, the real-time control provided by the AVM system enables the operator to increase both the utilization of each bus and the service level seen by each passenger. Also, the substantial amount of off-line AVM data can be used to improve the entire bus planning process.

(a) List all possible strategies that can be used by the supervisors in the control room of an advanced AVM system. Note that these strategies are carried out by instructing a driver or a group of drivers to follow certain actions.

(b) Outline all the potential benefits of AVM off-line data.

REFERENCES

Andreasson, I., 1976. A Method for the Analysis of Transit Networks, in M. Ruebens (ed.), *2nd European Congress on Operations Research*, North-Holland, Amsterdam.

Babin, A., M. Florian, L. James-Lefebvre, and H. Spiess, 1982. EMME/2: Interactive Graphic Method for Road and Transit Planning, *Transportation Research Record*, 866, 1–9.

Bartlett, T. E., 1957. An Algorithm for the Minimum Number of Transport Units to Maintain a Fixed Schedule, *Naval Research Logistics Quarterly*, 4, 139–149.

Blais, J.-Y., and J.-M. Rousseau, 1982. HASTUS: A Model for the Economic Evaluation of Drivers' Collective Agreements in Transit Companies, *INFOR*, 20, 3–15.

Blais, J.-Y., J. Lamont, and J.-M. Rousseau, 1990. The HASTUS Vehicle and Manpower Scheduling System at the Société de Transport de la Comminanté Vobaine de Montréal, *Interfaces* 20(1), 26–42.

Ceder, A., 1984. Bus Frequency Determination Using Passenger Count Data, *Transportation Research* A, 18(5/6), 439–453.

Ceder, A., 1986. Methods for Creating Bus Timetables, *Transportation Research* A, 21(1), 59–83.

Ceder, A., and H. Stern, 1981. Deficit Function Bus Scheduling with Deadheading Trip Insertions for Fleet Size Reduction, *Transportation Science*, 15, 338–363.

Ceder, A., and H. I. Stern, 1982. Graphical Person–Machine Interactive Approach for Bus Scheduling, *Transportation Research Record*, 857, 69–72.

Ceder, A., and H. Stern, 1984. *Optimal Transit Timetables for a Fixed Vehicle Fleet*, based on the 9th International Symposium on Transportation and Traffic Theory, VNU, Amsterdam, 331–355.

Ceder, A., and H. Stern, 1985. The Variable Trip Procedures Used in the Autobus Vehicle Scheduler, in J.-M. Rousseau (ed.), *Computer Scheduling of Public Transport 2*, North-Holland, Amsterdam, pp. 371–390.

Ceder, A., and Y. Israeli, 1992. Scheduling Consideration in Designing Transit Routes at the Network Level, *Computer-Aided Transit Scheduling*, Springer-Verlag, Berlin, pp. 113–136.

Chapleau, R., 1974. Réseaux de transport en commun: structure informatique et affectation, Ph.D. thesis, University of Montreal, Montreal, Quebec, Canada.

Chapleau, R., B. Allard, and M. Canova, 1982. *Madituc: un modèle de panification opérationnelle adapté aux enterprises de transport en commun de taille moyenne,*

Publication 265, Centre de recherche sur les transports, University of Montreal, Montreal, Quebec, Canada.

Chriqui, C., 1974. *Réseaux de transport en commun: les problèmes de cheminement et d'accès,* Ph.D. thesis, Dept. IRO, University of Montreal, Montreal, Quebec, Canada.

Chriqui, C., and P. Robillard, 1975. Common Bus Lines, *Transportation Science,* 9, 115–121.

Cohon, J. L., 1978. *Multiobjective Programming and Planning,* Academic Press, San Diego, CA.

Daduna, J. R., 1995. *Computer-Aided Transit Scheduling,* Lecture Notes in Economics and Mathematical Systems 410, Springer-Verlag, New York.

Daduna, J. R., and A. Wren (eds.), 1988. *Computer-Aided Transit Scheduling,* Lecture Notes in Economics and Mathematical Systems 308, Springer-Verlag, Berlin.

De Cea, J., J. P. Bunster, L. Zubieta, and M. Florian, 1988. Optimal Strategies and Optimal Routes in Public Transit Assignment Modes: An Empirical Comparison, *Traffic Engineering and Control,* 29(10), 520–526.

Desrochers, M., 1986. *La Fabrication d'horaires de travail pour les conducteurs d'autobus par une méthode de génération de colonnes,* Publication 470, Centre de Recherche sur les Transports, University of Montreal, Montreal, Quebec, Canada.

Desrochers, M., and J.-M. Rousseau (eds.), 1992. *Computer-Aided Transit Scheduling,* Lecture Notes in Economics and Mathematical Systems 386, Springer-Verlag, Berlin.

Desrochers, M., and F. Soumis, 1989. A Column Generation Approach to the Urban Transit Crew Scheduling Problem, *Transportation Science,* 23(1), 1–13.

Dial, R. B., 1967. Transit Pathfinder Algorithm, *Highway Research Record,* 205, 67–85.

Duckstein, L., and S. Opricovic, 1980. Multiobjective Optimization in River Basin Development. *Water Resources Research,* 16(1), 14–20.

Fearnside, K., and D. P. Draper, 1971. Public Transport Assignment: A New Approach, *Traffic Engineering and Control,* November, 298–299.

Florian, M., 1977. A Traffic Equilibrium Model of Travel by Car and Public Transit Modes, *Transportation Science,* 11(2), 166–179.

Florian, M., and H. Spiess, 1983. On Binary Mode Choice/Assignment Models, *Transportation Science,* 17(1), 32–47.

Florian, M., and H. Spiess, 1986. *Models and Techniques for the Planning of Urban and Regional Transportation Networks,* Publication 463, Centre de Recherche sur les Transports, University of Montreal, Montreal, Quebec, Canada.

Furth, P., 1986. Zonal Route Design for Transit Corridors, *Transportation Science,* 20(1), 1–12.

Furth, P., and N. H. M. Wilson,, 1981. Setting Frequencies on Bus Routes: Theory and Practice, *Transportation Research Record,* 818, 1–7.

General Motors, 1980. *Interactive Graphic Transit Design System (IGTDS) Demonstration Study: Bellevue, Washington, Final Report,* Vols. I and II (EP-80023A) General Motors Transportation System Center, Warren, MI.

Gertsbach, I., and Y. Gurevich, 1977. Constructing an Optimal Fleet for a Transportation Schedule, *Transportation Science,* 11, 20–36.

Goicoechea, A., D. R. Hansen, and L. Duckstein, 1982. *Multiobjective Decision Analysis with Engineering and Business Applications,* Wiley, New York.

Hasselström, D., 1981. Public Transportation Planning: A Mathematical Programming Approach, Ph.D. thesis, University of Gothenburg, Gothenburg, Sweden.

Israeli, Y., 1992. Transit Route and Scheduling Design at the Network Level, Doctoral dissertation, Technion, Israel Institute of Technology, Israel.

Israeli, Y., and A. Ceder, 1996. Public Transportation Assignment with Passenger Strategies for Overlapping Route Choice, *Transportation and Traffic Theory* (13th ISTTT), Elsevier Science and Pergamon Press, New York, pp. 561–588.

Jansson, K., and B. Ridderstolpe, 1992. A Method for the Route-Choice Problem in Public Transport Systems, *Transportation Science,* 26(3), 246–251.

Koutsopoulos, H., A. Odoni, and N. Wilson, 1985. Determination of Headway as a Function of Time-Varying Characteristics on a Transit Network, in J.-M. Rousseau (ed.), *Computer Scheduling of Public Transport 2,* North-Holland, Amsterdam, pp. 391–414.

Lampkin, W., and P. D. Saalmans, 1967. The Design of Routes, Service Frequencies and Schedules for a Municipal Bus Undertaking: A Case Study, *Operations Research Quarterly,* 18(4), 375–397.

Last, A., and S. E. Leak, 1976. Transept: A Bus Model, *Traffic Engineering and Control,* 17(1), 14–20.

Le Clerq, F., 1972. A Public Transport Assignment Method, *Traffic Engineering and Control,* 14(2), 91–96.

Lessard, R., M. Minoux, and J.-M. Rousseau, 1989. A New Approach to General Matching Problems Using Relaxation and Network Flow Subproblems, *Networks,* 19(4), 459–480.

Luedtke, L. K., 1985. Rucus II: A Review of System Capabilities, in J.-M. Rousseau (ed.), *Computer Scheduling of Public Transport 2,* North-Holland, Amsterdam, pp. 61–110.

Magnanti, T. L., and R. T. Wong, 1984. Network Design and Transportation Planning: Models and Algorithms, *Transportation Science,* 18(1), 1–55.

Mandle, C., 1980. Evaluation and Optimization of Urban Public Transportation Networks, *European Journal of Operational Research,* 5, 396–404.

Marguier, P., and A. Ceder, 1984. Passenger Waiting Strategies for Overlapping Bus Routes, *Transportation Science,* 18(3), 207–230.

Minieka, E., 1978. *Optimization Algorithms for Networks and Graphs,* Marcel Dekker, New York, Chapter 5.

Mohring, H., 1972. Optimization and Scale Economies in Urban Bus Transportation, *American Economic Review,* Vol. LXII, Number 4, 591–604.

Nguyen, S., and S. Pallottino, 1988. Equilibrium Traffic Assignment for Large Scale Transit Networks, *European Journal of Operational Research,* 37, 176–186.

Osleeb, J. P., and H. Moellering, 1976. TRANSPLAN: An Interactive Geographic Information System for the Design and Analysis of Urban Transit Systems, URISA 1976, *14th Annual Conference of the Urban and Regional Informational Systems Association,* pp. 212–225.

Pollock, S. M., M. H. Rothkopf, and A. Barnett (eds.), 1994. Operation Research and the Public Sector, in *Handbooks in Operations Research and Management Science,* Vol. 6, North-Holland, Amsterdam.

Rapp, M. H., P. Mattenberger, S. Piguet, and A. Robert-Grandpierre, 1976. Interactive Graphics System for Transit Route Optimization *Transportation Research Record,* 559, 73–88.

Rousseau, J.-M., R. Lessard, and J.-Y. Blais, 1985. Enhancements to the HASTUS Crew Scheduling Algorithm, in J.-M. Rousseau (ed.), *Computer Scheduling of Public Transport 2,* North-Holland, Amsterdam, pp. 255–268.

Rousseau, J.-M., J. Desrosiers, and Y. Dumas, 1995. Results Obtained with Crew-Opt, A Column Generation Method for Transit Crew Scheduling, in J. Paixao and J. R. Daduna (eds.), *Computer-Aided Transit Scheduling,* Springer-Verlag, Berlin.

Salzborn, F. J. M., 1972, Optimum Bus Scheduling, *Transportation Science,* 6, 137–148.

Schéele, C. E., 1977. *A Mathematical Programming Algorithm for Optimal Bus Frequencies,* Institute of Technology, University of Linköping, Linköping, Sweden.

Spiess, H., 1983. *On Optimal Route Choice Strategies in Transit Networks,* Publication 286, Centre de Recherche sur les Transports, Université de Montréal, Montreal, Quebec, Canada.

Spiess, H., and M. Florian, 1989. Optimal Strategies: A New Assignment Model for Transit Networks, *Transportation Research* B, 23(2), 83–102.

Stern, H. I., and A. Ceder, 1983. An Improved Lower Bound to the Minimum Fleet Size Problem, *Transportation Science,* 17, 471–477.

Syslo, M. M., N. Deo, and J. S. Kowalit, 1983. *Discrete Optimization Algorithms,* Prentice Hall, Upper Saddle River, NJ, Chapter 2.

Turnquist, M., 1979. Zone Scheduling of Urban Bus Routes, *Transportation Engineering Journal,* 105(1), 1–12.

UMTA/FHWA, 1977. *UTPS Reference Manual,* U.S. Department of Transportation, Washington, DC.

Ward, R. E., P. A. Durant and A. B. Hallman, 1981. A Problem Decomposition Approach to Scheduling the Drivers and Crews of Mass Transit Systems, in A. Wren (ed.), *Computer Scheduling of Public Transport: Urban Passenger Vehicle and Crew Scheduling,* North-Holland, Amsterdam, pp. 297–312.

Wilhelm, E., 1975. Overview of the Rucus Package Driver Run Cutting Program, preprints, in L. Bodin and D. Bergman (eds.), *Workshop on Automated Techniques for Scheduling of Vehicle Operators for Urban Public Transportation Services,* Chicago.

Wilson, N., and M. Banasiak, 1984. *The Role of Computer Models in Transit Service Allocation Decisions,* Report MA-11-0041, U.S. Department of Transportation, Washington, DC.

Wren, A., and J.-M. Rousseau, 1994. Bus Driver Scheduling, in J. Paixao and J. R. Daduna (eds.), *An Overview in Computer-Aided Transit Scheduling,* Springer-Verlag, Berlin.

Zeleny, M., 1973. Compromise Programming, in J. L. Cochrane and M. Zeleny (eds.), *Multiple Criteria Decision Making,* University of South Carolina Press, Columbia, SC.

Zeleny, M., 1974. A Concept of Compromise Solutions and the Method of the Displaced Ideal, *Computers and Operations Research,* 1(4), 479–496; *Linear Multiobjective Programming,* Springer-Verlag, New York.

CHAPTER 10

Airline Operations Research

CYNTHIA BARNHART and KALYAN T. TALLURI

10.A. INTRODUCTION

The goal of an airline company, as much as for any other business organization, is to maximize operating profits. This can be accomplished by making the right strategic decisions, but also by making sure that one implements that strategy properly, running the day-to-day operations efficiently (minimizing operating costs) and selling the airline's products at the right price (maximizing revenue). The latter two objectives are the main concern of operations research practice in the airline industry.

The airline business by nature is an operations-intensive business. Operating costs constitute a large portion of an airline's expenditure, providing many opportunities for efficiency. If one can, for example, reduce some of the flight operating costs (fuel, landing fees) by making better decisions on assigning aircraft types to flights, it will lead to an overall reduction in operating costs. On the other hand, by better matching aircraft sizes to the demand or by selling the right number of seats on a flight to the right fare class, one can maximize revenue. By building mathematical models that capture the relevant details of operational problems of this nature and using operations research techniques to solve the problem numerically, one can make these routine day-to-day operating decisions near-optimally and improve the profitability of the airline. In this chapter we survey airline operational areas where operations research techniques have had a big impact. The topics that we cover fall into two broad categories: (1) schedule planning and (2) revenue management.

10.B. SCHEDULE PLANNING

Schedule planning (Figure 10.1) is concerned with generating a schedule (a feasible plan of what cities to fly to and at what times) that has the most revenue potential and resolving a host of related issues involving aircraft

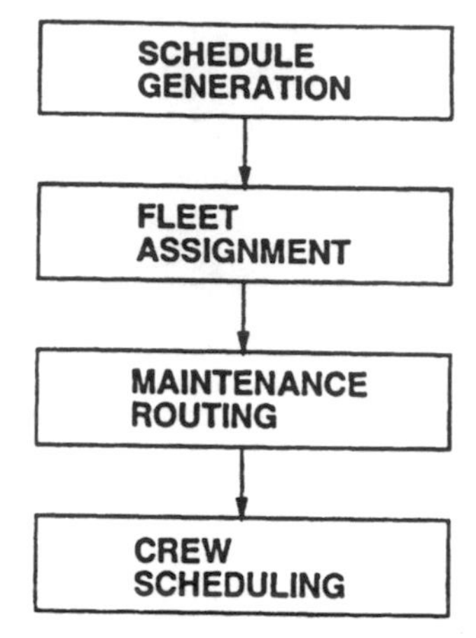

Figure 10.1 Schedule planning.

assignments and crew scheduling. For instance, after the schedule is fixed, the airline has to decide what aircraft type (such as 767, 727, etc.) will be assigned to each flight. This step of the schedule planning process is called *fleet assignment.* Then the airline has to decide the specific aircraft of that type, or *tail number,* in the airline's fleet to assign to each flight. This essentially is equivalent to deciding the routing of the aircraft. The primary motivation for this phase is that the aircraft be able to undergo planned periodic maintenance. This phase is called *periodic maintenance routing.* These two decisions involve the airline's equipment. The next step in schedule planning involves deciding who flies this equipment: that is, deciding how the crew (pilots, cabin crew) who fly these aircraft are assigned to the schedule. Crew costs make up a big part of the expenses (over $1 billion annually for major U.S. airlines), and by scheduling the crew properly and using their time efficiently, labor costs can be reduced. This phase is called *crew scheduling.* The crew of an airline usually work under some very strict regulatory agency guidelines and under restrictions imposed by labor agreements between the airline and its employees. Consequently, the objective is to find a crew schedule that satisfies all the constraints and at the same time makes efficient use of the crews.

The schedule planning process is usually completed a month or two in advance of the schedule inception date. Of course, one can formulate this entire schedule planning process as one giant model and try to solve the unified problem. Conceivably, such a joint formulation and optimization framework can lead to higher revenues. However, with current technology (computer hardware and optimization algorithms development), it seems unlikely that such a problem can be solved. This is the primary reason for breaking up the problem into sequential stages and solving each one individually. Even with this partitioning approach, operations research applications in the airline industry give output that is helpful to schedulers and planners. Clearly, as advances in technology occur, more of the problems will be modeled and solved simultaneously.

Schedule Implementation. We have to mention that planning is only half the story. Equally important is how the plans are implemented once the schedule becomes operational. The task of implementing this plan is the job of the system (or flight) controllers. The factors that make this difficult are weather disruptions, air traffic control delays, mechanical breakdowns of equipment, and crew cancellations and unavailability, all of which are present to some extent or the other on any given day. The goal of system controllers is to run operations as close to plan as possible with maximum on-time and revenue performance and minimal inconvenience to the customer. They are usually responsible for keeping track of weather patterns, planning the physical routes of flights (to go around bad-weather pockets, to minimize fuel consumption, to make up for delays), dispatching of aircraft, positioning of spare aircraft, coordinating spare parts for maintenance, coordinating flight and on-call crew, making decisions on delays and cancellations, among others. Because of the tight interlinking of flights in a schedule, even a small perturbation in the arrival or departure times of a flight can have repercussions for later flights throughout the system. If a particular aircraft breaks down, it may be better to cancel some other flight and switch the aircraft to minimize the overall disruption. If particular flights are delayed, other flights may be purposefully delayed so that delayed passengers can connect to these flights. All these decisions have to be made on-line using information that becomes available only a short time before when the decision has to be made.

In one of the few papers that describe some of the issues in this area, Jarrah et al. (1993) give models and solution experience at United Airlines. Before them, Teodorovic and Guberinic (1984) and Teodorovic and Stojkovic (1990) proposed some models to address these issues. It has to be said, however, that the use of decision support tools and operations research techniques is relatively scarce in the system control area. For this reason we do not have much to say on this important topic. However, the complexity of the problems, the large amount of information that has to be processed, and the speed at which decisions have to be made lead us to believe that this will be an active topic of research in the near future.

10.B.1. Schedule Generation

The basis for much of an airline's operations is the *schedule*. Different definitions of a schedule are possible, depending on how broad one chooses to make them. At its core, however, a schedule is a list comprising four things: origin city, destination city, time of departure (which roughly also determines the time of arrival), and flight frequency. The last quantity is the days of the week that this particular service is offered: for example, Monday to Friday, or Saturday and Sunday only.

The schedule is the primary product that an airline is selling. It affects every operational decision and has the biggest impact on profitability. It defines an airline's markets, its turf, so to speak, and is often a reflection of the airline's strengths, future plans, and ambitions. In the era before deregulation, the schedule would have been a stable entity modified, if at all, to accommodate seasonal changes in demand patterns. But that is no longer true. In a deregulated environment, a schedule for an airline is as much a competitive and marketing weapon as any other. Strategically, it is the airline's imperative to defend its traditional profit-making markets where it has a large market share, and improve its market share elsewhere, all the while keeping an eye open for new opportunities to expand. It is not unusual for an airline to make forays into an opponent's territory by increasing the frequency to the opponent's hub for a while, while the opponent may respond by increasing the frequency correspondingly in the first airline's home territory. If a particular market or hub is not profitable, the airline may decide to pull out of that market entirely. All these tend to make the schedule highly variable and unpredictable over the long term. However, in the short term, over a 3-month to one-year time frame at least, the schedule is relatively stable. The schedule announces the strategy of the airline for the time period and provides a period of time to develop that market and measure its impact.

A major constraint in developing the schedule is the size and composition of the airline's fleet. The number of markets and the frequency of service that the airline can offer are restricted by the number and types of aircraft that are in the airline's fleet. Not all types of aircraft can fly between all pairs of cities. A 737, for example, cannot fly between Los Angeles and Tokyo nonstop. Other considerations are the location of crew and maintenance bases. Some airports may be slot-constrained (currently, Chicago O'Hare, JFK, and La Guardia in the United States) and it may be prohibitively expensive for an airline to acquire slots to fly to these destinations. Also, for international flying, bilateral agreements and government allocations typically dictate who flies where.

For all its strategic and financial implications, operations research has had relatively little impact on the schedule generation process itself. Apart from using techniques such as passenger choice modeling and forecasting algorithms to predict passenger flows, no airline—as far as we know—uses a model that captures all the relevant factors and outputs a schedule. This may be partially because of the inherent strategic nature of the problem. It is very hard to capture in a mathematical model many of the objectives that go into schedule generation. Also, it is hard to design and solve a model that captures competitors' moves and strategies. Game theory may offer some insight there, but we don't know of any one using it in the airline industry. In this book schedule generation is also treated in the contexts of commercial freight transportation [Chapter 8 (Section 8.D.1)] and public-sector urban transportation [Chapter 9 (Section 9.D)].

Note that the schedule as we have defined it so far does not specify the specific aircraft flying each flight leg or even the type of aircraft. A broader definition of schedule generation would include these two succeeding stages of planning as part of the schedule generation process. We now describe these two stages in detail.

10.B.2 Fleet Assignment

Once the schedule is fixed, at least tentatively, the next step in the schedule generation process is fleet assignment. This is usually done around 3 months ahead of the start date of the schedule, with minor modifications as the day of operation approaches. Fleet assignment determines what type of aircraft (737L, 737S, F100, 767, etc.) are assigned to the flight legs. The capacities and operational characteristics of the different equipment types vary considerably. A Fokker 100 (F100) can seat around 100 passengers in coach, while a 767 can seat up to 220. On the other hand, a Fokker 100 has lower operating costs on short-haul routes, as it consumes less fuel and with its lesser weight is charged less by airports for landings and takeoffs. The motivation behind fleet assignment is to assign the equipment types to the flights so that the maximum passenger revenue is captured (by assigning larger planes to the high-demand flights) while simultaneously balancing the operating costs of such assignments. As equipment type assignments affect crew and maintenance, issues pertaining to them should also be taken into account. In this book, vehicle scheduling and assignment problems are also treated in the contexts of commercial freight transportation [Chapter 8 (Section 8.D.1)] and public-sector urban transportation [Chapter 9 (Section 9.E)].

Fleet assignment can be characterized as either daily or variable. Daily fleet assignment assumes a schedule that repeats every day of the week over the planning horizon and assigns equipment types such that every flight is assigned the same equipment every day. Variable fleet assignment does not assume a repeating schedule and does not necessarily assign the same equipment type to a flight every day. Daily fleet assignment makes operations—specifically, crew and maintenance scheduling—easier, while variable fleet assignment promises improvements in revenue by giving more flexibility to match demand and capacity. At the current time most U.S. airlines assign equipment types in a spirit that is closer to daily than variable fleet assignment. Of course, since most airlines do not have a full schedule on weekends, the assumption of a repeating schedule is rarely satisfied. Typically, assignments are done separately for the repeating weekday part of the schedule and for the weekend part of the schedule and the solutions are patched together.

The fleet assignment model has been widely applied in practice, in part because of the potentially large savings achieved with improved fleetings. For example, a 1.4% improvement in operating margin at American Airlines has been reported by Abara (1989); a $100 million per year savings in operating costs at Delta Airlines has been reported by Subramaniam et al. (1994); and significant savings have been reported by USAir (Dillon et al., 1993).

Fleet Assignment Network. The fleet assignment problem can be characterized as a multicommodity network flow problem with side constraints. Associated with each equipment type, or *fleet type,* is an underlying network (Figure 10.2). The nodes in each network represent a flight's origin location and departure time or its destination location and arrival plus *turn time.* Turn time is the minimum time needed to service an aircraft after its arrival at a station and before its departure from that station. Since servicing an aircraft involves refueling, cleaning, inspecting, and so on, turn times may vary by fleet type, by station, and by time of day.

The arcs represent either flight legs or ground connections. Each flight leg is represented by one arc from its departure node to its arrival node. The cost of that arc is the cost of flying the associated flight with the fleet type. Since exactly one fleet type must be assigned to a flight leg, each flight arc has a capacity of one.

Ground arcs connect adjacent nodes (the node representing the next point of time) at the same station, allowing aircraft to connect from one flight at a station to another at that same station. The set of ground arcs includes wrap-around (or overnight) arcs that connect the last node of the day with the first

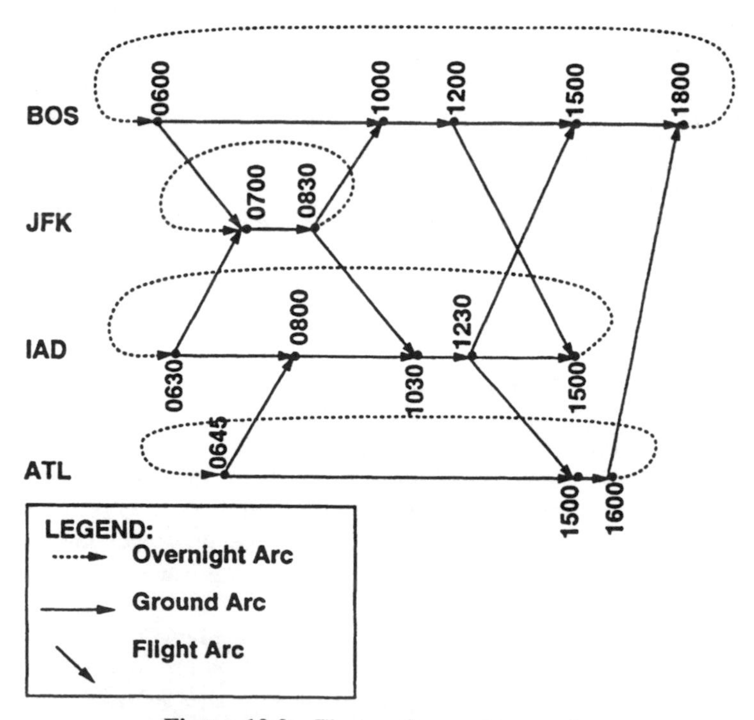

Figure 10.2 Fleet assignment network.

node of the day at that station. This arc models aircraft overnighting at a station. Ground arc costs are set to zero and capacities are unlimited since the number of aircraft on the ground at a station is not explicitly constrained.

Each fleet type has its own network because of differences in noise and gate restrictions for the various equipment types at certain stations, equipment operating limitations that vary by fleet type, and differences in the costs of different flight-fleet type assignments.

Fleet Assignment: Mathematical Model. We consider a *basic* domestic, daily fleet assignment model, similar to those proposed by Abara (1989), Daskin and Panayotopoulos (1989), Berge and Hopperstad (1993), Dillon et al. (1993), Subramaniam et al. (1994), and Hane et al. (1995). The basic model contains the following constraints:

- *Flight coverage and aircraft integrality.* Each flight must be assigned to exactly one aircraft type.
- *Aircraft balance.* The number of aircraft of a particular type flying into an airport, or station, must equal the number flying out.
- *Fleet size.* The number of assigned aircraft of a particular aircraft type must not exceed the number available.

Let the set of aircraft types be denoted by K, the number of aircraft of type $k \in K$ be denoted by $S(k)$, the sets of nodes and arcs in the fleet assignment network for fleet type $k \in K$ be denoted by $N(k)$ and $A(k)$, respectively, the set of all flights be denoted by F, the arc $j \in A(k)$ representing flight $f \in F$ be denoted by $a(k,f)$, the set of ground arcs for fleet type k be denoted by $G(k)$, the set of arcs out of node i for fleet type k be denoted by $O(k,i)$, the set of arcs into node i for fleet type k be denoted by $I(k,i)$, the decision variables and their costs be denoted by $x_{j,k}$ and $c_{j,k}$, respectively, for all $j \in A(k)$ and all $k \in K$. The decision variable $x_{j,k}$ represents a flow of type k aircraft on arc j.

$$\text{minimize} \sum_{j \in A(k)} \sum_{k \in K} c_{jk} x_{jk}$$

subject to

$$\sum_{k \in K} x_{a(k,f),k} = 1 \qquad \forall f \in F$$

$$\sum_{j \in O(k,i)} x_{jk} - \sum_{j \in I(k,i)} x_{jk} = 0 \qquad \forall\, i \in N(k), \quad \forall\, k \in K$$

$$\sum_{j \in C(k)} x_{jk} \leq S(k) \qquad \forall\, k \in K$$

$$x_{jk} \geq 0 \qquad \forall\, j \in A(k), \quad \forall\, k \in K$$

$$x_{a(k,f),k} \in \{0,1\} \qquad \forall f \in F, \quad \forall\, k \in K$$

The objective is to minimize the total fleet assignment cost. All flight legs must be flown and the repositioning of aircraft (i.e., flying the aircraft on a nonscheduled route to get the plane in position for a scheduled flight) is not allowed. By adjusting the objective function cost parameters, certain constraints that cannot be formulated easily or are specific to one fleet type can be enforced. For example, gate or noise restrictions might disallow aircraft of certain fleet types from landing at certain airports. These restrictions can be captured by giving an extremely high cost for the disallowed equipment type.

The first set of constraints in the basic model are the cover constraints that force each flight leg to be flown by exactly one fleet type. (Integrality of equipment types is enforced by the final set of constraints.) The second set of constraints are the aircraft balance constraints. These force the fleet assignment solution to be a flow circulation for each aircraft type, conserving flow at each node of the network. The third set of constraints are the fleet size constraints that count the number of aircraft of each type used in the solution. The number of aircraft are counted by selecting a point in time, referred to as the *count time,* and counting the number of aircraft of each type on the ground or in the air at that time. Since any fleet assignment is a flow circulation, counting aircraft at any single point in time is sufficient. Let $C(k)$ denote the subset of arcs $j \in A(k)$, including the count time; then aircraft are counted by summing the flow on all arcs $j \in C(k)$. The last two sets of constraints ensure integrality and nonnegativity of the number of aircraft assigned to a flight, on the ground, or overnighting at a station. Integrality restrictions are not explicitly included for equipment on the ground or overnighting since they are guaranteed by the last set of constraints.

Model Enhancements. Much of the fleet assignment research reports results for this basic model (or a slightly modified version of it). Although this model does not fully specify the fleet assignment problem, its solution provides valuable insights. Some of the additional constraints that may be included in an enhanced model are:

- *Maintenance requirements.* In the United States, the Federal Aviation Administration (FAA) requires different levels of maintenance checks. Details of maintenance requirements and industry practice are given in Feo and Bard (1989) and Gopalan and Talluri (1996). For the domestic fleet assignment, explicit satisfaction of maintenance constraints is impossible since fleet types, not specific aircraft, are assigned to flight legs. So to guide the process toward a maintenance-feasible solution, constraints may be added that bound the number of aircraft of different fleets that must be *available* for maintenance (i.e., an aircraft must be stationed at a maintenance location for a sufficient period of time to allow maintenance). A lower bound ensures that there are adequate maintenance opportunities for each fleet type, and an upper bound ensures that the number of aircraft available for maintenance does not exceed the capacity of the maintenance facility.

- *Gate restrictions.* The fleet assignment must allow for feasible gate assignments. Different jetways are required for different fleet types, and gate capacity may vary depending on the fleet assignment. For example, more flights can be accommodated at a station if narrow-body rather than wide-body aircraft are assigned to them.
- *Noise restrictions.* At airports, a variety of restrictions may be imposed to limit noise. The fleet assignment must satisfy these restrictions.
- *Crew considerations.* Ideally, we would like to minimize the combined fleeting and crew costs. However, this combined problem is currently computationally intractable. So, instead, a strategy is to enhance the fleet assignment model by including certain crew considerations. A discussion of possible enhancements is included in Clarke et al. (1994).

Fleet Assignment: Solution. Daskin and Panayotopoulos (1989) apply a Lagrangian relaxation approach to fleet assignment, while others have used classical mixed-integer programming techniques to solve the problem. Abara (1989), Subramaniam et al. (1994), and Hane et al. (1995), use mixed-integer programming techniques, available as commercial optimization software [CPLEX (Cplex, 1990) or OSL (IBM, 1990)]. Significant improvements in computational efficiency can be had by exploiting the special structure of the fleet assignment problem. Some of the techniques for achieving this are described in the paper by Hane et al. (1995).

10.B.3. Periodic Maintenance Routing

After assigning equipment types to flights, the next step is to assign the physical aircraft—called *tail* or *nose numbers* in the industry—to the flights. Maintenance plays a big role in determining the aircraft assignments. FAA regulations (in the United States or their equivalent in other countries) require that an airline do some periodic maintenance checks (called *transit checks*) on its aircraft after a certain number of hours of flying (currently 65 hours in the United States). These requirements are very strict and an aircraft will be grounded if it does not meet them. Airlines typically make sure that they meet these guidelines by requiring that their aircraft overnight at a maintenance base after at most 3 days of flying for some of the older equipment types and after at most 4 days of flying for some of the newer ones. In addition, some airlines require that aircraft go through a special maintenance base called the balance-check station, where they can undergo a specific type of maintenance called *balance check* once in awhile.

The assignment of aircraft to the flights is equivalent to determining how the planes are routed over the planning horizon. The routings during a day dictate where the plane spends the night. If the plane overnights at a maintenance base (an airline typically has maintenance bases at only a few cities), it has the opportunity to undergo maintenance. Therefore, the goal of the maintenance routing phase of the schedule generation process is to ensure that both the schedule and the fleet assignment allow for a routing *plan* that would let the aircraft overnight at the maintenance bases at periodic intervals. If the schedule or the fleet assignment does not allow for such a routing, either the fleet assignment or the schedule has to be modified.

In this section we survey aircraft routing models that are currently in use. These routing models are built upon the daily fleet assignment model [i.e., the fleeting (and the schedule) are assumed to be identical every day of the week]. In addition, in the routing model, the routings are also assumed to be repeated every day. We refer to this model as the *daily routing model.* Solution approaches for the daily routing model can be divided into two categories: one that uses graph-theoretic heuristics and one that uses mathematical programming models. We give brief descriptions of both below.

At this stage of our discussion, routing refers to "planned" routing. Its purpose is to assure the schedulers that the schedule and fleeting allows feasible routing to satisfy maintenance. In implementation, the routing is more ad hoc, with the planned routing serving only as a guide. So even though there may be short periods of maintenance opportunities during the day, during the longer layovers, for example, it is not very useful to "plan" for maintenance during those periods. For this reason, in these models we assume that maintenance opportunities are only during overnight stays at maintenance bases.

During a typical day, a plane starts at some station, makes a series of stops at airports and overnights at some other station, which may or may not be a maintenance station. Since planned maintenance is assumed to take place at night, it is useful to consider the day's activity only in terms of *lines of flying* (LOFs), which specify the origin at the start of the day and the destination at the end of the day for a particular airplane. LOFs are also sometimes referred to as *over-the-day routings* (Kabbani and Patty, 1992). The LOFs can easily be constructed from a fleeting by using simple rules such as first-in-first-out (FIFO) or last-in-first-out (LIFO). However, the LOFs so constructed need not satisfy the maintenance requirements.

To meet the requirement for transit checks, it is typical to require that every tail number spends a night at a maintenance station after at most k days of flying (k = 3,4) since its last overnight visit to a maintenance station. Assume that for every equipment type there is a single station where balance checks for that equipment type can be done and that this station is also a regular maintenance station.

Transit checks take a relatively short time to perform. We therefore assume that there is no real capacity constraint for doing the transit check maintenance. However, as balance checks require a considerable number of maintenance personnel and we are assuming that there is only one balance-check station for each aircraft type, there is a capacity restriction for each type of only one balance check per night. To provide equal utilization, it is typical of airlines that have the balance-check requirement to require that all the aircraft of one type pass through their balance-check station at regular periodic intervals. So a requirement could be that if there are n aircraft in the fleet of a particular equipment type, every one of the tail numbers of that equipment type should visit the balance-check station once every n days. As we mentioned before, not all airlines have a balance-check requirement. However, as it is very easy to modify both the heuristics and the math programming formulations described below to ignore this requirement, we will from now on assume that this requirement is present.

Let *k-day maintenance routing* denote a routing of tail numbers (of one equipment type) such that all the tail numbers visit a maintenance station after at most k days of flying without maintenance and every tail number spends at least one night at the balance-check station for that equipment type when it can undergo the balance check. We are mainly interested in $k = 3$ and $k = 4$. During operations the requirement is occasionally relaxed to stretch to 4 or 5 days (as this is still within FAA guidelines), allowing for some leeway in patching the routing over the weekends. For weekends when most airlines operate a restricted schedule, we will have to modify the daily routing plan. The LOFs over the weekends are arranged manually such that they fit seamlessly into the weekday maintenance schedule. Since gate, crew, and maintenance scheduling is much easier when the same LOFs occur daily, we will provide models that satisfy this requirement.

Graph-Theoretic Heuristics for Maintenance Routing. The overnighting stations and LOFs are represented by a directed graph $G = (V,E)$, where the vertices V represent the set of stations where the aircraft overnight and the arcs E represent the LOFs. G is called a *LOF graph.* The number of arcs will be equal to n, the number of aircraft in the fleet of the equipment type under consideration. The set of nodes V is partitioned into a set of nodes M, representing the maintenance stations, and $N = V \backslash M$, representing the nonmaintenance stations. We will refer to nodes in M as M nodes, or maintenance nodes, and nodes in N as N nodes, or nonmaintenance nodes.

Under the daily routing model, a k-day maintenance routing is an Euler tour (a tour along the arcs such that each arc is traversed exactly once) in G that includes at most $k - 1$ nodes of N in succession. Let $(i_1, i_2,\ldots,i_n, i_1)$ be the sequence of nodes in such an Euler tour. Note that nodes can be repeated in this sequence. We interpret the Euler tour in terms of maintenance routing as follows: At the beginning of day 1, position the n planes of the equipment type in the fleet at the nodes $i_1, i_2,\ldots,i_n$. Then the plane at node i_k goes to station i_{k+1}, following the Euler tour. Since the Euler tour passes through a maintenance node after $k - 1$ consecutive nonmaintenance nodes, the tail numbers overnight at a maintenance station after at most k days of flying without maintenance. In addition, since an Euler tour is connected by definition, every tail number passes through the balance-check station once every

n days. As we had mentioned before, the balance-check station is also a regular maintenance station. Hence if a tail number visits the balance-check station more than once in the n days, it is to undergo regular maintenance the other times.

Using this model, Talluri (1994) and Gopalan and Talluri (1994, 1996) have given algorithms for finding a 3-day maintenance routing, a 4-day maintenance routing, and a 3-day maintenance routing with additional restrictions. See also Talluri (1996) for some other issues pertaining to maintenance routing.

Mathematical Programming Approaches to Maintenance Routing. A *through* flight (also called a *direct* flight) is a pair of flight legs marketed under the same flight number. It offers a level of service to passengers that is in between a nonstop flight and a connection. It involves a stopover and hence is not as desirable as a nonstop flight, but on the other hand, does not have the passengers get off the plane and change aircraft at the hub as in a connection. The airline has to decide how to form through flights for maximum marketing effect.

Whenever a through flight is selected, we constrain the maintenance routing problem and make it more difficult to find a feasible routing. Since the pair of flight legs that constitute the through flight must be flown by the same aircraft, we provide less freedom at the routing phase for design of efficient routings that meet maintenance requirements. Hence addressing through flight selection and maintenance routing in separate modules might be myopic because a fixed set of through flights might constrain lines of flying so much that we are unable to meet maintenance requirements. To overcome this difficulty, Clarke et al. (1995) combine aircraft rotation and through-flight selection. As a starting point for their model, they work directly with the fleet-assigned flight network obtained following fleet assignment. This network (of a single equipment type) is an Eulerian directed graph (i.e., the indegree at each node equals the outdegree). Note that through-flight revenue really is derived from flight *sequence* (i.e., every pair of flights that are contiguous in the rotation represents a through-flight opportunity).

To incorporate maintenance and through-flight revenue, Clarke et al. (1995) deal with the line graph corresponding to the flight network. The line graph corresponding to an (underlying) graph is a graph containing a node for each arc in the underlying graph and an edge between nodes i and j if in the underlying graph, arc i's head and arc j's tail meet at a node. If the underlying graph is an Eulerian graph, there is a one-to-one correspondence between Euler tours in the underlying graph and Hamilton circuits (a circuit that visits every *node* exactly once) in the line graph. Incorporating maintenance and through-flight considerations in the rotation now amounts to solving a traveling salesman problem (TSP) with side constraints on the line graph corresponding to the Eulerian time-based network of flights. For details of the solution approach, see Barnhart et al. (1996).

10.B.4. Crew Scheduling

The next step in the schedule planning process is to decide how to assign crews to work activities. Typically, this is achieved in two steps. First, the *crew pairing problem* is solved to generate a set of work schedules, and second, the *crew assignment problem* is solved to assign specific crews to the work schedules.

The objective of the crew pairing problem is to find a set of work schedules that cover each flight the appropriate number of times and minimize total crew costs. These work schedules, called *pairings,* are comprised of a set of daily work activities, called *duty periods,* or *duties,* separated by rest periods. Every pairing is constructed so that a single crew can *legally* perform all the work activities it contains.

The crew pairing solution *does not* assign specific individuals to pairings. Instead, crew assignment accomplishes this. In crew assignment, pairings are combined with rest periods, vacations, training time, and so on, to create extended work schedules, typically spanning a period of about one month, that can be performed by an individual. With rostering, a practice commonly used in Europe, schedules are constructed for specific individuals, taking into consideration their particular needs or requests. Then a subset of schedules are selected such that total crew costs are minimized, each person is assigned to a schedule, and all pairings in the crew pairing problem solution are contained in the appropriate number of schedules. This is in contrast to bidline generation, a practice more commonly used in North America. With bidline generation, schedules are generated but not for specific individuals. The cost-minimizing subset of schedules is selected and individual employees reveal their relative preferences for these schedules through a *bidding* process. The airline then makes the specific assignment of schedule to employees based on individual priority rankings—rankings that are often related to seniority.

Tractability issues force the crew scheduling task to be broken into two stages. Solved separately, the crew pairing and assignment problems are each computationally challenging. In fact, for the crew assignment problem, problem size often makes finding optimal solutions impossible. Much of the literature (e.g., Nicoletti, 1975, Buhr, 1978, Moore et al., 1978, Tingley, 1979, Marchettini, 1980, Giafferri et al., 1982, Glanert, 1984, Byrne, 1988, and Sarra, 1988) focuses on heuristic solution methods. However, exact solution methods are presented in Ryan and Falkner (1988), Ryan (1992), and Gamache and Soumis (1993). In this book, crew scheduling is also treated in the contexts of commercial freight transportation [Chapter 8, (Section 8.D.1)] and public-sector urban transportation [Chapter 9, (Section 9.F)].

The focus of crew scheduling research has been the crew pairing problem, in part because of tractability issues associated with the crew assignment problem. In the next sections we describe models and methods for the crew pairing problem.

Crew Pairing Problem. As with daily fleet assignment, in crew pairing optimization we attempt to build pairings that can be repeated every day. Naturally, as the flight schedule does not repeat daily, this poses some difficulties. We therefore break up the problem into two stages. In the first stage, *daily* pairings (pairings that can be repeatedly daily) are constructed, and in the second stage, *weekly* pairings (pairings that can be repeatedly weekly) are constructed.

For the *daily* problem (the first stage), flights that are flown at least four times a week may be considered. Then some of the pairings generated can be flown daily, while others, called *broken pairings,* cannot because they contain flights that are flown only on selected days. This is the input to the second stage, the *weekly exceptions* problem. It consists of the flights in the broken pairings, and possibly some deadheads. (*Deadheads* are flights to which crew members are assigned as passengers. The purpose is to increase crew utilization by flying them to other locations where they may be assigned to flights earlier than if they had not been repositioned.) The objective of this second stage is to find the minimum cost set of weekly pairings that appropriately cover the flights in the broken pairings. Combined, the daily and weekly pairing solutions cover each flight the appropriate number of times.

The input for the weekly exceptions problem may be expanded to include flights from daily pairings. While this reduces the number of daily pairings and may result in increased operational complexity, the overall solution cost may be reduced significantly. The challenge is to find the appropriate balance between ease of operations (using daily pairings) and total cost (which is minimized by breaking all daily pairings and including all flights in the exceptions problem). Further details of the exceptions problem appear in Barnhart et al. (1996b).

The pilot and cabin crew pairing problems, although similar in structure, are treated separately. The cabin crew problem is more difficult to model and solve because of several reasons. The number of flight attendants assigned to a flight is a function of the number of passengers on that flight, while the crew size for the pilot problem is constant. Flight attendants can be assigned to several different aircraft types, unlike pilots, who are eligible to fly only one aircraft family type. This results in many more flights in the flight attendant problem than in the pilot problem. In addition, one flight attendant may work with several different flight attendants as he or she is assigned to various flights throughout the day. With pilots, a crew is not splitup; instead, each member of the crew flies the same pairing. This allows the pilot model to consider crews rather than individuals, reducing the number of variables in the model. A final difference is that the savings in the pilot problem is greater than in the cabin crew problem.

For these reasons and because much of the crew pairing literature pertains to the pilot problem (e.g., Arabeyre et al., 1969, Odier et al., 1983, Rannou, 1986, Lavoie et al., 1988, Gershkoff, 1989, Barutt and Hull, 1990, Desrosiers et al., 1991, Anbil et al., 1991a,b, 1993, Graves et al., 1993, Hoffman and Padberg, 1993, Barnhart et al., 1994, Vance et al., 1996), we restrict our discussion to the pilot crew pairing problem in the remainder of this section.

Crew Pairing Problem: Mathematical Model. The structure of *legal* duty periods and pairings are defined by some regulatory agency (the FAA in the United States) and by collective bargaining agreements between crews and airline management. For duty periods, restrictions may include limits on the maximum and minimum time that a crew is idle between consecutive flight legs; a limit on the maximum elapsed time from start to end; and a limit on the total number of hours of flying time, called *block* time. The cost of a duty period is typically expressed as the maximum of three quantities. The first quantity is the *flying time cost,* determined by multiplying the total flying time in the duty by a cost per unit time. The second cost is the *elapsed time cost,* expressed as the total elapsed time of the duty multiplied by a different (lower) cost per unit time. The third cost is the *minimum guarantee cost,* which is equal to some predetermined value. Duty costs, then, are the *maximum* of flying time, elapsed time, and minimum guarantee costs.

For pairings, regulatory and collective bargaining restrictions may include limits on the maximum number of duty periods in the pairing; a limit on the minimum number of hours of rest between duties—this limit may vary depending on the amount of flying time and previous rest received; and a limit on the maximum elapsed time, called *time away from base.* The cost of a pairing is also often expressed as the maximum of three quantities. The first quantity is the total cost of the duties contained in the pairing, the second is the cost associated with the pairing's total elapsed time, and the third is the minimum guarantee pay for the pairing.

Regulatory and collective bargaining restrictions may also restrict the amount of flying performed by crews at different crew bases. Crew base constraints, while relevant, often make the crew pairing problem difficult to solve. For this reason, crew base constraints are not included in many *basic* models, such as the one presented below.

Before describing the crew pairing problem formulation, we introduce the following notation. F is the set of flight segments and P is the set of pairings. The decision variables, denoted y_p, are equal to 1 if pairing p is in the solution, and 0 otherwise. A column p has a 1 in row i if flight i is flown by pairing p, and c_p is the cost of pairing p. The most common crew pairing models build on the set partitioning problem. Following is the *basic* formulation:

$$\text{minimize} \sum_{p \in P} c_p y_p$$

subject to

$$\sum_{p:i \in p} y_p = 1 \qquad i \in F$$

$$y_p \in \{0,1\} \qquad p \in P$$

Since every solution must assign one crew to each flight, a lower bound on total crew costs is the total flying time costs for the scheduled flights. Costs in excess of this lower bound are referred to as *penalty costs.* The objective, then, is to find a set of pairing that partitions the flights such that penalty costs are minimized.

This formulation requires the explicit enumeration of *all* pairings. Enumerating pairings can be difficult because of the numerous work rules that must be checked to ensure legality and because of the huge number of potential pairings. In fact, for most real instances, *explicit* enumeration of the constraint matrix is impractical. For example, a problem with several hundred flights typically has billions of pairings. To solve these problems, then, heuristic local optimization approaches or column generation methods (described in the next section) are used.

Crew Pairing Problem: Solution. Branch and bound is a *divide-and-conquer* algorithm used to solve integer problems to optimality. Applying branch and bound to the crew pairing problem is difficult because the model contains a huge number, possibly billions, of decision variables. Instead, a specialized branch-and-bound procedure, called branch and price (Barnhart et al., 1996d), is used to solve problems containing many decision variables. Branch-and-price solution approaches for solving crew pairing problems are described in Desrosiers et al. (1993) and Vance et al. (1996).

Like branch and bound, a branch-and-price procedure is a *smart* enumeration strategy in which a linear programming (LP) relaxation may be solved at each node of a branch-and-bound tree. The difference is that the huge constraint matrix requires the use of a specialized LP solution procedure, called *column generation.* With column generation, most variables, or columns, are not included in the constraint matrix. Instead, a subset of the columns are included and additional columns are added only if they may improve the solution (Ahuja et al., 1993). Adding variables to the constraint matrix is referred to as column generation, and generating such columns or determining that none exist is called the *pricing problem.* The repeated solution of the restricted master problem—the LP containing only a subset of its columns, followed by the modification of this restricted problem through the generation of columns—may continue until optimality is achieved.

As pointed out by Appelgren (1969), branch and price is not a simple extension of branch and bound. Traditional branching strategies based on variable dichotomy usually complicate the pricing problem, making it intractable or at least impractical to generate columns repeatedly. So new branching strategies must be developed. The branching rule most commonly used in solving crew pairing problems, referred to as *branch on follow-ons,* is derived from a general rule for set partitioning problems developed by Ryan and Foster (1981). Their rule has been applied successfully to a number of other applications, including binary cutting stock problems (Vance et al., 1994), generalized assignment problems (Savelsbergh, 1996), urban transit crew scheduling problems (Desrochers and Soumis, 1989), vehicle routing and scheduling problems (Desrosiers et al., 1993; Sol and Savelsbergh, 1994), and graph coloring problems (Mehrotra and Trick, 1993).

The column generation algorithm to solve the crew pairing LP involves the following steps:

Step 1. Solve the restricted master problem. Find the optimal solution to the current restricted master problem.

Step 2. Solve the pricing subproblem. Generate one or more columns that may improve the solution. If no columns are found, stop: the crew pairing problem LP relaxation is solved.

Step 3. Construct a new restricted master problem. Add the columns generated in solving the subproblem to the restricted master problem and return to step 1.

The solution and construction of the restricted master problem (steps 1 and 3) can be achieved using optimization software such as CPLEX (Cplex, 1990) or OSL (IBM, 1990). The solution of the pricing subproblem (step 2), however, should be tailored to exploit the network structure of the problem. The goal is to identify columns that may improve the solution without examining all variables. This often can be achieved by solving *multilabel shortest-path* problems on specially structured networks (Desrochers and Soumis, 1988). The networks take on different forms, depending on network structure, problem size, number of work rules, and so on. For domestic operations, a typical network has nodes representing the departure and arrival of each flight and an arc for each flight in the schedule. There are also *connection* arcs from the arrival node of a flight to the departure node of another flight if the two flights may be flown consecutively by the same crew. These connection arcs ensure that each crew pairing corresponds to a path in the network. However, every path does not represent a legal pairing.

Legal crew pairings correspond to the subset of network paths satisfying:

- The first and last nodes of the path represent the same crew base location.
- All work rules are satisfied by the path.

- Each flight appears at most once in the path. For the daily crew pairing problem, each flight in each pairing must be covered exactly once each day. A crew pairing containing n days therefore has n crews flying that pairing on any single day. If a multiday pairing contained the same flight more than once, the flight would be assigned more than one crew. The additional crews could be considered as deadheading crews. However, the decision to deadhead affects the pairing generation process since there are different costs associated with deadheading. In addition to the usual costs, deadheading costs include costs of assigning crews to airline seats that might otherwise generate revenue from paying passengers. For these reasons and others, most airlines do not allow deadheading in their domestic plans.

In contrast, the network associated with long-haul networks is typically *duty period* rather than *flight* based. That is, nodes represent the start or end of a *duty* and an arc is included in the network for each possible duty. *Connection arcs* between duties are included if two duties can be flown consecutively by the same crew. Due to limited number of connections for crews (a result of the sparsity of flights arriving and departing a station), it is often necessary to include potential deadhead flights to achieve a good, or even feasible, solution. Again, each crew pairing is represented by a network path, and only the subset of paths satisfying certain requirements represent pairings. In this case the requirements that must be satisfied are reduced significantly because the duty arcs guarantee satisfaction of many of the work rules.

Lavoie et al. (1988) and others found success in applying the multilabel shortest-path procedure to duty-based networks. This approach is especially successful when the number of duty periods is not excessive, as demonstrated by Barnhart et al. (1994), who used a duty-based network to solve long-haul crew problems containing about two to three times as many duties as flights. Vance et al. (1996) were also successful in using a duty network to solve a relatively small domestic daily problem.

Heuristic approaches, involving limited column generation at nodes of the branch-and-bound tree and exploration of a limited number of nodes in the tree have also been applied successfully (Anbil et al., 1991b, 1993; Barnhart et al., 1994; and Marsten, 1994). In the extreme case, column generation is restricted to the root node and a standard branch-and-bound algorithm is applied to the subset of the columns generated in solving the LP relaxation. In several cases (e.g., Anbil et al. and Barnhart et al.) this approach yields very good solutions to the integer programming problem. However, there is no guarantee that a good solution, or even a feasible solution, exists using only the columns generated to solve the initial LP relaxation, as demonstrated in the case described in Barnhart et al. (1996c).

10.C. REVENUE MANAGEMENT

The second broad category we cover in this chapter is *revenue management.* Airlines sell seats on the same plane in the same compartment with different restrictions on purchase and at different prices. Because of this pricing structure (which is almost universal in the airline industry), it becomes very important for airlines to manage the sales of their products carefully. If they reserve too many seats for high-fare passengers, they may fly out with many empty seats and forgo revenue, and if they sell too many seats to low-fare passengers, they will be turning away high-fare passengers in the later stages of the booking process and dilute their revenue. The objective of revenue management is to maximize overall revenue. It is estimated that practicing revenue management in controlling seat sales has brought in 1 to 5% additional revenues per year for some of the larger airlines (which can be as much as $300 to $500 million per year for the larger carriers). This revenue benefit is even more remarkable considering that it is obtained without any disruption to the operations of the airline, but simply by using information and optimization technology to the fullest extent.

Revenue management has emerged as one of the most crucial developments in airline operations research, and its use is spreading from the airline industry to other industries in the travel, hospitality, and broadcasting sectors. In this section we discuss the historical origins and motivation behind revenue management, the ticket booking process of airlines using central reservation systems, and the various elements that comprise revenue management optimization and control.

10.C.1. History of Revenue Management

The practice of revenue management was all but unknown (except for the overbooking aspect of it) in the airline industry until 1978, when the historical Airline Deregulation Act was passed. Prior to that, the U.S. Civil Aviation Board set the fares on each route and regulated competition in each market, allowing only a few carriers on any route. An airline made profits by managing its costs and making its operations efficient but had almost no say in the markets in which it would compete and the prices at which it would sell its product.

With deregulation, however, carriers were free to enter or exit any market and to set prices on all flights. Overnight, this changed the U.S. airline industry. Startups could now freely break into a new market and offer fares that reflect their lower operating costs and thereby drag business away from the larger established carriers. The larger carriers found benefit in stimulating demand by offering to sell a part of the capacity of a flight at lower fares, with restrictions on the purchase of the tickets and penalties for modifying the ticket. This way they could better match capacity to demand, filling flights that traditionally flew empty (e.g., Wednesday afternoon flights) with leisure travelers who are flexible as to schedule but sensitive to price. The lower fares were a big boon to the consumer and the period following deregulation saw a big increase in airline traffic.

The freedom to enter new markets and to set their own prices posed new challenges to many of the large carriers. Many startups and charter carriers entered markets traditionally dominated by one or two of the majors and started offering fares that were almost one-third of the standard prederegulation fares. Although the service and schedules of these small carriers did not match those of the majors, the attractiveness of the price diverted many leisure travelers to these low-cost alternatives. To preserve market share, the large carriers had to match the fares of the new entrants, but at the same time they did not want to see their most profitable customer, the relatively price-insensitive business passenger, buying the cheap tickets and diluting their revenue to such an extent that they would not make a profit on the flight.

So the large carriers came up with the solution—widely credited to Bob Crandall, then a vice president of marketing at American Airlines—which initially went by the name of *yield management* but is lately being referred to as *revenue management.* The idea behind revenue management is simple: First, segment the passengers into different classes based on booking characteristics and price sensitivity; and second, impose restrictions on the tickets for seats in the same cabin on the same flight, essentially making them different products. Each of these fare products is aimed at a different segment of the market and carries a different fare. The motivation for imposing the restrictions is to prevent a relatively price-insensitive business passenger for whom the higher fare class is intended from booking in a cheaper fare class intended for leisure travelers. Finally, the idea is to forecast the number of passengers expected to book in each fare class and control the number of seats sold in each fare class to maximize revenue. Over time, revenue management has come to include other auxiliary areas of the seat inventory control process, such as overbooking, point of sale, and group bookings.

10.C.2. Airline Booking Process

The benefit of market segmentation is well documented in marketing literature. The characteristics that distinguish revenue management from other market segmentation models are chiefly the perishable nature of the product, the fact that the physical product is essentially the same but made different by artificially imposed restrictions, the fact that the higher-fare customers typically tend to arrive closer to departure, the network nature of the product, and the mechanism used to control the sales. We now go into the details of each of these characteristics of airline revenue management.

Airlines start selling seats on flights up to a year before departure. But since the schedule is usually not confirmed more than 3 months ahead, most of the bookings that are made prior to 3 months before departure are tentative and tend to be group bookings by tour operators. Most regular-fare-class bookings come during the last 2 or 3 months before departure. Typically, leisure passengers are able to book earlier than business passengers, and the restrictions imposed on the cheaper fare classes try to exploit this inability of business passengers to book early.

The broad categories of leisure and business passengers are further segmented based on price sensitivity, trip purpose, and so on. We do not go into the details and motivation behind these market divisions but refer the reader to Belobaba (1987) for an economic and marketing analysis of passenger price behavior. For our purpose we assume that the bookings for seats in each cabin class (first, business, and coach) come in a number of booking classes (typically eight or more for coach, one or two for business, and one or two for first). These booking classes are represented by letters of the alphabet. Some industry standard booking classes are F for first class, J and C for business class, Y for full-fare coach, and M, B, K, H, Q, Z, and others for the discounted fare classes in coach. Z is usually reserved for the frequent-flyer class.

Each fare class in fact represents a collection of *fare codes* or *fare basis codes.* Each of these fare basis codes (with names such as QXE30) has specific published fares associated with them, and the requirements for booking under that fare basis code is encoded (somewhat cryptically) into the name of the fare code (e.g., the 30 in QXE30 represents a 30-day advance purchase requirement). The primary reason for this grouping is that many reservation systems could handle only a few fare categories (five to eight) per cabin, and this grouping of fare codes into fare classes allows the airline flexibility in posting a wider range of fares and to group the connecting traffic into fare codes.

A typical booking process goes as follows: The airline posts availability in each fare class to the reservation systems (something that looks like Flight 314: Y4 M4 B0 . . . , a notation that will be explained shortly), stating whether or not a fare class is open. If a fare class is open, the travel agent or the airline's own reservations agent can make a booking in that fare class. Within a fare class the agent quotes a fare for the itinerary based on one of the fare codes. If the restrictions and the fares are acceptable to the customer, a booking is made under that fare class. The fare code is recorded under the passenger name record and gives information about restrictions, penalties, and so on.

The bookings are also grouped—for forecasting purposes—based on fare classes rather than fare codes. From these historical data of bookings and fares, the objective of revenue management is to forecast how many bookings are expected to come in each fare class, and based on these forecasts, decide how many seats should be allowed to be sold in each fare class, or conversely, how many seats will be protected for each fare class from fare classes with lower average fares. Once this is decided, the allocations of the seats among the various fare classes is uploaded to the reservations systems and the control part begins.

For clarity, we will treat revenue management as a two-stage process, a decision-making part and a control part. The decision-making part takes the historical data on bookings on each fare class and makes a forecast of the bookings to come for a particular flight on a particular date. This forecast consists of a mean and a variance (from some distribution) of the number of bookings expected. Based on the forecast, the optimization part of the decision-making process makes an open/close decision for each fare class. How this open/close decision is implemented depends on the control part of revenue management (i.e., what kind of control is used by the reservation systems).

Until a few years ago the most widely used form of control was something called a *nested allocation control scheme.* In this control method, the fare classes for each compartment (cabin) are ordered in decreasing order of their average fares. The highest fare class always gets an allocation that is equal to the capacity of the cabin (or the capacity plus an overbooking pad). The one with the next-lower fare gets an allocation that is less than or equal to the highest, the one below that, an allocation less than or equal to the one above it, and so on. That is, the allocations are nested within each other in a decreasing order. For example, if 95 seats can be sold for the coach compartment on a flight, an example of a six-fare-class nested allocation is Y95 B80 M77 K40 Q20 Z0. Assume for a moment that lower-fare class bookings come before higher-fare class bookings. In this example, fare class Z gets a 0 allocation, or is closed out, while the allocations indicate that 20 seats can be sold in fare class Q, up to 40 seats can be sold in fare class K, and so on. The nestedness comes about this way: Say, if only 15 seats are sold in Q, it is possible to sell $40 - 15 = 25$ seats in K. If however, 20 seats get sold in Q (before any K seat is sold), only 20 seats are available to be sold in K. So at the minimum, it is always possible to sell at least $40 - 20 = 20$ seats in K if K passengers book before M, B, and Y passengers. This number is called the *protection* for fare class K. It is the number of seats protected for exclusive first use for that fare class (if lower-fare classes come before higher), from bookings for lower-fare classes. In the example above, Y has a protection of 15, B of 3, M of 37, K of 20, Q of 20, and Z of 0. The sum of the protections add up to the capacity. The nested allocations are the numbers 95 for Y, 80 for B, and so on. Note that these nested allocations are for use only within the airline and are considered confidential. These allocations are then posted on central reservation systems (CRSs) after converting them to open/close notices.

Periodically, the decision-making part of the airline's revenue management system makes the forecast and calculates these allocations to maximize the expected revenues over all such nested allocations and posts the allocations to its own reservation system (most large airlines have their own reservation systems) or to CRSs.

10.C.3. Central Reservation Systems

There are a number of CRSs currently in operation worldwide. Most are owned either by an individual airline (e.g., American Airline's SABRE or Continental Airline's System One) or by a consortium of airlines (Galileo is owned by United, USAir, British Airways, and others; AMADEUS by Lufthansa, Air France, and Iberia). The operation of these CRSs is governed by laws intended to prevent the airline or airlines owning the CRS to bias or direct bookings to their carriers. The CRS systems communicate with the reservation system of each airline to obtain the allocation decisions periodically. A travel agent makes a booking for an itinerary by sending a communication to the CRS to which the travel agency subscribes. The CRS in turn notifies the host reservation system of each one of the airlines involved in the itinerary about the booking. A small amount is charged for each one of these transactions.

If the host reservation system of an airline decides to close a booking class (because the number of seats sold in that class have reached the maximum allocated for that class, say), it sends a notification signal to all the external CRSs saying that a particular class on a particular flight is closed, and they in turn display the new availability in response to the travel agent's query. A mind-boggling number of such transactions go on every second of the day and the communication requirements are very demanding, usually requiring that the connect and transaction be completed in a second or less. For competitive purposes, so as not to reveal its inventory decisions to competing airlines, when the airline sends availability information to CRSs, it sends only open/close information for each fare class (represented as Y4 B4 M4 . . .).

A second aspect of the control process besides the nesting allocation scheme is at what level the control is being done. Three kinds of control are common in the airline industry. The simplest and most widely used control is at the leg level. Each flight leg is controlled independent of other legs in the network. This is clearly not a globally optimal approach since it ignores the network dependencies created by multileg itineraries. With the hub-and-spoke network topology of many of the large carriers, such connecting traffic forms a significant percentage of the airline's total traffic and one can conceivably obtain more revenue if one were to solve and control inventory by considering the network as a whole. Such a global optimization and control is called *network-level control.* A third form of control that is implemented by some airlines is segment control. A *segment* is a line of flight legs usually having the same flight number. An example of a segment is Flight 311, which has the route BOS-JFK-CLT-MIA. The same aircraft flies all these legs (typically) and the itineraries that span legs of this segment have the same number as the flight and are marketed as direct flights (same as the through flights referred to in Section 10.B.3). Some reservation systems are designed to handle segments, which is the motivation behind segment control.

The leg-level control was more or less the universal choice for revenue management until a few years ago. That was what most reservation systems could handle and if an airline wanted to do more sophisticated revenue management (i.e., a more global optimization approach), it had to find some way of mapping its solution to a nested allocation form. This led some airlines to develop new methods, called *virtual nesting methods.* They would try to map the fare classes of the itineraries to leg-level fare classes. An itinerary fare class can get mapped to a different leg-level fare class on each leg of the itinerary, allowing for somewhat better control on itinerary sales, but the approach is somewhat messy.

Many of the large airlines with sizable connecting traffic are now moving to network revenue management (or studying the move at least), and the nesting form of control is fundamentally inadequate to handle network level control. So for network level control, CRSs are adopting a new form of control called the *bid-price control* mechanism. In this form of network control, every leg compartment will have a real number called a bid price associated with it. A booking request for an itinerary fare class is accepted if every leg of the itinerary has adequate capacity and the fare of the booking request exceeds the sum of the bid prices for the legs in that itinerary. The airline has the responsibility of supplying and updating the bid prices. Since we are adding up the bid prices to make a decision for the itinerary, this form of control should properly be called *additive bid pricing.* In some ways this form of control on a leg is equivalent to a nested sequence of continuous infinite fare classes and hence is also referred to by some airlines as *continuous nesting.* It turns out, as shown in Talluri and van Ryzin (1995a) that additive bid pricing is a very good approximation to the optimal form of control for network revenue management.

This brings us to the end of our discussion of the current environment of airline revenue management. In what follows we describe the various elements that comprise the revenue management decision-making process and related issues.

10.C.4. Overbooking

Overbooking a flight is a practice that started much before revenue management but is closely identified with revenue management. Overbooking by itself is considered to contribute around 40% of the benefits of revenue management, according to a study based on American Airlines data (Smith et al., 1992).

Overbooking is the airline's response to passengers (or travel agents) making a booking and not showing up for the flight. Such no-shows happen for many reasons: Travel agents make speculative bookings or fictitious bookings to earn points in airline incentive schemes, passengers make duplicate bookings and do not bother to cancel, passengers miss connections, or passengers simply decide to cancel or postpone the trip. If the airlines did not overbook and, instead, closed out their bookings as soon as the number of bookings equaled the plane capacity, a significant number of seats on every flight would go empty. To counter this wastage (which does not benefit anyone), airlines accept more bookings than seats on a flight, a practice called *overbooking.* The number of seats to book above the physical capacity is the *overbooking pad.*

From historical data on no-shows and go-shows (passengers who show up at the airport and buy a ticket), a forecast is made and an overbooking pad is set to meet certain criteria. The trade-off here is that if the airline overbooks too little, a lot of seats may go empty, and if they overbook too much they will have to deny boarding to a large number of people, costing the airline both in compensation money and in loss of goodwill.

Until around 1978, overbooking was done rather surreptitiously by the airlines. They would not acknowledge it publicly but do it nevertheless, and if a passenger had to be bumped because of overbooking, they would make an arbitrary choice on who to remove. The passengers so involuntarily bumped would have to be paid an amount set by the Civil Aviation Board. The loss of goodwill from such involuntary bumpings is usually worth much more than the dollar amount. Then, around 1978, thanks mostly due to the idea—and tireless efforts to have the idea implemented—of an economist, Julian Simon, an auction-type scheme was finally adapted by most of the airlines.

The mechanism currently in use (at least among U.S. airlines) is to request volunteers 10 or 15 min before departure and only if there are not sufficient volunteers, deny someone a boarding. For both cases the compensation is typically one or two free round-trip tickets to anywhere in the continental United States for use anytime in the next one year. Involuntary denied boardings have shown a gradual decrease since the inception of this scheme, as the public has gotten used to the idea and practice.

Different airlines use different criteria to decide how many to overbook on a flight. Some common criteria: Always overbook by a fixed amount; overbook by the expected number of no-shows; assign a cost to each denied boarding and set the overbooking pad to maximize the expected revenue minus the cost; set the overbooking pad so that the probability of an oversale is equal to some acceptable target value; and similar ones.

Overbooking can be done jointly with revenue management, by including overbooking as another decision variable in revenue management formulations, or done separately by fixing an overbooking pad and then setting the allocations or calculating bid prices. The former model is more appropriate from the perspective of maximizing total revenue, while the latter is simpler to implement and more transparent.

10.C.5. Forecasting

There is extensive research on forecasting and it is not the intention of this brief section to survey forecasting techniques or methodologies as such. The techniques used (or reported to be used, as this is a secretive matter for many airlines) range from the standard time-series methods such as exponential smoothing, or Kalman filtering, to causal methods, and more recently, neural networks. The details on how each performs are generally not reported by the airline, since revenue management is used as a competitive tool, and therefore, confidentially is important. It is hard, therefore, to say which, if any, of the above-mentioned forecasting methods is suited for doing the kind of forecasting required for airline revenue management. We will in fact leave the subject of methodologies aside and concentrate on the forecasting requirements of revenue management and what data are typically available for making these forecasts. To know more about forecasting methodologies, we refer the reader to Taneja (1978) and Lee (1990).

We classify data requirements for forecasting into three categories, depending on what is being forecasted. The most important forecasted prediction is the distribution of the number of arrivals in each passenger booking class. The level at which this forecasting is done depends on whether the airline is controlling inventories at the market level [i.e., by itinerary (network revenue management)] or controlling at the leg-fare class level. The second group of variables that have to be forecasted are the no-show and cancellation probabilities of the arrivals. No-shows and cancellations are typically assumed to follow a binomial distribution and the probability of a cancellation or a no-show is forecasted. In the final category we group the forecasting requirements of some auxiliary variables. Among these are the utilization rates of groups, sell-up probabilities of the fare classes in the various markets, and the probability that a rejected booking will choose some other flight of the same airline (recapture rate). Very few airlines are actually known to forecast these, but the need exists.

10.C.6. Revenue Management on a Single Leg

The revenue management problem on a single leg arises directly in industries such as broadcasting and cruise lines. In the airline industry it is directly applicable to some small airlines that service hubs (commuter airlines) or do mostly point-to-point flying (Southwest, ValueJet, etc.), charter airlines, or more generally any part of an airline's network that has little connecting traffic on it.

At first glance, restricting attention to a single leg when the problem is more naturally defined on a network seems to be of limited interest. But historically, revenue management was practiced by controlling inventories on each leg in isolation, and that practice is very widespread even today. The reasons are partly limitations of CRSs, optimization technology, and lack of adequate information gathering. Moreover, when revenue management was introduced, seat inventory was controlled manually and considering the complex nature of the decision making, attention was restricted to one leg at a time. When CRSs were first designed, or when researchers first modeled the process and developed algorithms, it was therefore for the single-leg problem.

But the primary limitation for adaptation of network revenue management techniques was the capabilities of the CRSs. Many of them (even to this day) could display and control inventories one leg at a time. As a result, little information was retained about the itinerary flow of the passengers and their connecting information. Since inventory could not be controlled at the itinerary level anyway, solving the single-leg problem and setting booking limits for the fare classes on each leg in isolation seems more appropriate.

All this is changing, of course, as cheap computing power, new developments in CRSs, and better understanding of the network revenue management problem is motivating many of the larger airlines to abandon this single-leg approach and move to a globally optimal network solution to their revenue management. But the single-leg problem, because of the direct applications cited above, widespread usage, and its historical importance, retains its importance for the airline industry. It is on a single leg that the one can derive a clear-cut optimal algorithm for the revenue management problem (after making some minor assumptions), and one can gain a lot of insight into the nature of the problem by studying this solution. In addition, the complexity of solving the network revenue management problem in totality and the ease with which a good solution can be obtained for the single-leg problem, has led many researchers and practitioners to try to decompose the network problem into problems on a single leg and solve the leg problems optimally.

In this section we give two models of the problem on a single leg and state algorithms for each that solve the problem optimally under some assumptions or information requirements on passenger booking behavior.

Static Model. The first model assumes that bookings for the fare classes are such that the fare classes with lower fares arrive before the fare classes with the higher ones. Thus the fare classes can be ordered and nested based on the fares, and the order of arrivals conforms with the nesting. This is not such a bad assumption for bookings on a single leg, as most of the lower fare classes have advance purchase restrictions, and typically, the farther in advance one buys the ticket, the lower the fare (in addition, other restrictions, such as penalties for cancellations and alterations, are usually also imposed).

Under this assumption that lower comes before higher, the problem of setting the optimal booking limits has been solved independently by Curry (1990), Wollmer (1992), and Brumelle and McGill (1993).

We will have to formalize some definitions and notation before we describe the analytical solutions to this problem. Let F_i be the fare of fare class i, with the fare classes ordered such that $F_1 > F_2 > \cdots > F_k$, and the requests for different fare classes be independent of each other. Define B_i for class i as the booking limit for class i. That is, class i can book as long as the total bookings do not exceed B_i. Let N_i be the nested protection level for class i. That is, N_i seats are protected for class i from classes $i + 1$ upward. To give an example, a set of nested allocations Y95 B80 M77 K40 Q20 Z0 has the booking limit 95 for the Y fare class, 80 for the B fare class, and so on; and has nested protection levels of 15 for Y, 3 for B, 37 for M, 20 for K, and 20 for Q. Fare class 1 will have the highest fare and fare class n, the lowest fare.

Let $R_i(x)$ represent the maximum expected revenue for i fare classes when there are exactly the highest i fare classes available and the number of seats available is x. Then, as shown in Curry (1990), the optimal nested protection levels N_i^*, for $i = 1,\ldots,n$, satisfy

$$\frac{dR_i}{dB_i}(N_i^*) = F_i$$

Curry also shows that the optimal revenue function is concave with respect to the capacity. The optimal booking limits can be computed easily using either the equations above or using the dynamic programming algorithm of Wollemer (1992).

Belobaba (1987) has given a heuristic for setting the booking limits that is computationally somewhat faster than the algorithms to find the optimal booking limits. This heuristic is called the expected marginal seat revenue (EMSR) algorithm and works as follows: Let C be the capacity of the plane and let D_i be the random variable (usually taken as normally distributed) that represents demand for fare class i. Define Π_i as the protection limit for classes 1 through i. Note that $\Pi_i = N_1 + \cdots + N_i$, and $B_i = C - \Pi_i$. The protection level for fare class 1 (the highest fare class) is obtained by solving the equation

$$F_2 = F_1 \Pr[D_1 > \Pi_l]$$

Π_2 is obtained by solving the following two equations for π_2^1 and π_2^2.

$$F_3 = F_1 \Pr[D_1 > \pi_2^1]$$

and

$$F_3 = F_2 \Pr[D_2 > \pi_2^2]$$

and taking

$$\Pi_2 = \pi_2^1 + \pi_2^2$$

This procedure is repeated until all the Π_i's are calculated.

Dynamic Model. The second model breaks up time until departure into n discrete intervals and assumes that forecasts of passenger booking information is given in the form of P_i^j, the probability of a booking in fare class i arriving in period j. The intervals are assumed to be small enough that at most one booking can occur in one period. In every period, if a booking request occurs, an accept/deny decision has to be made. So the control also is assumed to be dynamic. The dynamic programming algorithm we give below for this problem is due to Lee and Hersh (1993) [a similar algorithm was discovered independently by Diamond and Stone (n.d.)].

As before, assume that F_i represents the fare of fare class i, the fare classes are ordered such that $F_1 > F_2 > \cdots > F_k$, and the requests for different fare classes are independent of each other. Assume for simplicity that requests come in for one seat at a time. Let P_i^n be the probability that a request for booking class i arrives in booking period n, for $i = 1,2,\ldots,k$. Since only one arrival can come in during each booking period, $\Sigma_{i=1}^k P_i^n \leq 1$ for all n. Let $P_0^n = 1 - \Sigma_{i=1}^k P_i^n$. Let booking period n be the current booking period and booking period 1 be the period prior to departure. Let s be the current seat capacity.

Let R_s^n be the value of the optimal expected revenue from period n onward (i.e., during periods $n, n-1, \ldots, 1$). During decision period n, if a request for booking class i comes in, the expected revenue if it is accepted is $F_i + R_{s-1}^{n-1}$, and if it is not accepted, it is R_s^{n-1}. Therefore, the optimal decision during period n is to accept a request for fare class i if

$$F_i + R_{s-1}^{n-1} \geq R_s^{n-1}$$

and deny otherwise.

The total expected revenue is therefore given by the following equations:

$$R_s^n = \sum_{i=1}^{k} P_i^n \max(F_i + R_{s-1}^{n-1}, R_s^{n-1}) \qquad \text{for } s > 0, \quad n > 0$$

This equation can be solved working backward and it gives a simple, easy-to-calculate method for calculating the optimal expected revenue and the optimal decision during each period. From this equation one can also prove certain properties of the expected marginal value of the seats. Define the expected marginal value of a seat in decision period n given booking capacity s as $\delta(n,s) = R_s^n - R_{s-1}^n$. For a fixed n, one would expect $\delta(n,s)$ to increase as s decreases (fewer seats for the same demand would mean that the value of the next seat to be sold should be higher). For a fixed s, one would expect $\delta(n,s)$ to decrease as n decreases (less time to sell the seats would make the value of the last seat less). These facts are summarized in the following theorem from Lee and Hersh (1993).

Theorem 1. For a given n, $\delta(n,s)$ is nonincreasing in s, and for a given s, $\delta(n,s)$ is nondecreasing in n.

The monotonocity properties of $\delta(n,s)$ imply:

1. For a fixed decision period n and for each booking class i, there exists a set of *critical booking capacities* $\hat{s}(i,n)$ such that booking class i requests are denied if the current capacity $s < \hat{s}(i,n)$ and accepted if $s \geq \hat{s}(i,n)$.
2. For a fixed capacity value s and for each booking class i, there exists a set of *critical decision periods* $\hat{n}(i,s)$ such that booking class i requests are denied if $n > \hat{n}(i,s)$ and accepted if $n \leq \hat{n}(i,s)$.

3. For a fixed booking capacity s and a decision period n, there exists a *critical booking class* $\hat{\imath}(n,s)$ such that a request for a seat in booking class i is denied if $i > \hat{\imath}(n,s)$, and accepted if $i \le \hat{\imath}(n,s)$.

Therefore, to make the accept/deny decisions, only a set of critical booking capacities or a set of critical decision periods need to be stored.

10.C.7. Network (O&D) Revenue Management

With the adaptation of hub-and-spoke route networks by most of the large carriers, the percentage of passengers who have itineraries involving two or more legs has increased tremendously. Smith et al. (1992) report that in 1980 around 10% of American Airlines traffic was connecting traffic, while by the mid-1980s this number has increased to more than 60%. This trend in traffic flow has a big impact on how airlines should do revenue management. New trade-offs have to be considered in deciding what bookings to accept and what to reject. Airline fares do not follow any particular pattern or logic. Under some conditions it may be more beneficial to the airline to reject a Y class through passenger and protect the seats for two local passengers on each one of the legs. Under some other conditions it may be better to reject the requests from local traffic for through bookings. It will all depend on the fare structure (which can be arbitrary), the capacity, and the network topology. In addition, the assumption that lower classes book before higher classes, although still somewhat true for any given itinerary, is rendered all but meaningless by the fact that the contribution of a high fare class two-leg itinerary to each of the legs may be less than the local fares in a lower fare class.

The complexity of solving the revenue management problem on a network becomes an order of magnitude more difficult than the leg case. The first thing that one has to resolve is what control is appropriate for network revenue management. Since a flight bank (the set of staggered flights coming in over an hour or so to connect to flights going out) in a hub-and-spoke network can contain up to 60 flights in and 60 flights out, it is not unusual to see a single flight carrying passengers with 40 different itineraries and 100 potential itineraries. Multiply this by the eight fare classes typical for each itinerary and it is clear that controlling itinerary sales by leg-level allocations (or even virtual nesting) becomes a horrendous task. Besides, many of these itinerary fare class combinations will have extremely small forecasted demands. To forecast them individually and protect seats for them will only lead to high unacceptable errors. So one has to move away from the protections/nested allocation framework, which is unsuitable for network level control.

The optimal form of control for network revenue management is a mechanism that makes an accept or reject decision for each itinerary request. Since at any point in time, if you are accepting an itinerary request with a certain fare you would be willing to accept a request for the same itinerary with a higher fare, this control can be in the form of a threshold value for each itinerary, where one would accept a booking request for an itinerary if the fare of that request exceeds the threshold value. Since any accept or deny decision changes the state of the system (an accept decision changes the capacity and forecast of bookings to come, while a deny decision will change only the forecast to come), it is important that the threshold values be updated very frequently.

The storage and computational requirements for maintaining and calculating threshold values for each itinerary are tremendous and practically infeasible. A network with 2000 flights per day can have more than 40,000 different itineraries. Storing the threshold values for the two or three compartments for each itinerary for 200 or more prior days places a significant burden on the airline's databases. But more important, calculating so many threshold values to any degree of optimality is next to impossible in the short overnight time frame that is allowed for updating the threshold values. As a compromise, a smaller set of threshold values, one for each leg compartment of the network, are calculated and the threshold value of an itinerary is taken to be the sum of the values for the legs in the itinerary. The threshold values associated with each leg compartment are called *bid prices,* and the control is called *additive bid-price control* or *bid-price control.*

The important question now is how good an approximation this bid-price control is to the optimal control policy for network revenue management. It turns out to be a very good approximation when the capacity is not too small. In Talluri and van Ryzin (1995a) it is shown that under the dynamic network revenue management model for a network, the expected revenue under a bid-price control scheme is asymptotically optimal as a function of capacity. It is important to remember though that the bid-price control mechanism can be double-edged if not implemented properly. Specifically, it is extremely important that bid prices be updated frequently. Otherwise, since the control provides no mechanism for limiting the number of sales in any fare class unless a new bid price is calculated, there is a danger that too many low-fare requests will be accepted.

For mathematical programming approaches to calculating the bid prices, refer to Williamson (1992). The idea is to use the dual information obtained from the optimal solution of the mathematical programming to obtain the bid prices. However, the resulting bid prices can often be quite far removed from the optimal bid prices (Talluri and van Ryzin, 1995b).

REFERENCES

Abara, J., 1989. Applying Integer Linear Programming to the Fleet Assignment Problem, *Interfaces, 19, July–August.*

Ahuja, R. K., T. L. Magnanti, and J. B. Orlin, 1993. *Network Flows: Theory, Algorithms, and Applications,* Prentice Hall, Upper Saddle River, NJ.

Anbil, R., E. Gelman, B. Patty, and R. Tanga, 1991a. Recent Advances in Crew-Pairing Optimization at American Airlines, *Interfaces,* 21, 62–74.

Anbil, R., E. L. Johnson, and R. Tanga, 1991b. A Global Optimization Approach to Crew Scheduling, *IBM Systems Journal,* 31, 71–78.

Anbil, R., C. Barnhart, L. Hatay, E. L. Johnson, and V. S. Ramakrishnan, 1993. Crew-Pairing Optimization at American Airlines Decision Technologies, in T. A. Cirani and R. C. Leachman (eds.), *Optimization in Industry,* Wiley, New York, pp. 31–36.

Appelgren, L. H., 1969. A Column Generation Algorithm for a Ship Scheduling Problem, *Transportation Science,* 3, 53–68.

Arabeyre, J. P., J. Fearnley, F. C. Steiger, and W. Teather, 1969. The Airline Crew Scheduling Problem: A Survey, *Transportation Science,* 3, 140–163.

Barnhart, C., E. L. Johnson, R. Anbil, and L. Hatay, 1994. A Column Generation Technique for the Long-Haul Crew Assignment Problem, in T. A. Cirani and R. C. Leachman (eds.), *Optimization in Industry II,* Wiley, New York.

Barnhart, C., N. Boland, L. W. Clarke, E. L. Johnson, G. L. Nemhauser, and R. G. Shenoi, 1996a. *String-Based Models and Methods for Aircraft Routing and Fleeting,* Center for Transportation Studies Working Paper, Massachusetts Institute of Technology, Cambridge, MA.

Barnhart, C., E. L. Johnson, G. L. Nemhauser, and P. H. Vance, 1996b. *Exceptions Crew Scheduling,* The Logistics Institute Working Paper, Georgia Institute of Technology, Atlanta, GA.

Barnhart, C., E. L. Johnson, G. L. Nemhauser, and P. H. Vance, 1996c. *Crew Pairing Optimization: Network Representation Effects,* The Logistics Institute Working Paper, Georgia Institute of Technology, Atlanta, GA.

Barnhart, C., E. L. Johnson, G. L. Nemhauser, M. W. P. Savelsbergh, and P. H. Vance, 1996d. Branch-and-Price: Column Generation for Solving Huge Integer Programs, Center for Transportation Studies Working Paper, Massachusetts Institute of Technology, Cambridge, MA.

Barutt, J., and T. Hull, 1990. Airline Crew Scheduling: Supercomputers and Algorithms, *SIAM News,* 23, 6.

Belobaba, P. P., 1987. Air Travel Demand and Airline Seat Inventory Management, Ph.D. thesis, Massachusetts Institute of Technology, Cambridge, MA.

Berge, M. A., and C. A. Hopperstad, 1993. Demand Drive Dispatch: A Method for Dynamic Aircraft Capacity Assignment, Models and Algorithms, *Operations Research,* 41(1).

Brumelle, S. L., and J. I. McGill, 1993. Airline Seat Allocation with Multiple Nested Fare Classes, *Operations Research,* 41, 127–137.

Buhr, J., 1978. Four Methods for Monthly Crew Assignment: A Comparison of Efficiency, *AGIFORS Symposium Proceedings,* 18, 403–430.

Byrne, J., 1988. A Preferential Bidding System for Technical Aircrew, *AFGIFORS Symposium Proceedings,* 28, 87–99.

Clarke, L. W., C. A. Hane, E. L. Johnson, and G. L. Nemhauser, 1994. *Maintenance and Crew Considerations in Fleet Assignment,* Technical Report COC-94-07, School of Industrial and Systems Engineering, Georgia Institute of Technology, Atlanta, GA.

Clarke, L. W., E. L. Johnson, G. L. Nemhauser, and Z. Zhu, 1995. *The Aircraft Rotation Problem,* LEC-95-03 Working Ppaer, Georgia Institute of Technology, Atlanta, GA.

Cplex, 1990. *Using the* CPLEX™ *Linear Optimizer,* CPlex Optimization, Inc., Incline Village, NV.

Curry, R. E., 1990. Optimum Seat Allocation with Fare Classes Nested by Origins and Destinations, *Transportation Science,* 24, 193–204.

Daskin, M. S., and N. D. Panayotopoulos. 1989. A Lagrangian Relaxation Approach to Assigning Aircraft to Routes in Hub and Spoke Networks, *Transportation Science,* 23.

Desrochers, M., and F. Soumis, 1988. A Generalized Permanent Labeling Algorithm for the Shortest Path Problem with Time Windows, *INFOR,* 26, 191–212.

Desrochers, M., and F. Soumis, 1989. A Column Generation Approach to the Urban Transit Crew Scheduling Problem, *Transportation Science,* 23, 1–13.

Desrosiers, J., Y. Dumas, M. Desrochers, F. Soumis, B. Sanso, and P. Trudeau, 1991. *A Breakthrough in Airline Crew Scheduling,* Report G-91-11, GERAD, Montreal, Quebec, Canada.

Desrosiers, J., Y. Dumas, M. M. Solomon, and F. Soumis, 1993. Time Constrained Routing and Scheduling, in *Vehicle Routing,* North-Holland, Amsterdam.

Diamond, M., and R. Stone, n.d. Dynamic Yield Management on a Leg, unpublished manuscript, Northwest Airlines, St. Paul, MN.

Dillon, J., R. Gopalan, S. Ramachandran, R. Rushmeir, and K. T. Talluri, 1993. Fleet Assignment at USAir, TIMS/ORSA Joint National Meeting, Phoenix, AZ.

Feo, T. A., and J. F. Bard, 1989. Flight Scheduling and Maintenance Base Planning, *Management Science,* 35(12), 1415–1432.

Hane, C. A., C. Barnhart, E. J. Johnson, R. E. Marsten, G. L. Nemhauser, and G. Sigismondi, 1995. The Fleet Assignment Problem: Solving a Large-Scale Integer Program, *Mathematical Programming,* 70, 211–232.

Gamache, M., and F. Soumis, 1993. *A Method for Optimally Solving the Rostering Problem,* les Cahier du GERAD, G-90-40, Ecole des Hautes Etudes Commerciales, Montreal, Quebec, Canada.

Gershkoff, I., 1989. Optimizing Flight Crew Schedules, *Interfaces,* 19, 29–43.

Giafferri, C., J. P. Hamon, and J. G. Lengline, 1982. Automatic Monthly Assignment of Medium-Haul Cabin Crew, *AGIFORS Symposium Proceedings,* 22, 69–95.

Glanert, W., 1984. A Timetable Approach to the Assignment of Pilots to Rotations, *AGIFORS Symposium Proceedings,* 24, 369–391.

Gopalan, R., and K. T. Talluri, 1994. *The Three-Day Aircraft Maintenance Problem with Red Eye Flights,* USAir Working Paper, USAir, Washington, DC.

Gopalan, R., and K. T. Talluri, 1996. The Aircraft Maintenance Routing Problem, Operations Management Working Paper, Universitat Pompeu Fabra, Barcelona, Spain.

Graves, G. W., R. D. McBride, I. Gershkoff, D. Anderson, and D. Mahidhara, 1993. Flight Crew Scheduling, *Management Science,* 39, 736–745.

IBM, 1990. *Optimization Subroutine Library, Guide and Reference,* IBM Corporation, Armonk, NY.

Hoffman, K. L., and M. Padberg, 1993. Solving Airline Crew-Scheduling Problems by Branch-and-Cut, *Management Science,* 39, 657–682.

Jarrah, A. I., G. Yu, N. Krishnamurthy, and A. Rakshit, 1993. A Decision Support Framework for Airline Flight Cancellations and Delays, *Transportation Science,* 27, 266–280.

Kabbani, N. M., and B. W. Patty, 1992. Aircraft Routing at American Airlines, *Proceedings of the 32nd Annual Symposium of AGIFORS,* Budapest, Hungary.

Lavoie S., M. Minoux, and E. Odier, 1988. A New Approach for Crew Pairing Problems by Column Generation with an Application to Air Transportation, *European Journal of Operational Research,* 35, 45–58.

Lee, A. O., 1990. Airline Reservations Forecasting: Probalistic and Statistical Models of the Booking Process, Ph.D. thesis, Massachusetts Institute of Technology, Cambridge, MA.

Lee, T. C, and M. Hersh, 1993. A Model for Dynamic Airline Seat Inventory Control with Multiple Seat Bookings, *Transportation Science,* 27, 252–265.

Marchettini, F., 1980. Automatic Monthly Cabin Crew Rostering Procedure, *AGIFORS Symposium Proceedings,* 20, 23–59.

Marsten, R., 1994. Crew Planning at Delta Airlines, Mathematical Programming Symposium XV presentation, Ann Arbor, MI.

Mehrotra, A., and M. A. Trick, 1996. A Clique Generation Approach to Graph Coloring, *INFORMS Journal on Computing,* 8, 344–354.

Moore, R., J. Evans, and H. Noo, 1978. Computerized Tailored Blocking, *AGIFORS Symposium Proceedings,* 18, 343–361.

Nicoletti, B., 1975. Automatic Crew Rostering, *Transportation Science,* 9, 33–42.

Odier, E., F. Lascaux, and H. Hie, 1983. Medium Haul Trip Pairing Optimization, *XXII AGIFORS Symposium,* 81–109.

Rannou, B. 1986. A New Approach to Crew Pairing Optimization, *XXVI AGIFORS Symposium Proceedings,* 153–167.

Ryan, D. M. 1992. The Solution of Massive Generalized Set Partitioning Problems in Air Crew Rostering, *Journal of the Operational Research Society,* 43, 459–467.

Ryan, D. M., and J. C. Falkner, 1988. On the Integer Properties of Scheduling Set Partitioning Models, *European Journal of Operational Research,* 35, 442–456.

Ryan, D. M., and B. A. Foster, 1981. An Integer Programming Approach to Scheduling, in A. Wren (ed.), *Computer Scheduling of Public Transport Urban Passenger Vehicle and Crew Scheduling,* North-Holland, Amsterdam, pp. 269–280.

Sarra, D., 1988. The Automatic Assignment Model, *AGIFORS Symposium Proceedings,* 28, 23–37.

Savelsbergh, M. W. P., 1996. A Branch-and-Price Algorithm for the Generalized Assignment Problem, The Logistics Institute Working Paper, Georgia Institute of Technology, Atlanta, GA.

Sol, M., and M. W. P. Savelsbergh, 1994. *A Branch-and Price Algorithm for the Pickup and Delivery Problem with Time Windows,* Report COC-94-06, Georgia Institute of Technology, Atlanta, GA.

Smith, B. C., J. F. Leimkuhler, and R. M. Darrow, 1992. Yield Management at American Airlines, *Interfaces,* January.

Subramanian, R., R. P. Scheff, J. D. Quillinan, D. S. Wiper, and R. E. Marsten, 1994. ColdStart: Fleet Assignment at Delta Airlines, *Interfaces,* 24, 104–120.

Talluri, K. T., 1994. *The Four-Day Aircraft Maintenance Problem,* USAir Working Paper, USAir, Washington, DC.

Talluri, K. T., 1996. Swapping Applications in a Daily Airline Fleet Assignment. *Transportation Science,* Vol. 30, No. 3, Aug. 1996, 220–236.

Talluri, K. T., and G. van Ryzin, 1995a. Bid-Price Control, *AGIFORS Reservations and Yield Management Conference Proceedings,* Washington, DC.

Talluri, K. T., and G. van Ryzin, 1995b. An Analysis of Bid-Price Control Revenue Management, presentation at the INFORMS Conference, New Orleans, LA.

Taneja, A. K., 1978. *Airline Traffic Forecasting,* Lexington Books, Lexington, MA.

Teodorovic, D., and S. Guberinic, 1984. Optimal Dispatching Strategy on an Airline Network After a Schedule Perturbation, *European Journal of Operational Research,* 15, 178–182.

Teodorovic, D., and G. Stojkovic, 1990. Model for Operational Daily Airline Scheduling, *Transportation Planning Technology,* 14, 273–285.

Tingley, G. A., 1979. Still Another Solution Method for the Monthly Aircrew Assignment Problem, *AGIFORS Symposium Proceedings,* 19, 143–203.

Vance, P. H., C. Barnhart, E. L. Johnson, and G. L. Nemhauser, 1994. Solving Binary Cutting Stock Problems by Column Generation and Branch-and-Bound, *Computational Optimization and Applications,* 3, 111–130.

Vance, P. H., C. Barnhart, E. L. Johnson, and G. L. Nemhauser, 1996. Airline Crew Scheduling: A New Formulation and Decomposition Algorithm, Center for Transportation Studies Working Paper, M.I.T., Cambridge, MA.

Williamson, E. L., 1992. Airline Network Seat Inventory Control: Methodologies and Revenue Impacts, Ph.D. thesis, Massachusetts Institute of Technology, Cambridge, MA.

Wollmer, R. D., 1992. An Airline Seat Management Model for a Single Leg Route When Fare Classes Book First, *Operations Research,* 40, 26–37.

CHAPTER 11

Environmental Planning for Electric Utilities

BENJAMIN F. HOBBS

11.A. INTRODUCTION

Planning and operating modern electric power systems involves several interlinked and complex tasks (Figure 11.1). Accomplishing each so that consumers receive power reliably at an acceptable economic and environmental cost is hugely difficult, for several reasons. First, the electric system itself encompasses an interconnected array of a large number of electrical machines and circuits. Maintaining acceptable voltages and frequency in such systems under rapidly changing circumstances is by itself a daunting task. Second, scheduling short-run generation and load management to minimize costs is complicated because of the sheer number of alternative schedules that are possible and uncertainties in equipment availability and demands (*loads*).

Finally, long-term planning involves sorting through a wide range of possible energy supply- and demand-side options and in-service dates, while keeping in mind the implications of each for short-term operations, cost and noncost criteria, and company competitiveness (Hirst and Goldman, 1991). Supply options include traditional hydroelectric, fossil steam, and nuclear generating units, along with increasingly popular combustion turbine, combined cycle, and renewable technologies (wind, photovoltaic, etc.). Demand-side management (DSM) options are increasingly important. They consist of utility-sponsored measures designed to alter the timing amount of energy demands. Examples include energy efficiency (e.g., efficient motors, high-efficiency lamps), sales promotion (such as security lighting and substitution of electricity for other fuels), shifting of loads from peak to off-peak periods (e.g., subsidization of thermal storage by customers), and even pricing reform (e.g., peak-load pricing).

RESOURCE AND EQUIPMENT PLANNING

RESOURCE PLANNING AND PRODUCTION COSTING (10 - 40 year horizon)

Given forecasts of loads, construction costs, fuel prices, & regulations, determine the mix of generator additions and retirements, power purchases and sales, and demand-side management. Risks and multiple objectives should be considered. Choosing emission controls and sites to comply with environmental laws a major consideration; also, utility regulators often encourage adoption of less polluting generation sources.

LONG RANGE FUEL PLANNING (10 - 20 years)

Given generating plants, find the least cost sources of fuel and schedule deliveries. Regulatory and environmental policy constraints (especially on air emissions) are important.

TRANSMISSION AND DISTRIBUTION PLANNING (5 - 15 years)

Given load forecasts and planned generation additions, design circuit additions that maintain reliability, minimize costs, and avoid possible environmental effects such as aesthetic impacts or exposure to EMF. Environmental considerations are key in route selection and substation siting.

RATE DESIGN AND DSM IMPLEMENTATION PLANNING (3 - 15 years)

Given base load forecasts and market opportunities, identify cost-effective DSM programs to target at particular markets, and design marketing and financing programs. Design rate structures for achieving load shape objectives such as valley-fill or peak shaving. Utility regulators often encourage utilities to consider environmental benefits of DSM programs.

MEDIUM TERM OPERATIONS PLANNING

MAINTENANCE AND PRODUCTION SCHEDULING (2 - 5 years)

Given load forecasts and available equipment, schedule interutility sales of energy and routine equipment maintenance to maintain reliability and minimize costs. Schedules can account for seasonal differences in emission restrictions, if any.

FUEL SCHEDULING (1 year)

Given long-term fuel contracts, schedule fuel deliveries to meet plant requirements.

SHORT RUN OPERATION

UNIT COMMITMENT (8 hours to 1 week)

Given load forecasts, decide when to start up and shut down generators so as to minimize costs and maintain reliability. Plant ramp rates and minimum down and up times must be respected, and fixed start-up costs must be considered. Unit commitment can be altered to decrease overall system emissions (usually air pollutants), or just those at certain times or from certain facilities.

DISPATCHING (1 to 10 minutes)

Given the load, schedule committed generators and load management to maintain voltages and frequencies, while minimizing cost and avoiding undue equipment stress. Air pollution often considered, as in unit commitment, usually via a cost penalty on emissions.

AUTOMATIC PROTECTION (Fractions of a second)

Design protection schemes to minimize damage to equipment and service interruptions resulting from faults and equipment failures.

Figure 11.1 Overview of utility planning and operations problems. (Adapted from Talukdar and Wu, 1981.)

Despite the power industry's compliance with the many environmental laws that have been enacted over the past decades, the public remains concerned with utility pollution and resource consumption. Consequently, environmental considerations come into play in many of the tasks of Figure 11.1. For instance, emissions are considered in real-time operations where there are economic penalties for SO_2 and NO_x emissions; this results in *emissions dispatch.* Emissions dispatch consists of altering the order in which plants are operated—or *dispatched* against load—so that cleaner plants generate more power and polluting plants generate less than they would if, instead, they were dispatched to minimize fuel costs. Environmental impacts are also important in intermediate-range fuel planning, especially because many power plants can switch between different fuels (e.g., coal and natural gas). Finally, environmental factors, in the form of public opinion and existing or potential legislation, influence capital budgeting decisions—in particular, decisions concerning what mix of supply- and demand-side "resources" to obtain, where to site them, and what capital-intensive pollution controls to install.

This chapter's subject is the use of models to integrate environmental concerns into electric resource planning and policymaking. The emphasis is on modeling of broad environment–cost trade-offs among different sources of electric energy and their interaction with long-term fuel choices and pollution control investments. The development and use of such methods is a natural application of the civil engineer's skills in modeling, economic analysis, and environmental engineering and planning. This type of modeling can be extended beyond just power systems to other energy systems, including combined heat-power systems and natural gas utilities (e.g., Gronheit, 1993; Fragniere and Haurie, 1995).

However, we should note that the modeling of long-run resource decisions is only the first step to including environmental factors in utility decisions. Environmental impacts usually depend on the location of emissions, where pollutants are transported, how they are transformed, and who or what is exposed to them. Hence environmental impacts are most credibly assessed by considering the possible location of generation and transmission facilities. Linear programming–based models for developing siting scenarios have been developed for closing the gap between resource planning models (which define resource mixes but not location) and pollutant transport and transformation models (which require that locations of pollutant releases be specified). These models indicate what areas within a large region would be most attractive for a new set of facilities, given the spatial distribution of demand, spatially varying environmental regulations and site availability, and the existing location of generation and transmission plant. Such models have been used successfully to analyze the effect of alternative environmental policies on distributions of emissions (e.g., Cohon et al., 1980). The interaction of location and regulation can lead to surprising conclusions. For instance, Meier (1979) showed that the requirement that all new coal plants meet strict

SO_2 emissions standards could result in worse exposure because the incentive to site plants far from populations would be diminished. Meanwhile, Hobbs and Meier (1979) found that stricter thermal pollution standards along coastlines could lead to deterioration of water quality for some rivers because inland plant siting would be encouraged.

Such siting scenario models are not considered further in this chapter. Neither will we address the many models used to include environmental considerations in decisions concerning system operations (i.e., unit commitment and dispatch, Figure 11.1), transmission line routing, and siting of individual power plants. Multiattribute decision analysis has long been used for siting and routing decisions (e.g., Economides and Sharifi, 1978; Hobbs, 1979; Keeney, 1980). Environmental attributes usually figure prominently in such analyses. Such multiattribute methods for eliciting value judgments can also be used to facilitate stakeholder input in resource planning (Hobbs and Meier, 1994). Recently, there has been intense interest in algorithms for including environmental penalties or constraints in short-run operations decisions, whose focus has been upon description of trade-offs between operating costs and either emissions or, less frequently, ambient pollutant concentrations (Gjengedal et al., 1992; Kuloor et al., 1992; Rose et al., 1994; Talaq et al., 1994). These various siting and operations models serve a very important role: to ensure that the environmental gains and cost savings identified by resource planning models are in fact achieved when plans are implemented.

The chapter is divided into three sections. Because environmental utility planning is an unfamiliar topic to most civil engineers, a review of external environmental costs of power generation and the evolution of environmental utility planning is presented in the first section. Formerly, environmental planning was merely a matter of complying with effluent limitations at individual generation facilities; now, however, environmental factors are an integral part of the power system planning process. The second section illustrates, through variations in a simple integrated resource planning (IRP) model, different approaches to environmental compliance planning. What distinguishes IRP from traditional power generation planning is its consideration of both energy supplies and demand modification under a range of economic, environmental, and other objectives. The model highlights the multiobjective approach to evaluating environmental effects, whose philosophy is to present information for use in negotiation and the political process (see Chapter 12).

The third section shows how planning for complying with pollution control laws can affect all aspects of the resource acquisition process. The focus is on planning under increasingly popular market-based schemes for environmental regulation. Under many such schemes, only total emissions matter and the utility must pay for the right to pollute. The U.S. acid rain control program is an example of such a system, and legislation has been proposed in the United States which would establish a similar market for CO_2 emission rights.

11.B. EXTERNAL ENVIRONMENTAL COSTS AND UTILITY RESOURCE PLANNING

11.B.1. What Are External Costs?

Economists divide the costs of generating, transmitting, and using electricity into two categories: internal and external costs. *Internal costs* are monetary expenditures by utilities and consumers for the inputs needed to produce and utilize power. For the utility, these costs show up on the income statement as expenses, taxes, and returns to the owner (i.e., what are termed *revenue requirements*). *External costs* are those costs that are not reflected in market prices but are, instead, borne by third parties. These costs have traditionally been ignored by utilities and consumers when they decide how much and by what means electricity is to be produced and consumed. The typical utility neither tallies these costs in its income statement nor considers them in determining what price to charge for electricity.

External costs can arise because residuals from production and consumption processes impose costs on others or because production facilities consume resources whose prices fail to reflect their true value to society. In addition, utility actions may also affect dispersion of pollutants from other sources: home weatherization's impact on indoor concentrations of radon is an example. Residuals may be either waste energy or material (Ortolano, 1984). Waste materials include gaseous, liquid, solid, or radioactive residuals. Waste energy includes heat, noise, or electromagnetic fields. Releases of such residuals may affect no one. But when they impose costs on other parties, the residuals are called *pollution*. Examples of such costs include human morbidity and mortality, discomfort, aesthetic impacts such as worsened visibility, effects upon plant and animal communities, damage to materials and equipment, and the expense of repairing and preventing these damages (Ottinger et al., 1990; ORNL and RFF, 1994).

But even if power production created no residuals, generation and transmission facilities can impose external costs by consuming or altering resources whose price does not reflect their actual value. For instance, the construction of hydroelectric dams changes the habitat available to wildlife at the dam site, and their operation alters downstream riparian ecosystems. The price the utility pays in the private market for the dam site is unlikely to reflect the worth of the wildlife there, while the utility may pay nothing for the ecologic damages its operation may cause downstream. Another example is the consumption by thermal power plants of water in locations where water rights are poorly defined; that consumption may preclude other uses of water whose value is not reflected in the price the utility pays for water. As a final example, transmission lines can impair scenic views and thus property values. The price the utility pays for a line right-of-way will omit the value that neighbors place on their views.

Table 11.1 lists some specific types of environmental impacts that production and use of electricity cause or are suspected to cause. The relative attention received by these impacts has shifted over time. Some that were little heard of a decade or two ago, such as electromagnetic fields (EMF), are major concerns today. Others receive less attention than they used to. For example, U.S. utility executives were quoted in the 1960s as saying that the biggest environmental problem facing utilities was the aesthetics of overhead low-voltage distribution lines—a minor concern today. In the late 1970s, nuclear safety was arguably the most salient issue. Presently, air pollution and potential greenhouse warming are major issues. The utility industry in the United States is responsible for roughly one-third of the nation's CO_2 emissions from fossil fuels and three-fourths of the total SO_2 emissions. Jones and Hanser (1991) estimate that air emissions are to blame for perhaps 90% of the external costs of power production, although such numbers are necessarily location and technology specific. Another important concern is the impact of hydropower dams upon recreation, wildlife, and in developing countries, the relocation and settlement of people.

If an industry such as the power sector imposes significant external costs on society, two distortions in society's allocation of resources can result. First, the price the industry charges for its product will be too low, as it will fail to reflect all the costs that society bears in producing it. Consequently, consumers will demand more of the product and too much will be produced. In the case of electric power, energy may be underpriced compared to its social cost; if so, buyers of power will have too little incentive to conserve (Woo et al., 1994). The second distortion—the focus of this chapter—is that utilities will tend to use too much of inputs and production processes that have high external costs and too little of those that are cleaner. For instance, if coal's external cost greatly exceeds that of natural gas, coal will look artificially cheap. Government environmental policies can be viewed as an attempt to correct these types of distortions.

11.B.2. Governmental Policies to Encourage Environmental Planning

Governments have responded to public concerns over the environmental costs of electricity in several ways. Generally, the tools that governments have to alter utility behavior can be divided into five general categories: taxation, subsidization, ownership, regulation, and on occasion, moral leadership (Ortolano, 1984). Through taxes and subsidies, governments can change the economic incentives facing the utility. A common means of doing so is by giving tax breaks for pollution control equipment. A less popular approach is to tax pollution in an attempt to make private costs the same as social costs. The early Nixon administration proposal to tax emissions of SO_2 is an example. Such taxes could motivate utilities to pollute or use resources at a level that is more optimal to society. Other instances of taxes and subsidies include tax-exempt pollution control bonds issued by governments and sub-

TABLE 11.1 Environmental External Costs from Electric Utility Production and Use[a]

Residual or Activity	Global Warming	Health, Safety	Aesthetics	Flora, Fauna	Materials, Property	Agricultural Productivity	Water Availability	Wilderness, Recreation
CO_2	×							
SO_2	M[b]	×	×	×	×	×		
NO_x	×(N_2O)	×	×	×	×	×		
Particulates		×	×					
Indoor air pollution		×						
Wastewater		×		×				
Cooling water				×			×	
EMF		×						
Ash, sludge		×	×					
Liquid and gaseous radionuclide emissions[c]		×			×	×		
Solid radioactive wastes		×						×
Occupation of land						×		×
Hydropower				×				×
Occupational hazards		×						

Source: Based on Ottinger et al. (1990).

[a]"×" designates concern over actual or possible negative impacts; in some cases, such as electromagnetic fields, the magnitude and even existence of these impacts is controversial.

[b]Possible positive impact.

[c]Both normal and accidental releases.

sidized research and development. A new example is the *access* or *wire charges* that are to be applied to retail electricity sales in several states, the proceeds of which will be used to subsidize DSM, renewable energy, and low-income energy assistance. Such charges are to replace utility provision of subsidies for these activities, which utilities will find difficult to sustain in a competitive energy market.

Public ownership is, in theory, another possibility, if such ownership would result in more recognition of the social costs of power. Yet public ownership does not necessarily lead to cleaner power, as Eastern Europe's experience shows, and there is no evidence that private power systematically imposes higher environmental costs in the United States. Governments might also attempt to use moral leadership to inspire or shame utilities into decreasing their environmental impacts, but its effectiveness is unlikely to last.

Regulation is the most common tool that governments wield. In the United States, this tool derives its legitimacy from the government's police power: the power to restrict or direct private actions for the welfare of the community. In the case of the federal government, this power is limited to certain enumerated but broad categories and is constrained by the federal constitution and Bill of Rights. The government's power to regulate private activity touches utilities in many ways.

1. *Permitting activities.* Examples include "certificates of convenience and necessity" for siting new facilities, limitations on facility emissions rates and types of control equipment that must be used, and requirements that certain minimum flows be maintained downstream of hydroplants.
2. *Creation of marketable rights to emit pollutants.* We classify this as a regulatory approach because it is an exercise of the government's police power. As an example of a pollution rights market, *offsets* are required in some air basins in the United States in which a new source of volatile organic compounds or NO_x must pay existing sources to lower their emissions by an equal or greater amount. Another example is the acid rain control provision of the U.S. Clean Air Act, which creates a system of SO_2 emission allowances that can be traded. Generally, utilities will not be able to emit any more SO_2 than they own allowances for.
3. *Requirements for environmental impact statements and consideration of these impacts in the acquisition and operation of resources.* Since 1970, the U.S. National Environmental Policy Act (NEPA) has required "environmental full disclosure" when federal activities, such as permitting of power plants, significantly affect the environment. More recently, several states have begun to require that utilities estimate external costs in dollar terms and include those costs in decisions to acquire new resources—even though the utility does not actually pay those costs. But a danger is that if environmental costs are considered only for new sources, total social costs will actually increase rather than decrease (Andrews, 1992a; Palmer et al., 1995). This can occur because utilities

would then favor extending the life of existing dirty units and would ignore cost-effective operating strategies to lower emissions, such as fuel switching and emissions dispatch. We illustrate this effect in Section 11.D.5.

4. *Regulation of the prices that private utilities charge and the services they offer.* In the United States, public utility commissions regulate the electric rates charged by private utilities, although deregulation is loosening their grip. In theory, government could set the price of electricity at a level that reflects not only the utility's expense of providing power, but also its external costs.
5. *"Portfolio" requirements for DSM and renewable energy.* Although rarely used now, some U.S. deregulation proposals include provisions that require that electricity providers maintain a minimum level of DSM or renewable energy sources in their resource mix. The more creative of these proposals allow utilities who have more than the minimum to sell excess DSM or renewable "credits" to utilities that would otherwise fail to reach the minimum; this encourages such investments to occur where they are most cost-effective.

Until the 1970s, permitting was the regulatory tool of choice. The evolution of air and water quality regulation illustrates this. Before the 1950s, regulation of water and air emissions was principally a state responsibility in the United States. The intent of early pollution regulations, such as "smoke" laws, were to eliminate the most obvious and offensive emissions. Since then, however, the scope of regulation—and the federal government's role in it—have grown. First, in the late 1950s and early 1960s, the federal government funded research and local control programs and established limited enforcement procedures. In the mid-1960s, the federal government forced states to adopt ambient air and water quality standards that were to be attained by forcing pollution sources to clean up. The federal role greatly expanded with the passage of the 1970 Clean Air Act and 1972 Federal Water Pollution Control Act. The federal government itself now defines ambient standards and has adopted, for example, emissions and control technology standards for new pollution sources.

Thus, until the 1970s, environmental planning was merely a matter of obtaining necessary permits and altering equipment design or operation to comply with a legal mandate. It was primarily a concern of the process engineers and lawyers whose job it was to bring individual facilities into compliance. Although the costs of building cooling towers or fish ladders could be significant, they were typically not a large factor in long-term resource planning. Generally, the only way in which environmental considerations affected resource acquisition was by increasing the cost of some resources.

The 1970s and 1980s dramatically altered this situation in the United States. The 1970s brought the following changes. First, the cost of new generating capacity grew rapidly in part because of new environmental and safety requirements. Second, environmental constraints became a major concern in choosing and siting resources. Legal restrictions and public concern lead to acrimonious debates over where, if anywhere, new facilities could be built. An example of such legal restrictions is the "no significant deterioration" provisions of the U.S. Clean Air Act. Third, utilities became obliged not only to comply with laws that regulated certain enumerated residuals, such as SO_2, but also to consider *all* the environmental and social impacts of its decisions. In the early 1970s, court decisions have made clear that NEPA mandates federal agencies, and thus the utilities they regulate, to balance all impacts carefully, not just those covered by existing statutory limits (*Calvert Cliffs' Coordinating Committee v. Atomic Energy Agency,* 449 F.2d 1109). These decisions motivated utilities to conduct, for example, power plant siting studies that weighted a dozen or more categories of environmental impacts in their evaluations (Hobbs, 1979).

This trend has been reinforced in 1980s and 1990s. Compliance costs continue to climb; of the capital cost of a new pulverized coal-fired power plant, roughly one-third is for environmental protection. One midwestern U.S. utility may have to invest more to comply the 1990 Clean Air Act Amendments than the entire book value of the firm. States have now implemented their own impact statement requirements, and many mandate specific numerical procedures for factoring environmental effects into resource decisions. Possible impacts that are not covered by existing statutes, such as global warming and electromagnetic fields, now influence choices among supply- and demand-side alternatives. For instance, the World Bank now makes consideration of environmental effects a prerequisite to approval of power-sector loans. Several utilities do planning in collaboration with consumer groups and environmentalists; environmental issues are a major concern to these new parties to the planning process.

A development in the last decade has been a shift away from command-and-control regulations, which specify how emissions are to be removed at each individual facility (see Chapter 15). Governments are beginning to adopt programs that establish economic incentives to clean up, but don't tell companies how to do it. Examples include tradable permits for water pollution control in the Denver, Colorado area, a program that allows refineries to buy and sell rights to use lead in their products, and the trading of air emissions credits in some areas of the United States that has been allowed since 1976. Elsewhere, the World Bank is encouraging adoption of such market-oriented environmental policies by borrower countries (Adamson et al., 1996). For such systems to work effectively, it is important that residuals can be monitored. Further, in the case of marketable pollution rights, regulators should encourage utilities to participate in the market and the market should be large enough so that anticompetitive behavior will not be a concern.

The 1990 U.S. Clean Air Act Amendments are perhaps the best known example of a market-based approach to pollution control (Torrens et al., 1992). In an effort to alleviate acid rain, Title IV of those amendments requires a reduction of over 50% in electric utility SO_2 emissions by the year 2000. Two innovative features of the act have irrevocably altered the nature of air quality compliance planning for utilities. These are:

1. *A national cap on SO_2 emissions.* This cap is implemented by allocating *emissions allowances,* measured in tons per year, to each generating unit by formulas specified in the act. In contrast, previous law only subjected units to an emissions rate cap, measured in pounds per million British thermal units (BTU) of heat input, and mandates for particular control technologies. Although these command-and-control rules remain in force for new generating units and where local pollution conditions require, the allowance system will motivate most large coal-fired units to lower their emissions further.
2. *The creation of a market in emissions allowances.* Generating units whose allowances exceed their SO_2 emissions will be able to sell their excess allowances to other units that have too few allowances.

These two provisions, in effect, create a tonnage cap for each utility, since emissions may be traded among the utility's generating units. They also create opportunities for profitable sales of excess allowances by utilities for whom pollution control is cheap.

These provisions affect utility planning profoundly. Because it is now a system's total tonnage of emissions that matters, and not so much emission rates at particular units, new options for compliance, including resource acquisition, are opened up. Among these are purchases of allowances from other utilities; DSM, which can decrease generation and thus emissions; retirement of dirtier units and replacement with cleaner resources; and emissions dispatch. Although the emission reductions that the new law requires are large, its market philosophy allows more flexibility than the old command-and-control approach in how to reduce emissions.

This flexibility means that utilities can lower their cost of compliance. But it also means that the environmental planning process has become more complex, as it is now inextricably linked to resource planning. *What* resources are acquired *when* has important implications for the cost of complying with the law. Consequently, the emissions impacts of different resources can alter their relative attractiveness drastically.

The most recent development has been the dramatic restructuring of the power industry in many parts of the world (Hunt and Shuttleworth, 1996). Privatization and breakups of utility monopolies have been widespread, motivated by the hope that markets will provide more effective incentives for cost minimization than government regulation. As a result, many utilities have shortened their time horizons, shied away from integrated resource planning and DSM, and are obtaining new power sources from independent power producers rather than building new plants themselves. In such an economic environment, new policy tools may be required to reconcile the commodity price reduction incentives produced by competition and other energy and environmental policy objectives. Market-based environmental regulations, such as emissions allowances, are quite compatible with a competitive generation services market. As another approach, some governments will also impose grid access fees on all customers to fund renewable energy sources and DSM.

Despite utility deregulation, environment–economy trade-offs will remain a public concern. However, their consideration will often be in different forums than traditional regulatory proceedings concerning the resource plans of individual utility monopolies. Examples of such forums include public hearings concerning facility siting; definition of regional pollution caps and allocations of reductions among pollution sources; determination of the level of "wires charges" and their disbursement among alternative DSM and renewable energy investments; and selection of portfolio targets in regulatory systems requiring a minimum amount of DSM or renewable energy. The modeling approach of this chapter can be just as useful for considering environment–economy trade-offs in those forums as for traditional resource planning.

In the next section we present a general resource planning model in which not just cost but also environmental effects are of concern. Then to illustrate its application, the model is used in Section 11.D to show how different types of environmental legislation can affect utility planning and operations decisions.

11.C. LP MODEL FOR INTEGRATED RESOURCES PLANNING

11.C.1. Multiobjective Framework: Cost–Emission Trade-offs

Multiobjective plots are a useful tool for understanding how coherent environmental compliance plans can be assembled. We use this device to illustrate how long-run resource acquisition decisions and short-run system operation strategies complement each other; both are needed to ensure an efficient outcome. Multiobjective plots can show how the options available to the utility affect important objectives, such as costs, rates, emissions, resource use, and financial indices. Figure 11.2 is a two-dimensional plot in which the only objectives are the present value of total power costs over a 20-year time horizon (the *Y* axis, in millions of Canadian dollars) and aggregate greenhouse gas (CO_2) emissions for the same time period (the *X* axis) from a study by BC Hydro, a Canadian utility (Meier, 1995). (This study actually had 12 objectives, but just these two are shown for simplicity.) In general, each point in such a plot stands for a distinct plan that includes a particular combination of supply sources, environmental controls, DSM programs, and rate design, along with a unique operating strategy. In Figure 11.2 each point label de-

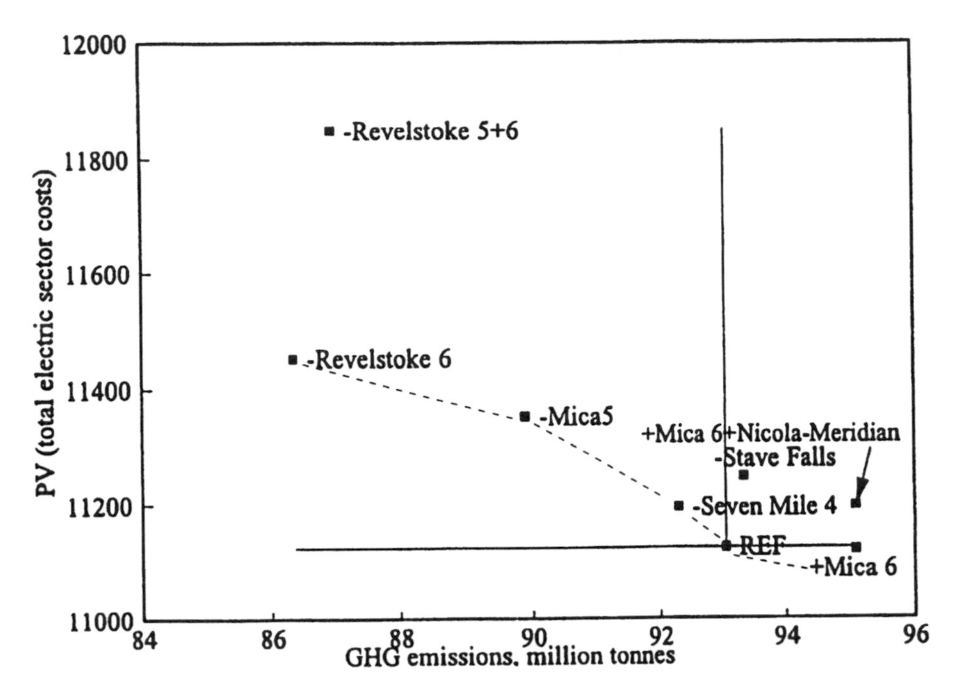

Figure 11.2 Emissions-internal cost trade-offs. (From Meier, 1995.)

scribes how the plan in question differs from a reference plan (REF), which consists of a mix of gas, hydropower, wood wastes, coal, and DSM additions. For instance, point −Mica5 indicates a portfolio that differs from REF in that generating unit Mica 5 has been subtracted.

We assume in Figure 11.2 that the costs and emissions resulting from a particular plan can be predicted with certainty; in reality, uncertain fuel prices, loads, and other factors mean that the location of each point should be described by a probability distribution or confidence interval. The points shown could also represent an average for each plan across a set of scenarios (Andrews and Conners, 1990).

Some plans in Figure 11.2 yield higher costs and higher emissions compared to other plants (e.g., -Stave Falls compared to -Seven Mile 4). Plans that are not dominated by any other plans are known as efficient or nondominated alternatives (e.g., points -Revelstoke 6 and REF), and are indicated by the dashed line. The only alternatives to a particular nondominated plan result in either worse costs or worse emissions, or both. In this case, if all a planner cares about is costs and CO_2, only plans in those sets are of interest (see Chapter 12 for further discussion of nondominated set generation methods, along with other techniques for multiobjective programming).

Another way to view the set of nondominated alternatives is as a total cost curve for emissions reductions. The slope of that curve is a supply curve for emissions reductions that gives the marginal cost of achieving each increment of reduction.

A major purpose of modeling is to define the nondominated set, as in Figure 11.2. It is then up to the interest groups in the political process to make the value judgments necessary to choose from among the alternatives in the set. This information can aid the negotiations by helping participants to focus on the key issues and trade-offs. This was the case in the BC Hydro case study (Meier, 1995), where examination of the trade-offs lead a diverse group of stakeholders to agree that BC Hydro should repower a natural gas–fueled plant in Vancouver (i.e., replace its boilers and generators). Previously, environmentalists had opposed the repowering because of the emissions that could result from greater use of the plant. However, repowering was found to be part of most of the nondominated portfolios, which showed that repowering actually helped reduce emissions for the system *as a whole.*

Environmental planning can be viewed as the process by which the utility defines and chooses from the possibilities in Figure 11.2. The government sets some rules as to which points are admissible and how the utility should weigh the trade-offs between the two objectives of minimizing internal costs and minimizing emissions. These rules might be very tight, giving the utility little discretion, or they may allow considerable flexibility. In Section 11.C.3 we present a simple integrated resource planning model that allows us to show how different rules affect the utility's planning process.

11.C.2. General Modeling Considerations

Before proceeding to the model itself, it is useful to point out that the solutions shown in Figure 11.2 can be generated in two ways: resource simulation or resource optimization. Resource simulation employs an operations, or *production costing,* model (Stoll et al., 1989; Wood and Wollenberg, 1995). The production costing model is used to estimate the cost and emissions resulting from predefined system designs—each being a combination of power plant capacities, DSM programs, and emissions controls. Emissions dispatch policies must also be prespecified, usually in the form of an economic penalty on emissions. The analyst can plot the results as in Figure 11.2, with each point representing a separate simulation of a distinct combination of system design and dispatch policy (e.g., Heslin and Hobbs, 1991). Then the nondominated alternatives can be identified visually or by the computer. Dominated alternatives (such as -Stave Falls) can then be eliminated.

It should be noted that the simulation approach itself involves some optimization in that estimates of production costs result from the optimization of system dispatch subject to demands and available generation capacity. However, resource capacities (e.g., MW of coal capacity or DSM) are not decision variables in production costing.

In contrast, the resource optimization approach uses an algorithm such as linear or dynamic programming to sort through the large number of possible combinations of generation plants, DSM programs, and emissions controls, along with dispatch policies in order to identify nondominated solutions.

Generally, if an IRP optimization model can calculate emissions, it can be used to generate nondominated alternatives by either of two methods: the constraint method and the weighting method (Cohon, 1978). Widely available IRP models such as EGEAS and PROVIEW/PROSCREEN have this capability (Stone and Webster Management Consultants, 1991; EDS, 1996).

A final general point about IRP modeling: In studying environment–cost trade-offs, it is critical to use a model of the entire generating system when considering options for lowering emissions. Simple analyses that study possible DSM, generation, or emissions control investments in isolation from each other may miss important interactions with operating strategies and other such investments.

As an example of such an interaction, installation of an emissions control (such as a SO_2 scrubber) at a generating unit will change the unit's variable cost and emission rate. This, in turn, will generally alter the dispatch order (describing which generators are to be operated most of the time and which are to be used only during peak demand periods), perhaps changing system costs and emissions in surprising ways. Moreover, stack gas controls generally exact a capacity penalty while increasing the plant's probability of outage, implying that power will have to made up from other generating units. Figure 11.3 illustrates the potential importance of this interaction (Huang and Hobbs, 1992). A hypothetical generating system consisting of 52 generating units is

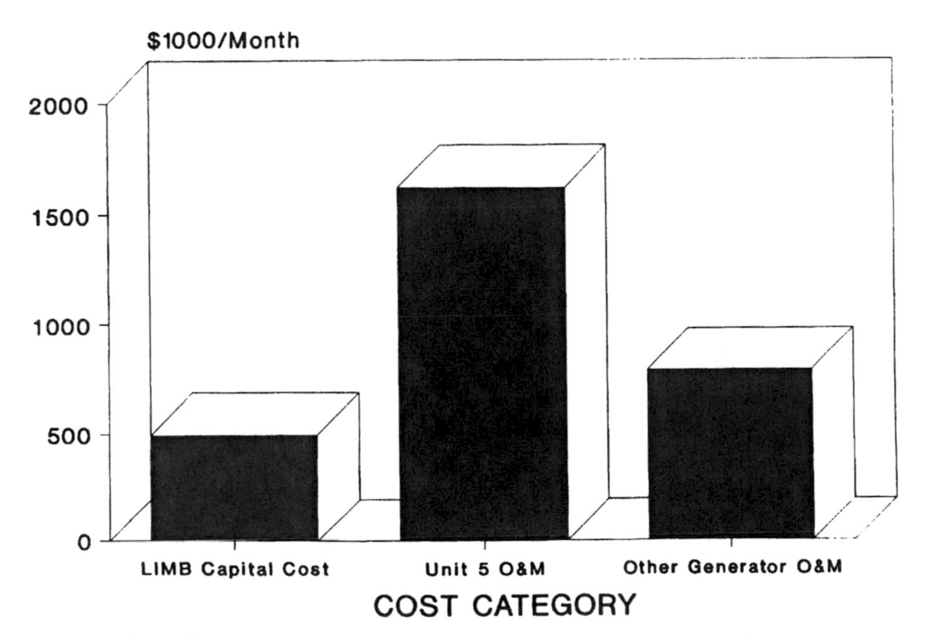

Figure 11.3 Effect of LIMB retrofit on generating unit costs and entire system costs.

dispatched to meet a mean load of 6300 MW. The figure shows the distribution of the incremental cost of retrofitting limestone injection multistage burners (LIMB) at one of the units, unit A. Fully one-third of the cost increase is due to the increased generation from units other than the one refitted. Those costs, and indeed the operating cost of the retrofitted unit itself, are difficult or impossible to estimate without a production costing analysis of the entire system.

As another example of these interactions, energy efficiency–type DSM can actually increase emissions, a possibility that is usually overlooked in simple benefit–cost analyses of individual DSM programs. This increase can occur because of the interaction of timing decisions for capacity expansion and system operating costs. Clarke and Parmelee (1994) document a situation in which energy efficiency, by postponing the addition of clean generation capacity (whose emissions are much less than existing plants), would actually increase emissions in New York State. A system-wide model, such as the linear program in Section 11.C.3, can account for such interactions.

11.C.3. Simplistic IRP Model for Environmental Planning

In this section we outline an integrated resource planning model for analyzing trade-offs between internal costs and environmental effects, such as those shown in Figure 11.2. Like any optimization model, this one includes one or more objective functions to be optimized (e.g., costs and emissions), a set of decision variables whose values are to be chosen by the utility (capacity and operation of each resource), and a set of technologic and economic constraints that must be respected (e.g., generation must meet demand, resource output cannot exceed resource capacity).

The model we present below is simplistic in that it considers just one year rather than all years over a time horizon. Also, reliability is ignored, as are nonlinearities such as scale economies in generation plant construction and heat rates that vary as a function of generator output. Important supply-side complications, such as transmission losses and limits, energy-limited units, and storage are omitted. Similarly, on the demand side, we disregard interactions among different DSM programs and do not consider programs that are dispatchable, such as interruptible rates. Most of these simplifications are avoided in more complex linear and dynamic programming models that utilities actually use to optimize the timing of resource additions over a time horizon (e.g., EGEAS and PROVIEW/PROSCREEN).

However, incorporation of these considerations here is unnecessary for our purposes, which are:

- To illustrate the basic trade-offs among environmental improvement measures and resource choices in long-run planning.

- To show how models can illuminate those trade-offs.
- To demonstrate how alternative government policies can direct or skew utility decisions. Some policies will yield dominated outcomes (i.e., plans and operating strategies for which there exist alternatives that would result in both lower emissions and costs). Other policies are more likely motivate the utility to choose nondominated mixes.

Another simplification made here is the assumption that environmental effects can be summarized in terms of total residuals, such as tons of SO_2. However, as pointed out earlier in this chapter, environmental impacts depend not just on the amount emitted, but also on the space- and time-dependent processes of transport, transformation, exposure, and physical and behavioral responses of affected populations. For instance, a ton of SO_2 emitted at one location may impose little cost (and might even be beneficial), while that same ton may cause great damage if released elsewhere. It is possible to embed simple representations of environmental processes in models such as the one we present here. For example, a few planning models tally the amount of sulfate deposited in sensitive watersheds in addition to total tons of SO_2 emitted (Ellis, 1988; Ottinger et al., 1990, Sec. VII). Usually, such models use coefficients to translate gaseous and liquid emissions linearly into concentrations or deposition rates elsewhere. Such coefficients are usually estimated from more complex water and air pollutant transport models (see, e.g., Chapters 2, 4, and 15). However, since most pollution control laws, including the new U.S. Clean Air Act, require utilities to consider emissions and not impacts, total emissions will remain a useful index for planners.

The model presented below focuses on generation operation and the acquisition of supply and DSM resources. However, there are also other options available to utilities for lowering net emissions. One is offsets, such as the purchase of SO_2 allowances or the planting of trees in Central America to take up CO_2, as some U.S. utilities are doing. Another is the adoption of rate policies that reflect external costs in the price of power (Woo et al., 1994). Our model also focuses on CO_2 as the environmental insult to be minimized; however, models that include multiple types of emissions and impacts are typically used for planning.

We use a linear programming (LP) model to optimize the mix of resource acquisition and generation dispatch. The model is based on the generation model of Turvey and Anderson (1977), modified to accommodate DSM programs (Lawrence, 1989). The formulation of any optimization model consists of a description of its decision variables, objective(s), and constraints. Lowercase letters designate decision variables, while capital letters define fixed parameters supplied by the analyst. Let the three basic decision variables of our model be:

x_i = generation capacity (MW) of supply resource i ($i = 1,2,...,I$). Supply resources can include not only utility-owned generation plants but also purchases from other utilities or independent generators of power. For existing plants, the value of this variable is fixed and is not altered in the LP.

y_{ift} = MW output of supply resource i during subperiod t ($t = 1,2,...,T$) using fuel f [$i = 1,2,...,I$, and $f = 1,2,...,F(i)$]. The 8760 hours in each year are divided into T demand periods. Period 1 represents peak demand, while the last period T is the lowest demand period. Commonly, these models include three to six periods; additional periods generally contribute little to model accuracy. Some units can choose among multiple fuel types f; in the application of our model below, a cofired generating unit could choose between using 100% coal and 85% coal/15% natural gas. In that case, the number of fuel types $F(i) = 2$.

z_k = 1 if DSM program k is fully implemented ($k = 1,2,...,K$). Intermediate values between 0 and 1 represent partial implementation.

There are two objectives. The first is the annual worth of capital and operating costs:

$$\text{minimize cost} = \sum_{i=1,2,...,I} CRF \cdot CX_i x_i + \sum_{k=1,2,...,K} CZ_k z_k + \sum_{i=1,2,...,I} \sum_{f=1,2,...,F(i)} \sum_{i=1,2,...,T} H_t CY_{ift} y_{ift} \qquad (11.1a)$$

The second objective is greenhouse gas emissions:

$$\text{minimize } CO_2 = \sum_{i=1,2,...,I} \sum_{f=1,2,...,F(i)} \sum_{t=1,2,...,T} H_t CO2_{ift} y_{ift} \qquad (11.1b)$$

Of course, in real applications, there are likely to be many more objectives. For instance, BC Hydro considered costs, price impacts, portfolio diversity, four types of air emissions (including a population-weighted metric), terrestrial ecosystem impacts, species effects, electromagnetic fields, land use, and employment effects (Meier, 1995).

The parameters in the objectives are as follows:

$CO2_{ift}$ = CO_2 emissions (tons/MWh) of generation from unit i using fuel f during subperiod t. This is the product of the unit's heat rate HR_{it} [in 10^6 BTU/kWh] and the CO_2 resulting from burning fuel (tons/10^6 BTU). The emissions from coal are roughly twice those from burning natural gas.

CRF = capital recovery factor (yr^{-1}) used to annualize capital costs.

CX_i = capital and other fixed costs (\$/MW) of building capacity of type i. In general, this parameter should include the present worth of the resource's fixed O&M costs, while deducting the resource's salvage value at the end of the planning period. For existing plants, this cost is omitted, as their capacity is fixed. For new plants, utilities must choose between low capital cost/high fuel cost plants, such as combustion turbines, and high capital cost/low fuel cost facilities, such as pulverized coal plants.

CY_{ift} = variable operating cost (\$/MWh) of supply type i using fuel type f during subperiod t. This cost includes fuel and any miscellaneous variable operating costs:

$$CY_{ift} = HR_{it}FC_{ift} + VO\&M_{it} \tag{11.2}$$

where HR_{it} is as defined above, FC_{ift} is the fuel cost (\$/10^6 Btu), and $VO\&M_{it}$ is the nonfuel variable O&M cost (\$/MWh). For some supply resources, variable costs may be the same in all t of a given year, but for others, costs can vary due to temperature-dependant heat rates or seasonal variations in fuel prices.

CZ_k = annual capital and other fixed costs (\$/year) of fully implementing DSM program k. These can include installation and administrative expenses, rebates to technology adopters, and postinstallation studies to confirm that energy savings occurred.

H_t = number of hours in time period t.

As is the case with any multiobjective mathematical program, trade-offs between the two objectives can be obtained by weighting and combining the two objectives and then solving the model several times for different values of the weights (Chapter 12). For example, a penalty PEN, in \$/ton, could be applied to the CO_2 emissions (11.1b), allowing that objective to be added to the cost objective (11.1a). If the values of all x_i and z_k are fixed, application of this penalty results in emissions dispatch (i.e., some reduction in emissions accompanied by an increase in other operating expenses). If, on the other hand, those capacities are also decision variables, the mix of resources will be chosen to minimize the utility's cost plus CO_2 penalties.

Demand, operating, and reliability constraints restrict which values of the decision variables can be chosen. Simple yet typical formulations of these constraints are given below. Consistent with standard mathematical programming notation, we list terms involving decision variables on the left-hand side of the equation, while constant terms are placed on the right.

- *Load must be met in each subperiod t:*

$$\sum_{i=1,2,\ldots,I}\ \sum_{f=1,2,\ldots,F(i)} y_{ift} + \sum_{k=1,2,\ldots,K} SAV_{kt} z_k \geq LOAD_t \qquad t = 1,2,\ldots,T \quad (11.3)$$

where $LOAD_t$ is the MW power demand during t, including transmission and distribution losses, and SAV_{kt} is the MW savings in period t resulting from fully implementing DSM program k. This constraint states that the sum of the MW output in subperiod t from all plants must equal or exceed the electricity demanded $LOAD_t$ at that time, as modified by any DSM programs. In spatial models, such as Meier (1979), the transmission grid is represented explicitly, and one load balance is defined for each node of the network for each t.

- *Generation must be no more than derated capacity for each resource in each t:*

$$\sum_{f=1,2,\ldots,F(i)} y_{ift} - (1 - FOR_i)x_i \leq 0 \qquad i = 1,2,\ldots,I, \quad t = 1,2,\ldots,T-1 \quad (11.4)$$

where FOR_i is the forced or unplanned outage "rate" (actually, the outage probability) of resource i. This constraint is generally unneeded for the time of lowest demand (subperiod T), as long as the annual energy constraint described next is also imposed. If the resource is an existing plant, x_i is fixed and its term is, instead, placed on the right side of (11.4). A more sophisticated version of this constraint would account for the fact that changing fuels alters plant capacity.

- *Generation must be no less than the must-run capacity for each resource in each t:*

$$\sum_{f=1,2,\ldots,F(i)} y_{ift} - MR_i x_i \geq 0 \qquad t = 1,2,\ldots,T \quad (11.5)$$

with MR_i being the minimum ("must run") level of output (as a fraction of total capacity). Minimum run levels result from, for example, efficiency considerations, safety (flame stability), and demands for process steam in cogeneration facilities.

- *Annual energy constraint for each resource:*

$$\sum_{t=1,2,\ldots,T}\ \sum_{f=1,2,\ldots,F(i)} H_t y_{ift} - 8760 CF_i x_i \leq 0 \qquad i = 1,2,\ldots,I \quad (11.6)$$

where CF_i is the maximum possible capacity factor for resource i (defined as the ratio of average output to capacity). This constraint is especially important for energy-limited resources such as hydroelectric power plants. CF_i also accounts for planned maintenance outages. Again, for existing resources, x_i is a constant and appears on the right side of the equation.

- *Reserve margin constraint:*

$$\sum_{i=1,2,\ldots,I} x_i + (1 + M) \sum_{k=1,2,\ldots,K} SAV_{k1} z_k \geq LOAD_1(1 + M) \quad (11.7)$$

where $LOAD_1$ is the yearly peak demand and M is the desired reserve margin. The reserve margin accounts for the fact that extra generation capacity must be constructed to ensure that demand can be met even when plant outages occur. This constraint can be viewed as a type of chance constraint.

- *Upper bound on new capacity:*

$$x_i \leq X_{i\text{MAX}} \qquad i = \text{new plants} \quad (11.8)$$

This constraint allows additions to be no more than a predetermined maximum size $X_{i\text{MAX}}$.

- *Upper bound on DSM programs:*

$$z_k \leq 1 \qquad k = 1,2,\ldots,K \quad (11.9)$$

- *Nonnegativity restrictions for all variables:*

$$x_i \geq 0 \qquad i = 1,2,\ldots,I$$

$$y_{ift} \geq 0 \qquad i = 1,2,\ldots,I, \quad f = 1,2,\ldots,F(i), \quad t = 1,2,\ldots,T$$

$$z_k \geq 0 \qquad k = 1,2,\ldots,K \quad (11.10)$$

Once the model is formulated and its parameter values estimated, it can be inserted into standard linear programming software and solved. The solution consists of the best values of the decision variables and the resulting total cost and emissions.

As an example, the model has been solved using the numerical assumptions in Tables 11.2, 11.3, and 11.4 under a minimum cost objective (i.e., the emissions objective is disregarded entirely). The data are for a hypothetical small utility in the year 2010. The utility's peak load in that year is projected to be 1050 MW. Its present generation mix consists of three coal units (totaling 500 MW), an oil-fired steam unit (150 MW), and natural gas–fired combustion turbines (200 MW). The options available for reducing emissions and meeting growing demands include emissions dispatch, fuel switching (in par-

TABLE 11.2 Miscellaneous Data for Linear Programming IRP Example

Parameter	Value[a]
Minimum reserve margin, M	15%
Capital recovery factor for plant investment CRF	12% per year
Load block widths H_t, t = 1,2,3,4,5	140, 580, 1950, 2545, 3545 hr
Load block heights $LOAD_t$, t = 1,2,3,4,5	1050, 950, 825, 525, 275 MW
Coal cost, CO_2 emissions	\$1.60/10^6 BTU, 221 lb/10^6 BTU
Natural gas cost, CO_2 emissions	\$3.80/10^6 BTU, 116 lb/10^6 BTU
Heavy fuel oil cost, CO_2 emissions	\$2.60/10^6 BTU, 168 lb/10^6 BTU

[a] 10^6 BTU.

ticular, cofiring natural gas at the second coal unit), acquiring supplies (new pulverized coal plants, gas-fired combined cycle, or combustion turbine capacity), and investing in DSM (energy efficiency and load controls). The costs of the DSM options are described using a *conservation supply curve* in which the per unit cost increases for larger programs. The resulting optimal supply solution (capacity additions and dispatch) is shown in Table 11.5. No DSM programs are added in this case, because their cost per MWh or MW saved exceeds the internal (if not environmental) cost of new supplies. The total cost is \$150.3 million per year, and 4,880,000 tons of CO_2 are emitted annually.

As pointed out, this model is simplistic in many ways. Some straightforward extensions of the model include:

- Incorporation of nonlinear cost functions
- Use of integer variables to represent discrete capacity and DSM additions
- Definition of a transshipment network representing spatially distributed generation and demands along with transmission (Makarova, 1966; Cohon et al., 1980; Meier, 1979).
- A dynamic formulation in which variables and coefficients are subscripted by year, and a set of objective function terms and constraints is defined for each year

Other complications can also be easily integrated into the model, including limited energy and storage units; nondispatchable units, such as wind turbines and cogenerators; and dispatchable DSM technologies (e.g., interruptible loads and water heater load controllers). Random generator outages can be modeled as an explicitly stochastic mathematical program (Bloom et al., 1984) rather than using the "derating" approach of constraint (11.4). The stochastic program is solved using Benders' decomposition, which divides the problem into a deterministic design model (which chooses trial values of the capacity variables) and a stochastic production costing model (which deter-

TABLE 11.3 Generating Unit Data for Linear Programming IRP Model

Plant, i	Type	Capacity, x_i (MW)	Must Run Capacity, MR_i (MW)	Heat Rate, HR_i (BTU/kWh)	Nonfuel Variable O&M, $VO\&M_{it}$ ($/MWh)	Capital Cost, CX_i ($/kW)	Forced Outage Rate, FOR_i	Maximum Capacity Factor, CF_i
A	Existing coal 1 (scrubbers)	200	80	10,000	9	n.a.[a]	0.05	0.85
B	Existing oil steam	150	0	10,200	3	n.a.	0.04	0.85
C	Existing combustion turbines	200	0	13,800	9	n.a.	0.035	0.8
D	Existing coal 2	150	60	10,400	5	n.a.	0.05	0.85
E	Existing coal 3	150	0	11,500	5	n.a.	0.05	0.85
F	New coal plant	500 (max.)	40% of capacity	10,000	5	2000	0.05	0.85
G	New combined cycle	500 (max.)	0	8,230	5	800	0.05	0.85
H	New combustion turbine	500 (max.)	0	13,800	8	550	0.035	0.8
I	Install 15% natural gas cofire capability, coal 2	n.a.	60, if installed	10,400	3	9	0.05	0.85

[a]n.a., not applicable.

TABLE 11.4 Demand-Side Program Characteristics for Linear Programming IRP Model

Program, k	Type	Load Decrease SAV_{kt} by Demand Period t	Cost[a]
A	Energy efficiency	40.0, 36.2, 31.4, 20.0, 10.5 MW	$55/MWh
B	Energy efficiency	40.0, 36.2, 31.4, 20.0, 10.5 MW	$65/MWh
C	Energy efficiency	40.0, 36.2, 31.4, 20.0, 10.5 MW	$75/MWh
D	Load control	40 MW peak only (no energy savings)	$96/peak kW/yr

[a]The objective function coefficient CZ_k is obtained by multiplying the per MWh cost by $\Sigma_t H_t SAV_{kt}$ for programs k = A, B, and C. For program D, total cost = $96/peak kW × peak kW savings (40,000 kW), or $3,840,000 per year.

mines the operating costs associated with trial system designs). Many or most of the features summarized in this paragraph are available in standard resource planning software.

An important modification for environmental planning would be consideration of control equipment retrofits that have a significant capital cost and possible impacts on plant capacity. These would be modeled with appropriate choices of capacity variables x_{ij}, where the subscript j represents the type of control. The generation variable y_{ifjt} would then also receive an extra subscript j, and it would be constrained by the capacity x_{ij}. An example of such a model which has been used by several utilities is EC-VIEW (EDS, 1996; based on Huang and Hobbs, 1994).

Because the LP model above is a multiobjective one, there is no unique solution—the best option depends on the relative importance of the cost and environmental objectives. To get a low value of one objective, a high value for the other must be accepted. This trade-off is the heart of the multiobjective problem. In a sense, the entire nondominated set of Figure 11.2 is the solution.

11.D. ALTERNATIVE APPROACHES TO ENVIRONMENTAL PLANNING

11.D.1. APPROACH

Government laws and regulations limit (1) which points in Figure 11.2 can be considered by the utility and (2) how the utility is to weigh the trade-offs between the objectives. In this section we use the model of Section 11.C to review how alternative government environmental policies affect utility planning. The policies we consider include traditional command-and-control regulations, constrained emissions, internalization of environmental costs via taxes or marketable emissions rights, and internalization of environmental costs via planning regulations.

The policies are compared in terms of whether they motivate the utility to choose an efficient (nondominated) plan and whether that plan also minimizes

TABLE 11.5 Cost-Minimizing Generation Solution for Linear Programming IRP Model

Plant, i	Type	Capacity, x_i (MW)	CO_2 (1000 tons)	Generation y_{ift}, During Demand Period t (MW)				
				$t = 1$	$t = 2$	$t = 3$	$t = 4$	$t = 5$
A	Existing coal 1 (scrubbers)	200	1392	190	190	190	190	76
B	Existing oil steam	150	438	144	144	144	50	0
C	Existing combustion turbines	200	217	193	193	68	0	0
D	Existing coal 2	150	1283	142.5	142.5	142.5	142.5	105.4
E	Existing coal 3	150	1366	142.5	142.5	142.5	142.5	93.6
F	New coal plant	0	0	0	0	0	0	0
G	New combined cycle	145.3	176	138	138	138	0	0
H	New combustion turbine	212.2	11	100	0	0	0	0
I	Install 15% natural gas cofire capability, coal 2	Not installed	0	0	0	0	0	0

total social cost. "Social" cost is defined for our purposes as the sum of internal costs plus emissions times the damage cost per ton of those emissions. If damages are, say, \$20 per ton for the emission in question, the least social cost plan for BC Hydro in Figure 11.2 is option REF. Only if the damage is \$50 per ton would a plan involving lower emissions (-Revelstoke 6) be selected in that case; the \$350 million increase in internal costs compared to REF would then be compensated by an equal decrease in CO_2 external costs.

The hypothetical utility described in Tables 11.2 to 11.4 is used to compare the policies. However, the exact generation and DSM cost assumptions do not affect our general conclusions concerning the relative efficiency of the different policies. Figure 11.4 shows the resource additions made in each of several plans, under the assumption that least (internal) cost dispatch is used to operate the system's power plants (i.e., zero emission penalty). Some plans in Figure 11.4 yield both high internal costs and high emissions (e.g., plan G, which includes a new coal plant). These plans are obviously less desirable than plans that have both lower costs and emissions (such as plan A). Here plans A, B, C, and D are the efficient or nondominated alternatives, which we connect by a line. Alternatives to a particular nondominated plan have either worse costs or worse emissions. If all a planner cares about is revenue requirements and CO_2, only nondominated plans are of interest (Burke et al., 1988; Andrews, 1992b; Gjengedal et al., 1992; Crousillat et al., 1993).

In Figure 11.5 we expand the options to include emissions dispatch and natural gas cofiring. As mentioned earlier, emissions dispatch is the operation

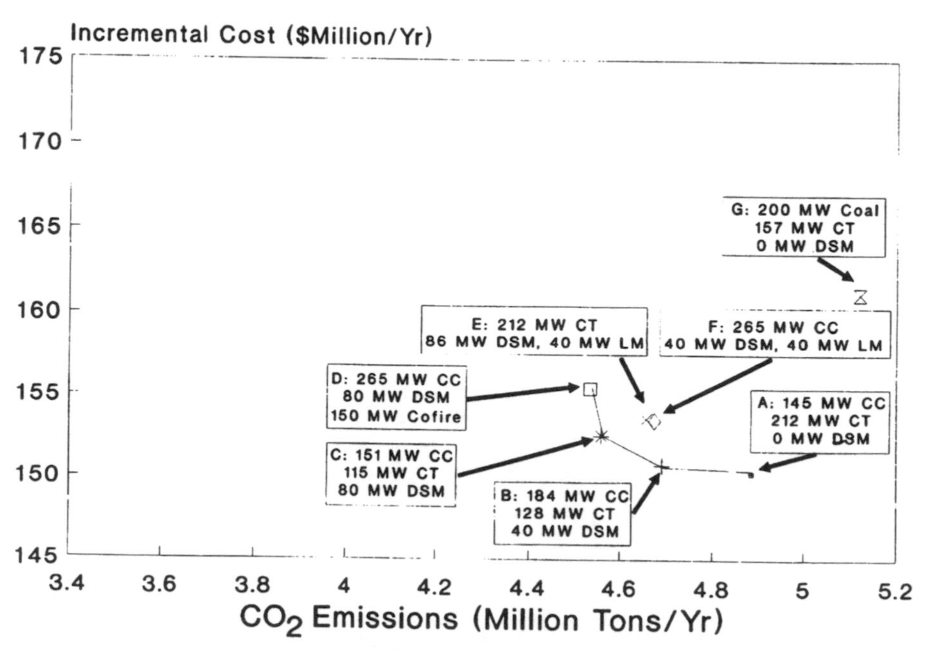

Figure 11.4 Trade-off plot: CO_2 versus cost for seven plans, least-cost dispatch assumed.

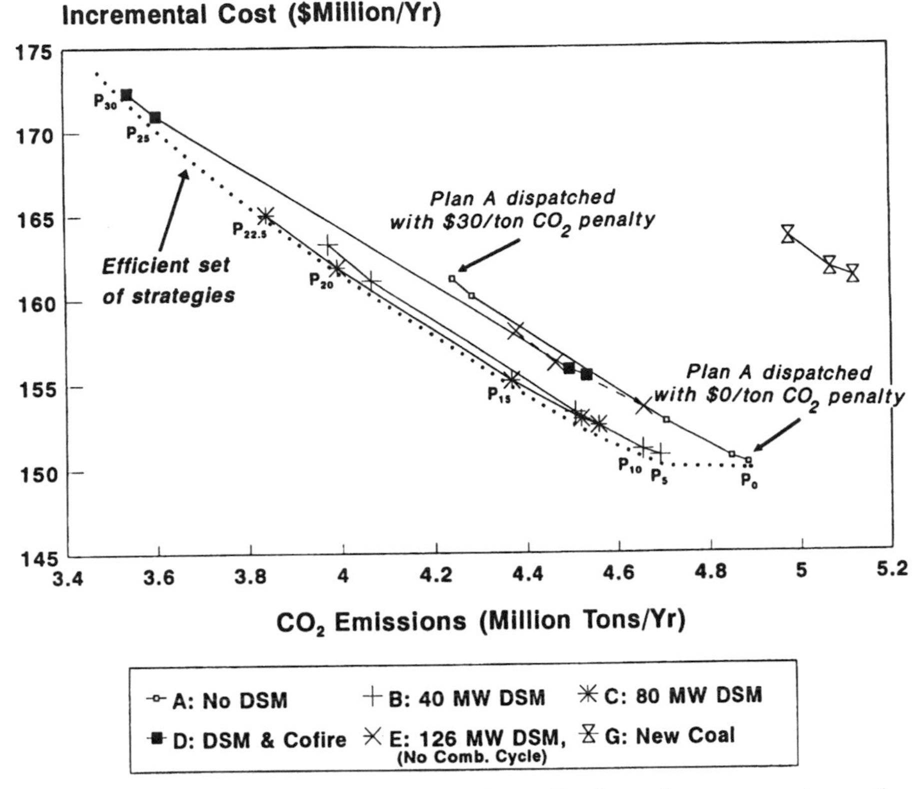

Figure 11.5 Identification of nondominated combinations of resource and operating strategies.

of a power system so that more generation is obtained from cleaner units than would be the case under least-cost dispatch. Emissions dispatch is usually implemented in practice by operating the system to minimize the sum of variable utility cost plus a penalty on emissions. In Figure 11.5, a range of penalties between $0 and $30 per ton were used to dispatch each plan's set of generating units; increasing the penalty results in lower emissions but higher costs. Solid lines connect points that represent different operating policies using the same set of facilities. The $0 dispatch points (which are the southeasternmost points for each plan) correspond to the points in Figure 11.5. Increasing degrees of emissions dispatch are represented by points to the northwest of the $0 point. For instance, the figure indicates how changing the CO_2 penalty from $0 per ton to $30 ton alters the cost and emissions resulting from plan A. One plan (plan D) also includes the operating option of burning natural gas in one of the coal units; this is done if the CO_2 penalty is sufficiently high to overcome the cost penalty of gas relative to coal, which happens near the northwestern end of the plan D curve.

Comparing Figures 11.4 and 11.5 we see that including operational strategies greatly increases the possibilities for decreasing emissions. Indeed, it turns out that every nondominated strategy except the least-cost point involves some degree of emissions dispatch. Several recent studies of actual utility systems confirm this fact (e.g., Heslin and Hobbs, 1991; Hess et al., 1992). In Figure 11.5 we label the nondominated points as P_0 through P_{30} and connect them with a dotted line. The subscript represents the CO_2 penalty at which a nondominated point minimizes the sum of utility costs plus CO_2 penalties. For example, P_{20} is the strategy that has the lowest value of utility cost plus $20 per ton times the CO_2 emissions.

As mentioned above, environmental planning could be viewed as the process by which the utility chooses from the possibilities in Figure 11.5. Government policy limits which points are admissible while providing incentives that guide the utility's choice from the feasible solutions. In the remainder of this section we discuss how, in general, alternative government environmental policies influence utility resource planning. The policies we consider include traditional command-and-control regulations, constrained emissions, internalization of environmental costs via taxes or marketable emissions rights, and internalization of environmental costs via planning regulations.

The policies are compared in terms of whether they motivate the utility to choose a nondominated plan and whether that plan also minimizes total social cost. *Social cost* is defined for our purposes as the sum of utility costs plus emissions of each type times the appropriate damage cost per ton for each type. As a hypothetical case, if damages are, say, $10 per ton for CO_2, the least social cost plan in Figure 11.5 is plan P_{10}, which is the point at which the marginal internal cost of reducing emissions further by moving to plan P_{15} exceeds $10 per ton. (We assume that changes in the expense of electricity do not induce consumers to alter the amount of power they consume. In reality, such rate feedback tends to decrease loads, which, in turn, lowers costs, emissions, and consumer surplus compared to the no-rate-feedback case (Rose et al., 1994). We also ignore emissions from energy sources that compete with electricity, which might increase if the price of electricity were raised (Hobbs, 1994).

11.D.2. Command-and-Control Regulations

Examples of command-and-control regulation include the U.S. New Source Performance Standards and local restrictions on fuel sulfur content. The effect of policies of this type would be to render some of the points in Figure 11.5 illegal. Only those plans that conform to the regulations can be considered. The utility is then free to choose from the permissible plans to minimize its internal cost; thus the emissions of those plans are disregarded at this stage.

For instance, say that a CO_2 law is passed that prohibits new coal plants or new fossil-fuel plants with heat rates worse than 9000 BTU/kWh. For our hypothetical utility, this eliminates all alternatives involving new combustion turbines and coal plants, leaving just the strategies represented by the solid line in Figure 11.6. The utility will choose plan $P_{C\&C}$ (command-and-control), which is the cheapest plan that excludes new plants of that type.

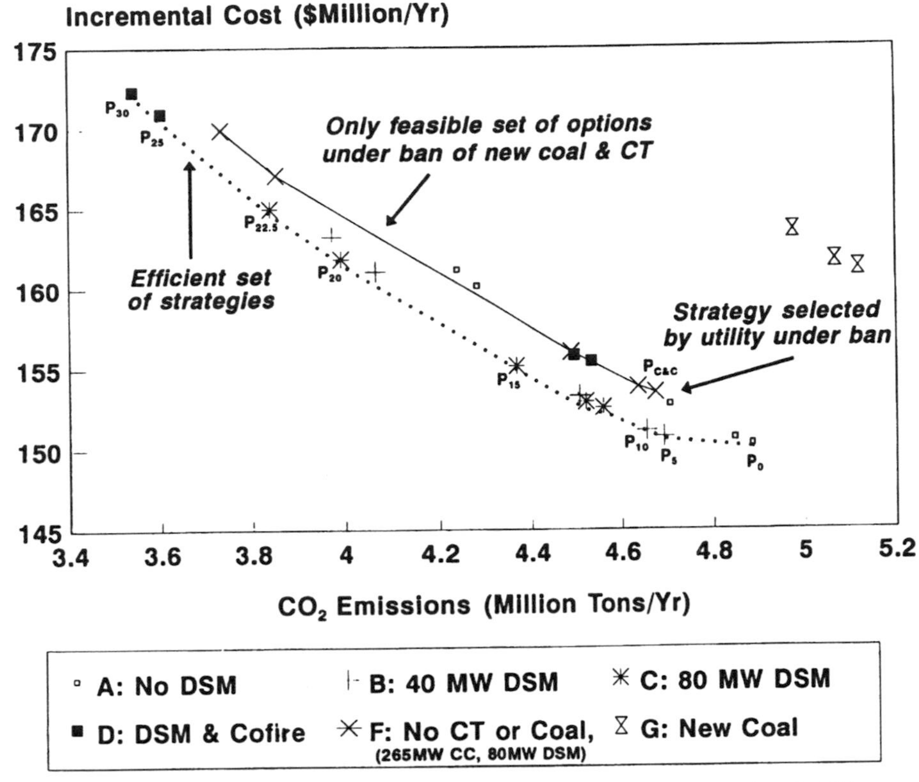

Figure 11.6 Inefficiency of a ban on new coal and combustion turbine facilities.

By focusing on individual supply resources, fuels, and emissions controls, the command-and-control approach ignores the environmental control benefits of system-wide strategies such as DSM and emissions dispatch. As a result, superior solutions that have lower costs and emissions may be eliminated prematurely. A number of such points exist southwest of $P_{C\&C}$. For instance, plan P_{10} in Figure 11.6 represents a combination of DSM programs and emissions dispatch together with some combustion turbine additions. The new turbines lower the system's cost but also make the plan illegal in this case. The utility could compensate for the higher emissions rate of the turbines through dispatch and DSM, but command-and-control regulations do not permit consideration of that strategy. The result is that the utility will choose $P_{C\&C}$ instead, which unfortunately gives both higher costs *and* more emissions than point P_{10}.

An unfortunate consequence of command-and-control regulation is that even if the environmental damages of CO_2 are high, the utility has no incentive to use dispatch and DSM to reduce emissions beyond what the law requires. This is true even if the cost of doing so is small compared to the damages those emissions cause. Thus the utility's choice under command-and-control regulation is unlikely to minimize total social cost.

11.D.3. Emissions Cap

Another philosophy of environmental regulation, which involves imposing a cap on total emissions, has been called *environmental least-cost utility planning* (Brick and Edgar, 1990). This approach specifies the problem as follows: The utility is to minimize its internal cost, subject to a constraint on overall emissions. For instance, Figure 11.7 shows the effect of imposing of a cap of 4.4 million tons per year of CO_2, 10% below the least-cost level. It would motivate the utility to choose point P_{15}, as all the lower-cost points to the right of the constraint are rendered infeasible by the cap.

The advantage of emissions caps over the command-and-control approach is that the utility can freely choose from any combination of supply resources, emissions controls, DSM programs, and dispatch strategies to meet the constraint. Unlike command-and-control regulations, a nondominated point will always result from this process, at least in theory. If the cap happens to be set at the least social cost point, this policy is also optimal.

A variation of the cap approach is to constrain the system's average emissions *rate,* measured in terms of kg/kWh or lb/MBTU of heat input. Con-

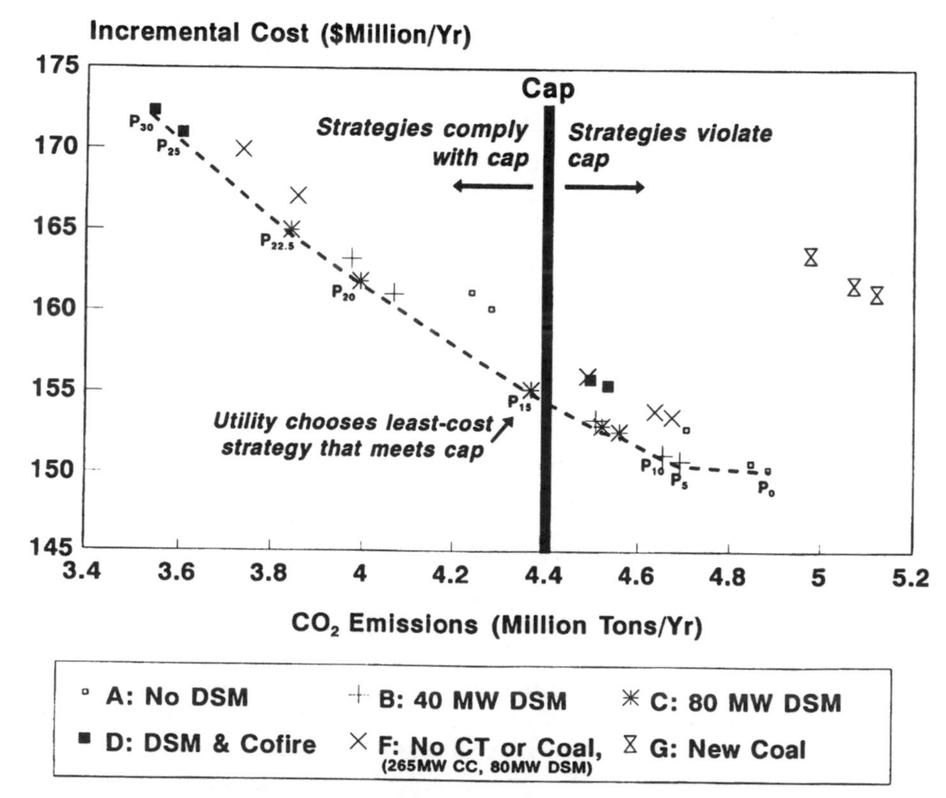

Figure 11.7 Effect of emissions cap upon plan choice.

straining a rate rather than total emissions can yield operating difficulties (Hess et al., 1992). Dominated solutions can also result (Leppitsch and Hobbs, 1996), especially if the limit is imposed only on the emissions rate of a subset of units or if there exist significant power trading opportunities with other utilities. No incentive is given for DSM or to shift generation to less sensitive areas by, for instance, power purchases.

11.D.4. Market-Based Systems: Taxes and Emissions Rights/Allowances

External environmental costs can be internalized by imposing emissions taxes or by creating marketable rights to pollute. The effect of taxes and marketable rights is to convert external costs of emissions into an internal cost from the utility's perspective. For instance, let us assume that a CO_2 tax *PEN* of \$15 per ton is levied on our hypothetical utility's emissions. The utility will minimize its costs by minimizing "Cost," the sum of its capital, fuel, and variable costs [equation (11.1a), shown as the *Y* axis of the figures] plus \$15 $\times$ CO_2, where CO_2 is its emissions [equation (11.1b), the *X* axis]. This process is shown in Figure 11.8, in which isoquants of the quantity "cost + \$15 $\times$ CO_2" are shown as straight lines running northwest to southeast. Isoquants lying to the southwest represent lower costs. The solution lying on the lowest such isoquant is the plan that minimizes the utility's total cost. In Figure 11.8, this is point P_{15}. No point has a lower total cost when the CO_2 penalty of \$15 per ton has been included.

The same decision process results if, instead of a tax, the utility must secure emissions allowances whose market price is \$15 per ton. Even if the utility is granted more than enough allowances to cover its emissions (as some are under the 1990 U.S. Clean Air Act SO_2 provisions), the fact that it could sell them at that price means that emissions should still be valued at \$15 per ton when making investment and operating decisions.

Like the emissions cap approach but unlike command-and-control regulations, the strategy chosen is in theory nondominated because the benefits of *all* options for reducing emissions, including DSM and dispatch, are recognized. If the tax penalty/allowance price happens to equal the marginal damage of emissions, the solution is also socially optimal.

11.D.5. Adders and Other Requirements for Considering External Costs

Another means of internalizing environmental costs is for the government to require that the utility estimate and consider external costs when making resource acquisition or operation decisions. Environmental impact statements are examples of such requirements. Several states have gone further by specifying particular numerical "adders" to be included in the decision calculus. The utility does not actually pay these costs, unlike the tax or emissions

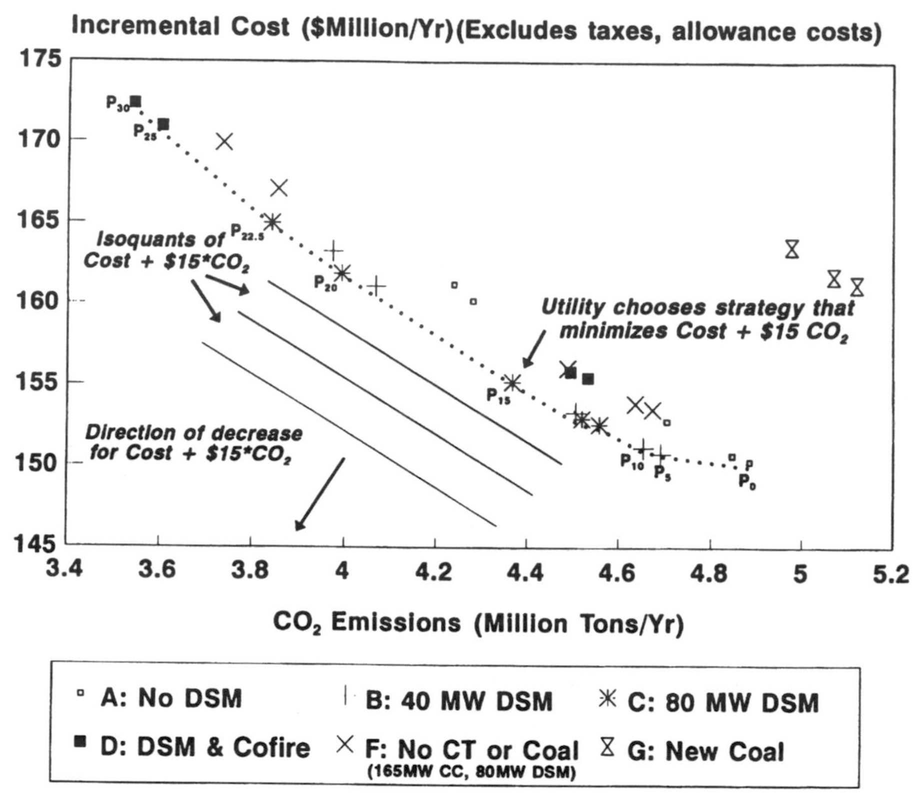

Figure 11.8 Effect of CO_2 tax of $15 per ton upon plan choice.

allowance systems; it is merely forced by regulation to make decisions *as if* it does. The most common version of this requirement in the United States applies only to decisions concerning resource acquisition (Wiel, 1991). For our hypothetical utility, this would mean that the "cost," including adders, of new combustion turbines, combined cycle facilities, and coal plants would be increased relative to energy conservation. In theory, however, adders could also be extended to dispatch and pricing decisions (Bernow et al., 1991; Busch and Krause, 1993).

If the utility is forced to consider external costs in all its resource operation and procurement decisions, the effect would be the same as an emissions tax (Figure 11.8). A nondominated plan would be chosen, and if the estimated external cost was an accurate estimate of actual damages, the least social cost plan would be achieved.

This happy outcome is unlikely to occur, however, if external costs are only factored into resource procurement decisions and not into operation. The environmental costs of capacity expansion and DSM programs would be considered, but the utility would dispatch its resources to minimize its own internal cost. This is inefficient because, as we pointed out earlier, nondominated alternatives almost always include some degree of emissions dispatch. These inconsistent incentives can lead to inefficient decisions, such as the adoption of expensive pollution controls when changes in dispatch order would accomplish the same emissions reductions at less cost. Thus, just like command-and-control regulations, this policy can result in the choice of a dominated alternative over a superior one (Andrews, 1992a; Palmer et al., 1995).

As another example of such an inefficiency, uneconomic life extension of coal units might be encouraged because such decisions would not be subject to the adders system. Consider three resource options that might be compared by our hypothetical utility:

- Construction of 145 MW of combined cycle capacity
- A load control program that clips 40 MW off the system peak, paired with 86 MW of energy efficiency-type DSM
- Repowering of a 145-MW coal-fired unit that would otherwise be retired

Each option is paired with 212 MW of new combustion turbines. The amount of each resource is chosen so that the system achieves a 15% reserve margin.

Table 11.6 summarizes the costs and emissions of each resource based on how they would be dispatched in our hypothetical system. Under least-cost dispatch, the combined cycle unit would be cycled (capacity factor = 0.29), while the repowered coal unit would instead be base loaded (capacity factor = 0.85). In a typical resource "screening" study, the costs and emissions directly associated with each resource would be tallied and compared, often without rigorous consideration of the system-wide effects. In particular, the different resources might be compared on a $ per MWh basis, with the following definitions of the $ numerator and MWh denominator:

- The $ numerator would be either the utility cost (which excludes CO_2 costs) or total societal cost (which includes those costs) directly associated with the investment and operation of the new resource. Changes in the dispatch and resulting variable costs of existing resources are ignored.
- The MWh denominator would be the anticipated energy output of just the resource (if a new supply) or the energy savings (if DSM). Again, we disregard changes in the operation of existing resources.

For the sake of argument, we assume in the societal cost calculation that the external cost of CO_2 emissions is $30 per ton but that no emissions dispatch takes place. Under these assumptions, repowering has the lowest cost ($ per MWh) from the utility's perspective, costing $43 per MWh. But from society's point of view, the CO_2 penalty causes repowering's cost to jump to $76 per MWh; in that case, the DSM program is the least expensive, costing $70 per MWh. In contrast, the combined cycle plant is apparently the most

TABLE 11.6 Per-Unit Utility and Societal Cost Calculations: Three-Resource Comparison

Resource Option	Annualized Capital Cost ($/yr)	Annual Variable Cost ($/yr)	CO_2 Produced (tons/yr)	Energy Produced or Saved (MWh/yr)	Utility Cost of Option ($/MWh)	Social Cost of Option (at $30/ton) ($/MWh)
145-MW combined cycle, gas fired	$13,900,000	$13,400,000	176,000	368,000	$74.2	$88.5
126-MW DSM	$27,000,000	$0	0	385,000	$70.1	$70.1
Repowering of 145-MW coal plant	$24,100,000	$22,700,000	1,195,000	1,080,000	$43.3	$75.7

costly program from either perspective, with a utility cost of $74 per MWh and a social cost of $89 per MWh.

Yet these results are misleading because those resources are used very differently. Table 11.7 presents the resulting costs and emissions for the *entire system.* A comparison of the total system results reveals that the combined cycle unit yields the *lowest* system-wide cost for the utility, mainly because of its dispatching flexibility. The DSM option is most expensive. So, in the absence of an adder or tax upon CO_2 emissions, the utility would recommend construction of the combined cycle unit.

But if a $30 adder were applied to CO_2 emissions from just new resources, and not repowering of existing units, the decision would be different. Regulators would conclude that DSM was definitely cheaper—whether by the $/MWh or total social cost figure (which includes a $30 per ton CO_2 adder) (see the last columns of Tables 11.6 and 11.7). As a result, the regulators would prohibit construction of the combined cycle plant. In that case the utility would decide instead to extend the life of the coal unit, since that decision is not subject to the adders system and has the next-lowest utility system cost. If that happens, Table 11.7 shows that the effect of the adders policy is adoption of an option (repowering) that yields both higher costs *and* higher emissions than would have been the case if there was no adders policy (which would instead result in selection of a combined cycle unit). Although this example is contrived, it does illustrate the possibility that adders applied only to new resource additions can make matters worse from both an economic and environmental perspective.

11.D.6. Effect of Price-Sensitive Demand

In the foregoing comparisons of environmental planning approaches, we assumed that changes in electricity prices would not significantly alter the consumers' use of other goods and services that also have external costs. This allowed us to focus on the utility's emissions. However, because electricity is in competition with other fuels, this assumption may be inaccurate. For instance, if considering external costs causes changes in resource choices or dispatch policies that increase electricity rates, less electricity will be demanded. This may be beneficial from a societal perspective if the social cost of providing electricity exceeds its price to the consumer. However, if consumers substitute other fuels for electricity, the net environmental effects of resource choices might be considerably different than just the changes in the utility's emissions (Hobbs, 1994). Indeed, a policy designed to lower utility emissions could, in theory, increase emissions from all energy sources combined.

This problem arises because the external environmental costs of competing fuels, such as natural gas, oil, and wood, are not accounted for their market prices or by the consumer. For example, increases in electric rates might motivate more "bypass" by industrial customers, who would generate their

TABLE 11.7 Total Utility System Cost and Societal Cost Calculations: Three-Resource Comparison

Resource Option	Utility Cost for Entire System ($/yr)	System CO_2 Produced (tons/yr)	Adjusted System Cost (Including CC CO_2 costs) ($/yr)	System Societal Cost (at $30/ton) ($/yr)
145-MW combined cycle, gas fired	150,300,000	4,880,000	155,600,000	296,700,000
126-MW DSM	153,600,000	4,660,000	153,600,000	293,400,000
Repowering of 145-MW coal plant	152,000,000	5,090,000	152,000,000[a]	304,700,000

[a]Excludes coal plant CO_2 because those emissions are not subject to the CO_2 adder.

own power using natural gas rather than take power off the grid. This could yield a net increase in regional NO_x and CO_2 emissions if the customer's equipment has higher emissions rates than the utility's power plants. Another example is the encouragement that rate hikes would give to wood burning in areas that already have problems with particulates and volatile organic compounds (Hobbs, 1994).

The implication of competition with other polluting fuels is that plots such as those in this chapter might not tell the full story about environmental impacts. As a consequence, our conclusions about the ability of the various policies to achieve the social least-cost solution may be misleading or even dead wrong. We advise utility planners to be aware of the potential importance of this issue. In circumstances in which there is significant competition among fuels and where inclusion of externalities will alter rates, planners should attempt to estimate the effects of those rate increases upon consumer fuel choices using demand forecasting models or comprehensive energy market models that account for price elasticities and fuel substitution. Then, using information on the environmental effects of various fuels (Ottinger et al., 1990), a rough estimate of the increase in emissions due to use of competing fuels can be obtained and its potential significance assessed (as in Hobbs, 1994).

11.E. CONCLUSIONS

Environmental considerations are important in utility planning and operations. In this chapter we presented a simple optimization model for examining the implications of long-run supply- and demand-side resource choices for environment–economic trade-offs. This general type of model is widely used within the industry. The reader is urged to examine any of a number of examples of the application of this approach (e.g., Andrews and Conners, 1990; Northwest Power Planning Council, 1991; Andrews, 1992b; Busch and Krause, 1993; Meier, 1995; Palmer et al., 1995). Other models for utility environmental planning address more detailed implementation issues, especially facility siting and short-term operations.

At least with respect to pollutants that mix uniformly over broad geographic areas, such as SO_2, NO_x, greenhouse gases, and some air toxics, the analyses of this chapter show that market-based environmental regulation offers the greatest potential for minimizing the expense of achieving emission reductions. By contrast, externality adders provide limited near-term benefits, no direct incentive to reduce emission control costs, and under some conditions, can worsen both utility costs and environmental impacts.

ACKNOWLEDGMENTS

Partial support was provided by the National Regulatory Research Institute (NRRI), sponsored by the National Association of Regulatory Utility Commissioners (NARUC). The author benefited from collaboration with K. Rose and P. Centolella and comments by P. Meier. Portions of Sections 11.C and 11.D are reprinted from Hobbs and Centolella (1995) with kind permission from Elsevier Science, Ltd., The Boulevard, Langford Lane, Kidlington, 0X5 1Gb, UK. An earlier version of that material appeared in the *1992 Summer Study Proceedings* of the American Council for an Energy Efficient Economy, and is reprinted with permission. Opinions expressed are the author's and not necessarily those of NRRI or NARUC.

EXERCISES

11.1. Imagine that you are a utility planner for a combined electric-natural gas utility. Your objective might be to minimize the cost of providing energy services (heat, light, etc.) to consumers and to consider environmental impacts appropriately. Consider energy end uses for which electricity and natural gas compete, such as water heating and space conditioning. How might the net environmental effects of using electricity versus natural gas for these uses be estimated and compared? With what policies might a utility regulator encourage use of the least cost fuel from society's point of view? Or should electricity and gas be allowed to compete freely without regulatory interference?

11.2. Consider the formulation of the water and air quality models of Chapters 2, 4, and 15. How might constraints on ambient concentrations of, say, SO_2 be incorporated into the resource planning model of this chapter?

11.3. Set up and solve a LP model for the following problem, using any convenient package. In the year 2010, the peak demand in country X is expected to be 1200 MW, with a load factor of 0.6. (Approximate the load-duration curve with a two-block representation, with block $t = 1$ lasting $H_1 = 1752$ hr/yr and load $LOAD_1 = 1200$ MW. This information, together with the load factor—defined as the ratio of the average to peak load—is sufficient to define $LOAD_2$.) The existing plants and potential new generating units have the following characteristics:

Plant	Capacity (MW)	CF_i	FOR_i	HR_i (10^6 BTU/MWh)	FC_i (\$/10^6 BTU)	CX_i (\$/MW)
Existing oil steam	400	0.7	0.1	10	5	—
Existing diesel	100	0.8	0.2	13	7	—
Existing hydro	50	0.35	0	0	0	—
New coal steam	Decision variable	0.65	0.15	10.5	2	1,200,000
New combustion turbine	Decision variable	0.8	0.1	12	7	400,000

A 20% reserve margin must be maintained. Assume a capital recovery factor of 0.09 per year. Assume that each fuel has the emissions rate shown in Table 11.2. Assume that the diesel unit burns oil. Calculate the emissions resulting from solving the model above assuming that the objective is to minimize the sum of capital and operating (fuel) costs. Then solve the model again, assuming a tax of $15 per ton for CO_2. How does the solution change (costs, emissions, and plant additions)?

REFERENCES

Adamson, S., R. Bates, R. Laslett, and A. Potoschig, 1996. *Energy Use, Air Pollution, and Environmental Policy in Krakow: Can Economic Incentives Really Help?* Technical Paper 308, Energy Series, World Bank, Washington, DC.

Andrews, C. J., 1992a. The Marginality of Regulating Marginal Investments: Why We Need a Systemic Perspective on Environmental Externality Adders, *Energy Policy,* 20(5), 450–463.

Andrews, C. J., 1992b. Spurring Inventiveness by Analyzing Tradeoffs: A Public Look at New England's Electricity Alternatives, *Environmental Impact Assessment Review,* 12, 185–210.

Andrews, C. J., and S. R. Conners, 1990. Cost-Effective Emissions Reductions: Combining New England's Options into a Coordinated Strategy, Energy Laboratory, Massachusetts Institute of Technology, Cambridge, Mass., presented at the 12th Annual North American Conference of the International Association for Energy Economics, Ottawa, October 2.

Bernow, S., B. Biewald, and D. Marron, 1991. Full-Cost Dispatch: Incorporating External Costs in Power System Operations, *Electricity Journal,* March 20–33.

Bloom, J. A., M. Caramanis, and L. Charny, 1984. Long-Range Generation Planning Using Generalized Benders' Decomposition: Implementation and Experience, *Operations Research,* 32, 290–313.

Brick, S., and G. Edgar, 1990. Blunting Risk with Caution: The Next Step for Least-Cost Planning, *Electricity Journal,* July, 56–63.

Burke, W. J., et al., 1988. Trade Off Methods in System Planning, *IEEE Transactions on Power Systems,* 3(3), 1284–1290.

Busch, J. F., and F. L. Krause, 1993. Environmental Externality Surcharges in Power System Planning: A Case Study of New England, *IEEE Transactions on Power Systems,* 8(3), 789–795.

Clarke, D., and J. Parmelee, 1994. DSM and System Emissions in New York State, in *Proceedings of the 3rd International Energy Efficiency and DSM Conference: Charting the Future,* Synergic Resources Corp., Bala Cynwyd, PA, pp. 131–143.

Cohon, J., 1978. *Multiobjective Programming and Planning,* Academic Press, San Diego, CA.

Cohon, J., et al. 1980. Application of a Multiobjective Facility Location Model to Power Plant Siting in a Six-State Region of the U.S., *Computers and Operations Research,* 7, 107–123.

Crousillat, E. O., et al., 1993. Conflicting Objectives and Risk in Power Systems Planning, *IEEE Transactions on Power Systems,* 8(3), 887–893.

Economides, S., and M. Sharifi, 1978. Environmental Optimization of Power Lines, *J. of the Environmental Engineering Division, ASCE,* 104(4), 675–684; discussions by B. F. Hobbs, R. L. Church, and T. J. Clifford, 105(4), 794–795.

1996. *PROVIEW/PROSCREEN.* Electronic Data Systems, Utilities Division, Atlanta, GA.

Ellis, J. H., 1988. Acid Rain Control Strategies, *Environmental Science and Technology,* 22(11), 1248–1255.

Fragniere, E., and A. Haurie, 1995. MARKAL-Geneva: A Model to Assess Energy-Environmental Choices for a Swiss Canton, *Proceedings, IEA Energy Technical Systems Analysis Programme/Annex V,* 5th Workshop, Gothenburg, Sweden, Netherlands Energy Research Foundation.

Gjengedal, T., O. Hansen, and S. Johansen, 1992. Qualitative Approach to Economic-Environmental Dispatch; Treatment of Multiple Pollutants, *IEEE Transactions on Energy Conversion,* 7(3), 367–373.

Gronheit, P. E., 1993. Modeling CHP Within a National Power System, *Energy Policy,* 21(4), 420–428.

Heslin, J. S., and B. F. Hobbs, 1991. A Probabilistic Production Costing Analysis of SO_2 Emissions Reduction Strategies for Ohio: Emissions, Cost, and Employment Tradeoffs, *Journal of the Air and Waste Management Association,* 41(7), 956–966.

Hess, S. W., et al., 1992. Planning System Operations to Meet NO_x Constraints, *IEEE Computer Applications in Power,* 5, 10.

Hirst, E., and C. Goldman, 1991. Creating the Future: Integrated Resource Planning for Electric Utilities, *Annual Review of Energy,* 16, 91–121.

Hobbs, B. F., 1979. *Analytical Multiobjective Decision Methods for Power Plant Siting: A Review of Theory and Applications, NUREG/CR-1687,* Brookhaven National Laboratory, Upton, NY; prepared for the U.S. Nuclear Regulatory Commission.

Hobbs, B. F., 1994. Emission-Cost Tradeoffs and Rate Feedback for Electric Utilities, *ASCE Journal of Energy Engineering,* 120(3), 103–121.

Hobbs, B. F., and P. Centolella, 1995. Environmental Policies and Their Effects on Utility Planning and Operations, *Energy,* 20(4), 255–271.

Hobbs, B. F., and P. Meier, 1979. An Analysis of Water Resources Constraints on Power Plant Siting in the Mid-Atlantic States, *Water Resources Bulletin,* 15(6), 1666–1676.

Hobbs, B. F., and P. Meier, 1994. Multicriteria Methods for Resource Planning: An Experimental Comparison, *IEEE Transactions on Power Systems,* 9(4), 1811–1817.

Huang, W., and B. F. Hobbs, 1992. Estimation of Marginal System Costs and Emissions of Changes in Generating Unit Characteristics, *IEEE Transactions on Power Systems,* 7(3), 1251–1258.

Huang, W., and B. F. Hobbs, 1994. Optimal SO_2 Compliance Planning Using Probabilistic Production Costing and Generalized Benders' Decomposition, *IEEE Transactions on Power Systems,* 9(1), 174–180.

Hunt, S., and G. Shuttleworth, 1996. *Competition and Choice in Electricity,* Wiley, New York.

Jones, D. E., and P. Hanser, 1991. *Environmental Externalities: An Overview of Theory and Practice,* Report EPRI CU/EN 7294, Electric Power Research Institute, Palo Alto, CA.

Keeney, R. L., 1980. *Siting Energy Facilities,* Academic Press, San Diego, CA.

Kuloor, S., G. S. Hope, and O. P. Malik, 1992. Environmentally Constrained Unit Commitment, *IEE Proceedings C,* 139(2), 122–128.

Lawrence, D., 1989. Integrated Demand-Side Management into Least-Cost Resource Planning, *Demand-Side Management Strategies for the 90s,* EPRI CU-6367, Electric Power Research Institute, Palo Alto, CA, Chapter 40.

Leppitsch, M. J., and B. F. Hobbs, 1996. The Effect of NO_x Regulations on Emissions Dispatch: A Probabilistic Production Costing Analysis, *IEEE Transactions on Power Systems,* 11(4), 1711–1717.

Makarova, A. A., 1966. Optimum Power Systems Expansion Using Mathematical Models, *Proceedings, World Power Conference,* Paper 124, Vol. IIA(2), p. 1911.

Meier, P. M., 1979. Long-Range Regional Power Plant Siting Model, *Journal of the Energy Division, ASCE,* 105(1), 117–135.

Meier, P. M., 1995. *Resource Trade-off Decision Analysis for BC Hydro's 1995 Integrated Electricity Plan,* prepared for the BC Hydro Planning Integration and Consultation Department by IDEA, Inc., Washington, DC.

Northwest Power Planning Council, 1991. *1991 Northwest Conservation and Electric Power Plan,* Portland, OR.

ORNL and RFF (Oak Ridge National Laboratory and Resources for the Future), 1994. *Estimating Fuel Cycle Externalities: Analytical Methods and Issues,* Report 2, McGraw-Hill/Utility Data Institute, Washington, DC.

Ortolano, L., 1984. *Environmental Planning and Decision Making,* Wiley, New York.

Ottinger, R. L., et al., 1990. *Environmental Costs of Electricity,* Oceana Publications, New York.

Palmer, K. L., A. J. Krupnick, and H. Dowlatabadi, 1995. Social Cost of Electricity in Maryland: Effects on Pollution, Investment, and Prices, *Energy Journal,* 16(1), 1–26.

Rose, K., P. A. Centolella, and B. F. Hobbs, 1994. *Public Utility Commission Treatment of Environmental Externalities,* Report NRRI-94-10, National Regulatory Research Institute, Columbus, OH.

Stoll, H. G., et al., 1989. *Least Cost Electric Utility Planning,* Wiley, New York.

Stone and Webster, 1991. *EGEAS: Electric Generation Expansion Analysis System,* Version 7, Stone and Webster Management Consultants, Englewood, CO.

Talaq, J. H., F. El-Hawary, and M. E. El-Hawary, 1994. A Summary of Environmental/Economic Dispatch Algorithms, *IEEE Transactions on Power Systems,* 9(3), 1508–1516.

Talukdar, S. N., and F. F. Wu, 1981. Computer-Aided Dispatch for Electric Power Systems, *Proceedings of the IEEE,* 69(10), 1212–1231.

Torrens, I. M., J. E. Cichanowicz, and J. B. Platt, 1992. The 1990 Clean Air Act Amendments: Overview, Utility Industry Responses, and Strategic Implications, *Annual Review of Energy and the Environment,* 17, 211–233.

Turvey, R., and D. Anderson, 1977. *Electricity Economics: Essays and Case Studies,* Johns Hopkins University, Baltimore, MD, Chapter 13.

Wiel, S., 1991. The New Environmental Accounting: A Status Report, *Electricity Journal,* 4(9), 46–54.

Woo, C. K., B. F. Hobbs, R. Orans, R. Pupp, and B. Horii, 1994. Emission Costs, Customer Bypass, and Efficient Pricing of Electricity, *Energy Journal,* 15(3), 43–54.

Wood, A. J., and B. W. Wollenberg, 1995. *Power Generation Operation and Control,* 2nd ed., Wiley, New York.

CHAPTER 12

Multiobjective Methods

JARED COHON and KRISTINA ROTHLEY

12.A. INTRODUCTION

Multiple criteria decision making (MCDM) and multiobjective programming (MOP) are vibrant areas of research and application. Multiobjective problems appear in virtually every field and with a wide variety of contexts. As an analytical challenge, multiobjective problems are especially attractive to researchers for their integration of human decision making and political dimensions with economic and physical aspects. There is, as a result, a wide variety of MCDM and MOP problem-solving methods.

We present here four established and widely used techniques: two MCDM methods (the simple multiattribute rating technique and the analytical hierarchy process) and two MOP methods (the weighting method and the constraint method). Interested readers can explore these and other techniques further in the many good texts and monographs that have been published in the last 20 years. Zeleny (1982) provides a somewhat dated but very good overview of the entire multiobjective field. Chankong and Haimes (1983) provide a comprehensive and rigorous treatment. Cohon (1978) focuses on MOP and its applications at the mathematical level found in this volume. Steuer (1986) covers similar ground but with a more theoretical treatment and with much more detail on the multiobjective simplex algorithm. Keeney and Raiffa (1976) is the award-winning source of the main ideas in multiattribute value and utility theory and provides complete coverage of the theory and techniques of this area. The analytical hierarchy process came to the fore somewhat later than the books mentioned above, but the method has spawned many papers and a few books. Saaty (1977, 1990, 1994), the creator of AHP, provides a detailed and comprehensive account of the technique and his views on decision making.

The proceedings of the international conferences on MCDM are helpful for keeping up with the latest multiobjective techniques. Held nominally every other year—the most recent were in Hagen, Germany, in June 1995 and Capetown, South Africa, in January 1997—these conferences have provided an excellent way to gauge what is happening in MCDM and MOP. Proceedings have been published more or less every other year since 1972. Another good source is the relatively new *Journal of Multiple Criteria Decision Analysis.*

We conclude the chapter with example applications that highlight the utility of these methods. Elsewhere in this book, problem formulations are presented that have multiobjective features. A multiobjective problem is discussed in water supply engineering in Chapter 1 (Section 1.H), and a multiobjective trade-off problem in water quality management in Chapter 2 (Section 2.D.3). The air pollution models in Chapters 4 and 11 (Section 11.C.1) are formulated in a multiobjective context, as are hazardous waste facility location problems in Chapter 6 (Section 6.D.3). Finally, a multiobjective transportation route design problem appears in Chapter 9 (Section 9.C.3).

12.B. NEED FOR MULTIOBJECTIVE TECHNIQUES

It is a rare and blessed event when we are faced with a decision for which we need consider only one goal. Not that any variety of decision problems is necessarily easy. Just ask the engineer tasked with designing a bridge span with the single design goal of minimizing cost. Such a project requires an enormous investment of time, labor, and complex analysis. As challenging as such problems are, however, single-objective problems are generally more easily addressed than are their multiobjective counterparts.

Given the complexity of the world and the variety of interests that may have a stake in most decisions, it is more often the case that we are asked to make decisions that will satisfy several, potentially conflicting, interests. The importance and magnitude of a decision and the extent of its impact may be a loose indicator of the number of objectives that must be considered. For example, when selecting a trucking route for distribution of, say, consumer goods, it may be sufficient to assume that cost is the only objective that needs to be considered (although even in this case, one can easily think of other objectives, e.g., time and the probability of major delays). However, when determining the alignment of a new highway, we are dealing with a decision that has a potentially large impact on many people. Here, air quality, increased noise impacts, dislocation of structures and people, and filling of wetlands all represent potential objectives in addition to cost. Furthermore, it is unlikely that any highway alignment can optimize all the objectives simultaneously, as the objectives may conflict. For example, the least-cost alignment, which will tend to be a straight line, may traverse many wetlands and be close to many residential areas. Avoiding those impacts would require a longer, and therefore more expensive, alignment. Multiobjective techniques have been developed to deal with decision-making problems such as this.

Multiobjective decision problems are ubiquitous, but civil and environmental engineering may represent areas in which multiple, conflicting objectives are especially common. Environmental and natural resource management problems are, by nature, multiobjective, pitting environmental quality against economic considerations. Water resources problems almost always exhibit conflicts between upstream and downstream interests, as well as other objectives. Infrastructure systems are public works that necessarily bring forth multiple constituencies and stakeholders whose objectives are usually in conflict. Even structural problems, which may appear to be amenable to a relatively narrow, technical analysis, are often multiobjective. The trade-off between safety and cost in structural engineering is a classic multiobjective problem.

12.B.1. Benefit–Cost Analysis and Multiobjective Problems

Prior to the advent of multiobjective analysis, the standard approach to dealing with multiple objectives was to identify a single fundamental objective and then to convert all other objectives into this single currency. This is the basic logic behind *benefit–cost analysis,* which has been the dominant analytical tool for civil and environmental problems for 60 years. Drawbacks and benefits associated with an alternative may exist in many forms. But if all drawbacks can be converted into monetary costs and all benefits can be converted into monetary gains, we can look for the solution that maximizes the overall net benefit function. Benefit–cost analysis has been and continues to be applied successfully in many situations, but there are problems with benefit–cost analysis that limit its utility.

The fundamental difficulty becomes clear as soon as we try to apply benefit–cost analysis in a multiobjective situation. There are a large number of desirable and important objectives that are not easily translated into dollars: for example, human health and life, environmental impacts, convenience, and aesthetic quality. These objectives may rank just as highly as economic benefits, yet cannot be compared directly with profits because they cannot be converted into common units. Even when they can be converted, the value judgments necessary to perform the conversions may be unacceptable or premature.

Consider the case of human health impacts from a new fossil-fueled power plant to be sited by a utility. Suppose that there is a quantifiable risk of additional lung cancer due to the air emissions from the plant. Considering these impacts in a benefit–cost analysis would require a monetary value to be associated with this increased risk of lung cancer. One approach to doing this is to estimate the cost of treating lung cancer and the loss of income due to disability and premature death. This strictly economic interpretation of health impacts does not capture the loss of quality of life, pain and suffering, or the inescapably political dimensions of trading off facility costs and benefits against lives. These represent noneconomic dimensions of value that can only be weighed against economic benefits by value judgments articulated by decision makers. However, it may be undesirable to elicit and incorporate value judgments as part of the analysis. Decision makers may be unwilling or unable to describe their preferences exactly. This passes the burden on to the analysts, who should avoid incorporating their own preferences into a decision solution. Furthermore, decision makers should know the full range of alternatives and understand the trade-offs among the objectives implied by each alternative before making their selection. Multiobjective techniques recognize these difficulties and provide mechanisms for dealing with them in practice.

12.B.2. What Multiobjective Techniques Can Do

Based on the discussion above, it is clear that there is a need for quantitative approaches to handle decision making with more than one objective. The decision-making support techniques presented in this chapter were devised to meet that need. With these methods it is possible to state and solve multiobjective decision problems. There is no requirement to convert objectives into a common currency, nor must objectives be combined into a single function. Instead, it is possible to capture with multiobjective techniques the realistic situation that decisions must be made to balance multiple, conflicting objectives, each of which may have a unique currency. The elimination of a need for a common currency has an associated implicit benefit. Analysts and decision makers are able to focus their efforts on creating an accurate and complete representation of the decision at hand. No objectives need to ignored because they cannot be converted into a common currency, nor do value judgments have to be introduced before decision makers develop insight into the alternatives, their choices, and the implied trade-offs. The exercise of creating and revising a decision model that includes all the relevant objectives can prove to be very informative. The formal approach of MO methods could be said to help to organize and define the problem.

12.C. SETTING THE DECISION CONTEXT: DECISION MAKERS, STAKEHOLDERS, AND OBJECTIVES

Perhaps the most difficult and important part of problem solving is understanding and setting the decision-making context. By this we mean identifying the decision makers, the other people with a stake in the outcome, and the objectives, and agreeing on and understanding how a decision will be reached. Similar to the creation of free-body diagrams in engineering mechanics, without the accurate execution of this initial step, it is impossible to obtain an accurate solution. This may seem obvious, but it is remarkable how infrequently enough time is spent on setting the decision context. Perhaps this should not be surprising. We, the analysts, are, after all, engineers, operations researchers, and economists who are trained to deal with problems within contexts set by others. The fact is, however, that big and complex problems are rarely so well ordered. To be effective, we *have* to deal with decision contexts. Multiobjective techniques were created to deal explicitly with conflict and to contribute to setting the decision context.

Solving multiobjective problems requires at least two sets of information that help to define the decision context: (1) the set of objectives and (2) the set of preferences that indicate the relative importance of the objectives, whether these are used explicitly by the analytical technique (preference-oriented techniques) or not (generating techniques). Before we begin, therefore, we must know what the objectives are and whose objectives and preferences we are interested in.

12.C.1. Whose Objectives?

Open and inclusive decision-making processes represent one of the most important recent developments in the design and management of civil structures and environmental and natural resource problems. Decision processes are "open and inclusive" when they give standing and voice to parties that have a stake in the outcome. Several recent papers have called for such processes (e.g., Ruckelshaus, 1995, for environmental problems and the National Research Council, 1995, for infrastructure systems). Other papers have presented actual cases in which open decision making was pursued (e.g., Ridgley and Rijsberman, 1992, in water resources, which is discussed in detail in Section 12.G).

There is no definitive process for identifying stakeholders or for devising ways to include them in decision making. All problems have history, and understanding their history will usually lead to insight into what's at stake and who cares about it. The Rio Colorado in Argentina offers a good and early example. A project based at MIT in the early 1970s developed optimization and simulation models to support decisions about reservoirs, power plants, and irrigation areas in this then-undeveloped river basin (Major and Lenton, 1979). The waters of the Rio Colorado had been the subject of years of conflict among the five provinces along the river. Consultation with federal government officials and a review of documents related to past disputes and agreements revealed the central conflicts between upstream and downstream provinces and between water use for hydroelectric energy generation and irrigation. This lead to appropriate multiobjective models (Cohon and Marks, 1973; Major and Lenton, 1979).

Today, problems of the magnitude of the Rio Colorado are characterized by many more stakeholders: environmental groups, commercial interests, and local governments, as well as provincial and national governments and multilateral agencies such as the United Nations and the World Bank. Situations are more complicated, but consultation and review of relevant past work are still reliable techniques. Public meetings early in the design process can also help to identify and to engage stakeholders.

Considerable effort and time may be required to structure an appropriate decision context when there is a large number of stakeholders. There are many possible ways to do this, and the proper structure must be influenced by the political and social context in which decisions will be made. The case of the Rhine River delta in the Netherlands, discussed in Section 12.G, provides one example.

12.C.2. What Are the Objectives?

Specification of the objectives is a crucial step toward any problem solution. The objectives represent what is important to decision makers and stakeholders, and they are, therefore, the basis for evaluating alternatives. Proper care and time at this stage are wise investments. As with the identification of stakeholders, there is no universally approved method for the specification of objectives. Keeney (1988) has provided the most valuable and authoritative guidance. Objectives must have certain characteristics to support the analysis. They must be meaningful in the sense of capturing what is truly important to decision makers and stakeholders, and to the extent possible they should represent ends rather than means to an end. For example, a preliminary analysis of the Rio Colorado problem (Cohon and Marks, 1973) used a measure of distributional equity of water as an objective. This was hard to explain to decision makers and did not contribute to their understanding or resolution of the conflict. Objectives must also be specific enough to permit quantification and measurement. For example, maximizing environmental quality is insufficiently specific, but maximizing dissolved oxygen in a water body is something that we can measure and that represents a good indicator of one important dimension of water quality. As a set, the objectives should capture all of the issues important to all stakeholders, and double-counting should be avoided. It is important to keep in mind that the number and selection of the objectives determine the final set of options. If an objective were to be added or deleted, an entirely new set of options may emerge. This reinforces the importance of accurate definition of objectives.

12.C.3. Objective Hierarchy

Some problems have a very large number of objectives. The *objective hierarchy* has been created as a way to elicit objectives from stakeholders and to organize them in a systematic fashion (see Edwards, 1977; Edwards and Newman, 1982). The hierarchical structure allows for the representation of the logical relationships of the objectives to each other. It may be argued that the creation of the objective hierarchy may be the most valuable step in any multiobjective process. Not only are decision makers and interest groups provided the opportunity to voice their interests, but the creation of a complete and satisfactory objective hierarchy guarantees the inclusion of all named objectives in the evaluation process. Furthermore, the overall integrity and credibility of the decision-making process is boosted, which improves the likelihood of determining and implementing an agreed-upon solution. The objective hierarchy is also helpful in communicating the considerations and solution process to persons outside the immediate decision-making group. There is no single objective hierarchy applied to all decision problems. Instead, a new hierarchy is created for every new decision problem and revised with the inclusion of every new interest group.

The case of the Rhine River delta, discussed later, offers a good example of a large objective hierarchy created for a real decision problem. We present here a simple, hypothetical example taken from a multiobjective software program called VISA (Visual Thinking, 1995) to demonstrate the concepts. Consider the problem of choosing one of three alternative automobiles: a Porsche, a Land Rover, and a Metro. As these are very different cars, we anticipate that no car will be the single best alternative on all features that we may find important in a car. Therefore, we cannot identify our preferred solution immediately and, instead, structure this decision as a multiple-criteria decision problem. To begin, we identify four car features that we feel are important to make our final selection: total cost, space, comfort, and speed. These four features form the first level of the objective hierarchy (Figure 12.1a).

We evaluate our four initial car objectives, as described in Section 12.C.2, to be sure that they are sufficiently specific to capture our real issues about choosing a car. Space and comfort pass the test. For space, the passenger compartment cubic feet estimate provided by the manufacturers is a sufficient measure, assuming that we are concerned only about passenger space and not other possible measures, such as trunk space. For comfort, we decide that our only interest is the cushioning in the driver's seat. Each driver's seat will be scored on a comfort scale of 1 to 10. However, total cost and speed need more work. Total cost will be divided into three subobjectives—capital cost, running cost, and maintenance—as we expect that the values for these subobjectives will vary greatly between the cars, and we consider these subobjectives to have different relative importance. Speed is broken down into two subobjectives: acceleration and top speed. Again, we anticipate that the values for these subobjectives vary greatly among cars. Being "speed demons," we also place a high value on top speed relative to acceleration and would like to have this preference reflected in the hierarchy. The updated hierarchy is shown in Figure 12.1b. Notice how the construction of the hierarchy preserves the relationships among objectives and subobjectives. There is no limit to the number of layers in a hierarchy. For example, capital cost, which is itself a subobjective of total cost, could have its own sub-subobjectives: say, capital cost of the standard model and capital cost of the options. This additional specification may be important to keep separate what is essential and what is, well, frivolous. Having long ago come to peace with our propensity for frivolity, we have decided not to go further with the hierarchy. We revisit this example when we discuss the various solution techniques.

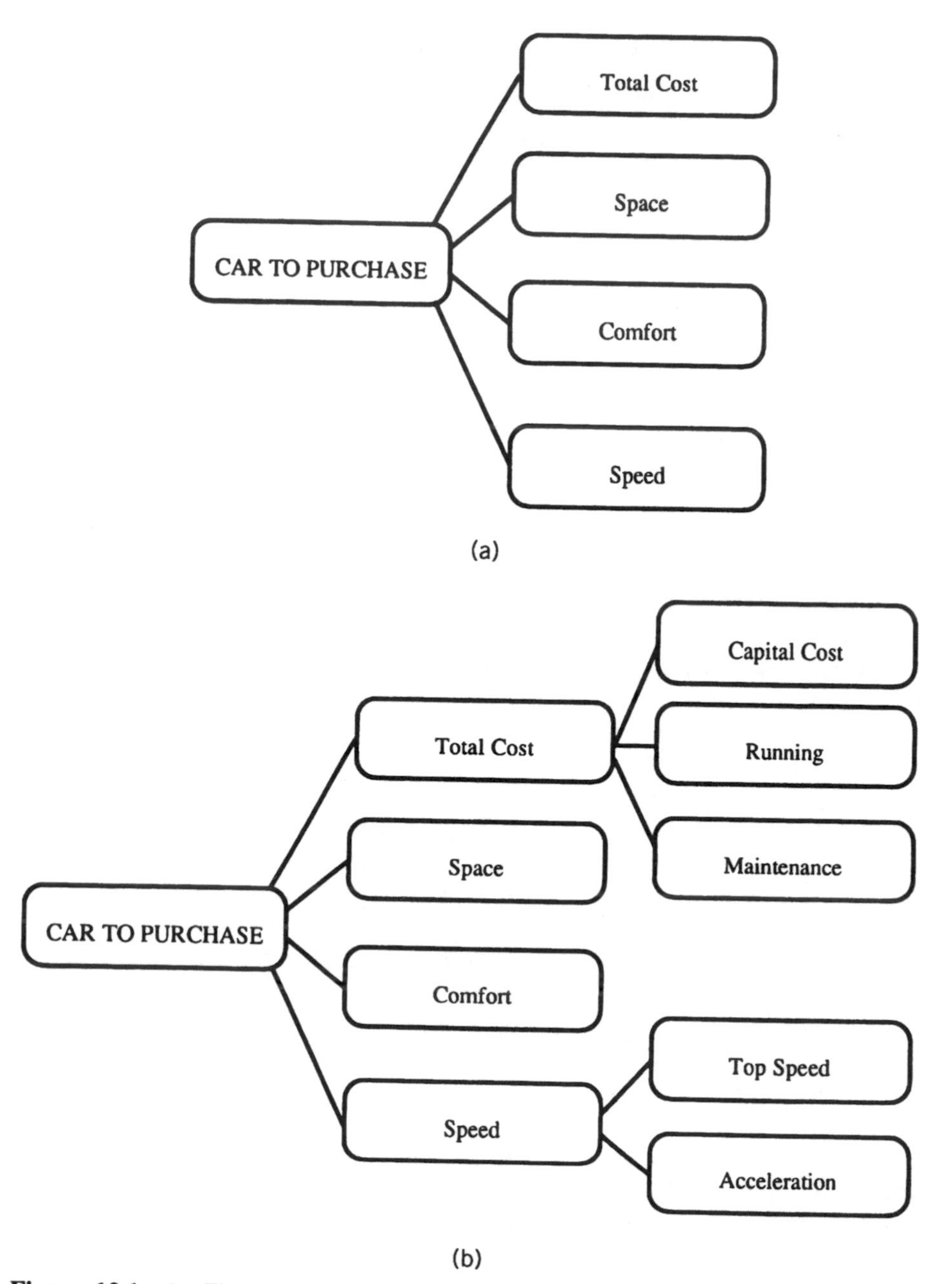

Figure 12.1 (a) First level of an objective hierarchy for the car selection problem; (b) complete objective hierarchy for the car selection problem.

Treatment of Preferences	Nature of the alternatives Predefined Set	Not Predefined
Explicitly quantified and used in the analysis	MCDM • SMART • AHP	Preference-oriented MOP techniques • Goal Programming
Not quantified, expressed after the anaylsis	Seat of the pants decision making	MOP Generating Methods

Figure 12.2 Matrix of the multiobjective methods.

12.D. TYPES OF MULTIOBJECTIVE DECISION PROBLEMS AND SOLUTION

12.D.1. Nature of the Alternatives

Before we begin to solve a decision problem, it is important to determine the sort of problem that we have. Multiobjective decision problems can be classified into two general categories: (1) problems for which the potential alternatives are predefined, and (2) problems for which the alternatives are not predefined. To understand this distinction, consider the selection of one of six potential power plant locations versus the selection of the capacity for a power plant. In the first instance we are simply selecting from a list of known alternatives. In the second case we are determining the optimal value for a variable that may take on any value within some range defined by the design constraints. Assigning a decision problem to one of these categories is not always obvious. Virtually any problem can be stated such that it falls into either category. The proper treatment may depend on the level of detail or the point in the decision process at which decision support analysis is started. As an example, consider the decision problem of locating a dam along a river. It may be more efficient and intuitive to consider the potential location alternatives as a set of discrete locations evenly spaced along the length of the river rather than as the set of all continuous numbers that can take on any value within the range that matches the length of the river.

There are separate multiobjective methods to deal with these two different categories of problems (columns of Figure 12.2). Methods to deal with decisions with known alternatives are referred to as *multiple-criteria decision making* (MCDM) methods.[1] The term *multiobjective programming* (MOP) is used to describe methods for solving decision problems in which the alter-

[1]The term *multiple-criteria decision making* is also used to refer to the complete set of analytical techniques used to handle problems with multiple objectives, both those with predefined alternative solutions and those without. For the purposes of this chapter, we use this term to describe only those techniques designed to solve decision problems for which the alternatives are predefined.

natives are not predefined. In this chapter we describe methods for dealing with each kind of problem. The analytical hierarchy process (AHP) and the simple multiattribute rating technique (SMART) are presented as forms of MCDM techniques. Multiobjective linear programming is presented as a form of MOP techniques.

12.D.2. Treatment of Preferences

Some multiobjective techniques require the solicitation and incorporation of decision makers' preferences as part of the problem analysis. For other techniques, decision makers' preferences are not considered explicitly until completion of the analysis phase of the problem solution. This distinction provides further means for categorizing multiobjective techniques (rows of Figure 12.2). It may seem natural that preferences should be quantified and incorporated into the decision-making process early on, but there are many good reasons why we would like to defer the solicitation of the decision maker's preferences until completion of the analysis phase of the problem solution. We discuss these reasons when we present the MOP methods.

Methods for which preferences are required for the analysis include SMART, AHP, and goal programming. Methods for which preferences are not used and for which the alternatives are not predefined are called *generating methods.* We refer to the case of decision making for predefined alternatives in which preferences are not elicited until completion of the analysis as "seat-of-the-pants decision making" because of the lack of formal, quantitative analysis of preferences.

12.E. MULTIPLE-CRITERIA DECISION MAKING

As described previously, multiple-criteria decision making (MCDM) refers to decision problems for which the alternatives are predefined. Here we present two widely used quantitative MCDM decision support methods: the analytical hierarchy process (AHP) (Saaty, 1977, 1994) and the simple multiattribute rating technique (SMART) (Edwards, 1971). Each has been applied successfully to a wide variety of decision problems in many disciplines. The significant differences between these techniques lay in the assumptions made about the type of information required from decision makers and the style of questioning with which this information is extracted. Neither method is clearly more appropriate for all types of decision contexts. Rather, the best way to select an MCDM technique is to have experience with the techniques, to be familiar with the decision maker, and to test the validity of the assumptions. The details of these differences are highlighted in the descriptions of the methods.

It is important to note that for both of the MCDM methods described here, a decision maker's preferences are an integral part of the solution process

(Figure 12.2). This explicit incorporation of decision-maker preferences is not required for the MOP techniques described later. It is also important to note that software packages are available that incorporate the logic and analysis of AHP and SMART. Perhaps the most significant benefit of these software packages is that they increase the accessibility of MCDM to unfamiliar users. They also increase the efficiency of the methods. The visual displays of the information improve the decision maker's understanding of the problem. Finally, these packages automate the process of sensitivity analysis, which could arguably be the most important part of the process. References for the software are included with the MCDM method descriptions. However, given the frequency with which new software is published, a quick review of the most current offerings is recommended. The newsletter of the organization INFORMS publishes software reviews that frequently include decision analysis software.

12.E.1. Analytical Hierarchy Process

The analytical hierarchy process is a mix of quantitative and qualitative decision support methodologies created to solve decision problems with predefined alternative solutions. Keeping in the theme of methods described in this chapter, the AHP methodology proceeds without the conversion of objectives into a common unit or the creation of unitless indices. Instead, through AHP it is possible to order an entire set of alternatives from knowledge of only pairwise preferences (Ridgley and Rijsberman, 1992). The AHP has gained widespread use across many disciplines. The broad appeal of AHP may be derived both from its ease of use and the excellent software packages available to assist in the application of AHP to decision problems.

12.E.2. Using AHP

We will use our car example and a decision support software package called Hipre (Systems Analysis Laboratory, 1993) to demonstrate the AHP method. The AHP employs *pairwise comparisons* to determine the relative weights or priorities of the decision maker for the objectives and ultimately the alternatives. The AHP starts with a completed objective hierarchy. For each pair of objectives on the same branch of every level of the objective hierarchy, the decision maker is asked to indicate the intensity of his or her preferences for one objective with respect to another in the form of a positive number. For example, the analyst would ask "please give me a number that indicates the relative importance to you of top speed verses acceleration" and the decision maker may say "top speed is three times as important to me as acceleration." Although the comparison values can be any positive number, a scale of 1 to 9 is typically used in AHP. For the car problem, pairwise comparisons would be made for top speed versus acceleration, comfort versus speed, space versus speed, total cost versus comfort, running versus maintenance, and so on. The notation used for these comparisons is a_{ij}, where the value of a_{ij} is the degree to which i is preferred to j. These pairwise comparisons can be represented as a matrix A with 1's along the diagonals.

Pairwise comparisons are also made for every pair of alternatives with respect to the bottom-level criteria on every branch. In the car problem the Porsche would be compared with the Land Rover, the Land Rover would be compared with the Metro, and the Metro would be compared with the Porsche on capital cost, running cost, maintenance, space, comfort, top speed, and acceleration. The pairwise comparison values represent the decision maker's preferences, but the basis for the values can be objective (as opposed to subjective) data if available. For example, when assessing the pairwise comparisons for the alternative cars relative to the top speed objective, the comparison values can be the ratios of the top speeds themselves.

To make this more clear, we show a sample of pairwise comparisons for the car example. Table 12.1 presents the performance of each car on each objective. In Table 12.2(a) the a_{ij} values for the pairwise comparisons of cars are shown with respect to the bottom-level objectives. These pairwise comparisons were calculated as ratios of the scores of cars on the bottom-level objectives. Tables 12.2(b), (c), and (d) show the A matrices for each branch of the objective hierarchy. These present our particular value judgments [e.g., that total cost is twice as important as comfort; see Table 12.2(d)].

Before we can calculate the weights on the objectives from the pairwise comparisons, we must consider the consistency of each matrix A. *Consistency* refers to the congruity of the comparisons. For example, if a decision maker has expressed his or her preferences in a perfectly consistent manner, the following would hold true: If objective 1 is preferred twofold to objective 2, and objective 2 is preferred threefold to objective 3, objective 1 is preferred sixfold to objective 3. Mathematically,

$$a_{13} = a_{12}a_{23} \tag{12.1}$$

or in this case,

TABLE 12.1 Performance of the Car Alternatives on the Bottom-Level Objectives of the Hierarchy

Car	Capital Cost	Running	Maintenance	Space	Comfort	Acceleration	Top Speed
Porsche	50,000	60	100	15	10	6	160
Land Rover	35,000	50	70	20	8	11	90
Metro	20,000	25	45	12	6	10	110

TABLE 12.2 Pairwise Comparisons
(a) Performance of Cars on Bottom-Level Objectives of the Hierarchy

Capital Cost

	Porsche	Land Rover	Metro
Porsche	1.0	0.7	0.4
Land Rover	1.4	1.0	0.6
Metro	2.5	1.8	1.0

Maintenance

	Porsche	Land Rover	Metro
Porsche	1.0	0.7	0.5
Land Rover	1.4	1.0	0.6
Metro	2.2	1.6	1.0

Comfort

	Porsche	Land Rover	Metro
Porsche	1.0	1.3	1.7
Land Rover	0.8	1.0	1.3
Metro	0.6	0.8	1.0

Top Speed

	Porsche	Land Rover	Metro
Porsche	1.0	1.8	1.5
Land Rover	0.6	1.0	0.8
Metro	0.7	1.2	1.0

Running

	Porsche	Land Rover	Metro
Porsche	1.0	0.8	0.4
Land Rover	1.2	1.0	0.5
Metro	2.4	2.0	1.0

Space

	Porsche	Land Rover	Metro
Porsche	1.0	0.8	1.3
Land Rover	1.3	1.0	1.7
Metro	0.8	0.6	1.0

Acceleration

	Porsche	Land Rover	Metro
Porsche	1.0	1.8	1.7
Land Rover	0.5	1.0	1.1
Metro	0.6	0.9	1.0

TABLE 12.2 ***(Continued)***

(b) Cost Objectives

	Capital Cost	Running	Maintenance
Capital cost	1.0	0.5	0.7
Running	2.0	1.0	1.2
Maintenance	1.4	0.8	1.0

(c) Speed Objectives

	Acceleration	Top Speed
Acceleration	1	0.5
Top speed	2	1

(d) Highest-Level Objectives

	Total Cost	Space	Comfort	Speed
Total cost	1.0	2.5	2.0	0.5
Space	0.4	1.0	0.6	0.4
Comfort	0.5	1.8	1.0	0.7
Speed	2.2	2.5	1.4	1.0

$$a_{13} = (2)(3) = 6 \tag{12.2}$$

Stated generally,

$$a_{ik} = a_{ij}a_{jk} \tag{12.3}$$

When consistency holds for all pairwise comparisons, equation (12.3) is true for all i, j, and k.

If a set of pairwise comparisons is perfectly consistent, to find the decision maker's weights for the objectives, we would simultaneously solve the set of equations

$$w_i = \frac{1}{n} \sum_{j=1}^{n} a_{ij} w_j \qquad i = 1,\ldots,n \tag{12.4}$$

where n is the number of objectives and w_i is the weight on objective i. For example, using the pairwise comparisons for the speed objectives, we form the equations

$$w_1 = (\tfrac{1}{2})(1w_1 + 0.5w_2) \qquad (12.5a)$$

$$w_2 = (\tfrac{1}{2})(2w_1 + 1w_2) \qquad (12.5a)$$

From these equations

$$w_1 = \tfrac{1}{2}w_2 \tag{12.6}$$

Any pair of weights such that equation (12.6) holds [e.g., $(w_1, w_2) = (1,2)$] will express the proper preferences for acceleration and top speed.

It is usually the case, however, that the decision maker's preferences are not perfectly consistent [equation (12.3) does not hold for all i, j, and k]. In this case the set of equations in (12.4) would have no solution. We can still find the weights by introducing another degree of freedom to (12.4). We do this by replacing n with a variable λ_{max}:

$$w_i = \frac{1}{\lambda_{max}} \sum_{j=1}^{n} a_{ij} w_j \qquad i = 1,\ldots,n \tag{12.7}$$

The system of weight equations now becomes an eigenvalue problem where the vector of weights **w** is the principal right eigenfactor of matrix A and λ_{max} is its eigenvalue. Saaty (1977) showed that the components of **w** computed this way provide a good estimate of the weights (Ridgley and Rijsberman, 1992).

The eigenvalue approach allows AHP to proceed in the face of inconsistency, but it is important to know *how* inconsistent a decision maker is. There may be degrees of inconsistency that are considered intolerable. The eigenvalue of matrix A, λ_{max}, provides an estimate of the consistency of matrix A. Comparing equations (12.4) and (12.7), we see that the closer λ_{max} is to n, the more consistent is matrix A. A consistency index (CI) has been developed to capture this relationship:

$$\text{CI} = \frac{\lambda_{max} - n}{n - 1} \tag{12.8}$$

The CI for matrix A is compared to the CI calculated from a matrix of the same size as A whose elements are all chosen randomly. This random index can be used as a basis for determining whether or not the pairwise comparisons that comprise matrix A are sufficiently consistent (Saaty, 1990, Chap. 3). When using an AHP software package [e.g., Hipre or Expert Choice (Expert Choice, Inc., 1993)], this consistency evaluation is normally calculated automatically and the user is notified if the pairwise comparisons are too inconsistent.

After all pairwise comparisons for all A matrices are determined to be sufficiently consistent, the system of equations (12.7) is used to calculate the weights for each objective. These weights are then used to calculate the "score" for each alternative. Starting with an alternative, its weight is multiplied by the objective's weight immediately above it, working upward through the hierarchy until all the products have been calculated. The result is a score for each of the alternatives for each of the lowest-level criteria. The scores for each alternative are summed to create a total score for each alternative. The final scores for each car alternative are shown graphically in Figure 12.3. Based on the current set of pairwise preferences, the Porsche has the overall highest score, followed closely by the Metro and then the Land Rover. Therefore, our car purchase should be the Porsche.

One could stop at this point or proceed to a sensitivity analysis on the results, a step that is often not given the attention it deserves. It is important to remember that through sensitivity analysis we are able to understand the robustness of the results and to identify the parts of the problem to which the results are most sensitive. The software package that we have used to solve our example has an automated sensitivity analysis function that makes this process very easy and intuitive. A typical sensitivity analysis is shown in Figure 12.4. The vertical line (between 0.1 and 0.2) indicates the current weight placed on space, one of the criteria. The slanted lines in the graph represent the total scores for each car as a function of the weight placed on space. Notice that for the range of space weights from 0 through almost 0.4, the Porsche has the highest score. However, if the weight placed on space

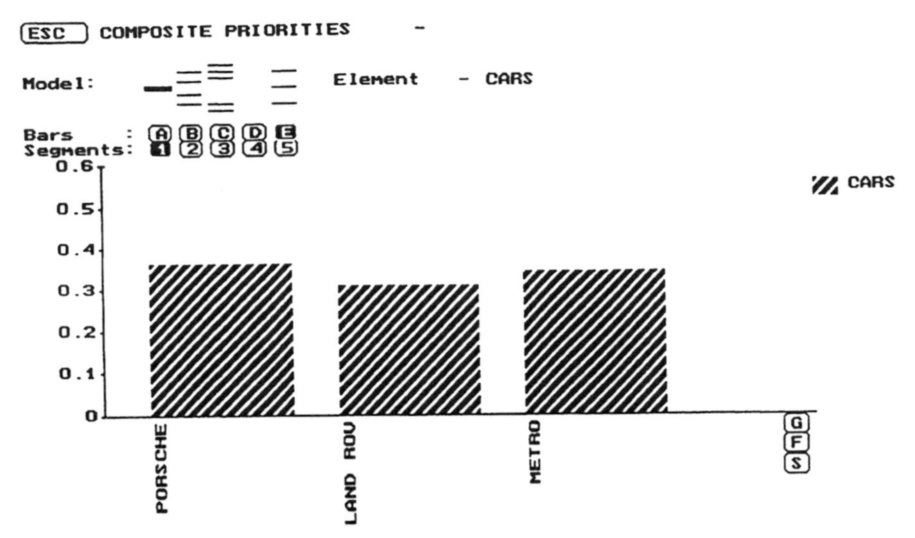

Figure 12.3 AHP scores for the car selection problem. (R.P. Hamalainen and H. Lauri: HIPRE 3+ Decision Support Software (v.3.13 / 1996). System Analysis Laboratory, Helsinki University of Technology. Distributed by YVA Inc.)

were higher than 0.4, the Land Rover would have the highest score. We proceed this way for each sensitivity analysis panel. For this example it seems that the weights would need to change significantly before the results would change. Therefore, we can be confident about our choice of the Porsche.

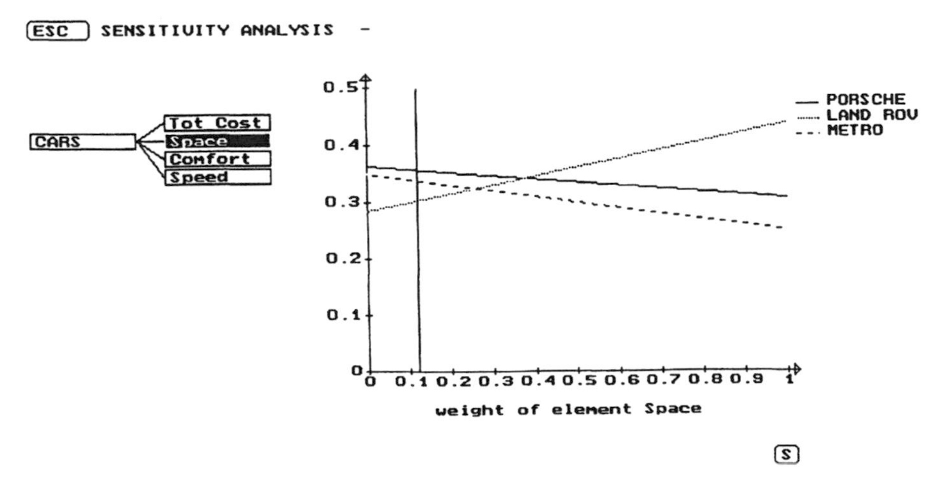

Figure 12.4 Sensitivity analysis for the space objective. (R.P. Hamalainen and H. Lauri: HIPRE 3+ Decision Support Software (v.3.13 / 1996). Systems Analysis Laboratory, Helsinki University of Technology. Distributed by YVA Inc.)

12.E.3. Simple Multiattribute Rating Technique

SMART is a simplified version of multiattribute utility (or value) theory, a general framework for representing decision makers' preferences. The best reference for multiattribute utility theory is Keeney and Raiffa (1976). A value function is a mathematical function in which the objectives are the arguments

$$v(Z_1,\ldots,Z_p) \tag{12.9}$$

The value function produces a single number from a set of objective values and so converts a multiobjective problem into a single-dimensional one. The particular form of the value function, v, depends on the decision makers' preferences and how they change with different combinations of objective values. In general, v can take on any form, but there are certain behavioral assumptions that restrict it. For example, monotonicity is often assumed. A monotonic value function, v, never goes down when Z_k goes up (i.e., more of a good objective, such as benefits, is always preferred). Of course, the direction would be reversed for an undesirable objective such as cost.

Most of multiattribute utility and value theory is about estimating the value function. Once we have v, it is a simple matter to identify the preferred alternative. Using the value function, v, each alternative is scored. The best alternative is then the alternative with the highest score.

The distinction between value functions and utility functions is related to uncertainty and whether uncertainty plays a role. Value functions apply to deterministic situations. Utility functions are value functions that include decision makers' attitudes toward risk as well as their relative preferences for the conflicting objectives. Our focus is on value functions.

SMART is a methodology that uses a particular form of the value function called an *additive value function.* This function is of the form

$$v(Z_1,\ldots,Z_p) = \sum_{j=1}^{p} \lambda_j v_j(Z_j) \tag{12.10}$$

where Z_j is an alternative's raw score on objective j, p the total number of objectives, λ_j the relative weight on objective j, and v_j an individual value function for objective j. By using an additive value function, SMART assumes that the decision makers' preferences for each objective are independent, in a certain way, from the other objectives (see Keeney and Raiffa, 1979; Goodwin and Wright, 1991). As was the case with AHP, SMART assumes that the alternatives have already been identified and that an objective hierarchy has already been created. Each alternative must also be scored according to each objective. This can be accomplished either by directly assigning values for each alternative for each objective (called *direct assessment*) or by creating a value function for each objective (described below). When using direct assessment, either subjective (the decision maker's opinions) or objective (the scores of each alternative on each objective, such as top speed) information can be used to determine the scores. While either scoring method is permitted by SMART, some software packages that execute the SMART methodology do not allow for both scoring methods. For example, Hipre allows either direct assessment or the creation of value functions, while VISA allows only direct assessment. Finally, SMART requires the decision maker's relative preferences for the objectives. Once again we refer to our car example to demonstrate this method and rely on Hipre to perform the analysis and provide the graphics. We demonstrate the use of value functions to provide the scoring for the alternatives (rather than direct assessment).

The first step of SMART is to identify the individual value functions (the v_j's) for each objective. An individual value function captures a decision maker's preferences for different amounts of an objective. Let's consider a trivial but easily understood case: your preferences for apples. We'll assume that you like apples and that more of them is better than fewer of them (you can store them). Now, consider how much you'd prefer two apples to one, three apples to two, four apples to three, and so on. If you are like most people, the more apples you already have, the less valuable the next one would be (i.e., your value function would show decreasing marginal value). This is one typical characteristic of an individual value function that we try to capture.

The *midvalue splitting technique* (also called the *bisection method*) is one technique for deriving a decision maker's individual value functions (Keeney and Raiffa, 1979; Goodwin and Wright, 1991). The basic logic of the bisection method is a series of steps involving questions for the decision maker. These steps are:

Step 1. Set the range for the objective j values. Let's call the worst objective score for the range Z^0 and the best possible objective score for the range Z^1. Set $v_j(Z^0) = 0$ and $v_j(Z^1) = 1$. (Note that the scale of the value function is arbitrary. Here we use a scale of 0 to 1.)

Step 2. Ask the decision maker for $Z^{0.5}$, the value of Z such that the decision maker is indifferent between moving from level Z^0 of the objective to level $Z^{0.5}$ and moving from level $Z^{0.5}$ of the objective to level Z^1. Set $v_j(Z^{0.5}) = \frac{1}{2}$.

Step 3. Find $Z^{0.25}$ and $Z^{0.75}$ using the same procedure as in step 2, where $Z^{0.25}$ is the midvalue of $[Z^0, Z^{0.5}]$ and $Z^{0.75}$ is the midvalue of $[Z^{0.5}, Z^1]$.

Step 4. Continue this process until enough points have been obtained to estimate a curve for v_j.

As an example, consider the capital cost objective in the car-buying problem. First, set Z^0 = \$50,000, Z^1 = \$20,000, v_j(\$50,000) = 0, and v_j(\$20,000) = 1 (points *A* and *E* in Figure 12.5a). Notice that because cost is a "bad" rather than a "good," the highest cost is given the lowest value. Next, ask the decision maker for a car price such that he or she is indifferent between

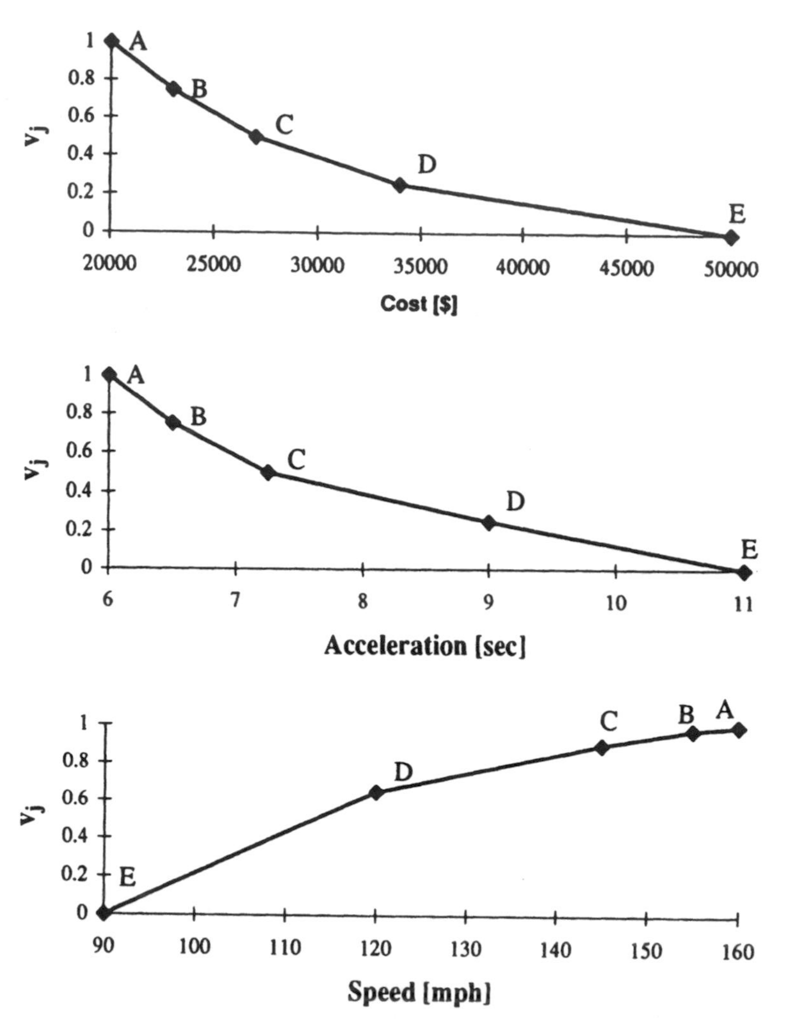

Figure 12.5 Value function for (a) capital cost; (b) acceleration; (c) top speed.

moving from \$20,000 to this price and moving from this price to \$50,000. The decision maker gives a price of \$27,000. Set $v_j(\$27,000) = 0.5$ (point C in Figure 12.5a), which means in preference terms that the decision maker feels equally bad between a cost increase of \$7,000, from \$20,000 to \$27,000, as he or she does when cost increases from \$27,000 to \$50,000. We could interpret this as sticker-shock and post-sticker-shock. Next, ask the decision maker for a car price such that they are indifferent between moving from \$20,000 to this price and moving from this price to \$27,000. The decision maker gives a price of \$23,000. Set $v_j(\$23,000) = 0.75$ (point B in Figure 12.5a). Finally, ask the decision maker for a car price such that he or she is indifferent between moving from \$27,000 to this price and moving from this price to \$50,000. The decision maker gives a price of \$34,000. Set $v_j(\$34,000)$

= 0.25 (point D in Figure 12.5a). We now have a value function for capital cost. We repeat this process for every other lowest-level objective in the hierarchy (see Figure 12.5b and c for other examples).

After finding the v_j values for each objective, the next step is to determine the decision maker's weights [the λ_j values in (12.10)] for each set of objectives for each branch on the objective hierarchy. Let's say that we are trying to find the relative weights for objectives 1 and 2. First, since only the relative values of the weights matter, we can set

$$\lambda_1 + \lambda_2 = 1 \tag{12.11}$$

We ask the decision maker for two alternatives [two sets of scores (Z_1^1, Z_2^1) and (Z_1^2, Z_2^2)] for which they are indifferent [i.e., $v(Z_1^1, Z_2^1) = v(Z_2^2, Z_2^2)$]. Using equation (12.10), we write

$$\lambda_1 v_1(Z_1^1) + \lambda_2 v_2(Z_2^1) = \lambda_1 v_1(Z_1^2) + \lambda_2 v_2(Z_2^2) \tag{12.12}$$

We can read $v_1(Z_1^1)$, $v_2(Z_2^1)$, $v_1(Z_1^2)$, and $v_2(Z_2^2)$ off the value function graphs and then solve (12.11) and (12.12) simultaneously to find λ_1 and λ_2. This is generalizable to any number of weights in a particular set, but more indifference points are needed (two for three weights, three for four weights, etc.)

For our car problem, we determine three sets of relative weights: For the total cost subobjectives, for the speed subobjectives, and for the subobjectives of the overall cars objective. To illustrate the method, we determine the weights for the subobjectives of the speed objective. Say that our decision maker has stated that they are indifferent between having a car with level 8 acceleration and level 150 top speed and a car, all else being equal, with level 7 acceleration and level 130 top speed. In the style of equation (12.12) we write

$$\lambda_1 v_1(8) + \lambda_2 v_2(150) = \lambda_1 v_1(7) + \lambda_2 v_2(130) \tag{12.13}$$

Using our value functions for these objectives (Figures 12.5b and c) we rewrite (12.13) as

$$\lambda_1(0.4) + \lambda_2(0.92) = \lambda_1(0.55) + \lambda_2(0.7) \tag{12.14}$$

and solving (12.13) and (12.14) simultaneously, we obtain

$$(\lambda_1, \lambda_2) = (0.59, 0.41) \tag{12.15}$$

Next, we calculate the overall score for each alternative. For each bottom-level objective we replace the raw objective score with the value (the v_j value) for the alternative. Then working from bottom to top, for each node we calculate its score as the weighted sum of its branches, using the appropriate

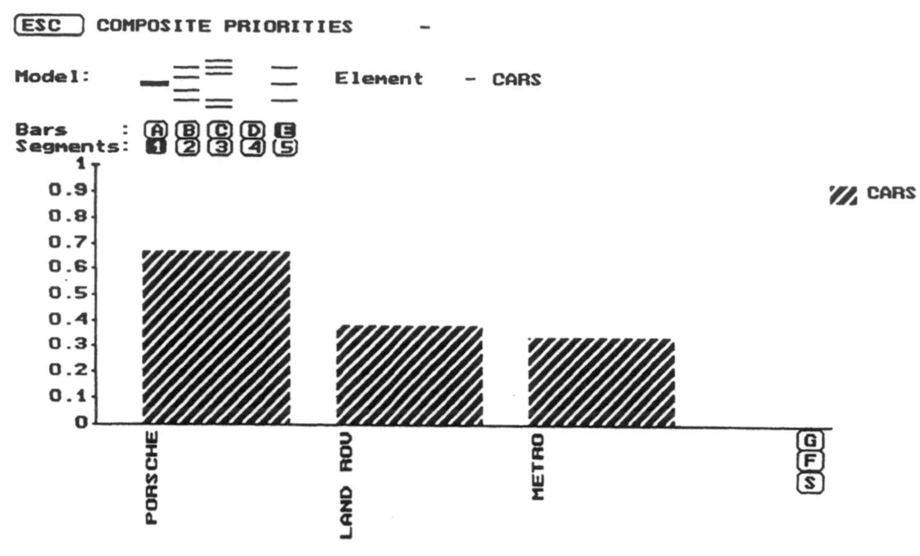

Figure 12.6 Smart scores for the car selection problem. (R.P. Hamalainen and H. Lauri: HIPRE 3+ Decision Support Software (v.3.13 / 1996). Systems Analysis Laboratory, Helsinki University of Technology. Distributed by YVA Inc.)

λ_j's for the weights. Figure 12.6 shows the overall score for each car. Again we have chosen the Porsche.

The final step is sensitivity analysis. We use Hipre to generate our sensitivity screens (see the example in Figure 12.7). The interpretation for these

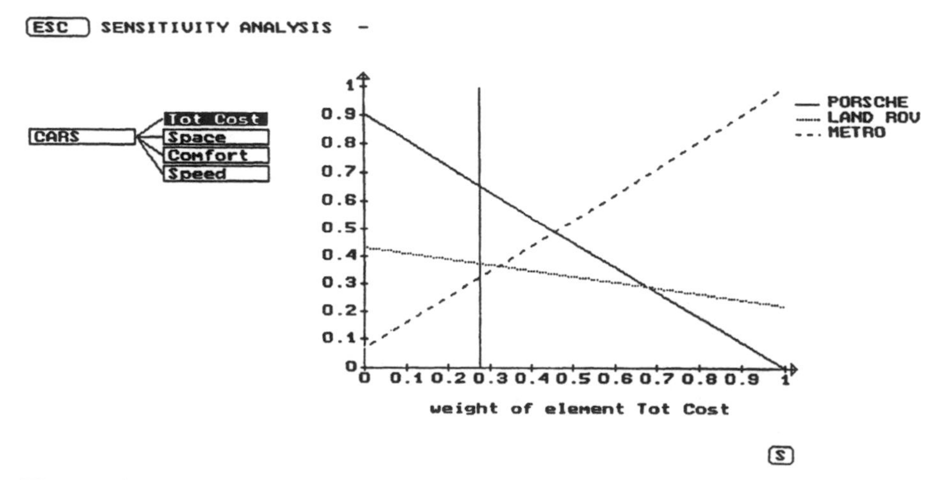

Figure 12.7 Sensitivity analysis for the highest-level objectives. (R.P. Hamalainen and H. Lauri: HIPRE 3+ Decision Support Software (v.3.13 / 1996). Systems Analysis Laboratory, Helsinki University of Technology. Distributed by YVA Inc.)

screens is identical to that for the sensitivity screens generated by the AHP method, discussed in Section 12.E.2.

12.F. MULTIOBJECTIVE PROGRAMMING

Multiobjective programming (MOP) is one subdiscipline among a wide variety of quantitative decision support techniques within the realm of mathematical programming (MP) (Figure 12.8). Its roots can be found in the early work of Kuhn and Tucker (1951) and Koopmans (1951). MOP was not really considered a separate specialty, however, until a 1972 conference in South Carolina (Cochrane and Zeleny, 1973) which brought together most of the active researchers in the then-emerging field.

MOP can be thought of as a set of methodologies for generating a preferred solution or range of efficient solutions to a decision problem. The idea of "generating" solutions is important because this is the basic distinction between MOP and its MCDM cousins (Figure 12.8). MCDM methods are used to identify the optimal solution for a decision problem from a set of *predefined* alternative solutions. MOP methods are invoked when the alternatives are *not* predefined.

An MOP problem is any decision problem that can be stated in the following format:

minimize (or maximize) some set of objectives

subject to

some set of constraints

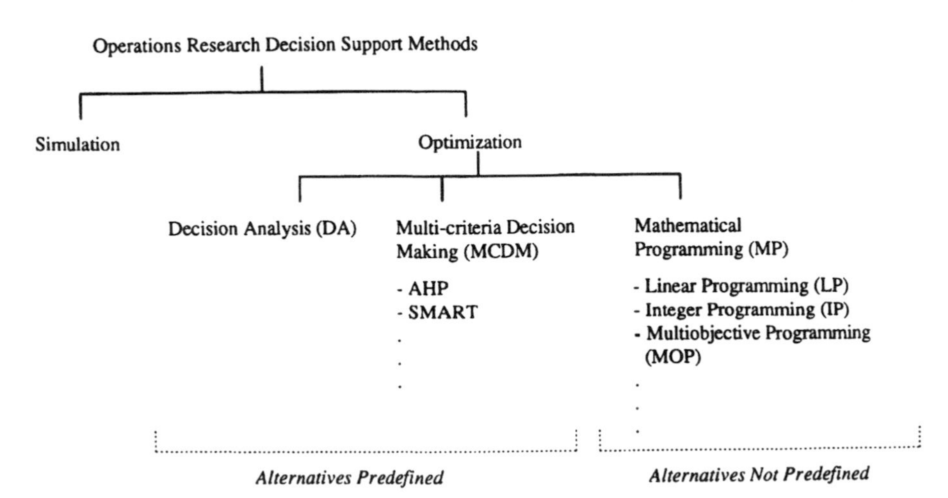

Figure 12.8 Categories of decision support methods.

The objective function(s) and constraints are mathematical functions of a set of decision variables and parameters. The form (e.g., linear versus nonlinear) of these equations determines the particular variety of the MOP. This apparently simple format can be and has been applied to an enormously broad range of decision applications, from highly technical design problems in water resources systems to behavior in animal ecology.

For the MOP methods covered in this chapter, all objective functions and constraint functions are assumed to be linear. Therefore, to be more specific, the techniques described in this chapter may be referred to as multiobjective linear programming (MOLP) techniques. MOLP problems must meet the following conditions:

1. All objective functions and constraints must be linear. This linearity has two implications. First, the functions exhibit *constant returns to scale.* Therefore, given some objective function or constraint function (f) of two decision variables (x_1 and x_2),

$$cf(x_1 + x_2) = f(cx_1 + cx_2) \tag{12.16}$$

   Second, the functions exhibit *separability:*

$$f(x_1, x_2) = f_1(x_1) + f_2(x_2) \tag{12.17}$$

   The effect or contribution of a variable does not depend on the value of another variable.
2. All decision variables are continuous. Integer programming (IP) techniques exist for handling cases for which the decision variables must be constrained to integer values.
3. All decision variables must be nonnegative. This is an artifact of the algorithm used to solve LPs, and the condition does not impose real limitations, as there are techniques for getting around this assumption.

To use MOLP solution techniques, these assumptions must be met. For nonlinear objective functions and constraint functions, it is necessary to use some form of nonlinear programming technique (Chankong and Haimes, 1983). However, in certain cases of nonlinear objective functions, a linear approximation coupled with MOLP techniques is adequate.

The general formulation of an MOLP is

$$\text{minimize (or maximize) } Z_1, Z_2, \ldots, Z_p \tag{12.18a}$$

subject to

$$\sum_{j=1}^{n} a_{ij}x_j \leq b_i \qquad i = 1,...,m \tag{12.18b}$$

$$x_j \geq 0 \qquad j = 1,...,n \tag{12.18c}$$

where $Z_1, Z_2, ...Z_p$ are the objective functions, p the number of objectives, m the number of constraints, n the number of decision variables, x_j the jth decision variable, b_i the limiting value of resource i, and the a_{ij}'s are the coefficients of the constraint equations. Each objective function has the form

$$Z_k = \sum_{j=1}^{n} c_{kj}x_j \qquad k = 1,...,p \tag{12.19}$$

12.F.1. Solving an MOP

Nondominated set. The most important concept in MOP is the notion of the *nondominated set.* It is, in fact, this set of alternatives that we seek as solutions to a multiobjective decision problem. Therefore, we explain this concept even before presenting the solution techniques for MOLPs. The nondominated set is the collection of alternatives that represent potential compromise solutions among the objectives. The result of the application of an MOLP technique to a decision problem is the nondominated set for the problem, and it is from this subset of potential solutions that the final, preferred decision is chosen by the decision makers.

This concept is most easily understood with an example (borrowed from Cohon, 1978). Consider a typical MOLP problem for which we have two objectives, two decision variables, and four constraints. We write these statements mathematically in the style of equations (12.18a), (12.18b), and (12.18c) as follows:

$$\text{maximize } Z_1 = 5x_1 - 2x_2 \tag{12.20a}$$

$$Z_2 = -x_1 + 4x_2 \tag{12.20b}$$

subject to

$$-x_1 + x_2 \leq 3 \tag{12.20c}$$

$$x_1 \leq 6 \tag{12.20d}$$

$$x_1 + x_2 \leq 8 \tag{12.20f}$$

$$x_2 \leq 4 \tag{12.20g}$$

$$x_1, x_2 \geq 0 \tag{12.20h}$$

where the variables are defined similarly as for equations (12.18a), (12.18b), and (12.18c).

The constraint equations define the set of feasible combinations of decision variables for the problem. Because we have only two decision variables, we can plot these constraint equations on a graph as straight lines using the decision variables x_1 and x_2 as the axes (Figure 12.9a). For example, the line passing through points B and C in Figure 12.9a corresponds to the constraint (20d). The shaded region indicates the *feasible set* of (x_1, x_2) pairs in *decision space.* The points at which two constraints intersect (e.g., point B along the x_1 axis) are called the *extreme points.* Each extreme point's coordinates in decision space along with the value that it yields for the two objective variables Z_1 and Z_2, are given in Table 12.3.

Again, because we have only two objectives, we can plot each feasible solution onto a new graph using objectives Z_1 and Z_2 as the axes (Figure 12.9b). This representation is referred to as the *objective space* for the problem. It is important to note that the shaded region indicating the feasible objective value combinations in objective space correspond to the same set of feasible solutions as those in the shaded region in decision space. One is a mapping of the other, with the objectives acting as the mapping functions. The extreme points identified in decision space correspond to those similarly labeled in objective space.

Next we identify the nondominated set. Consider point N in Figure 12.9b. The solution that corresponds to point D gives an equal level of objective Z_1 but gives a higher level of Z_2. Point D is said to *dominate* point N. In fact, all points within the area indicated by the dashed lines are said to dominate point N because they all yield higher levels of both objectives (or more of one objective and the same amount of the other). Using similar logic, we can show that for all points inside the boundaries of the feasible region in objective space, there is at least one point along the "northeast" boundary of the feasible region that dominates each of the inside points. We also see that for all points along the northeast boundary, there are no points that dominate them. These points are said to be nondominated[2] and the collection of these points is called the nondominated set (Figure 12.9b, borders BC, DC, and ED, the darkened borders). Stated formally and more generally, say that $\bar{x}$ is a particular feasible set of values for the decision variables. A solution, $\bar{x}^*$, is nondominated if it is feasible and if there is no other feasible solution $\bar{x}$ such that

$$Z_k(\bar{x}) \geq Z_k(\bar{x}^*) \qquad k = 1,\ldots,p \tag{12.21}$$

where p is the number of objectives, and with at least one of these inequalities satisfied as a strict inequality (assuming all objectives are to be maximized).

[2]The term *nondominated* is equivalent to the terms *noninferior, Pareto optimal,* and *efficient.*

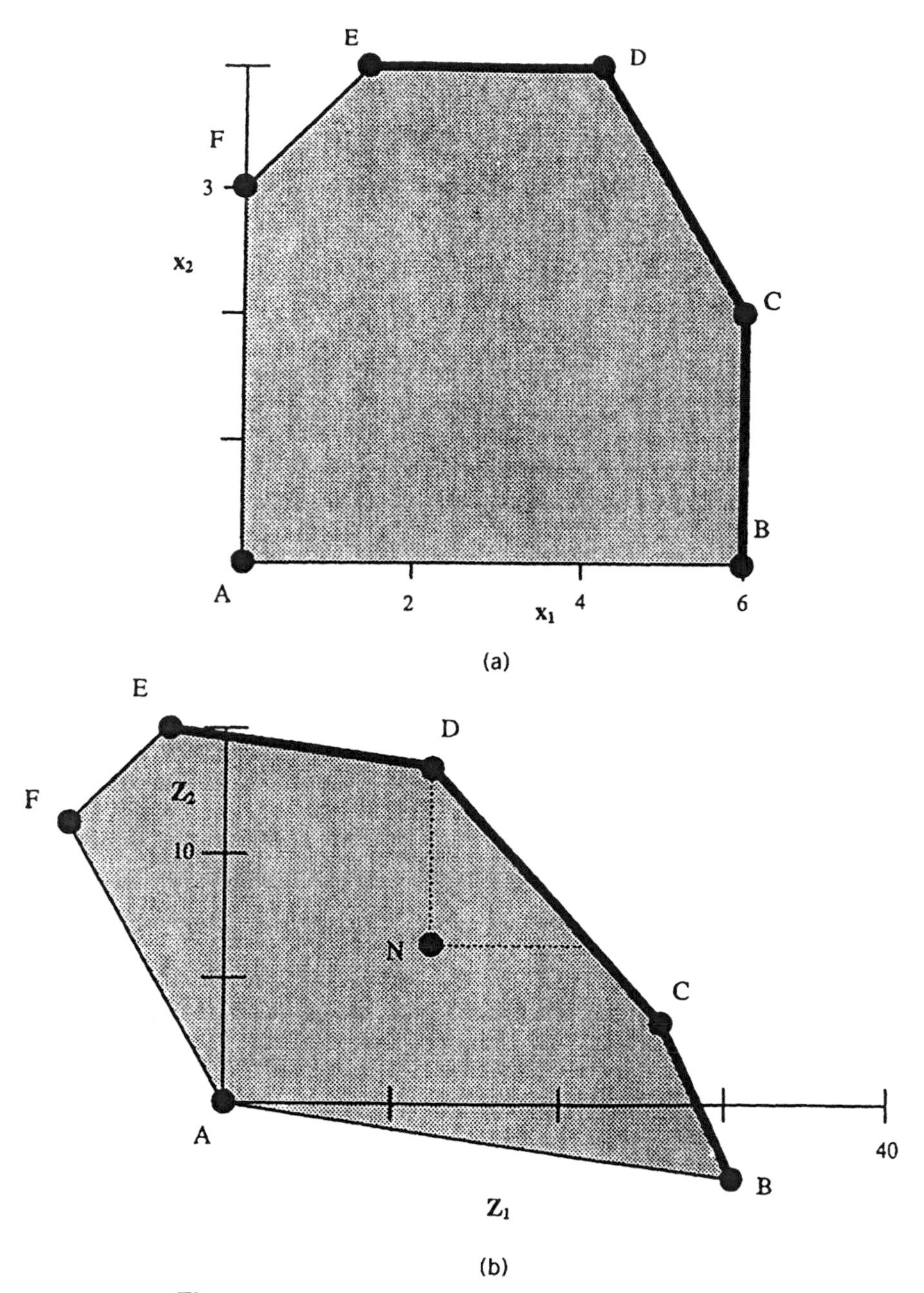

Figure 12.9 (a) Decision space; (b) objective space.

TABLE 12.3 Decision Variable Values and Objective Values for the Extreme Points in Figure 12.9

Extreme Point	x_1	x_2	Z_1	Z_2
A	0	0	0	0
B	6	0	30	−6
C	6	2	26	2
D	4	4	12	12
E	1	4	−3	15
F	0	3	−6	12

The solution to an MOLP, then, is not a single solution but, instead, is a set of nondominated points. Each point along the nondominated set in objective space has an associated point in decision space that shows the particular combination of decision variable values yielding that solution. Therefore, we can also locate the set of nondominated feasible solutions in decision space (Figure 12.9a, borders *BC*, *DC*, and *ED*, the darkened borders). Note, however, that the notion of dominance as defined in (12.21) has meaning only in objective space, so the nondominated set in decision space need not lie along the northeast boundary of the feasible region.

Solution Techniques: Preference-Oriented Techniques and Generating Methods. Unlike other areas of mathematical programming in which there are one or a few solution techniques (e.g., the simplex algorithm of linear programming), MOP is a collection of several techniques, each with its own approach to dealing with the information needs of decision makers and the elicitation of the decision maker's preferences. It is convenient to consider MOP methods to belong to one of two categories: generating techniques and preference-oriented methods (Figure 12.2). As the name suggests, preference-oriented methods involve the use of the decision-maker's relative preferences for the objectives as part of the solution method. Depending on the preference-oriented approach taken, these preferences are solicited either prior to the initiation of the solution method or during execution of the method. In both cases the decision maker's preferences are directly incorporated into the solution process. Therefore, solutions produced by preference-oriented techniques are dependent on the preferences expressed by the decision maker. But there may be situations in which we would like to defer the procurement of these preferences. The decision makers may be unable to state their preferences exactly or may simply not want to reveal their preferences to the analyst. The more desirable scenario would be to present the decision maker with the set of nondominated choices, determined independent of a priori preferences. Then the decision makers could consider their relative preferences for the objectives and select the final solution with the benefit of knowing their choices, as represented by the nondominated set. Preference-oriented methods such as goal programming are widely studied and applied (Zeleny, 1982; Chankong and Haimes, 1983). But given the limited space, we have chosen to focus on generating methods.

Generating techniques allow the analyst to solve for the set of nondominated solutions in the absence of knowledge about the decision maker's preferences. The result of the application of a generating technique to an MOLP is the set of nondominated solutions for the problem (described above). It is only upon the completion of the analysis phase of the problem solution that the decision maker's preferences are obtained. If this set is small enough, the decision maker may be able to select a final solution without quantitative support. However, if there are a large number of alternatives or a large number of objectives, one of the MCDM methods may be employed. We present two generating techniques for MOLPs: the weighting method and the constraint method (Cohon, 1978). Other major generating techniques not covered here are the NISE method (Cohon, 1978), the multiobjective simplex method (Steuer, 1986), and multiobjective integer programming techniques (Rasmussen, 1986).

12.F.2. Weighting Method

Recall that the goal of an MOLP method is to identify the nondominated set of alternative solutions. For the two-dimensional example described above, it was relatively straightforward to define the nondominated set exactly. However, as the complexity of a decision problem increases, the ease with which the nondominated set is identified disappears. As the number of constraints increases, it becomes more complicated to solve for all the extreme points. If there are more than three objectives, it is no longer possible to see the nondominated set graphically. Instead, we must turn to mathematical approximation techniques, such as the weighting method and the constraint method.

The weighting method is used to approximate the nondominated set through the identification of extreme points along the nondominated surface. The basic strategy is to transform the multiobjective problem into a series of single objective problems that can be solved using the well-known simplex algorithm for single objective problems. An approximation of the nondominated set is formed by "connecting" the extreme points identified.

To explain the weighting method, we return to our two objective example. As originally stated, the decision problem has two separate objectives, Z_1 and Z_2. But suppose that we were to hypothesize the relative weight between Z_1 and Z_2. The two objective equations could then be combined into a single objective function as follows:

$$\text{maximize } Z = Z_1 + wZ_2 \tag{12.22}$$

where w represents the weight on Z_2 relative to Z_1. The solution to this single objective problem would be the optimal solution for a decision maker whose relative preference for these objectives was represented accurately by w. However, the weight here is used merely as a solution device; we do not necessarily ascribe any preference relevance to it. Rewriting this equation in the standard form of a line gives us

$$Z_2 = -\frac{1}{w}Z_1 + \frac{Z}{w} \tag{12.23}$$

We see that this objective equation can be graphed as a line in objective space where the slope of the line is $-1/w$ and the intercept along the Z_2 axis is Z/w. Figure 12.10 shows the objective-space representation of our two-objective problem with a sample set of contours in the form of (12.23) for which $w = 0.5$ and Z is varied from 5 to 20.

The solution to the maximization problem (12.22) can be found graphically by "pushing" a contour of slope $-1/w$ as far to the northeast as possible until the line just touches the boundary of the feasible region. In this example, that solution occurs at extreme point B. Mathematically, the problem can be solved as a single objective linear program with the simplex algorithm to get the same solution.

In this two-dimensional form it is possible to visualize that for decision problems with strictly linear equations, the solutions to the weighted problem (12.22) will always occur at extreme points. Furthermore, as long as $w \geq 0$, the solution will be on the northeast boundary of the feasible region (i.e., in the nondominated set). Consider the two extremes $w = 0$ and $w = \infty$, which produce solutions B and E, respectively (Figure 12.11). All other nonnegative values of w will produce solutions between E and B. The weighting method uses these observations to find extreme points on the nondominated set. As the relative weight of the objectives, w, is varied, the single objective maximizing solution [the solution to (12.22)] moves from extreme point to extreme point. Therefore, by systematically varying this weight, we can locate nondominated extreme points. By connecting the extreme points with straight lines, an approximation of the nondominated surface is formed. The interval between the weights will change the resolution at which extreme points are located and hence the accuracy of the approximation of the nondominated surface. For example, if only two weights, $w = \infty$ and $w = 0$, were used, only two extreme points, the solution that maximizes Z_2 and the solution that maximizes Z_1, would be identified. The approximation of the nondominated surface would be just the straight line that connects these two points (Figure 12.12a). As we solve more problems with different weights, the number of identified extreme points increases and the approximation improves (Figure 12.12b).

The weighting method logic used to approximate the nondominated surface in two dimensions (i.e., for two objectives) can be generalized to handle

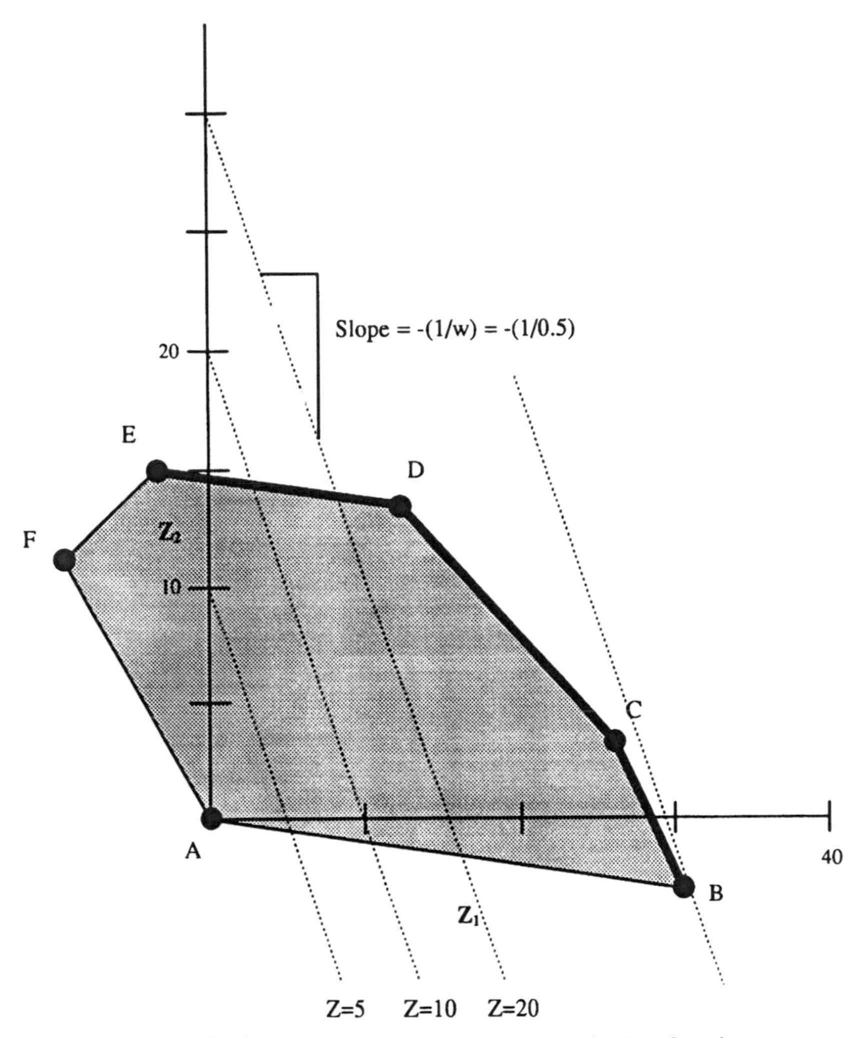

Figure 12.10 Contours for the weighted objective function.

problems with more than two objectives. The weighted single-objective function for any number of objectives is

$$\text{maximize} \sum_{j=1}^{p} w_j Z_j(x_1, \ldots, x_n) \tag{12.24}$$

where p is the total number of objectives, n the total number of decision variables, and w_j the weight on objective j. Just as for the two-dimensional case, the p objectives are combined into a single weighted objective function in the form of (12.24), the weights are systematically varied, and the single-

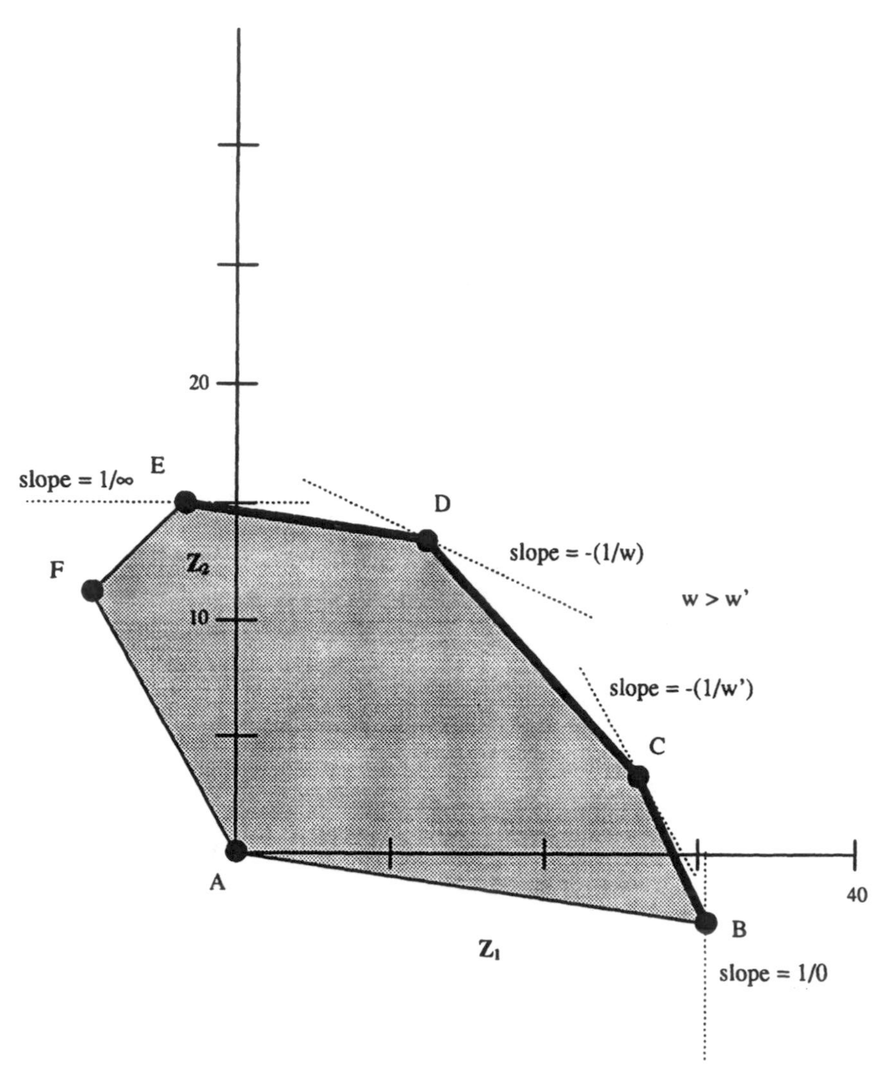

Figure 12.11 Extreme points located by varying relative weights of the objectives.

objective problem is solved using simplex to identify extreme points on the nondominated surface.

The strength of the weighting method is its ability to locate extreme points. While all points along the nondominated set are nondominated and therefore worthy of consideration, the extreme points may be of the most interest to the decision maker. It is at these points that the *rate of exchange* or trade-off between the objectives changes. These rates of exchange may be of importance to the decision maker. For example, for our two-objective example, the trade-off between objectives 2 and 1 along the border *DC* is −0.71 (the slope of border *DC*, Figure 12.9b). That is, for every unit of Z_1 gained, 0.71 unit

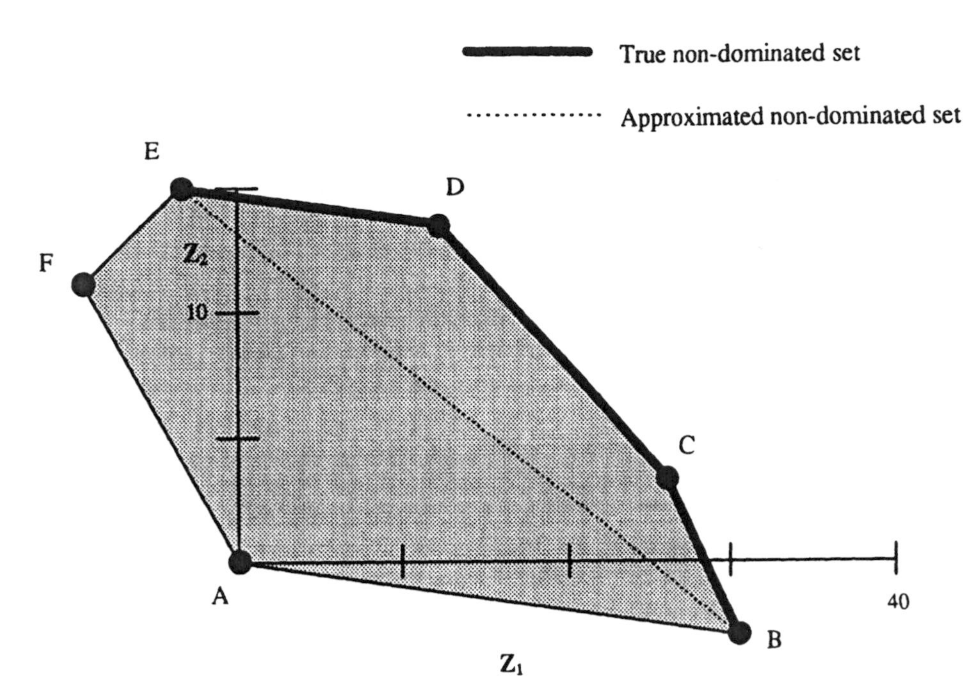

(a)

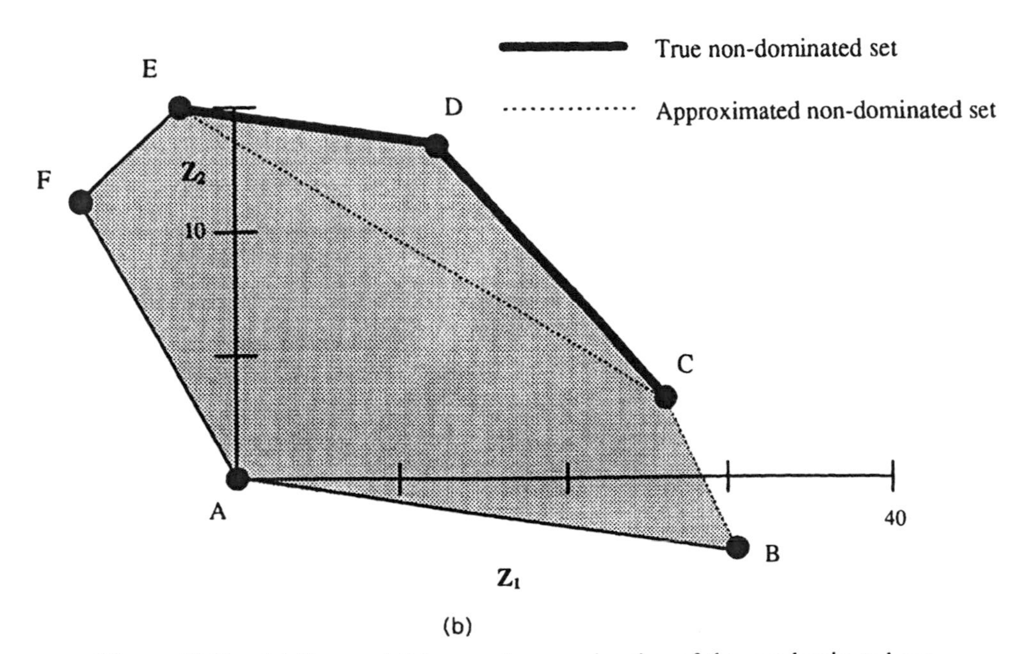

(b)

Figure 12.12 (a) First and (b) second approximation of the nondominated set.

of Z_2 is lost. Along the border *CB* the rate of exchange is -2. The decision maker may be interested in having a high level of Z_1 and be willing to trade 0.71 unit of Z_2 for each unit of Z_1 but not be willing to trade 2 units of Z_2 for a unit of Z_1. Therefore, the decision maker's preferred alternative will be the solution corresponding to extreme point *C*, where he or she can have the highest level of Z_1 while maintaining an exchange rate of -0.71.

The major drawback to the weighting method lies in its inefficiency. While the interval between the weights influences the resolution with which extreme points are identified, there is no way to control explicitly resolution of the approximation of the nondominated surface. Consider the nondominated surface shown in Figure 12.13. To find all the extreme points on this surface, particularly points *B* and *C*, it would be necessary to vary *w* gradually, changing it by very small increments and thereby increasing the number of single-objective problems that must be solved. It is also difficult to control the region of the nondominated surface on which extreme points are found. Referring back to Figure 12.13, say that the decision maker was only interested in solutions that heavily favored objective 1. Then the large number of problems solved to identify points *A*, *B*, and *C* are a waste of time. These difficulties with the weighting method are greatly magnified in problems with $p > 3$ when we can't "watch" as the approximation of the nondominated set unfolds. The constraint method has certain advantages over the weighting method in this regard.

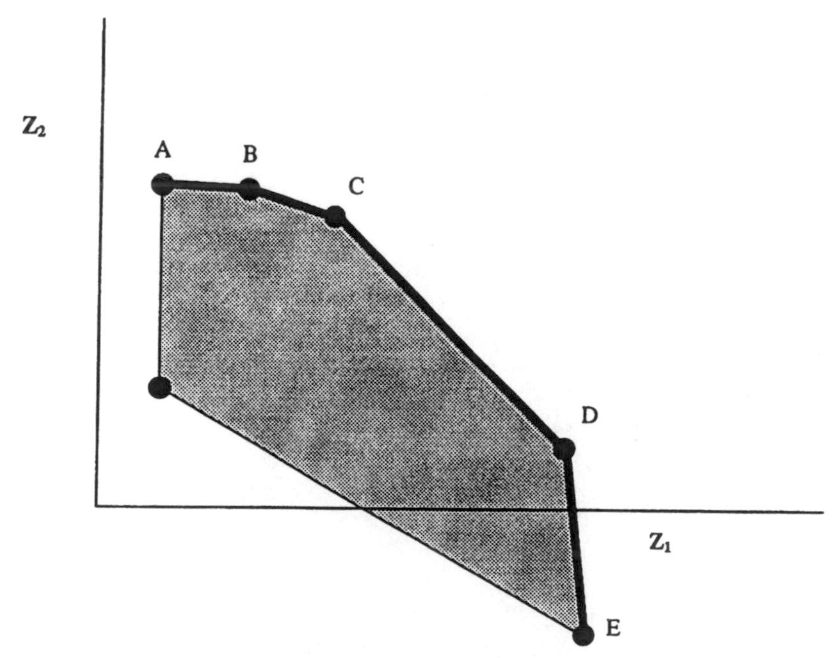

Figure 12.13 Clumped extreme points on the nondominated set.

Care must also be exercised when using the weighting method to solve multiobjective integer problems. Integer programming is used to solve problems for which the decision variables are not continuous variables but, instead, are constrained to integer values (e.g., the knapsack problem). If all extreme points are located along a concave surface, as is the case for mathematical programs with continuous variables, the weighting method can find all of the extreme points. However, for an integer program, the alternatives of the nondominated set do not necessarily fall along a concave surface. The weighting method can be used to solve integer programs but at the risk of "missing" some extreme points. This issue, referred to as the *duality gap,* is addressed in more detail in the nuclear waste management example (Section 12.H).

12.F.3. Constraint Method

The constraint method is another technique for generating an approximation of the nondominated set. Again, the basic strategy is to transform a multiobjective problem into a series of single-objective problems that can be solved using the simplex algorithm for single-objective problems. But unlike the weighting method that locates only extreme points, the constraint method can find points anywhere along the nondominated surface; in general, the constraint method will *not* find the extreme points of the original MOP. The constraint method offers the advantage of better control over exploration of the nondominated set.

The logic of the constraint method can be shown graphically. Figure 12.14 is a reproduction of the feasible region defined by the original constraints of our example problem. Again, the nondominated set is indicated by the dark-

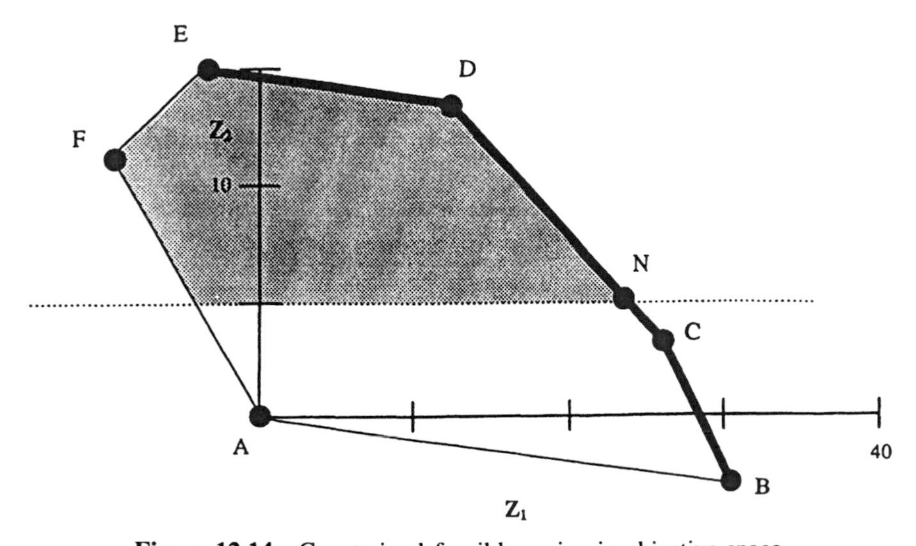

Figure 12.14 Constrained feasible region in objective space.

ened border. If the decision problem was to find the solution that maximized Z_1 alone, the result would be point B, the point within the feasible region that yields the highest value of Z_1. Now we add a new constraint as follows:

$$Z_2 \geq 5 \tag{12.25}$$

and again solve the problem to find the solution that maximizes Z_1. The new constraint reduces the size of the feasible region as indicated in Figure 12.14. The new Z_1-maximizing solution given the reduced feasible region is point N. Notice that point N lies on the nondominated set of the original problem. Notice also that the right-hand-side value of the new constraint (12.25) determined the location of point N along the nondominated border. To find other points along the nondominated surface, the right-hand-side of (12.25) is changed and the problem is resolved. By connecting these points, an approximation to the nondominated surface is formed.

The constraint method process can be generalized for problems with any number of objectives. To make the constraint method clear, it is stated in a series of steps for a multiobjective problem with p objectives.

Step 1. Solve p individual maximization problems to find the optimal solution for each of the individual p objectives.

Step 2. Compute the value of each of the p objectives for each of the p individual optimal solutions. In this way the potential range of values for each of the p objectives is determined. The maximum possible value is the individual objective-maximizing solution. The minimum possible value is the minimum value for that objective found when maximizing the other $p - 1$ objectives individually.

Step 3. For each objective and its range of potential values, select a desired level of resolution and divide the range into the number of intervals determined by this level of resolution. These interval values will be used as the right-hand-side values for the constraints that will be formed for each objective.

Step 4. Select a single objective to be maximized. Transform the remaining $p - 1$ objectives into constraints of the form

$$Z_k \geq L_k \tag{12.26}$$

and add these new $p - 1$ constraints to the original set of constraints, where L_k represents the right-side values that will be varied.

Step 5. Solve the constrained problem set up in step 4 for every combination of right-hand-side values determined in step 3. These solutions form the approximation for the nondominated surface.

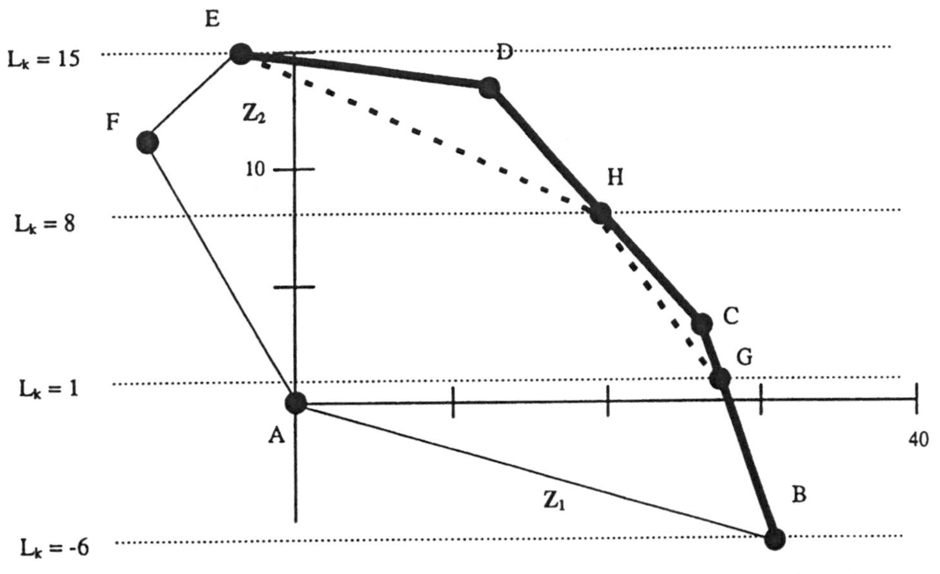

Figure 12.15 Constraint method applied to approximate the nondominated set.

There is an interesting and rich theoretical connection between the weighting and constraint methods, in the context of the fundamental theory of optimization; see Chankong and Haimes (1983) for a thorough discussion.

We apply the constraint method to our two-objective example problem. First Z_1 and Z_2 are maximized individually (points B and E on Figure 12.15, respectively). Next the value of each of the two objectives for each of the p individual optimal solutions is computed (Table 12.4, extreme points B and E). Next, the range of potential values for each objective is determined. In a problem with two objectives, Z_2 achieves its minimum in the noninferior set

TABLE 12.4 Right-Hand-Side Constraint Values Used to Estimate the Nondominated Set

Point	Right-Hand-Side Constraint Value, L_k	x_1	x_2	Z_1	Z_2	
B	−6	6	0	30	−6	(the Z_1 maximizing solution)
G	1	6	1.75	26.6	6	
H	8	4.8	3.2	17.6	8	
E	15	4	4	−3	15	(the Z_2 maximizing solution)

at the solution that maximizes Z_1, and Z_1 achieves its minimum in the noninferior set at the solution that maximizes Z_2 (Beeson, 1971). In this example the range for Z_1 is from -6 to 30 and the range for Z_2 is from -6 to 15. Next, a level of resolution of 3 units is chosen (arbitrarily) for objective Z_2. The range of Z_2 is divided up into three intervals, each 7 units wide. Z_1 is chosen to be maximized (again, arbitrary). A constraint is added for Z_2:

$$Z_2 \geq L_k \tag{12.27}$$

The constrained problem is then solved for each intermediate right-hand-side value in the Z_2 value range, producing points G and H (Table 12.4 and Figure 12.15). The solutions generated are connected to form the approximation to the nondominated set (Figure 12.15).

The strength of the constraint method is its ability to specifically control the resolution of the approximation of the nondominated set. Its major weakness is its inability to locate extreme points, the value of which was described previously.

12.G. APPLICATION OF MCDM TO WATER RESOURCES PLANNING

Ridgley and Rijsberman (1992) published a study that captures and demonstrates many of the key points and methods presented in this chapter. The Rhine River delta at Rotterdam and their approach to supporting decision making about its management are fine examples of decision context, objective hierarchy, and the use of AHP. The significant similarities that exist between AHP and SMART make this example case also instructive for those interested in using SMART. All of the following is adapted from Ridgley and Rijsberman.

The Haringvliet estuary is a part of the Rhine River delta near Rotterdam in the Netherlands (Figure 12.16). Because of flooding from the North Sea, sluice gates were built in the 1960s. In addition to providing flood protection, however, the structures created significant changes in the local ecology by converting a tidal estuary into a freshwater body with greatly increased sedimentation rates. In 1990, reflecting the shift in attitude toward ecology and water resources, the Dutch government began to explore alternatives and measures to restore the ecological health of the delta in the region highlighted in Figure 12.16. In particular, the government wanted to consider new operating rules for the Haringvliet Dam at the mouth to the North Sea and removal of contaminated sediments by dredging.

Deciding how best to manage a dam and whether to dredge in a biophysical setting like this is a challenging problem, requiring sophisticated hydrodynamic and ecological models. But the Rhine River delta case is especially complicated because of development and socioeconomic changes induced by the conversion of the area to a nontidal, freshwater body. With the control

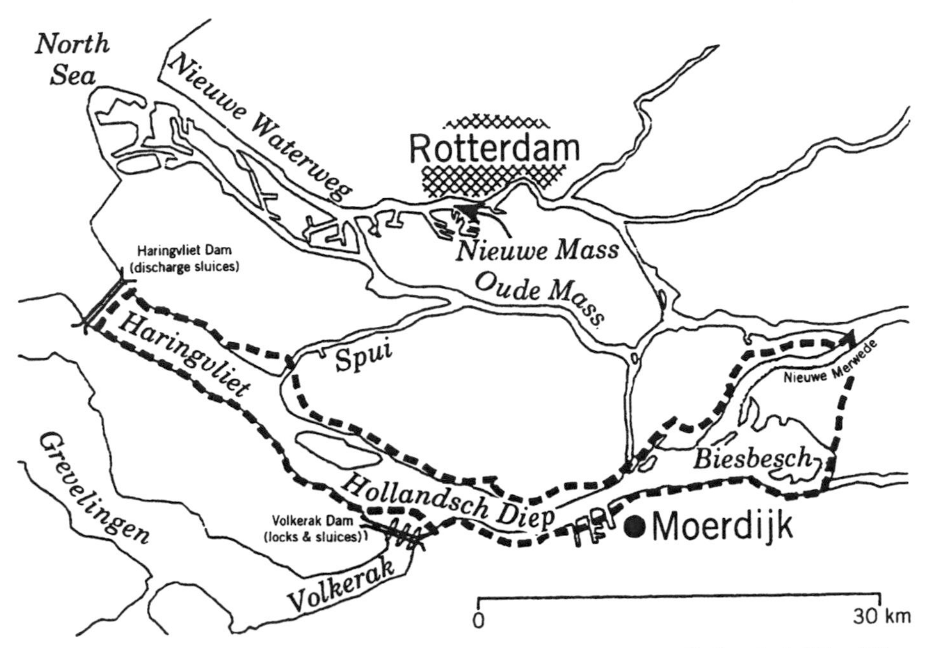

Figure 12.16 Northern Rhine delta, showing the Haringvliet–Hollandsch Diep Biesbosch (HHB) system (delimited) and part of the broader study region. (From Ridgley and Rijsberman, 1992.)

provided by the Haringvliet Dam, the delta had become over the ensuing 30 years an important source of water for agricultural, municipal, and industrial uses, a heavily used recreational resource, and an area with its own biological values, but based on a freshwater wetland ecology. Reverting back to the conditions of the 1960s before the water structures were built would necessarily have a significant impact on all that had since developed. The determination of the best alternative in this case is clearly a problem with multiple stakeholders and multiple, conflicting objectives.

The Dutch government created an interesting structure (Figure 12.17) to deal with the problem. The final decision-making authority rested with the Consultative Committee, which included high-level officials of all relevant government agencies. The committee was charged with choosing a management plan. Representatives of these same agencies joined representatives of stakeholders on the coordinating group, the organizational point where all the interests and analysis came together and conflicts were resolved. The biophysical and socioeconomic complexity of the problem shows up in the 10 work groups that were formed to analyze in detail the many dimensions of the problem. Notice that the stakeholders were represented on the work groups as well, giving interested parties a voice in the detailed analyses as well as in the evaluation of the results.

Consultative Committee
  Government officials
  Role: decision-making authority

⇕

Coordinating Group
  Government officials and stakeholders
  Role: communication and consensus building

⇕

Policy Analysis Team
  Dutch Water Agency
  Role: coordinate work groups and integrate their results

⇕

Ten Work Groups
  Each comprised of people from the Dutch Water Agency, consultants, and stakeholders
  Role: detailed analysis of a specific dimension of the problem
  The 10 work groups were:
    Hydrology and Geomorphology
    Water Quality and Ecology
    Drinking Water Supply
    Flooding and Safety
    Recreation
    Shipping
    Infrastructure
    Sediment Quality and Removal
    Agricultural Water Supply
    Policy Analysis

Figure 12.17 Decision making and analytical organization for the Rhine river delta problem. (Adapted from Figure 2, Ridgley and Rijsberman, 1992, p. 1097.)

We made the distinction earlier between problems with a discrete number of predefined alternatives, which lend themselves to analysis by MCDM techniques, and problems with an infinite number of possible alternatives, which require MOP. This case is an example of the former, but it demonstrates an interesting fact: The problem type is largely a matter of how one chooses to treat it. As Ridgley and Rijsberman point out (p. 1098), there are an infinity of operating schemes for the Haringvliet sluice gates and an infinity of sediment removal plans (and, therefore, an infinite number of combinations of the two). In this case, however, the planners could reasonably choose five individual options for sluice gate management (e.g., no change from the current scheme and, at the other extreme, storm-surge protection only, which provides for maximum tidal fluctuation) and five for sediment removal (e.g., removal of sediments at various quality levels). The alternatives then consisted of the 25 possible combinations of these five options for each of the two management dimensions. Preliminary analysis allowed the 25 alternatives to be reduced to seven, which then became the focus of detailed analysis, evaluation, and decision making.

Our earlier discussion emphasized the importance of the objectives and how time and effort invested early in a project to identify an appropriate set of objectives will pay off in a better evaluation and result. This particular case study is one of the finest demonstrations of this point. Almost four months of this 14-month-long study were devoted to the creation of an objective hierarchy. And what a hierarchy it is! We have reproduced here as Figure 12.18 and Table 12.5 the final hierarchy as presented in Ridgley and Rijsberman (1992). There are three top-level objectives: ecology, human uses, and administration. Each of these is further broken down into subobjectives (e.g., fishing, recreation, etc. under human uses), sub-subobjectives (e.g., beach-based recreation, etc. under recreation) and sub-sub-subobjectives (e.g., Haringvliet, etc. under beach-based recreation). There is a total of 68 bottom-level objectives: 31 under ecology, 20 under human uses, and 17 under administration. This objective hierarchy was the last in a series of seven hierarchies that evolved during those four months of consensus building and revision to achieve the desirable and necessary characteristics of objectives that were discussed earlier.

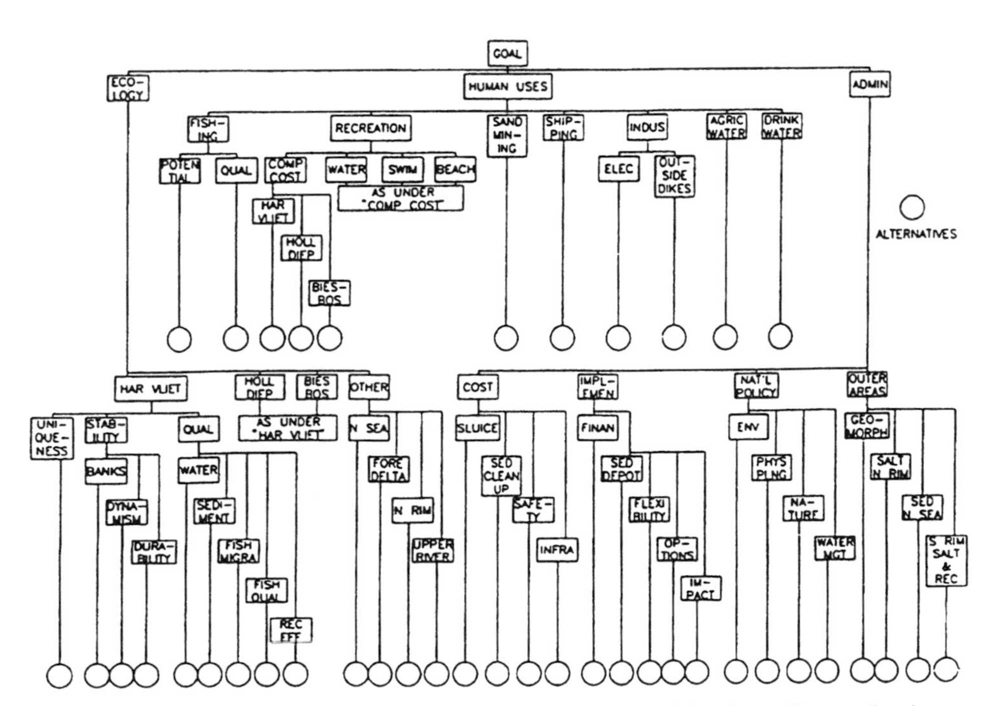

Figure 12.18 Seventh and final objective hierarchy used in the policy evaluation. (From Ridgley and Rijsberman, 1992.)

TABLE 12.5 Definitions of Criteria in the Final Value Tree

Name	Definition
ECOLOGY	Ecological dimension
HAR VLIET	Effects pertaining to Haringvliet
UNIQUENESS	Expected floral/faunal uniqueness in W. Europe
STABILITY	Expected durability without further interventions
BANKS	Bank erosion/stability with water-level changes
DYNAMISM	Duration and frequency of tidal/water-level changes
DURABILITY	Durability based on vegetative successional stages
QUAL	Achievement of area's ecological development potential
WATER	Degree of water pollution
SEDIMENT	Degree of contamination of the benthic zone
FISH MIGRA	Degree to which North Sea fish can pass in and out
FISH QUAL	Types and suitability of fish for human consumption
REC EFF	Recreational impacts: inversely proportional to ecol. quality
HOLL DIEP	Effects pertaining to Hollansch Diep
BIES BOS	Effects pertaining to the Biesbosch
OTHER	Effects pertaining to areas outside the planning area
N SEA	Pollution and sediment effects on North Sea, esp. fisheries
FOREDELTA	Sedimentation in foredelta and transport towards N Sea
N RIM	Water quality along North Rim: Rotterdam harbor, H. IJssel
UPPER RIVER	Effects on fish migration into upper rivers
HUMAN USES	Socioeconomic and cultural dimension
FISHING	Effects on fisheries
POTENTIAL	Fisheries potential: fish type and amount
QUALITY	Fish quality; suitability for human consumption
RECREATION	Effects pertaining to recreation
COMP COST	Cost of compensating measures
WATER	Attractiveness for boating and water recreation
SWIM	Quality of water for swimming
BEACH	Attractiveness and suitability for beach recreation
SAND MINING	Effects on sand rinsing and transfer in Haringvliet
SHIPPING	Effects on shipping (transit, nautical dredging, etc.)
INDUS	Effects on industry outside the dikes
ELEC	Effects on cooling-water supply for power plants
OUTSIDE DIKES	Cost of compensating measures for industry
AGRIC WATER	Compensating costs for agricultural water supply
DRINK WATER	Compensating costs for drinking water supply
ADMIN	Administrative and institutional dimensions
COST	Cost of measures constituting a policy option
SLUICE	Costs of adapting Haringvliet sluices and operations
SED CLEANUP	Costs of removing contaminated sediments
SAFETY	Costs of flood-protection measures
INFRA	Costs of other infrastructure (e.g., bank protection)
IMPLEMENT	Administrative feasibility of implementation
FINAN	Possibilities (ease, sources) of financing
SED DEPOT	Flexibility in dredge-spoil storage options
FLEXIBILITY	Degree to which the policy reduces planning flexibility
OPTIONS	Degree to which the policy creates new opportunities

TABLE 12.5 ***Continued***

Name	Definition
IMPACT	Degree to which existing functions are disturbed
NAT'L POLICY	Compatibility with existing national policies
ENV	National environmental policy ("*NMP+*")
PHYS PLNG	Spatial Planning Policy ("*4e Nota*")
NATURE	Nature policy (*Naturbeleidsplan*)
WATER MGT	Water management ("*3e Nota*")
OUTER AREAS	Implications for areas outside the planning area
GEOMORPH	Geomorphic effects in lower rivers area
SALT N RIM	Saliniation along North Rim (e.g., Rotterdam harbor)
SED N SEA	Sediment/contamination transport into North Sea
S RIM SALT & REC	Recreation and saliniation in southern delta mouth

Source: Ridgley and Rijsberman (1992).

Each of the seven alternatives was analyzed and evaluated in terms of the 68 bottom-level objectives. This large task was done by the work groups. For example, the work groups had to estimate the impact of an alternative, say, maximum tidal range for the sluice gate operation and removal of all sediments worse than a specified quality, on fishing (number of each species and quality), shipping, agricultural water use, bank erosion, cost of flood protection, and every one of the other 68 bottom-level objectives. This is a large analytical challenge that required sophisticated computer models and extensive data collection and management.

The results of these extensive analyses were seven different values (one for each alternative) for each of the 68 bottom-level objectives. Arriving at a decision required value judgments to be made about the relative importance of each group of objectives, starting from the bottom of the hierarchy and moving to the top. For example, the relative weights on fishing "potential" (quality and species of fish) and quality had to be determined in order to arrive at an overall score on fishing. A similar weight determination had to be applied to every other group of bottom-level objectives. Then relative weights had to be determined on fishing, recreation, sand mining, and the rest of the human uses, and a similar weight determination process had to be followed for all of the other groups of objectives one level up. Finally, relative weights were determined for ecology, human uses, and administration.

The analytical hierarchy process (AHP) was used in this case study to determine the relative weights on the objectives. Recall that AHP uses comparisons of pairs of objectives to arrive at weights. For example, at the highest level of the hierarchy, decision makers would consider the relative importance of ecology versus human uses, ecology versus administration, and human uses versus administration. AHP's combining procedure results in a set of weights. Ridgley and Rijsberman used a 1990 version of Expert Choice (see also Expert Choice, 1993) and found its ease of use and interface to have contributed to the success of the project. We have used AHP on Hipre (Systems Analysis Laboratory, 1993) and on VISA (Visual Thinking, 1995) and found them also to be very easy to use. With all of these software packages, the user merely builds the hierarchy and specifies the alternatives; the package takes care of the rest.

The result of the analysis of the Rhine River delta problem was the selection of an intermediate operating scheme for the sluice gates and a decision to study sediment removal further because of a perceived lack of sufficient information. This is a good result for such a complicated problem, and the use of MCDM can be justifiably considered to have been a major factor in coming to a successful resolution.

12.H. APPLICATION OF MOP TO NUCLEAR WASTE MANAGEMENT

There are 119 commercial nuclear reactors in the United States, which by the end of 1995 had generated 32,000 metric tons of spent nuclear fuel. They will continue to generate spent fuel at the rate of 2000 metric tons per year (NWTRB, 1996). What to do with the spent fuel is one of the major unresolved environmental problems in the United States.

The following passage from a recent report from the Nuclear Waste Technical Review Board (NWTRB) summarizes the history and current status of nuclear waste management in the United States.

> The Nuclear Waste Policy Act of 1982, as amended, established a statutory basis for managing the nation's civilian (or commercially produced) spent nuclear fuel. The law established a process for siting, developing, licensing, and constructing an underground repository for the *permanent disposal* of that waste. Utilities were given the primary responsibility for storing spent fuel until it is accepted by the Department of Energy (DOE) for disposal at the repository—originally expected to begin operating in 1998. Since then, however, the repository operation schedule has been delayed several times, and according to testimony submitted to the U.S. Senate by the Secretary of Energy in December 1995, repository operations may be delayed again, perhaps until 2015. These delays, along with the absence of a federal centralized storage facility, similarly delay the prospect of federal acceptance and removal of the spent fuel from utility sites. As a result, much more commercial spent nuclear fuel will require *temporary storage* for much longer periods of time than originally were anticipated. (NWTRB, 1996, p. vii)

Currently, almost all of the 32,000 metric tons of spent fuel are stored in special spent fuel pools—pools of treated water designed to prevent the still highly radioactive material from initiating a chain reaction such as occurs in a controlled way inside a reactor—located on the site of the reactors that generated the spent fuel. The utilities that own the reactors want the U.S.

Department of Energy (DOE) to take possession of the fuel by 1998 or as soon as possible thereafter, as was originally intended by the law. Some environmental groups and local communities also would like to see the spent fuel removed from the reactors and stored in centralized locations until a permanent repository opens. The question, of course, is where to put it? To which, we might add, how do you get it there?

The problem of choosing where to site one or more temporary storage facilities is a complicated one. Finding suitable sites has all of the problems associated with any "noxious" facility, in particular the NIMBY ("not in my backyard") problem. In addition, a major issue is the shipment of nuclear waste from reactors to the storage facilities, and clearly, where the facility is sited will unavoidably affect shipping routes. Thus the spent fuel storage problem requires that three questions be answered simultaneously:

- Where should storage facilities be sited?
- If there is more than one facility, to which facility should each reactor send its waste?
- Over which routes should waste be shipped from reactors to storage sites?

As we will see, this is an intricate analytical problem. The nuclear waste storage problem is also extremely complicated from a decision-making perspective. The President and Congress have decision-making authority, and there are a huge number of stakeholders: communities (and the governments of the states in which they are located) that have reactors and/or storage sites and/or through which waste might be shipped, the utilities and their rate payers, and local and national environmental groups. The following analysis, which is drawn from ReVelle et al. (1991), focuses on two objectives that capture much of the conflict inherent in nuclear waste storage and shipping decisions: shipping cost and shipping risk.

The nuclear waste storage siting and shipping problem was formulated by ReVelle et al. (1991) as a two-objective zero–one integer programming problem:

$$\text{Minimize } Z_1 = \sum_{i=1}^{m} \sum_{j=1}^{n} t_i p_{ij} x_{ij} \tag{12.28a}$$

$$\text{Minimize } Z_2 = \sum_{i=1}^{m} \sum_{j=1}^{n} t_i d_{ij} x_{ij} \tag{12.28b}$$

subject to

$$\sum_{j=1}^{n} x_{ij} = 1 \qquad i = 1,\ldots,m \tag{12.28c}$$

$$y_j \geq x_{ij} \qquad i = 1,\ldots,m, \quad j = 1,\ldots,n \tag{12.28d}$$

$$\sum_{j=1}^{m} y_j = q \tag{12.28e}$$

$$x_{ij}, y_j = 0,1 \qquad i = 1,\ldots,m, \quad j = 1,\ldots,n \tag{12.28f}$$

where

$$x_{ij} = \begin{cases} 1 & \text{if the spent fuel at reactor } i \\ & \text{is assigned to the storage facility at } j \\ 0 & \text{otherwise} \end{cases}$$

$$y_j = \begin{cases} 1 & \text{if a storage facility is established at potential site } j \\ 0 & \text{otherwise} \end{cases}$$

t_i = metric tons of spent fuel at reactor i

p_{ij} = number of people living along the shipping route from reactor i to potential storage site j

d_{ij} = distance in miles of the shipping route from reactor i to potential storage site j

q = number of storage facilities that must be established

Equation (12.28a) is a representation of the risk associated with shipping spent fuel, measured here as the number of people living along shipping routes weighted by the amount of waste shipped along that route. This "people-tons" objective is not an explicit risk measure. It ignores, for example, the probability of radioactive releases and the probability of exposure in the event of releases. Still, it seems to be a relevant characterization of, at least, public perception of risk—the more tons moving past more people, the higher the perceived risk —which, for this problem, may be more determining than actual risk.

Equation (12.28b) is an objective related to shipping cost. Here, again, a surrogate—"ton-miles"—was used, with similar logic: The more waste shipped over longer routes, the higher the cost.

The first constraint, Equation (12.28c), requires that the waste at every reactor be assigned to one storage facility. The constraints in Equation (12.28d) ensure that assignments can only be made to sites where a storage facility will be established (i.e., $x_{ij} = 1$ only if $y_j = 1$). Constraint (12.28e) requires that exactly q storage facilities be opened.

Nondominated solutions were obtained for this problem by using the weighting method, discussed in Section 12.F.2. Before presenting the results, however, there are two interesting aspects of this particular problem that must be noted. First, notice that the formulation in Equations (12.28a)–(12.28f) only determined explicitly the answers to the first two questions posed earlier (i.e., storage location and reactor-storage assignment). The third question of shipping routes is not explicitly part of this problem, but it *is* being decided. Here's how it works (see ReVelle et al., 1991 for details):

- Weights are chosen for the two objectives.
- The weights are used in a shortest-path algorithm to determine the optimal path (in a weighted objective function sense) from reactor i to storage site j.
- The distances of each of these optimal paths, the populations living along them, and the weights used in the first step are used in the weighted objective function of problem (12.28a)–(12.28e).

Thus, although the model does not route explicitly, it is actually choosing predetermined (and optimal) routes when it makes reactor assignments.

The second aspect relates to the special nature of this and all other multiobjective integer programming (IP) problems. Because IPs are, by definition, discrete, their nondominated sets consist of unconnected points (i.e., compare Figure 12.19 with the nondominated set for the linear program in Figure

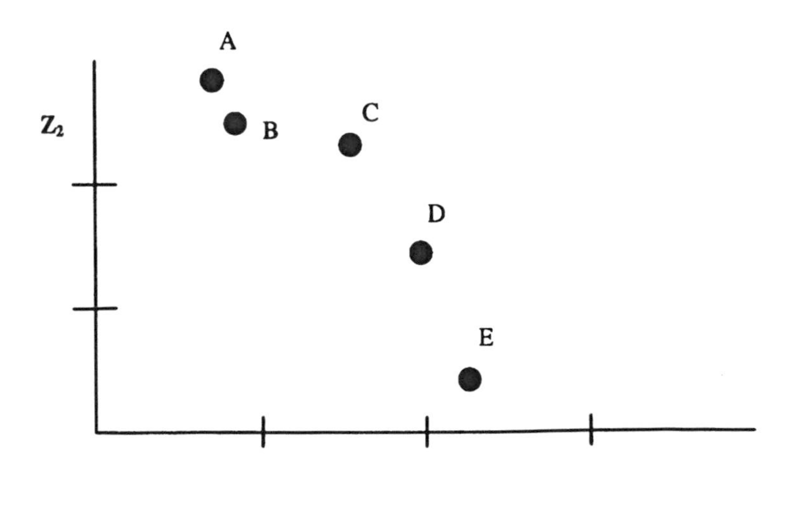

Figure 12.19 Nondominated set for a hypothetical two-objective integer program (IP).

12.9b). The nondominated set for this hypothetical problem is the set of points A, B, C, D, and E and *not* the line segments that connect them.

The discrete nature of multiobjective IPs creates an interesting and practically important problem called a *duality gap* (see Cohon, 1978). Notice that point B in Figure 12.19 lies below the line segment that would connect points A and C. This is emphasized in Figure 12.20, which also demonstrates the central point: There is no weighted objective function that would produce point B as the optimal solution of a weighted problem. Thus, solutions like B cannot be found with the weighting method. Although the constraint method may find such solutions, the computational burden in an IP context may be prohibitive. Duality gaps have provided much of the impetus for the research in multiobjective IP (Rasmussen, 1986). In the following application, the weighting method was used, knowing that potential solutions in the duality gaps could not be identified.

Figure 12.21 shows the location of the 119 commercial reactors in the United States as of the end of 1995. Note that the great majority of the reactors are located in the eastern United States, east of the Mississippi River. This was the region studied by ReVelle et al. (1991) using a subset of the reactors and data from 1980 and focusing on three potential storage locations at Morris Plains (Illinois), West Valley (New York), and Barnwell (South Carolina), all of which are existing nuclear facilities no longer being used. Transportation of spent fuel was assumed to be done by truck on the network of interstate and U.S. highways.

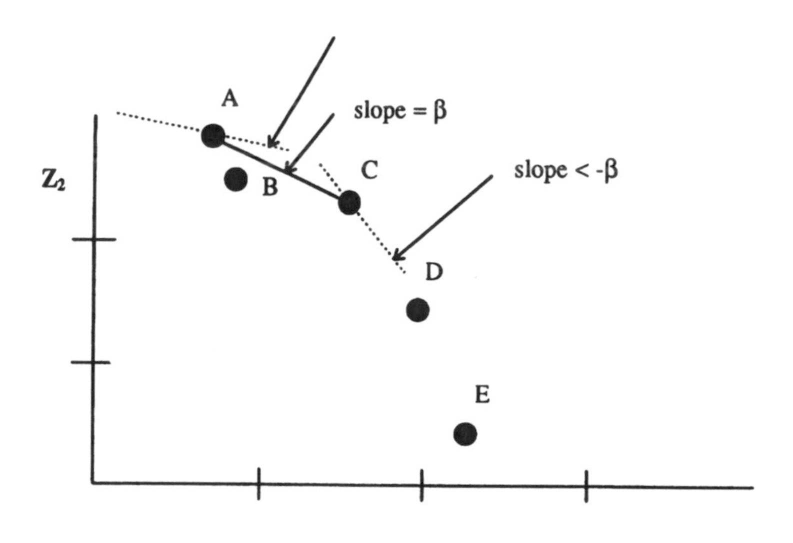

Figure 12.20 Duality gap.

Figure 12.21 Location of the 119 commercial nuclear reactors in the United States as of the end of 1995. (From NWTRB, 1996, p. 9.)

The formulation in (12.28a)–(12.28f) with $q = 2$ (i.e., choose two of the possible storage sites) was used with the weighting method to generate the nondominated set shown in Figure 12.22. Note that due to the discrete nature of the problem, the line segments in Figure 12.22 do not represent nondominated solutions; only the points do. Point A in Figure 12.22 is the lowest people-tons solution (i.e., it is the solution that minimizes Z_1). The solution corresponding to this point is shown in Figure 12.23. Of the three potential storage sites, the two selected are Barnwell (South Carolina) and Morris Plains (Illinois). Figure 12.23 also shows the assignments of the reactors and the storage sites to which they ship. We see that the reactors in Illinois and Michigan ship to Morris Plains (Illinois) and all of the rest ship to Barnwell in South Carolina. Not shown in Figure 12.23 are shipping routes, only because the picture gets too messy.

As we move away from A along the trade-off curve, the cost objective is decreased at the expense of the people-tons objective, which goes up. Notice the unlabeled points between A and B in Figure 12.22. These correspond to new nondominated solutions in which the storage sites and assignments are the same as at A and as shown in Figure 12.23. These solutions differ from A and from each other in the routes chosen to ship waste from reactors to storage sites. As we move from A to B, the routes that avoid population concentrations and are, therefore, longer are replaced by shorter routes that have higher populations.

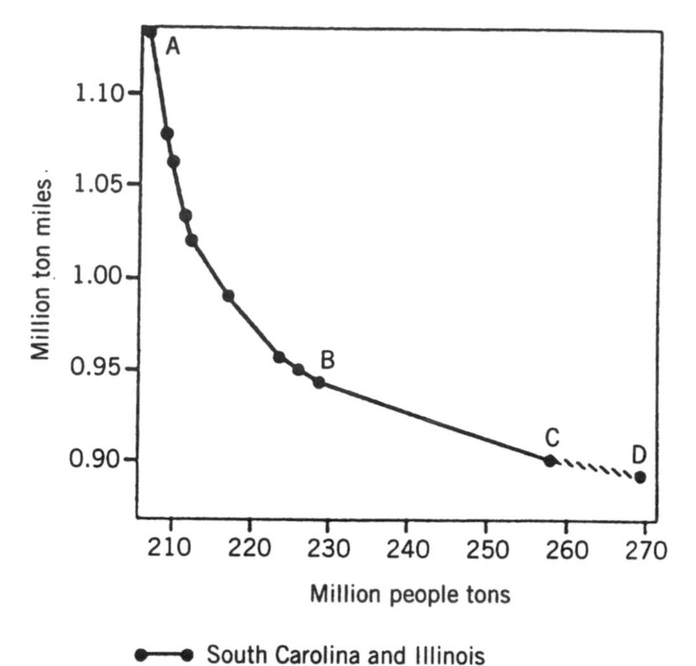

Figure 12.22 Approximated nondominated set. (Line segments between points for presentation purposes only.) (From ReVelle et al., 1991.)

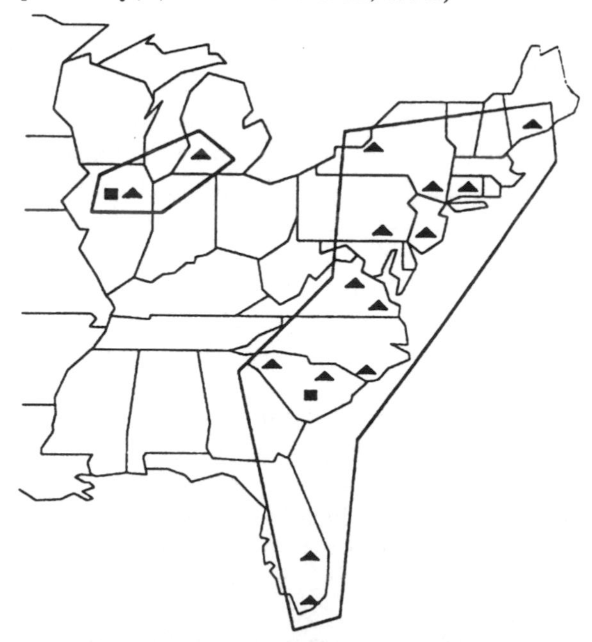

Figure 12.23 Reactor assignments and spent fuel storage sites corresponding to portion *A* up to (but not including) *B* in Figure 12.22. Squares represent storage sites, triangles represent reactors. (From ReVelle et al., 1991.)

Moving further along the trade-off curve in Figure 12.22 toward ever-lower cost (and higher people-tons) takes us to *B*, where there is a shift in reactor assignments as well as routes. Figure 12.24 shows that at *B* the reactor in upstate New York reassigns from Barnwell (South Carolina) to Morris Plains (Illinois). Moving further along the trade-off curve we get to *C* (Figure 12.25) and, finally, the least-cost solution at *D*, where not only have routes and assignments changed, but the set of storage sites has changed to West Valley (New York) and Barnwell (South Carolina).

Unlike the Rhine River delta case, where there was a well-defined decision-making structure, the problem of nuclear waste management in the United States is characterized by a poorly defined and very complicated decision-making situation. Nonetheless, MOP applied in the manner shown here can play a valuable role, even in such complicated and poorly structured problems.

12.I. CONCLUSIONS

As analysts, we strive to solve decision problems in a systematic, sensible, repeatable, and defensible manner. This becomes a tall order to fill as the complexity of a decision problem and the number of stakeholders increase. Each party with an interest in the outcome of a decision may advocate a unique set of desirable objectives.

Multiobjective programming and multiple-criteria decision making techniques can be used to solve decision problems with more than one objective. The significant feature of MOP and MCDM is that there is no need to convert objectives into a common currency. Analysts and decision makers are able to focus their efforts on creating an accurate and complete representation of the decision problem. Using MOP, optimal alternative solutions are generated independent of stakeholders' preferences for the importance of the various criteria. The trade-offs between the objectives for the alternative solutions are explicitly quantified as a result of the analysis. In other words, the decision makers can see exactly what they must give up and what they get with each alternative solution. Using MCDM, the decision makers can invoke their preferences for the various objectives to identify a final preferred solution. Again, MCDM proceeds without conversion of the objectives into a common currency.

It is worth restating that real-world decision problems are not always readily classified between those appropriate for MOP analysis and those appropriate for MCDM analysis. In fact, it is possible to state almost any decision problem in the format appropriate for either MOP or MCDM methods. The choice of the right analysis method can be based on the analyst's prior experiences with the techniques and/or the decision makers' desired results from the analysis. If resources permit, it may be desirable to compare the results of two or more methods.

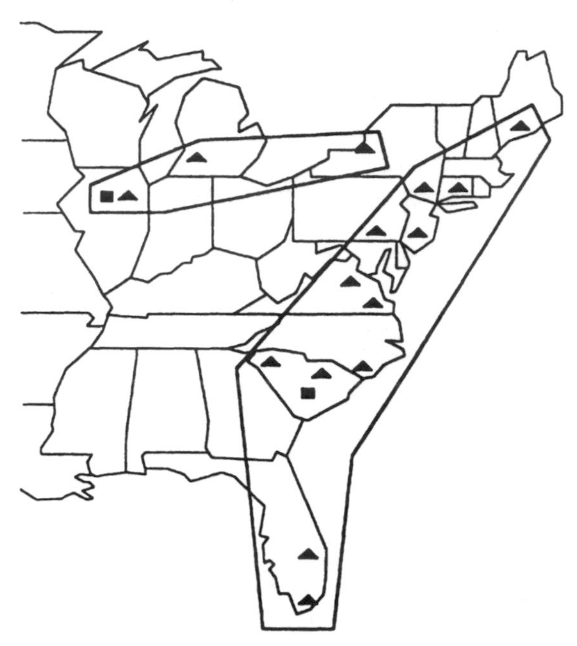

Figure 12.24 Reactor assignments and spent fuel storage sites corresponding to portion of trade-off curve from B up to (but not including) C in Figure 12.22. (From ReVelle et al., 1991.)

Figure 12.25 Reactors assignments and spent fuel storage sites corresponding to portion of trade-off curve from C to D in Figure 12.22. Squares represent storage sites, triangles represent reactors. (From ReVelle et al., 1991.)

REFERENCES

Beeson, R., 1971. Optimization with Respect to Multiple Criteria, Ph.D. thesis, University of Southern California, Los Angeles.

Chankong, V., and Y. Y. Haimes, 1983. *Multiobjective Decision Making: Theory and Methodology,* North-Holland, New York.

Cochrane, J., and M. Zeleny, 1973. *Multiple Criteria Decision Making,* University of South Carolina Press, Columbia, SC.

Cohon, J., 1978. *Multiobjective Programming and Planning,* Academic Press, San Diego, CA.

Cohon, J., and D. Marks, 1973. Multiobjective Screening Models and Water Resources Investment, *Water Resources Research,* 9(4), 826–836.

Edwards, W., 1971. Social Utilities, *Engineering Economist,* Summer Symposium Series 6.

Edwards, W., 1977. How to Use Multiattribute Utility Measurement for Social Decision Making, *IEEE Transactions on Systems, Man and Cybernetics,* 7(5), 326–340.

Edwards, W., and J. R. Newman, 1982. *Multiattribute Evaluation,* Sage, Newbury Hills, CA.

Expert Choice, 1993. *Expert Choice, Version 8,* Expert Choice, Inc., Pittsburgh, PA.

Goodwin, P., and G. Wright, 1991. *Decision Analysis for Management Judgement,* Wiley, New York.

Keeney, R., 1988. Structuring Objectives for Problems of Public Interest, *Operations Research,* 36(3), 396–405.

Keeney, R., and H. Raiffa, 1976. *Decisions with Multiple Objectives,* Wiley, New York. (Reprinted in paperback in 1993 by Cambridge University Press.)

Koopmans, T. C., 1951. *Activity Analysis of Production,* Wiley, New York.

Kuhn, H., and A. Tucker, 1951. Nonlinear Programming, in J. Neyman (ed.), *Proceedings of the 2nd Berkeley Symposium on Mathematical Statistics and Probability,* University of California Press, Berkeley, CA, pp. 481–492.

Major, D., and R. Lenton, 1979. *Applied Water Resource Systems Planning,* Prentice Hall, Upper Saddle River, NJ.

National Research Council, 1995. *Measuring and Improving Infrastructure Performance,* National Academy Press, Washington, DC.

NWTRB, 1996. *Disposal and Storage of Spent Nuclear Fuel: Finding the Right Balance,* Nuclear Waste Technical Review Board, Arlington, VA.

Rasmussen, L. M., 1986. Zero–One Programming with Multiple Criteria, *European Journal of Operational Research,* 26, 83–95.

ReVelle, C., J. Cohon, and D. Shobrys, 1991. Simultaneous Siting and Routing in the Disposal of Hazardous Wastes, *Transportation Science,* 25(2), 138–145.

Ridgley, M., and F. Rijsberman, 1992. Multicriteria Evaluation in a Policy Analysis of a Rhine Estuary, *Water Resources Bulletin,* 28(6), 1095–1110.

Ruckelshaus, W. D., 1995. Stopping the Pendulum, speech given at the Environmental Law Institute, Washington, DC, October 18.

Saaty, T. L., 1977. A Scaling Method for Priorities in Hierarchical Structures, *Journal of Mathematical Psychology,* 15, 234–281.

Saaty, T. L., 1988. *Decision Making for Leaders,* RWS Publications, Pittsburgh, PA.

Saaty, T. L., 1990. *The Analytical Hierarchy Process,* McGraw-Hill, New York.

Saaty, T. L., 1994. How to Make a Decision: The Analytical Hierarchy Process, *Interfaces,* 24(6), 19–43.

Steuer, R., 1986. *Multiple Criteria Optimization: Theory, Computation, and Application,* Wiley, New York.

Systems Analysis Laboratory, 1993. *Hipre 3+, Decision Support Software,* Helsinki University of Technology, Espoo, Finland.

Visual Thinking, 1995. *VISA: Visual Interactive Sensitivity Analysis,* Visual Thinking International Limited, Mississauga, Ontario, Canada.

Zeleny, M., 1982. *Multiple Criteria Decision Making,* McGraw-Hill, New York.

CHAPTER 13

Cost Allocation in Water Resources

JAMES HEANEY

13.A. INTRODUCTION

The purpose of this chapter is to provide an overview of the cost allocation problem in water resources. Cost allocation is required whenever a project deals with multiple purposes and/or groups. Also, the water resources field is very capital intensive (i.e., a large capital investment is required to generate revenue). Thus assignment of these capital costs among the various purposes and groups is a very important issue. In the next section we provide a broad introduction to the interrelated issues of efficiency and equity. Then conventional methods of cost allocation in the water resources field are reviewed. Related work in cooperative *n*-person game theory and cost allocation is described. Then approaches used by accountants to allocate costs are summarized.

In the latter part of the chapter we summarize the concepts for cost allocation. Then the important question of the context in which cost allocation takes place is described. Each cost allocation problem has its unique history and set of agreements as to what constitutes a "fair" division of costs. These concepts are described and selected applications are presented. Finally, summary and conclusions are included.

13.B. EFFICIENCY AND EQUITY

Major advances have been made in increasing the economic efficiency of water resource projects by taking advantage of some or all of the following (Heaney and Dickinson, 1982):

1. Economies of scale in production and distribution facilities
2. The assimilative capacity of the receiving environment
3. Excess capacity in existing facilities
4. Multipurpose opportunities

5. Multigroup cooperation

The results of the economic analysis will typically indicate that the best overall solution is a complex blend of management options. Unfortunately, the increasing sophistication of the optimal economic solution results in a very complex cost allocation problem, since it is necessary to divide the costs of this project among many purposes and groups. Thus the search for improved economic efficiency exacerbates the financial analysis of how these costs should be divided. Methods from cooperative n-person game theory can provide some help in sorting through this complex efficiency-equity question. Also, conventional cost allocation procedures are reviewed to describe how the problem has been solved in practice.

Young (1994a) presents a thorough review of the broad concept of equity and how it relates to the cost allocation problem. Young (1994a) defines key terms as follows:

- An *allocation* problem occurs whenever a bundle of resources, rights, burdens, or costs is temporarily held in common by a group of persons and must be allotted to them individually.
- An allocation is an assignment of the objects to specific individuals.
- An allocation is a decision about who gets a good or who bears a burden and is usually decided by a group or an institution acting on behalf of the group.
- Exchange involves many voluntary, decentralized transactions and can occur only after the goods and burdens have been allocated.
- Allocation comes first, exchange afterward.

With regard to *equity,* Young (1994a) stresses that it is a complex issue: "Equity is a complex idea that resists simple formulations. It is strongly shaped by cultural factors, by precedent, and by the specific types of goods and burdens being distributed. To understand what equity means in a given situation, we must therefore look at the contextual details." Young (1994a) analyzes the following seven cases to demonstrate the wide diversity of problems that have complex equity implications:

1. Demobilization of soldiers from the U.S. Army at the end of World War II
2. Allocation of kidneys among transplant patients
3. Apportionment of representation among political parties and states
4. Adjudication of conflicting property claims
5. Allocation of benefits and costs among participants in joint enterprises
6. Distribution of tax burden

7. Division of inheritances

The equity problem of most direct interest to the water resources field is the fifth category listed above, although some of the equity issues in water resources are related to tax burdens.

As Young (1994a) points out, the study of equity has close ties to the axiomatic method in mathematics. By postulating one or more axioms of equity, one can evaluate whether a proposed solution meets these conditions. Unfortunately, when several seemingly reasonable axioms are combined, the result may be that no method satisfies these seemingly reasonable conditions, (e.g., Arrow's impossibility theorem; Arrow, 1963). A second limitation of the axiomatic method is that it can oversimplify real problems by not accounting properly for the actual, complex context within which cost allocation problems exist. By focusing on the axiomatic method, we may lose the reality of the problem. Thus the preferred strategy is a blend of the axiomatic method and a thorough understanding of the problem context. Alternatively, equity is not necessarily an ethical or moral judgment, but rather, what society deems to be *appropriate,* which is a combination of principles and precedent (Young, 1994a).

Allocation rules usually exhibit one or a combination of three concepts of equity (Young, 1994a):

- *Parity.* Claimants are treated equally.
- *Proportionality.* Differences among claimants are acknowledged and the goods divided in proportion to these differences.
- *Priority.* The person with the greatest claim to the good gets it.

Students of economics have detailed knowledge of the exchange process. However, until recently, economics has been virtually silent on the allocation process. By assuming that all property has been divided, they sidestep the issue of how common property should be allocated (Young, 1994a). In the water resources field in particular, much of the resources are held in common; thus the allocation issue is of paramount importance. A contemporary example is the World Wide Web and related communications, which will have a profound effect on the future of the world, depending on how these resources are allocated.

Elsewhere in this book there are examples of problem formulations that deal with cost allocation, efficiency, and equity. In Chapter 1 (Section 1.H) a linear programming model for a multipurpose reservoir is presented. Chapter 2 (Section 2.D.3) deals with the problem of assigning treatment efficiencies to multiple dischargers into a single stream. Problems relating to effluent charges and transferable discharge permits are treated in Chapter 15.

13.C. COST ALLOCATION IN WATER RESOURCES

13.C.1. Early Debate at TVA

The earliest reported literature on the cost allocation problem in water resources is in a book by Ransmeier (1942), who reports on the results of several years of debate on how the costs of the Tennessee Valley Authority (TVA) should be divided among (1) flood control, (2) navigation, (3) fertilizer production, (4) national defense, and (5) development of power (Straffin and Heaney, 1981). This debate was important because it represented the first time that public water projects would compete with private water development. In particular, there was strong concern that multipurpose public water projects could outcompete the existing private hydropower development because a significant part of the total cost could be assigned to other purposes.

The outcome of these deliberations was five criteria for cost allocation (Ransmeier, 1942):

1. The method should have a reasonable logical basis. It should not result in charging any objective with a greater investment than the fair capitalized value of the annual benefit of this objective to the consumer. It should not result in charging any objective with a greater investment than would suffice for its development at an alternative single-purpose site. Finally, it should not charge any two or more objectives with a greater investment than would suffice for alternate dual-purpose or multiple-purpose improvement.
2. The method should not by unduly complex.
3. The method should be workable.
4. The method should be flexible.
5. The method should apportion to all purposes present at a multiple-purpose enterprise a share in the overall economy of the operation.

The conditions described in criterion 1 can be represented mathematically as follows:

$$x(i) \leq \min[b(i),c(i)] \qquad \forall i \in N \tag{13.1}$$

where $x(i)$ is the cost allocated to group i, $b(i)$ the benefit of group i, $c(i)$ the alternative cost if group i acts independently, N the set of all groups, equal to $\{1,2,...,i,...,n\}$, and

$$\sum_{i \in S} x(i) \leq c(S) \qquad \forall S \subset N \tag{13.2}$$

where $c(S)$ is the alternative cost if subset S acts independently and S is any subset of the master set N.

The TVA group could not agree on whether it was essential that total costs be paid, or

$$\sum_{i \in N} x(i) = c(N) \tag{13.3}$$

Economists insisted on using marginal costs to optimize efficiency, while the remainder of the group insisted that it is reasonable that the sum of the allocations equal the total cost.

13.C.2. Cost Allocation in the Federal Water Agencies

Large-scale federally sponsored water development after World War II brought attention to the need to develop cost allocation methods for water projects. The separable costs, remaining benefits (SCRB) method developed out of this initiative (Federal Interagency River Basin Committee, 1950). Separable costs are defined as

$$sc(i) = c(N) - c[(N) - \{i\}] \qquad \forall i \in N \tag{13.4}$$

where $sc(i)$ is the separable cost to group i, $c(N)$ the total cost for grand coalition of n groups, and $c[(N - \{i\}]$ the total cost for the grand coalition with group i excluded.

Alternatively, separable costs are the incremental costs of adding group i as the last member of the grand coalition of N members. If economies of scale exist, the lowest incremental costs occur when a group joins last. Thus joining last is assumed to be the most favored position from a cost-allocation point of view. If group i cannot pay $sc(i)$, they will have to be subsidized by the other groups.

Each group is assigned its separable costs. The remaining nonseparable costs (NSC) that must be assigned are

$$\text{NSC} = c(N) - \sum_{i \in N} sc(i) \tag{13.5}$$

Each group's share of NSC is based on the ratio of their remaining benefits to the total remaining benefits, or

$$\beta(i) = \frac{\min[(b(i),c(i)] - sc(i)}{\sum_{i \in N} \{[\min(b(i),c(i))] - sc(i)\}} \tag{13.6}$$

where $\beta(i)$ is the prorating factor using the separable costs, remaining benefits method. Unlike the TVA committee, the Federal Interagency River Basin Committee (1950) required that the total cost be paid, or

$$\sum_{i \in N} \beta(i) = 1.0 \tag{13.7}$$

Thus, using the SCRB method, the total charge to the ith group is

$$x(i) = sc(i) + \beta(i)(\text{NSC}) \tag{13.8}$$

13.C.3. Cost Allocation in the Water Supply Industry

Beecher et al. (1991) provide a current summary of cost allocation methods used in the urban water supply field. The decision areas and principal considerations in cost allocation are summarized in Table 13.1. As you can see from perusal of the table, there are many facets to the cost allocation problem.

13.D. GAME THEORY METHODS AND COST ALLOCATION

13.D.1. Cooperative *N*-Person Game Theory and Cost Allocation

Independent of the water resources field, game theorists were pondering the same questions of efficiency and equity using cooperative N-person game theory. A good current summary of this topic can be found in Owen (1995) or Aumann and Hart (1994). Moulin (1995) presents a current summary of applications of game theory to economics. Young (1994b) provides a nice explanation of game theory and cost allocation. We will focus on n-person cooperative games in characteristic function form. Game theory is also used in Chapter 14 (Section 14.E.4) in the context of regional wastewater treatment.

Three types of cooperative games are presented in the literature. Games in which the players in the coalition seek to maximize profits or net revenues are called *value games.* Games in which the coalition of players seeks to minimize costs are called *cost games.* Cost games can be converted to *savings games* by measuring savings relative to the cost of not participating in a coalition. Thus savings are zero if a player chooses not to participate. These three categories are interrelated and lead to equivalent results. To keep the notation as simple as possible, only cost games are discussed. The following example is used to help the reader understand these concepts. Ransmeier (1942) describes the negotiations at TVA regarding how costs should be allocated for a three-purpose reservoir. Straffin and Heaney (1981) provide more details on this example. The problem is to assign the costs of the reservoir among the following three purposes: (1) navigation, (2) flood control, and (3) hydropower. The costs for each of the coalitions are

$$c(1) = \$163{,}520 \qquad c(2) = \$140{,}826 \qquad c(3) = \$250{,}096$$

$$c(12) = \$301{,}607 \qquad c(13) = \$378{,}821 \qquad c(23) = \$367{,}730$$

$$c(123) = \$412{,}584$$

TABLE 13.1 Cost Allocation and Rate Design for Water Utilities: Decision Areas and Principal Considerations

Decision Area	Principal Considerations
Identification of revenue requirement	Capital investments/rate base
	Return on rate base
	Operation and maintenance expenses
	Depreciation
	Taxes
Cost functionalization	Source development
	Pumping
	Transmission
	Treatment
	Storage
	Distribution
	Nontraditional supply
Cost classification	Customer costs
	Capacity (demand) costs
	Commodity (operating) costs
Cost allocation	Functional cost
	Commodity demand
	Base-extra capacity
	Embedded direct
	Fully distributed
	Marginal/incremental
Cost assignment	Residential
	Commercial
	Industrial
	Wholesale
	Institutional
	Public authorities
	Fire protection
Rate design	Flat rates
	Fixture rates
	Uniform rates
	Decreasing block rates
	Increasing block pricing
	Seasonal rates
	Excess use charges
	Indoor/outdoor rates
	Lifeline rates
	Sliding scale pricing
	Scarcity pricing
	Spatial pricing
Tariff design	Customer charges
	Capacity (demand) charges
	Commodity (operating) charges
	Dedicated-capacity charges
	Capital contributions
	Fire protection charges
	Ancillary charges

Source: Beecher et al. (1991).

An N-person game (N,c) in characteristic function form consists of a set $N = 1,2,...,n$ of players along with the characteristic function c, which assigns the real number $c(S)$ to each nonempty subset S of players. Cost games are subadditive; that is,

$$c(S) + c(T) \geq c(S \cup T) \qquad S \cap T = \emptyset, \quad S,T \subset N \tag{13.9}$$

where $\emptyset$ is the empty set.

For a three-person game, we need to check:

$$c(1) + c(2) \geq c(12) \quad \text{or} \quad 304{,}346 \geq 301{,}607 \quad \text{OK}$$

$$c(1) + c(3) \geq c(13) \quad \text{or} \quad 413{,}616 \geq 378{,}821 \quad \text{OK}$$

$$c(2) + c(3) \geq c(23) \quad \text{or} \quad 390{,}922 \geq 367{,}730 \quad \text{OK}$$

$$c(1) + c(23) \geq c(123) \quad \text{or} \quad 531{,}250 \geq 412{,}584 \quad \text{OK}$$

$$c(2) + c(13) \geq c(123) \quad \text{or} \quad 519{,}647 \geq 412{,}584 \quad \text{OK}$$

$$c(3) + c(12) \geq c(123) \quad \text{or} \quad 551{,}703 \geq 412{,}584 \quad \text{OK}$$

If the engineering economic analysis has been done correctly, subadditivity should be satisfied automatically. In the engineering economic analysis, all coalition possibilities are considered. The characteristic function is defined as the minimum cost alternative for all the options. For example,

$$c(12) = \min[c(1) + c(2), c(12)'] \tag{13.10}$$

where $c(12)'$ is the project cost for a joint venture. Thus the worst that can happen is that there are no savings from the joint venture. In this case

$$c(12) = c(1) + c(2) \tag{13.11}$$

Such coalitions are said to be *inessential* (i.e., nothing is gained).

Interestingly, game theorists in the 1930s developed virtually the same axioms of fairness as did the TVA group working completely independently (Heaney and Dickinson, 1982). Game theorists enumerate three general axioms that a fair solution to a cost game should satisfy. The first criterion is that the costs assigned to the ith group, $x(i)$, must not exceed their costs if they act independently; that is,

$$x(i) \leq c(N) \qquad \forall i \in N \tag{13.12}$$

The second criterion states that the total cost, $c(N)$, must be apportioned among the n groups; that is,

$$\sum x(i) = c(N) \tag{13.13}$$

All solutions satisfying conditions (13.12) and (13.13) are called *imputations.* The third criterion extends the first criterion by insisting that the cost assigned to each group, S, be no more than the costs they would incur in any coalition, S, contained in N:

$$\sum x(i) \leq c(S) \qquad \forall S \subset N \tag{13.14}$$

All solutions satisfying conditions (13.12)–(13.14) constitute the *core* of the game. The core is a very important notion in cost allocation. The core represents the set of *fair* solutions in the sense defined above. Thus proposed cost-sharing vectors, $\mathbf{x} = (x_1, x_2, \ldots, x_n)$ should fall in the core. For the TVA example, the core conditions are:

$$\begin{aligned} x(1) \qquad\qquad\qquad\qquad &\leq 163{,}520 \\ x(2) \qquad\qquad &\leq 140{,}826 \\ x(3) &\leq 250{,}096 \\ x(1) + x(2) \qquad\qquad &\leq 301{,}607 \\ x(1) \qquad\qquad + x(3) &\leq 378{,}821 \\ x(2) + x(3) &\leq 367{,}730 \\ x(1) + x(2) + x(3) &= 412{,}584 \end{aligned} \tag{13.15}$$

Properties of the Core. For two-person games, the core always exists. For example, consider a subset of the three-purpose TVA example, purposes 1 and 3. The core is shown below.

$$\begin{aligned} x(1) \qquad\qquad &\leq 163{,}520 \\ x(3) &\leq 250{,}096 \\ x(1) + x(3) &= 378{,}821 \end{aligned} \tag{13.16}$$

The bounds on the charges to $x(1)$ and $x(3)$ are

$$\begin{aligned} 128{,}725 \leq x(1) \leq 163{,}520 \\ 215{,}301 \leq x(3) \leq 250{,}096 \end{aligned} \tag{13.17}$$

The lower bounds on the charges to $x(1)$ are the difference between the total cost and the maximum charge for $x(3)$, and vice versa. The core for this two-purpose game is shown in Figure 13.1. By plotting the three conditions, we see that the core is a straight line with the upper and lower bounds indicated

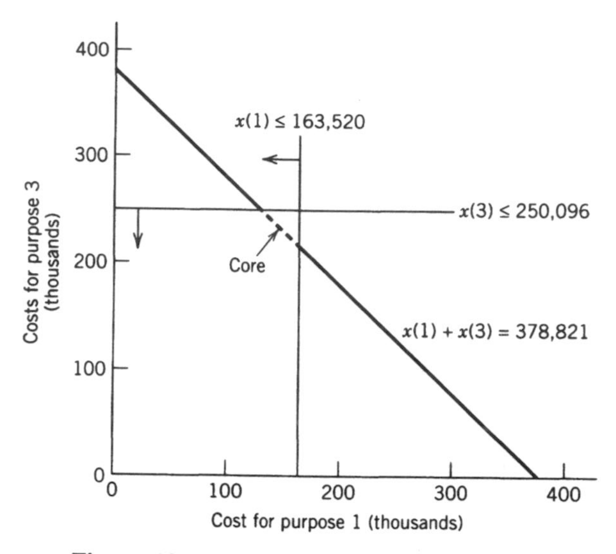

Figure 13.1 Core of two-person cost game.

above. If the game is inessential, the core is a single point. For two-person games, the core always exists and the cost allocation is relatively simple.

For three- or more-person games, the core conditions become much more complex. Three types of cores can be described: convex, nonconvex, and empty. A *convex* core is one that satisfies the following conditions:

$$c(S) + c(T) \geq c(S \cup T) + c(S \cap T) \qquad \text{for } S \cap T \neq \emptyset \tag{13.18}$$

For a three-person cost game, we need to check the following conditions in addition to the subadditivity conditions. The results for the TVA cost game are also presented.

$$c(12) + c(13) \geq c(123) + c(1) \quad \text{or} \quad 680{,}428 \geq 576{,}104 \qquad \text{OK}$$

$$c(12) + c(23) \geq c(123) + c(2) \quad \text{or} \quad 669{,}337 \geq 553{,}410 \qquad \text{OK}$$

$$c(13) + c(23) \geq c(123) + c(3) \quad \text{or} \quad 746{,}551 \geq 662{,}680 \qquad \text{OK}$$

Thus the example game is convex. Convexity is a highly desirable property of cost games. It means that the savings in the game are relatively large and it will be relatively easy to find a cost allocation vector that will satisfy all participants.

The second case is that the core is *nonconvex* but it is not empty. If the cost of the grand coalition in our example is raised from \$412,584 to \$515,000, the convexity conditions are no longer satisfied since

$$c(13) + c(23) \geq c(123) + c(3) \quad \text{or} \quad 746{,}551 \geq 765{,}096 \qquad \text{No}$$

As the cost of $c(123)$ increases, the grand coalition becomes less desirable and the core shrinks in size. Thus it will be harder to find an acceptable vector of assigned costs that satisfies the core conditions.

The worst case is that the core is *empty.* In this case no cost allocation satisfies the core conditions. This can happen even though the cost game is subadditive. Cost games with nonempty cores are inherently unstable since we cannot offer a solution that does not leave at least one purpose or group worse off than they would be if they simply formed their subcoalition.

The core of three-person cost games can be plotted using isometric paper with equilateral triangles as the unit cells. Isometric paper can be used for cases where the sum of the three values is a constant. The core conditions for plotting the example on isometric paper are described below.

$$\begin{aligned} 44{,}854 &\leq x(1) \leq 163{,}520 \\ 33{,}763 &\leq x(2) \leq 140{,}826 \\ 110{,}977 &\leq x(3) \leq 250{,}096 \\ \textstyle\sum x(i) &= 412{,}584 \end{aligned} \tag{13.19}$$

The upper bounds are obtained directly from the core bounds. The lower limits are obtained as follows:

$$x_i(\min) = c(123) - c(jk) \tag{13.20}$$

For example,

$$x_1(\min) = c(123) - c(23) = 412{,}584 - 367{,}730 = 44{,}854.$$

Figure 13.2 shows the core for this TVA cost game. The core is relatively large. For convex games, the nominal bounds shown above turn out to be the actual core bounds. For three-person games, the core geometry will vary as the values of $c(123)$ vary. A cost game can be converted to a savings game by using the no cooperation situation as the base case. Furthermore, the savings game can be normalized so that the total savings equal 1. The geometry of the core for this example, shown in Figure 13.3 in normalized savings form, illustrates how the core shrinks as $c(123)$ increases. If the cost game is nonconvex, at least one of the upper and/or lower nominal bounds of the core are outside the core. Using the TVA example, with $c(123) = \$515{,}000$, the nominal core upper bound on $x(3)$ is \$250,096, while the upper core bound on $x(3) = \$231{,}551$. Thus the maximum cost assessment on purpose 3 must not exceed the upper core bound of \$231,551, not the nominal upper bound

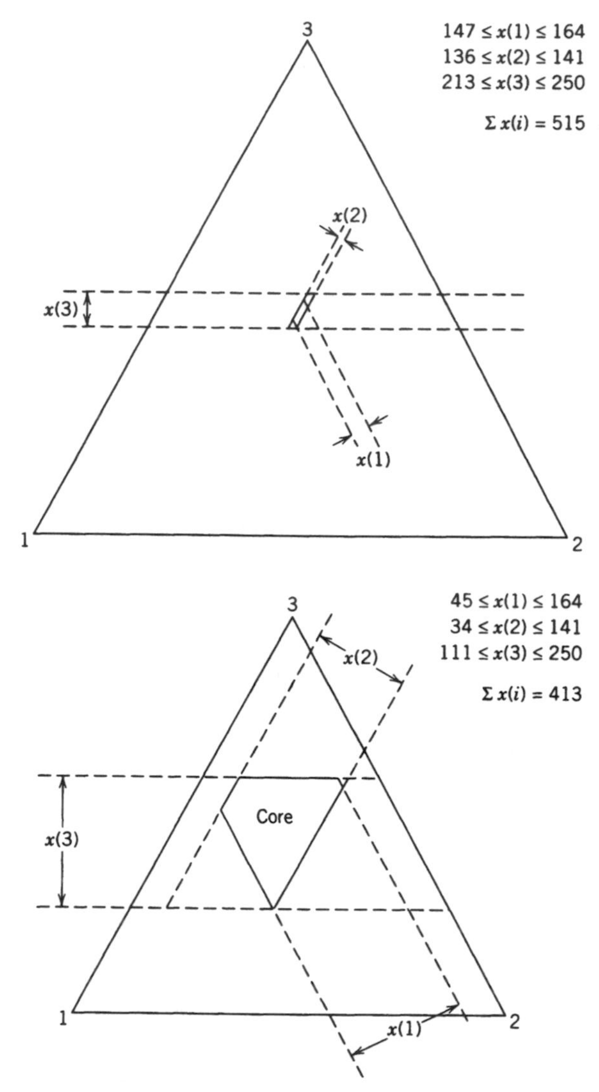

Figure 13.2 Cores of three-person cost games with $c(N) = 412{,}584$ and 515,000.

of \$250,096. The actual core bounds can be read directly from the isometric plot. They also can be found directly by solving the following $2n$ linear programs:

$$\text{maximize or minimize } Z = x(i)$$

subject to

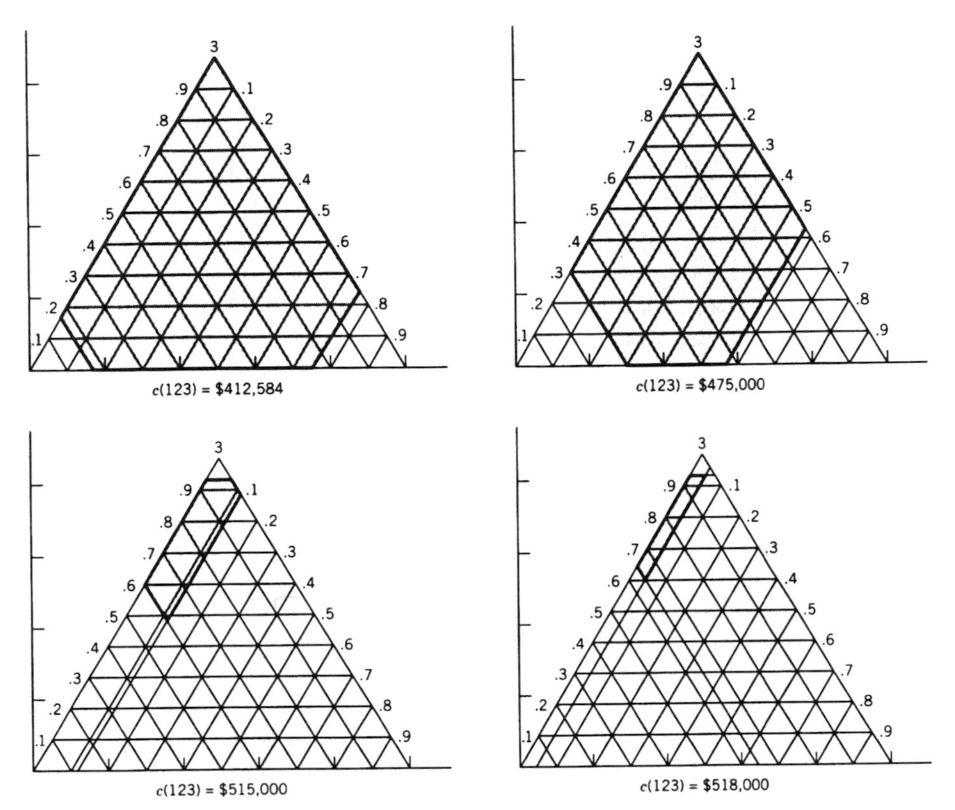

Figure 13.3 Effects of $c(N)$ on the size of the core.

$$\begin{aligned} \sum x(i) &\leq c(i) && \forall i \in N \\ \sum x(i) &\leq c(S) && \forall S \subset N \\ \sum x(i) &= c(N) && \\ x(i) &\geq 0 && \forall i \in N \end{aligned} \tag{13.21}$$

These $2n$ linear programs are simple to solve since only the objective function needs to be changed. LP solvers are built in to contemporary spreadsheets. All calculations in this chapter were done using Excel.

The core for the three-person TVA cost game with $c(N) = 515{,}000$ is shown in Figure 13.2. By comparison with the original cost game with $c(N) = 412{,}584$, the core is barely visible at the same scale. This very small core tells us several things, including the following:

1. The savings are relatively small.

2. The required minimum charges to each purpose are relatively large. For example, if $c(N) = 515{,}000$, the sum of the minimum charges is \$496,842. Thus the remaining costs to be apportioned are only \$18,158, whereas they were \$222,990 with $c(N) = \$412{,}584$.
3. There is a high probability that conventional cost allocation formulas will yield a solution that is not in the core.
4. Even a modest added charge to account for the transactions costs associated with a cooperative solution would probably absorb all the potential savings. Alternatively, it may not be worth the hassle of forming a three-person coalition if the savings are relatively minor.

13.D.2. Unique Solution Notions from Game Theory

Shapley Value. The two most popular unique solution notions from game theory are the Shapley value and the nucleolus. In proposing the concept of the *Shapley value,* Shapley (1953) argues that each participant should pay the incremental cost of adding him to the coalition; that is, the incremental charge for the ith player is

$$c(S) - c(s - [I]) \tag{13.22}$$

However, we don't know in which order the coalitions will form. In the three-person case, six coalition formation sequences exist:

123	213	312
132	231	321

Shapley (1953) argues that in the absence of other information on coalition formation sequences, assume that each sequence above is equally likely. Then the probability that city i will be the last one to join an S-member coalition is $\alpha_i(S)$, where

$$\alpha_i(S) = \frac{(s - 1)!\ (n - s)!}{n!} \tag{13.23}$$

Thus each player pays the incremental cost that he or she brings to coalitions expected out over all coalition formation sequences; that is,

$$\phi_i = \sum_{S \subseteq N} \alpha_i(S)[(c)S) - c(s - \{i\})] \tag{13.24}$$

Consider the TVA example with the following coalition costs:

$$c(1) = 163{,}520 \qquad c(2) = 140{,}826 \qquad c(3) = 250{,}096$$
$$c(12) = 301{,}607 \qquad c(13) = 378{,}821 \qquad c(23) = 367{,}730$$
$$c(123) = 412{,}584$$

The Shapley value is calculated for this problem is calculated in Table 13.2. The sum of the Shapley values equals \$412,584, the total cost of the project.

If the core is convex, the Shapley value is in the center of the core. However, for nonconvex games, the Shapley value may fall outside the core—a very undesirable property.

Nucleolus. The *nucleolus* maximizes the minimum savings of any coalition (Schmeidler, 1969). The measure of the happiness of coalition S, called e_s, is the savings realized by the members of S from cooperation. Let e_1 denote the minimum savings of any S. Then

$$e_1 = c(S) - \sum_{i \in S} x_i \tag{13.25}$$

The nucleolus is found by solving a sequence of linear programs. The initial LP is (Hallefjord et al., 1995)

$$\text{LP1:} \qquad \text{maximize } e_1$$

subject to

$$\begin{aligned} e_1 + \sum_{i \in S} x_i &\le c(S) \qquad \forall S \subset N \\ \sum_{i \in N} x_i &= c(N) \end{aligned} \tag{13.26}$$

TABLE 13.2 Calculation of the Shapley Value for the TVA Game

	Incremental Cost			
Sequence	$dc(1)$	$dc(2)$	$dc(3)$	Probability
123	163,520	138,087	110,977	$\frac{1}{6}$
132	163,520	33,763	215,301	$\frac{1}{6}$
213	160,781	140,826	110,977	$\frac{1}{6}$
231	45,214	140,826	226,544	$\frac{1}{6}$
312	128,725	33,763	250,096	$\frac{1}{6}$
321	45,214	117,274	250,096	$\frac{1}{6}$
SV(i)	117,829	100,757	193,998	

Let $y(S)$ be the dual variables associated with the S member coalitions and let (e_1^*, x^*) be an optimal solution to LP1 and $y(S)$ the corresponding dual solution. If LP1 has a unique optimal solution with the dual variables positive for the single coalitions, this initial optimal solution is the nucleolus. If not, we have to find the smallest $e_2(x)$ among the $x \in X$ with $e_1(x) = x_1$ for every optimal solution to LP1.

The new LP to be solved is (Maschler 1994)

$$\text{LP2:} \qquad \text{maximize } e_2$$

subject to

$$\begin{aligned} e_2 + \sum_{i \in S} x_i &\le c(S) && \forall S \subset N{:}y^*(S) = 0 \\ \sum_{i \in N} x_i &= c(N) && \\ \sum_{i \in S} x_i &= v(S) - e_1^* && \forall S \subset N{:}y^*(S) > 0 \end{aligned} \tag{13.27}$$

This process continues until the LP problem has a unique solution. This procedure for calculating the nucleolus can be tedious. The nucleolus is a very attractive unique solution notion. If a core exists, the nucleolus is in the core and occupies a central place within the core. The nucleolus is also attractive from a negotiating point of view as a "fair" solution since it provides the largest minimum payoff to players in the game. Maschler (1994) provides a thorough summary of the nucleolus.

Computation of the nucleolus for the original and modified TVA games is shown in Tables 13.3 and 13.4. For the original convex cost game with $c(123) = \$412{,}584$, a unique solution is found to LP1 as shown in Table 13.3. The minimum excess is \$47,286 and each of the three players receives this excess. Correspondingly, the associated dual variables for $c(1)$, $c(2)$, and $c(3)$ are positive. The nucleolus for the modified TVA game with $c(123) = 515{,}000$ is calculated in Table 13.4. The solution to the first LP is an excess of 2323.5. The dual variables are positive for the $c(2)$ and $c(13)$ coalitions. The solution to the LP2 fixes $c(2)$ and $c(13)$ and resolves the LP. The new minimum excess is 7737.25 and the nucleolus has been found.

As the size of the game becomes larger, it gets increasingly tedious to calculate the nucleolus because of the need to make significant adjustments in the LP at each iteration. Despite this limitation, the nucleolus provides a valuable benchmark as to a "fair" solution.

Six-City Water Supply Game and Monotonicity. Young (1994b) describes cost allocation for a proposed joint water supply project to serve six Swedish cities. The characteristic function costs for each of the 26 essential coalitions

TABLE 13.3 Computation of the Nucleolus for Original TVA Game, $c(N)$ = 412,584 (LP1)[a]

	Value							
	$116,234	$93,540	$202,810	$47,286[b]				Dual
Coalition	$x(1)$	$x(2)$	$x(3)$	e	LHS	Inequal.	RHS 1	Var. > 0
1	1			1	$163,105	<=	$163,520	Yes
2		1		1	140,826	<=	140,826	Yes
3			1	1	250,096	<=	250,096	Yes
12	1	1		1	257,060	<=	301,607	No
13	1		1	1	366,330	<=	378,821	No
23		1	1	1	343,636	<=	367,730	No
123	1	1	1		412,584	=	412,584	

[a]LP solved using Microsoft Excel solver.
[b]$47,286 = max. Z.

TABLE 13.4 Computation of the Nucleolus for Revised TVA Game, $c(N)$ = \$515,000

(a) LP1[a]

	Value							
	\$160,781	\$138,502.5	\$215,716.5	\$2323.5[b]				Dual
Coalition	$x(1)$	$x(2)$	$x(3)$	e	LHS	Inequal.	RHS 1	Var. > 0
1	1			1	\$163,105	<=	\$163,520	No
2		1		1	140,826	<=	140,826	Yes
3			1	1	218,040	<=	250,096	No
12	1	1		1	301,607	<=	301,607	No
13	1		1	1	378,821	<=	378,821	Yes
23		1	1	1	356,543	<=	367,730	No
123	1	1	1		515,000	=	515,000	

(b) LP2

	Value								
	\$155,367.3	\$138,502.5	\$221,130.25	\$7737.25[c]					
Coalition	$x(1)$	$x(2)$	$x(3)$	e	LHS	Inequal.	RHS 1	Excess	RHS 2
1	1			1	\$163,104.5	<=	\$163,520		\$163,520
2		1		1	138,502.5	=	140,826	2323.5	138,502.5
3			1	1	218,040.0	<=	250,096		250,096
12	1	1		1	301,607.0	<=	301,607		301,607
13	1		1	1	378,821.0	=	378,821	2323.5	376,497.5
23		1	1	1	356,542.5	<=	367,730		367,730
123	1	1	1		515,000.0	=	515,000		515,000

[a] Result of LP 1: Binding constraints on coalitions 2 and 13. Thus $x(2)$ = \$138,502.5, $x(1) + x(3)$ = \$378,821.

[b] \$2323.5 = max. Z.

[c] \$7737.25 = max. Z.

for this six-city game are shown in Table 13.5. The results of this evaluation are shown in Table 13.6. Three solutions are presented: the alternative cost avoided (ACA) method, the minimum cost, remaining savings (MCRS) method, and the nucleolus. Each of the three solutions falls within the lower and upper core bounds. However, only the nucleolus falls within the core. The ACA and MCRS methods result in negative excesses for coalitions HKL and AHKL. However, Young (1994b) points out a serious flaw in the nucleolus; it may not satisfy *monotonicity.*

The values calculated for the characteristic function are typically preliminary cost estimates. Thus the participants are negotiating using provisional numbers. These estimates will typically differ from the final cost estimates. The participants may decide to accept the principle of how the actual costs will be allocated. The monotonicity rule simply states that if the actual costs turn out to be lower than the estimated costs, no one's costs should go up. Conversely, if the final costs turn out to be higher than the estimated costs, no one's costs should decrease. An allocation rule ϕ is said to be *monotonic in the aggregate* if for any set of projects N and any two cost functions c and c' on N (Young, 1994b),

$$c'(N) \geq c(N) \quad \text{and} \quad c'(S) = c(S) \qquad \forall S \subset N \tag{13.28}$$

implies that

$$\phi_i(c') \geq \phi_i(c) \qquad \forall i \in N$$

For the Swedish water supply cost game, assume that there is a cost overrun

TABLE 13.5 Cost of Alternative Water Supply Systems (Millions of Swedish Crowns)

Coalition	Group	Cost	Coalition	Group	Cost
1	*A*	21.95	14	*HKL*	27.26
2	*H*	17.08	15	*HKM*	42.55
3	*K*	10.91	16	*LMT*	51.46
4	*L*	15.88	17	*AHKL*	48.95
5	*M*	20.81	18	*AHKM*	60.25
6	*T*	21.98	19	*HKMT*	59.35
7	*AH*	34.69	20	*HLMT*	64.41
8	*HK*	22.96	21	*KLMT*	56.61
9	*HL*	25.00	22	*AHKMT*	77.42
10	*LM*	31.10	23	*AHLMT*	83.00
11	*MT*	39.41	24	*AKLMT*	73.97
12	*AHK*	40.74	25	*HKLMT*	66.46
13	*AHL*	43.22	26	*AHKLMT*	83.82

Source: (Young 1994b).

TABLE 13.6 Summary of Results for the Six-City Swedish Water Supply Problem

Item	City						Total	Minimum Excess	In Core?	Objecting Coalitions
	A	*H*	*K*	*L*	*M*	*T*				
Stand-alone cost	21.95	17.08	10.91	15.88	20.81	21.98				
Core minimum cost	17.36	9.85	1.19	6.4	12.89	14.06				
Core maximum cost	21.95	17.08	10.91	14.14	20.81	21.98				
ACA	19.54	13.28	5.62	10.9	16.66	17.82	83.82	−2.54	No	*HKL*, *AHKL*
MCRS	19.61	13.39	5.94	10.19	16.76	17.93	83.82	−2.26	No	*HKL*, *AHKL*
Nucleolus	20.34	12.06	5.00	8.61	18.32	19.49	83.82	1.6	Yes	None

Source: Adapted from Young (1994b).

of 4 million Swedish crowns. Resolving the problem with this overrun yields the allocation of this cost overrun shown in Table 13.7. The results show that neither the nucleolus nor the ACA method satisfies monotonicity. Despite an added cost of 4 million Swedish crowns, the cost allocated to city *K* decreases.

13.E. ACCOUNTING PERSPECTIVE ON COST ALLOCATION

13.E.1. Activity-Based Costing

In this section we describe a relatively new technique in accounting called activity-based costing (ABC). ABC is a recently developed private-sector innovation designed to provide more accurate and meaningful benefit and cost information for internal decision making. Traditional cost and management accounting systems may not provide correct signals as to the "true" cost of producing outputs in multiproduct organizations. This gap has led accountants and engineers to develop new cost estimating procedures that provide much more accurate estimates of these costs. These cost data are vital for getting meaningful information for decision support systems. Without such information, the results of sophisticated simulation and optimization methods can be quite misleading and counterproductive. (Heaney et al., 1991).

13.E.2. Previous Efforts to Assign or Allocate Costs

The fundamental issue is how to assign or allocate costs that are common to more than one product or output to these products and to assign costs among beneficiaries. The popular name for this problem is the cost allocation problem. If the costs are truly joint, the allocation basis is usually felt to be arbitrary. Cooper and Kaplan (1991) distinguish between cost allocation and cost assignment in management accounting. They use cost allocation to refer to the apportionment of costs that are truly joint. Such apportionment of cost is done using an arbitrary basis. Cost assignment, on the other hand, refers to the case where a logical basis exists to apportion these costs. In the water resources field, the accepted terminology is that costs are allocated among purposes and shared among groups (U.S. Interagency Committee on Water Resources, 1958). However we define these costs, the fundamental problem

TABLE 13.7 Allocation of a Cost Overrun of 4 Million Swedish Crowns by the ACA and Nucleolus Methods

	A	*H*	*K*	*L*	*M*	*T*
Nucleolus	0.41	1.19	−0.49	1.19	0.84	0.84
ACA	1.88	0.91	−0.16	0.07	0.65	0.65

Source: Young (1994b).

is to establish a reasonable basis for assigning costs as accurately as appropriate for the problem under study.

Although the cooperative NPGT approach provides some additional insights to solving the cost allocation problem from a theoretical point of view, its major limitation is that the concept is based on comparisons with alternatives not pursued. Thus we run into major accounting problems of needing to keep track of what others would have done if the alternative selected had not been pursued. Although such calculations are feasible, they do impose a serious computational burden. Thus this approach has not gained widespread acceptance in water resources in particular or accounting in general (Kaplan, 1982). Activity-based accounting is a positive step in obtaining the desired information to assign costs in a more rigorous manner within feasible accounting systems.

13.E.3. Management Accounting and ABC

If there is only one product, all costs are legitimately assignable to that product. However, most public- and private-sector organizations produce multiple products. As Johnson and Kaplan (1987) point out, labor was the dominant cost in traditional organizations. However, labor and other direct costs have declined in relative importance. Instead, indirect costs now are a much larger share of total costs.

Traditional management accounting systems are designed primarily for *external* purposes (e.g., to value inventory for financial and tax statements). Such an approach does not necessarily provide managers with the accurate and timely information they need for *internal* purposes to promote operating efficiencies and measure product cost. Seriously distorted cost analysis can lead managers to choose a losing competitive strategy by dropping highly profitable products or by expanding commitments to complex, unprofitable lines (Kaplan, 1982). Other reports, such as profit and loss statements, budgets, and margin analyses, have become too aggregated, too late, and too one-dimensional to be of use to operating managers (McNair et al., 1990). Under these conditions, distorted internal evaluations eventually cause deterioration of the firm's external performance measures. It has been suggested that the decline of U.S. competitiveness over the past decade, both in domestic models and abroad, is due in part to the inadequacy of traditional internal decision-making models (Johnson and Kaplan, 1987).

Much of the distortion of internal performance information stems from seemingly sensible procedures developed decades ago, when the nature of production was much different. Traditional managerial accounting procedures best serve companies with narrow product lines, through which direct labor and materials can easily be traced. Distortions from assigning indirect costs are minor under these conditions, making the expense of collecting and processing data for more sophisticated assignment methods unattractive.

Many companies no longer fit the foregoing description. New production technologies, intensified global competition, and the explosion of marketing, distribution, and other support expenses have moved these firms to rethink their approach to cost analysis. The plummeting cost of information technology also reshaped their efforts—payroll hours and material purchasing records are no longer viewed as the only reasonable databases that a firm can maintain for strategic uses (Cooper and Kaplan, 1991).

ABC can provide firms with better product costs and a better understanding of the impacts of pricing and product mix decisions on their profitability. However, there is no guarantee that the additional benefits of creating and operating an ABC system will exceed its costs. ABC can benefit:

1. *Firms that produce a broad range of products* (Drury, 1990). Cost distortions occur in virtually all organizations producing and selling multiple products or services (Cooper and Kaplan, 1991).
2. *Firms with heavy overhead and/or support services expenses* (Cooper and Kaplan, 1991). Traditionally, these expenses are often "treated like peanut butter" and spread indiscriminately across a firm's output. The larger these expenses, the greater the possibility of distortion. ABC relates overhead expenses to the products that cause them.

13.E.4. ABC Methodology

The first step in creating any product costing system is generally the collection of accurate data on direct labor and material consumption. The designer of an ABC system must proceed further by examining the demand on support services made by particular products. Three rules should guide this process (Cooper and Kaplan, 1991):

1. Focus on the most expensive support activities.
2. Focus on activities whose consumption varies significantly between products.
3. Focus on activities whose demand patterns are uncorrelated with traditional allocation measures.

The goal is to "unbundle" all indirect expenses and trace them back to the individual products or product families that caused them. Virtually all of a company's activities exist to support the production and delivery of the company's products; they should therefore be considered product costs. Even administrative expenses should be allocated, especially if they vary across product lines. For example, the portion of a company's legal expenses caused by a certain product could be a function of that product's risk of liability, risk of environmental damage, or antitrust concerns (Cooper and Kaplan, 1991).

The process of tracing costs from resources to activities to products relies heavily on the judgment of the system designer, unlike the highly formalized procedures of traditional cost allocation. Unlike a system that serves the needs of financial reporting, an ABC system does not need to be precise to the penny.

13.E.5. Activity-Based Costing in Public Sector

Decision making in the public sector is often confounded by difficulties unfamiliar to the private sector. There has been trouble measuring outputs, with few empirical production functions expressing the technical relationships between factor inputs and product outputs (Milliman, 1972). External performance evaluation is challenged by the lack of market-type response to a firm's pricing policies. Frequently, no conceptual framework other than the cumbersome political process exists to inform a public-sector firm of the value of its outputs (Mushkin and Bird, 1972).

An important objective of developing an ABC system for a public-sector organization is to improve its efficiency. A major justification of public systems is that they produce public goods the cost of which cannot easily be assigned to individuals or groups. However, many public-sector services are hybrids of public and private goods. Without the aid of computers, it was more reasonable to view public projects such as water resources as public goods and not to attempt to assign responsibility to individual taxpayers. The principles of benefit–cost analysis for federal water resources projects included a single national economic efficiency goal that the benefits, to whomsoever they may accrue, should exceed the costs, to whomsoever they may accrue. Thus, funding for such projects came from the federal government under the premise that the benefits were widespread in nature.

With modern computer systems, our ability to track these benefits and costs has grown to levels never imagined in precomputer days. Also, contemporary water resources management is no longer dominated by the federal government, so cost-sharing of initial and operating costs is more critical. Indeed, in recent years, user fees have become increasingly popular as a way to assign costs to beneficiaries directly. Of course, the feasibility of assigning costs and benefits is heavily dependent on the quality of the databases and associated decision support systems.

Recent evidence suggests that despite the difficulties discussed above, private-sector evaluation systems have found applications in the public sector. Coursey and Bozeman (1990) investigated the influence of a firm's "publicness" on the types of strategic decisions encountered by managers within that firm. In an analysis of a mail survey of 210 managers in 39 organizations in the United States, Coursey and Bozeman found a small difference in the nature of the decisions confronting public-, versus private-sector, managers. In addition, publicness did not account for differences in smoothness, including processing time and internal disagreement over strategic decisions (Coursey and Bozeman, 1990).

The tendency to apply traditional private-sector performance measures, and the lack of such measures that capture the complexities of public-sector performance, is reminiscent of the development of activity-based costing. Mushkin and Bird (1972) concluded that "within such larger service categories . . . regarded by some as public goods, there are numerous subservices that provide essentially private benefits." For example, the police who handle crowds at ballgames and direct traffic around large shopping centers are providing services to identifiable beneficiaries even though the larger service of preventing crime may perhaps best be considered a public good (Mushkin and Bird, 1972). This observation fits the theme of activity-based costing remarkably well; the police force deploys its resources in a range of activities, from crowd control to traffic operations. Arguable, the end products of this system are the individual packages of police activities "consumed" by each citizen within their jurisdiction. As highlighted by Mushkin and Bird (1972), the proportion of certain activities consumed varies from citizen to citizen. By relating resources to activities, and activities to products, the single issue of police protection is unbundled into operational components.

Water management provided to the public is arguably similar to Mushkin and Bird's (1972) police protection example. Different citizens may require different proportions of the subservices of water management, making the final product as varied as the citizenry. An activity-based decision support system would address the relationship between the products received by the district's constituents and the activities consumed in the product's delivery.

Anderson (1993) has applied ABC to the problem of estimating the cost of filling potholes and removing snow from streets in Indianapolis, Indiana. The city of Indianapolis is considering privatizing certain functions that traditionally have been performed by the public sector. However, when management tried to determine the actual cost of repairing a pothole, they found that the accounting system did not provide this information. Working with the employees affected, ABC was used to develop an accurate estimate of this unit cost. The city then issued a request for bids wherein their own utility had to bid in competition with private-sector competitors on who could provide this service in the most cost-effective way. In the process of setting up the ABC system, the city employees discovered several significant ways to reduce their costs. The end result was that they were awarded the contract to continue to repair potholes.

13.F. SUMMARY OF METHODS AND COSTS

13.F.1. Cost Allocation Methods

James and Lee (1971) summarize cost allocation methods in water resources as the problem of selecting the appropriate cost allocation method from the following 18 options. A general expression for the cost allocation problem for the ith participant is

$$x(i) = x(i)_{\text{min}} + \beta(i)[c(N) - \sum_{i \in N} x(i)_{\text{min}}] \tag{13.29}$$

Let

$$RC = \text{remaining costs} = c(N) - \sum_{i \in N} x(i)_{\text{min}} \tag{13.30}$$

Then equation 13.29 can be expressed as

$$x(i) = x(i)_{\text{min}} + \beta(i)(RC) \tag{13.31}$$

Methods for Assigning Minimum Costs. James and Lee (1971) present three options for calculating $x(i)_{\text{min}}$:

$$\begin{aligned} &\text{a. Let } x(i)_{\text{min}} = 0, \text{ or} \\ &\text{b. Let } x(i)_{\text{min}} = \text{direct or specific costs, or} \\ &\text{c. Let } x(i)_{\text{min}} = \text{separable costs.} \end{aligned} \tag{13.32}$$

Specific or direct costs are those components of total costs that can be identified as being directly assignable to one or several groups. For example, the canal from the reservoir to the irrigation district is directly assignable to this user.

Another way to define minimum costs is based on the core of an n-person game (Heaney and Dickinson, 1982). The minimum cost to participant i in an n-person game which is in the core of the game is the solution to the following linear program:

$$\text{minimize } Z = x(i)$$

subject to

$$\begin{aligned} x(i) &\leq c(i) && \forall i \in N \\ \sum_{i \in N} x(i) &= c(N) && \forall S \subset N \\ x(i) &\geq 0 && \forall i \in N \end{aligned} \tag{13.33}$$

This definition of the minimum assignable cost, while less intuitive, is very important. It is the minimum cost that can be assigned to participant i which is stable in the sense that no other individual or subgroup of N can object that they are subsidizing participant i. For coalitions with only two members, the separable costs and the core minimum cost are identical. However, for coalitions with three or more members, the core minimum can differ from the separable cost. The core minimum is a more robust definition of the minimum cost.

Thus, in addition to James and Lee's three methods for defining the minimum costs, we can add a fourth method, the minimum assignable cost as defined by the core of the game.

Methods for Assigning the Apportionment Factor, $\beta(i)$ James and Lee (1971) enumerate six ways that the remaining costs can be apportioned:

1. Equally among the n cost centers:

$$\beta(i) = \frac{1}{n} \qquad \forall i \in N \tag{13.34}$$

2. Proportional to the use the participant makes of the facilities based on a single measure of use, such as volume, flow rate, or BOD load:

$$\beta(i) = \frac{Q(i)}{\Sigma_{i \in N} Q(i)} \qquad \forall i \in N \tag{13.35}$$

   where $Q(i)$ is the measure of use. An immediate problem with a single measure of use is that the cost of facilities is usually a function of several factors. For a given measure of use, cost is usually based on a combination of peak and average use. The design size of the system is based on projected peak use, while operating cost is based on expected average use.

3. Entirely to the highest priority participant up to the limit of the benefit that this participant receives. Thus if we have strict ordering of the priorities of the n participants, the proportion of the remaining costs chargeable to the ith participant is

$$\beta(i) = \min \left[1 - \sum_{i=1}^{s-1} \beta(i), \frac{\min(B(i), C(i)) - x(i)_{\min}}{RC} \right] \tag{13.36}$$

   where $B(i)$ represents the benefits for coalition i, and $C(i)$ represents the costs for coalition i.

4. Proportional to the benefit in excess of assigned minimum cost.
5. Proportional to the excess cost to provide the service by some alternative means. This is called the *alternative cost avoided* (ACA) *method.*
6. Proportional to the minimum of options 4 and 5.

Thus for options 4, 5, and 6,

$$\beta(i) = \frac{B'(i) - x(i)_{\min}}{RC} \tag{13.37}$$

where β_i is the gross benefit, alternative cost, or the minimum of gross benefit and alternative cost. Consideration of the core provides at least one additional way to define $\beta(i)$.

Heaney and Dickinson (1982) proposed the minimum cost, remaining savings method. This method prorates costs between the minimum and maximum charges to $x(i)$ based on the lower and upper bounds in the core. The lower bounds can be calculated based on the linear program described above. The upper bound can be determined simply by replacing minimum $x(i)$ with maximum $x(i)$ as the objective function in the same problem. Thus

$$\beta(i) = \frac{x(i)_{\max} - x(i)_{\min}}{\sum_{i \in N} [x(i)_{\max} - x(i)_{\min}]} \tag{13.38}$$

where $x(i)_{\max}$ and $x(i)_{\min}$ are the upper and lower bounds of the core for the ith participant.

Overall, we have described four ways to calculate the minimum cost assignment and seven ways to apportion the remaining cost, a total of 28 options.

13.F.2. Cost of Environmental Regulations

Heaney et al. (1979) estimated the nationwide cost of controlling combined sewer overflows to be about $10 billion. Previous estimates had costs ranging as high as $400 billion. A major source of uncertainty in these estimates is the wide variability in reported unit costs for facilities. For example, a nationwide assessment of the cost of street sweeping revealed unit costs ranging from $0.28 to $28,000 per curb mile. The lower cost estimate is based on fuel costs only, while the higher estimate includes much of the overhead costs of the city. Because of similar uncertainty in estimating the economic impact of drinking water regulations, the U.S. Environmental Protection Agency has sponsored several studies directed at determining the "true" costs of added regulations (Raucher et al., 1995). Clark and Males (1986a,b) summarize the development of a simulation model that includes a spatial costing model. As far as the author knows, this is the only model in the water resources field that does cost allocation as an integral part of a modeling package. Sample results of this activity are shown in Figure 13.4.

13.F.3. Cost Allocation Along a Network

Networks are the single most important component of the cost of water resource systems. Thus the problem of allocating the cost of a network is of

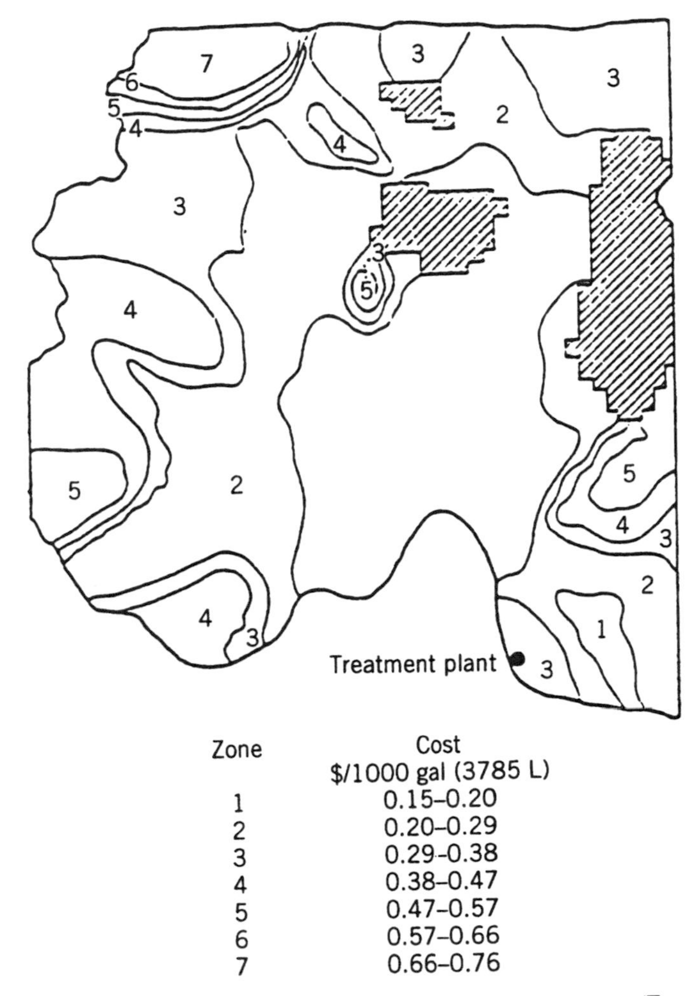

Figure 13.4 Cost contours for Cincinnati Water Works service area. (From Clark and Males, 1986b.)

prime importance. Clark and Males (1986a,b) developed a spatial costing procedure for allocating the cost of a water distribution network among service areas. The objective of this analysis is to account for spatial cost differences for serving customers based on the length of pipe needed to serve them and the pumpage required due to pressure requirements based on head losses and differences in elevation. For example, the customer(s) located on a hill would require extra pumping. With contemporary hydraulic analysis software such as EPANET, it is relatively simple to evaluate the flows and pressures in a water network. However, other than the one effort by Clark and Males (1986a,b), no commonly acceptable methods are available.

Cooperative game theory provides valuable insights for this problem. The problem of allocating costs along a water network is similar to the problem

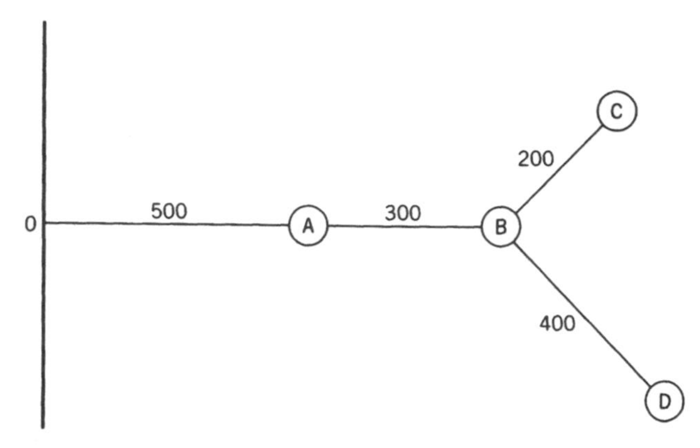

Figure 13.5 Cost of connecting four houses to an existing trunk power line. (From Young 1996.)

of allocating costs along other types of networks, such as transportation and communications networks (Granot and Granot, 1992). Young (1994b) describes how such network games can be decomposed into simple subproblems. We use his example of a power line shown in Figure 13.5. This power line connects four houses (*A*,*B*,*C*,*D*) to the existing trunk power line. In this simplest of cases, each of the demands is considered to be equal. In this case the cost assignment algorithm is simply to divide the total cost of each segment equally among the users. Thus the cost allocation for this simple game is as shown in Table 13.8.

Of course, Young's example of a power line could easily be adapted to apportion the cost of a water network. The procedure would need to be refined to apportion the cost according to the relative sizes of the flows of each user.

An interesting contextual extension of this problem is to revise it so that the original system was built to only serve houses *A*, *B*, and *C*. After the contract has been signed, house *D* approaches *A*, *B*, and *C* regarding joining the group. One could simply let *D* join and assess him as shown above. Another option would be to treat this as a two-person game with *A*, *B*, *C*

TABLE 13.8 Decomposition of Electrical Line Costs

Cost Element	House *A*	House *B*	House *C*	House *D*	Segment Cost
OA	125	125	125	125	500
AB		100	100	100	300
BC			200		200
BD				400	400
Charge	125	225	425	625	1400

Source: Young (1994b).

representing one person and D representing the other person. Thus we would have the following two-person game:

$$c(ABC) = 1000 \qquad c(D) = 1200 \qquad c(ABCD) = 1400$$

In this case the cost allocation would be $x(ABC) = 600$ and $x(D) = 800$. Thus D would pay an additional 175. Thus whether the basic units have formed binding coalitions can have a significant impact on the outcome of the cost allocation.

13.F.4. Cost Allocation Among Products

Heaney (1979) used the simple Shapley value to allocate the cost of a wastewater treatment plant among various pollutants. This problem was motivated by a study that sought to determine whether it was cheaper to remove nutrients from the wastewater treatment plant (WWTP) or from treating urban runoff. The problem is complicated by the fact that a treatment facility removes numerous constituents simultaneously. For example, a WWTP designed to remove biochemical oxygen demand (BOD) and suspended solids (SS) will also remove nitrogen (N) and phosphorus (P) in the process. Thus the cost of removing nutrients depends on whether they are the primary product or a by-product of the design. The players in this game are the pollutants that will share in the costs. Let's assume that the four pollutants are BOD, SS, N, and P. The required pollutant removals are shown in Table 13.9. The removal efficiencies for the 12 wastewater treatment systems and their annual costs are shown in Table 13.10. Using Tables 13.9 and 13.10, the subset of feasible treatment systems is determined and the characteristic function is the minimum cost for the feasible systems, as shown in Table 13.11.

The cost of any system is based on the *limiting pollutant* (i.e., one of the pollutant standards dictates the sizing of the plant). Thus we have three different characteristic function values: 3.27, 5.37, and 7.43.

Littlechild and Owen (1973) have shown that the Shapley value may be represented in this special case by

TABLE 13.9 Required Pollutant Removals for a 20-MGD WWTP

		Concentration (mg/L)	
Pollutant	Group	Influent	Effluent
BOD	1	210	10
SS	2	230	5
Nitrogen	3	11	0.5
Phosphorus	4	20	2

TABLE 13.10 Removal Efficiencies for 12 Wastewater Treatment Systems

	Proportion Removed				Annual Cost[a]		
System	BOD	SS	N	P	*a*	*b*	Cost (millions)
1	0.38	0.57	0	0.18	0.27	0.62	$ 1.73
2	0.52	0.78	0	0.82	0.37	0.71	3.10
3	0.79	0.74	0.1	0.27	0.38	0.67	2.83
4	0.88	0.87	0.1	0.82	0.49	0.75	4.63
5	0.9	0.91	0.15	0.36	0.41	0.69	3.24
6	0.93	0.93	0.15	0.82	0.48	0.72	4.15
7	0.95	0.91	0.9	0.27	0.44	0.67	3.27
8	0.95	0.91	0.95	0.27	0.64	0.71	5.37
9	0.95	0.96	0.9	0.91	0.57	0.71	4.78
10	0.98	0.98	0	0.91	1.27	0.74	11.66
11	0.98	0.98	0.95	0.95	0.86	0.72	7.43
12	0.99	0.98	0.95	0.95	1.11	0.71	9.31

Source: EPA-600/9-76-014, 1976.
[a] 1984 dollars using $C = aQ^b$.

TABLE 13.11 Determination of the Least Cost Solutions for the Coalitions

Coalition	Feasible Systems	Optimal System	Annual Cost (millions)
1	7–12	7	$3.27
2	10–12	11	7.43
3	11,12	11	7.43
4	8,11,12	8	5.37
12	10–12	11	7.43
13	8,11,12	8	5.37
14	11,12	11	7.43
23	11,12	11	7.43
24	11,12	11	7.43
34	11,12	11	7.43
123	11,12	11	7.43
124	11,12	11	7.43
134	11,12	11	7.43
234	11,12	11	7.43
1234	11,12	11	7.43

$$SV_i = SV_{i-1} + \frac{c_i - c_{i-1}}{r_i} \tag{13.39}$$

where c_i is the cost associated with group i, c_{i-1} the cost associated with group $i - 1$, SV_i the Shapley value cost assigned to group i, SV_{i-1} the Shapley value cost assigned to group $i - 1$, and r_i the number of players of type i. The procedure for doing the calculations is described below. First, arrange the players from smallest to largest costs, as shown in Table 13.12.

In general, there are m group types. Each group type has n players associated with it. In this pollutant removal cost game, group types 1 and 2 have only one member associated with them, while group 3 has two members. Thus, for this game,

TABLE 13.12 Coalitions Listed in Increasing Order of Costs

Coalition	Feasible Systems	Least Cost System	Annual Cost (millions)	Group Type
1	7–12	7	$3.27	1
4	8,11,12	8	5.37	2
2	10–12	11	7.43	3
3	11,12	11	7.43	3

$$r_1 = 1 + 1 + 2 = 4$$
$$r_2 = 1 + 2 = 3 \tag{13.40}$$
$$r_3 = 2$$

The simple Shapley value assigned costs can now be calculated as

$$SV_1 = \frac{c_1}{r_1} = \$0.8175$$
$$SV_4 = SV_1 + \frac{c_4 - c_1}{r_2} = \$1.5175$$
$$SV_2 = SV_4 + \frac{c_2 - c_4}{r_3} = \$2.5475 \tag{13.41}$$
$$SV_3 = SV_2 = 2.5475$$

The total assigned cost is \$7.43, which is the required amount.

The original application of this simplified Shapley value was to set landing fees for the Birmingham Airport in England. The game consisted of 11 types of aircraft and 13,572 landings or players. The charging algorithm is as follows (Littlechild and Thompson, 1977): Divide the total cost of catering for the smallest aircraft equally among the number of landings of all aircraft. Then divide the incremental cost of catering for the second smallest type of aircraft among the landings of all but the smallest type of aircraft. Continue in this manner until the incremental cost of the largest type of aircraft is divided equally among the number of landings made by the largest aircraft type.

13.G. CONTEXT FOR CALCULATIONS

13.G.1. Context for Cost Allocation

The examples presented typically seen in the game theory literature assume a very simple design case (e.g., a brand new system). The more typical case is that we are considering a change in an existing system. For every cost allocation problem it is essential to define clearly the *context* within which the calculations will be done. There is no single correct way to determine and allocate costs. Next we present some guidelines and examples regarding the importance of the problem context.

13.G.2. System Development Charges for Water, Wastewater, and Stormwater Facilities

Nelson (1995) describes various ways to determine system development charges (SDCs) for water, wastewater, and stormwater systems. SDCs are one-time charges paid by new development to finance the construction of public facilities needed to serve it. Numerous legal challenges to SDCs have been made on the basis that these charges are simply another form of taxation. Thus cities must prove that these assessments are for actual services rendered and are not just general taxes. According to Nelson (1995), the 14th Amendment to the Constitution requires that laws treat similarly situated persons equally. There are two basic elements of equal protection:

1. The SDC must not apply arbitrarily to some classes of development but not to others.
2. The fee must be related to the public purpose.

Based on legal considerations, SDCs should comply with the *rational nexus test* as follows (Nelson, 1995):

1. A connection is established between the new development and the new or expanded facilities required to accommodate such development. This establishes the basis for public policy.
2. The cost of the expanded facilities to serve these new developments needs to be quantified. These costs may be determined using *Banberry factors.*
3. The cost apportioned to the new development should be based on the benefit it reasonably receives.

Banberry Rules. According to Nelson (1995), the Banberry court ruling requires that the following seven factors be considered to determine the proportionate share of cost to be borne by new development [*Banberry Development Company* v. *South Jordan City* (631 P.2d 899, Utah 1981)]:

1. The cost of existing facilities.
2. The means by which existing facilities have been financed.
3. The extent to which new development has already contributed to the cost of providing existing excess capacity.
4. The extent to which existing development will, in the future, contribute to the cost of providing existing facilities used community wide for nonoccupants of new development.

5. The extent to which new development should receive credit for providing at its cost, facilities the community has provided in the past without charge to other development in the service area.
6. Extraordinary costs incurred in serving new development.
7. The time–price differential inherent in fair comparisons of amounts of money paid at different times.

Each of these seven considerations is discussed below.

To estimate the cost of existing facilities, detailed engineering estimates are needed of the present value of the facilities. A critical component of this valuation is proper depreciation of the existing facilities. This should reflect actual loss in value and not be a simple depreciation formula used for other purposes.

The financing of existing facilities must be determined to assign costs properly. This requires detailed record keeping, including an evaluation of the extent to which existing facilities have been financed from revenues other than utility charges. Some of the infrastructure may have been financed by grants from the federal government. Moreau (1996) shows the capital outlays for water and wastewater facilities in the United States from 1965 to 1990. Total expenditures in 1987 dollars ranged from 8 to 13 billion per year (see Figure 13.6). Federal grants for wastewater facilities increased dramatically in the 1970s from less than \$1 billion per year to over \$6 billion per year in

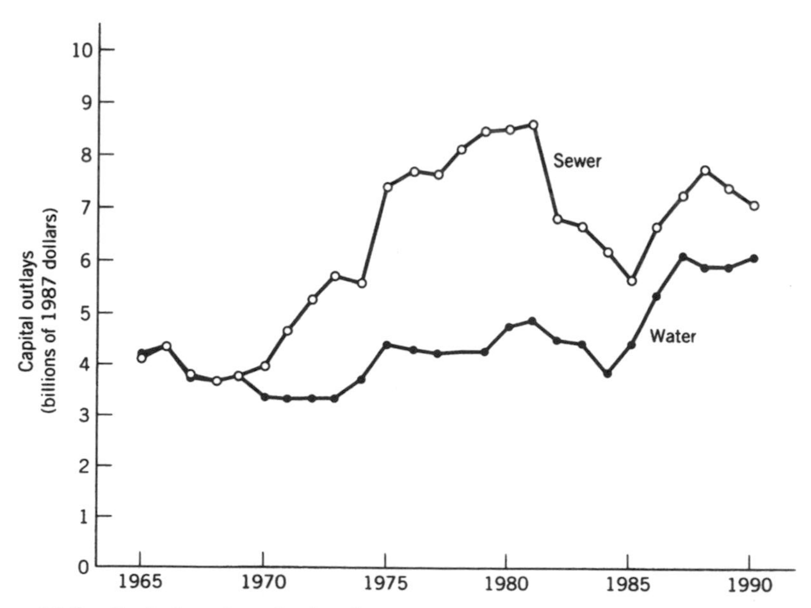

Figure 13.6 Capital outlays by local government for water and wastewater facilities in constant 1987 dollars. (From Moreau, 1987.)

the late 1970s (see Figure 13.7). During the 1970s, the federal government paid 75% of the cost of wastewater treatment plants. Thus about one-half of the local expenditures for wastewater treatment plants came from federal grants. Federal participation in financing local wastewater infrastructure has declined dramatically since 1980, and cities no longer receive significant funding from the federal government. Some judgment needs to be made as to how these federal grants are apportioned among existing and new development.

New development may have already paid for part of the infrastructure if the infrastructure has been financed by property taxes. This accounting needs to be included in the evaluation. The present value of future payments by new development also needs to be included in the analysis.

New facilities are typically required to dedicate part of their land to the community for utility easements, open space, road right-of-way, and so on. The value of these contributions should be credited. However, debate exists as to the value of this dedicated land. Should it be valued under the assumption that it otherwise would have been buildable, or should it be given a lower value? This can be a complex question. Assigned costs may vary based on different levels of service (e.g., communities may charge more to service water customers who live on a hill and require extra pumping) (Raucher et al., 1995).

Finally, the present value of the costs is determined. This last step is straightforward once the component costs have been determined.

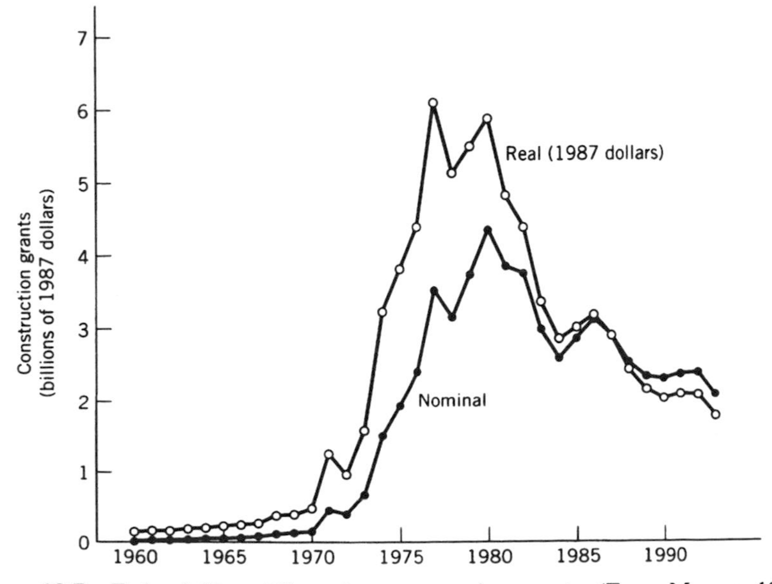

Figure 13.7 Federal Clean Water Act construction grants. (From Moreau 1987.)

Applying the* Rational Nexus *Principles. As Nelson (1995) states, the heart of the rational nexus test is the *proportionate share,* defined as "that portion of the cost of existing and future system improvements that is reasonably related to the demands of new developments." The SDC is calculated for each service area in the community and for specified levels of service. Capital costs are estimated on a per unit basis (e.g., dollars per gallon for water and wastewater). Multiple measures can be used, such as percent imperviousness, slope of land, and so on, for stormwater. The estimated average and maximum demand for each development must be determined since cost depends on both of these activity measures.

How SDCs Are Calculated in the United States. Nelson (1995) summarizes several methods used to estimate SDCs in the United States. He applies six of these methods to a case study and comes up with the assigned costs shown in Table 13.13. Perusal of the wide variability in assigned cost from \$620 to \$9365 reveals the importance of the assumptions made in assigning costs. Overall, Nelson (1995) feels that the total cost attribution method is probably the most accurate.

13.G.3. Calculating Characteristic Function Values

The core for an essential two-person cooperative game always exists and is a line. For a three-person game, the core may exist and be convex or nonconvex, or it may not exist at all. It is necessary to evaluate the core constraints to determine its bounds. For more than three-person games the problem becomes increasingly complex. The number of characteristic function values to be calculated is $2^n - 1$. Thus the number of combinations becomes unwieldy as n grows. Fortunately, it is possible to reduce the computational burden.

Ng and Heaney (1989) present a graph-theoretic method for efficiently enumerating the cost of water resources alternatives. They point out the lim-

TABLE 13.13 SDCs Calculated Using Six Different Methods

Method	SDC/ERU[a]
Growth-related cost allocation method	\$9365.22
Recoupment value method	620.50
Replacement cost method	812.71
Marginal cost method	4474.74
Average cost method	1706.29
Total cost attribution method	2388.91

Source: Nelson (1995).

[a]SDC/ERU, system development charges/equivalent residential unit.

itations of using optimization techniques that permit partial enumeration of the set of alternatives. These partial enumeration techniques may not be very robust in the sense of realistically depicting the actual problem. Thus one ends up with an "optimal" solution to an overly simplified problem. For example, Miles and Heaney (1988) were able to find a "better than optimal" solution to a storm drainage network design problem using a heuristic. This approach found a better solution than two previous studies that used dynamic programming. The reason they were able to find a better solution is that they represented the hydraulics more accurately than did the dynamic programming methods. Another limitation of optimization methods is that they only present the best solution. An analyst would like to be aware of nearly optimal solutions that may have other more attractive features. With regard to cost allocation, the optimal solution may prescribe a single grand coalition with six members, whereas the second best solution may be simply to pair off the six members into three groups and avoid a complex cost allocation problem with associated higher transaction costs.

The simple water network analyzed by Ng and Heaney (1989) is shown in Figure 13.8. Water is to be delivered from a single source, S, to three communities. The cost of the pipeline is estimated using

$$C = aQ^bL \tag{13.42}$$

where C is the total annual cost of the pipeline (\$/yr), Q = quantity of flow, L = length of pipeline, and a, b = parameters, $0 < b < 1$. An efficient total enumeration procedure was developed. The options for the three-city case are presented in Figure 13.9 for this three-node digraph. The associated characteristic function values for the alternative pathways are shown in Figure 13.10. A total of 15 characteristic function values need to be calculated. However, only 8 of these calculations are independent. Figure 13.11 and Table 13.14 show a summary of the computational effort for three digraphs. For the five-node digraph, a total of 109 calculations are required to find the 31 optimal solutions. However, only 19 of these 109 calculations are independent. Thus the actual computational burden is much less.

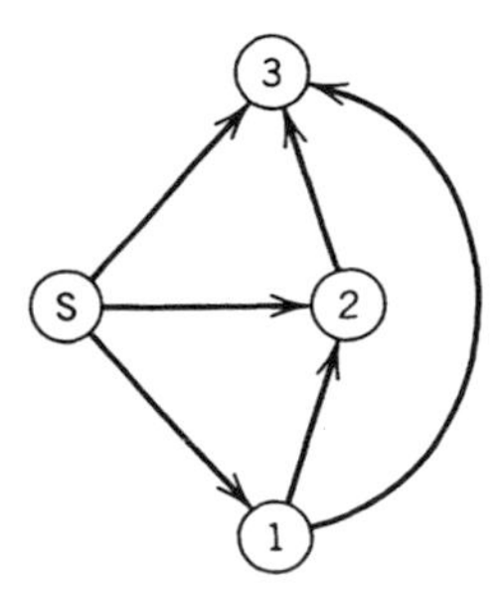

Figure 13.8 Digraph representing a regional water network for three users. (From Ng and Heaney, 1988.)

```
Distance: L(i,j) is the distance in feet from i to j
L(S,1)-    17000  L(S,3)-    30250  L(1,3)-    19670
L(S,2)-    26000  L(1,2)-    13100  L(2,3)-    15500
Demand: Q(i) is the demand in mgd for user i
             Q(1)-        1    Q(2)-        6    Q(3)-        3
Cost Function: C-a(Q b)L          a-       38       b-     0.51
-----------------------------------------------------------------
                   Calculations with total enumeration procedure
C(i..j) [x]-Cost of network [x] for i..j ; C(i..j)-least cost
for i..j
-----------------------------------------------------------------
 Number
       1 C(1)[S1]-              646000
       2 C(2)[S2]-             2463849
       3 C(3)[S3]-             2012986
         -------------------------------------------------------
       4 C(12)[S1,12]-         2984141          min C(12)- 2984141
     1+2 C(12)[S1^2]-          3109849
         -------------------------------------------------------
       5 C(13)[S1,13]-         2618976          min C(13)- 2613976
     1+3 C(13)[S1^3]-          2658986
         -------------------------------------------------------
       6 C(23)[S1,23]-         4061294          min C(23)- 4061294
     2+3 C(23)[S2^3]-          4476835
         -------------------------------------------------------
       7 C(123)[S1,12,23]-  4648439
     3+4 C(123)[S1,12;S3]-  4997127
       8 C(123)[S1,12,13]-  4640756             min C(123)-4640756
     1+5 C(123)[S1,13;S2]-  5082825
     2+6 C(123)[S1;S2,23]-  4707294
   1+2+3 C(123)[S1;S2;S3]-  5122835
         -------------------------------------------------------
         Sort C(123) in ascending order
             Path                   Cost      Rank % savings
         C(123)[S1,12,13]-  4640756           1       9.41
         C(123)[S1,12,13]-  4648439           2       9.25
         C(123)[S1;S2,23]-  4707294           3       8.11
         C(123)[S1,12;S3]-  4997126           4       2.45
         C(123)[S1,13;S2]-  5082824           5       0.78
         C(123)[S1;S2;S3]-  5122835           6       0.00
```

Figure 13.9 Total enumeration procedure for three-node digraph. (From Ng and Heaney, 1988.)

Iteration i	i-Node Subdigraphs	Spanning Directed Trees for i-node Subdigraph	Are Spanning Directed Trees Essential?
i - 1	(S, 1)	S → 1	Yes
	(S, 2)	 S → 2	Yes
	(S, 3)	S → 3	Yes
i - 1	(S,1,2)	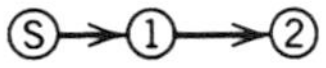 S → 1 → 2	Yes
		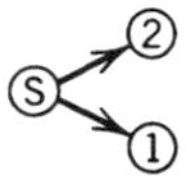 S → 2, S → 1	No
	(S,1,3)	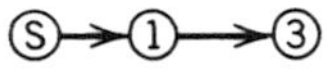 S → 1 → 3	Yes
		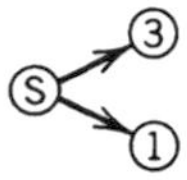 S → 3, S → 1	No
	(S,2,3)	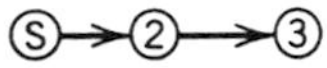 S → 2 → 3	Yes
		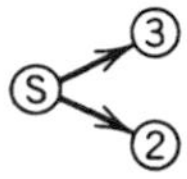 S → 3, S → 2	No
i - 3	(S,1,2,3)	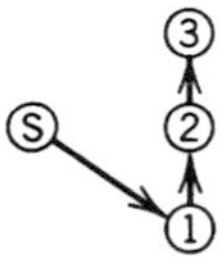 S → 1 → 2 → 3	Yes
		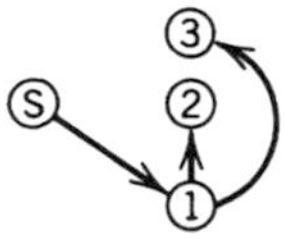 S → 1 → 2, 1 → 3	Yes
		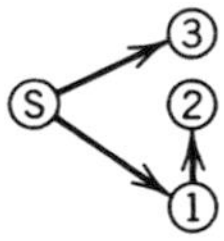 S → 3, S → 1 → 2	No
		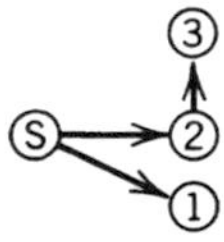 S → 2 → 3, S → 1	No
		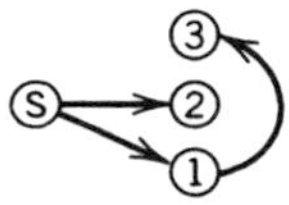 S → 2, S → 1 → 3	No
		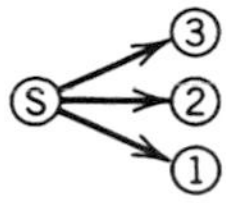 S → 3, S → 2, S → 1	No

Figure 13.10 Efficiency analysis of a three-user water supply network with nonlinear cost functions using a spreadsheet. (From Ng and Heaney, 1988.)

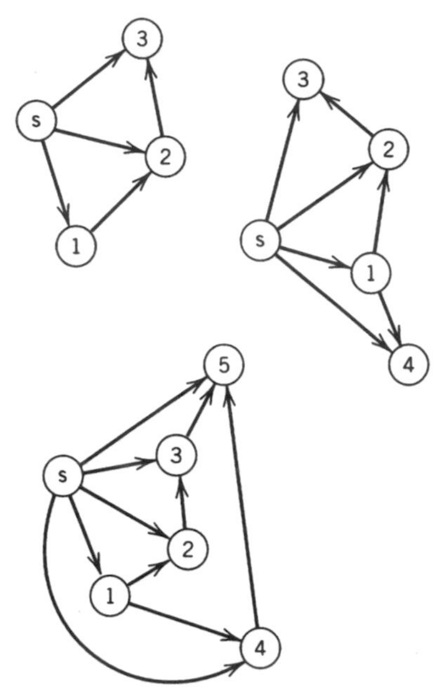

Figure 13.11 Three-, four-, and five-node digraphs. (From Ng and Heaney, 1989.)

13.G.4. Tracking the Nearly Optimal Solutions

As mentioned above, a major advantage of total enumeration or using heuristics is that the good as well as the optimal solutions can be saved and reviewed. The bottom portion of Figure 13.9 shows the six alternatives ranked from most desirable to least desirable. The optimal solution results in 9.41% savings. However, two other solutions provide 9.26% and 8.11% savings, respectively. The option with 8.11% savings results in a separate pipe for city 1 and a shared pipe system for cities 2 and 3. This solution may be best from

TABLE 13.14 Summary of Computational Effort for Digraphs Shown in Figure 13.10

Digraph	$2^n - 1$ Optimal Solutions	Number of Spanning Directed Trees	Number of Inessential Spanning Directed Trees	Number of Calculations to Find $2^n - 1$ Optimal Solutions	Number of Independent Calculations to Find $2^n - 1$ Optimal Solutions
Three-node	7	4	3	12	6 (50%)
Four-node	15	8	7	33	10 (30%)
Five-node	31	24	22	109	19 (17%)

Source: Ng and Heaney (1989).

the joint consideration of efficiency and equity because it results in very good savings and a straightforward cost allocation since the largest coalition is only 2.

Recent advances in heuristics using genetic algorithms, tabu search, and simulated annealing offer much promise for more robust approaches to evaluating good as well as optimal solutions (Reeves, 1993). Dandy et al. (1996) show how genetic algorithms can provide superior solutions to nonlinear programming approaches to optimizing water distribution networks. Thus these more robust heuristics provide an excellent way to generate the good solutions as well as the optimal solution.

A feasible option to finding the characteristic function values for all options is to calculate only the characteristic function values for specified S member coalitions. For a given problem the engineer would typically evaluate only a subset of the available combinations. This subset can be used to approximate the core. Another reason for looking at only a subset of the combinations is that the interested parties would only be interested in a small subset of the possible S member coalitions. They wouldn't view it as realistic to have to negotiate over a sample set that is quite large. For example, if $N = 10$, the maximum number of coalitions is 1023, an unwieldy number. Thus we may only need to evaluate a small subset of these coalitions.

13.G.5. Effect of Transactions Costs

Transactions costs are incurred as part of cooperative solutions. As the size of the group grows, transactions costs would be expected to increase even more rapidly due to multiple jurisdictions, growing administration costs, more complex environmental impacts, and so on (Heaney, 1983). These costs are seldom counted as part of the characteristic function costs. Figure 13.12 shows hypothetical curves of savings and transactions costs as a function of the size of the system (Heaney, 1982). Savings tend to increase at a decreasing rate as the coalition size grows, resulting in the concave shape. On the other hand, transactions costs would tend to increase at an increasing rate as coalition size increases. At least two justifications can be given for the nonlinear nature of transaction costs. First, the number of coalitions to consider, k, is

$$k = 2^n - 1 \tag{13.43}$$

where n is the size of the grand coalition. The number of orderings to consider, o, can be calculated using

$$o = n! \tag{13.44}$$

For a modest-size problem with six coalitions, $k = 63$ and $o = 720$. Given these rapidly increasing transactions costs, it may be better to stop short of the regional optimum, as dictated by calculating total project costs without

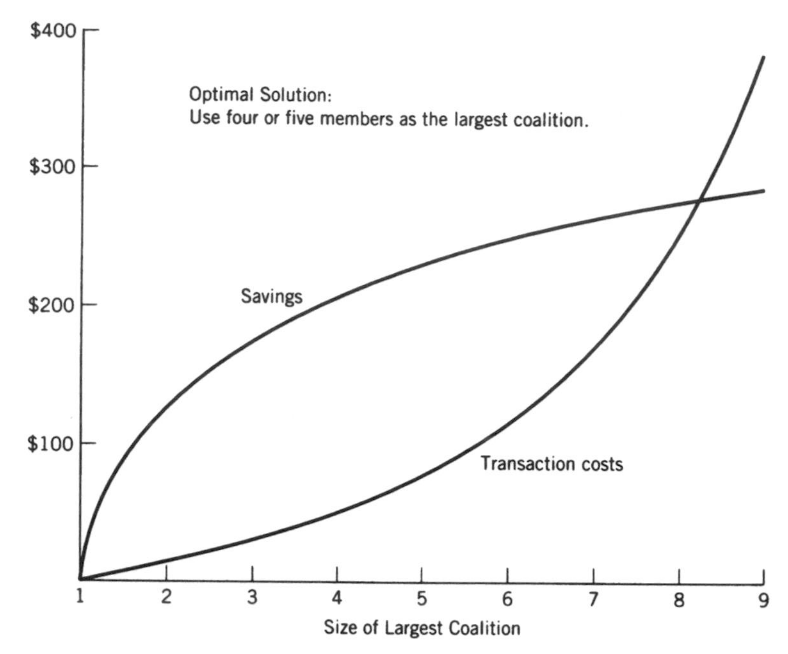

Figure 13.12 Savings-Transactions costs for hypothetical regional problems. (Heaney 1983.)

transactions costs. Heaney (1982) gives an example of this approach applied to an eight-city regional wastewater treatment plant example.

13.G.6. Cooperation or Competition: Who Gets to Go First?

A key question in determining the characteristic functions for some games is who gets to go first. The TVA example can be used to illustrate this point. The TVA example consists of three purposes: (1) navigation, (2) flood control, and (3) hydropower. If the three purposes cooperate, a single dam will be built and they will share the costs. But how do we calculate the cost of building a dam to serve a single purpose such as flood control? Do they get to pick their best site along the river, or should they get to calculate their cost after the other two purposes have built their system? Depending on the answer to this question, the game will be either cooperative or competitive. For the game to end up cooperative, it is essential that coalition $N–S$ be given preference over coalition S in deciding who gets to go first.

13.G.7. Appropriate Measure of Use

Cost allocation methods typically assign costs among purposes and groups based on a single or multiple measure of use. In general accounting, labor hours are often used as the allocation basis. In the water resources field, there does not usually exist a single measure of use that can be used to assign costs accurately. The capital costs of a facility are typically based on peak usage, whereas operating costs are based on average usage. Cities use a variety of measures of use in actual practice. Accurate measures of the cost of a project can be obtained from the cost estimating program of an engineering analysis. Unfortunately, such models are not carried over from the design phase to the operating phase. Rather, they are replaced by a much more simplistic accounting model that does not reflect these costs accurately. This problem can be resolved by linking engineering process models and accounting models using activity-based costing described elsewhere in this chapter.

13.H. COMPUTER MODELS FOR COST ALLOCATION

Heaney (1979) included a detailed description of how cost allocation could be incorporated into cost allocation done as part of the evaluation of urban stormwater management analyses. Clark and Males (1986a) include spatial cost allocation in their model of water distribution systems. Nelson (1995) describes simple spreadsheet procedures for estimated the impact of new developments. However, it is fair to say that no active cost allocation models exist in water resources. A recent review of water resources models does not mention any such programs (Wurbs, 1994).

13.I. SUMMARY AND CONCLUSIONS

The purpose of this chapter was to describe various ways to address the cost allocation problem. Conventional practices in the water resources field were described. Then related methods from cooperative *n*-person game theory were presented. Next, recently developed methods from the field of accounting are outlined. Then these methods were summarized and applications to selected problems in cost allocation were described.

The importance of the context within which cost allocation takes place is stressed by reviewing current practices for establishing system development charges for new developments. Methods for calculating characteristic function values are presented to reduce the computational burden. Also, methods for tracking good solutions as well as the optimal solution are described. Participants may be willing to sacrifice a small gain in economic efficiency for a simpler cost allocation procedure. New institutions must be created to implement cooperative solutions. The associated transactions costs can offset the gains from a cooperative solution. Finally, the effects of the measures of use on the final solution are presented.

Overall, the cost allocation problem is complex. Practitioners have developed simple procedures for performing this task. However, the stakes are becoming higher as more competition enters the water resources field. It is essential that we fully understand the nature of and methods for properly answering the seemingly simple question, "What does it cost"?

REFERENCES

Anderson, M. B., 1993. Using Activity-Based Costing for Efficiency and Quality, *Government Finance Review,* June, 7–9.

Arrow, K. J., 1963. *Social Choice and Individual Values,* 2nd ed., Wiley, New York.

Aumann, R. J., and S. Hart (ed.), 1994. *Handbook of Game Theory,* Vols. I and II, Elsevier, New York.

Beecher, J. A., P. C. Mann, and J. R. Landers, 1991. *Cost Allocation and Rate Design for Water Utilities,* NRRI 90-17, National Regulatory Research Institute, Ohio State University, Columbus, OH.

Clark, R. M., and R. M. Males, 1986a. Simulating Cost and Quality in Water Distribution, *Journal of Water Resources Planning and Management,* 111(4), 454–466.

Clark, R. M., and R. M. Males, 1986b. Developing and Applying the Water Supply Simulation Model, *Journal of the American Water Works Association,* 78(8), 61–65.

Cooper, R., and R. S. Kaplan, 1991. *The Design of Cost Management Systems,* Prentice Hall, Upper Saddle River, NJ.

Coursey, D., and B. Bozeman, 1990. Decision Making in Public and Private Organizations: A Test of Alternative Concepts of "Publicness," *Public Administration Review,* 50, September–October, 525–535.

Dandy, G. C., A. R. Simpson, and L. J. Murphy, 1996. An Improved Genetic Algorithm for Pipe Network Optimization, *Water Resources Research,* 32(2), 449–458.

Drury, C., 1990. Product Costing in the 1990's, *Accountancy,* 105, May, 122–126.

Federal Interagency River Basin Committee, 1950. *Proposed Practices for Economic Analysis of River Basin Projects,* Report, Washington, DC.

Granot, D., and F. Granot, 1992. On Some Network Flow Games, *Mathematics of Operations Research,* 17, 792–841.

Hallefjord, A., R. Helming, and K. Jornsten, 1995. Computing the Nucleolus When the Characteristic Function is Given Implicitly: A Constraint Generation Approach, *International Journal of Game Theory,* 24, 357–372.

Heaney, J. P., 1979. *Economic/Financial Analysis of Urban Water Quality Management Problems, Final Report,* EPA R802411-024, U.S. Environmental Protection Agency, Cincinnati, OH.

Heaney, J. P., 1982. Urban Wastewater Management Planning. in S. J. Brams, W. F. Lucas, and P. D. Straffin, Jr. (eds.), *Modules in Applied Mathematics,* Vol. 2, *Political and Related Models,* Springer-Verlag, New York, pp. 98–108.

Heaney, J. P., 1983. Coalition Formation and the Size of Regional Pollution Control Systems, in E. F. Joeres and M. H. Davis (eds.), *Land Economics Monograph 6,* University of Wisconsin Press, Madison, WI.

Heaney, J. P., and R. E. Dickinson, 1982. Methods for Apportioning the Cost of a Water Resource Project, *Water Resources Research,* 18(3).

Heaney, J. P., et al., 1979. Nationwide Cost of Controlling Combined Sewer Overflows and Urban Stormwater Discharges, *Journal of the Water Pollution Control Federation,* 51(8).

Heaney, J. P., N. C. Simpson, and K. R. Coates, 1991. Activity-Based Costing for Performance-Based Management of a Water Management District Final Report to the South Florida Water Management District, Florida Water Resources Research Center, University of Florida, Gainesville.

James, L. D., and R. R. Lee, 1971. *Economics of Water Resources Planning,* McGraw-Hill, New York.

Johnson, H. J., and R. S. Kaplan, 1987. *Relevance Lost: The Rise and Fall of Management Accounting,* Harvard Business School Press, Boston.

Kaplan, R. S., 1982. *Advanced Management Accounting,* Prentice Hall, Upper Saddle River, NJ.

Littlechild, S. C., and G. Owen, 1973. A Simple Expression for the Shapley Value in a Special Case, *Management Science,* 20, 370–372.

Littlechild, S. C., and G. F. Thompson, 1977. Aircraft Landing Fees: A Game Theory Approach, *Bell Journal of Economics,* 8, 186–204.

Maschler, M., 1994. The Bargaining Set, Kernel and Nucleolus, in R. J. Aumann and S. Hart (eds.), *Handbook of Game Theory,* Vol. 1, Elsevier Science Publishers, New York, Chapter 18.

McNair, C. J., R. L. Lynch, and K. F. Cross, 1990. Do Financial and Nonfinancial Performance Measures Have to Agree? *Management Accounting,* 72, November, 28–36.

Miles, S. M., and J. P. Heaney, 1988. Better Than Optimal Method for Designing Drainage Systems, *Journal of Water Resources Planning and Management,* 114(5), 477–495.

Milliman, J. W., 1972. Beneficiary Charges: Toward a Unified Theory, in *Public Prices for Public Products,* The Urban Institute, Washington, DC, pp. 27–33.

Moreau, D. H., 1996. Principles of Planning and Financing for Water Resources in the United States, in L. Mays (ed.), *Water Resources Handbook,* McGraw-Hill, New York, Chapter 4.

Moulin, H., 1995. *Cooperative Economics: A Game-Theoretic Approach,* Princeton University Press, Princeton, NJ.

Mushkin, S. J., and R. M. Bird, 1972. Public Prices: An Overview, in *Public Prices for Public Products,* The Urban Institute, Washington, DC, pp. 3–25.

Nelson, A. C., 1995. *System Development Charges for Water, Wastewater, and Stormwater Facilities,* CRC Press, Boca Raton, FL.

Ng, E., and J. P. Heaney, 1989. Efficient Total Enumeration of Water Resources Alternatives, *Water Resources Research,* 25(4), 583–590.

Owen, G., 1995. *Game Theory,* 3rd ed., Academic Press, New York.

Ransmeier, J. S., 1942. *The Tennessee Valley Authority,* Vanderbilt University Press, Nashville, TN.

Raucher, R. S., et al., 1995. *Estimating the Cost of Compliance with Drinking Water Standards: A User's Guide,* AWWA Research Foundation, Denver, CO.

Reeves, C. R. (ed.), 1993. *Modern Heuristic Techniques for Combinatorial Problems,* Halsted Press, New York.

Schmeidler, D., 1969. The Nucleolus of a Characteristic Function Game, *SIAM Journal on Applied Mathematics,* 17, 1163–1170.

Shapley, L. S., 1953. A Value for *n*-Person Games, in H. W. Kuhn and A. W. Tucker (eds.), *Contributions to the Theory of Games,* Vol. II, Annals of Mathematics Studies 28, Princeton University Press, Princeton, NJ.

Straffin, P., and J. P. Heaney, 1981. Game Theory and the Tennessee Valley Authority, *International Journal of Game Theory,* 10(1).

U.S. Interagency Committee on Water Resources, 1958. *Proposed Practices for Economic Analysis of River Basin Projects,* Washington, DC.

Wurbs, R. A., 1994. *Computer Models for Water Resources Planning and Management,* IWR Report 94-NDS-7, U.S. Army Corps of Engineers, Fort Belvoir, VA.

Young, H. P. (ed.), 1985. *Cost Allocation Methods: Principles and Practices,* Elsevier, Amsterdam.

Young, H. P., 1994a. *Equity in Theory and Practice,* Princeton University Press, Princeton, NJ.

Young, H. P., N. Odaka, and T. Hashimoro, 1982. Cost Allocation in Water Resources Development. *Water Resources Research,* 18(3), 463–475.

Young, H. P., 1994b. Cost Allocation. Chapter 34 in Aumann, R. J. and S. Hart, Eds. Handbook of Game Theory with Economic Applications. Elsevier, Amsterdam, 1193–1235.

CHAPTER 14

Siting Regional Environmental Facilities

EARL WHITLATCH

14.A. INTRODUCTION

In this chapter we address the concept of regionalization in civil and environmental engineering facilities and projects and focus on models for siting regional wastewater treatment facilities.

14.B. REGIONALIZATION

Regionalization has been defined as the "integration or coordination of the physical, economic, social, informational, or personnel structure of water resource projects to better achieve national, regional, and local societal objectives and constraints" (Whitlatch and ReVelle, 1990). Although the definition emphasizes such water projects as water quality control, water supply, flood control, navigation, hydroelectric generation, commercial fisheries, and water-based recreation, it applies equally to other civil and environmental engineering areas, such as solid and hazardous waste, air pollution, transportation, energy, and structures. Most of the mathematical programming models in this book are illustrative of the regional approach wherein a high degree of coordination among physical and economic factors is achieved.

However, Table 14.1 indicates that regionalization can occur to a varying degree over a wide range of system functions. For example, a low degree of regionalization which is still very appropriate and valuable can be illustrated by joint purchase agreements among rural water supply systems for materials, equipment, or services. Indeed, for each specific system there is an appropriate degree of regionalization that maximizes total societal well-being. The goal of the engineering planning process is to identify and arrive at this appropriate level. Although it is important to recognize that systems can be overregionalized, as illustrated by uneconomic overextension of water sys-

TABLE 14.1 Degrees of Regionalization and Generalized Examples

Degree of Regionalization	Physical/Economic/ Social/Informational/ Personnel Structure	Generalized Example
Over (O)	Overly integrated/ coordinated	Excessive capacity; overextension; strong negative externalities; bureaucratic inefficiency, waste; not representative of public served
Complete (C)	Completely integrated/ coordinated; multipurpose	Federal autonomous water agencies; river basin authorities; unique federal or state water plans, projects; uniquely coordinated multiparty water projects
High (H)	Highly integrated/ coordinated; special-purpose	Most federal or state water plans, projects; most international treaty projects; most external support agency plans, projects; general-purpose river basin organizations
Intermediate (I)	Closely integrated/ coordinated; special-purpose	Wholesale metropolitan water and sewer services; water and wastewater service districts; municipal combined water or wastewater treatment plants; municipal capacity sharing of reservoirs; interconnection of municipal water distribution systems; interstate compacts for water allocation; judicial allocation of river flow; special-purpose river basin organizations
Low (L)	Loosely integrated/ coordinated; special-purpose	Centralized operation of multiple, small water/wastewater treatment plants; joint purchase agreements for materials, equipment, or services; shared laboratory facilities; purchase of services from large companies; joint emergency management plans; federal and state planning grants; federal construction grants; federal/ state data collection programs, software development, and material standards
Zero (Z)	Not integrated/ coordinated	Separate facilities, plans, operations
Under (U)	Underintegrated/ coordinated	Uncaptured economies of scale; overlooked system interactions; duplicative services; strong negative externalities; short-range planning

Source: Whitlatch and ReVelle (1990).

tems, the existence of excessive system capacity, or prevalence of inefficient bureaucracies, it is fair to say that the most common situation in practice is more likely to be a system with some degree of underregionalization.

Advantages of the regional approach include cost reduction through economies of scale, consideration of a wider range of economic, environmental, and social impacts (and externalities) as well as planning alternatives, greater operational reliability of facilities, a higher-quality product, easier financing, longer planning horizon, larger geographic area for site selection, greater public participation, and more political power. Disadvantages include increased planning costs, potentially larger project impacts, greater consequences of failure, need for more highly skilled professionals, political/jurisdictional opposition, and perceived inequities in cost sharing. Current worldwide trends that would tend to promote the regional approach are demand for a higher-quality product (e.g., water and wastewater treatment standards), greater computational abilities, increased urbanization, tighter fiscal resources, and crisis response (Whitlatch and ReVelle, 1990).

14.C. ECONOMIES OF SCALE IN WASTEWATER SYSTEMS

The present worth total cost (capital cost plus operation, maintenance, and replacement cost) of most component facilities for wastewater systems can be expressed as

$$C = \alpha Q^{\beta} \tag{14.1}$$

where C is the present worth total cost ($), Q the wastewater design flow (million gallons per day, *mgd*), α the empirical coefficient, and β the empirical coefficient. If $(0 < \beta < 1)$, the cost function is said to demonstrate *economies of scale* and has a strictly concave shape. The function holds if cost versus design flow plots as a straight line on log-log paper. Table 14.2 gives the total cost functions for conventional activated sludge treatment plants, gravity interceptor sewers, force (pressure) main sewers, pumping stations, and package (small manufactured) treatment plants. Since in some cases the cost data do not plot exactly as straight lines on log-log paper, segmented straight-line equations are provided in the form

$$C = \alpha Q^{\beta(Q)} \tag{14.2}$$

and should be used for greater accuracy. However, in most cases, a single equation that very closely approximates the cost data is also provided. The latter is useful in summarizing the economy of scale factor, β. For example, interceptor sewers demonstrate greatest economies of scale ($\beta = 0.52900$), while pumping stations demonstrate the least ($\beta = 0.75699$). The presence of economies of scale argues for the combined treatment of wastewater in one or a few plants rather than in many separate plants.

TABLE 14.2 Equations for Total Cost of Various Components of Wastewater Systems

Item and Data Source	Flow Range (mgd)	Cost Equation[a]
Secondary treatment plants (activated sludge) (Eilers and Smith, 1971)	1.0–2.0	$C = 1{,}838{,}377Q^{0.35364}$
	2.0–25.0	$1{,}622{,}888Q^{0.53351}$
	5.0–20.0	$1{,}334{,}286Q^{0.65517}$
	20.0–100.0	$980{,}502Q^{0.75801}$
	1.0–100.0 (approx.)	$1{,}500{,}000Q^{0.65052}$
Interceptor sewers (Smith, 1971)	0.10–0.50	$149{,}653Q^{0.53088}$
	0.50–2.5	$154{,}697Q^{0.57870}$
	2.5–100.0	$165{,}346Q^{0.50604}$
	0.10–100.0 (approx.)	$148{,}748Q^{0.52900}$
Force mains (Smith, 1971)	0.10–0.3	$92{,}609Q^{0.49544}$
	0.30–1.0	$98{,}228Q^{0.54427}$
	1.0–5.0	$98{,}228Q^{0.58505}$
	5.0–100.0	$94{,}100Q^{0.61173}$
	0.10–100.0 (approx.)	$103{,}153Q^{0.56632}$
Pumping stations (Smith, 1971)	>0.1	$414{,}387Q^{0.75699}$
Package treatment plants (McMichael, 1971)	0.001–0.005	$131{,}308Q^{0.13865}$
	0.005–0.02	$282{,}910Q^{0.28352}$
	0.02–0.06	$486{,}526Q^{0.42211}$
	0.06–0.2	$733{,}115Q^{0.56784}$
	0.2–0.5	$878{,}493Q^{0.68025}$
	0.001–0.5 (approx.)	None

Source: Whitlatch (1973).

[a] January 1971 dollars; present worth total cost (C) given as ($) for treatment plants and pumping stations, and ($/mile) for conveyance; ENR sewerage cost index = 150.6 for treatment plants and 157.4 for sewer lines.

However, Figure 14.1 illustrates that there exists a trade-off between the cost of transporting wastewater to a regional facility and the savings attained through economies of scale in regional treatment. An appropriate degree of regionalization generally exists that will minimize total societal costs, here represented by the sum of treatment and transport cost. Note that to determine the minimum total cost for a given number of regional plants, optimal plant locations and the best pattern of wastewater flow must be found.

Figure 14.2 shows the impact of economies of scale on the well-known relationship between cost and efficiency of removal of a pollutant such as BOD. The unit cost of removal, measured either by dollars per unit of flow or per pound of pollutant removed, will be lower for a larger plant. Thus economies of scale will be important even when efficiencies of removal are allowed to vary in seeking to meet a water quality goal at least cost.

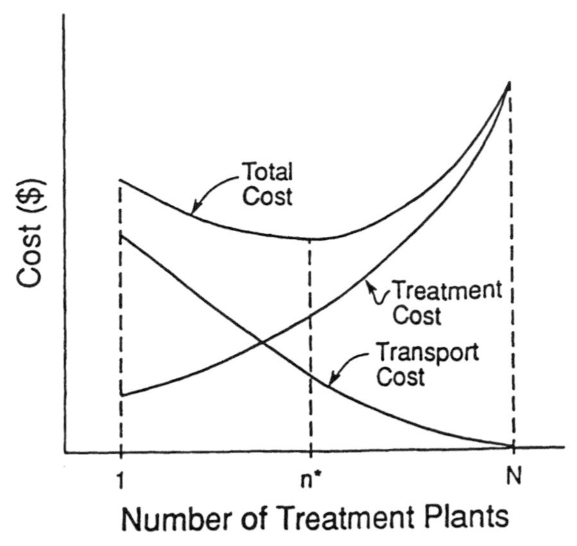

Figure 14.1 Determination of optimal number and location of treatment plants. (From Whitlatch and ReVelle, 1990.)

14.D. COST SAVINGS IN REGIONAL WASTEWATER SYSTEMS

It is useful to break the discussion of regional wastewater systems into two problems. The phase I problem can be summarized as

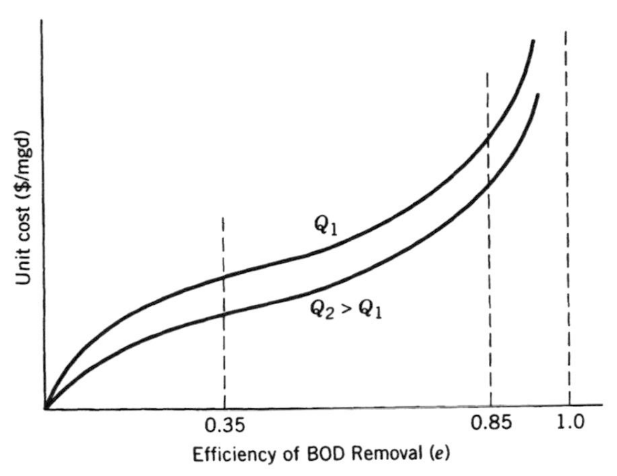

Figure 14.2 Variation in unit cost of treatment with plant size and efficiency of removal.

$$\text{minimize } Z = \sum_i C_i(x_i) + \sum_i \sum_j C_{ij}(x_{ij}) \tag{14.3}$$

subject to

(1) all waste flows treated to a fixed removal efficiency level (14.4)

(2) continuity of flow and BOD mass is maintained

where x_i is the flow treated at site i (mgd), x_{ij} the flow transported from site i to site j (mgd), $C_i(x_i)$ the present worth total treatment cost at site i ($), and $C_{ij}(x_{ij})$ the present worth total transport cost in moving waste flow from site i to site j ($). Treatment and transport cost must include capital, operation, maintenance, and replacement cost. The cost data in Table 14.2 therefore apply to the phase I problem when secondary treatment is required everywhere.

The phase II problem aims to determine the optimal degree of regionalization, pattern of wastewater transport, and efficiency of removal at each active treatment plant to minimize cost while meeting prescribed water quality goals:

$$\text{minimize } Z = \sum_i C_i(x_i,e_i) + \sum_i \sum_j C_{ij}(x_{ij}) \tag{14.5}$$

subject to

(1) water quality goals are met everywhere (14.6)

(2) continuity of flow and BOD mass is maintained

where e_i is the efficiency of removal of pollutant at site i, and $C_i(x_i,e_i)$ is the present worth total treatment cost at site i ($). In certain cases, further conditions are placed on removal efficiencies at the various sites in the phase II problem to meet equity goals (e.g., uniform treatment, zoned uniform treatment). The phase II problem allows much greater flexibility in treatment options than does the phase I problem, but it is also more difficult to formulate and solve. Detailed models for both the phase I and phase II problems are developed in the sections that follow.

At this point it is useful to ask what the order-of-magnitude savings from wastewater regionalization might be. Table 14.3 summarizes actual case studies wherein optimization models have been applied and compared to at-source treatment. Case-based average savings for the phase I problem is 21.7% which is quite high. For the one case study involving the phase II problem, savings are even greater, about 35%. Clearly, it would be advantageous to have realistic models and efficient solution procedures for both problems.

Table 14.4 highlights cost savings possible from consideration of a wide range of regional alternatives rather than focusing only on at-source effluent

TABLE 14.3 Cost Savings Possible from Regional Wastewater Treatment Plants

Author(s)	Geographic Area	Number of Sources/ Potential Regional Sites	Optimal Number of Plants	Planning Horizon (yr)	Cost Without Regionalization ($ millions)	Cost with Regionalization ($ millions)	Percent Savings
		Phase I Case: Effluent Standard					
Maryland Environmental Service (1971)	Patuxent R., MD	4/1	1	10	16.1	14.4	10.6
		4/1	1	30	35.7	27.5	22.9
Deininger and Su (1971)	Hypothetical	5/4	1	—	3.46	3.08	11.0
Converse (1972)	Merrimack R., NH/VT	18/18	4	NA	8.86/yr	5.56/yr	37.2
Wanielista and Bauer (1972)	Little Econlockahatchee R., FL	11/1	2	20	30.3	27.5	9.1
McConagha and Converse (1973)	Upper Conn. valley region, VT/NH	7/NA	NA	NA	0.197	0.149	24.7
Joeres et al. (1974)	Dane County, WI	12/12	4	20	29.2	25.5	12.7
Whitlatch and ReVelle (1976)	Delaware estuary, PA/NJ/DE	22/9	14	11	260.4	179.8	30.9

TABLE 14.3 (*Continued*)

Author(s)	Geographic Area	Number of Sources/ Potential Regional Sites	Optimal Number of Plants	Planning Horizon (yr)	Cost Without Regionalization ($ millions)	Cost with Regionalization ($ millions)	Percent Savings
Phase I Case: Effluent Standard (*Continued*)							
Whitlatch (1978)	Olentangy R., OH	2/1	1	20	8.3	4.7	43.4
Leighton and Shoemaker (1984)	Suffolk County, NY	50/23	14	0	661.7	566.8	14.3
						Case-based average:	21.7
Phase II Case: Water Quality Standard							
Graves et al. (1970, 1972)[a]	Delaware estuary, PA/NJ/DE	22/9	15	11	53.5	33.3[a]	37.8[a]
Whitlatch and ReVelle (1976)	Delaware estuary, PA/NJ/DE	22/9	16	11	65.9	43.0	34.8

[a]Includes the option of bypass piping to remote estuary sections for water quality improvement.

TABLE 14.4 Cost Savings Possible from Various Regional Water Quality Management Alternatives for the Delaware Estuary

Management Alternative[a]	Total Cost
Required secondary, at source	$260,414,688
Uniform treatment, at source	192,781,753
Required secondary, regional (phase I)	179,836,696
Least cost, at source with primary treatment required	65,870,270
Least cost, at source without primary treatment required	54,064,103
Least cost, at source without primary treatment required (Graves et al. 1970)	53,495,000
Least cost, regional (phase II) with primary treatment required	42,975,555
Least cost, regional (phase II) without primary treatment required	39,637,015[b]
Least cost, regional (phase II) without primary treatment required and with bypass piping	≈33,300,000[b]
Least cost, regional without primary treatment required and with bypass piping (Graves et al., 1970)	33,300,000

Source: Whitlatch and ReVelle (1976).

[a]Results are from Whitlatch and ReVelle (1976) unless noted.

[b]Cost given may possibly be reduced further through application of heuristic procedures.

standards. For the Delaware estuary, requiring secondary treatment at the source costs $260,614,688. By allowing regionalization but still requiring secondary treatment (phase I problem), cost is reduced by 30.9%, to $179,836,696. By allowing regionalization and plant efficiencies to vary while meeting imposed water quality standards (phase II problem) and requiring at least primary treatment, costs are reduced by 83.5%, to $42,975,555. Permitting bypass piping of wastewater reduces cost further. Although complete implementation of all options may not be acceptable, there are clear benefits to considering a wide range of alternatives in a regional format.

14.E. SITING MODELS FOR MEETING AN EFFLUENT STANDARD

The Federal Water Pollution Control Act of 1972 required that all publicly owned wastewater treatment plants provide secondary treatment by July 1, 1977, and best practicable control technology currently available (BPT) by July 1, 1983. All industrial plants were to provide BPT by July 1, 1977, and best available control technology economically achievable (BAT) by July 1, 1983. Technologies were subsequently specified, effectively establishing an effluent standard for wastewater discharges. Associated cost functions for these treatment levels, as well as others (e.g., tertiary treatment), demonstrate economies of scale just as do those illustrated in Table 14.2 for secondary treatment.

14.E.1. Network Configuration, Nonlinear Programming

A particularly clear formulation of the regional wastewater treatment plant siting problem results from a network viewpoint first described by Deininger and Su (1971). Figure 14.3 shows a hypothetical set of source flows (large circles), transport paths (arcs), transshipment points (small circles), and potential regional treatment plant sites (triangles). As drawn, only one source of wastewater (the northeast source) can accommodate a treatment plant (for either at-source treatment or regional treatment); all other sources must send their flow elsewhere for treatment. In the general case, more sources could accommodate treatment plants. For the given network there can be from one to five active treatment plant sites. A further condition is that if either or both of the two plant sites along the small river are active, tertiary treatment must be provided. Along the larger river (three potential sites), BPT is adequate. Cost functions that demonstrate economies of scale are known for each potential treatment plant site, and these functions include the cost of an outfall pipeline and any associated pumping cost for discharge to the river.

Solid lines, or arcs, indicate possible flow direction, which can be restricted by the engineer or left flexible. Dashed lines from potential treatment plant sites to an artificial effluent node are purely for convenience in formulating

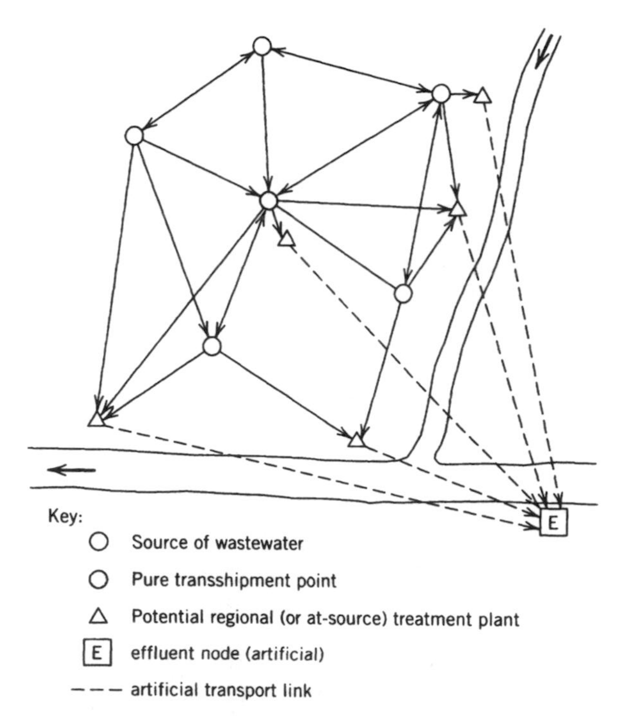

Figure 14.3 Network representation of the regional wastewater siting problem. (Adapted from Deininger and Su, 1971.)

the mathematical programming model. It is imagined that flow going to any regional treatment plant site is transferred to the artificial effluent node, and the cost of such transfer is the cost of treatment at the regional plant. Constraints in the model will ensure that source wasteflows are sent through the network via allowable paths, ultimately reaching the artificial effluent node, thereby providing treatment to the proper effluent quality level. Assume that source, transshipment, and regional plant site nodes are numbered consecutively.

Define

x_{ij} = wastewater flow sent from node i to node j (mgd)

$f_{ij}(x_{ij})$ = function expressing total cost of transport or treatment when flow is sent from node i to node j ($, present worth)

a_i = amount of wastewater generated by source node i (mgd)

r = number of regional sites

s = number of source nodes

t = number of transshipment nodes

The objective function is to

$$\text{minimize } Z = \sum_i \sum_j f_{ij}(x_{ij}) \tag{14.7}$$

which represents the sum of many concave cost functions, including the cost of treatment (and sewer outfall) when flow occurs along the path from a regional plant site to the artificial effluent node.

Constraints are designed to maintain continuity of flow. At source nodes:

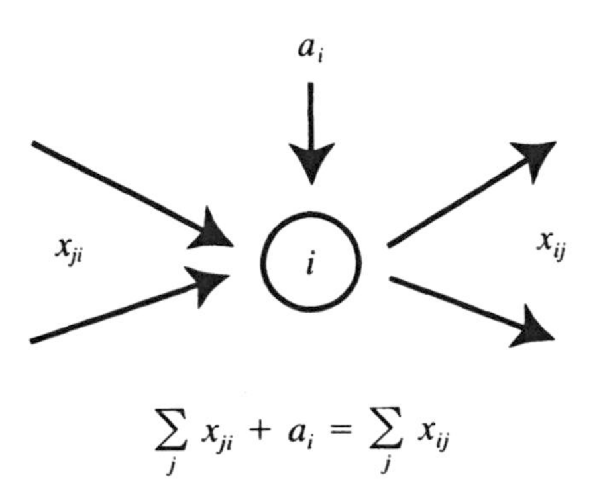

$$\sum_j x_{ji} + a_i = \sum_j x_{ij}$$

Rearranging and multiplying by (−1) yields

$$\sum_j x_{ij} - \sum_j x_{ji} = a_i \qquad i = 1,2,\ldots,s \tag{14.8}$$

At pure transshipment nodes,

x_{ji} i x_{ij}

$$\sum_j x_{ij} - \sum_j x_{ji} = 0 \qquad i = s + 1, s + 2, \ldots, s + t \tag{14.9}$$

At potential regional treatment plant sites,

x_{ji} i x_{iE}

$$x_{iE} - \sum_j x_{ji} = 0 \qquad i = s + t + 1, s + t + 2, \ldots, s + t + r \tag{14.10}$$

At the artificial effluent node,

x_{jE} E

$$\sum_j x_{jE} = \sum_i a_i \tag{14.11}$$

with

$$x_{ij} \geq 0 \qquad \text{all } (i,j)$$

If any flow paths or treatment plants have capacity limitations, the appropriate flow variables can be bounded in the constraints.

Model (14.7)–(14.11) is nonlinear and difficult to solve. In general, existing solution algorithms suffer from long computation times and/or termination at locally optimal solutions. One method, a ranking extreme point algorithm, takes advantage of the fact that the optimal solution can be shown to lie at an extreme point of the convex set formed by the linear equations (14.8)–(14.11). However, the method may have to enumerate all extreme points before the answer is obtained (Deininger and Su, 1973). For large problems, this may be computationally prohibitive. Another shortcoming of the nonlinear programming approach is that fixed costs cannot be handled easily. However, the model is appealing from the standpoint that it is a straightforward mathematical representation of the physical and economic problem.

14.E.2. Network Configuration, Mixed Integer Programming

Mixed-integer programming models have the strong advantage of being naturally amenable to the inclusion of fixed costs. Figure 14.4 shows how a fixed-cost concave cost curve can be approximated by a series of straight lines, each having a "fixed-cost" intercept on the y axis. For proper representation of the cost function, each of the variables, x_{ijk} (k = 1,2,3,...) must be mutually exclusive. For example, if the optimal solution were to lie between the bounds b_{ij1} and b_{ij2}, variable x_{ij2} would be in the solution and equal to the optimal solution. The value of the objective function would be well approximated between the two bounds by $C = f_{ij2} + c_{ij2}\, x_{ij2}$. The other variables would be zero because none could produce a lower cost in the range between b_{ij1} and b_{ij2}. Similarly, if the optimal solution were to lie between zero and b_{ij1}, x_{ij1} would be equal to the optimal solution, all other variables would equal zero, and cost would be approximated by $C = f_{ij1} + c_{ij1}\, x_{ij1}$. However, the model must be formulated to ensure that when x_{ijk} is active, it may not exceed its upper bound, and fixed cost f_{ijk} is added to the total cost.

Although many mixed-integer programming models have been put forward for the regional wastewater treatment plant siting problem, one of the clearest and most direct is presented by Joeres et al. (1974). The formulation is for a network representation similar to that presented in Figure 14.3 but with some

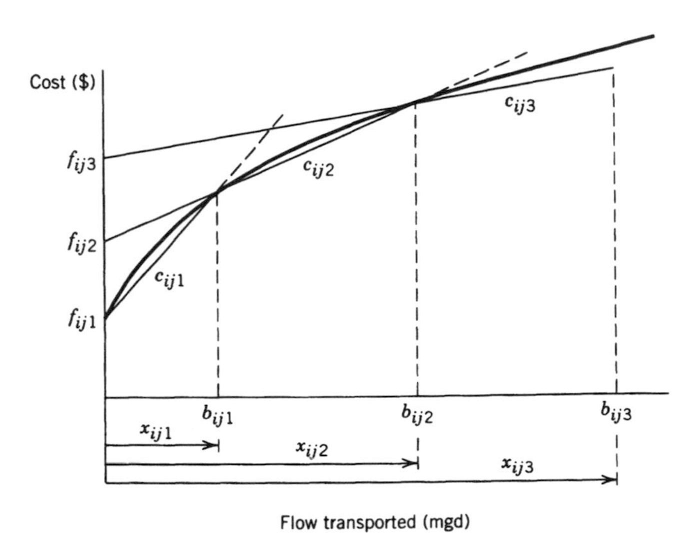

Figure 14.4 Piecewise approximation of fixed-cost, concave cost curve.

minor network flow modifications that should become evident in development of the constraints. Define

x_{ijk} = flow transported from node i to node j, as represented by variable k in the approximation of transport cost (mgd)

x_{ik} = flow treated at node i, as represented by variable k in the approximation of treatment cost (mgd)

c_{ijk} = unit cost slope of the line associated with variable k in the approximation of transport cost ($/mgd)

c_{ik} = unit cost slope of the line associated with variable k in the approximation of treatment cost ($/mgd)

y_{ijk} = (0,1) dummy variable associated with f_{ijk}

z_{ik} = (0,1) dummy variable associated with f_{ik}

f_{ijk} = intercept fixed cost associated with flow transported from node i to node j, as represented by variable k ($)

f_{ik} = intercept fixed cost associated with flow treated at node i, as represented by variable k ($)

b_{ijk} = upper bound on variable x_{ijk}

b_{ik} = upper bound on variable x_{ik}

p = k index of the last approximation variable for x_{ik} or x_{ijk}

The objective function becomes

$$\text{minimize } Z = \sum_i \sum_j \sum_k c_{ijk} x_{ijk} + \sum_i \sum_j \sum_k f_{ijk} y_{ijk} + \sum_i \sum_k c_{ik} x_{ik} + \sum_i \sum_k f_{ik} z_{ik} \quad (14.12)$$

wherein the first two summations account for transport cost, and the third and fourth account for treatment cost.

A general expression for continuity of flow at each node can be written based on the following nodal flows:

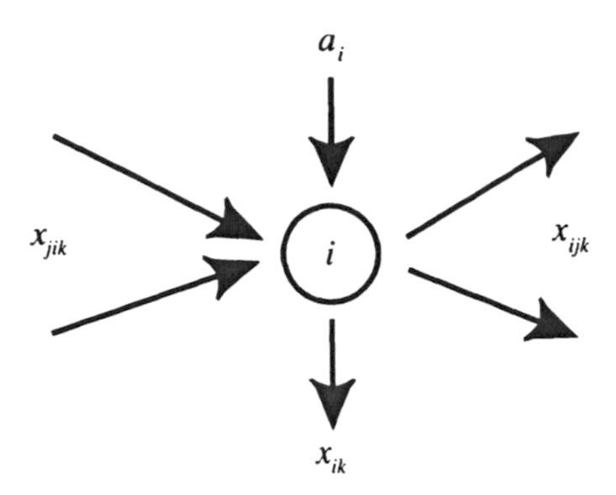

$$a_i + \sum_j \sum_k x_{jik} - \sum_j \sum_k x_{ijk} - \sum_k x_{ik} = 0 \qquad \text{all } i \qquad (14.13)$$

where $a_i = 0$ at pure transshipment nodes and potential treatment plant sites, x_{ik} terms are omitted for any node at which treatment cannot be given, and certain flow terms x_{ijk} and x_{jik} are omitted depending on network-allowable flows into or out of a particular node. For example, the constraint requires that a source node that cannot serve as an at-source or regional plant site must send its flow, a_i, into the network for treatment elsewhere. By requiring continuity of flow at all nodes, waste flow contributions from sources will be passed along the network until treatment is eventually given (at the sites most economical overall).

A potential treatment plant site may have capacity limitation, in which case b_{ip}, the upper bound on x_{ip}, is set equal to the capacity limitation. Therefore, it is not necessary to write specific constraints for capacity limitations. Instead, constraints are written that bound all variables to the range established in approximating the respective cost functions:

$$x_{ijk} \leq b_{ijk} y_{ijk} \qquad \text{all } (i,j,k) \qquad (14.14)$$

$$x_{ik} \leq b_{ik} z_{ik} \qquad \text{all } (i,k) \qquad (14.15)$$

Consider constraint (14.14): If x_{ijk} is active (nonzero), y_{ijk}, the (0,1) variable, must be equal to 1, which simultaneously causes x_{ijk} to be bounded by b_{ijk} and fixed cost, f_{ijk}, to be added to the objective function, as intended. However, note that if x_{ijk} is equal to zero, the constraint does not compel y_{ijk} to equal zero as is desired; the latter could be equal to 1 and still satisfy the constraint. This will not happen in the final solution, however, since cost would not be minimized until y_{ijk} is assigned a zero value. Conversely, if y_{ijk} is equal to zero, x_{ijk} is restricted to be zero, as desired. If y_{ijk} equals 1, x_{ijk} is bounded by b_{ijk} as required, but could equal zero. Again, the latter situation will not be present in the final solution since cost is to be minimized and there is no reason to keep y_{ijk} equal to 1 if x_{ijk} is equal to zero. Constraint (14.15) operates on x_{ik} in the same fashion.

Because capacity limitations usually exist on potential regional treatment plant sites and transport paths, constraints must be added to ensure that multiple treatment facilities are not placed at one site nor multiple flow paths used between two nodes:

$$\sum_k y_{jik} + \sum_k y_{ijk} \leq 1 \qquad \text{all } (i,j) \tag{14.16}$$

$$\sum_k z_{ik} \leq 1 \qquad \text{all } i \tag{14.17}$$

Constraint (14.16) says that, at most, only one pipeline will be constructed between any two nodes, requiring flow to be in one direction only. The constraint could be broken into two parts, thereby limiting flow to, at most, one pipeline in each direction; but as both a practical and an economic matter, flow would be in one direction only. Constraint (14.17) limits the number of treatment plants to one at a site.

Finally,

$$\begin{aligned} x_{ijk} &\geq 0 \qquad y_{ijk} = (0,1) \qquad \text{all } (i,j,k) \\ x_{ik} &\geq 0 \qquad z_{ik} = (0,1) \qquad \text{all } (i,k) \end{aligned} \tag{14.18}$$

Model (14.12)–(14.18) can be solved using any commercially available mixed-integer programming software package. Many, if not most, of these packages utilize a branch-and-bound solution algorithm. First, the entire model is solved without integer restrictions as a linear program, but with the additional constraints that y_{ijk} and z_{ik} be less than or equal to 1. Many of the y_{ijk} and z_{ik} will turn out to be zero or 1, but some will not. Branching then occurs by setting a selected noninteger y_{ijk} or z_{ik} equal to zero and 1, respectively, in two alternative linear programming models. The usual branch-and-bound procedure then continues.

Computational experience with the mixed-integer programming regional wastewater treatment problem leads one to be very cautious. A Wisconsin case study involving 12 sources and four transshipment points resulted in a total of 136 constraints having 96 continuous and 96 integer variables. Computer runtime was excessive. By making engineering observations and simplifications, a model involving 81 constraints having 49 continuous and 47 integer variables was solved in a reasonable amount of time (Joeres et al., 1974). A larger case study on Long Island, New York, which involved 73 nodes and 167 arcs, resulted in a model with almost 200 integer and 200 continuous variables. A solution was unable to be obtained. Again, through engineering and mathematical observations, modifications were made to obtain a solution under the rather restrictive assumption of an uncapacitated network (Leighton and Shoemaker, 1984).

Despite its drawbacks, a mixed-integer programming model is probably the method of choice for most engineering applications involving a medium number of sources and flow paths arranged in a spatial network pattern. Commercial software is usually accessible, and only one, or at most two, cost segments have been found adequate to give a good approximation for the fixed-cost concave cost function. In most cases, engineering and economic observations can be used to simplify the network. Further, the method produces many closely optimal solutions through the branch-and-bound algorithm, and these can be used to provide flexibility in system design and acceptance.

14.E.3. Linear Configuration, Linear Programming

Figure 14.5 shows a chain, or linear configuration, of waste flow sources and potential regional wastewater treatment plants. At-source treatment of waste flow is possible, as is transport to a regional plant for treatment. A very efficient solution approach can be used for the linear configuration case when no current treatment is provided, no bypassed plants are allowed, a fixed level of treatment is required, and water quality constraints are not present. Further, it is assumed that the fixed-charge concave cost function for treatment can be well approximated by the fixed charge itself and one straight-line segment (no approximation is needed for transport costs in the method). Under these assumptions, a linear programming model can be derived for the regional wastewater treatment plant location problem. Rather than taking a transshipment approach as was adopted in the two previous models, the current model utilizes the literature on fixed-charge plant location models found within the field of location systems analysis (Zhu and ReVelle, 1988). In this book, fixed-charge formulations for problems are also presented in Chapter 1 (Section 1.F; water supply), Chapter 5 (Section 5.C; solid waste), and Chapter 8 (Section 8.B.1; location of freight terminal facilities).

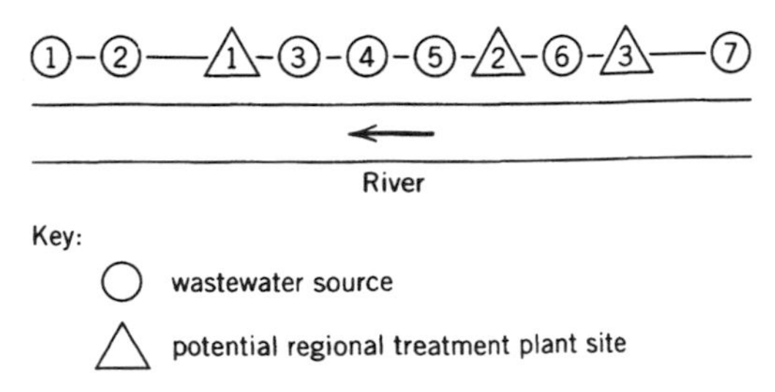

Figure 14.5 Linear configuration of waste flow sources and potential regional treatment plants.

In particular, the method is inspired by the mixed-integer programming model for the uncapacitated plant location problem, as suggested by Balinski (1965):

$$\text{minimize } Z = \sum_{i=1}^{m} \sum_{j=1}^{n} c_{ij}x_{ij} + \sum_{i=1}^{m} f_i y_i \tag{14.19}$$

subject to

$$\sum_{i=1}^{m} x_{ij} = 1 \qquad (j = 1,2,\ldots,n) \tag{14.20}$$

$$x_{ij} \leq y_i \qquad (i = 1,2,\ldots,m; \quad j = 1,2,\ldots,n) \tag{14.21}$$

with $x_{ij} \geq 0$ and $y_i = (0,1)$, where x_{ij} is the fraction of the demand at j supplied by plant i; c_{ij} the cost of supplying all of the demand at j from plant i, including transportation and production costs; y_i an integer variable equal to 1 if plant i is open and 0 otherwise; and f_i the fixed cost of opening plant i. Constraints (14.20) require that all demand points be entirely supplied. Constraints (14.21) require that a demand point be supplied only by those plants that are open. It has been shown that if this model is solved as a linear programming problem (i.e., without including integer restrictions on y_i), the resulting solution is very likely to be all-integer. This derives from the fact that there generally are many more demand points than supply points, and the supply points are uncapacitated (not limited in size). The result is that demand points invariably are supplied entirely by one plant (x_{ij} = 0 or 1). Because noninteger y_i can occur only in association with fractional x_{ij} [through constraints (14.21)] and the latter are rare, an all-integer solution is the result. This observation and model format is the basis of the regional siting model developed by Zhu and ReVelle (1988).

Define

x_{ij} = (0,1) variable equal to 1 if source i sends its waste flow to regional plant j for treatment and zero otherwise ($i = 1,2,\ldots,n$; $j = 0,1,2,\ldots,m$; where x_{i0} indicates presence or absence of at-source treatment at source i

y_j = (0,1) variable equal to 1 if regional plant j is open and zero otherwise ($j = 1,2,\ldots,m$)

f_j = fixed cost for opening regional treatment plant j ($j = 1,2,\ldots,m$)

k_j = slope of the fixed-cost straight-line segment at regional plant j

c_{ij} = total additional (incremental, or marginal) cost of treatment and transport incurred when sending waste flow from source i to regional plant j for treatment (c_{i0} is the at-source treatment cost at source i)

d_{ik} = distance from source i to the next adjacent node, either another source or a regional treatment plant

Q_i = quantity of waste flow at source i

m = number of potential regional treatment plant sites

n = number of waste flow sources

Note that the cost of at-source treatment, c_{i0}, does not have to be approximated; it is evaluated exactly as the relevant fixed cost plus variable (concave) cost to treat the known flow, Q_i. To evaluate c_{ij}, the incremental cost of treatment and transport incurred when sending waste flow from source i to regional plant j for treatment, consider a portion of the linear configuration as illustrated in Figure 14.6. Because plant bypasses are not allowed, source can join the regional treatment plant j only if waste flow from source $(i + 1)$ has already been sent to the regional plant. The total cost for transporting and treating the flows Q_{i+1} and Q_i at the regional plant is

$$\alpha Q_i^\beta d_{i,i+1} + \alpha(Q_i + Q_{i+1})^\beta d_{i+1,j} + k_j(Q_i + Q_{i+1}) + f_j$$

The cost of transporting and treating only Q_{i+1} at the regional plant is

$$\alpha Q_{i+1}^\beta d_{i+1,j} + f_j + k_j Q_{i+1}$$

Therefore, the additional cost of sending waste flow from source i to the regional plant is the difference between the two expressions, or

$$c_{ij} = \alpha Q_i^\beta d_{i,i+1} + \alpha[(Q_i + Q_{i+1})^\beta - Q_{i+1}^\beta]d_{i+1,j} + k_j Q_i \qquad (14.22)$$

Note that c_{ij} includes only transport costs and a linear term associated with the approximated treatment cost function; it does not include the fixed cost of opening the regional plant because this has already been counted in the

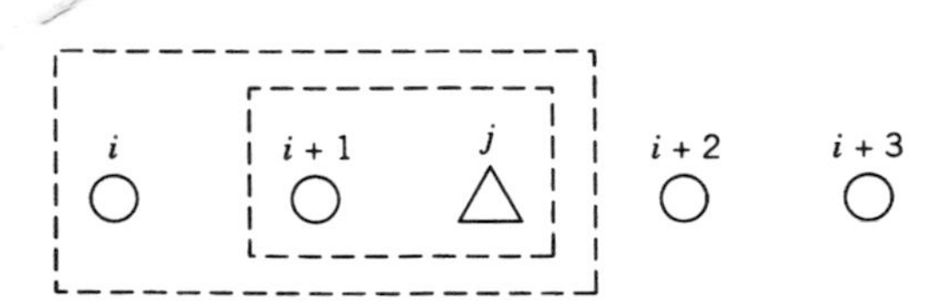

Figure 14.6 Portion of linear considered for regionalization.

"additional cost" of sending the waste flow from source $(i + 1)$ to the regional plant, which is a necessary precursor to sending source i. It is also observed that from a variable-cost standpoint, it does not matter whether or not sources $(i + 2)$ and $(i + 3)$ are contributing to regional plant, j because the variable cost of treatment at the regional plant is approximated by one straight-line segment; the additional variable cost of adding flow Q_i would remain the same, k_jQ_i (discussion of the additional complication in this case of whether or not the fixed cost of opening the regional plant has already been counted will be delayed until after the constraints have been written). Given the manner of computation of the incremental costs, the total cost of a regional coalition can be found by adding the fixed cost plus all incremental costs for contributing sources. The model is

$$\text{minimize } Z = \sum_{i=1}^{n} \sum_{j=0}^{m} c_{ij}x_{ij} + \sum_{j=1}^{m} f_j y_j \tag{14.23}$$

subject to

$$\sum_{j=0}^{m} x_{ij} = 1 \qquad i = 1,2,\ldots,n \tag{14.24}$$

$$x_{cj} \leq y_j \qquad j = 1,2,\ldots,m \tag{14.25}$$

$$x_{aj} \leq x_{bj} \qquad j = 1,2,\ldots,m \tag{14.26}$$

with

$$x_{ij} = (0,1) \qquad i = 1,2,\ldots,n, \quad j = 0,1,2,\ldots,m \tag{14.27}$$

$$y_j = (0,1) \qquad j = 1,2,\ldots,m \tag{14.28}$$

Constraints (14.24) require that a source send its waste flow to a regional plant or provide at-source treatment ($x_{i0} = 1$). Constraints (14.25) ensure that waste flows are not sent to regional plants that are not open; the indicator variable for fixed cost at a regional plant must be greater than or equal to the larger of the indicator variables for waste flow sources closest to j on each side. If either one or both of these sources contribute to the regional plant, the plant must be open. This resolves the dilemma posed earlier as to how to handle fixed cost when calculating additional cost; fixed cost is added once, and it does not matter which source is first to cause a regional plant to open. Constraints (14.26) constitute the sequential priority of contributions to a regional plant. Sources a and b are on the same side of the regional plant, and

a is next closest to j after b. Although not stated, this constraint must be written for all sequential pairs of sources on each side of j that are eligible to contribute to j. Simply stated, source a cannot contribute to regional plant j unless source b, one step closer to j, is also contributing to j. The objective function incorporates the idea of adding the additional costs associated with each source that is contributing to a given regional treatment plant, or treating at-source.

As written, the model is an integer program. However, because the model has essentially the same form as the uncapacitated plant location model of Balinski, all-integer solutions can be expected if it is solved as a linear program [by omitting conditions (14.27) and (14.28)]. In over 50 trials in different problem sizes, all-integer solutions have resulted. Further, in the worst case, the number of variables that might possibly terminate fractional is expected to be few. In this case, resolution using the branch-and-bound algorithm should be quite rapid (Zhu and ReVelle, 1988).

14.E.4. Cost Sharing

It has been noted previously that the development of equitable cost-sharing allocations among regional participants has been a stumbling block when trying to implement optimal regional configurations. Briefly, if there are n participants in a regional plan, each participant and every subgroup of participants must be better off in the regional plan than they would be when acting alone. If any are not, they will not cooperate in the regional plan, and it must be abandoned. Analyzing such a situation becomes complex due to the many combinations of subgroups that are possible. Fortunately, much work on this type of problem has been carried out in a field known as n-person cooperative game theory.

The purpose of this section is fourfold: (1) to provide a brief background to n-person cooperative game theory; (2) to present some conventional cost-sharing approaches; (3) to illustrate how the dual of the linear programming regional siting model developed in the preceding section can be used to help find acceptable cost-sharing arrangements in regional systems, and; (4) to derive a linear programming cost-sharing model that has general applicability.

n-*Person Cooperative Game Theory.* The current presentation of n-person cooperative game theory will follow the format used by Heaney and Dickinson (1982) in describing the application of game theory to the field of water resources. Games wherein players seek to minimize costs are called *cost games.* An n-person cost game consists of a set, $N = 1,2,\ldots,n$, of players and a characteristic function, c, which assigns a cost, $c(S)$, to each subset (or coalition), S, of players. For the regional wastewater treatment problem, $c(S)$ is the cost of the optimal regional plan designed to serve only the set of players, S, in the coalition. There are three general criteria that a "fair" cost game must satisfy. If a cost game is not fair, some or all players will drop out. First, the cost, $x(j)$, assigned to any player j must be no more than that player's cost if the player acted independently, that is,

$$x(j) \leq c(j) \qquad j = 1,2,...,n \tag{14.29}$$

The second criterion is that the total cost, $c(N)$, of the regional system serving all players be apportioned among the N players:

$$\sum_{j=1}^{n} x(j) = c(N) \tag{14.30}$$

Solutions satisfying (14.29)–(14.30) are called *imputations.* The third criterion expands the first by saying that the sum of costs assigned to any subset, or coalition, S, of players cannot exceed the cost that would be incurred if the players acted within the subset:

$$\sum x(j) \leq c(S) \tag{14.31}$$

where the summation occurs over all players, j, in the subset, S, and the condition must hold for all possible subsets. This ensures that subsets of players will not drop out and act on their own. Conditions (14.29)–(14.31) are called *rationality conditions,* and their solutions constitute the *core* of the game. However, in a given problem situation, there is no guarantee that the core exists; there may be no cost allocation that can satisfy all the conditions. In general, the greater the cost savings from players acting as a whole rather than individually, the greater the chance that a core exists. In this book a more extensive discussion of game theory in the context of cost allocation decisions related to water supply is presented in Chapter 13 (Section 13.D).

Conventional Cost Allocation Approaches. It is useful to test a few of the many cost allocation procedures that have been suggested. A more extensive discussion of possible procedures is available in James and Lee (1971). A numerical example incorporating economies of scale is the following (Giglio and Wrightington, 1972):

$$Q_1 = 10 \qquad Q_2 = 20 \qquad Q_3 = 50$$

$$c(1) = 20 \qquad c(2) = 32 \qquad c(3) = 48$$

$$c(1,2) = 45 \qquad c(2,3) = 60 \qquad c(1,3) = 60 \qquad c(1,2,3) = 78$$

One simple cost-sharing approach is to charge players based on the flow or pollution contributed. Assume that flow is directly proportional to pollution amount. Thus player 1 pays $x(1) = (10/80)(78) = 9.75$; similarly, $x(2) = 19.5$; $x(3) = 48.75$. The rationality conditions are violated since player 3 pays more in the grand coalition than by going it alone, and players in the coalition (2,3) pay more than they would by forming the coalition. Even though the cost allocation formula seems equitable (everyone pays according to the amount of flow or pollutant contributed), it leads to an allocation that is not "fair" in the game theory sense. The grand coalition would not hold together as players 2 and 3 try to leave.

Another approach is to base cost sharing on single plant costs with a rebate proportional to flow or pollutant contributed. Here, the total amount paid, 100, exceeds the cost of the regional system by 22. The 22 is apportioned back to the players according to their flow. Thus player 1 pays $20 - (10/80)(22) = 17.25$; similarly; $x(2) = 26.5$; $x(3) = 34.25$. All players are now better off than going it alone, but again coalition (2,3) produces a lower cost than the sum of the required payments by the two potential partners. The grand coalition again fails.

The separable cost, remaining benefit (SCRB) method is widely used to allocate costs of multiple-purpose reservoir projects. Separable cost assigned to a player is the cost that the player would add if joining an already formed coalition of all other players. The difference between the cost of the grand coalition and the sum if all assigned separable costs is then allocated to the players on the basis of their proportional share of total benefits remaining after separable costs are deducted from the benefits of each player. As applied to the regional wastewater treatment plant problem, benefits must be interpreted as waste flow or pollution treated, and separable costs cannot be deducted from these benefits because units are not commensurate. Separable costs assigned to each of the three players in the example are:

$$x(1)_{sep} = c(1,2,3) - c(2,3) = 78 - 60 = 18$$

$$x(2)_{sep} = c(1,2,3) - c(1,3) = 78 - 60 = 18$$

$$x(3)_{sep} = c(1,2,3) - c(1,2) = 78 - 45 = 33$$

The total is 69, leaving $78 - 69 = 9$ units to be allocated on the basis of remaining benefits, which in this case are assumed to be equal to the waste flows treated. Thus

$$x(1) = 18 + (10/80)(9) = 19.125$$

$$x(2) = 18 + (20/80)(9) = 20.25$$

$$x(3) = 33 + (50/80)(9) = 38.625$$

These assignments are better than each player going it alone and better than the cost of any coalition of two that can be formed. The grand coalition is stable in this example, but it cannot be expected that the method will work in all cases. Basically, the method does not consider all coalitions that may form in larger problems and will therefore fail in certain cases.

Bargaining LP Model When the Regional Authority Is a Player. A novel viewpoint of the bargaining game is to include the regional authority as a player (Giglio and Wrightington, 1972). The authority will collect from each player an amount equal to the player's cost of going it alone [i.e., $c(j)$]. Because the regional plan is so economical, the authority will have an excess of funds to build the regional system. Out of the excess funds, the authority gives back to each player a savings amount, s_j, large enough that each player finds that joining the regional system is better than joining any coalition, or subset of players. For this to be the case, the sum of savings paid to each player in any potential coalition, i, must be at least as great as the "payoff," p_i, the players would get by joining the coalition as opposed to going it alone. The values for p_i are known because the optimal design and corresponding cost for each coalition can be found and compared to the total cost of the go-it-alone case. Note that the cost to individual j of joining the regional system will never be greater than $c(j)$, so the self-treatment "coalition" can be ignored. Of course, being a player itself, the authority will want to minimize the total savings given back. Define

s_j = savings amount received by player j ($j = 1,2,\ldots,n$)

p_i = payoff obtained by coalition i as compared to a go-it-alone strategy ($i = 1,2,\ldots,m$)

a_{ij} = indicator coefficient equal to 1 if player j is a member of coalition i, and zero otherwise

$c(N)$ = cost of the grand coalition

$c(j)$ = cost of player j treating his own waste

The model is

$$\text{minimize } S = \sum_{j=1}^{n} s_j = S_{\min} \tag{14.32}$$

subject to

$$\begin{aligned} a_{11}s_1 + a_{12}s_2 + \cdots + a_{1n}s_n &\geq p_1 \\ \vdots \qquad \vdots \qquad\qquad \vdots \quad & \quad \vdots \\ a_{m1}s_1 + a_{m2}s_2 + \cdots + a_{mn}s_n &\geq p_m \end{aligned} \tag{14.33}$$

with

$$s_j \geq 0 \qquad j = 1,2,\ldots,n \tag{14.34}$$

There will always be a solution to the linear program, since there is no explicit upper bound on the total savings payment. However, if the total payment becomes too large, the authority will not have enough money left to build the regional system; that is,

$$c(N) > \sum_{j=1}^{n} c(j) - S_{\min} \qquad \text{(case a)}$$

In this case the grand coalition cannot be maintained without an external subsidy. Since the regional plan is the least expensive of all plans, there is good reason to try to subsidize its creation. If

$$c(N) < \sum_{j=1}^{n} c(j) - S_{\min} \qquad \text{(case b)}$$

the authority has funds left over after distributing savings and may choose to distribute even more savings to players or place the excess into a fund that might be used to subsidize regional systems in the future. In any case, there is a good economic argument that can be made for subsidies to bring about the creation of regional systems. It is also possible for funds remaining with the authority to just equal the cost of the regional system, $c(N)$.

In all three cases, the linear program may produce a unique savings distribution plan or it may produce alternative optimal solutions. If the latter, an infinite number of solutions can be produced through a convex combination of the alternative optimal solutions. It may be desirable to consider a large number of alternative payout plans, since each will be seen somewhat differently by the players involved (i.e., there are many sociopolitical considerations that are not included in the economics-oriented savings allocation model).

The chief drawback of the bargaining LP model is that the constraints are formed by considering all possible coalitions that might occur. Even for medium-sized problems, this could be a very large number. For n players, the number of possible coalitions is given by

$$\sum_{r=2}^{n-1} \binom{n}{r} = \sum_{r=2}^{n-1} \frac{n!}{r!\,(n-r)!} \tag{14.35}$$

where at-source treatment ($r = 1$) and the grand coalition ($r = n$) are excluded. For each coalition the optimal regional plan must be determined in order to compute the right-hand-side values, p_i. The combined computational burden makes the bargaining model useful only for cases involving a small number of players. The advantage of the model is that it clearly states the fundamental cost allocation problem.

Dual of the Balinski Plant Siting Model. Sources sending their waste flows to a common regional plant for treatment constitute a natural subset of N, and as noted, there are very many such possible subsets (coalitions). It is difficult to find a cost allocation scheme that discourages all such subsets from forming and results in stability of the grand coalition. Rather unexpectedly, however, it is possible to use the linear configuration linear programming regional siting model, presented previously, to derive such a cost allocation. The derivation is most clearly illustrated using the original Balinski plant siting model for a small example problem composed of two potential plant sites and three demand points. The siting model becomes (Zhu and ReVelle, 1989)

$$\text{minimize } Z = c_{11}x_{11} + c_{12}x_{12} + c_{13}x_{13} + c_{21}x_{21} + c_{22}x_{22} + c_{23}x_{23} + f_1y_1 + f_2y_2 \tag{14.36}$$

subject to

$$\begin{array}{llllllllll}
\text{(a)} & x_{11} & & & x_{21} & & & & = 1 & (u_1) \\
\text{(b)} & & x_{12} & & & x_{22} & & & = 1 & (u_2) \\
\text{(c)} & & & x_{13} & & & x_{23} & & = 1 & (u_3) \\
\text{(d)} & x_{11} & & & & & & -y_1 & \le 0 & (-v_{11}) \\
\text{(e)} & & x_{12} & & & & & -y_1 & \le 0 & (-v_{12}) \\
\text{(f)} & & & x_{13} & & & & -y_1 & \le 0 & (-v_{13}) \\
\text{(g)} & & & & x_{21} & & & -y_2 & \le 0 & (-v_{21}) \\
\text{(h)} & & & & & x_{22} & & -y_2 & \le 0 & (-v_{22}) \\
\text{(i)} & & & & & & x_{23} & -y_2 & \le 0 & (-v_{23})
\end{array} \tag{14.37}$$

with

$$x_{ij} \ge 0 \quad \text{all } (i,j) \qquad y_i \ge 0 \quad \text{all } i \tag{14.38}$$

The dual variables for each constraint are named to the right of the constraint, and negative dual variables have a negative sign denoting this. The dual problem is

$$\text{maximize } Z' = u_1 + u_2 + u_3 \tag{14.39}$$

subject to

$$\begin{array}{llllllllll}
\text{(a)} & u_1 & & & -v_{11} & & & & & \leq c_{11} \\
\text{(b)} & & u_2 & & & -v_{12} & & & & \leq c_{12} \\
\text{(c)} & & & u_3 & & & -v_{13} & & & \leq c_{13} \\
\text{(d)} & u_1 & & & & & & -v_{21} & & \leq c_{21} \\
\text{(e)} & & u_2 & & & & & & -v_{22} & \leq c_{22} \\
\text{(f)} & & & u_3 & & & & & & -v_{23} \leq c_{23} \\
\text{(g)} & & & & v_{11} & +v_{12} & +v_{13} & & & \leq f_1 \\
\text{(h)} & & & & & & & v_{21} & +v_{22} & +v_{23} \leq f_2
\end{array} \tag{14.40}$$

with

$$u_j \text{ unrestricted for all } j \qquad v_{ij} \geq 0 \qquad \text{all } (i,j) \tag{14.41}$$

The general form of the dual of the Balinski model is

$$\text{maximize } Z' = \sum_{j=1}^{m} u_j \tag{14.42}$$

subject to

$$u_j \leq c_{ij} + v_{ij} \qquad \text{all } (i,j) \tag{14.43}$$

$$\sum_{j=1}^{m} v_{ij} \leq f_i \qquad (i = 1,2,\ldots,n) \tag{14.44}$$

with

$$u_j \text{ unrestricted } (j = 1,2,\ldots,m) \qquad v_{ij} \geq 0 \qquad \text{all } (i,j) \tag{14.45}$$

From constraint (14.43), the term u_j can be interpreted as the payment assigned to demand point j and is composed of two parts: c_{ij}, the cost of transportation from i to j plus the production cost at i, and v_{ij}, a portion of the cost of opening plant i. Constraint set (14.43) sets an upper bound on the payment by each demand point, and constraint (14.44) says that the total of the assigned fixed costs at plant i cannot exceed that incurred.

The constraints of the dual also establish the core conditions of the cost game. For example, by combining constraints (14.40) a, b, c with g; and d, e, f with h, respectively, we obtain

$$u_1 + u_2 + u_3 \leq c_{11} + c_{12} + c_{13} + f_1$$

$$u_1 + u_2 + u_3 \leq c_{21} + c_{22} + c_{23} + f_2$$

which restricts the assigned charges to be less than the costs incurred if all players were to be assigned to plants 1 and 2, respectively.

Similarly, assigned charges can be restricted to be less than the costs incurred in any coalition by combining the appropriate constraints. For example, if users 1 and 2 wished to be supplied from plant 1 and user 3 wanted to be supplied by plant 2, the related rationality constraints would be

$$u_1 + u_2 \leq c_{11} + c_{12} + f_1$$

$$u_3 \leq c_{23} + f_2$$

The first condition can be produced by combining constraints (14.40) a, b with g (and recognizing that $v_{13} = 0$). The second condition can be obtained from constraint (14.40) f (and recognizing that $v_{23} = f_2$).

The dual can be interpreted as maximizing the total amount of charges collected, subject to not charging more to any subset of demand points than they would have to pay in that subset. The dual is a cost allocation model.

14.F. SITING MODELS FOR MEETING WATER QUALITY GOALS

When water quality standards are not met with BPT and BAT levels of treatment at all municipal and industrial plants, respectively, a water body is termed *water quality limited.* Higher levels of treatment must be given to meet the water quality standard, and removal efficiencies at each source and potential regional plant become additional variables. This makes the optimization model considerably more complicated to write and harder to solve. Not only must constraints be written to describe water quality but also to ensure continuity of pollutant mass at every node.

In addition, many model assumptions made in the case of siting regional facilities having a fixed level of treatment now become invalid. For example, both split flows at sources and plant bypasses become potentially worthwhile options; by distributing effluents more evenly along a stream or estuary, greater use is made of assimilative capacity, and less treatment is required.

Relatively few regional siting models have been developed to meet water quality goals. Two of these, both developed for the case of a linear configuration along an estuary, are presented in this section. Water quality management models involving optimal treatment efficiencies are also discussed in this book in Chapters 2 (Section 2.D) and 15.

14.F.1. Linear Configuration, Nonlinear Programming

A nonlinear programming formulation of the regional wastewater treatment problem has been developed for the Delaware estuary near Philadelphia by Graves et al. (1970, 1972). In addition to allowing for regional treatment plants, plant bypasses, and split flows, the model allows *bypass piping,* whereby direct shipment of waste effluents is permitted from either regional, industrial, or at-source municipal plants to any section of the estuary for discharge. Bypass piping allows maximum use of the assimilative capacity of the water body consistent with overall cost minimization.

Define

q_{ij} = flow from source j to regional treatment plant i (mgd)

r_{ij} = flow from regional plant j to estuary section i (mgd)

s_{ij} = flow from source j to estuary section i (mgd)

h_i = BOD concentration of effluent from source i (lb/mg)

e_i = BOD removal efficiency at regional treatment plant i

g_i = water quality improvement goal for estuary section i (mg/L)

a_{ij} = transfer coefficient showing the change in DO in estuary section i due to a change in BOD loading in section j (mg/L per lb/day)

S = set of sources

R = set of regional plants

E = set of estuary sections

With the convention that the present level of any variable be denoted by a bar, such as $\bar{e}_i$, the objective function is

$$\begin{aligned}\text{minimize } Z = &\sum_{j\in S} \mathrm{TC}_j\left(\sum_{i\in R} q_{ij} + \sum_{i\in E} s_{ij}, h_j\right)\\ &+ \sum_{j\in R} \mathrm{TC}_j\left(\sum_{i\in E} r_{ij}, e_j\right)\\ &+ \sum_{i\in R}\sum_{j\in S} \mathrm{PC}_{ij}(q_{ij}) + \sum_{i\in E}\sum_{j\in R} \mathrm{PC}_{ij}(r_{ij})\\ &+ \sum_{i\in E}\sum_{j\in S} \mathrm{PC}_{ij}(s_{ij})\end{aligned} \tag{14.46}$$

The constraint set for flow balance at a source is

$$\sum_{i \in R} q_{ij} + \sum_{i \in E} s_{ij} = \sum_{i \in R} \bar{q}_{ij} + \sum_{i \in E} \bar{s}_{ij} \qquad j \in S \tag{14.47}$$

and similarly, for a regional plant,

$$\sum_{j \in S} q_{ij} - \sum_{j \in E} r_{ji} = 0 \qquad i \in R \tag{14.48}$$

The constraint set for water quality is more complicated. It is

$$\begin{aligned} a_{ii} \Bigg\{ & \sum_{j \in S} (\bar{s}_{ij}\bar{h}_j - s_{ij}h_j) + \sum_{j \in R} \Bigg[(1 - \bar{e}_j) \Bigg(\sum_{k \in S} \bar{q}_{jk}\bar{h}_k \Bigg) \frac{\bar{r}_{ij}}{\Sigma_{i \in E} \bar{r}_{ij}} \\ & - (1 - e_j) \Bigg(\sum_{k \in S} q_{jk}h_k \Bigg) \frac{r_{ij}}{\Sigma_{i \in E} r_{ij}} \Bigg] \Bigg\} + \sum_{\substack{l \in E \\ l \neq i}} a_{il} \Bigg[\sum_{j \in S} (\bar{s}_{lj}\bar{h}_j - s_{lj}h_j) \\ & + \sum_{j \in R} \Bigg[(1 - \bar{e}_j) \Bigg(\sum_{k \in S} \bar{q}_{jk}\bar{h}_k \Bigg) \frac{\bar{r}_{lj}}{\Sigma_{l \in E} \bar{r}_{lj}} \\ & - (1 - e_j) \Bigg(\sum_{k \in S} q_{jk}h_k \Bigg) \frac{r_{lj}}{\Sigma_{l \in E} r_{lj}} \Bigg] \Bigg\} \geq g_i \qquad i \in E \end{aligned} \tag{14.49}$$

where the first three terms in equation (14.49) account for water quality improvement caused by changes in waste discharge directly into estuary section i, and the last three terms account for changes in water quality in section i caused by changes in waste discharge in other estuary sections.

Note that the cost of treatment at a source is defined in terms of source flow and final pollutant concentration, h_i, at the source, while cost of treatment at a regional plant is defined in terms of removal efficiency, e_i, at the regional plant. A formal definition of removal efficiency is not needed in the constraint set. The model assumes that pollution exists and that increased treatment is needed to meet the water quality standard.

The formulation is highly nonlinear, both in the objective function and constraint set. Cross-product terms as well as ratios of variables are present in each. To solve the formulation, a modified gradient method is used. This technique requires periodic evaluation of the derivatives of the objective function and constraints and the solution of an associated local linear programming problem. Computational difficulties are encountered in both of these steps, and special mathematical refinements are necessary to achieve results. Specifically, infinite derivatives of the cost function occur when flows are initially zero or very low, and column generation is necessary to shorten the solution times for the local linear programs. The large range of parameters in the constraint matrix causes the scaling of numbers to be difficult in attempting to reduce round-off error propagation.

Since the gradient method guarantees only a locally optimal solution, a procedure has to be developed to avoid obviously poor solutions. The method orders variables into priority classes. Initially, only variables in the highest priority class are allowed to be in the solution and the problem is solved. Using the results given in previous runs, lower-priority class variables are successively allowed to enter the solution. The best ordering of variable entry was (1) regional treatment with discharge to the adjacent estuary section, (2) at-source treatment with discharge to the adjacent estuary section, (3) flow from regional plants to any section, and (4) flow from sources to any section. This procedure substantially increased the computation time necessary to achieve a solution (Graves et al., 1970, 1972).

The manner of accounting for transport costs in the model would seem to indicate that the solution obtained will only be a locally optimal one. In defining the transport variable, q_{ij}, as flow from any source j to any regional plant i, no allowance is made for the combining of flows from many sources in a common pipeline for transport to the regional plant. Since economies of scale in transport are even larger than those in treatment, neglect of combined flows in the model would bias the results toward a decentralized system. Justification for direct piping is perhaps stronger in the case of bypass piping to estuary sections, but even here, gains may be had in a consolidated system. As the model stands, manual calculation of the effect of consolidation of pipelines must be carried out after results are obtained from the algorithm.

The method was applied to a case example in the Delaware estuary having 22 industrial and 22 domestic waste sources, with 9 potential regional treatment sites. The estuary was divided into 30 sections. This resulted in a problem having approximately 2000 variables and 80 constraints. A solution time of about 10 minutes using an IBM 360/91 is reported for the case of best variable entry order and initial starting solution. As shown in Table 14.4, permitting bypass piping lowers system costs by about $6,000,000, 15.8% below the pure regionalization option.

Noting the unrealistic treatment of flows, the nonlinear programming model has been extended through the inclusion of additional nodes representing pipe junctions (Graves, 1972). However, this tends to increase the computational burden, and experience with the revised model has not been reported.

14.F.2. Linear Configuration, Heuristic Algorithm

All the models so far for the siting of regional wastewater treatment plants rely heavily on formal mathematical programming formulations and numerical solution algorithms. A somewhat different approach is to apply heuristic siting procedures from the field of location theory while involving the water quality engineer more directly in the solution procedure (Whitlatch and ReVelle, 1976). An incremental approach is taken wherein small changes are made in the regional configuration to test whether an improvement in the objective function occurs. Four basic types of incremental changes involved in the algorithm are illustrated, followed by a general summary of the procedure.

The *Maranzana best plant substitution* (M-B) *technique,* patterned after the work of Maranzana (1964), is designed to determine the best location for a regional plant within its current partition, or coalition. Figure 14.7 illustrates the method. It is first necessary to calculate the total cost for the initial configuration, where total cost is composed of transport and treatment cost. Because the piping configuration and location of regional plants are known, piping cost can be found by simple evaluation of the relevant cost functions, while treatment cost can be found using a standard at-source linear programming water quality optimization model. Assuming that the plants are located along an estuary, the linear programming model can be developed based on the following definitions:

x_{jk} = amount of BOD removed at plant k in estuary section j (lb/day)

a_{ij} = transfer coefficient showing the change in dissolved oxygen in estuary section i due to a change in BOD loading in estuary section j (mg/L per lb/day)

$c_{jk}(x_{jk})$ = cost of removal function for plant k in estuary section j ($)

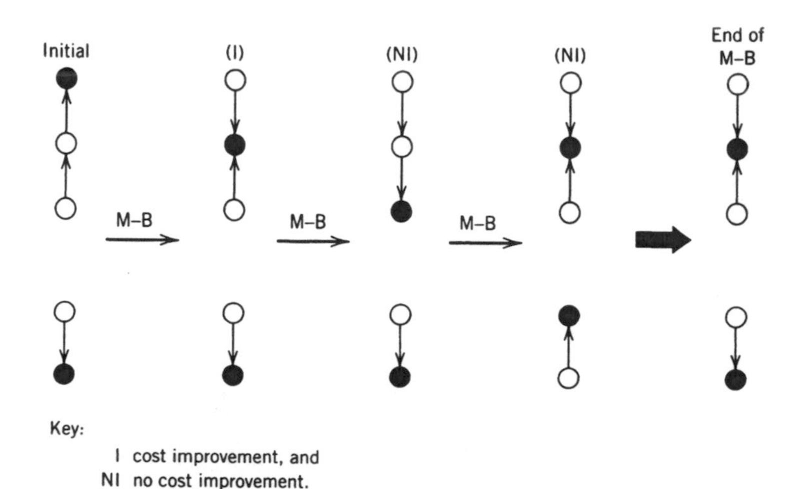

Figure 14.7 Example of the M-B plant substitution technique. All sites are sources and also potential regional plants.

n = number of estuary sections
n_j = number of plants located in estuary section j
b_{jk} = upper limit of BOD that can be removed at plant k in estuary section j (lb/day)
g_i = required increase in DO to meet the water quality standard in estuary section i (mg/L)

The linear programming model solved at every iteration of the procedure is

$$\text{minimize } Z = \sum_{j=1}^{n} \sum_{k=1}^{n_j} c_{jk}(x_{jk}) \tag{14.50}$$

subject to

$$\sum_{j=1}^{n} \sum_{k=1}^{n_j} a_{ij} x_{jk} \geq g_i \qquad i = 1,2,\ldots,n \tag{14.51}$$

$$x_{jk} \leq b_{jk} \qquad j = 1,2,\ldots,n, \quad k = 1,2,\ldots,n_j \tag{14.52}$$

with

$$x_{jk} \geq 0 \qquad j = 1,2,\ldots,n, \quad k = 1,2,\ldots,n_j \tag{14.53}$$

Each term of the cost function is strictly convex, allowing it to be approximated by piecewise segments and the entire problem to be solved using any standard linear programming code. The LP model is based on the work of Sobel (1965). If the plants were located along a river, the LP model of Revelle et al. (1967, 1968) could be used.

Knowing the total cost of the initial configuration, a new location is tested for the regional plant in the top coalition, as shown in Figure 14.7. Again, total cost is found using a simple costing program for transport cost and a linear programming model for treatment cost. Since the location and discharge point for the regional plant has changed, both transport and treatment costs will change. An improvement in total cost is indicated in the illustration. The new configuration is therefore the best so far, and a further incremental change (M-B substitution) is made in the location of the regional plant in the top coalition. No improvement occurs, so the best location for the regional plant in the top coalition has been found. Next, the lower coalition is tested to see whether the regional plant should be moved. Again, no improvement results, and all possibilities have been tested for both coalitions. The M-B technique terminates with the best configuration found.

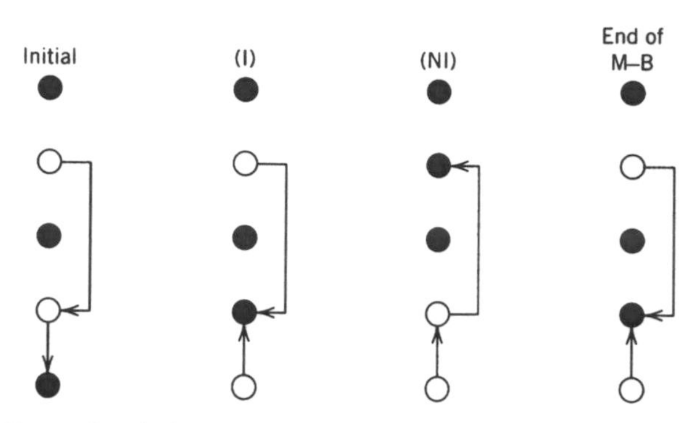

Figure 14.8 Example of plant bypasses in the M-B technique. All sites are sources and also potential regional plants.

A case involving plant bypass produces no further complications for the M-B technique. Thus, if the situation were as shown in Figure 14.8, the concept of a partition still holds and the M-B step would seek to find the best location of the regional plant for the partition consisting of the second, fourth, and fifth sites. The sources that self-treat are each a separate partition in which no M-B steps are possible. In two steps all possible plant substitutions are made and the best location for the regional plant is found. Plant bypasses commonly occur if some sources are already giving a high level of treatment.

Another incremental step is the *Maranzana arc substitution* (M-A) *technique,* illustrated in Figure 14.9. The initial configuration shown is that resulting from the M-B step of Figure 14.7. Sources at the extreme end of partitions are tested to see whether they would be served more economically by regional plants in immediately adjacent partitions. The first source tested results in no improvement in total cost. However, the second source tested

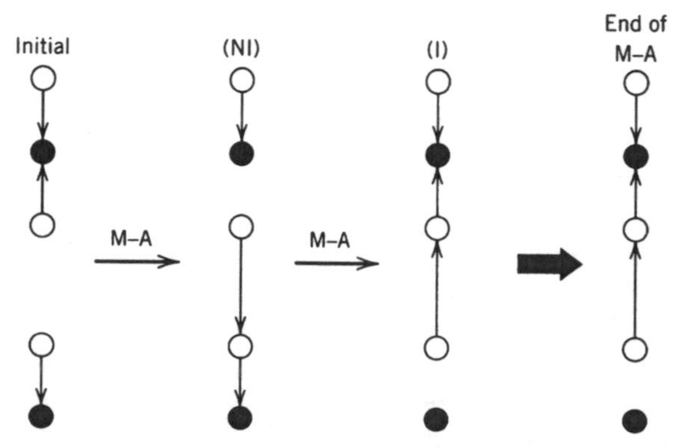

Figure 14.9 Example of the Maranzana arc substitution technique. All sites are sources and also potential regional plants.

does produce an improvement and the resulting configuration is the best given by the M-A technique. It should be noted that if a source at the end of a partition does reassign, exposing another source within that partition, that source also becomes eligible for reassignment. The process can continue if improvements are found at each step until the regional plant itself is reached for that partition. Since any reassignment modifies both partitions, as well as water quality conditions in the problem, the M-B technique must be returned to at some stage of the ensuing algorithm. It is more efficient, however, to have intervening procedures before returning to the M-B technique.

Note that neither the M-B nor M-A techniques change the *number* of active treatment plants in the regional configuration. They merely change the location of such plants or the configuration of contributing sources. In contrast, the SEPARATE and COMBINE techniques are designed specifically to change the number of active plants. The origin of the two procedures can be found in the location theory literature, specifically in the works of Kuehn and Hamburger (1963) and Feldman et al. (1966). The SEPARATE step is illustrated in Figure 14.10. Any waste flow source currently at the end of a partition is eligible to be "added" in the SEPARATE step. Thus, in the upper partition, the separation of the most remote waste flow source and its consequent self-treatment results in an improvement in total cost. This exposes another source of waste flow for the SEPARATE procedure. When this is carried out, no further improvement is gained. There are no other candidate sources in the system, and the procedure terminates with the best configuration found to date.

The COMBINE technique allows consolidations of regional plants between adjacent partitions. An example is provided in Figure 14.11. Initially, an attempt is made to combine the upper partition's regional plant into that serving waste flow sources of the middle partition. An improvement results, so this configuration is used to make further trial combinations. No further improvements occur and the configuration is adopted as the best so far.

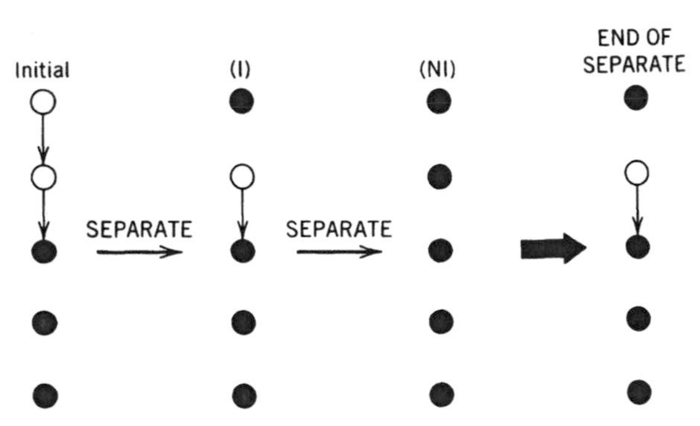

Figure 14.10 Example of the SEPARATE procedure. All sites are sources and also potential regional plants.

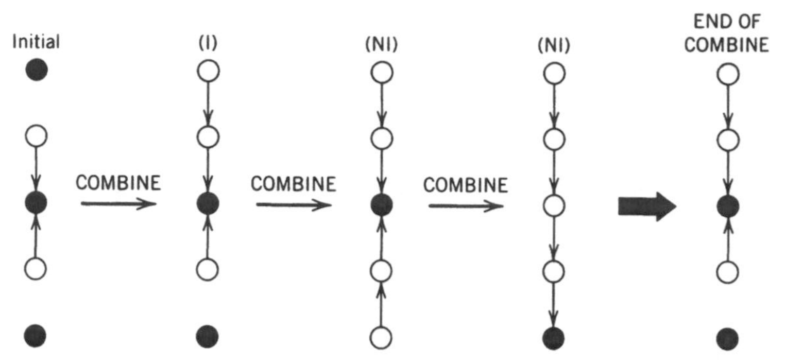

Figure 14.11 Example of the COMBINE technique. All sites are sources and also potential regional plants.

The four incremental procedures can be arranged sequentially to solve the regionalization problem for a fixed treatment level (phase I problem) or to meet a water quality standard at minimum total cost (phase II problem). Both problems were solved for the Delaware estuary near Philadelphia (Whitlatch and ReVelle, 1976). A flowchart of the solution procedure for the phase II problem is shown in Figure 14.12. The first step is to select the desired form of the linear programming model used to calculate treatment cost at each stage of the algorithm. This may be purely a least cost LP model, as given in equations (14.50)–(14.53), or a special variation such as uniform treatment, zoned uniform treatment, or other format. The second step is to use the LP formulation to calculate a baseline cost assuming all at-source treatment. This can be compared to the total cost of the initial regional plant configuration in the next step to indicate how worthwhile the initial regional configuration is.

The initial configuration can be selected by the engineer but is assumed in the flowchart to be the result of similar heuristic steps applied to the case of required secondary treatment (phase I, where $e = 0.85$). If the linear program selected previously does not call for uniform treatment, a uniform treatment LP model (LPUNI) is now run to determine the level of uniform treatment, e^U, required in the regional configuration to meet water quality standards. If e^U is less than 0.85, the right branch of the flowchart is taken where the first algorithmic steps are M-B followed by SEPARATE. If e^U is greater than 0.85, the left branch is taken and the first steps are M-B followed by COMBINE. The rationale for this, verified in application of the procedure to the Delaware estuary, is that if e^U is less than 0.85, it indicates that treatment cost is not as significant in the phase II problem as it was in the phase I problem. Therefore, it is anticipated that the degree of regionalization in phase II will not be as great as in phase I because the need for treatment cost savings is not as great. Conversely, if e^U is more than 0.85, treatment costs assume greater importance in the phase II problem, and greater regionalization should occur

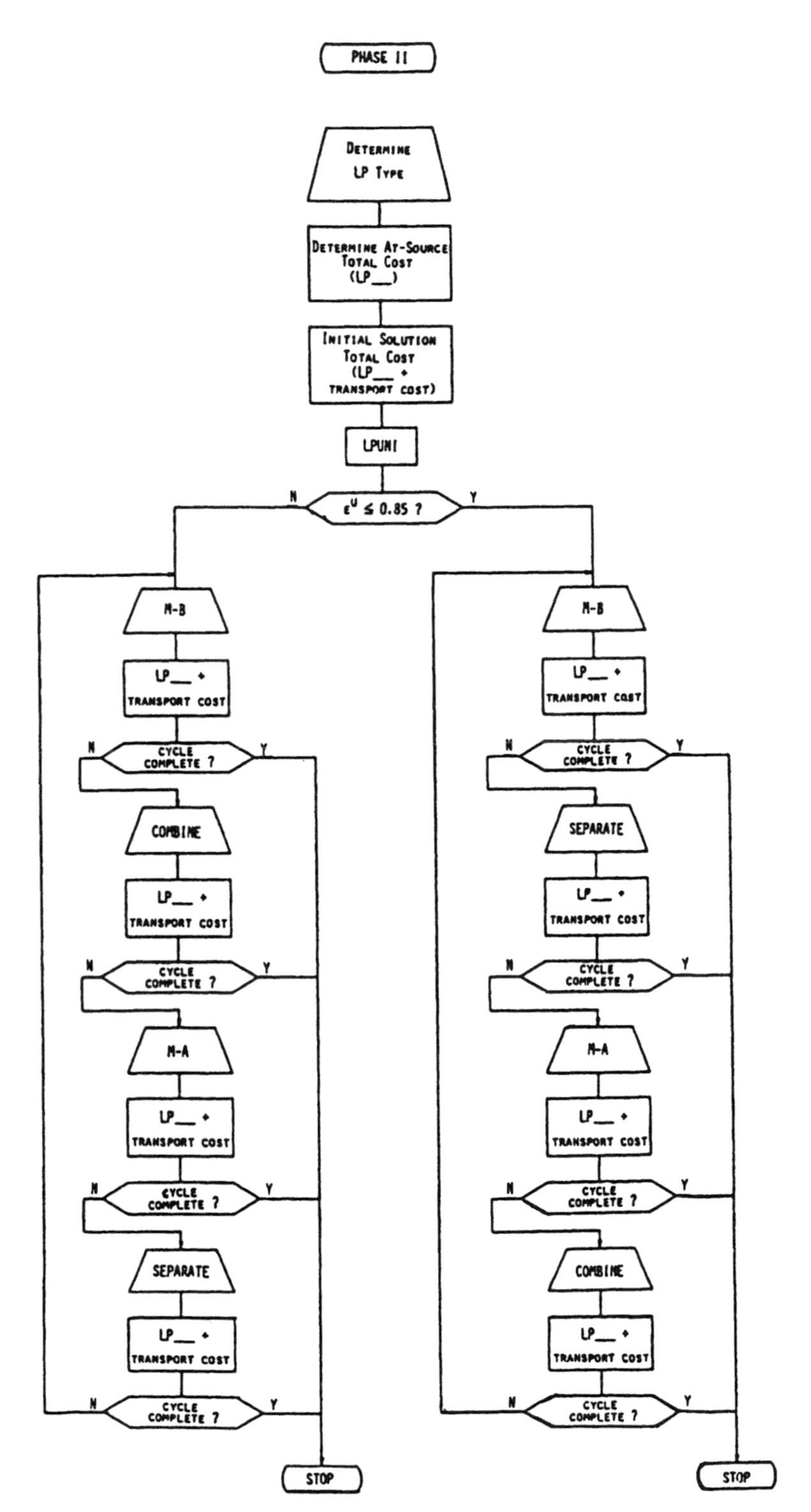

Figure 14.12 Flowchart of heuristic algorithm for regionalization to meet a water quality goal. Rectangles indicate computer processes, trapezoids indicate user-controlled processes.

to reduce treatment costs through economies of scale. The algorithm terminates only when one complete cycle of the heuristic procedure results in no improvement in total cost.

The heuristic algorithm was applied to the Delaware estuary with good results. The regional configuration resulting from solving the phase I problem was used to initiate the phase II algorithm. A total of only 38 incremental steps in phase II were needed to arrive at the optimal solution, which is reported in Table 14.4. The solution was the same as that found by Graves et al. (1970, 1972), for the same problem but without the option of bypass piping to remote estuary sections.

Advantages of the heuristic approach are that (1) the water quality engineer can control the algorithmic steps, allowing flexibility and engineering judgment to be applied; (2) the procedure builds insight for the user through accurate feedback on system behavior as incremental changes are made; (3) a great many closely optimal solutions are generated; and (4) the method is not complex mathematically and is guaranteed to produce at least locally optimal solutions. Disadvantages include (1) the possibility that a great many incremental steps might be necessary, and (2) the danger that only a locally optimal solution might be found.

14.G. SUMMARY AND CONCLUSIONS

It is obvious that a rich set of optimization models have been developed for siting regional wastewater treatment facilities. Under a given set of circumstances, any one of the models presented may be most appropriate; certainly all are instructional. Selection of a model will depend on the problem setting, background of the engineer, and available hardware and software computer facilities. Data on possible cost savings are impressive and should provide a constant incentive to investigate regional alternatives. The nonmonetary advantages of regionalization are also numerous, and several widespread economic and technological trends should serve to promote the regionalization alternative. Primary among these are increased funding limitations and the more widespread availability of computer hardware and software. The greatest need is to make existing models more accessible and regional approaches more familiar to practitioners.

EXERCISES

14.1. Estimate the value of the exponent, β, in $C = \alpha Q^{\beta}$, the cost equation for gravity interceptor sewers, using basic hydraulic principles as expressed by Manning's equation:

$$v = \frac{1.486}{n} r^{2/3} s^{1/2}$$

where v is the velocity (ft/sec); n an empirical coefficient, constant for a given pipe material; r the hydraulic radius (ft), = (area)/(wetted perimeter) = A/p; and s = slope (unitless). Assume that the sewer flows full and that cost varies as amount of materials used. Find b assuming that wall thickness:

(a) Is a constant with sewer diameter, D.

(b) Varies linearly with diameter.

(c) Varies as $D^{0.5}$.

(d) Varies as found from actual manufacturer specifications in your geographic area (cite the source).

14.2. Determine the break-even distance, d, for transport of sewage from source 2 to source 1 by gravity interceptor sewer for regional treatment. Use the segmented cost equations for greater accuracy, then the approximate cost equation. What is the percentage error for break-even distance in this case?

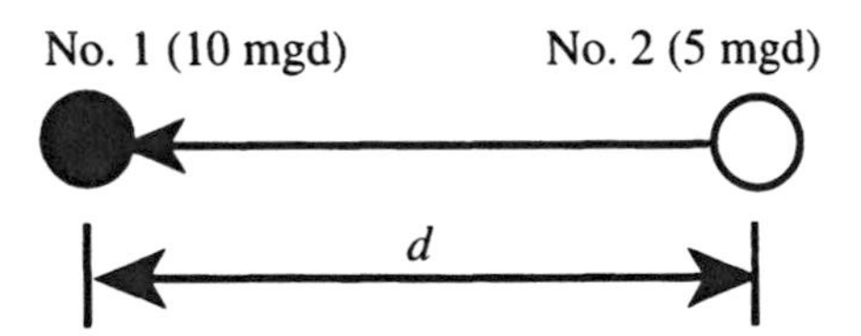

14.3. Expand the nonlinear programming model, equations (14.7)–(14.11) by allowing for capacity expansion. Define:

x_{ijk} = wastewater flow sent from node i to node j at the end of time increment k (mgd)

k = index of five-year time increments (k = 1,2,...,6; 30-year planning period)

a_{ik} = amount of wastewater generated by source node i at the end of time increment k (mgd)

y_{ijk} = increase in capacity between node i and node j at the beginning of time increment k (mgd)

$f_{ijk}(y_{ijk})$ = total present worth cost of providing an increase in capacity between node i and node j at the beginning of time increment k ($)

Sketch the capacity expansion diagram (capacity versus time) using appropriate labels. Write the new capacity expansion model. How much larger is the new model than the original?

14.4. Expand the mixed-integer programming model, equations (14.12)–(14.18), by allowing for capacity expansion at fixed time intervals. Define your decision variables and explain briefly the purpose of each constraint in your model.

14.5. Consider a linear configuration of three sources of waste flow. Each source can serve as a regional plant location (do not use triangles) or treat its wastes at-source. Assuming no bypass piping nor split flows, sketch out all possible treatment configurations (give dark shading to source nodes to indicate active treatment plant sites). Verify that the number of configurations you have found satisfies *Moivre's formula* (Deininger and Su, 1971):

$$A(n) = \frac{1}{2^n\sqrt{5}}\left[(3 + \sqrt{5})^n - (3 - \sqrt{5})^n\right]$$

Evaluate the formula for $n = 1, 2, 3, 4, 5, 6, 10$.

14.6. It is desired to place some equity conditions on the bargaining LP model, equations (14.32)–(14.34). Write out the general form of the constraint(s) or other changes that must be made to the model so as to meet each of the following conditions; explain why each condition might be equitable or inequitable, depending on the viewpoint of the participant; and state what effect the modification will have on the total savings given back to the players.

(a) The savings payment to each player must be the same.

(b) The unit savings (dollars saved per unit of flow) for each player must be the same.

(c) The final unit cost of treatment (dollars per unit of flow) must be the same for each player.

(d) Given a fixed total savings payment, minimize the largest "overpayment" for any coalition (i.e., total savings received in excess of the payoff gained by forming the coalition).

REFERENCES

Balinski, M. L., 1965. Integer Programming: Methods, Uses, Computation, *Management Science*, 12(3), November, 253–313.

Converse, A. O., 1972. Optimum Number and Location of Treatment Plants, *Journal of the Water Pollution Control Federation,* 44(8), 1629–1636.

Deininger, R. A., and S. Y. Su, 1971. Regional Waste Water Treatment Systems, *Proceedings, Annual and National Environmental Engineering Meeting, American Society of Civil Engineers,* St. Louis, MO, October 18–22, Meeting Preprint 1536.

Deininger, R. A., and S. Y. Su, 1973. Modelling Regional Waste Water Treatment Systems, *Water Research,* 7(4), 633–646.

Eilers, R. G., and R. Smith, 1971. *Wastewater Treatment Plant Cost Estimating Program,* Advanced Waste Treatment Research Laboratory, U.S. Environmental Protection Agency, Cincinnati, OH, April.

Feldman, E., F. A. Lehrer, and T. L. Ray, 1966. Warehouse Location Under Continuous Economies of Scale, *Management Science,* 12(9), 670–684.

Giglio, R. J., and R. Wrightington, 1972. Methods for Apportioning Costs Among Participants in Regional Systems, *Water Resources Research,* 8(5), 1133–1144.

Graves, G. W., 1972. *Extensions of Mathematical Programming for Regional Water Quality Management,* U.S. EPA Water Pollution Control Research Series, 16110-EGQ-04/72, U.S. Government Printing Office, Washington, DC, April.

Graves, G. W., A. B. Whinston, and G. B. Hatfield, 1970. *Mathematical Programming for Regional Water Quality Management,* Federal Water Quality Administration Water Pollution Control Research Series 16110 FPX, U.S. Government Printing Office, Washington, DC, August.

Graves, G. W., G. B. Hatfield, and A. B. Whinston, 1972. Mathematical Programming for Regional Water Quality Management, *Water Resources Research,* 8(2), 273–290.

Heaney, J. P., and R. E. Dickinson, 1982. Methods for Apportioning the Cost of a Water Resource Project, *Water Resources Research,* 18(3), 476–482.

James, L. D., and R. R. Lee, 1971. *Economics of Water Resources Planning,* McGraw-Hill, New York.

Joeres, E. F., J. Dressler, C.-C. Cho, and C. H. Falkner, 1974. Planning Methodology for the Design of Regional Waste Water Treatment Systems, *Water Resources Research,* 10(4), 643–649.

Kuehn, A. A., and M. J. Hamburger, 1963. A Heuristic Program for Locating Warehouses, *Management Science,* 9(4), 643–666.

Leighton, J. P., and C. A. Shoemaker, 1984. An Integer Programming Analysis of the Regionalization of Large Wastewater Treatment and Collection Systems, *Water Resources Research,* 20(6), 671–681.

Maranzana, F. E., 1964. On the Location of Supply Points to Minimize Transport Costs, *Operational Research Quarterly,* 15(3), 261–270.

Maryland Environmental Service, 1971. Preliminary Proposal and Feasibility Study for a Central Patuxent Regional Wastewater Treatment Plant, unpublished report, Maryland Environmental Service, Annapolis, MD, June 4.

McConagha, D. L., and A. O. Converse, 1973. Design and Cost Allocation Algorithm for Waste Treatment Systems, *Journal of the Water Pollution Control Federation,* 45(12), 2558–2566.

McMichael, W. F., 1971. *Package Treatment Plant Costs,* U.S. Government Memorandum, Advanced Waste Treatment Laboratory, U.S. Environmental Protection Agency, Washington, DC, March 24.

ReVelle, C. S., D. P. Loucks, and W. R. Lynn, 1967. A Management Model for Water Quality Control, *Journal of the Water Pollution Control Federation,* 39(7), 1164–1183.

ReVelle, C. S., D. P. Loucks, and W. R. Lynn, 1968. Linear Programming Applied to Water Quality Management, *Water Resources Research,* 4(1), 1–9.

Smith, R. 1971. Cost-Effectiveness Task Force: Economics of Consolidating Sewage Treatment Plants by Means of Interceptor Sewers and Force Mains, (U.S. Government Memorandum, Advanced Waste Treatment Laboratory, U.S. Environmental Protection Agency, March 10.

Sobel, M. J., 1965. Water Quality Improvement Programming Problems, *Water Resources Research,* 1(4), 477–487.

Wanielista, M. P., and C. S. Bauer, 1972. Centralization of Waste Treatment Facilities, *Journal of the Water Pollution Control Federation,* 44(12), 2229–2238.

Whitlatch, E. E., Jr., 1973. Optimal Siting of Regional Wastewater Treatment Plants, Ph.D. dissertation, Johns Hopkins University, Baltimore, MD.

Whitlatch, E., 1978. *Expert Testimony in Ohio* v. *U.S. EPA,* U.S. District Court, Columbus, Ohio, Case C-2-76-704 and Case C-2-76-780, March 29.

Whitlatch, E. E., Jr., and C. S. ReVelle, 1976. Designing Regionalized Waste Water Treatment Systems, *Water Resources Research,* 12(4), 581–591.

Whitlatch, E. E., and C. S. ReVelle, 1990. Regionalization in Water Resource Projects, *Water International,* 15(2), 70–79.

Zhu, Z. P., and C. ReVelle, 1988. A Siting Model for Regional Wastewater Treatment Systems: The Chain Configuration Case, *Water Resources Research,* 24(1), 137–144.

Zhu, Z. P., and C. ReVelle, 1989. Adaptation of the Plant Location Model for Regional Environmental Facilities and Cost Allocation Strategy, *Annals of Operations Research,* 18, February, 345–366.

CHAPTER 15

Effluent Charges and Transferable Discharge Permits

E. DOWNEY BRILL

15.A. INTRODUCTION

Mathematical optimization models can be used to simulate the outcome of implementing an environmental management strategy based on economic incentives; these incentives could be in the form of effluent charges or waste discharge permits that can be bought or sold. Incentives-based management approaches are intended to lead to reductions in waste discharges for a given environmental system such that the ambient standard is met in a cost-effective way. Simulation studies of these strategies can be used in carrying out policy analysis at the national, state, or regional level or as part of a planning process for a given environmental system.

Quantitative modeling methods also offer the potential to assist in future studies of innovative management strategies. Management approaches continue to become more complex as our understanding of environmental problems increases. For instance, increased attention is being directed to the multimedia environmental impacts of municipal and industrial activity. Full life-cycle effects of products are also being examined more closely. Innovative combinations of charges, transferable permits, and direct regulations will probably be needed as we design management strategies to respond to a deeper understanding of such complexities. In the following two sections we review effluent charges and transferable permits. Mathematical formulations and illustrations for example systems are provided. In the last section we consider future challenges and the industrial ecology paradigm.

15.B. ANALYZING EFFLUENT CHARGE PROGRAMS

15.B.1. Effluent Charges: Economically Efficient Management Strategy

An effluent charge is a charge, or tax, assessed by a regulatory agency on waste discharged by an industry or municipality. A charge is intended to provide an economic incentive for the discharger to reduce the waste load to an acceptable level. The discharger would be expected to increase the level of waste reduction as long as the incremental cost of doing so is less than the incremental cost of paying the charge.

Effluent charges were investigated by economists and policy analysts extensively in the 1960s and 1970s because in theory the approach would tend to produce cost-effective waste management solutions (Johnson, 1967; Kneese and Bower, 1968). In particular, the application of a unit charge uniformly to a group of dischargers would produce a least-cost solution for meeting a target for total waste reduction by the group.

Two other chapters of this book deal with topics relating to the interactions among groups involved in wastewater discharges. Methodology for the allocation of costs among a group is discussed in Chapter 13, and the siting of regional facilities to handle the wastewaters generated by several different sources is covered in Chapter 14.

Response of a Single Discharger to an Effluent Charge. How a discharger responds to an effluent charge can be seen by considering an example with the units of the discharge, reduction of the discharge, cost, and effluent charge given in detail. Consider a discharger with an initial waste discharge level of t kg/day of pollutant P. (Assume steady-state conditions for the purposes of the examples of this chapter unless specified otherwise.) The total cost of reducing the discharge level is given by TC, which is a function of the waste reduction level, r, in kg/day. TC represents all capital, operating, and other costs as well as any anticipated salvage values evaluated using the principles of engineering economics. Although TC would usually be expressed in present-value dollars for a fixed time period or as an annual cost, here a daily cost is used for simplicity in considering examples (i.e., the units of TC are $/day). The graph of TC($r$) given in Figure 15.1 is typical of total

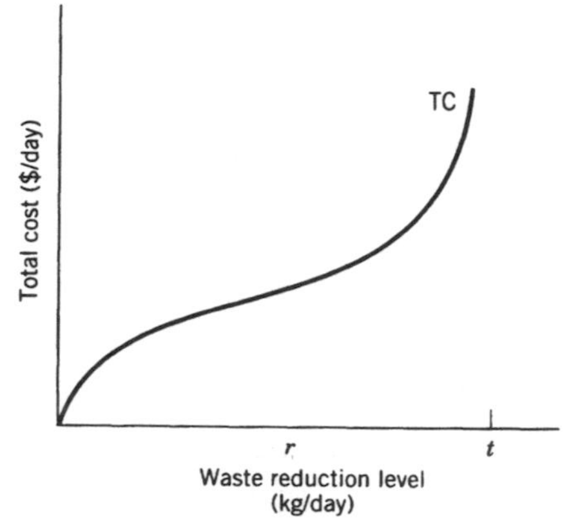

Figure 15.1 Total cost of waste reduction.

cost curves for dischargers where the incremental cost of removing waste increases sharply for very high levels of r.

Ideally, TC represents the net effects on cost of all actions for the given municipality or industry, including the application of waste treatment processes, changes in production processes, changes in raw materials, and changes in product mixes. In practice, the boundaries of an economic analysis are quite important. Steady state is assumed here, implying that the overall production level of the discharger remains unchanged. In practice, this assumption should be investigated and the boundaries of the economic analysis chosen carefully. This point is illustrated by an example in Section 15.B.3.

The incremental cost, or marginal cost (MC), is the change in TC per unit change in the level of waste reduction. The value of MC is a function of r, as shown for the example in Figure 15.2. The units of the MC function are \$/day per kg/day, or \$/kg of waste reduction. If TC(r) is expressed mathematically and has the appropriate properties, MC(r) can be obtained by differentiating TC(r). Again, a key feature of the MC of waste reduction is that it typically rises steeply for very high levels of waste reduction.

Now consider the application to the discharger of an effluent charge. One type of charge is a constant unit charge, \$$c$/kg, applied to the amount of waste discharge measured in kilograms. The total charge, TCH, assessed to the discharger for one day would be

$$\mathrm{TCH}(r) = c(t - r) \tag{15.1}$$

since $(t - r)$ would be the amount actually discharged during one day (i.e., the initial discharge level minus the waste reduction level). If the discharger provides no waste reduction, the charge payment would be ct. By increasing the waste reduction level the discharger is able to reduce the charge payment. The total savings, TS in \$/day, that would result is given by the maximum

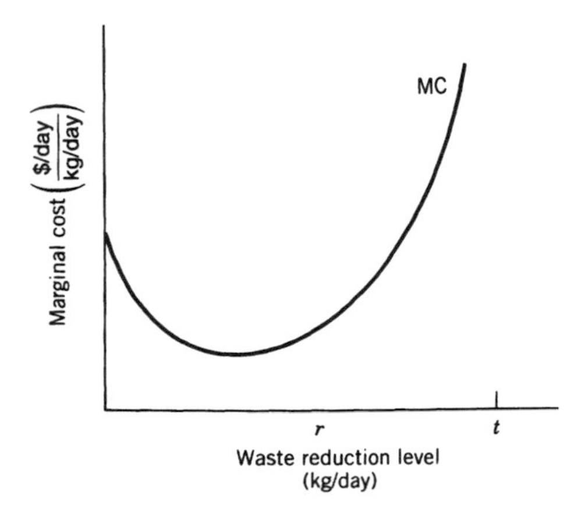

Figure 15.2 Marginal cost of waste reduction.

possible charge payment (ct) minus the charge payment after waste reduction [$c(t - r)$]:

$$\mathrm{TS}(r) = ct - c(t - r) \tag{15.2}$$

or

$$\mathrm{TS}(r) = cr \tag{15.3}$$

Thus the total savings that results from waste reduction is given by the unit charge multiplied by the waste reduction.

The total savings function, TS(r), is shown in Figure 15.3. The maximum possible level of waste reduction is t, the initial level of discharge. As the discharger increases the level of reduction, r, the total savings increases linearly. This savings, measured in \$/day, provides an economic incentive for the discharger to increase the level of waste reduction.

The incremental savings, or marginal savings (MS), is the value of the additional savings for an additional unit of waste reduction. It is given by the rate of change in the TS function. In this case the MS(r) is given by the derivative of the TS function given in Equation 3:

$$\mathrm{MS}(r) = c \tag{15.4}$$

Thus, for every additional kg/day of waste reduction the discharger would save \$$c$/day. The units of the MS are \$/day per kg/day, or \$/kg. Given that the unit charge is \$$c$/kg, it is expected that the MS, the savings associated with one additional kilogram of waste reduction, would also be \$$c$/kg.

The MS function, MS(r), is a horizontal line with value c, as shown in Figure 15.4. The figure also shows the marginal cost function, MC(r). By

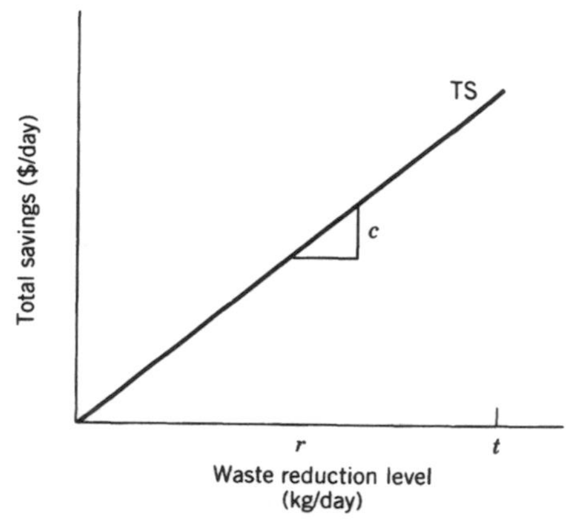

Figure 15.3 Total savings from waste reduction.

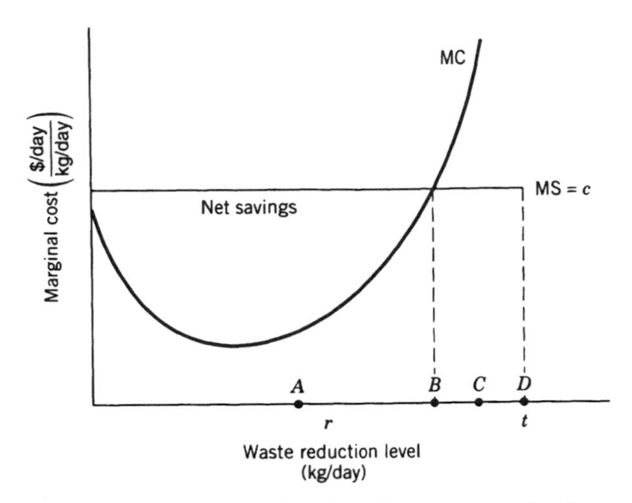

Figure 15.4 Marginal savings from waste reduction.

providing an additional kg/day of waste reduction, the discharger would save an additional $\$c$/day. It would, however, incur an incremental cost per day as well. If the discharger were to decide to reduce waste to level A in Figure 15.4, the marginal savings would be greater than the marginal cost, and it would be economically advantageous for the discharger to provide additional waste reduction. If the discharger were to decide to reduce waste to level C in the figure, it would be economical to reduce the level of waste reduction. By reducing waste discharge to level B, the discharger would obtain the most economically attractive solution.

The cost of waste reduction for the given discharger would be given by the value of the TC function, or the area under the MC curve in Figure 15.4, up to point B. The MS would be the area under the MS function up to point B, and the difference between the two areas would be the net savings associated with the waste reduction. The area under the MS function after point B would be the MS not earned by waste reduction. In other words, since waste would not be reduced beyond point B, the waste discharged would be represented by D minus B, and the charge payment would be represented by the area under the MS curve between points B and D [i.e., $C(D - B)$]. This area, in general, would be given by $c(t - r)$ using the notation from equation (15.1).

The economic model suggests that the individual discharger would respond to the incentive provided by an effluent charge of value c by reducing waste to level B in Figure 15.4. Depending on the shape of the MC curve, it is possible that the optimal solution for a discharger would be the boundary solution ($r = 0$). This possibility would need to be examined. In practice, a minimal standard for waste reduction might be in effect. Such a standard would then be the lower bound for r. The upper bound on r, given by t, would probably not be an optimal boundary condition given a steeply rising MC curve for high values of r since the value of c would have to be extremely high.

One approach to designing a management program would be to apply such an effluent charge to a set of dischargers—for instance, along a river or in an airshed. Each discharger would be expected to reduce its discharge by choosing its own economically optimal solution given its individual MC function. The effluent charge program would be designed with the expectation that waste discharges would be reduced to an acceptable level given the applicable environmental standards.

Application of a Single Effluent Charge to a Group of Dischargers in a Special Case Involving a Conservative Pollutant. One reason that effluent charges are potentially attractive as a management strategy is that the approach has the potential to be very efficient economically. This feature is demonstrated by considering an example with three dischargers along a river. Suppose that each of the discharges of pollutant P, measured in kg/day, mixes with the river water. Pollutant P is assumed to be conservative (i.e., the concentration is affected only by dilution and does not change through other mechanisms, such as decay or settling). Furthermore, the flow in the river is constant and much greater than the flow of the wastewater discharges. The highest concentration of P in the river occurs downstream of the last discharger. This concentration is obtained by summing the three discharge rates and calculating the effect of the dilution by the river flow. Dilution processes are described in detail in Chapter 2 (Section 2.C.1) of this book.

Assume that a water quality standard (expressed as a concentration of P) is not met downstream of the last discharger and that it is necessary to reduce the aggregate discharge so that it would be. Since P is conservative, the total allowable discharge level can be calculated, taking into account dilution by the river flow, such that the standard would just be met. Furthermore, for the purpose of meeting the standard downstream it does not matter what level of waste reduction each discharger provides. Since the total allowable discharge level is known, the total waste reduction required to meet the standard can also be calculated. The total reduction required is given by the total of the three discharge levels minus the total allowable discharge level.

The three graphs shown in Figure 15.5 illustrate an effluent charge solution where the same unit charge, c, is applied to all dischargers. The three MC curves are different, and the expected waste reduction level is shown for each of the three dischargers. By increasing (or decreasing) the value of c it is possible to increase (or decrease) the waste reduction level for each discharger, and therefore to increase (or decrease) the total waste reduction (or mathematical optimization). Using trial and error, the value of c can be varied until the solution would provide exactly the total reduction required to meet the water quality standard.

A key feature of this solution is that, in theory, it is the most economically efficient solution. It would provide the total level of waste reduction required

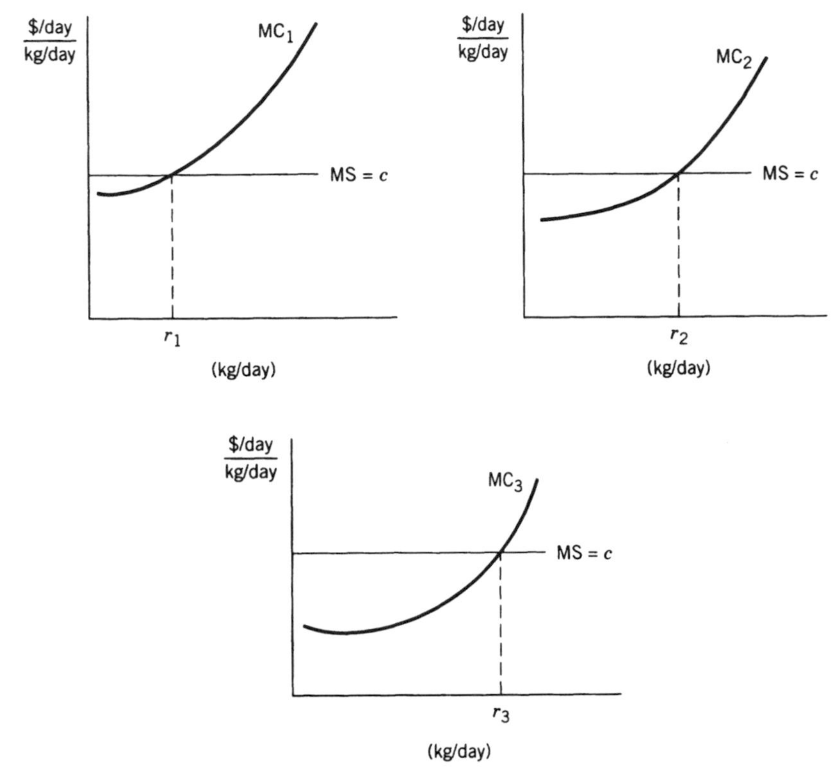

Figure 15.5 Effluent charge solution for three dischargers.

to meet the ambient standard at the least overall cost. One way to confirm this result is to consider any variation from this solution. Pick any of the three discharges and assume that it might be better to decrease its level of waste reduction by x kg/day. The reduction in cost for that discharger would be the area under the MC curve to the left of the original solution for a distance of x kg/day. Such an area is shown in Figure 15.6 for discharger 3. Since the total level of waste reduction must be maintained at the level of the original solution to met the standard, at least one of the other dischargers must increase its level of waste reduction. Using discharger 1 for illustration purposes in Figure 15.6, it can be seen that the incremental cost of increasing the waste reduction level is given by the area under its MC curve. Furthermore, the area showing the increase in cost to discharger 1 must be greater than the area showing the decrease in cost for discharger 3. This result holds since the value of c, and the value of the MC at the intersection of the MS and MC curves, is the same for each discharger. In general, for any variation from the original solution, the decrease in cost associated with decreasing the waste reduction level for any discharger will be more than offset by the increase in cost for another discharger. Thus the original solution must be the least cost solution.

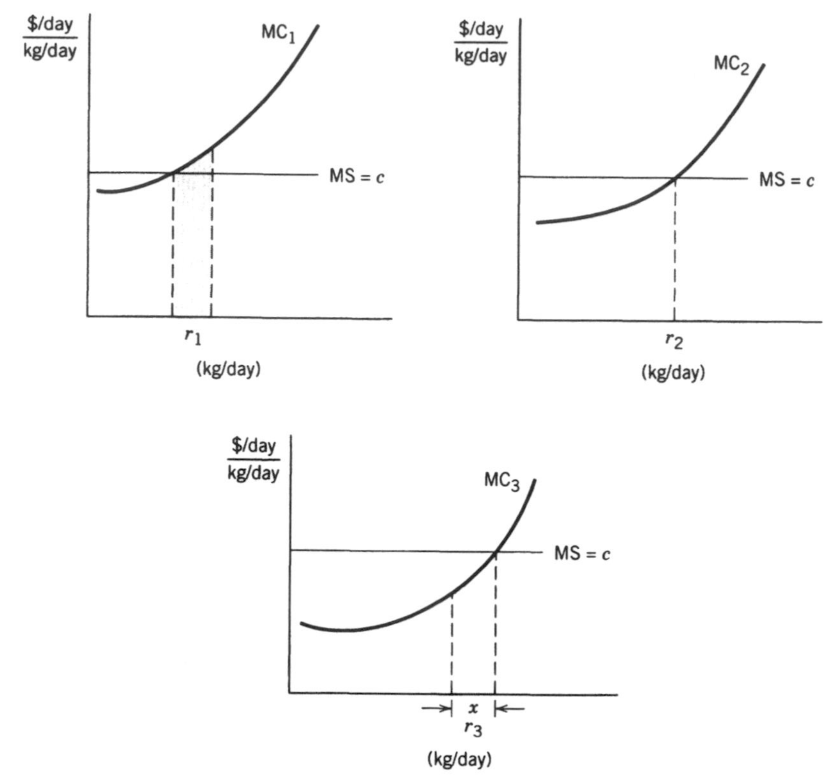

Figure 15.6 Effluent charge solution for three dischargers is a least cost solution.

Mathematical Optimization to Obtain the Least Cost Solution. The graphical result, that the solution obtained using a single unit charge is least cost, can also be obtained mathematically. The sum of the three TC function can be minimized subject to the requirement that the total waste reduction level is met. Assuming that the functions are continuous and continuously differentiable, the Lagrangian can be used to obtain the overall least cost solution.

The problem can be formulated as

$$\text{minimize total cost} = \sum_{j=1}^{3} \text{TC}_j(r_j) \tag{15.5}$$

subject to

$$\sum_{j=1}^{3} r_j = f \tag{15.6}$$

The objective function is to minimize the total cost, and the constraint ensures that the total waste reduction level is f kg/day, the value that will just meet the water quality standard. For this example, lower and upper bounds for each r are omitted since none of these bounds is in effect in the solution. In a case where the MC curves vary more, one of the dischargers might be at a bound, particularly if there are lower limits on waste reduction levels or if charges are not assessed once a specified upper limit is reached for waste reduction. Either upper or lower bounds can be added to the formulation using inequality constraints.

As shown, the formulation can be solved using the Lagrangian approach as done below. If inequality constraints are added, the Kuhn–Tucker conditions can be used to solve such problems. Alternatively, the Lagrangian method can be applied while checking boundary conditions and setting r for individual dischargers at a bound as appropriate.

The Lagrangian is formed and solved as follows:

$$\text{minimize } L(r_1, r_2, r_3, \lambda) = \sum_{j=1}^{3} \mathrm{TC}_j(r_j) - \lambda \left(\sum_{j=1}^{3} r_j - f \right) \tag{15.7}$$

subject to

$$(1)\ \frac{\partial L}{\partial r_1} = \frac{\partial \mathrm{TC}_1(r_1)}{\partial r_1} - \lambda = 0 \tag{15.8}$$

$$(2)\ \frac{\partial L}{\partial r_2} = \frac{\partial \mathrm{TC}_2(r_2)}{\partial r_2} - \lambda = 0 \tag{15.9}$$

$$(3)\ \frac{\partial L}{\partial r_3} = \frac{\partial \mathrm{TC}_3(r_3)}{\partial r_2} - \lambda = 0 \tag{15.10}$$

$$(4)\ \frac{\partial L}{\partial \lambda} = -\left(\sum_{j=1}^{3} r_j - f \right) = 0 \tag{15.11}$$

Rewriting (15.1), (15.2), and (15.3), we obtain

$$\frac{\partial \mathrm{TC}_1}{\partial r_1} = \lambda \tag{15.12}$$

$$\frac{\partial \mathrm{TC}_2}{\partial r_2} = \lambda \tag{15.13}$$

$$\frac{\partial \mathrm{TC}_3}{\partial r_3} = \lambda \tag{15.14}$$

Since

$$\frac{\mathrm{TC}_j(r_j)}{\partial r_j} = \mathrm{MC}_j(r_j) \tag{15.15}$$

$$\mathrm{MC}_1(r_1) = \mathrm{MC}_2(r_2) = \mathrm{MC}_3(r_3) = \lambda \tag{15.16}$$

or

$$\mathrm{MC}_1 = \mathrm{MC}_2 = \mathrm{MC}_3 \tag{15.17}$$

Optimality, the Equal Marginal Cost Condition, and Effluent Charges. The key feature of the optimal solution, whether obtained graphically or using the Lagrangian, is that the MC values are equal in a least cost solution. This important characteristic of a least cost solution—for the problem considered—will be used later in this chapter and is referred to as the *equal marginal cost condition.*

An important characteristic of effluent charge programs is that the approach would tend to be very efficient economically. The application of a single unit charge assessed to all dischargers would provide an incentive for the individual dischargers to reduce wastes such that the equal marginal cost condition would be met. If the dischargers act to minimize their individual costs, the water quality standard would be met at least overall cost.

Many water quality and air quality management problems, of course, are more complex. Many pollutants, for instance, are not conservative. One common case is where a pollutant decays exponentially. In such cases the impact of a given discharge into a river depends on both the location of the discharge and the location in the river where the water quality is checked. If a standard is imposed all along the river, the impact of the discharge will change along it. If there are multiple dischargers, the relative impacts of different dischargers will vary when checked at a given downstream location. Furthermore, the effect of an upstream discharger may be much greater at an upstream location. As a result, its impact at an upstream location might be greater than the combined effects of that discharger and a downstream discharger on water quality farther downstream. Thus the critical, or binding, water quality checkpoint may be upstream of some dischargers.

Another example of a complicating factor is the situation where the stream conditions may change. For instance, a tributary may increase stream flow so that the dilution effect downstream of the tributary decreases the impact of all discharges. Even though there may be dischargers downstream of the tributary, the limiting water quality check points may lie upstream of the tributary. A more thorough discussion of impact coefficients in water quality models is provided in Chapter 2 (Section 2.C.5), where a multireach–multidischarge model is presented for biochemical oxygen demand, a nonconservative pollutant with exponential decay, and dissolved oxygen concentration in a freshwater stream. Optimization models for siting wastewater treatment plants are presented in Chapter 14 (Section 14.F). The use of optimization models for water supply engineering is covered in Chapter 1.

In cases where the impacts of the dischargers are location dependent, the equal marginal cost condition as developed does not apply, and therefore a single effluent charge, will not generally produce a least cost solution. Variations of the equal marginal cost condition and of effluent charge programs, however, can be developed for special cases. For instance, Exercise 4 deals with the special case where there is a set of dischargers along a river, as in the case considered above, but where the impacts on a single critical downstream water quality checkpoint are different.

The simple case with a conservative pollutant considered above does, however, suggest why the effluent charge approach tends to be relatively economically efficient even if it is not optimal for more complex systems. The dischargers with relatively low costs would provide relatively high levels of waste reduction to avoid paying effluent charges. Dischargers with relatively high costs would elect to pay relatively more effluent charges instead of providing as much waste reduction. Thus the overall level of waste reduction would be carried out at least cost. Since the solution would not take the location effects into account, however, it would not be expected to provide a least cost solution for meeting an ambient standard.

15.B.2. Effluent Charges Applied to More Complex Systems

Effluent charge programs could be designed for particular water bodies or airsheds. Such an approach could also be applied broadly, even nationally. In the 1960s it was suggested that the U.S. Congress consider applying a national effluent charge on some discharges, such as conventional wastewaters that exert a biochemical oxygen demand (BOD) on receiving water bodies. The idea was that assessing a charge per pound of BOD discharge nationwide might be a desirable strategy for reducing overall pollution in an economically efficient way. The examples considered below, however, are for a given water body with multiple discharges.

Examination of Effluent Charges for the Case of the Delaware Estuary. In one of the first such studies for a realistic water resources system, discharge information and cost estimates for 44 dischargers, along with a water quality model, were used to simulate the application of effluent charges to meet a given water quality standard for the Delaware Estuary (Johnson, 1967). The waste loads were measured in lb/day of BOD. The impact of a given combination of waste discharges on water quality was evaluated using a matrix of impact coefficients. Each coefficient provides the impact of a unit discharge (lb/day of BOD) in section i on dissolved oxygen (mg/L) in section j of the estuary (Thomann, 1972).

One computational aspect of simulating the application of an effluent charge program is estimating the response of a given discharger to a given charge. In Section 15.B.1, continuous MC curves were used for the examples. In practice it is unlikely that such information is available. In the Delaware case, several points had been estimated for the TC curve for each discharger,

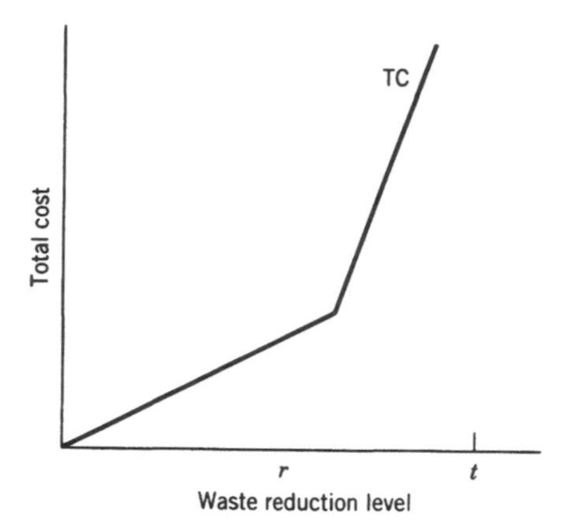

Figure 15.7 Piecewise linear total cost function.

and a piecewise linear cost function had been used in optimization studies (Thomann, 1972).

A piecewise TC function is shown in Figure 15.7. The method used to estimate the MC functions was to use the slope of each segment to form a step function, as shown in Figure 15.8. One result of this method of approximation is that a unit charge, which produces a horizontal MS function, would be expected to lead to a waste reduction level at the end of one of the steps. In Figure 15.8, for instance, the expected solution would be point *A*, *B*, or *C*. For the discharger represented, a range of values of the unit charge (up to c_1) would produce point A, a higher range (from c_1 up to c_2) would produce *B*, and a yet higher range (c_2 or above) would produce point *C*.

In a later study using the same data, the MC curves were estimated by using a continuous, piecewise linear fashion. As can be seen by examining Figure 15.9, such an approach allows for a continuous set of unit charges that would be estimated to produce a continuous set of waste reduction levels

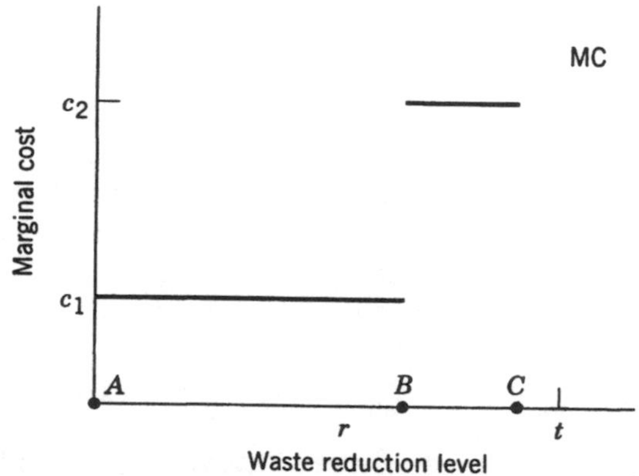

Figure 15.8 Step function approximation of marginal cost function.

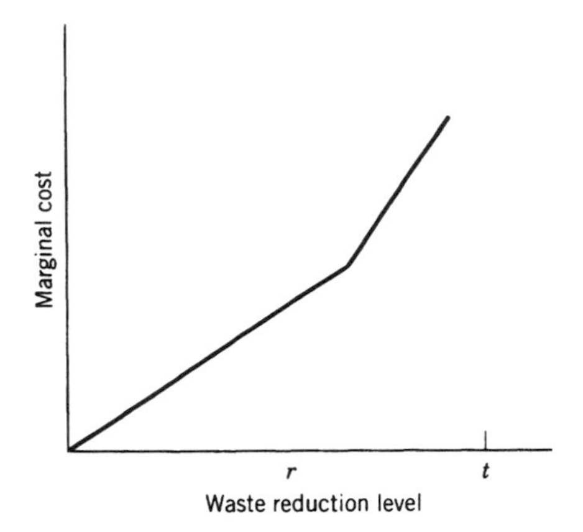

Figure 15.9 Piecewise linear approximation of marginal cost function.

(Brill et al., 1979). Either of the two approaches to approximating the MC curves can be used in an application, depending on the judgment of the analyst. The second approach is assumed for the discussions and examples used below.

Given a set of approximations of the MC curves, the result of applying a single unit charge, \$$c$/lb of BOD, to all dischargers for a system such as the Delaware Estuary can be obtained using the approach described in Section 15.B.1.:

1. A value is selected for c.
2. The MS curve is determined given that value of c.
3. The intersection of the MC and MS functions is determined for each discharger, providing the level of waste reduction, r_j, that each discharger j would undertake.
4. The set of waste reductions, r_j, are then multiplied by the set of impact coefficients, a_{ij}, to determine the improvement in DO for each section i as a result of each r_j. The improvement is summed for all 44 dischargers, giving the total improvement in each section.
5. The improvement in each section i is then compared to the improvement required to meet the water quality standard. If the standard is not met in all sections, the value of c must be increased. If the standard is exceeded, the value of c may be decreased. By trial and error the value of c can be changed until the standard is met within a given tolerance.
6. The cost of the final solution can be obtained by simply summing the values of $\text{TC}(r_j)$ for $j = 1,...,444$.
7. The charge payment can be estimated for each discharger by multiplying the unit charge by the waste discharge $[c(t_j - r_j)]$. The sum of the

charge payments can be obtained by summing the individual charge payments for $j = 1,...,44$.

This method can be used to obtain the single effluent charge solution that is most cost-effective in meeting a given standard. The reason is that the total cost and the level of water quality both increase (or at least are monotonically nondecreasing) as the value of c increases. That is, for a higher value of c, each discharger will provide a higher level of waste reduction and incur a higher cost.

The specific solution procedure used in an application would depend on the purpose of the analysis. To obtain just one solution (for one water quality standard), it might be desirable to use a more efficient search than trial and error. For instance, consecutive bisection could be used to search values of c between a lower bound and an upper bound. (Bounds would be provided by the lowest and highest MC values over all dischargers.) Often in carrying out an analysis of management strategies, however, it is important to consider a set of solutions that would meet a range of water quality standards. The method outlined above could be modified to change the value of c incrementally and to record the value of the standard that would be met for each value. Each such solution would then provide information that could be used in choosing both a water quality standard and a management strategy. Regardless of the actual solution approach, the calculations are straightforward vector and matrix operations.

Comparison of Effluent Charge and Direct Regulation Solutions. Using the approach described above, simulated solutions were obtained for the Delaware case for a range of values of a single unit charge applied to all 44 dischargers (Brill et al., 1979). The solutions obtained are quite economical compared to other solutions for meeting a particular water quality standard. Results were obtained for a variety of management strategies for the same water quality standard and under identical conditions.

The overall least cost solution was obtained using linear programming. [Linear programming models for water quality management are described in detail in Chapter 2 (Section 2.D).] The sum of all waste reduction costs was minimized subject to a set of constraints used to ensure that the water quality standard would be met throughout the estuary. Impact coefficients were used to sum the impacts of all reductions of BOD discharges on the improvement in DO in each segment of the estuary. As shown in Table 15.1, the least cost (LC) solution would require expenditures of $67 million (present value cost for the planning period) for the conditions considered.

This solution provides important cost information for benchmark purposes, but implementing it directly is likely to be impractical because of inequities. The waste reduction levels that would be required from the 44 discharger vary widely throughout the range from 20 to 90%. The inequities associated with this strategy were apparent in the first modeling study of the Delaware (Thom-

TABLE 15.1 Cost and Unit Charges for Direct Regulation and Effluent Change Programs

	Cost ($ millions)	Unit Change ($/lb BOD)
Least cost	67	
UT	195	
ZUT	108	
SECH	87	0.036
ZECH	76	(0, 0.04, 0)

ann, 1972). One of the primary purposes of such a model, however, is to provide just this sort of insight about a given problem. This particular result, an efficient but inequitable solution, has been observed for other environmental systems and has provided part of the motivation for the ongoing research effort to identify cost-effective management strategies, including those based on economic incentives. After a brief discussion of alternative direct regulation management programs, a comparison between the direct regulation and effluent charge programs is provided.

In the original study of the Delaware estuary, the analysts also considered a more equitable strategy. In that management approach each discharger would provide the same percent waste reduction; the required reduction percentage was determined such that the DO standard would be met. This solution can be obtained by varying the uniform percentage, using an approach similar to that described above for varying the single unit effluent charge, until the DO standard is just met. For the case considered here, the solution, called the *uniform treatment* (UT) *solution,* would cost $195 million. Although this solution is judged to be equitable, there is a severe cost penalty in comparison to the LC solution.

In the Delaware study this trade-off led the study participants to seek alternative programs and prompted the design of a third alternative. Called a *zoned uniform treatment* (ZUT) *solution,* the dischargers were divided into groups by geographical zones. Within each zone the dischargers are required to provide the same percent removal, but the removal level can vary from zone to zone. A linear programming model can be solved to obtain the optimal ZUT solution; constraints can be added to the least cost model to require the same percent removal by each discharger in a group. A typical solution is shown in Table 15.1 for three groups of dischargers defined as part of the original Delaware study. It can be observed that the extra flexibility offered by the ZUT approach results in a considerable saving compared to the UT solution; the total cost would be reduced from $195 million to $108 million.

As a side comment, cost and equity are often considered two major objectives in devising an effective management strategy, and trade-offs between them can sometimes be analyzed using modeling methods. In addition to the approach described above, the trade-off can be probed further for direct regulation programs using formal multiobjective methods (Brill et al., 1976a).

As shown in Table 15.1, the *single effluent charge* (SECH) *solution* would cost just $87 million. Although more costly than the benchmark LC solution, the total cost appears very attractive in comparison to the ZUT and UT solutions. Furthermore, the effluent charge program appears equitable in that it would apply the same unit charge to all dischargers. The solution for this example would require a unit charge of approximately $0.04 per pound of BOD. As discussed in Section 15.B.1, the effluent charge approach tends to be cost-effective because the dischargers with high costs would tend to pay the charge, whereas those with low costs would provide waste reduction. The SECH approach does not provide a least cost solution because the impacts of the discharges vary with their locations.

15.B.3. Equity Aspects of Effluent Charge Programs

The single effluent charge program discussed above is equitable in that it treats all dischargers in a similar way. It would be possible in theory to implement a least cost solution using effluent charges after solving a least cost model. Such an approach, however, would require that the set of TC curves (or MC curves) for all dischargers would be used to calculate the set of individual unit charges needed to produce the required set of waste reduction levels. Such unit charges would vary widely and would probably be viewed as highly inequitable. For instance, for the Delaware example, the unit charges would vary from $0.012 to $0.52 per pound of BOD. The single effluent charge program, in contrast, is similar to the UT program in that it treats all dischargers the same.

It is also possible to define a *zoned effluent charge* (ZECH) *program* by applying the same unit charge to each discharger in a designated group. The optimal solution can be obtained by varying the unit charge in each zone incrementally or by trial and error to obtain a good solution. A formal algorithm can also be used under certain conditions (Brill et al., 1976b). The cost of a simulated solution for the example problem is $76 million. As shown in Table 15.1, the cost of the SECH program is much less than the cost of the UT program, so the decrease from the SECH to the ZECH solution is not nearly as great as that from the UT to the ZUT solution.

Equity concerns discussed so far address the issue of treating equals among dischargers equally. One way to treat equals equally in a direct regulation program is to require equal percent waste reduction requirements. A way to treat equals equally in an effluent charge program is to apply the same unit charge to each discharger. Another major equity concern associated with effluent charge programs, however, is the need for the municipalities and industries to pay the charges for the wastes that are discharged. As discussed in Section 15.B.1, the charge payment for an individual discharger would be given by $c(t - r)$, the unit charge multiplied by the amount of waste discharged. The charge payment for an individual discharger is shown in Figure 15.4. The total charge payments for a given effluent charge program can be estimated by summing the payments for all dischargers, as described above in step 7 of the method used for the Delaware example.

As shown in Table 15.2, the total charge payments can be very large compared to the total cost of waste reduction. For example, the total charge payments for the SECH program would be $100 million. These payments represent an income transfer from the dischargers to the governmental agency assessing the charge. They are not an economic cost from the point of view of society as a whole since no resources are expended. (The funds are available to the agency instead of the dischargers.) The distribution of income, however, represents a major equity issue. Certainly, the fact that they would need to pay $100 million to the agency would be a major concern to the municipalities and industries making the payments (and possibly to the agency as well).

When the $100 million in charge payments is added to the $87 million in total cost, the total financial burden on the dischargers would be $187 million, which is nearly the same as the total cost of the UT solution. Thus, in comparison to the UT approach, society would save considerable resources, but the dischargers would not benefit directly. In making the same comparison of the ZECH and ZUT programs, it can be seen that the total cost of the SECH program is less costly ($76 versus $108 million), but the financial burden is greater ($149 versus $108 million). From the point of view of the municipalities and industries, the effluent charge approach adds an extra burden on top of the considerable burden of paying the costs of reducing wastes; largely for this reason, dischargers have not received the approach well historically. Furthermore, there have been concerns that charges may be punitive if they are applied to the residual discharges after municipalities or industries have provided high levels of waste removal (U.S. Congress, 1970, p. 175).

One concern with charges is that the financial burden would be a significant part of the cost of industrial activities. In such a case the industrial production levels might change, and there could be adverse impacts on individual firms

TABLE 15.2 Cost, Charge Payments, and Total Financial Burden for Direct Regulation and Effluent Charge Programs

	Cost ($ millions)	Charge Payments ($ millions)	Total Financial Burden ($ millions)
LC	67	n.a.[a]	67
UT	195	n.a.	195
ZUT	108	n.a.	108
SECH	87	100	187
ZECH	76	73	149

[a]n.a., not applicable.

or on the regional economy. In the first major empirical study of effluent charges, it was shown for the Delaware example that this would not be likely under the assumptions at that time (Johnson, 1967). The method of analysis was to calculate the economic burden for each industry as a percentage of the value of the output of each firm. One caution is that there could always be a marginal firm that would not remain economically feasible under an effluent charge, or perhaps any other, pollution control program.

Effluent Charge Schedules. One way to ameliorate the financial burden on dischargers from the charge payments is to modify the effluent charge function (Brill et al., 1979). One simple way to do this would be to eliminate the charge for some portion of the discharge. Since the cost increases steeply for very high levels of waste reduction, it might be practical to allow, for instance, 10% of the waste to be discharged without penalty. This change would redefine the MS function so that it would end at 90% waste reduction. There would be an incentive for the discharger to reduce waste up the 90% level but not beyond. Also, for any given level of reduction, such as point B in Figure 15.4, the charge payment would be eliminated beyond the 90% level.

Another way to reduce the charge payments would be to apply a unit charge that would decrease as the level of waste reduction increases. One example (from Brill et al., 1979) is a unit charge that would decrease linearly with the waste reduction level expressed as a percentage, $e(e = r/t$ using the notation from Section 15.B.1):

$$c = a - be \tag{15.18}$$

It can be shown that the actual charge payments can be reduced considerably by applying such a charge function. This approach is the subject of Exercise 15.5.

Either of these approaches can be used to reduce the charge payments for a SECH or ZECH program. They can also be combined to reduce the charges further. Nevertheless, the dischargers would continue to have to pay some charges and the distribution of income issue remains as an equity concern.

15.B.4. Other Characteristics of Effluent Charges Programs

As illustrated above, effluent charge programs can be analyzed using quantitative models. The discussion has focused on cost, equity among dischargers, equity for the dischargers as a group because of the need for substantial transfer payments, and meeting a given water quality standard. Other issues are also important, including the certainty of meeting the ambient standard, the need for information to implement and administer such a program, the incentive provided for technological innovation, and the implications of nonsteady-state conditions.

One concern on the part of regulatory agencies and environmental groups is the uncertainty of the dischargers' responses to a charge. Even if the cost information is known, how will the dischargers respond if they behave strategically given anticipated changes in regulations or are faced with major capital investments? Even more fundamental is the need for accurate estimates of the cost curves before a reasonable value can be selected, for instance, for a unit charge that will lead to a solution that meets the ambient standard. Such cost information is notoriously difficult to obtain, yet its absence may lead to programs that fail to meet the standard. Dischargers may not even know this information themselves prior to actually designing processes or facilities to provide the waste reduction. Furthermore, it would be impractical to implement a unit charge, to wait several years to assess the result, and then to adjust the charge if the desired result is not obtained. Some processes and facilities are literally cast in concrete and cannot be altered without incurring extra costs. Thus there is considerable uncertainty about meeting the water quality standard even under steady-state conditions.

Real-world conditions, however, are not steady state. Municipalities may experience growth, and industries may expand. If an effluent charge program is in effect, the amount of wastes discharged would increase, possibly leading to violations of a standard. On the other hand, costs could decrease. If a MC curve decreases for a discharger, the same unit charge would lead to a higher level of waste reduction. In fact, an advantage of the effluent charge approach is that the incentive for technological innovations that would reduce waste is greater than that under a direct regulation program; the dischargers have the ongoing opportunity to decrease the charge payments if a less costly method can be devised for reducing the discharge. Thus changes resulting from growth and technological innovation would offset each other, and the net effect of changes would be difficult to predict.

15.B.5. Summary Assessment of Effluent Charges

Effluent charges offer the potential for designing programs that in theory would be cost-efficient. As shown in this section, models can be used to analyze the potential costs, required unit charges, and charge payments. Trade-offs between cost and equity issues can be examined.

Other programs attributes are equally important. While the application of an effluent charge would provide an ongoing incentive for technological innovation and other ways to reduce waste discharges, there would be considerable uncertainty with respect to meeting an ambient water quality standard. Lack of information about costs and growth in populations or industrial activities could lead to increased levels of discharge and possible violations of ambient standards over time. In general, uncertainty about dischargers' behavior when facing an effluent charge translates into uncertainty about the levels of waste reduction and the resulting improvements achieved in water quality (Rose-Ackerman, 1973). Although much of the discussion is in the context of water quality management, the same general observations apply to air quality management.

Two of the major concerns about effluent charges, the requirement for charge payments and uncertainties about meeting ambient standards, have lead to increased attention to another incentives-based management approach, transferable discharge permits. This approach is discussed in the next section.

15.C. ANALYZING TRANSFERABLE DISCHARGE PERMIT PROGRAMS

15.C.1. Transferable Discharge Permits: A More Practical Incentives-Based Approach?

Under a transferable discharge permit program, a discharger is allocated a discharge permit and then allowed to transfer (i.e., to sell) it to another discharger (Dales, 1968; Joeres and David, 1983). Similarly, it could buy another permit from a discharger willing to sell one. The approach would tend to be economically efficient since a discharger with low waste reduction costs would be able to sell a permit to another discharger with relatively high costs. Higher levels of waste reduction would be carried out by dischargers with low costs, and lower levels of reduction would be carried out by those with high costs, leading to lower overall costs to achieve a given level of aggregate waste reduction. Furthermore, each discharger would have an incentive to develop new ways to reduce wastes since any unneeded permits could be sold. Permits could be traded, of course, only within a particular region corresponding to the particular water body or airshed being managed.

Various names for the approach that have been used include pollution rights, marketable effluent permits, and transferable discharge allocations. The concept has been examined for applications to both air pollution control (e.g., Tietenberg, 1974, 1980) and water pollution control (e.g., David et al., 1980; Eheart, 1980). Moreover, versions of the concept are currently being tested in practice. For instance, permits can now be exchanged in a national market in the United States for SO_2 emissions, and various states are experimenting with transferable permits for water quality control.

There are two important differences between the transferable permit and effluent charge approaches. In the effluent charge approach a charge is set that provides an incentive for a municipality or industry to reduce waste discharged. The response of the discharger, as noted above, is then observed—leading to considerable uncertainty about the impact on the environment. In contrast, under the transferable discharge permit approach, the amount of discharge is regulated. The value of the permits when bought and sold might then be observed. Thus, under the charge approach the price of discharges is set and the discharge rates are observed. Under the transferable permits approach the aggregate amount of the discharges is set and the price is observed. The uncertainty about the level of discharge and the impact on the environment would be greatly reduced compared to the effluent charge approach.

Another key feature of the transferable discharge permit approach is that the permits may be allocated initially without charge (as nontransferable permits would under a direct regulation program, for instance). Since all transfers would be voluntary and would be expected only if the positions of all trading parties improved, no inequity would be introduced by a transfer. All payments for permits would be voluntary payments from one discharger to another. In comparison to the effluent charge approach, for the dischargers as a group there would be no need for the extra financial burden of the charge payments to the regulatory agency. Thus there is a major difference between the two incentives-based approaches with respect to this important equity issue.

Alternatively, the transferable discharge permit approach could be implemented by selling all permits initially. The concept of transferable permits has been applied to a wide range of problems involving the management of public resources. In many cases it has been customary for governmental agencies to sell permits for the use of publicly owned resources. Examples are rights for commercial fishing, airport landings, offshore drilling for oil, grazing farm animals, or harvesting forestland. If transferable discharge permits were sold initially, there would be an income transfer from the dischargers to the regulatory agency, but it would be an option selected as a matter of public policy.

There are many variations of the transferable discharge permit concept. These variations can be examined with respect to cost, equity, and environmental quality using models such as those discussed in the preceding section. Such studies may be used for designing a program for a particular region or for developing regulatory policies.

Application of Transferable Discharge Permits to a Special Case Involving a Conservative Pollutant. One special case that illustrates the concept is illustrated by Figure 15.10. The three discharges along a river are the same as those considered in Section 15.B.1. Given a water quality standard, the total allowable discharge of pollutant P can be determined since P is conservative and the flow in the river is constant.

If the aggregate allowable discharge is d, one way to manage the water quality is to allocate a total of d discharge permits to the three dischargers in proportion to their initial levels of waste production, t_j for $j = 1, 2$, and 3. This solution is shown in Figure 15.10 by points A, B, and C. This solution could be implemented as a uniform treatment direct regulation program.

By allowing the permits to be transferred, however, the dischargers could reduce their individual costs and therefore the total program cost. Assuming steady-state conditions for this example, Discharger 1, for instance, would be willing to buy an additional permit for a cost up to c_1. Discharger 3 would

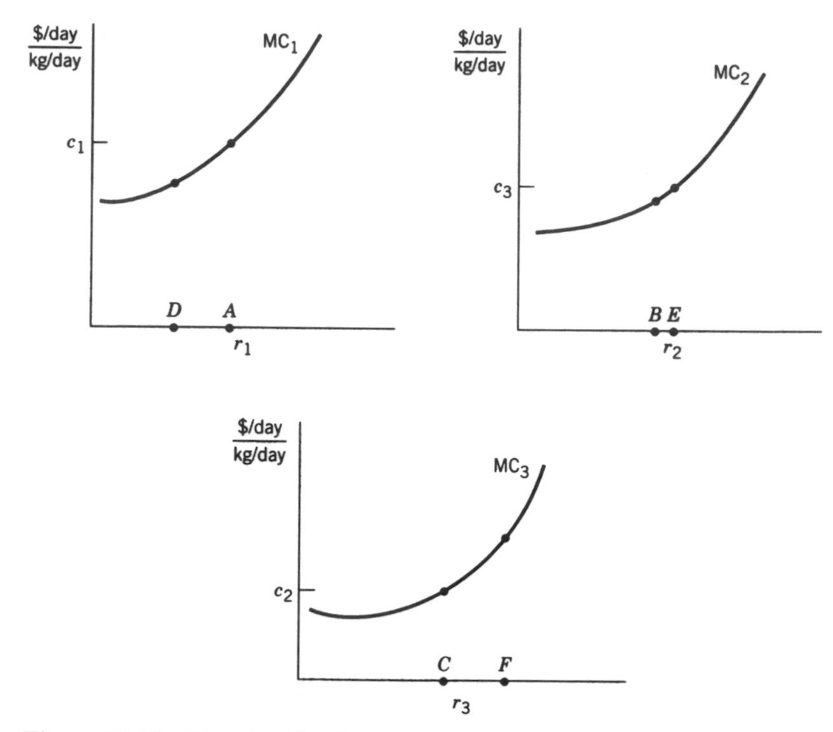

Figure 15.10 Transferable discharge permit solution for three dischargers.

be willing to sell a permit for a price above c_2. The dischargers could continue to buy and sell permits in this fashion until an equilibrium is reached at price c_3. The three waste reduction levels for the equilibrium solution are points *D*, *E*, and *F* in Figure 15.10.

An important feature of this solution is that the equal marginal cost condition would be met, and as discussed in Section 15.B.1 it would be a least cost solution for the case considered. Both the single effluent charge and the transferable discharge permit approaches would tend to be economically efficient for the special case where only the total amount of discharge is being regulated. By using transferable discharge permits, however, the total discharge is directly regulated. As long as no new permits are issued, the water quality standard would be met regardless of changing economic conditions or population growth.

An important observation is that the same solution could be obtained if the regulatory agency sold the permits initially (e.g., David et al., 1980). The agency, for instance, could ask each discharger to provide a list of how many permits it would buy for each in a range of prices. Assuming the dischargers behave according to the economic model, the bids would reflect their costs for reducing wastes. For an additional permit a discharger would be willing to pay up to the MC of removing that unit of waste. Thus the agency could select an overall market price, and initial allocation of permits, that would meet the standard. The same solution would be obtained as in Figure 15.10. The equal marginal cost condition would be met, and the solution would be least cost.

As discussed in Section 15.B.1, this simple case suggests why the transferable discharge permit approach tends to be relatively economically efficient even if it does not result in the exact least cost solution for more complex systems. Dischargers with relatively high waste abatement costs would tend to buy permits, and those with low costs would tend to sell them. This result is based on a simple economic model of behavior, however, and does not reflect strategic decisions by dischargers who may anticipate future changes in their needs or regulations or may take into account competitive positions with respect to other dischargers.

Other complications may arise with respect to meeting an environmental standard if the impact of a discharge depends on its location. In this case, if permits are transferred, the impact on a water body or airshed may change, and an ambient standard might be violated. This issue is considered further in the next section.

15.C.2. Illustrations of Transferable Discharge Permit Programs

Transferable Permits for Managing the Total Phosphorus Load in a Lake. One early illustration of the transferable permit approach was in the context of phosphorus management in Lake Michigan (David et al., 1980). Since the transport and fate of phosphorus is quite complicated in a complex water system, early regulatory efforts focused on controlling the total waste load. Using this approach and assuming steady-state conditions, the application is the same as that discussed in Section 15.C.1. The use of transferable discharge permits would be expected to provide a least cost solution for meeting a given target for total waste discharges.

This approach was examined for 53 sources of phosphorus in the portion of the Lake Michigan basin within Wisconsin. These sources were under orders from the Wisconsin Department of Natural Resources to reduce discharges of phosphorus. A simulated market solution was provided to illustrate the concept, and it was estimated that the cost savings would be approximately $750,000 per year over the direct regulation approach.

Transferable Permits for Management of BOD, a Nonconservative Pollutant, in a River System. The application of transferable discharge permits to a nonconservative case is more complicated because the effects of discharges vary, depending on their locations (Eheart, 1980). A simplified view of BOD management would be to design a program to meet a total target for the sum of BOD discharges. For instance, a uniform treatment (UT) program could be designed for a given water quality standard, and the solution would provide a total allowable discharge. Permits for BOD could be allocated according to this solution. If trading were then allowed, the expected market solution would meet the given target on total discharges at least overall cost (just as in the conservative pollutant case).

Without knowing in advance the outcome of trading, however, the locations of the discharges would not be known. Thus, after trading, the water quality standard for DO might be exceeded or might be violated. Since the standard would probably not be met exactly, the final solution would probably not be a least cost solution for meeting the given standard. If the standard were exceeded, additional permits could have been allocated. If the standard were not met, fewer permits should have been allocated. Different ways to analyze and deal with this uncertainty in meeting a water quality standard are considered in the following three subsections.

For different versions of the transferable discharge permit approach for BOD control, simulated market solutions are very cost effective (e.g., Eheart, 1980) because of the aggregate cost savings that result from trading. Depending on the details of implementation, the simulated solutions are identical or similar to simulated SECH solutions. They are typically much less costly from a societal perspective than simulated uniform treatment solutions.

Simulation of DO Level Resulting from a BOD Permit Market Solution. One step that can be taken is to simulate a BOD permit market and then to evaluate the water quality impact of the solution obtained. The market solutions can be simulated for a total discharge level in a fashion similar to that used for the single effluent charge case. For a given total discharge level, the final distribution of permits would be assumed to be the solution that meets the equal marginal cost condition. The water quality standard for DO that would be met for the final distribution of permits can be evaluated using a water quality model. For example, impact coefficients, such as those derived for the general water quality management model in Chapter 2 (Section 2.D.1) could be used to calculate the total impacts at specified water quality checkpoints.

In one study using published data for the Delaware estuary and Willamette River (Brill et al., 1984), the uniform treatment program was used to determine an initial allocation of BOD permits. The total BOD load in this case is simply the sum of the individual permits that would be allocated under the UT program. The result of trading was then simulated and the water quality impacts determined. Three cases with different DO standards were examined for each water system. It was shown that in five of the six cases the simulated solution would meet the DO standard; in some cases the standard would be exceeded. In the one case where the standard would be violated, the violation would be only 0.1 mg/L.

Evaluation of Worst Possible Violation of DO Standard Under Transferable BOD Permits. A second step in the analysis is needed, however, since it is possible, in general, to have more serious violations of the ambient standard as a result of trades. Given the availability of a water quality model, it should be very easy to calculate the worst possible impact on DO that could occur as a result of trading. For instance, for given discharge locations and water quality checkpoints, each checkpoint can be examined to see the worst-possible DO level. If impact coefficients are known, the worst location for a discharge has the highest impact coefficient. If all the permits are assumed to be exercised at that location, the worst possible DO level can be calculated for that checkpoint. By making a similar analysis of each checkpoint, the maximum possible violation of a DO standard can be calculated. It is even possible to take into account a DO standard that varies along the waterway in carrying out these calculations.

When this type of analysis was done for the Delaware Estuary and Willamette River examples, it was shown that the results differed dramatically for the two cases (Brill et al., 1984). For each of the three Delaware cases the worst possible solutions would violate the standard greatly—violations on the order of 3.0 mg/L. The same study also showed, using probabilistic simulations, that it would be very likely that if there were permit trades, there would be serious violations of the DO standard in each case. For the three Willamette cases, the maximum violation would be 1.2 mg/L, regardless of the final distribution of permits.

Additional Safeguards to Prevent Violations of the DO Standard Under Transferable BOD Permits. If a worst-case analysis shows that there could not be a serious violation of the DO standard regardless of the distribution of permits, a UT program could be designed and implemented with the addition of a provision that trading would be allowed. In a case where there are potential violations of the standard, additional safeguards can be implemented as part of the water quality management program. Such safeguards can be designed and implemented using methods of analysis similar to those discussed above.

One simple approach would be to vary the total number of permits and to calculate the worst possible violation of the DO standard that could occur under design conditions, using the method described above. The total number of permits issued could then be selected so that the standard would not be violated regardless of the final distribution. Although this approach might work well in some cases, it would often be very conservative and therefore increase the cost considerably. A likely benefit, however, would be water quality levels higher than the standard.

Another approach to ensuring that the water quality standard would be met is to require that all trades be approved by the regulatory agency. The permits could be distributed initially, for instance, using a UT program such that the standard would be meet. Only proposed trades that would not lead to a violation of the standard would be approved. This approach would require the agency to carry out water quality modeling studies, but once a model is developed in an initial study it can readily be used by the dischargers as well as the agency to examine permit transfers.

If a proposed trade would lead to a violation of the standard, the discharge allowed by the permit could be reduced. Using the water quality model, the agency could determine the waste discharge level that could be allowed such that the standard would be met. For instance, consider such a case where there is only one critical water quality checkpoint, say $j = 9$, and impact coefficients, $a_{1,9}$ and $a_{2,9}$, are available for dischargers 1 and 2. If discharger 1 is initially allocated p_1 permits (kg/day of BOD), the impact, b(mg/L of DO), at location 9 would be

$$b = p_1 a_{1,9} \tag{15.19}$$

If these permits are transferred to location 2 and $a_{2,9} > a_{1,9}$, the permits would have to be revalued so that the impact would be no greater than b at location 9. The new value of the permits, p_2, could be calculated as follows:

$$p_2 a_{2,9} = b \tag{15.20}$$

or

$$p_2 a_{2,9} = p_1 a_{1,9} \tag{15.21}$$

$$p_2 = \frac{b}{a_{2,9}} \tag{15.22}$$

or

$$p_2 = \frac{p_1 a_{1,9}}{a_{2,9}} = p_1 \left(\frac{a_{1,9}}{a_{2,9}} \right) \tag{15.23}$$

Equation (15.21) shows that the impact of each permit holding would be the same at location 9, the critical water quality checkpoint. After a transfer it would be necessary to confirm that there would not be a violation at a different location; if there were, the permit would have to be revalued further.

Other types of restrictions can also be placed on the transfer of permits to ensure that the water quality standard would not be violated (Brill et al., 1984). One restriction would be to limit the transfer of permits such that a permit would have to be exercised within a geographical zone. These zones could be designed to minimize the effects of changes in the locations of the discharges. Another restriction would be to decrease the value of any permit automatically when transferred using a uniform (for equity) revaluation factor. This approach could be used to gradually decrease the value of total waste discharges to the water system.

Another approach to dealing with location effects is to define permits in terms of the impact on DO at a given critical location (Eheart, 1980). Such permits are called dissolved oxygen deficit contribution (DODC) permits. The value of a permit in terms of the allowable BOD discharge would have to be calculated using a water quality model. For instance, if discharger 1 in the example above owns a permit to contribute b mg/L to the DO deficit at location 9, the permit would allow it to discharge p_1 kg/day of BOD. If the permit were sold to discharger 2, the permit would allow a waste load of p_2 kg/day of BOD, as shown in equation (15.23). Thus a DODC permit with value b would represent different allowable BOD loads according to the following relationship:

$$b = p_1 a_{1,9} = p_2 a_{2,9} \tag{15.24}$$

or

$$\frac{p_1}{p_2} = \frac{a_{2,9}}{a_{1,9}} \tag{15.25}$$

For example, if discharger 1 has twice the impact per kg/day of BOD discharge as discharger 2, it would be allowed half the discharge.

The DODC approach is most appropriate for a water body (or airshed) where there are location-dependent effects of waste discharges and where there is one distinct area where aggregate impacts are critical. The economic efficiency, equity, and water quality aspects of this approach are the subject of Exercise 15.7.

15.C.3. Summary of Considerations in Designing a Transferable Discharge Permit Program

The transferable discharge permit approach has attractive properties in comparison to effluent charges. It would tend to be economically efficient and to offer a way to avoid the income transfers and uncertainty of meeting the environmental quality standard associated with effluent charges. The information requirements are also fewer since cost information is not needed to implement a transferable permit program.

The discussion has focused on using models (1) to investigate cost, equity, and water quality impacts of alternative permit programs, and (2) to support the implementation of a permit program if permits need to be revalued if traded. Aspects of permit programs that have been considered include the definition of the basis of the permit (e.g., BOD discharge or DO impact) and possible limits on transfers. Although water quality examples are used, the concepts apply directly to air quality as well.

There are numerous other issues that should be considered in a study of a transferable permit program (Joeres and David, 1983). For instance, what should be the life of a permit? If the life is too long, there would not be an opportunity for the regulatory agency to upgrade the ambient standard over time. If the life is too short, dischargers would not have adequate time to plan an optimal response that might include building new facilities. Some have suggested that the initial allocation include permits with different terms (e.g., David et al., 1980).

Another issue to be considered is the likelihood of strategic behavior by the dischargers that would prevent the effective functioning of a permit market. A related issue is to investigate the alternative of selling permits initially instead of distributing them free of charge (see, e.g., Eheart, 1980). Dischargers could be required to bid for permits, thus revealing their demand for permits. The agency could then determine how many permits would be sold a given price and to which dischargers. Using a water quality model, the agency could select a price and sell permits such that the initial allocation would just meet the standard. If it wished, the agency could also hold some permits in reserve as a cushion against uncertainties in water quality modeling or to accommodate new dischargers. If permits are sold initially, there are alternative procedures for auctioning them. Two major considerations are the economic efficiency of the process and the likelihood of strategic behavior by dischargers (e.g., Lyon, 1982).

Most studies have examined the implementation of a permit program for one pollutant for a set of steady-state design conditions. If multiple pollutants are being managed for the same set of dischargers, it would be important to evaluate technical and cost interrelationships (Lence et al., 1988). It may be cost-effective to design permits to vary the allowable discharge over time to reflect changes in the assimilative capacity of a waterway (O'Neil, 1983; Eheart et al., 1987).

15.D. FUTURE CHALLENGES

In this chapter we described how quantitative models of environmental systems can be used to examine incentives-based management strategies. Models can be used to examine these strategies as part of a policy analysis. For example, it is useful to observe the equity as well as efficiency aspects of effluent charges for illustrative systems. The practical advantages of transferable discharge permits can also be seen by considering examples. The cost savings in comparison to direct regulation approaches can be evaluated along with the information requirements for implementation and the uncertainties with respect to meeting the ambient standard.

Models can also be used to support the implementation of a program. For instance, the worst possible water quality impacts that would result from allowing permits to be exchanged can be calculated. Models can also be used to examine proposed transfers and to revalue permits if desired. Potential complexities associated with implementing an incentives-based program are numerous and have received attention in the literature.

In this chapter we have illustrated that models may be very helpful for environmental policy analysis or program design, by aiding analysts and decision makers in gaining insights and understanding. The problems are much too complex, however, to expect to capture all aspects within a single model that can be used to obtain the final, optimal solution. Rather, by using an array of models in an iterative fashion, such models can be used as tools to support decision making and the development of creative solutions.

One major challenge is to continue to develop management strategies that reflect our increasing understanding of the complexity of environmental systems. Management approaches will be sought that deal with multimedia environmental impacts of municipal and industrial activities. For example, solid waste management practices that include additional recycling may lead to increased air pollution from additional use of collection and transport vehicles. Analyses of the full lifecycle effects of products, including ultimate disposal, might lead to new designs that make reuse of materials and components more practical.

One concept that may be used as we envision future management approaches is called industrial ecology (e.g., Graedel and Allenby, 1995). The notion is that society should attempt to design its activities so that the "waste" from one activity serves—to the extent possible—as a "food" for another. In this way, masses would continuously circulate or recycle in complex patterns, and residual discharges to the environment—while not eliminated—would be minimized.

Quantitative modeling methods offer the potential to assist in future policy studies of innovative management strategies. Figure 15.11d provides an illustration of the elaborate feedback links in an ecological model of industrial and municipal activity (Brill et al., 1992). This model is considerably more complicated than forerunner models (Figure 15.11a–c) of environmental systems used to develop regulatory approaches. The earliest models focused on end-of-pipe treatment. Evolutionary models have dealt with recycle flows within a firm or even recycled products. The industrial ecology model considers more complex combinations of mass flows. The optimal design of a product would reflect the values of any residuals from producing it as well as the product itself, ultimately, as foods for other products or processes.

The management challenge is to devise strategies that promote innovative products and processes (and resulting mass flows) that maximize the use of wastes as foods and minimize the adverse environmental impacts. These strategies may be complex and include combinations of effluent charges (or the analogous subsidies; Kneese and Bower, 1968), transferable permits, and direct regulations. Models such as those suggested in this chapter can be used to develop and evaluate such combinations. A wide array of issues would need to be examined, including economic efficiency, equity, environmental impacts, information requirements, and incentives for innovation.

EXERCISES

15.1. Consider a discharger with a total cost curve given by

$$\mathrm{TC}(r) = 2r^2 + 6r$$

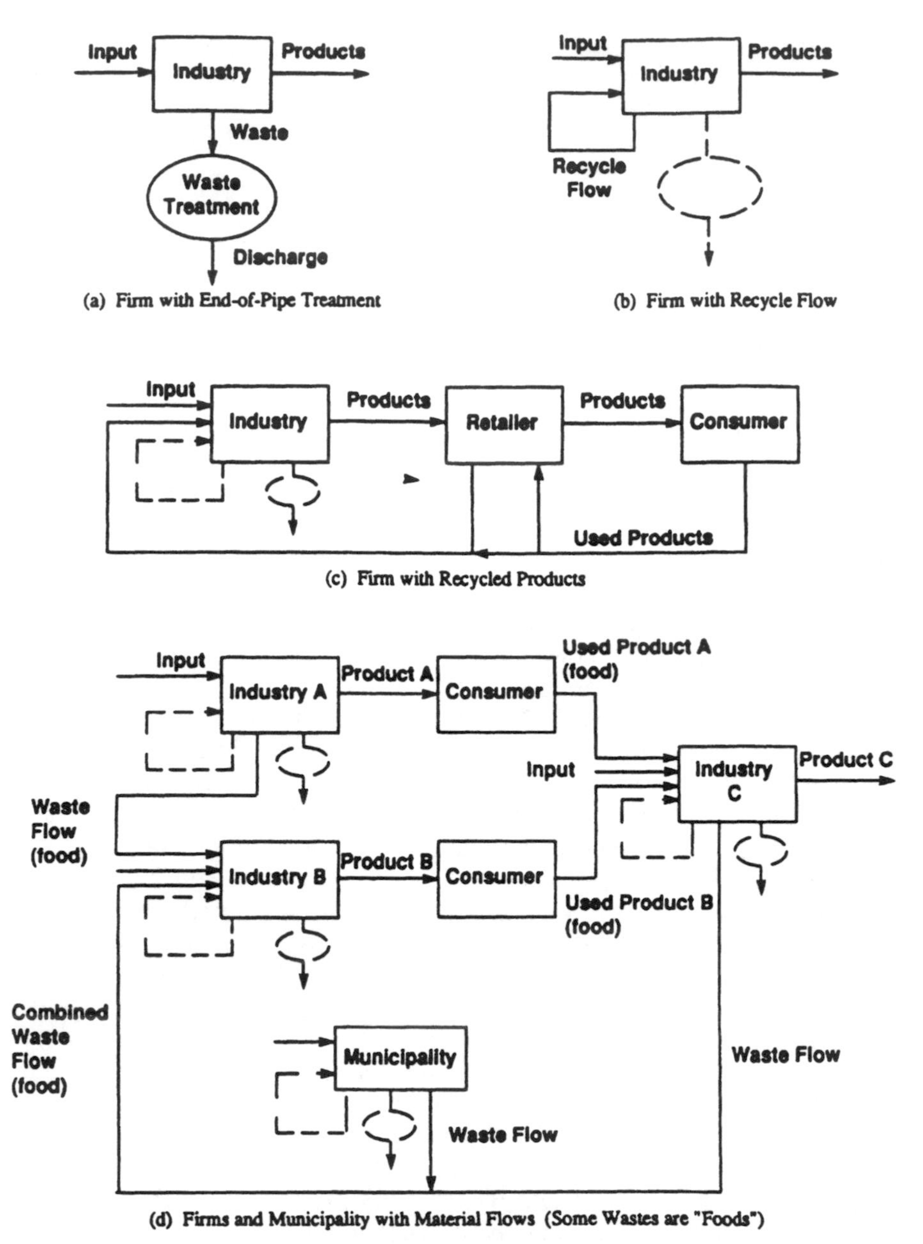

Figure 15.11 Hierarchy of systems for managing materials flows.

where the units of TC are thousands of dollars per day and the units of r are kg/day. Show how to calculate the value of a unit charge, $\$c$/kg, that would produce any given value of r from a lower value of zero to an upper value of t.

15.2. For the same discharger considered in Exercise 15.1, let c_1 be the unit charge that would provide an incentive for the discharger to reduce waste r_1 kg/day (i.e., c_1 corresponds to the MS curve that would intersect the MC curve at r_1). Show mathematically that the solution, r_1, minimizes the sum of the waste reduction cost plus the charge payments for the discharger.

15.3. Consider the application of the Lagrangian shown in Section 15.B.1. Develop a formulation that includes lower bounds for the three dischargers ($r_1 \geq b_1$, etc.). Write the Lagrangian and the Kuhn–Tucker conditions for an optimum. Describe the result. When would a lower bound be tight in an optimal solution?

15.4. The Lagrangian formulation shown in Section 15.B.1 applies to a river system with a set of dischargers where the pollutant is conservative and there is one downstream water quality checkpoint. Consider a similar system where there also is one downstream location where the water quality standard is critical. In this case, however, the pollutant is not conservative and the effects of the discharges are location dependent. Specifically, the impact of each discharger's waste reduction (the impact on improving water quality at the critical location) is determined by multiplying its value of r by an impact coefficient, a. The values of a_i are different for the different dischargers i.

(a) Develop a mathematical formulation similar to that used in Section 15.B.1 to obtain the least cost solution, assuming that boundary conditions are not needed for each r.

(b) Solve the formulation using the Lagrangian.

(c) Is it possible, in general, to select a single unit charge ($/kg of BOD) that could be applied to all dischargers to produce the least cost solution? Why?

(d) Describe the set of different unit charges (all measured in $/kg of BOD) that could be used to produce a least cost solution.

(e) How might this solution be viewed as being inequitable? How might it be viewed as being equitable?

(f) How can the equal marginal cost condition discussed in Section 15.B.1 for a least cost solution be modified to apply to such cases?

(g) Suppose that a unit charge is defined on the basis of the impact at the critical location (dollars per mg/L of DO deficit at the critical location). Evaluate an effluent charge program where this type of unit charge is applied to all dischargers.

15.5. Consider the unit charge function defined in equation (15.18).

(a) Derive the MS function as a function of r for a given discharger.

(b) Compare the value of the MS to the unit charge for a given value of r.

(c) Discuss the reason the charge payment would be expected to decrease if the unit charge function is applied to a discharger instead of a constant unit charge.

(d) How could the optimal values of a and b be obtained if the objective is to produce a given waste reduction level, r, such that the charge payments would be minimized. What are the optimal values?

(e) Now suppose that the same charge schedule is to be applied to a set of dischargers along an estuary such as the Delaware. How could the optimal values of a and b be determined such that the BOD discharges would be reduced so that a DO standard is met and the total charge payments would be minimized?

15.6. Equation (15.23) shows the revaluation of a transferable discharge permit to ensure that the water quality standard is not violated at location 9 using impact coefficients for the example shown.

(a) Suppose that the regulations in such a waterway require nondegradation throughout the river. How would each permit be revalued if transferred?

(b) Under what conditions would it be possible to increase the value of a permit if it is transferred? How could impact coefficients be used to ensure that there is no degradation in water quality at any location in the river if such increases are allowed?

15.7. Evaluate the DODC permit approach for a water body. Assume steady-state design conditions and that trading takes place according to the market model.

(a) Assume that the program is defined using one critical water quality checkpoint. How might permits be initially distributed?

(b) How economically efficient will the market solution be? Use a graphical or mathematical approach to justify your conclusion.

(c) Evaluate the DODC approach with respect to equity.

(d) Evaluate the DODC approach with respect to the certainty of meeting the water quality standard.

(e) Compare this permit program to an effluent charge program where the unit charge for a unit of discharge is based on the impact at the critical water quality checkpoint [see part (g) of Exercise 15.4].

15.8. Consider the network shown in Figure 15.11 used to illustrate the industrial ecology concept. Describe an example for a given set of products or activities where an innovative set of mass flows and processes might be used to allow "wastes" to be "foods" such that environmental impacts are decreased. Also discuss how incentives-based regulations, or combinations of regulations, might be used to promote the implementation of this innovative combination of mass flows.

REFERENCES

Brill, E. D., Jr., J. C. Liebman, and C. S. ReVelle, 1976a. Equity Measures for Exploring Water Quality Management Alternatives, *Water Resources Research,* 12, 845–851.

Brill, E. D., Jr., J. C. Liebman, and C. S. ReVelle, 1976b. Evaluating Environmental Quality Management Programs in Which Dischargers Are Grouped, *Journal of Applied Mathematical Modelling,* 1(2), 77–82.

Brill, E. D., Jr., C. S. ReVelle, and J. C. Liebman, 1979. Alternative Effluent Charge Functions: Cost, Financial Burden, and Punitive Effects, *Water Resources Research,* 15, 993–1000.

Brill, E. D., Jr., J. W. Eheart, S. R. Kshirsagar, and B. J. Lence, 1984. Water Quality Impacts of Biochemical Oxygen Demand Under Transferable Discharger Permit Programs, *Water Resources Research,* 20, 445–455.

Brill, E. D., Jr., G. W. Roberts, S. Roberts, K. Smith, M. S. Soroos, L. Pietrafesa, Q. W. Lindsey, and R. Lea, 1992. Industrial Ecology, Appendix C, *Report of the Blue Ribbon Advisory Commission, Environmental Research, Training, Policy* and *Extension Complex,* North Carolina State University, Raleigh, NC.

Dales, J. H., 1968. *Pollution, Property, and Prices,* University of Toronto Press, Toronto, Ontario, Canada.

David, M., W. Eheart, E. Joeres, and E. David, 1980. Marketable Permits for the Control of Phosphorus Effluent into Lake Michigan, *Water Resources Research,* 16, 262–270.

Eheart, J. W., 1980. Cost Efficiency of Transferable Discharge Permits for the Control of BOD Discharges, *Water Resources Research,* 16, 980–989.

Eheart, J. W., E. D. Brill, Jr., B. J. Lence, J. D. Kilgore, and J. G. Uber, 1987. Cost Efficiency of Time-Varying Discharge Permit Programs for Water Quality Management, *Water Resources Research,* 23, 245–251.

Graedel, T. E., and B. R. Allenby, 1995. *Industrial Ecology,* Prentice Hall, Upper Saddle River, NJ.

Joeres, E. F., and M. H. David (eds.), 1983. *Buying a Better Environment: Cost-Effective Regulation Through Permit Trading,* Land Economics Monograph 6, University of Wisconsin Press, Madison, WI.

Johnson, E. L., 1967. A Study in the Economics of Water Quality Management, *Water Resources Research,* 3(2), 291–305.

Kneese, A. V., and B. T. Bower, 1968. *Managing Water Quality: Economics, Technology, Institutions,* Johns Hopkins University Press, Baltimore, MD.

Lence, B. J., J. W. Eheart, and E. D. Brill, Jr., 1988. Cost Efficiency of Transferable Discharge Permit Markets for Control of Multiple Pollutants, *Water Resources Research,* 24(7), 897–905.

Lyon, R. M., 1982. Auctions and Alternative Procedures for Allocating Pollution Rights, *Land Economics,* 58, 16–32.

O'Neil, W. B., 1983. The Regulation of Water Pollution Permit Trading Under Conditions of Varying Streamflow and Temperature, in E. F. Joeres and M. H. David (eds.), *Buying A Better Environment: Cost-Effective Regulation Through Permit*

Trading, Land Economics Monograph 6, University of Wisconsin Press, Madison, WI.

Rose-Ackerman, S., 1973. Effluent Charges: A Critique, *Canadian Journal of Economics,* 6(4), 512–528.

Thomann, R. V., 1972. *Systems Analysis and Water Quality Management,* McGraw-Hill, New York.

Tietenberg, T. H., 1974. The Design of Property Rights for Air-Pollution Control, *Public Policy,* 22, 275–292.

Tietenberg, T. H., 1980. Transferable Discharge Permits and the Control of Stationary Source Air Pollution: A Survey and Syntheses, *Land Economics,* 56(4), 391–416.

U.S. Congress, 1970. Hearings before the Subcommittee on Air and Water Pollution of the Committee of Public Works 91st Congress, 2nd Session, April 20, 21, and 27.

CHAPTER 16

Applications of Optimization Techniques to Structural Design

TIMOTHY L. JACOBS

16.A. INTRODUCTION

The concept of developing the optimal structural design for a given project has been the goal of structural engineers, architects, and master builders since the time of the ancient Egyptians and the early civilizations of Greece. Structural optimization focuses on determining the most cost-effective structure while satisfying functionality and safety requirements. This basic concept is reflected in many of the structural design codes used throughout the past 30 years.

The choice of an efficient structural design can have a significant impact on cost. Although a marginal cost difference between two beams may appear to be insignificant, the repeated use of the more efficient beam throughout a large structure can result in sizable cost savings. Formulating a structural design problem in terms of an optimization model is a natural extension of the design process itself. The design process focuses on creating the most efficient and safe structure possible. As an optimization problem, the objective is to maximize the efficiency of the structural system. In most cases this involves minimizing either the weight or the overall cost of the structural system. This objective is subject to numerous constraints defining the safe physical limits of the construction materials used and the functionality requirements of the structure. These constraints often include limits on the allowable stresses and deflections expected within the structural elements that make up the structural system. In general, the constraints represent the behavioral and safety aspects of the system, while the objective represents the goals of the designer.

In recent years much research has been devoted to the development and implementation of structural optimization techniques. In many cases the primary objective has been to minimize the total structural weight, thereby minimizing the cost of the structural material (Hager and Balling, 1988; Amir and Hasegawa, 1989). In actuality, optimal structural design is a multiobjective process involving trade-offs between design cost and design reliability (ASCE, 1972; Jacobs, 1991). Presently, the issue of reliability is treated implicitly by meeting current design code requirements within the optimization model (Hager and Balling, 1988).

Optimal structural design is a difficult combinatorial problem when considering the discrete nature of available design sections (Templeman, 1988). Several recent papers have focused on the combinatorial problems that result when considering only standard structural sections within the optimization model. Hager and Balling (1988) have proposed an approach for reducing the number of possible discrete design solutions considered by creating a feasible envelope that includes the structural sections that meet code stress requirements. A modified branch-and-bound procedure is then employed to determine the optimal combination of structural sections comprising the structure.

A sequential search interval technique has been proposed for solving nonlinear discrete structural optimization problems (Amir and Hasegawa, 1989). Although this approach works well in determining design solutions for nonlinear structural problems, global optimality cannot be guaranteed. Optimal control theory has been applied to determine the optimal cross-section characteristics of beams using a segment-wise linear model (Goh and Wang, 1988). The optimal design of frames subject to nonlinear elastic bending has been considered by Nakamura and Takewaki (1989) to develop an overall structural flexibility factor. Templeman (1988) presents a review of many of the structural optimization methods and algorithms currently available.

In this chapter we present a brief introduction to the subject of structural optimization and some of the methodologies and algorithms used to optimize the design of structural systems. To help the reader appreciate the basic concepts, motivations, and problems involved in formulating and solving structural optimization problems, we begin by considering the optimal design of a three-bar truss. Following this example, we examine many of the more commonly used structural optimization methods and illustrate the use of these methods with examples.

16.B. BASIC CONCEPTS

To begin our exploration of the field of structural optimization, consider the problem of determining the most efficient members for the statically determinate three-bar truss shown in Figure 16.1. For the first part of this example, we consider only the stresses in the truss members in the design process. From statics we can determine the maximum loads that will exist in each of the truss members (Table 16.1). To express this problem as a formal optimization problem, let the decision variable A_i represent the cross-sectional area of member i. Using an allowable stress approach to design the truss, the

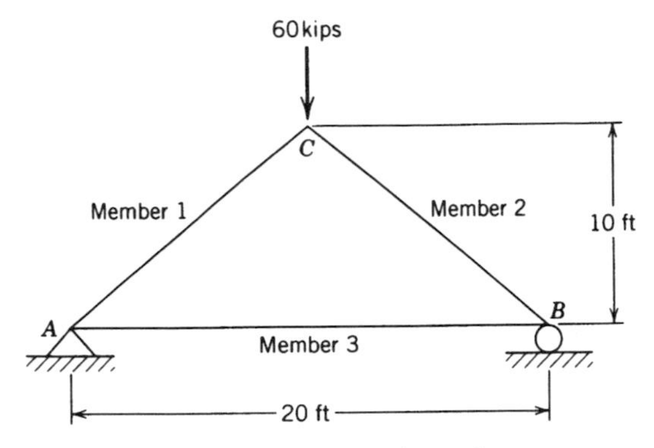

Figure 16.1 Statistically determinate truss.

governing design constraint stipulates that the stress in any member of the truss cannot exceed a specified maximum. Mathematically, this is expressed as

$$\sigma_a \geq \frac{F_i}{A_i} \tag{16.1}$$

where σ_a and F_i represent the allowable stress and the maximum expected force in member i, respectively. For this example assume that the allowable stress equals 0.6 of the yield stress of the construction material in tension or compression. For steel with a yield stress of 36 ksi, the allowable stress would equal approximately 21.6 ksi. By using an allowable stress approach, we assume that the structural behavior will always remain linear and elastic. In addition, we are incorporating a margin of safety into the model. Therefore, the constraint set for this formulation will be linear. Rewriting equation (16.1) as a constraint in standard form yields

$$\sigma_a A_i \geq F_i \qquad \forall i \tag{16.2}$$

One common objective for this type of optimization problem is minimization of the total weight of the structure. The total weight of a truss consisting of n members is defined by

TABLE 16.1 Member Forces for Three-Bar Truss

Truss Member	Member Force (kips)
1	42.43 (compression)
2	42.43 (compression)
3	30.00 (tension)

$$W = \sum_{i=1}^{n} \gamma l_i A_i \tag{16.3}$$

where γ is the specific weight of the construction material and l_i equals the length of each truss member. The complete model formulation for this example is now presented as

$$\text{minimize } W = \sum_{i=1}^{n} \gamma l_i A_i \tag{16.4}$$

subject to

$$\sigma_a A_i \geq F_i \qquad \forall i \in 1,2,\ldots,n \tag{16.5}$$

$$A_i \geq 0 \qquad \forall i \in 1,2,\ldots,n \tag{16.6}$$

Equation (16.6) stipulates that all the decision variables must be positive.

The formulation above is linear and can be solved using any commercially available linear programming package. Table 16.2 presents the optimal design for the three-bar truss assuming that the allowable stress equals 21.6 ksi ($0.6\sigma_y$), the specific weight of the construction material is assumed to equal 488 (lb/ft^3), and using the member lengths shown in Figure 16.1 and the member forces presented in Table 16.1. Recognizing that the truss is symmetrically loaded, members 1 and 2 will be the same at optimality. Therefore, this problem can also be solved graphically as shown in Figure 16.2.

Although this problem is extremely simple, it provides a useful starting point for our study of structural optimization. By considering stress as the only design criteria, the resulting constraint set consists of only lower bounds on the cross-sectional area of each truss member. Therefore, the global optimal solution to this problem can be found by individually solving each constraint algebraically as an equality. This is due to the fact that the truss is statically determinate and no serviceability considerations, such as limiting the truss deflection, are included. Therefore, the design variables are independent of one another. However, this example is unrealistic in two ways.

TABLE 16.2 Optimal Solution for Three-Bar Truss $W^* = 282$ lb

Truss Member	Optimal Cross-Sectional Area (in^2)
1	1.96
2	1.96
3	1.39

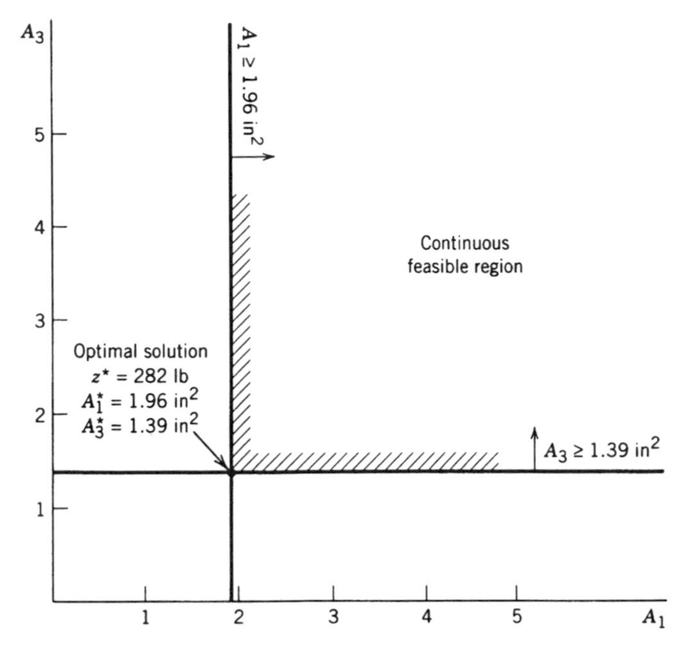

Figure 16.2 Graphical solution for three-bar truss.

First, it is unlikely that stress would be the only design criteria considered. In reality, the engineer will probably consider deflections in the structure during the design process. Second, it is unlikely that the engineer will have ready access to custom-fabricated structural members. In most cases the engineer will select the truss members from a list of standard sections available. For the optimization model, this means that the feasible region will consist of discrete rather than continuous solutions. Otherwise, the fabrication costs associated with using nonstandard sections could increase the total cost of the structure dramatically.

To illustrate the impact of these additional design criteria, again consider the three-bar truss example. First consider the addition of a constraint limiting the vertical deflection of joint C. Although several methods are available for determining the deflection of joints within a truss, for this example we will use virtual work (Chen, 1995). Using virtual work, the deflection at joint C, δ_c, is determined by

$$\delta_c = \sum_{i=1}^{n} \frac{P_i p_i l_i}{A_i E_i} \tag{16.7}$$

where P_i represents the actual force in member i, p_i the virtual load in member i due to a unit load at joint C, and E_i represents Young's modulus of elasticity for the construction material. As a constraint, equation (16.7) becomes

TABLE 16.3 Data for Displacement Constraint

Member	P_i (kips)	p_i	l_i (ft)	E_i (ksi)
1	−42.4	−0.707	14.1	30,000.0
2	30.0	0.5	20	30,000.0
3	−42.4	−0.707	14.1	30,000.0

$$\sum_{i=1}^{n} \frac{P_i p_i l_i}{A_i E_i} \leq \delta_c \qquad \forall c \tag{16.8}$$

Table 16.3 presents the actual and virtual loads for determining the vertical deflection of joint C for the three-bar truss. Substituting the information in Table 16.3 into equation (16.8) yields the following constraint limiting the total vertical deflection of joint C:

$$\frac{0.336}{A_1} + \frac{0.12}{A_3} \leq \delta_c \tag{16.9}$$

Recall that the cross-sectional areas of members 1 and 2 are equal. Figure

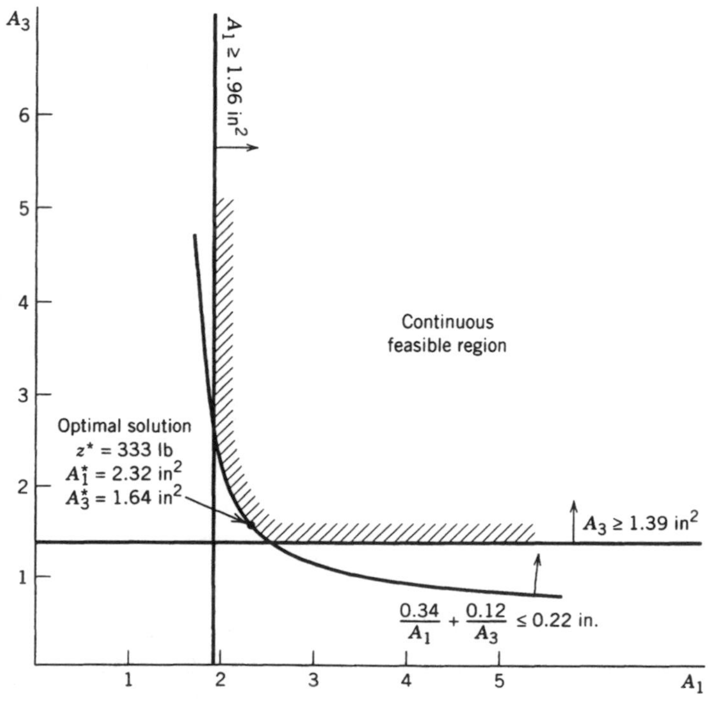

Figure 16.3 Three-bar truss with deflection constraint.

16.3 presents a graphical solution to the three-bar truss example assuming that the maximum allowable deflection downward for joint C equals 0.22 in. From Figure 16.3, we see that the addition of a constraint limiting the deflection makes the optimization problem nonlinear. For this example, the optimal solution lies along the nonlinear deflection constraint. In addition, we see that we can no longer determine the optimal solution simply by solving the individual constraint equations as equalities. This is due to the fact that the deflection at any point in the structure is a function of the geometric and material characteristics of all members in the truss, resulting in a constraint that is a function of both decision variables.

We can further complicate the problem by stipulating that the truss members be selected from a list of available sections. To accomplish this, the optimization model must be formulated as a binary model in which the decision variables take on values of 0 or 1 and specify the selection of a specific section for each member in the truss. Let X_{ij} represent the binary decision variable such that

$$X_{ij} = \begin{cases} 1 & \text{if section } j \text{ is selected for member } i \\ 0 & \text{otherwise} \end{cases} \tag{16.10}$$

Using this definition, the selection of each truss member is defined by

$$A_i = \sum_{j=1}^{m} A_{ij} X_{ij} \qquad \forall i \in 1,2,\ldots,n \tag{16.11}$$

where A_{ij} represents the cross-sectional area of standard section j available for use as member i. In general form the complete model formulation is presented as

$$\text{minimize } W = \sum_{i=1}^{n} \gamma l_i A_i \tag{16.12}$$

subject to

$$\sigma_a A_i \geq F_i \qquad \forall i \in 1,2,\ldots,n \tag{16.13}$$

$$\sum_{i=1}^{n} \frac{P_i p_i l_i}{A_i E_i} \leq \delta_k \qquad \forall k \in \text{specific points} \tag{16.14}$$

$$A_i - \sum_{j=1}^{m} A_{ij} X_{ij} = 0 \qquad \forall i \in 1,2,\ldots,n \tag{16.15}$$

$$A_i \geq 0 \qquad \forall i \in 1,2,\ldots,n \tag{16.16}$$

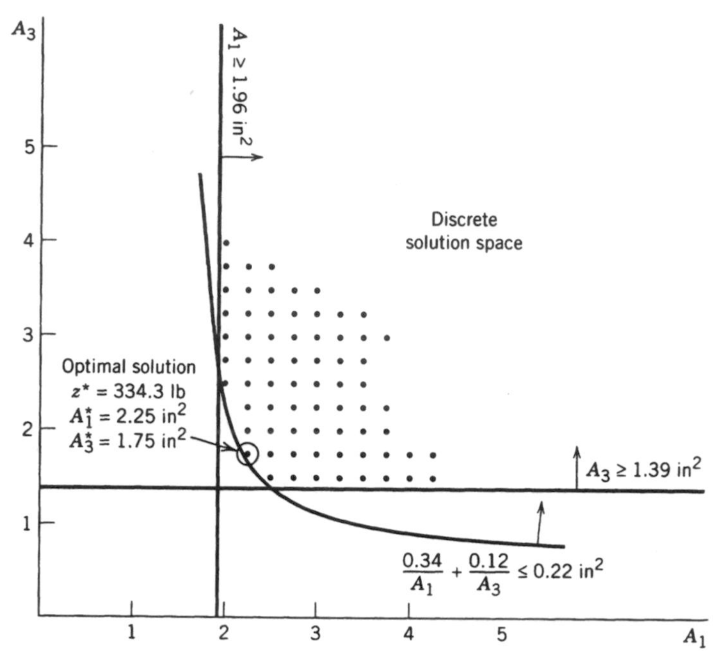

Figure 16.4 Three-bar truss solution with discrete solution space.

$$X_{ij} = \begin{cases} 1 & \text{if section } j \text{ is selected for member } i \\ 0 & \text{otherwise} \end{cases} \qquad \forall(i,j) \qquad (16.17)$$

This problem is now nonlinear and discrete. For the three-bar truss, assume that the material supplier can furnish truss members with cross-sectional areas between 0.5 and 2.25 in^2 in increments of 0.25 in^2. Figure 16.4 presents a graphical solution of this problem. Notice how the feasible solution space is no longer continuous, but consists of discrete solutions.

The final formulation of this problem can also be solved using a nonlinear optimization code and a branch-and-bound algorithm to determine the optimal selection of the discrete truss members. Table 16.4 compares all three solutions as determined using a commercially available optimization software

TABLE 16.4 Comparison of Three-Bar Truss Solutions

Model Parameter	Original Formulation	With Deflection Constraint	With Discrete Solution Space
A_1 (in^2)	1.96	2.32	2.25
A_2 (in^2)	1.96	2.32	2.25
A_3 (in^2)	1.39	1.64	1.75
W^* (lb)	281.92	333.11	334.27

package. Notice how the optimal value of the objective function increases as the formulation becomes more constrained.

Although this example is very simple, it illustrates many of the mathematical characteristics common to structural optimization problems. First we see that structural optimization problems that are statically determinate and consider elastic stress as the only design criterion are relatively easy to formulate and solve algebraically or as simple linear programs. However, if we also consider behavioral constraints such as maximum allowable deflections, the optimization model becomes nonlinear. In addition, serviceability requirements, such as deflection constraints, are typically functions of more than one of the variables (i.e., the cross-sectional areas of the truss members). Finally, if we stipulate that the structural members must be selected from a discrete group of available members, the problem must be formulated as a binary or discrete optimization model. In general, structural optimization problems have three types of constraints: (1) design constraints, (2) behavioral constraints, and (3) feasibility constraints.

Design constraints typically relate the structural loads to the structure's capacity to carry the specified loads. The structure's capacity to carry the specified loads are often written in terms of design parameters such as a member's cross-sectional area or moment of inertia. In the example presented earlier, the design constraints related the loads in truss members due to the external loading and the cross-sectional area of each member using an allowable stress. Several other design criteria are available and could have been used to develop the design constraints for this example. For statically determinate structures, these constraints are usually linear. However, for indeterminate structural systems the design constraints are often nonlinear.

The behavioral constraints relate the structural behavioral requirements to the design variables. These constraints are typically nonlinear even for linear and elastic and statically determinate systems. This characteristic is illustrated by the deflection limit in the example. Other examples of behavioral constraints include thermal expansion considerations, limits on the dynamic behavior of the system, and limits on the amount of displacement of one or more of the system's supports.

Feasibility constraints define the feasible region for the problem. The feasibility constraints relate the design variables to structural sections that are actually available. Most books on structural optimization do not consider this type of constraint. However, this constraint is very important and can have a profound impact on the computational effort required to determine the optimal solution. For example, eight discrete sections were available for each member of the three-bar truss in the example presented above. Using binary variables to define the selection of each member results in a total of 24 decision variables. This translates to 2^{24} or 16,777,216 possible solutions to the three-bar truss problem. However, we were able to linearize the model formulation and solve the problem using a technique known as branch and bound. This technique uses a "divide and conquer" scheme to efficiently prune the tree of possible solutions and usually results in a substantial savings in the computational effort required to determine the global optimal solution. For more information on the branch-and-bound method, see ReVelle et al. (1997). In addition, we assumed that members 1 and 2 were always the same, thereby further reducing the number of possible solutions to 2^{16} (or 65,536).

Some model formulations do not require a feasibility constraint due to the fact that the design variables can be continuous. Reinforced concrete beams and columns are examples of structural members that can be custom fabricated. However, structural engineers often use discrete dimensions to minimize potential errors due to field conditions.

16.C. OPTIMAL DESIGN OF INDETERMINATE STRUCTURES

Many structural systems encountered in practice, such as frames and multiple-span beams, are indeterminate and cannot be analyzed using statics. For these types of structures, the analysis must incorporate the deflection of the structure in determining the member forces. For indeterminate structures, we find that many of the design constraints are often nonlinear. To illustrate this, consider an example of the indeterminate three-bar truss shown in Figure 16.5. A similar problem was presented by Kirsch (1981) and Ossenbruggen (1984). For this example, assume that the allowable stresses in the members are 30 ksi in tension and 18 ksi in compression. For illustrative purposes we assume that the number of design variables is reduced to two by making members 1 and 3 the same size in the final design. This assumption will allow us to solve the problem graphically. In an actual design, the engineer might not include this restriction and would therefore solve the problem using three design variables. In addition, only one loading condition is considered. To formulate the mathematical constraints for this problem, we must determine

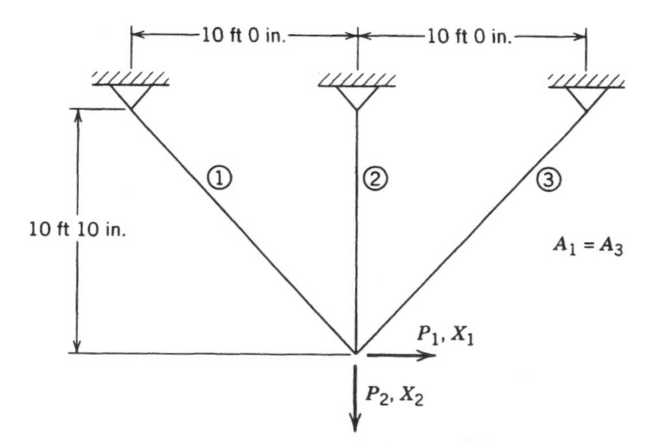

Figure 16.5 Three-bar indeterminant truss.

the stress in each member as a function of the design variables. As before, the design variables for this example represent the cross-sectional areas of truss members 1 and 2. Using the stiffness method to analyze the forces in each truss member (see Meyers, 1983) yields the following relationship:

$$\begin{Bmatrix} P_1 \\ P_2 \end{Bmatrix} = \frac{E}{\sqrt{2}\,L} \begin{bmatrix} A_1 & 0 \\ 0 & A_1 + \sqrt{2}\,A_2 \end{bmatrix} \begin{Bmatrix} \delta_1 \\ \delta_2 \end{Bmatrix} \tag{16.18}$$

where A_1 and A_2 represent the cross-sectional areas of members 1 and 2, respectively, E represents Young's modulus of elasticity, and L is the length of the members. Solving for the displacements for the case where P_1 and P_2 equal 20 kips yields

$$\begin{Bmatrix} \sigma_1 \\ \delta_2 \end{Bmatrix} = \frac{200\sqrt{2}}{E} \begin{bmatrix} \dfrac{1}{A_1} \\ \dfrac{1}{A_1 + \sqrt{2}\,A_2} \end{bmatrix} \tag{16.19}$$

Backsubstituting into the local stiffness relationships yields the following member axial forces:

$$\begin{Bmatrix} p_1 \\ p_2 \\ p_3 \end{Bmatrix} = \begin{bmatrix} 14.14 + \dfrac{14.14A_1}{A_1 + \sqrt{2}\,A_2} \\ \dfrac{20\sqrt{2}\,A_2}{A_1 + \sqrt{2}\,A_2} \\ 14.14 - \dfrac{14.14A_1}{A_1 + \sqrt{2}\,A_2} \end{bmatrix} \tag{16.20}$$

and the following member stresses as functions of the cross-sectional areas of each member in the system:

$$\begin{Bmatrix} \sigma_1 \\ \sigma_2 \\ \sigma_3 \end{Bmatrix} = \begin{bmatrix} \dfrac{14.14}{A_1} + \dfrac{14.14}{A_1 + \sqrt{2}\,A_2} \\ \dfrac{20\sqrt{2}}{A_1 + \sqrt{2}\,A_2} \\ \dfrac{14.14}{A_1} - \dfrac{14.14}{A_1 + \sqrt{2}\,A_2} \end{bmatrix} \tag{16.21}$$

Using an allowable stress design criteria and ignoring deflection considerations yields the following continuous nonlinear optimization model that minimizes the total volume of material used:

$$\text{minimize } Z = 339.4A_1 + 120A_2 \tag{16.22}$$

subject to

$$\frac{14.14}{A_1} + \frac{14.14}{A_1 + \sqrt{2}\,A_2} \leq 30$$

$$\frac{20\sqrt{2}}{A_1 + \sqrt{2}\,A_2} \leq 30 \tag{16.23}$$

$$\frac{14.14}{A_1} - \frac{14.14}{A + \sqrt{2}\,A_2} \leq 18$$

$$A_1, A_2 \geq 0 \tag{16.24}$$

The objective function for this formulation is in terms of volume (in^3). Recall that the allowable stress for tension members differs from that used for compression members. Figure 16.6 presents a graphical solution to this nonlinear optimization model. From this example one can easily imagine the cumbersome nature of more complicated indeterminate structures. Most nonlinear

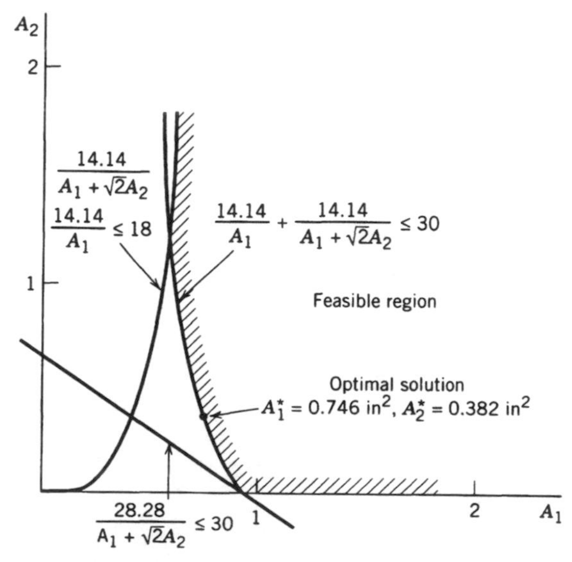

Figure 16.6 Graphical solution for indeterminant truss.

optimization codes are not capable of solving general nonlinear problems. Therefore, many times the analyst must use a heuristic or iterative algorithm to solve the optimization model. However, this example is simple enough to be solved using a commercially available nonlinear solver. The optimal solution to this example using AMPL and MINOS (Fourer et al., 1993) is $A^* = \langle 0.746, 0.382 \rangle$ with an objective value of $Z^* = 299$ in^3.

Another interesting example formulation for an indeterminate structural system is the optimal design of the supporting structure for the two suspended walkways shown in Figure 16.7. A problem similar to this example was presented by Jewell (1986). For this problem, assume that each floor beam is to carry a section of the walkway measuring 50 ft long by 10 ft wide. The design loading for the walkways is 150 lb/ft^2 and includes both the dead and live loads. The deflection of the walkways due to elongation of the supports must be limited. The upper walkway must not deflect more than 0.12 in. and the lower walkway deflection must not exceed 0.22 in. The allowable stress in each of the supporting members is 24 ksi and the modulus of elasticity for the steel supports is 29,000 ksi. For this example we must determine the optimal size of the steel hangers that support the walkways.

To solve this problem, we must first formulate the optimization problem in general terms. To begin, the stress in each of the hangers must not exceed 24 ksi. Mathematically, this constraint is written as

$$\sigma_i \leq 24 \qquad \forall i \tag{16.25}$$

where i represents each of the members supporting the walkways. The be-

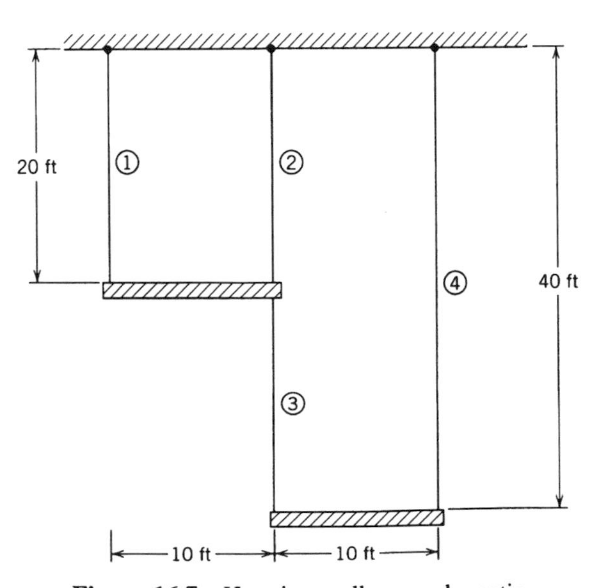

Figure 16.7 Hanging walkway schematic.

havioral constraints required limit the amount of deflection in each walkway. Mathematically, these are written as

$$d_1, d_2 \leq 0.12 \tag{16.26}$$

$$d_4 \leq 0.22 \tag{16.27}$$

$$d_1 - d_2 = 0 \tag{16.28}$$

$$d_2 + d_3 - d_4 = 0 \tag{16.29}$$

where d_i represents the deflection of member i due to elongation. To formulate a complete model, each of the constraints must be formulated in terms of design variables. For this problem the design variables are the cross-sectional areas of the walkway hangers. The tension in each member, T_i, is related to the elongation of each member by the relationship

$$T_i = \frac{E_i A_i}{L_i} d_i \tag{16.30}$$

where E_i, A_i, and L_i represent the modulus of elasticity, cross-sectional area, and length of the support members, respectively. From this relationship the stress in each support is defined by

$$\sigma_i = \frac{E_i}{L_i} d_i \tag{16.31}$$

In matrix form, the tension forces and stresses in each member are defined by

$$\begin{bmatrix} T_1 \\ T_2 \\ T_3 \\ T_4 \end{bmatrix} = E \begin{bmatrix} \frac{A_1}{L_1} & 0 & 0 & 0 \\ 0 & \frac{A_2}{L_2} & 0 & 0 \\ 0 & 0 & \frac{A_3}{L_3} & 0 \\ 0 & 0 & 0 & \frac{A_4}{L_4} \end{bmatrix} \begin{bmatrix} d_1 \\ d_2 \\ d_3 \\ d_4 \end{bmatrix} \tag{16.32}$$

and

$$\begin{bmatrix} \sigma_1 \\ \sigma_2 \\ \sigma_3 \\ \sigma_4 \end{bmatrix} = E \begin{bmatrix} \frac{1}{L_1} & 0 & 0 & 0 \\ 0 & \frac{1}{L_2} & 0 & 0 \\ 0 & 0 & \frac{1}{L_3} & 0 \\ 0 & 0 & 0 & \frac{1}{L_4} \end{bmatrix} \begin{bmatrix} d_1 \\ d_2 \\ d_3 \\ d_4 \end{bmatrix} \tag{16.33}$$

For equilibrium of the structure, the following relationships must hold:

$$75 = T_1 + T_2 - T_3 \tag{16.34}$$

$$75 = T_3 + T_4 \tag{16.35}$$

In terms of the member displacements, the equilibrium equations yields

$$\begin{bmatrix} 75 \\ 75 \end{bmatrix} = \frac{E}{20} \begin{bmatrix} A_1 & A_2 & -A_3 & 0 \\ 0 & 0 & A_3 & A_4/2 \end{bmatrix} \begin{bmatrix} d_1 \\ d_2 \\ d_3 \\ d_4 \end{bmatrix} \tag{16.36}$$

More conveniently, the equilibrium relationship can be expressed in terms of the displacements of the individual floor beams. This yields

$$\begin{bmatrix} 75 \\ 75 \end{bmatrix} = \frac{E}{20} \begin{bmatrix} A_1 & A_2 & -A_3 & 0 \\ 0 & 0 & A_3 & A_4/2 \end{bmatrix} \begin{bmatrix} 1 & 0 \\ 1 & 0 \\ -1 & 1 \\ 0 & 1 \end{bmatrix} \begin{bmatrix} d_a \\ d_b \end{bmatrix} \tag{16.37}$$

or

$$\begin{bmatrix} 75 \\ 75 \end{bmatrix} = \frac{E}{20} \begin{bmatrix} A_1 + A_2 + A_3 & -A_3 \\ -A_3 & A_3 + A_4/2 \end{bmatrix} \begin{bmatrix} d_a \\ d_b \end{bmatrix} \tag{16.38}$$

where d_a and d_b represent the displacements of the two walkways. Solving for the displacement allows us to write the behavioral constraints that limit the displacement of the supports. Mathematically, this set of constraints is expressed as

$$\frac{20(12)}{E}\begin{bmatrix} A_1 + A_2 + A_3 & -A_3 \\ -A_3 & A_3 + A_4/2 \end{bmatrix}^{-1}\begin{bmatrix} 75 \\ 75 \end{bmatrix} \leq \begin{bmatrix} 0.12 \\ 0.22 \end{bmatrix} \tag{16.39}$$

The next set of constraints for this problem define the capacity of each of the support members in terms of the allowable stress. Recalling that the force in each support member is the product of the stress in that member and the cross-sectional area of the member, the stress in each member is defined by

$$\begin{bmatrix} 75 \\ 75 \end{bmatrix} = \begin{bmatrix} A_1 & A_2 & -A_3 & 0 \\ 0 & 0 & A_3 & A_4/2 \end{bmatrix}\begin{bmatrix} \sigma_1 \\ \sigma_2 \\ \sigma_3 \\ \sigma_4 \end{bmatrix} \tag{16.40}$$

and

$$\begin{bmatrix} \sigma_1 \\ \sigma_2 \\ \sigma_3 \\ \sigma_4 \end{bmatrix} \leq 24 \tag{16.41}$$

The final constraint for this problem restricts the cross-sectional area of each member to be a positive number. Mathematically, this is expressed as

$$A_1, A_2, A_3, A_4 \geq 0 \tag{16.42}$$

One objective of this problem is to minimize the volume of steel used in the supporting members. This can be accomplished by minimizing the sum of the cross-sectional areas of the supports. Mathematically, this is represented by

$$\text{minimize } Z = \sum_{i=1}^{4} L_i A_i \tag{16.43}$$

Other objective functions might include fabrication costs and be nonlinear. The solution to this problem can be found using a number of iterative search procedures. Assuming an initial feasible solution, the actual deflection is calculated by solving equation (16.38) and then comparing to the allowable deflection. The stresses in the hangers are then determined using equation (16.33) and compared to the allowable stress. Once the feasibility of the solution is confirmed, the cross-sectional areas of the hangers are reduced using an iterative relationship and the process is repeated. The algorithm is repeated until the change in the design variables is small for any two successive iterations or the solution becomes infeasible. Although this strategy is straightforward and easy to implement, it does not guarantee that the global optimal solution will be found.

Now that we have seen some of the basic ideas and requirements used to formulate structural problems, let's consider some of the more popular optimization approaches.

16.D. OPTIMIZATION APPROACHES

Structural optimization techniques or approaches typically fall into one of three categories: (1) traditional approaches, (2) optimality criteria approaches, and (3) mathematical approaches. In the following sections each of these approaches is introduced and illustrated by example.

16.D.1. Traditional Approaches

Traditional approaches to structural optimization usually attempt to minimize the total structural weight based on a series of mathematical constraints that reflect behavior of the structure. Using this approach, the optimal solution is determined by preselecting the constraints that are assumed to control the design. Often, the design engineer will resort to a simple iterative or trial-and-error approach to determine the most efficient design. In reality, this process is not a formal optimization approach. However, it is the most common approach to structural design used in practice today and it typically results in reasonably efficient designs. In many cases this type of strategy can be automated and incorporated into commercially available software packages.

To illustrate this methodology, consider the task of designing the simply supported steel beam shown in Figure 16.8. This beam is subject to both a uniform load throughout its span and a concentrated load at its center. Assume that the beam is supported laterally throughout its length. Therefore, we can ignore the possibility of lateral buckling in the design process. Since it is

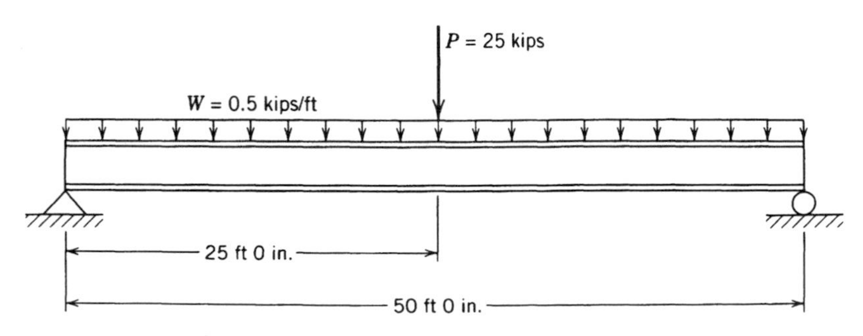

Figure 16.8 Simply supported steel beam.

simply supported, the structure can be analyzed using simple statics. The design objective is to determine the most efficient or lightest-weight standard W section (I-beam) that will carry the loads. Assume that the yield stress of the steel is 36 ksi and the allowable stress is 66% of the yield stress. The maximum allowable deflection at the center of the beam is 1.8 in. down. The design process is as follows:

Step 1. Determine the maximum moments experienced by the structure (neglecting the weight of the beam)

$$M_{\text{max}} = \frac{wl^2}{8} + \frac{Pl}{4} = 468.8 \text{ kip-ft} \tag{16.44}$$

Step 2. Estimate the size of the beam required, using the section modulus:

$$f_a = 0.66 f_y = 0.66(36^{\text{ksi}}) \simeq 23.76^{\text{ksi}} \tag{16.45}$$

$$f_a \geq \frac{M_{\text{max}}}{S_{\text{req}}} \tag{16.46}$$

Rearranging equation (16.46) yields

$$S \geq \frac{M_{\text{max}}}{f_a} = \frac{468.8^{\text{kip-ft}}(12 \text{ in./ft})}{23.76^{\text{ksi}}} = 237 \text{ in}^3 \tag{16.47}$$

Referring to the standard section tables published by AISC, we select the lightest W section that has a section modulus, S, of more than 237 in^3. In this case we will try a W27 × 94. This represents a wide-flange steel section that is approximately 27 in. deep and weighs 94 lb per linear foot. The properties of this standard section are:

$$I = 3270 \text{ in}^4$$

$$S = 243 \text{ in}^3$$

Step 3. Check the structural deflections at the midpoint (where it is a maximum). In this step we include the weight of the beam. Using superposition, we know the maximum deflection occurs at the center of the beam and is given by

$$\delta = \frac{Pl^3}{48EI} + \frac{5wl^4}{384EI} = 1.99 \text{ in.} \tag{16.48}$$

where E, Young's modulus of elasticity, is approximately 30,000 ksi for A36 steel. For this problem we see that the deflection is larger than the allowable deflection. Therefore, we must return to step 2 and select a larger steel section.

This time let's try a W30 × 99, which is slightly deeper and heavier and has a moment of inertia of 3970 in^4 and a section modulus of 269 in^3. The deflection, including the weight of the beam, equals 1.65 in., which satisfies the constraint stipulated in the problem description. Therefore, we use a W30 × 99 for this design.

Although this example is very simple, it illustrates the trial-and-error approach traditionally used to determine the most economically efficient design. The same approach can be used for concrete, timber, or composite structural systems. Design engineers with a great deal of experience are often able to select nearly optimal size sections or design dimensions on the first iteration. This allows them to converge to a good design rather quickly.

For large and complicated structural systems, the iterative procedure is likely to include a reanalysis of the structural system during each iteration. From the example presented here, we see that there is no reanalysis required for a statically determinate structure.

16.D.2. Optimality Criteria Approaches

Optimality criterion approaches assume that some criterion related to the behavior of the structural system is satisfied at optimality. For example, the structural engineer might feel that the deflection of the structural system at a specific location will control the overall design. In this case the optimality criterion would assume that the deflection at that location would equal the maximum allowable deflection at optimality. Two other examples include fully stressed design in which each member is subjected to its limiting stress at optimality and buckling stability. Optimality criteria methods utilize an iterative algorithm to converge to a superior and more efficient design. As a result, these methods are commonly referred to as *indirect methods.*

Optimality criteria methods rely on the behavioral characteristics of the structure and approximate the optimal values of the design variables in any successive iteration by modifying the previous approximation. These methods usually result in specialized but computationally efficient algorithms. The resulting iterative procedures are simple to implement but provide no guarantee that they will converge to the optimal solution. The key to these techniques lies in their use of a recurrence relationship that dictates the values of the design variables for each iteration. Although several optimality criteria have been developed and implemented, only two of the most common are discussed here. These are the stress ratio (or zero-order) and first-order approximation procedures.

The *stress-ratio procedure* approximates the optimal value of each successive design variable, X_i^{n+1}, using the redesign rule

$$X_i^{n+1} = X_i^n \frac{\sigma_i^n}{\sigma_i^U} \tag{16.49}$$

where X_i^n is the value of design variable i in the preceding iteration, σ_i^n the stress in member i in iteration n, and σ_i^U the maximum allowable stress for member i. For this iterative procedure, the structural members are resized (increased or decreased) proportional to the ratio of the computed stresses and the maximum allowable stresses in each member. The iterative process is started with an initial guess of the optimal values of the design variables and continues until the solution converges. Although this procedure is simple and easily implemented, there is no guarantee that the algorithm will determine the global optimal solution or that it will converge at all.

To illustrate this method, consider the mathematical formulation for the indeterminate three-bar truss presented previously [equations (16.22)–(16.24)]. For this formulation the design variables are the cross-sectional areas of the truss members. Recall that the left-hand side of equation (18.23) defines the stresses in each of the members for the three-bar indeterminate truss presented earlier. The recurrence or redesign rules used for this example are defined by

$$A_1^{n+1} = A_1^n \max\left(\frac{\sigma_1^n}{\sigma_1^U}, \frac{\sigma_3^n}{\sigma_3^U}\right) \tag{16.50}$$

$$A_2^{n+1} = A_2^n \frac{\sigma_2^n}{\sigma_2^U} \tag{16.51}$$

For this example, an initial cross-sectional area of 2 in^2 was assumed for each member. Table 16.5 presents the values for each iteration in the procedure. The results show that the algorithm converges slowly to yield a structure with only two members (members 1 and 3). The solution shown in Table 16.5 represents the fully stressed design. However, the final design is suboptimal with an objective function value, Z^*, of 319.2 in^3 as compared to the global optimal solution of 299 in^3 presented earlier. The underlying assumption of this method is that the members will be fully stressed at optimality. If this assumed optimality criterion is not correct, the method will converge to a suboptimal solution.

The term *zero-order approximation* stems from the fact that the exact constraint surface is approximated by a plane normal to the *i*th axis in the feasible region. Several modifications of the basic stress ratio procedure can be applied to improve the rate of convergence. One such strategy adds an exponent or overrelaxation term to the ratio term. In this case the redesign rule becomes

TABLE 16.5 Results of Iterative Algorithm for Indeterminate Three-Bar Truss

n	A_1	A_2	σ_1	σ_2	σ_3
0	2.0	2.0	10.0	5.86	−4.14
1	0.667	0.391	32.82	23.20	−9.61
2	0.729	0.302	31.62	24.46	−7.16
3	0.769	0.246	31.06	25.32	−5.74
4	0.796	0.208	30.75	25.95	−4.79
5	0.816	0.180	30.56	26.43	−4.12
6	0.831	0.158	30.43	26.81	−3.62
7	0.843	0.142	30.34	27.12	−3.22
8	0.852	0.128	30.28	27.37	−2.91
9	0.860	0.117	30.23	27.58	−2.65
10	0.867	0.107	30.20	27.76	−2.43
⋮					
500	0.941	0.003	30.00	—	−0.06

$$X_i^{n+1} = X_i^n \left(\frac{\sigma_i^n}{\sigma_i^U} \right)^{\nu} \tag{16.52}$$

where ν represents the overrelaxation factor, which is greater than 1. Another iterative relationship, developed by Melosh (1970), is based on a quadratic fit extrapolation. Once the iterative procedure has stabilized over three successive approximations, a quadratic relationship is fit through the values X_i^n, X_i^{n+1}, and X_i^{n+2}. The new redesign rule is then given by

$$X_i^{n+3} = X_i^{n+1} + 2 \left\{ \frac{(X_i^n \quad X_i^{n+1})(X_i^{n+2} - X_i^{n+1})}{X_i^n - 2X_i^{n+1} + X_i^{n+2}} \right\} \tag{16.53}$$

Another optimality criteria approach uses a Taylor series expansion to approximate the final stress in each member in the structure. To begin, assume that the member stresses at iteration $n + 1$ are approximated by

$$\sigma^{n+1} \approx \sigma^n + \nabla\sigma_X^n \{X^{n+1} - X^n\} \tag{16.54}$$

where $\nabla\sigma_x^n$ represents the changes in member stresses with respect to changes in the design variable. Ultimately, the stresses in the members should converge to the allowable stresses. Mathematically, this is represented by

$$\sigma_a \approx \sigma^n + \nabla\sigma_X^n \{X^{n+1} - X^n\} \tag{16.55}$$

where σ_a is the allowable stress for the design material. For an individual structural member, the allowable stress is approximated by

$$\sigma_{a_i} \approx \sigma_i^n + \frac{\partial \sigma_i^n}{\partial X_i} \{X^{n+1} - X^n\} \tag{16.56}$$

Solving for X_i^{n+1} yields the following redesign rule:

$$X_i^{n+1} = X_i^n + \frac{\sigma_a - \sigma_i^n}{(\partial \sigma_i / \partial X_i)^n} \tag{16.57}$$

This redesign rule can be further simplified by evaluating the change in stress with respect to a change in the design variable of the member. To accomplish this, assume that we are dealing with a truss and let the stress in any structural member, σ_i, be defined by

$$\sigma_i = \frac{P_i}{X_i} \tag{16.58}$$

where P_i represents the axial force in member i and X_i is the cross-sectional area of the member. Taking the derivative of equation 16.58 with respect to X_i yields the following relationship:

$$\frac{\partial \sigma_i}{\partial X_i} = -\frac{P_i}{X_i^2} + \frac{1}{X_i} \frac{\partial P_i}{\partial X_i} \tag{16.59}$$

where $\partial P_i / \partial X_i$ represents the redistribution of the member forces due to a change in the design variable or, in this case, the cross-sectional area of the truss member. For statically determinate structures, the $\partial P_i / \partial X_i$ equals zero, thereby eliminating the second term from equation (16.59). Using this assumption, we can derive the redesign rule for determining the fully stressed design solution of a statically determinate truss:

$$X_i^{n+1} = X_i^n + \frac{\sigma_i^n \quad \sigma_a}{P_i} (X_i^n)^2 \tag{16.60}$$

To illustrate this method, consider again the three-bar statically determinate truss presented at the beginning of the chapter. For this example we assume that the truss members will be fully stressed at optimality and that no consideration is given to deflection. In addition, the cross-sectional areas for each member are assumed to be continuous. The redesign rules for each truss member are

$$A_1^{n+1} = A_1^n + \frac{\sigma_1^n - 21.6 \text{ ksi}}{42.43 \text{ kips}} (A_1^n)^2 \tag{16.61}$$

and

$$A_3^{n+1} = A_3^n + \frac{\sigma_3^n - 21.6 \text{ ksi}}{30 \text{ kips}} (A_3^n)^2 \tag{16.62}$$

where A_1^n and A_3^n are the cross-sectional areas of the truss members for each iteration n. Table 16.6 presents the results of applying equations (16.61) and (16.62) iteratively using A_1^0 and A_3^0 equal to 2.5 in^2 and 2.0 in^2, respectively. Remember, due to the symmetry of the truss, members 1 and 2 will have the same cross-sectional area. Although this methodology seems to work very well, it is often prone to problems. One problem is that the method tends to diverge if the initial estimates of the design variables are not reasonably close to the optimal solution. To illustrate this point, again consider the previous example in which the original estimates of A_1^0 and A_2^0 both equal 4.0 in^2. In this case the solution become unrealistic in the first iteration. In addition, this method becomes much more complicated and tedious for indeterminate structural systems. In the case of an indeterminate structural system, we must consider the redistribution of the loading due to changes in the member sizes. To evaluate more complex structures, the approximation of the allowable stress is commonly assessed in matrix form. This leads to other problems with this method:

1. This method assumes that the stress constraints are active at optimality.
2. The matrix of partial derivatives must be evaluated for each iteration.
3. This method requires the solution of simultaneous equations for each iteration.

The two methods presented here are limited by the fact that they consider only stress in determining the optimal design. At no time did we consider any behavioral constraints or the value of the objective function. In fact, these methods completely ignore the objective function throughout the solution process. This omission can lead to solutions that in no way reflect the true optimal solution, as was seen for the stress ratio illustration. In addition, it is clear that these methods cannot be adapted to consider structural systems in which other conditions, such as a maximum allowable deflection, control. To

TABLE 16.6 Results of First-Order Approximation for Statically Determinate Truss

n	A_1^n (in^2)	A_3^n (in^2)	σ_1 (ksi)	σ_3 (ksi)
0	2.5	1.5	16.97	15.00
1	1.82	1.12	23.31	26.79
2	1.95	1.34	21.76	22.44
3	1.96	1.39	21.65	21.58
4	1.97	1.39	21.6	21.60

consider such adaptation, another methodology is needed. One candidate is Lagrange multipliers, discussed later in the chapter.

16.D.3. Mathematical Programming Approaches

Mathematical programming approaches to structural optimization include techniques such as linear programming, nonlinear programming, and dynamic programming. These approaches require the optimization problem to be formulated in terms of mathematical inequalities that define the feasible solution space. Mathematical programming techniques are general enough to consider a variety of objective functions and constraints.

All three versions of the first example presented at the beginning of this chapter are illustrations of model formulations that were solved as linear, nonlinear, or mixed-integer programming methods. Mathematical programming approaches are appropriate when the constraint equations can be written as relatively simple mathematical inequalities. In many cases this limits the usefulness of applying standard mathematical approaches. The walkway design formulation presented earlier is an excellent illustration of this point. In addition, the problem gets much more complicated as the size of the structural system increases.

For the indeterminate truss example, we solved for the member stresses in terms of the cross-sectional areas of each member in the truss. Therefore, the design constraints were written in terms of an allowable stress and the cross-sectional areas of the structural members. One can easily imagine the amount of computational work that would be required to formulate the design constraints as functions of the design variables in an indeterminate system consisting of 100 or 1000 structural members. For indeterminate structural systems that have more than a few members, it is often more convenient to simplify the problem by stipulating a governing criterion that will define the optimal solution. Such optimality criteria approaches were discussed in the preceding section.

16.E. LINEAR PROGRAMMING APPLICATIONS

Several realistic structural optimization problems can be posed easily and accurately as linear problems. One example of this is the optimal design of a prestressed concrete beam. Consider the optimal design of the simply supported prestressed concrete beam shown in Figure 16.9. This problem was first presented in graphical form by Magnel in 1950. For this example, the only design variables will be the prestressing force and the tendon eccentricity. Assume that the beam is prestressed by a straight cable or tendon that applies a force of F_i at an eccentricity of $-e_0$. The moment applied to the beam due to the prestressing equals $-F_i e_0$. Taking the x axis as the centroid axis, the stresses due to the prestress cable are

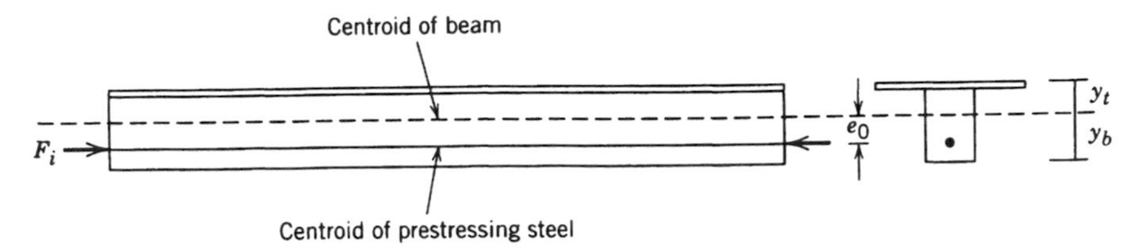

Figure 16.9 Schematic of simply supported prestressed concrete beam.

$$\left\{\frac{F_i}{A_c} - F_i e_0 \frac{y_t}{I}\right\} \tag{16.63}$$

at the top fiber of the beam and

$$\left\{\frac{F_i}{A_c} + F_i e_0 \frac{y_b}{I}\right\} \tag{16.64}$$

at the bottom fiber. A_c represents the cross-sectional area of the concrete, I the moment of inertia of the cross section, and y_t and y_b the distance from the centroid to the extreme fibers at the top and bottom of the beam, respectively.

The self-weight of the beam defines the initial stresses at the top and bottom fiber of the beam at the instant of prestress. Mathematically, these stresses equal

$$\left\{\frac{F_i}{A_c} - F_i e_0 \frac{y'}{I}\right\} + \frac{M_{\min} y_t}{I} \tag{16.65}$$

at the top fiber and

$$\left\{\frac{F_i}{A_c} + F_i e_0 \frac{y_b}{I}\right\} - \frac{M_{\min} y_b}{I} \tag{16.66}$$

at the bottom fiber, where $M_{\min}$ represents the moment due to the distributed weight of the beam.

By the time the live loads are added to the beam, the prestressing in the beam will have suffered some losses, primarily due to creep in the concrete and steel. Therefore, the remaining prestress in the beam will be a fraction of F_i. After accounting for the losses in the prestressing, the final stresses in the beam equal

$$\eta \left\{ \frac{F_i}{A_c} - F_i e_0 \frac{y_t}{I} \right\} + \frac{M_{\max} y_t}{I} \tag{16.67}$$

for the top fiber and

$$\eta \left\{ \frac{F_i}{A_c} + F_i e_0 \frac{y_b}{I} \right\} - \frac{M_{\max} y_b}{I} \tag{16.68}$$

for the bottom fiber, where η represents the remaining prestress in the beam. $M_{\max}$ represents the maximum moment in the beam due to both the dead and live load. For the beam to function safely, it cannot be overstressed at any time in its design life. Let σ_t and σ_c represent the allowable tensile stress and the allowable compressive stress for concrete, respectively. Note that for this problem we assume σ_t is a negative value and that tension in the concrete should be numerically smaller than the allowable tension stress. Using these stress limits, the design conditions that must be satisfied can be written as the four inequalities listed below.

1. Initial stress in the top fiber:

$$\left\{ \frac{F_i}{A_c} - F_i e_0 \frac{y^t}{I} \right\} + \frac{M_{\min} y_t}{I} \geq \sigma_t \tag{16.69}$$

2. Initial stress in the bottom fiber:

$$\left\{ \frac{F_i}{A_c} + F_i e_0 \frac{y_b}{I} \right\} - \frac{M_{\min} y_b}{I} \leq \sigma_c \tag{16.70}$$

3. The final stress in the top fiber:

$$\eta \left\{ \frac{F_i}{A_c} - F_i e_0 \frac{y_t}{I} \right\} + \frac{M_{\max} y_t}{I} \leq \sigma_c \tag{16.71}$$

4. The final stress in the bottom fiber:

$$\eta \left\{ \frac{F_i}{A_c} + F_i e_0 \frac{y_b}{I} \right\} - \frac{M_{\max} y_b}{I} \geq \sigma_t \tag{16.72}$$

To simplify these constraints, we introduce the following convenient relationships:

$$Z_t = \frac{I}{y_t} \tag{16.73}$$

$$Z_b = \frac{I}{y_b} \tag{16.74}$$

$$k_t = -\frac{I}{A_c y_b} \tag{16.75}$$

$$k_b = \frac{I}{A_c y_t} \tag{16.76}$$

Using these definitions, equations (16.69)–(16.72) can be rewritten as the following linear constraints:

$$e_0 - \frac{1}{F_i}(M_{\min} - \sigma_t Z_t) \le k_b \tag{16.77}$$

$$e_0 - \frac{1}{F_i}(M_{\min} + \sigma_c Z_b) \le k_t \tag{16.78}$$

$$e_0 - \frac{1}{\eta F_i}(M_{\max} - \sigma_c Z_t) \ge k_b \tag{16.79}$$

$$e_0 - \frac{1}{\eta F_i}(M_{\max} + \sigma_t Z_b) \ge k_t \tag{16.80}$$

Two additional constraints are required to define the feasible region for this problem completely. They include a limit on the eccentricity of the prestressing cable and nonnegativity. Mathematically, these constraints are written as

$$e_0 \le y_b \tag{16.81}$$

$$F_i,\ e_0 \ge 0 \tag{16.82}$$

The objective for this design problem is to minimize the amount of prestressing steel required by the beam. This is accomplished by minimizing the prestressing force, F_i, or by maximizing its reciprocal. The complete model formulation is now presented as

$$\text{maximize } S = \frac{1}{F_i} \tag{16.83}$$

subject to

$$e_0 - \frac{1}{F_i}(M_{\min} - \sigma_t Z_t) \leq k_b \tag{16.84}$$

$$e_0 - \frac{1}{F_i}(M_{\min} + \sigma_c Z_b) \leq k_t \tag{16.85}$$

$$e_0 - \frac{1}{\eta F_i}(M_{\max} - \sigma_c Z_t) \geq k_b \tag{16.86}$$

$$e_0 - \frac{1}{\eta F_i}(M_{\max} + \sigma_t Z_b) \geq k_t \tag{16.87}$$

$$e_0 \leq y_b \tag{16.88}$$

$$F_i, e_0 \geq 0 \tag{16.89}$$

Figure 16.10 presents an illustrative graphical solution to this linear formulation.

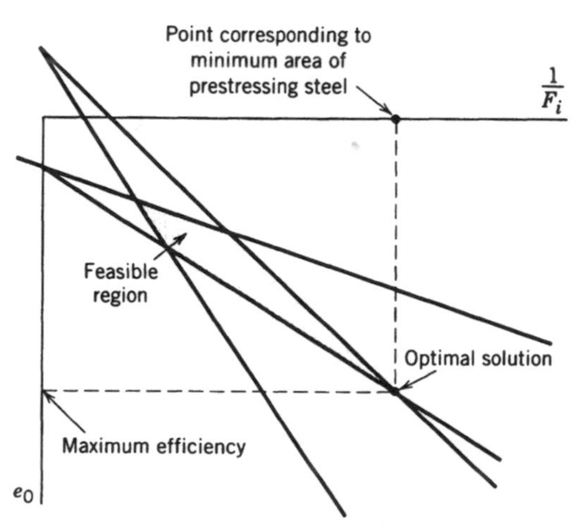

Figure 16.10 Solution for prestressed concrete beam example.

Another problem that can be posed as a linear programming problem is the plastic design of a steel portal frame. Consider the portal frame subject to both a vertical and horizontal loading shown in Figure 16.11. For this example, we will assume that the columns will be identical and have a plastic moment of $M_c = \sigma_y Z_c$, where σ_y represents the yield stress of the steel and Z_c represents the plastic section modulus of the column. Similarly, the plastic moment of the beam is given by $M_b = \sigma_y Z_b$. The frame will not collapse as long as the internal work done at the plastic hinges within the frame exceeds the external work done by the applied loads. Therefore, our optimization model must consider all the possible failure mechanisms of the frame. Figure 16.12 shows the six possible failure mechanisms of the frame (Chen, 1995).

Relating the internal and external work for each of the six failure mechanisms yields the following design constraints:

$$\text{Mechanism 1:} \quad 4\sigma_y Z_b \geq V\frac{l^b}{2} \tag{16.90}$$

$$\text{Mechanism 2:} \quad 2\sigma_y Z_c + 2\sigma_y Z_b \geq V\frac{l^b}{2} \tag{16.91}$$

$$\text{Mechanism 3:} \quad 2\sigma_y Z_c + 2\sigma_y Z_b \geq Hl_c \tag{16.92}$$

$$\text{Mechanism 4:} \quad 4\sigma_y Z_b \geq Hl_c \tag{16.93}$$

$$\text{Mechanism 5:} \quad 2\sigma_y Z_c + 4\sigma_y Z_b \geq V\frac{l^b}{2} + Hl_c \tag{16.94}$$

$$\text{Mechanism 6:} \quad 4\sigma_y Z_c + 2\sigma_y Z_b \geq V\frac{l^b}{2} + Hl_c \tag{16.95}$$

Of course, we still must restrict the plastic section modulus for each member in the frame to a positive value:

$$Z_c, Z_b \geq 0 \tag{16.96}$$

One possible objective for this formulation is to minimize the sum of the plastic section moduli for the frame members. This objective is expressed as

$$\text{minimize } Z = 2Z_c + Z_b \tag{16.97}$$

For vertical and horizontal loads of 25 kips and 10 kips, respectively, and a yield stress of 36 ksi, the optimal values for Z_c and Z_b are 20.8 in^3. The value

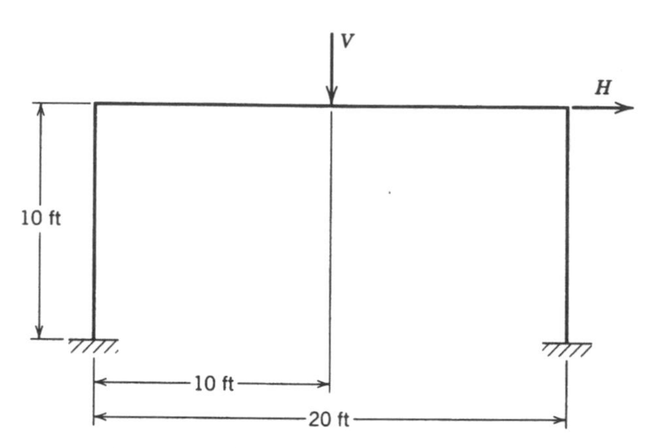

Figure 16.11 Fixed-base portal frame.

of the objective function is 62.5 in^3. Figure 16.13 presents a graphical solution to this problem.

By using plastic analysis. we are able to analyze the indeterminate frame relatively easily and formulate its design as a linear programming model. Plastic analysis assumes that the structure is made of a ductile material that can undergo deformations beyond the elastic limit and that the structural deflections are small enough to ignore the effect of the loads on the deformed geometry of the structure (second-order effects) (Chen, 1995). Three conditions must be satisfied for any plastic hinge solution: equilibrium, mechanism, and plastic moment. A mechanism condition occurs when a sufficient number of plastic hinges form to allow the structure to deform or collapse. The plastic moment simply represents a structural member's plastic strength. Equilibrium requires that the algebraic sum of all external forces and reactions equal zero and all internal forces balance. When all three conditions are satisfied, the limiting load required to form the mechanism is called the *collapse load.* Collapse loads for simple structures such as portal frames and beams are easy to determine. It was relatively easy to identify the six possible collapse mechanisms for the portal frame example. However, identifying all the possible collapse mechanisms for more complex structural systems is difficult.

Once all the collapse mechanisms have been identified, it is relatively easy to relate the internal and external work as inequalities to form the problem constraints. Because the fully plastic moment for any given structural section is a constant, the work relationships or problem constraints are linear. For a more general overview of plastic analysis, see Chen (1995).

16.F. NONLINEAR APPLICATIONS

As was shown in the first three examples presented in this chapter, many structural optimization problems are nonlinear. This is especially true for

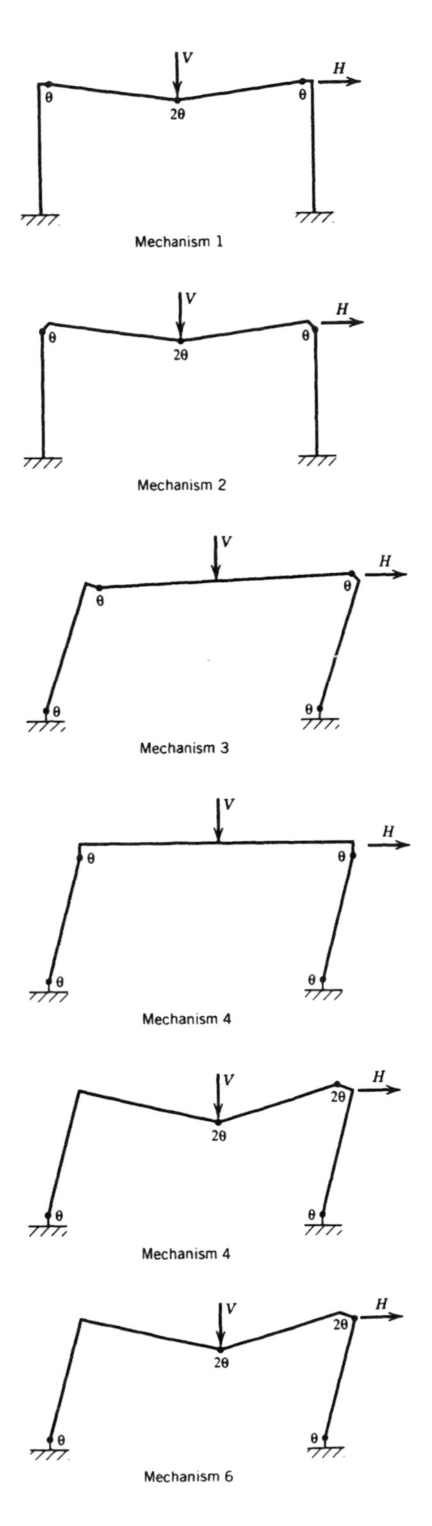

Figure 16.12 Collapse mechanisms for portal frame example.

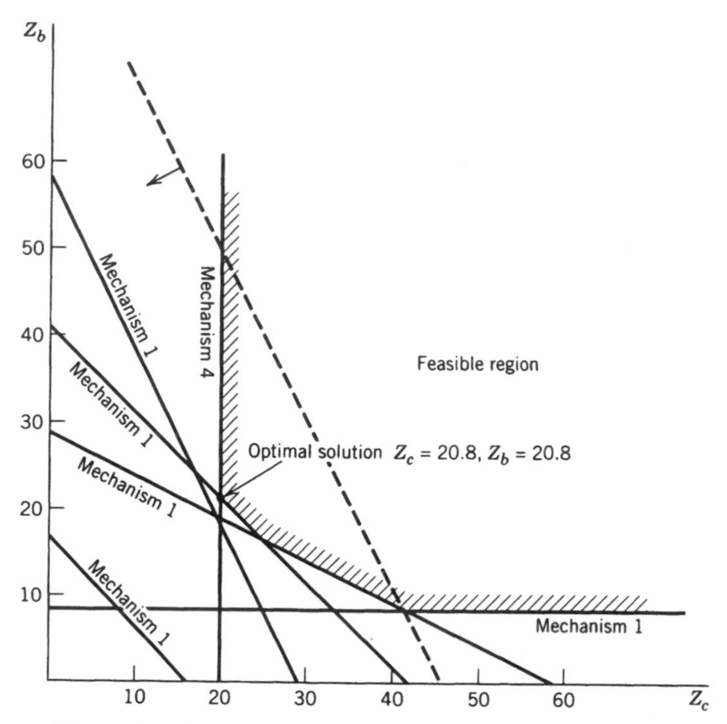

Figure 16.13 Graphical solution to portal frame example.

structural systems that are indeterminate or are subjected to behavioral constraints such as deflections at specific locations. A significant amount of research has been done in the area of operations research to develop and evaluate reliable methods for solving nonlinear problems. However, most of the methods developed in the past deal with model formulations that have nonlinear objective functions subject to a number of linear constraint equations. Most structural optimization problems can be formulated with linear objective functions such as the minimization of total weight. However, the constraint set is often nonlinear. One of the most common methods for dealing with nonlinear optimization problems is the use of Lagrange multipliers. Lagrange multipliers basically transform the constrained problem into a single unconstrained equation.

To begin our study of the use of Lagrange multipliers, consider an optimization problem with a single active constraint:

$$\text{minimize } Z = c(X) \tag{16.98}$$

subject to

$$g(X) = b \tag{16.99}$$

where X is the vector of n design variables. By introducing a new variable called the *Lagrange multiplier*, the optimization problem can be written as an unconstrained optimization problem. The unconstrained optimization problem is called the *Lagrange function* and is defined mathematically as

$$L(X,\lambda) = c(X) + \lambda\{g(X) - b\} \tag{16.100}$$

The necessary conditions for a local minimum or critical point are

$$\frac{\partial L}{\partial X} = 0 \tag{16.101}$$

and

$$\frac{\partial L}{\partial \lambda} = 0 \tag{16.102}$$

To determine if the critical point is a global maximum or minimum, the following four theorems are needed (Ossenbruggen, 1984):

1. If the objective function $f(X)$ is a convex function over a convex constraint set $g(X) \leq b$, any local minimum of $f(X)$ satisfying the constraint set is the global minimum.
2. If the objective function $f(X)$ is a concave function over a convex constraint set $g(X) \leq b$, the global minimum will occur at one or more of the extreme points of the constraint set.
3. If the objective function $f(X)$ is a concave function over a convex constraint set $g(X) \leq b$, any local maximum of $f(X)$ satisfying the constraint set is the global maximum.
4. If the objective function $f(X)$ is a convex function over a convex constraint set of $g(X) \leq b$, the global maximum will occur at one or more of the extreme points of the constraint set.

To illustrate the application of Lagrange multipliers to structural optimization problems, again consider the statically determinate three-bar truss presented at the beginning of this chapter. If we assume a priori that the deflection constraint is binding at optimality, the optimization model is written as a model with a single active constraint:

$$\text{minimize } Z = 339.4A_1 + 240A_3 \tag{16.103}$$

subject to

$$\frac{0.34}{A_1} + \frac{0.12}{A_3} = 0.22 \tag{16.104}$$

For this example the objective function is expressed in terms of cubic inches. The Lagrange function is written as

$$\text{minimize } L = 339.4A_1 + 240A_3 - \lambda\left(\frac{0.34}{A_1} + \frac{0.12}{A_3} - 0.22\right) \tag{16.105}$$

Standard calculus methods can be used to determine the local minimum of the Lagrange function. The necessary conditions for optimality are

$$\frac{\partial L}{\partial A_1} = \frac{\partial L}{\partial A_3} = \frac{\partial L}{\partial \lambda} = 0 \tag{16.106}$$

Enforcing these conditions of optimality results in the following set of simultaneous equations:

$$\frac{\partial L}{\partial A_1} = 339.4 + \frac{0.34\lambda}{A_1^2} = 0 \tag{16.107}$$

$$\frac{\partial L}{\partial A_3} = 240 + \frac{0.12\lambda}{A_3^2} = 0 \tag{16.108}$$

$$\frac{\partial L}{\partial \lambda} = \frac{0.34}{A_1} + \frac{0.08}{A_3} - 0.22 = 0 \tag{16.109}$$

Solving for the design variables, A_1 and A_3, and the Lagrange multiplier, λ, yields the critical point:

$$A_1^* = 2.32 \text{ in}^2$$

$$A_3^* = 1.64 \text{ in}^2$$

$$\lambda = -5375.8 \text{ in}^3/\text{in.}$$

At this point, we can only conclude that the critical point is a local minimum. The next step is to determine if the critical point represents the global optimal solution to the problem. To make this determination, we must consider the four theorems presented earlier. The original objective function for this problem was linear and therefore can be considered a convex function. Recalling Figure 16.3, we find that the deflection constraint is also convex.

Therefore, from Theorem 1 we can conclude that our solution to this problem represents the global optimal solution.

In many cases it is not possible to visualize the constraint set. In such cases, the Hessian matrix can be used to determine if the constraint equation is convex. In general, the Hessian matrix is defined as

$$H = \begin{bmatrix} \frac{\partial^2 F}{\partial x_1^2} & \frac{\partial^2 F}{\partial x_1 \, \partial x_2} & \cdots & \frac{\partial^2 F}{\partial x_1 \, \partial x_n} \\ \frac{\partial^2 F}{\partial x_2 \, \partial x_1} & \frac{\partial^2 F}{\partial x_2^2} & \cdots & \frac{\partial^2 F}{\partial x_2 \, \partial x_n} \\ \cdot & \cdot & \cdot & \cdot \\ \cdot & \cdot & \cdot & \cdot \\ \frac{\partial^2 F}{\partial x_n \, \partial x_1} & \frac{\partial^2 F}{\partial x_n \, \partial x_2} & \cdots & \frac{\partial^2 F}{\partial x_n^2} \end{bmatrix} \tag{16.110}$$

where F represents a constraint function or the objective function. The Hessian matrix is always symmetric about its major diagonal. If the Hessian matrix is positive definite, the function F, is a convex function. Mathematically, the sufficient conditions to determine if the Hessian matrix is positive definite are:

1. All second derivatives of the function, $(\partial^2 F / \partial x_i \, \partial x_j)$, exist.
2. All the principal minors are positive.

For our example, the Hessian matrix is defined by

$$H = \begin{bmatrix} 0.68A_1^{-3} & 0 \\ 0 & 0.16A_3^{-3} \end{bmatrix} \tag{16.111}$$

The corresponding principal minors are:

$$p_1 = 0.68A_1^{-3} \geq 0 \tag{16.112}$$

and

$$p_2 = 0.109A_1^{-3}A_3^{-3} \geq 0 \tag{16.113}$$

for all values of A_1 and A_3 greater than or equal to zero. Therefore, we can conclude that the function is convex and the optimal solution found using Lagrange multipliers is truly a global minimum. Recall that negative values for the cross-sectional area of the structural members are infeasible and not considered. For further explanation of this concept, the reader is referred to Smith et al. (1983), Ossenbruggen (1984), and ReVelle et al. (1997).

The Lagrange multiplier is the dual variable. For this example it represents the marginal value of increasing the maximum allowable displacement. Put another way, the value of the dual at optimality tells us how much the objective function will change for a unit change in the right-hand side of the constraint equation limiting the deflection. For this example we must express the value of the dual variable in terms of units of volume per inch, which yields a value for λ equal to -5375.8 in^3/in. This means that if we increased the maximum allowable deflection by 0.01 in., the total volume of the three truss members would decrease by 53.8 in^3, or 15.2 lb. However, we must remember that the dual variable is only valid for small changes in the right-hand side of the constraint. For large changes the problem may become infeasible or another constraint may become binding. To get a feeling for this concept, think about how the deflection constraint would change on the graphical solution to this problem as the right-hand side is increased.

Now let us consider the case where the constraint is expressed as an inequality. In this case, the model is expressed as

$$\text{minimize } f(x) \tag{16.114}$$

subject to

$$g(x) \leq b \tag{16.115}$$

As with linear programming models, we can express the constraint set as a set of equalities by adding slack variables. However, since we will need to be able to differentiate the constraint equations, the slack variables will take the form of s^2. The model formulation may now be presented as

$$\text{minimize } f(x) \tag{16.116}$$

subject to

$$g(x) + s^2 = b \tag{16.117}$$

The corresponding Lagrangian equation is defined by

$$L(x) = f(x) - \lambda(g(x) + s^2 - b) \tag{16.118}$$

From this point forward, the problem is solved as before. ReVelle et al. (1997)

present an excellent example of this type of Lagrange multiplier problem for the design of a reinforced concrete beam.

Although Lagrange multipliers provide a nice method for solving nonlinear optimization models, it is easy to imagine how cumbersome this approach would become for relatively large problems. For such cases we need to consider other more efficient methods for solving nonlinear problems. One popular method for solving nonlinear problems is to employ a gradient or hill-climbing approach. For this approach an iterative equation is used. One form of the iterative equation is defined by

$$X_i^{k+1} = X_i^k + h\nabla f(X)_{x_i} \tag{16.119}$$

where $\nabla f(X)_{x_i}$ represents the gradient of the function $f(X)_i$ with respect to the decision variable and h is the step size. The unit gradient is defined as

$$\nabla f(X)_{x_i} = \frac{\partial f(X)_i / \partial x_i}{\sqrt{\sum_{j=1}^{n} (\partial f(X)_j / \partial x_j)^2}} \tag{16.120}$$

where n represents the number of decision variables involved in the function $f(X)$.

Choosing the appropriate step size for use in equation (16.119) often has a profound impact on how well the iterative approach works. If the step size remains constant, the algorithm will tend to oscillate around the optimal or a local optimal solution. In most cases it is beneficial to decrease the step size systematically as the algorithm progresses. The rate and frequency in which the step size is decreased is up to the person solving the problem. To illustrate the use of this method, consider the following simple unconstrained optimization model:

$$\text{maximize } Z = 2x_1x_2 + 2x_2 - x_1^2 - 2x_2^2 \tag{16.121}$$

The derivatives for this problem are defined by

$$\frac{\partial Z}{\partial x_1} = 2x_2 - 2x_1 \tag{16.122}$$

and

$$\frac{\partial Z}{\partial x_2} = 2x_1 + 2 - 4x_2 \tag{16.123}$$

Using $X = \langle 0.1, 0.1 \rangle$ as a starting point, Figure 16.14 illustrates the convergence of the algorithm to the optimal solution of $X^* = \langle 1.0, 1.0 \rangle$. The initial step size for this example was 0.5 and was decreased by 20% each iteration.

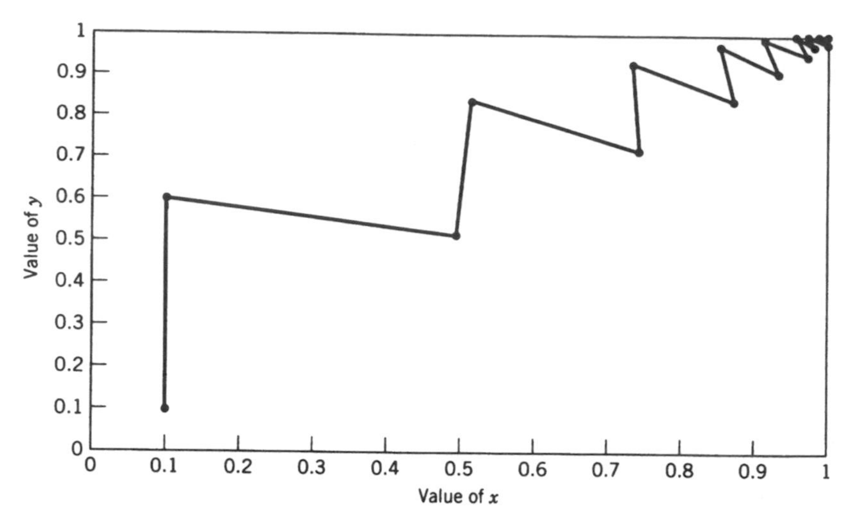

Figure 16.14 Search path for gradient example.

This iterative approach can easily be adapted for use with more difficult and complex problems. Gradient search algorithms are very useful for problems that have nonlinear objective functions. One thing to remember is that for each iteration the feasibility of the new solution must be checked. This can be time consuming and slow down the algorithm. In addition, the quality of the final solution can be improved by reducing the step size each iteration. However, this too will slow down the process.

One area in which a gradient search is very useful is for determining better designs for large, complex structural systems. In these cases the objective function can be nonlinear. In addition, complex structural systems often require repeated analyses as the design converges. Once the structure has been analyzed for an initial design, the design variable can be modified using Equation 119. Using the new design variables, the structure is again analyzed. This process is repeated until one or more of the design constraints is violated or another stopping criterion is met. The user should experiment with a number of step sizes and refinement schemes when using a gradient search approach.

Another iterative algorithm for solving nonlinear optimization problems is to use a succession of linear approximations to the original nonlinear model. This is different from piecewise linearization of a constraint or objective function in that we express the nonlinear equation in terms of a Taylor series. To begin our study of this approach, consider the following nonlinear model:

$$\text{maximize } f(X) \tag{16.124}$$

subject to

$$g(X) \leq 0 \tag{16.125}$$

where both $g(X)$ and $f(X)$ are assumed to be nonlinear equations and X is the vector of design variables. Let us first consider the constraint equation $g(X)$. The constraint equation, $g(X)$, can be approximated using a Taylor's series as

$$g_i(X) = g_i(X^0) + \sum_{j=1}^{m} \frac{\partial g_i}{\partial x_j}(x_j - x_j^0) \leq 0 \tag{16.126}$$

If we let $a_{ij} = \partial g_i / \partial x_j$ evaluated at X^0 and simplify equation (16.126), we have

$$\sum_{j=1}^{m} a_{ij} x_j \leq \sum_{j=1}^{m} a_{ij} x_j^0 - g_i(X^0) \tag{16.127}$$

The first thing that we notice is that everything on the right-hand side of equation (16.127) is a constant. Therefore, equation (16.127) is essentially a linear approximation of the original nonlinear constraint equation $g(X)$. If we approximate all the nonlinear equations in this manner, we find that the original nonlinear formulation is approximated by a standard linear program. The resulting nonlinear model is now approximated by the following linear program:

$$\text{maximize } Z = \sum_{j=1}^{m} c_j x_j + f(X^0) - \sum_{j=1}^{m} c_j x_j^0 \tag{16.128}$$

subject to

$$\sum_{j=1}^{m} a_{ij} x_j \leq \sum_{j=1}^{m} a_{ij} x_j^0 - g_i(X^0) \qquad \forall i \tag{16.129}$$

$$x_j^0 - \delta_j \leq x_j \leq x_j^0 + \delta_j \qquad \forall j \tag{16.130}$$

where c_j represents the partial derivative of the function $f(X)$ with respect to x_j evaluated at x_j^0. In addition to the linear approximation of the objective function and constraint equation, we must restrict the value of x_j to be near x_j^0. This restriction stems from the fact that we are using a truncated Taylor series to *approximate* the original nonlinear equation. Consequently, if x_j is allowed to vary substantially from x_j^0, the accuracy of the approximation is compromised. In practice, it is common to reduce δ_j by 30 to 50% at each iteration. Bradley et al. (1977) outline the following steps for implementing this algorithm.

Step 1. Let $X^0 = (x_1^0, x_2^0, \ldots, x_n^0)$ represent any candidate solution the problem. The solution selected is usually feasible or near feasible. Set $k = 0$.

Step 2. Evaluate the partial derivatives of the objective function and constraint equations (c_i and a_{ij}) at x_j^k.

Step 3. Formulate and solve the resulting linear-approximation problem defined by equations (16.128)–(16.130). Let $X^{k+1} = (x_1^{k+1}, x_2^{k+1}, ..., x_n^{k+1})$ equal the optimal solution. Increment k by 1 and return to step 2.

This procedure is repeated until the optimal solutions for two successive approximations are similar. Other stopping criteria, such as the maximum number of iterations, may also be used.

To illustrate the application of this method to a structural optimization problem, again consider the three-bar truss subject to stress and deflection constraints. Unlike previous examples, we will assume that all the constraints are inequalities. Restating the problem formulation in a slightly different form, we have

$$\text{minimize } Z = 339.4A_1 + 240A_3 \tag{16.131}$$

subject to

$$0.12A_1 + 0.34A_3 - 0.22A_1A_3 \leq 0 \tag{16.132}$$

$$A_1 \geq 1.96 \tag{16.133}$$

$$A_3 \geq 1.39 \tag{16.134}$$

To begin the algorithm, assume that $A_1 = A_3 = 4$ and $\delta = 2$ for both design variables. The partial derivatives for the nonlinear deflection constraint $[g(A)]$ are defined by

$$a_{11} = \frac{\partial g}{\partial A_1} = 0.12 - 0.22A_3 \tag{16.135}$$

and

$$a_{12} = \frac{\partial g}{\partial A_3} = 0.34 - 0.22A_1 \tag{16.136}$$

For the first approximation we evaluate the partial derivatives at A_1 and A_3 equal to 4 and find that $a_{11} = -0.76$ and $a_{12} = -0.54$. Using these values, a new constraint is formulated using Equation 129:

$$0.76A_1 + 0.54A_3 \geq 3.52 \tag{16.137}$$

Therefore, the linear-approximation model is defined by

$$\text{minimize } Z = 339.4A_1 + 240A_3 \tag{16.138}$$

subject to

$$0.76A_1 + 0.54A_3 \geq 3.52 \tag{16.139}$$

$$A_1 \geq 1.96 \tag{16.140}$$

$$A_3 = 1.39 \tag{16.141}$$

$$2 \leq A_1 \leq 6 \tag{16.142}$$

$$2 \leq A_3 \leq 6 \tag{16.143}$$

The optimal solution to this linear approximation is $A_1^* = 2.0$ and $A_3^* = 3.7$ with an objective function value of $Z^* = 1567.7$ in^3. This solution is then used as the evaluation point for the next iteration. The second iteration yields the following linear approximation, where $\delta = 1$, $A_1 = 2.0$, and $A_3 = 3.7$:

$$\text{minimize } Z = 339.4A_1 + 240A_3 \tag{16.144}$$

subject to

$$0.695A_1 + 0.1A_3 \geq 1.63 \qquad (16.145)(16.146)$$

$$A_1 \geq 1.96 \tag{16.147}$$

$$A_3 \geq 1.39 \tag{16.148}$$

$$1 \leq A_1 \leq 3 \tag{16.149}$$

$$2.7 \leq A_3 \leq 4.7$$

which has an optimal solution of $A_1^* = 1.96$, $A_3^* = 2.70$, and $Z^* = 1314.2$ in^3. Table 16.7 presents a summary of the results using this iterative approach.

Figure 16.15 presents a graphical representation of the successive solutions for each approximation to the original nonlinear problem. The user must be careful of numerical roundoff when using this method. A quick check of the original deflection constraint for the solution to iteration 5 proves that it is slightly infeasible. This is a result of roundoff and the fact that we are approximating the nonlinear constraint over a limited range with a linear equation. This behavior is not uncommon for this method. After several more iterations, the method does converge to a solution that is very near the optimal.

TABLE 16.7 Results of Method of Approximate Programming

Iteration	δ	A_1 (in^2)	A_3 (in^2)	Z (lb)	Z (in^3)
0	2	4	4	654.5	2317.6
1	1	1.96	2.70	371.1	1314.2
2	0.5	2.38	2.32	385.6	1365.3
3	0.25	2.14	2.07	345.3	1222.7
4	0.125	2.13	1.95	336.1	1190.2
5	0.625	2.17	1.88	335.8	1189.2
6	0.03125	2.19	1.85	335.4	1187.6
7	0.01563	2.20	1.83	335.2	1186.8

The reader should note that the linear constraints are not affected by the approximation and remain the same throughout this procedure. This method is a generalized extension to the Frank–Wolfe algorithm. For more information concerning this algorithm, the interested reader should see Bradley et al. (1977).

16.G. STOCHASTIC MODELS FOR STRUCTURAL OPTIMIZATION

In each of the techniques discussed above and those discussed by Templeman, the loading conditions are considered to be deterministic. However, the actual

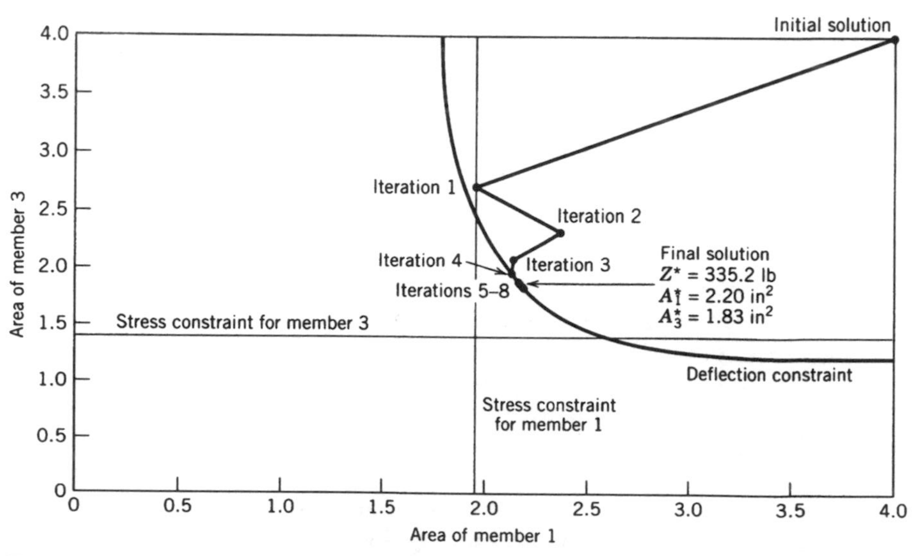

Figure 16.15 Successive solutions for linear approximation to nonlinear truss formulation.

loads experienced by an existing structure throughout its lifetime are actually probabilistic in nature and can be addressed in terms of structural reliability.

Structural reliability provides the engineer with a methodology for making a probabilistic assessment of design performance by considering structural design loads and material properties as random processes with predefined distributions. In recent years structural engineers have attempted to incorporate the probabilistic behavior of material, loads, and resistance into structural design and analysis (Bloom, 1984). Since the introduction of structural reliability by Freudenthal (1947), significant progress has been made and structural reliability concepts have been implemented in many design codes and procedures (Yao and MacGregor, 1982).

The literature is replete with papers addressing structural reliability topics, including fatigue, design, analysis, and optimal design under random loading conditions and damage analysis (Yao, 1974; Frangopol, 1985; Yao et al., 1986; Frangopol and Nakib, 1986). Structural reliability concepts have been applied to a variety of topics within structural engineering (Yao, 1978), including structural design, analysis, and damage assessment (Yao, 1974; Yao et al., 1986). Frangopol (1985) has applied the concepts of classical reliability in a structural optimization framework by specifying bounds on structural weight and reliability indices. In fact, the determination of a system's reliability is itself an optimization problem (Ang and Tang, 1984).

One method for incorporating the uncertainties inherent to the structural design process is the use of chance constraints (Jacobs, 1991). In the following example we illustrate the use of chance constraints by formulating and solving a mixed-integer optimization model for the design of a steel frame conditioned on probabilistic structural loads. This technique can be used to evaluate the likely performance of structural designs subjected to random loadings. In addition, if information concerning variabilities in material properties, cross-sectional properties, and workmanship are available, the chance constraint can be expanded to incorporate these design parameters as random variables.

Chance constrained programming is an optimization method in which the constraints are expressed as probability statements (Charnes et al., 1958; Wagner, 1975). The chance constraints ensure that each design constraint is realized with a minimum probability. This is accomplished by expressing design parameters as functions of random variables with established probability distributions. This method has been applied extensively in water resources to determine the optimal design and operation of reservoirs conditioned on stochastic streamflows and to manage stream-aquifer systems (ReVelle et al., 1969; Houck, 1979; Hantush and Marino, 1989). The chance-constrained optimization model for a structural system may be expressed as

$$\text{minimize } Z = \sum_{i=1}^{n} \gamma_i A_i l_i \tag{16.150}$$

subject to

$$P(\sigma_f \le \sigma_y)_i \ge \alpha \qquad \forall i \tag{16.151}$$

$$P(\sigma_v \le \sigma_y)_i \ge \beta \qquad \forall i \tag{16.152}$$

$$P(\Delta \le \Delta_a)_i \ge \xi \qquad \forall i \tag{16.153}$$

where

γ_i = specific weight of the material in structural member i

A_i = cross-sectional area of structural section i

l_i = length of structural section i

σ_f = maximum flexural stress within the structural section

σ_v = maximum shear stress within the structural section

σ_y = yield stress of the structural section material

Δ = maximum deflection of the structural section

Δ_a = maximum allowable deflection

$P(\cdot)_i$ = probability function for structural section i

α, β, ξ = minimum reliabilities specified for the optimal design

The objective in this formulation is to minimize the total weight of the structure by minimizing the cross-sectional areas of each structural member within the structure. By minimizing the total weight, the engineer in turn minimizes the material costs associated with the design. Fabrication and construction costs are not included in this model formulation.

The set of constraints for this model formulation represents the failure criteria for each structural element within a probabilistic framework in which the structure's loading is modeled as a random variable. The constraints impose lower bounds on the likelihood or chance that the failure criteria are exceeded for each section within the structure. By specifying a cumulative distribution function for the loading, each chance constraint may be expressed as a deterministic equivalent. To demonstrate this process, consider the one-bay one-story frame subject to a random horizontal loading shown in Figure 16.16.

For this example, design considerations will be based on the maximum moment. Shear and deflection constraints will be neglected. Using the stiffness method (Meyers, 1983), the frame can be analyzed in terms of the random load, W (kN/m). The local member forces in terms of the random loading are found to be

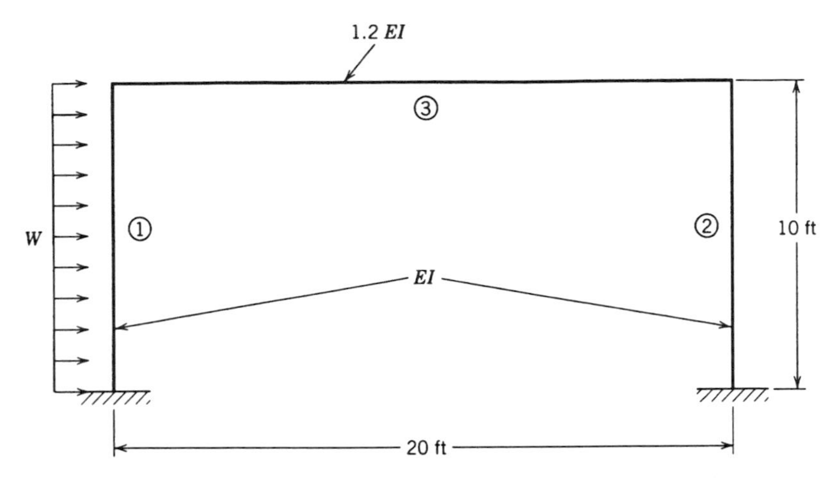

Figure 16.16 Steel frame for chance constraint example.

$$r^1 = \langle -8.98W, 7.53W, -10.81W, 32.85W \rangle \tag{16.154}$$

$$r^2 = \langle 8.98W, 10.14W, -8.98W, 17.22W \rangle \tag{16.155}$$

$$r^3 = \langle -2.89W, -7.53W, 2.89W, -10.14W \rangle \tag{16.156}$$

For each of these local member forces, the first and third terms represent the shear force (kN) at the ends of the member. The second and fourth terms represent the end moment (kN · m) for the members. In this case the maximum moments are found to coincide with a local end moment for each element within the structure. Assuming elastic bending and symmetric cross sections, the maximum stress within each member can be determined using the flexural formula. Substituting the maximum bending moment for member 1 as a function of the random load yields

$$\sigma_{f\,\max} = \frac{M_{1,\max}}{S_{1,\text{req}}} = \frac{32.85W}{S_{1,\text{req}}} \tag{16.157}$$

where σ_f is the flexural stress in the member and $S_{1,\text{req}}$ is the required section modulus. The chance constraint for flexural stress in member 1 can now be stated as

$$P\left(\frac{32.85W\ (\text{kN}-\text{m})}{S_{1,\text{req}}\ (\text{m}^3)} \le \sigma_y\right)_1 \ge \alpha_1 \tag{16.158}$$

Simplifying this expression to isolate the random loading yields

$$P(W \leq 0.0304\sigma_y S_{1,\text{req}})_1 \geq \alpha_1 \tag{16.159}$$

Because W is a random variable, the left-hand side of the equation represents the cumulative distribution function (CDF) of the structure's loading. Denoting $F(\cdot)$ as the CDF of the structure's loading, equation (16.159) may now be restated as

$$F(0.0304\sigma_y S_{1,\text{req}})_1 \geq \alpha_1 \tag{16.160}$$

Letting $F^{-1}(\cdot)$ represent the inverse function of the loading CDF, equation (16.160) can be restated as a deterministic equivalent constraint in standard form:

$$0.0304\sigma_y S_{1,\text{req}} - F^{-1}(\alpha_1) \geq 0.0 \tag{16.161}$$

If the negative inverse CDF is convex, it can be approximated using piecewise linearization and the simplex algorithm can be used to determine an optimal solution to the resulting linear problem. Because it is unlikely that reliabilities of less than 0.5 would be considered, only the upper half of the loading CDF need be linearly approximated. Similarly, the flexural chance constraints for the remaining two members can be stated as deterministic equivalent constraints.

$$0.0580\sigma_y S_{2,\text{req}} - F^{-1}(\alpha_2) \geq 0.0 \tag{16.162}$$

and

$$0.0990\sigma_y S_{3,\text{req}} - F^{-1}(\alpha_3) \geq 0.0 \tag{16.163}$$

Equations (16.161), (16.162), and (16.163) relate the required section modulus for each structural member to the probabilistic distribution of the load. To complete the model formulation, the values of each section modulus must be restricted to available standard steel sections.

To restrict the possible design solutions to combinations of available standard steel sections, three additional constraints are needed for each structural member. The decision to include a specific structural section in the design can be modeled using binary (0–1) decision variables. Formally, the decision to select a specific structural section as part of the overall design can be represented by the variable x_{ij} such that

$$x_{ij} = \begin{cases} 1 & \text{if structural section } j \text{ is selected for member } i \\ 0 & \text{otherwise.} \end{cases} \tag{16.164}$$

The optimal design for the structure is completely specified by the vector $\mathbf{X} = (x_{11}, x_{12}, \ldots, x_{nm})$ and consists of $n \times m$ binary variables. Let S_{ij} be defined as the section modulus of structural section j for member i. Using the binary decision variable, the section modulus for the structure is defined as

$$S_{i,\text{req}} - \sum_{j=1}^{m} S_{ij} x_{ij} = 0.0 \qquad \forall i \tag{16.165}$$

where m is the number of available structural sections.

A similar constraint is needed to define the cross-sectional area of the structural member. Let A_{ij} be defined as the cross-sectional area of section j to be used for member i. The cross-sectional area of the section to be used in the structure may now be defined as

$$A_{i,\text{req}} - \sum_{j=1}^{m} A_{ij} x_{ij} = 0.0 \qquad \forall i \tag{16.166}$$

Finally, a constraint is needed to ensure that no more than one available section is chosen for any structural member. This may be expressed as

$$\sum_{j=1}^{m} x_{ij} = 1.0 \qquad \forall i \tag{16.167}$$

Collectively, these constraints enforce the discrete nature of selecting structural members for design from a finite number of available standard structural sections. For this example, 51 AISC standard W sections were selected as possible structural members. To ensure that assumptions made in analyzing the structural member forces are met, two constraints are added to guarantee that the section modulus for member 3 is 1.2 times that of members 1 and 2. These constraints can be expressed as

$$1.2 S_{1,\text{req}} - S_{3,\text{req}} \leq 0.0 \tag{16.168}$$

$$S_{1,\text{req}} - S_{2,\text{req}} = 0.0 \tag{16.169}$$

The complete deterministic equivalent mathematical formulation is defined by

$$\text{minimize } Z = \sum_{i=1}^{n} \gamma_i A_i l_i \tag{16.170}$$

subject to

$$0.0304\sigma_y S_{1,\text{req}} - F^{-1}(\alpha_1) \geq 0.0 \tag{16.171}$$

$$0.0580\sigma_y S_{2,\text{req}} - F^{-1}(\alpha_2) \geq 0.0 \tag{16.172}$$

$$0.0990\sigma_y S_{3,\text{req}} - F^{-1}(\alpha_3) \geq 0.0 \tag{16.173}$$

$$S_{i,\text{req}} - \sum_{j=1}^{m} S_{ij}x_{ij} = 0.0 \qquad \forall i \tag{16.174}$$

$$A_{i,\text{req}} - \sum_{j=1}^{m} A_{ij}x_{ij} = 0.0 \qquad \forall i \tag{16.175}$$

$$1 - \sum_{j=1}^{m} x_{ij} = 0.0 \qquad \forall i \tag{16.176}$$

$$1.2S_1 - S_3 \leq 0.0 \tag{16.177}$$

$$S_1 - S_2 = 0.0 \tag{16.178}$$

$$x_{ij} = \begin{cases} 1 & \text{if structural section } j \text{ is selected for member } i \\ 0 & \text{otherwise.} \end{cases} \tag{16.179}$$

Using piecewise linear segments to approximate the negative inverse CDF of the loading, this formulation may be solved as a deterministic mixed-integer program.

To investigate the trade-offs between total structural weight and the probability of failure, the model was solved for α values ranging from 0.9 to 0.99999. To compare the model results to designs obtained by conventional methods, the frame was designed using the AISC allowable stress and LRFD methods. Table 16.8 presents the results for this example. Figure 16.17 represents the trade-off relationship between design weight and the probability of failure. Also noted are the design weights using allowable stress and LRFD design criteria. Due to the discrete nature of the available structural sections, the trade-off relationship between the weight and probability of failure is not continuous.

Assuming that design weight reflects cost, the trade-off relationship illustrates that decreasing the probability of failure from 0.0123 to 0.00042 results in a 19.9% increase in the structure's total weight. However, the additional increase in structural weight associated with achieving a lower probability of failure is substantial. To reduce the probability of failure from 0.00029 to 0.000001 results in a 102.6% increase in the total structural weight. This result demonstrates that the probability of failure can be reduced significantly at a moderate cost, but as the probability of failure approaches zero, the design weight increases rapidly and the associated cost becomes prohibitive.

TABLE 16.8 Chance-Constraint Model Results

Total Weight	Probability of Failure	Member 1	Member 2	Member 3
4903.9	0.0123	W12 × 26	W12 × 26	W14 × 30
5085.4	0.0033	W16 × 26	W16 × 26	W16 × 31
5421.7	0.00292	W12 × 30	W12 × 30	W16 × 31
5706.3	0.00061	W14 × 30	W14 × 30	W18 × 35
5878.5	0.00042	W16 × 31	W16 × 31	W18 × 35
7053.2	0.00029	W18 × 35	W18 × 35	W21 × 44
7507.3	0.00021	W16 × 40	W16 × 40	W21 × 44
8022.0	0.000162	W18 × 40	W18 × 40	W21 × 50
8839.1	0.00005	W21 × 44	W21 × 44	W24 × 55
10413.2	0.00004	W18 × 55	W18 × 55	W24 × 62
13561.5	0.000006	W24 × 68	W24 × 68	W24 × 84
14287.9	0.000001	W24 × 76	W24 × 76	W24 × 84

This optimization model is economical and computationally feasible to solve. The example presented here contained 102 (0–1) binary variables. Although the model is a mixed-integer program and the number of possible solutions increases exponentially with each additional decision variable, solutions for more complex frames can be obtained on a real-time basis by reducing the size of the model formulation through the application of a few practical assumptions (Garey and Johnson, 1979). For larger more complex

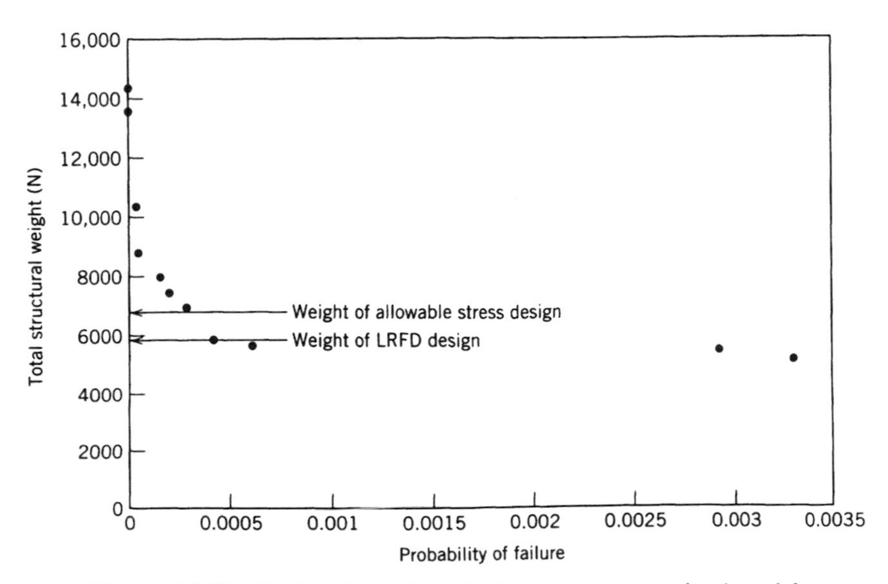

Figure 16.17 Trade-off relationship for chance-constrained model.

frames, the size of the optimization model can be reduced by assuming that all beams are the same size and columns within each story are the same.

This type of model provides engineers with a method for incorporating loading condition uncertainties directly into the design optimization process. By utilizing the probabilistic distribution of the loading conditions, the complete range of likely loading magnitudes for a specific type of loading is considered. One advantage of this approach is that the probabilistic optimization model has the same number of variables and only slightly more constraints as that of a conventional deterministic optimization model. In addition, the model can be used to evaluate the trade-offs between design alternatives and the associated probability of failure. This approach has also been used to evaluate the dynamic control of frames subject to random excitations (Zadoks and Jacobs, 1991).

16.H. APPLICATION OF STOCHASTIC METHODS TO STRUCTURAL OPTIMIZATION

Structural optimization problems can also be solved using a family of optimization procedures known as *stochastic optimization algorithms*. Stochastic optimization algorithms use a randomized approach to determine the candidates for the optimal solution. Stochastic algorithms include routines such as Monte Carlo simulation, simulated annealing, and genetic algorithms. One problem with all these methods is that they cannot guarantee that the best solution found will be anywhere close to the true global optimal solution.

16.H.1. Monte Carlo Simulation

Monte Carlo simulation is the most straightforward of the stochastic algorithms. Monte Carlo simulation determines a nearly optimal solution by randomly generating and comparing a large number of independent solutions. Table 16.9 presents the results obtained using Monte Carlo simulation to solve

TABLE 16.9 Monte Carlo Simulation Results

Sample Size	A_1^* (in^2)	A_3^* (in^2)	Z^* (lb)
10	2.01	8.11	742.5
50	2.83	2.80	461.3
100	2.6	1.53	353.1
1,000	2.13	2.06	343.2
5,000	2.43	1.61	342.2
10,000	2.18	1.86	334.7
20,000	2.33	1.60	332.0
100,000	2.34	1.58	331.5
500,000	2.27	1.66	330.6

the simple truss problem subject to both stress and deflection constraints. As illustrated by the results, the Monte Carlo results do not find the true global optimal solution. However, if the number of iterations is sufficiently large, the Monte Carlo simulation will usually determine a relatively good solution to the problem. In many cases this may be sufficient and no additional refinement is needed. Due to its random nature, Monte Carlo simulation does not search the solution space in a "smart" way and can be very inefficient. Incidentally, the example presented above is only improved by 0.08 lb using 5 million Monte Carlo iterations.

One of the factors that adds to the inefficiency of all stochastic methods is the evaluation of solution feasibility. For the Monte Carlo results presented here, the feasibility of each potentially improved solution must be evaluated. Computationally, this takes time and as the complexity of the constraint set and structural configuration increases, the computational effort needed to evaluate the constraint set increases rapidly. Furthermore, as the problem becomes more constrained, the number of feasible solutions found decreases. This can lead to solutions that are substantially inferior and in no way close to the true global optimum.

However, for difficult problems Monte Carlo simulation has been used effectively. One application by Zadoks and Jacobs (1991) used Monte Carlo simulation to determine the optimal feedback controller parameters for structural frames subjected to probabilistic dynamic loadings. In this work, both displacement and velocity feedback control parameters were determined to limit the maximum defection of steel frames subject to probabilistic loading magnitudes. The results of this work clearly defined a relationship between the energy needed to control the structure's deflection and the reliability as defined by the probability that the defections will not exceed a specified maximum.

In another application, Monte Carlo simulation was used to determine the most probable failure scenarios for multistory frames (Pulido et al., 1992). In this work Monte Carlo simulation incorporating variance reduction techniques is combined with a nonlinear finite element formulation to identify the most likely sequence of beam and joint failures leading to the collapse of the structure. In addition, Monte Carlo methods are dependent on the quality of the random number generator used in the algorithm. For the results presented here, the uniform distribution random number generator included in the standard C^{++} library was used. A multitude of information exists on the use of Monte Carlo simulation, and the interested reader should consult texts such as Ang and Tang (1984).

16.H.2. Simulated Annealing

Another stochastic algorithm often used to solve messy nonlinear problems is simulated annealing. Simulated annealing is a search technique similar to a descent algorithm that allows the occasional acceptance of an inferior solution in an effort to avoid being trapped in a local optimum. This technique is motivated by an analogy to the annealing process observed in the formation of solids. During the annealing process, a material is heated to a temperature that allows free movement of the atoms. The temperature is then carefully (slowly) cooled until the material forms a nearly perfect crystal. Simulated annealing uses this same process to transform a poor unordered solution into a highly optimized solution. As with all stochastic search techniques, locating the global optimum is not guaranteed. However, by adding a stochastic element to the search process, simulated annealing attempts to avoid convergence to local optimums. The stochastic feature of simulated annealing simply allows the acceptance of an inferior solution with a probability that decreases as the algorithm iterates. One advantage of using simulated annealing is that it allows the incorporation of hill-climbing methods, gradient search procedures, and heuristics to guide the search process. Rutenbar (1989) presents an excellent overview of simulated annealing.

16.H.3. Genetic Algorithms

Genetic algorithms (GA) provide a robust framework for determining the optimal solution to nonlinear optimization problems in which the objective function is not well understood and the solution space is filled with local optima. Using a process analogous to natural selection, the algorithm searches the solution space for a better design. This method can also be used to develop more efficient structural designs. Goldberg (1989) presents a summary of using a genetic algorithm to design a truss.

16.I. SUMMARY

This chapter has presented only a brief introduction to the topic of structural optimization and a summary of some of the more common techniques available to structural engineers for developing efficient designs. The interested reader is encouraged to investigate this topic further using the references cited in this chapter.

REFERENCES

Amir, H. M., and T. Hasegawa, 1989. Nonlinear Mixed-Discrete Structural Optimization, *Journal of Structural Engineering,* 115(3), 626–646.

Ang, A. H.-S., and W. H. Tang, 1984. *Probability Concepts in Engineering Planning and Design,* Vol. 2, *Reliability,* Wiley, New York.

ASCE, 1972. Structural Safety: A Literature Review, *Journal of the Structural Division, ASCE,* 98(4), 845–884.

Bloom, J. M., 1984. Probabilistic Fracture Mechanics: A State of the Art Review, *Advances in Probabilistic Fracture Mechanics,* 92, 1–19.

Bradley, S. P., A. C. Hax, and T. L. Magnanti, 1977. *Applied Mathematical Programming,* Addison-Wesley, Reading, MA.

Charnes, A., W. W. Cooper, G. H. Symonds, 1958. Cost Horizons and Certainty Equivalents: An Approach to Stochastic Programming of Heating Oil, *Management Science,* 4(3), 235–263.

Chen, W. F., 1995. *The Civil Engineering Handbook,* CRC Press, Boca Raton, FL.

Fourer, R., D. M. Gay, and B. W. Kernighan, 1993. *AMPL: A Modeling Language for Mathematical Programming,* Boyd & Fraser, Danvers, MA.

Frangopol, D. M., 1985. Structural Optimization Using Reliability Concepts, *Journal of Structural Engineering,* 111(11), 2288–2301.

Frangopol, D. M., and R. Nakib, 1986. Analysis and Optimal Design of Nondeterministic Structures Under Random Loads, *Proceedings of the 9th Conference on Electronic Computation, Committee on Electronics Computation Structural Division, ASCE,* pp. 483–493.

Freudenthal, A. M., 1947. Safety of Structures, *Transactions of ASCE,* 112, 125–180.

Garey, M. R., and D. S. Johnson, 1979. *Computers and Intractability: A Guide to the Theory of NP-Completeness,* W.H. Freeman, San Francisco.

Goh, C. J., and C. M. Wang, 1988. Optimization of Segment-Wise Linear Structures via Optimal Control Theory, *Computers and Structures,* 30(6), 1367–1373.

Goldberg, D. E., 1989. *Genetic Algorithms in Search, Optimization, and Machine Learning,* Addison-Wesley, Reading, MA.

Hager, K., and R. Balling, 1988. New Approach for Discrete Structural Optimization, *Journal of Structural Engineering,* 114(5), 1120–1134.

Hantush, M. S., and M. A. Marino, 1989. Chance Constrained Model for Management of Stream-Aquifer System, *Journal of Water Resources Planning and Management,* 115(3), 278–298.

Houck, M. H., 1979. A Chance Constrained Optimization Model for Reservoir Design and Operation, *Water Resources Research,* 15(5), 1011–1016.

Jacobs, T. L., 1991. Chance-Constrained Optimization Model for Structural Design, *Journal of Structural Engineering,* 117(1), 100–110.

Jewell, T. K., 1986. *A Systems Approach to Civil Engineering Planning and Design,* Harper & Row, New York.

Kirsch, U., 1981. *Optimum Structural Design,* Wiley, New York.

Magnel, G., 1950. *Prestressed Concrete,* 2nd Edition, Concrete Publications, London.

Melosh, R. J., 1970. *Structural Analysis, Frailty Evaluation and Design:* Vol. 1. *Safer Theoretical Basis, AFFDL-TR-70-15,* July.

Meyers, V. J., 1983. *Matrix Analysis of Structures,* Harper & Row, New York.

Nakamura, T., and I. Takewaki, 1989. Ductility Design via Optimum Design of Nonlinear Elastic Frames, *Journal of Structural Engineering,* 115(3).

Ossenbruggen, P. J., 1984. *Systems Analysis for Civil Engineers,* Wiley, New York.

Pulido, J. E., T. L. Jacobs, and E. C. Prates de Lima, 1992. Structural Reliability Using Monte-Carlo Simulation with Variance Reduction Techniques on Elastic–Plastic Structures, *Journal of Computers and Structures,* 43(3), 419–431.

ReVelle, C., E. Joeres, and W. Kirby, 1969. The Linear Decision Rule in Reservoir Management and Design: 1. Development of the Stochastic Model, *Water Resources Research,* 5(4), 767–777.

ReVelle, C. S., E. E. Whitlach, and J. R. Wright, 1997. *Civil and Environmental Systems Engineering,* Prentice Hall, Upper Saddle River, NJ.

Rutenbar, R. A., 1989. Simulated Annealing Algorithms: An Overview, *IEEE Circuits and Devices,* January.

Smith, A. A., E. Hinton, and R. W. Lewis, 1983. *Civil Engineering Systems: Analysis and Design,* Wiley, New York.

Templeman, A. B., 1988. Discrete Optimum Structural Design, *Computers and Structures,* 30(3), 511–518.

Wagner, H. M., 1975. *Principles of Operations Research,* Prentice Hall, Upper Saddle River, NJ.

Yao, J. T. P., 1974. Fatigue Reliability and Design, *Journal of the Structural Division, ASCE,* 100(9), 1827–1836.

Yao, J. T. P., 1978. A General Note on Structural Reliability, *Nuclear Structural Engineering,* 50(2), 145–147.

Yao, J. T. P., and J. G. MacGregor, 1982. Structural Reliability Theory and Design Codes, *ASCE Structures Congress,* New Orleans, LA.

Yao, J. T. P., F. Kozin, Y.-K. Wen, J.-N. Yang, G. I. Schueller, and O. Ditlevsen, 1986. Stochastic Fatigue, Fracture and Damage Analysis, *Structural Safety,* 3, 231–267.

Zadoks, R. I., and T. L. Jacobs, 1991. Optimal Determination of Feedback Controller Parameters Conditioned on Probabilistic Dynamic Loadings, *Mechatronics,* 1(3), 321–338.

INDEX

Printed in the United States
100002LV00001B/1-2/A

9 780471 128168